Methods in Enzymology

Volume 334
HYPERTHERMOPHILIC ENZYMES
Part C

METHODS IN ENZYMOLOGY

EDITORS-IN-CHIEF

John N. Abelson   Melvin I. Simon

DIVISION OF BIOLOGY
CALIFORNIA INSTITUTE OF TECHNOLOGY
PASADENA, CALIFORNIA

FOUNDING EDITORS

Sidney P. Colowick and Nathan O. Kaplan

Methods in Enzymology

Volume 334

Hyperthermophilic Enzymes
Part C

EDITED BY

Michael W. W. Adams

THE UNIVERSITY OF GEORGIA
ATHENS, GEORGIA

Robert M. Kelly

NORTH CAROLINA STATE UNIVERSITY
RALEIGH, NORTH CAROLINA

ACADEMIC PRESS
San Diego London Boston New York Sydney Tokyo Toronto

This book is printed on acid-free paper.

Copyright © 2001 by ACADEMIC PRESS

All Rights Reserved.
No part of this publication may be reproduced or transmitted in any form or by any means, electronic or mechanical, including photocopy, recording, or any information storage and retrieval system, without permission in writing from the Publisher.

The appearance of the code at the bottom of the first page of a chapter in this book indicates the Publisher's consent that copies of the chapter may be made for personal or internal use of specific clients. This consent is given on the condition, however, that the copier pay the stated per copy fee through the Copyright Clearance Center, Inc. (222 Rosewood Drive, Danvers, Massachusetts 01923), for copying beyond that permitted by Sections 107 or 108 of the U.S. Copyright Law. This consent does not extend to other kinds of copying, such as copying for general distribution, for advertising or promotional purposes, for creating new collective works, or for resale. Copy fees for pre-2000 chapters are as shown on the title pages. If no fee code appears on the title page, the copy fee is the same as for current chapters. /00 $35.00

Explicit permission from Academic Press is not required to reproduce a maximum of two figures or tables from an Academic Press chapter in another scientific or research publication provided that the material has not been credited to another source and that full credit to the Academic Press chapter is given.

Academic Press
A Harcourt Science and Technology Company
525 B Street, Suite 1900, San Diego, California 92101-4495, USA
http://www.academicpress.com

Academic Press
Harcourt Place, 32 Jamestown Road, London NW1 7BY, UK
http://www.academicpress.com

International Standard Book Number: 0-12-182235-4

PRINTED IN THE UNITED STATES OF AMERICA
01 02 03 04 05 06 07 SB 9 8 7 6 5 4 3 2 1

Table of Contents

CONTRIBUTORS TO VOLUME 334 ix
PREFACE xiii
VOLUMES IN SERIES xv

Section I. Redox and Thiol-Dependent Proteins

1. Ferredoxin and Related Enzymes from *Sulfolobus* — TOSHIO IWASAKI AND TAIRO OSHIMA — 3

2. Ferredoxin from *Thermotoga maritima* — REINHARD STERNER — 23

3. Ferredoxin from *Pyrococcus furiosus* — CHULHWAN KIM, PHILLIP S. BRERETON, MARC F. J. M. VERHAGEN, AND MICHAEL W. W. ADAMS — 30

4. Ferredoxin:NADP Oxidoreductase from *Pyrococcus furiosus* — KESEN MA AND MICHAEL W. W. ADAMS — 40

5. Rubredoxin from *Pyrococcus furiosus* — FRANCIS E. JENNEY, JR. AND MICHAEL W. W. ADAMS — 45

6. NAD(P)H:rubredoxin Oxidoreductase from *Pyrococcus furiosus* — KESEN MA AND MICHAEL W. W. ADAMS — 55

7. Protein Disulfide Oxidoreductase from *Pyrococcus furiosus*: Biochemical Properties — SIMONETTA BARTOLUCCI, DONATELLA DE PASCALE, AND MOSÈ ROSSI — 62

8. Protein Disulfide Oxidoreductase from *Pyrococcus furiosus*: Structural Properties — BIN REN AND RUDOLF LADENSTEIN — 74

Section II. Nucleic Acid Modifying Enzymes

9. DNA Polymerases from Hyperthermophiles — HOLLY H. HOGREFE, JANICE CLINE, AMY LOVEJOY, AND KIRK B. NIELSON — 91

10. Archaeal Histones and Nucleosomes	KATHLEEN SANDMAN, KATHRYN A. BAILEY, SUZETTE L. PEREIRA, DIVYA SOARES, WEN-TYNG LI, AND JOHN N. REEVE	116
11. DNA-Binding Proteins Sac7d and Sso7d from *Sulfolobus*	STEPHEN P. EDMONDSON AND JOHN W. SHRIVER	129
12. Reverse Gyrases from Bacteria and Archaea	ANNE-CÉCILE DÉCLAIS, CLAIRE BOUTHIER DE LA TOUR, AND MICHEL DUGUET	146
13. DNA Gyrase from *Thermotoga maritima*	OLIVIER GUIPAUD AND PATRICK FORTERRE	162
14. DNA Topoisomerases VI from Hyperthermophilic Archaea	CHANTAL BOCS, CYRIL BUHLER, PATRICK FORTERRE, AND AGNÈS BERGERAT	172
15. Topoisomerase V from *Methanopyrus kandleri*	ALEXEI I. SLESAREV, GALINA I. BELOVA, JAMES A. LAKE, AND SERGEI A. KOZYAVKIN	179
16. pGT5 Replication Initiator Protein Rep75 from *Pyrococcus abyssi*	STÉPHANIE MARSIN AND PATRICK FORTERRE	193
17. Stability and Manipulation of DNA at Extreme Temperatures	EVELYNE MARGUET AND PATRICK FORTERRE	205
18. Ribonucleotide Reductase from *Pyrococcus furiosus*	MARC FONTECAVE	215
19. Preparation of Components of Archaeal Transcription Preinitiation Complex	YAKOV KORKHIN, OTIS LITTLEFIELD, PAMLEA J. NELSON, STEPHEN D. BELL, AND PAUL B. SIGLER	227
20. Methylguanine Methyltransferase from *Thermococcus kodakaraensis* KOD1	MASAHIRO TAKAGI, YASUSHI KAI, AND TADAYUKI IMANAKA	239
21. DNA Polymerases from Euryarchaeota	YOSHIZUMI ISHINO AND SONOKO ISHINO	249
22. RecA/Rad51 Homolog from *Thermococcus kodakaraensis* KOD1	NAEEM RASHID, MASAAKI MORIKAWA, SHIGENORI KANAYA, HARUYUKI ATOMI, AND TADAYUKI IMANAKA	261
23. Hyperthermophilic Inteins	FRANCINE B. PERLER	270

Section III. Biophysical and Biochemical Aspects of Protein Stability

24. Assaying Activity and Assessing Thermostability of Hyperthermophilic Enzymes	ROY M. DANIEL AND MICHAEL J. DANSON	283
25. Chaperonin from *Thermococcus kodakaraensis* KOD1	SHINSUKE FUJIWARA, MASAHIRO TAKAGI, AND TADAYUKI IMANAKA	293
26. Organic Solutes from Thermophiles and Hyperthermophiles	HELENA SANTOS AND MILTON S. DA COSTA	302
27. Pressure Effects on Activity and Stability of Hyperthermophilic Enzymes	MICHAEL M. C. SUN AND DOUGLAS S. CLARK	316
28. Thermodynamic Analysis of Hyperthermostable Oligomeric Proteins	JAN BACKMANN AND GÜNTER SCHÄFER	328
29. Dynamics and Thermodynamics of Hyperthermophilic Proteins by Hydrogen Exchange	S. WALTER ENGLANDER AND REUBEN HILLER	342
30. Nuclear Magnetic Resonance Analysis of Hyperthermophile Ferredoxins	GERD N. LA MAR	351
31. Calorimetric Analyses of Hyperthermophile Proteins	JOHN W. SHRIVER, WILLIAM PETERS, NICHOLAS SZARY, ANDREW CLARK, AND STEPHEN EDMONDSON	389
32. Crystallographic Analyses of Hyperthermophilic Proteins	DOUGLAS C. REES	423
33. Thermostability of Proteins from *Thermotoga maritima*	RAINER JAENICKE AND GERALD BÖHM	438
34. Structural Basis of Thermostability in Hyperthermophilic Proteins, or "There's More Than One Way to Skin a Cat"	GREGORY A. PETSKO	469

AUTHOR INDEX 479

SUBJECT INDEX 511

Contributors to Volume 334

Article numbers are in parentheses following the names of contributors.
Affiliations listed are current.

MICHAEL W. W. ADAMS (3, 4, 5, 6), *Department of Biochemistry and Molecular Biology, University of Georgia, Athens, Georgia 30602*

HARUYUKI ATOMI (22), *Department of Synthetic Chemistry and Biological Chemistry, Kyoto University Graduate School of Engineering, Kyoto 606-8501, Japan*

JAN BACKMANN (28), *Dienst Ultrastruktuur, Vrije Universiteit Brussel, Sint-Genesius-Rode B-1640, Belgium*

KATHRYN A. BAILEY (10), *Department of Microbiology, Ohio State University, Columbus, Ohio 43210*

SIMONETTA BARTOLUCCI (7), *Department of Organic and Biological Chemistry, University of Naples Federico II, Naples 80134, Italy*

STEPHEN D. BELL (19), *Wellcome Trust and Cancer Research Campaign Institute of Cancer and Developmental Biology, Cambridge CB2 1QR, United Kingdom and Department of Zoology, University of Cambridge, Cambridge CB2 3EJ, United Kingdom*

GALINA I. BELOVA (15), *M. M. Shemyakin and Yu.A. Ovchinnikov Institute of Bioorganic Chemistry, Russian Academy of Sciences, Moscow 117871, Russia*

AGNÈS BERGERAT (14), *Institut de Génétique et Microbiologie, Université de Paris Sud, Orsay Cedex 91405, France*

CHANTAL BOCS (14), *Institut de Génétique et Microbiologie, Université de Paris Sud, Orsay Cedex 91405, France*

GERALD BÖHM (33), *Institut für Biotechnologie, Martin-Luther Universität Halle-Wittenberg, Halle, Saale D-06120, Germany*

CLAIRE BOUTHIER DE LA TOUR (12), *Laboratoire d'Enzymologie des Acides Nucléiques, Institut de Génétique et Microbiologie, Université de Paris Sud, Centre d'Orsay, Orsay Cedex 91405, France*

PHILLIP S. BRERETON (3), *Department of Biochemistry and Molecular Biology, University of Georgia, Athens, Georgia 30602*

CYRIL BUHLER (14), *Institut de Génétique et Microbiologie, Université de Paris Sud, Orsay Cedex 91405, France*

ANDREW CLARK (31), *Department of Biochemistry and Molecular Biology, School of Medicine, Southern Illinois University, Carbondale, Illinois 62901-4413*

DOUGLAS S. CLARK (27), *Department of Chemical Engineering, University of California, Berkeley, California 94720*

JANICE CLINE (9), *Stratagene, La Jolla, California 92037*

MILTON S. DA COSTA (26), *Departamento de Bioquímica and Centro de Neurociências de Coimbra, Universidade de Coimbra, Coimbra 3000, Portugal*

ROY M. DANIEL (24), *Department of Biological Sciences, University of Waikato, Hamilton, New Zealand*

MICHAEL J. DANSON (24), *Department of Biology and Biochemistry, University of Bath, Bath BA2 7AY, United Kingdom*

ANNE-CÉCILE DÉCLAIS (12), *Department of Biochemistry, CRC Nucleic Acid Structure Research Group, University of Dundee, Dundee DD1 5EH, United Kingdom*

DONATELLA DE PASCALE (7), *Department of Organic and Biological Chemistry, University of Naples Federico II, Naples 80134, Italy*

MICHEL DUGUET (12), *Laboratoire d'Enzymologie des Acides Nucléiques, Institut de Génétique et Microbiologie, Université de Paris Sud, Centre d'Orsay, Orsay Cedex 91405, France*

STEPHEN P. EDMONDSON (11, 31), *Department of Biochemistry and Molecular Biology, School of Medicine, Southern Illinois University, Carbondale, Illinois 62901-4413*

S. WALTER ENGLANDER (29), *Johnson Research Foundation, Department of Biochemistry and Biophysics, University of Pennsylvania, Philadelphia, Pennsylvania 19109-6059*

MARC FONTECAVE (18), *Laboratoire de Chimie et Biochimie des Centres Rédox Biologiques, DBMS-CEA, CNRS, Université Joseph Fourier, Grenoble Cedex 9, 38054, France*

PATRICK FORTERRE (13, 14, 16, 17), *Institut de Génétique et Microbiologie, Université de Paris Sud, Orsay Cedex 91405, France*

SHINSUKE FUJIWARA (25), *Department of Biotechnology, Osaka University Graduate School of Engineering, Osaka 565-0871, Japan*

OLIVIER GUIPAUD (13), *Institut de Génétique et Microbiologie, Université de Paris Sud, Orsay Cedex 91405, France*

REUBEN HILLER (29), *Johnson Research Foundation, Department of Biochemistry and Biophysics, University of Pennsylvania, Philadelphia, Pennsylvania 19109-6059*

HOLLY H. HOGREFE (9), *Stratagene, La Jolla, California 92037*

TADAYUKI IMANAKA (20, 22, 25), *Department of Synthetic Chemistry and Biological Chemistry, Kyoto University Graduate School of Engineering, Kyoto 606-8501, Japan*

SONOKO ISHINO (21), *Department of Molecular Biology, Biomolecular Engineering Research Institute, Osaka 565-0874, Japan*

YOSHIZUMI ISHINO (21), *Department of Molecular Biology, Biomolecular Engineering Research Institute, Osaka 565-0874, Japan*

TOSHIO IWASAKI (1), *Department of Biochemistry and Molecular Biology, Nippon Medical School, Tokyo 113-8602, Japan*

RAINER JAENICKE (33), *Institut für Biophysik und Physikalische Biochemie, Universität Regensburg, Regensburg D-93040, Germany*

FRANCIS E. JENNEY, JR. (5), *Department of Biochemistry and Molecular Biology, University of Georgia, Athens, Georgia 30602*

YASUSHI KAI (20), *Departments of Biotechnology and Applied Chemistry, Osaka University Graduate School of Engineering, Osaka 565-0871, Japan*

SHIGENORI KANAYA (22), *Department of Material and Life Science, Osaka University Graduate School of Engineering, Osaka 565-0871, Japan*

CHULHWAN KIM (3), *Department of Biochemistry and Molecular Biology, University of Georgia, Athens, Georgia 30602*

YAKOV KORKHIN (19), *Department of Molecular Biophysics and Biochemistry and the Howard Hughes Medical Institute, Yale University, New Haven, Connecticut 06511*

SERGEI A. KOZYAVKIN (15), *Fidelity Systems, Inc., Gaithersburg, Maryland 20879*

RUDOLF LADENSTEIN (8), *Center for Structural Biochemistry, Karolinska Institutet NOVUM, Huddinge S-14157, Sweden*

JAMES A. LAKE (15), *Molecular Biology Institute and MCD Biology, University of California, Los Angeles, California 90095*

GERD N. LA MAR (30), *Department of Chemistry, University of California, Davis, California 95616*

WEN-TYNG LI (10), *Department of Microbiology, Ohio State University, Columbus, Ohio 43210*

OTIS LITTLEFIELD (19), *Department of Molecular Biophysics and Biochemistry and the Howard Hughes Medical Institute, Yale University, New Haven, Connecticut 06511*

AMY LOVEJOY (9), *Stratagene, La Jolla, California 92037*

KESEN MA (4, 6), *Department of Biology, University of Waterloo, Waterloo, Ontario N2L 3G1, Canada*

EVELYNE MARGUET (17), *Institut de Génétique et Microbiologie, Université de Paris Sud, Orsay Cedex 91405, France*

STÉPHANIE MARSIN (16), *Institut de Génétique et Microbiologie, Université de Paris Sud, Orsay Cedex 91405, France*

MASAAKI MORIKAWA (22), *Department of Material and Life Science, Osaka University Graduate School of Engineering, Osaka 565-0871, Japan*

PAMLEA J. NELSON (19), *Department of Molecular Biophysics and Biochemistry and the Howard Hughes Medical Institute, Yale University, New Haven, Connecticut 06511*

KIRK B. NIELSON (9), *Stratagene, La Jolla, California 92037*

TAIRO OSHIMA (1), *Department of Molecular Biology, Tokyo University of Pharmacy and Life Science, Tokyo 192-0392, Japan*

SUZETTE L. PEREIRA (10), *Department of Microbiology, Ohio State University, Columbus, Ohio 43210*

FRANCINE B. PERLER (23), *New England BioLabs, Inc., Beverly, Massachusetts 01915*

WILLIAM PETERS (31), *Department of Biochemistry and Molecular Biology, School of Medicine, Southern Illinois University, Carbondale, Illinois 62901-4413*

GREGORY A. PETSKO (34), *Departments of Biochemistry and Chemistry and Rosenstiel Basic Medical Sciences Research Center, Brandeis University, Waltham, Massachusetts 02454-9110*

NAEEM RASHID (22), *Department of Synthetic Chemistry and Biological Chemistry, Kyoto University Graduate School of Engineering, Kyoto 606-8501, Japan*

DOUGLAS C. REES (32), *Howard Hughes Medical Institute, Division of Chemistry and Chemical Engineering, California Institute of Technology, Pasadena, California 91125*

JOHN N. REEVE (10), *Department of Microbiology, Ohio State University, Columbus, Ohio 43210*

BIN REN (8), *Center for Structural Biochemistry, Karolinska Institutet NOVUM, Huddinge S-14157, Sweden*

MOSÈ ROSSI (7), *Department of Organic and Biological Chemistry, University of Naples Federico II, Naples 80134, Italy*

KATHLEEN SANDMAN (10), *Department of Microbiology, Ohio State University, Columbus, Ohio 43210*

HELENA SANTOS (26), *Instituto de Tecnologia Química e Biológica, Universidade Nova de Lisboa, Oeiras 2780-156, Portugal*

GÜNTER SCHÄFER (28), *Institut für Biochemie, Medizinische Universität zu Lübeck, Lübeck D-23538, Germany*

JOHN W. SHRIVER (11, 31), *Department of Biochemistry and Molecular Biology, School of Medicine, Southern Illinois University, Carbondale, Illinois 62901-4413*

PAUL B. SIGLER (19), *Department of Molecular Biophysics and Biochemistry and the Howard Hughes Medical Institute, Yale University, New Haven, Connecticut 06511*

ALEXEI I. SLESAREV (15), *Laboratory of Gene Bioengineering, M. M. Shemyakin and Yu.A. Ovchinnikov Institute of Bioorganic Chemistry, Russian Academy of Sciences, Moscow 117871, Russia*

DIVYA SOARES (10), *Department of Microbiology, Ohio State University, Columbus, Ohio 43210*

REINHARD STERNER (2), *Universität zu Köln, Institut für Biochemie, Köln D-50674, Germany*

MICHAEL M. C. SUN (27), *Department of Chemical Engineering, University of California, Berkeley, California 94720*

NICHOLAS SZARY (31), *Department of Biochemistry and Molecular Biology, School of Medicine, Southern Illinois University, Carbondale, Illinois 62901-4413*

MASAHIRO TAKAGI (20, 25), *Department of Biotechnology, Osaka University Graduate School of Engineering, Osaka 565-0871, Japan*

MARC F. J. M. VERHAGEN (3), *Department of Biochemistry and Molecular Biology, University of Georgia, Athens, Georgia 30602*

Preface

More than thirty years ago, the pioneering work of Thomas Brock of the University of Wisconsin on the microbiology of hot springs in Yellowstone National Park alerted the scientific community to the existence of microorganisms with optimal growth temperatures of 70°C and even higher. In the early 1980s, the known thermal limits of life were expanded by the seminal work of Karl Stetter and colleagues at the University of Regensburg, who isolated from a marine volcanic vent the first microorganisms that could grow at, and even above, the normal boiling point of water. Subsequent work by Stetter and several other groups have led to the discovery in a variety of geothermal biotopes of more than twenty different genera that can grow optimally at or above 80°C. Such organisms are now termed *hyperthermophiles*.

Initial efforts to explore the enzymology of hyperthermophiles were impeded by the difficulty of culturing the organisms on a scale large enough to allow the purification of specific proteins in sufficient quantities for characterization. This often meant processing hundreds of liters of nearly boiling fermentation media under anaerobic conditions. In addition, relatively low biomass yields were typically obtained. Nevertheless, the first "hyperthermophilic enzymes" were purified in the late 1980s. It was demonstrated that they are, indeed, extremely stable at high temperatures, that this is an intrinsic property, and that they exhibit no or very low activity at temperatures below the growth conditions of the organism from which they were obtained. At that time it was difficult to imagine how quickly the tools of molecular biology would make such a dramatic impact on the world of hyperthermophiles. In fact, it was unexpected that the recombinant forms of hyperthermophilic enzymes would, to a large extent, correctly achieve their active conformation in mesophilic hosts grown some 70°C below the enzyme's source organism's normal growth temperature. This approach provided a much-needed alternative to large-scale hyperthermophile cultivation. With the ever-expanding list of genomes from hyperthermophiles that have been or are being sequenced, molecular biology provides universal access to a treasure chest of known and putative proteins endowed with unprecedented levels of thermostability.

In Volumes 330, 331, and 334 of *Methods in Enzymology*, a set of protocols has been assembled that for the first time describe the methods involved in studying the biochemistry and biophysics of enzymes and proteins from hyperthermophilic microorganisms. As is evident from the various chapters, hyperthermophilic counterparts to a range of previously studied but less thermostable enzymes exist. In addition, the volumes include descriptions of many novel enzymes that were first identified and, in most cases, are still limited to hyperthermophilic organisms. Also included in these volumes are genomic analyses from selected hyperthermophiles that provide some perspective on what remains to be investigated in terms of hyperthermophilic enzymology. Specific chapters address the basis for extreme levels of thermostability and special considerations that must be taken into account in defining experimentally the biochemical and biophysical features of hyperthermophilic enzymes.

There are many individuals whose pioneering efforts laid the basis for the work discussed in these volumes. None was more important than the late Holger Jannasch of Woods Hole Oceanographic Institute. His innovation and inspiration opened a new field of microbiology in deep-sea hydrothermal vents and provided the research world access to a biotope of great scientific and technological promise. Holger will be remembered in many ways, and it is a fitting tribute that the first genome of a hyperthermophile to be sequenced should bear his name: *Methanocaldococcus jannaschii*. We wish to recognize Holger's pioneering efforts by dedicating these volumes to him.

MICHAEL W. W. ADAMS
ROBERT M. KELLY

METHODS IN ENZYMOLOGY

VOLUME I. Preparation and Assay of Enzymes
Edited by SIDNEY P. COLOWICK AND NATHAN O. KAPLAN

VOLUME II. Preparation and Assay of Enzymes
Edited by SIDNEY P. COLOWICK AND NATHAN O. KAPLAN

VOLUME III. Preparation and Assay of Substrates
Edited by SIDNEY P. COLOWICK AND NATHAN O. KAPLAN

VOLUME IV. Special Techniques for the Enzymologist
Edited by SIDNEY P. COLOWICK AND NATHAN O. KAPLAN

VOLUME V. Preparation and Assay of Enzymes
Edited by SIDNEY P. COLOWICK AND NATHAN O. KAPLAN

VOLUME VI. Preparation and Assay of Enzymes (*Continued*)
Preparation and Assay of Substrates
Special Techniques
Edited by SIDNEY P. COLOWICK AND NATHAN O. KAPLAN

VOLUME VII. Cumulative Subject Index
Edited by SIDNEY P. COLOWICK AND NATHAN O. KAPLAN

VOLUME VIII. Complex Carbohydrates
Edited by ELIZABETH F. NEUFELD AND VICTOR GINSBURG

VOLUME IX. Carbohydrate Metabolism
Edited by WILLIS A. WOOD

VOLUME X. Oxidation and Phosphorylation
Edited by RONALD W. ESTABROOK AND MAYNARD E. PULLMAN

VOLUME XI. Enzyme Structure
Edited by C. H. W. HIRS

VOLUME XII. Nucleic Acids (Parts A and B)
Edited by LAWRENCE GROSSMAN AND KIVIE MOLDAVE

VOLUME XIII. Citric Acid Cycle
Edited by J. M. LOWENSTEIN

VOLUME XIV. Lipids
Edited by J. M. LOWENSTEIN

VOLUME XV. Steroids and Terpenoids
Edited by RAYMOND B. CLAYTON

VOLUME XVI. Fast Reactions
Edited by KENNETH KUSTIN

VOLUME XVII. Metabolism of Amino Acids and Amines (Parts A and B)
Edited by HERBERT TABOR AND CELIA WHITE TABOR

VOLUME XVIII. Vitamins and Coenzymes (Parts A, B, and C)
Edited by DONALD B. MCCORMICK AND LEMUEL D. WRIGHT

VOLUME XIX. Proteolytic Enzymes
Edited by GERTRUDE E. PERLMANN AND LASZLO LORAND

VOLUME XX. Nucleic Acids and Protein Synthesis (Part C)
Edited by KIVIE MOLDAVE AND LAWRENCE GROSSMAN

VOLUME XXI. Nucleic Acids (Part D)
Edited by LAWRENCE GROSSMAN AND KIVIE MOLDAVE

VOLUME XXII. Enzyme Purification and Related Techniques
Edited by WILLIAM B. JAKOBY

VOLUME XXIII. Photosynthesis (Part A)
Edited by ANTHONY SAN PIETRO

VOLUME XXIV. Photosynthesis and Nitrogen Fixation (Part B)
Edited by ANTHONY SAN PIETRO

VOLUME XXV. Enzyme Structure (Part B)
Edited by C. H. W. HIRS AND SERGE N. TIMASHEFF

VOLUME XXVI. Enzyme Structure (Part C)
Edited by C. H. W. HIRS AND SERGE N. TIMASHEFF

VOLUME XXVII. Enzyme Structure (Part D)
Edited by C. H. W. HIRS AND SERGE N. TIMASHEFF

VOLUME XXVIII. Complex Carbohydrates (Part B)
Edited by VICTOR GINSBURG

VOLUME XXIX. Nucleic Acids and Protein Synthesis (Part E)
Edited by LAWRENCE GROSSMAN AND KIVIE MOLDAVE

VOLUME XXX. Nucleic Acids and Protein Synthesis (Part F)
Edited by KIVIE MOLDAVE AND LAWRENCE GROSSMAN

VOLUME XXXI. Biomembranes (Part A)
Edited by SIDNEY FLEISCHER AND LESTER PACKER

VOLUME XXXII. Biomembranes (Part B)
Edited by SIDNEY FLEISCHER AND LESTER PACKER

VOLUME XXXIII. Cumulative Subject Index Volumes I-XXX
Edited by MARTHA G. DENNIS AND EDWARD A. DENNIS

VOLUME XXXIV. Affinity Techniques (Enzyme Purification: Part B)
Edited by WILLIAM B. JAKOBY AND MEIR WILCHEK

VOLUME XXXV. Lipids (Part B)
Edited by JOHN M. LOWENSTEIN

VOLUME XXXVI. Hormone Action (Part A: Steroid Hormones)
Edited by BERT W. O'MALLEY AND JOEL G. HARDMAN

VOLUME XXXVII. Hormone Action (Part B: Peptide Hormones)
Edited by BERT W. O'MALLEY AND JOEL G. HARDMAN

VOLUME XXXVIII. Hormone Action (Part C: Cyclic Nucleotides)
Edited by JOEL G. HARDMAN AND BERT W. O'MALLEY

VOLUME XXXIX. Hormone Action (Part D: Isolated Cells, Tissues, and Organ Systems)
Edited by JOEL G. HARDMAN AND BERT W. O'MALLEY

VOLUME XL. Hormone Action (Part E: Nuclear Structure and Function)
Edited by BERT W. O'MALLEY AND JOEL G. HARDMAN

VOLUME XLI. Carbohydrate Metabolism (Part B)
Edited by W. A. WOOD

VOLUME XLII. Carbohydrate Metabolism (Part C)
Edited by W. A. WOOD

VOLUME XLIII. Antibiotics
Edited by JOHN H. HASH

VOLUME XLIV. Immobilized Enzymes
Edited by KLAUS MOSBACH

VOLUME XLV. Proteolytic Enzymes (Part B)
Edited by LASZLO LORAND

VOLUME XLVI. Affinity Labeling
Edited by WILLIAM B. JAKOBY AND MEIR WILCHEK

VOLUME XLVII. Enzyme Structure (Part E)
Edited by C. H. W. HIRS AND SERGE N. TIMASHEFF

VOLUME XLVIII. Enzyme Structure (Part F)
Edited by C. H. W. HIRS AND SERGE N. TIMASHEFF

VOLUME XLIX. Enzyme Structure (Part G)
Edited by C. H. W. HIRS AND SERGE N. TIMASHEFF

VOLUME L. Complex Carbohydrates (Part C)
Edited by VICTOR GINSBURG

VOLUME LI. Purine and Pyrimidine Nucleotide Metabolism
Edited by PATRICIA A. HOFFEE AND MARY ELLEN JONES

VOLUME LII. Biomembranes (Part C: Biological Oxidations)
Edited by SIDNEY FLEISCHER AND LESTER PACKER

VOLUME LIII. Biomembranes (Part D: Biological Oxidations)
Edited by SIDNEY FLEISCHER AND LESTER PACKER

VOLUME LIV. Biomembranes (Part E: Biological Oxidations)
Edited by SIDNEY FLEISCHER AND LESTER PACKER

VOLUME LV. Biomembranes (Part F: Bioenergetics)
Edited by SIDNEY FLEISCHER AND LESTER PACKER

VOLUME LVI. Biomembranes (Part G: Bioenergetics)
Edited by SIDNEY FLEISCHER AND LESTER PACKER

VOLUME LVII. Bioluminescence and Chemiluminescence
Edited by MARLENE A. DELUCA

VOLUME LVIII. Cell Culture
Edited by WILLIAM B. JAKOBY AND IRA PASTAN

VOLUME LIX. Nucleic Acids and Protein Synthesis (Part G)
Edited by KIVIE MOLDAVE AND LAWRENCE GROSSMAN

VOLUME LX. Nucleic Acids and Protein Synthesis (Part H)
Edited by KIVIE MOLDAVE AND LAWRENCE GROSSMAN

VOLUME 61. Enzyme Structure (Part H)
Edited by C. H. W. HIRS AND SERGE N. TIMASHEFF

VOLUME 62. Vitamins and Coenzymes (Part D)
Edited by DONALD B. MCCORMICK AND LEMUEL D. WRIGHT

VOLUME 63. Enzyme Kinetics and Mechanism (Part A: Initial Rate and Inhibitor Methods)
Edited by DANIEL L. PURICH

VOLUME 64. Enzyme Kinetics and Mechanism (Part B: Isotopic Probes and Complex Enzyme Systems)
Edited by DANIEL L. PURICH

VOLUME 65. Nucleic Acids (Part I)
Edited by LAWRENCE GROSSMAN AND KIVIE MOLDAVE

VOLUME 66. Vitamins and Coenzymes (Part E)
Edited by DONALD B. MCCORMICK AND LEMUEL D. WRIGHT

VOLUME 67. Vitamins and Coenzymes (Part F)
Edited by DONALD B. MCCORMICK AND LEMUEL D. WRIGHT

VOLUME 68. Recombinant DNA
Edited by RAY WU

VOLUME 69. Photosynthesis and Nitrogen Fixation (Part C)
Edited by ANTHONY SAN PIETRO

VOLUME 70. Immunochemical Techniques (Part A)
Edited by HELEN VAN VUNAKIS AND JOHN J. LANGONE

VOLUME 71. Lipids (Part C)
Edited by JOHN M. LOWENSTEIN

VOLUME 72. Lipids (Part D)
Edited by JOHN M. LOWENSTEIN

VOLUME 73. Immunochemical Techniques (Part B)
Edited by JOHN J. LANGONE AND HELEN VAN VUNAKIS

VOLUME 74. Immunochemical Techniques (Part C)
Edited by JOHN J. LANGONE AND HELEN VAN VUNAKIS

VOLUME 75. Cumulative Subject Index Volumes XXXI, XXXII, XXXIV–LX
Edited by EDWARD A. DENNIS AND MARTHA G. DENNIS

VOLUME 76. Hemoglobins
Edited by ERALDO ANTONINI, LUIGI ROSSI-BERNARDI, AND EMILIA CHIANCONE

VOLUME 77. Detoxication and Drug Metabolism
Edited by WILLIAM B. JAKOBY

VOLUME 78. Interferons (Part A)
Edited by SIDNEY PESTKA

VOLUME 79. Interferons (Part B)
Edited by SIDNEY PESTKA

VOLUME 80. Proteolytic Enzymes (Part C)
Edited by LASZLO LORAND

VOLUME 81. Biomembranes (Part H: Visual Pigments and Purple Membranes, I)
Edited by LESTER PACKER

VOLUME 82. Structural and Contractile Proteins (Part A: Extracellular Matrix)
Edited by LEON W. CUNNINGHAM AND DIXIE W. FREDERIKSEN

VOLUME 83. Complex Carbohydrates (Part D)
Edited by VICTOR GINSBURG

VOLUME 84. Immunochemical Techniques (Part D: Selected Immunoassays)
Edited by JOHN J. LANGONE AND HELEN VAN VUNAKIS

VOLUME 85. Structural and Contractile Proteins (Part B: The Contractile Apparatus and the Cytoskeleton)
Edited by DIXIE W. FREDERIKSEN AND LEON W. CUNNINGHAM

VOLUME 86. Prostaglandins and Arachidonate Metabolites
Edited by WILLIAM E. M. LANDS AND WILLIAM L. SMITH

VOLUME 87. Enzyme Kinetics and Mechanism (Part C: Intermediates, Stereochemistry, and Rate Studies)
Edited by DANIEL L. PURICH

VOLUME 88. Biomembranes (Part I: Visual Pigments and Purple Membranes, II)
Edited by LESTER PACKER

VOLUME 89. Carbohydrate Metabolism (Part D)
Edited by WILLIS A. WOOD

VOLUME 90. Carbohydrate Metabolism (Part E)
Edited by WILLIS A. WOOD

VOLUME 91. Enzyme Structure (Part I)
Edited by C. H. W. HIRS AND SERGE N. TIMASHEFF

VOLUME 92. Immunochemical Techniques (Part E: Monoclonal Antibodies and General Immunoassay Methods)
Edited by JOHN J. LANGONE AND HELEN VAN VUNAKIS

VOLUME 93. Immunochemical Techniques (Part F: Conventional Antibodies, Fc Receptors, and Cytotoxicity)
Edited by JOHN J. LANGONE AND HELEN VAN VUNAKIS

VOLUME 94. Polyamines
Edited by HERBERT TABOR AND CELIA WHITE TABOR

VOLUME 95. Cumulative Subject Index Volumes 61–74, 76–80
Edited by EDWARD A. DENNIS AND MARTHA G. DENNIS

VOLUME 96. Biomembranes [Part J: Membrane Biogenesis: Assembly and Targeting (General Methods; Eukaryotes)]
Edited by SIDNEY FLEISCHER AND BECCA FLEISCHER

VOLUME 97. Biomembranes [Part K: Membrane Biogenesis: Assembly and Targeting (Prokaryotes, Mitochondria, and Chloroplasts)]
Edited by SIDNEY FLEISCHER AND BECCA FLEISCHER

VOLUME 98. Biomembranes (Part L: Membrane Biogenesis: Processing and Recycling)
Edited by SIDNEY FLEISCHER AND BECCA FLEISCHER

VOLUME 99. Hormone Action (Part F: Protein Kinases)
Edited by JACKIE D. CORBIN AND JOEL G. HARDMAN

VOLUME 100. Recombinant DNA (Part B)
Edited by RAY WU, LAWRENCE GROSSMAN, AND KIVIE MOLDAVE

VOLUME 101. Recombinant DNA (Part C)
Edited by RAY WU, LAWRENCE GROSSMAN, AND KIVIE MOLDAVE

VOLUME 102. Hormone Action (Part G: Calmodulin and Calcium-Binding Proteins)
Edited by ANTHONY R. MEANS AND BERT W. O'MALLEY

VOLUME 103. Hormone Action (Part H: Neuroendocrine Peptides)
Edited by P. MICHAEL CONN

VOLUME 104. Enzyme Purification and Related Techniques (Part C)
Edited by WILLIAM B. JAKOBY

VOLUME 105. Oxygen Radicals in Biological Systems
Edited by LESTER PACKER

VOLUME 106. Posttranslational Modifications (Part A)
Edited by FINN WOLD AND KIVIE MOLDAVE

VOLUME 107. Posttranslational Modifications (Part B)
Edited by FINN WOLD AND KIVIE MOLDAVE

VOLUME 108. Immunochemical Techniques (Part G: Separation and Characterization of Lymphoid Cells)
Edited by GIOVANNI DI SABATO, JOHN J. LANGONE, AND HELEN VAN VUNAKIS

VOLUME 109. Hormone Action (Part I: Peptide Hormones)
Edited by LUTZ BIRNBAUMER AND BERT W. O'MALLEY

VOLUME 110. Steroids and Isoprenoids (Part A)
Edited by JOHN H. LAW AND HANS C. RILLING

VOLUME 111. Steroids and Isoprenoids (Part B)
Edited by JOHN H. LAW AND HANS C. RILLING

VOLUME 112. Drug and Enzyme Targeting (Part A)
Edited by KENNETH J. WIDDER AND RALPH GREEN

VOLUME 113. Glutamate, Glutamine, Glutathione, and Related Compounds
Edited by ALTON MEISTER

VOLUME 114. Diffraction Methods for Biological Macromolecules (Part A)
Edited by HAROLD W. WYCKOFF, C. H. W. HIRS, AND SERGE N. TIMASHEFF

VOLUME 115. Diffraction Methods for Biological Macromolecules (Part B)
Edited by HAROLD W. WYCKOFF, C. H. W. HIRS, AND SERGE N. TIMASHEFF

VOLUME 116. Immunochemical Techniques (Part H: Effectors and Mediators of Lymphoid Cell Functions)
Edited by GIOVANNI DI SABATO, JOHN J. LANGONE, AND HELEN VAN VUNAKIS

VOLUME 117. Enzyme Structure (Part J)
Edited by C. H. W. HIRS AND SERGE N. TIMASHEFF

VOLUME 118. Plant Molecular Biology
Edited by ARTHUR WEISSBACH AND HERBERT WEISSBACH

VOLUME 119. Interferons (Part C)
Edited by SIDNEY PESTKA

VOLUME 120. Cumulative Subject Index Volumes 81–94, 96–101

VOLUME 121. Immunochemical Techniques (Part I: Hybridoma Technology and Monoclonal Antibodies)
Edited by JOHN J. LANGONE AND HELEN VAN VUNAKIS

VOLUME 122. Vitamins and Coenzymes (Part G)
Edited by FRANK CHYTIL AND DONALD B. MCCORMICK

VOLUME 123. Vitamins and Coenzymes (Part H)
Edited by FRANK CHYTIL AND DONALD B. MCCORMICK

VOLUME 124. Hormone Action (Part J: Neuroendocrine Peptides)
Edited by P. MICHAEL CONN

VOLUME 125. Biomembranes (Part M: Transport in Bacteria, Mitochondria, and Chloroplasts: General Approaches and Transport Systems)
Edited by SIDNEY FLEISCHER AND BECCA FLEISCHER

VOLUME 126. Biomembranes (Part N: Transport in Bacteria, Mitochondria, and Chloroplasts: Protonmotive Force)
Edited by SIDNEY FLEISCHER AND BECCA FLEISCHER

VOLUME 127. Biomembranes (Part O: Protons and Water: Structure and Translocation)
Edited by LESTER PACKER

VOLUME 128. Plasma Lipoproteins (Part A: Preparation, Structure, and Molecular Biology)
Edited by JERE P. SEGREST AND JOHN J. ALBERS

VOLUME 129. Plasma Lipoproteins (Part B: Characterization, Cell Biology, and Metabolism)
Edited by JOHN J. ALBERS AND JERE P. SEGREST

VOLUME 130. Enzyme Structure (Part K)
Edited by C. H. W. HIRS AND SERGE N. TIMASHEFF

VOLUME 131. Enzyme Structure (Part L)
Edited by C. H. W. HIRS AND SERGE N. TIMASHEFF

VOLUME 132. Immunochemical Techniques (Part J: Phagocytosis and Cell-Mediated Cytotoxicity)
Edited by GIOVANNI DI SABATO AND JOHANNES EVERSE

VOLUME 133. Bioluminescence and Chemiluminescence (Part B)
Edited by MARLENE DELUCA AND WILLIAM D. MCELROY

VOLUME 134. Structural and Contractile Proteins (Part C: The Contractile Apparatus and the Cytoskeleton)
Edited by RICHARD B. VALLEE

VOLUME 135. Immobilized Enzymes and Cells (Part B)
Edited by KLAUS MOSBACH

VOLUME 136. Immobilized Enzymes and Cells (Part C)
Edited by KLAUS MOSBACH

VOLUME 137. Immobilized Enzymes and Cells (Part D)
Edited by KLAUS MOSBACH

VOLUME 138. Complex Carbohydrates (Part E)
Edited by VICTOR GINSBURG

VOLUME 139. Cellular Regulators (Part A: Calcium- and Calmodulin-Binding Proteins)
Edited by ANTHONY R. MEANS AND P. MICHAEL CONN

VOLUME 140. Cumulative Subject Index Volumes 102–119, 121–134

VOLUME 141. Cellular Regulators (Part B: Calcium and Lipids)
Edited by P. MICHAEL CONN AND ANTHONY R. MEANS

VOLUME 142. Metabolism of Aromatic Amino Acids and Amines
Edited by SEYMOUR KAUFMAN

VOLUME 143. Sulfur and Sulfur Amino Acids
Edited by WILLIAM B. JAKOBY AND OWEN GRIFFITH

VOLUME 144. Structural and Contractile Proteins (Part D: Extracellular Matrix)
Edited by LEON W. CUNNINGHAM

VOLUME 145. Structural and Contractile Proteins (Part E: Extracellular Matrix)
Edited by LEON W. CUNNINGHAM

VOLUME 146. Peptide Growth Factors (Part A)
Edited by DAVID BARNES AND DAVID A. SIRBASKU

VOLUME 147. Peptide Growth Factors (Part B)
Edited by DAVID BARNES AND DAVID A. SIRBASKU

VOLUME 148. Plant Cell Membranes
Edited by LESTER PACKER AND ROLAND DOUCE

VOLUME 149. Drug and Enzyme Targeting (Part B)
Edited by RALPH GREEN AND KENNETH J. WIDDER

VOLUME 150. Immunochemical Techniques (Part K: *In Vitro* Models of B and T Cell Functions and Lymphoid Cell Receptors)
Edited by GIOVANNI DI SABATO

VOLUME 151. Molecular Genetics of Mammalian Cells
Edited by MICHAEL M. GOTTESMAN

VOLUME 152. Guide to Molecular Cloning Techniques
Edited by SHELBY L. BERGER AND ALAN R. KIMMEL

VOLUME 153. Recombinant DNA (Part D)
Edited by RAY WU AND LAWRENCE GROSSMAN

VOLUME 154. Recombinant DNA (Part E)
Edited by RAY WU AND LAWRENCE GROSSMAN

VOLUME 155. Recombinant DNA (Part F)
Edited by RAY WU

VOLUME 156. Biomembranes (Part P: ATP-Driven Pumps and Related Transport: The Na, K-Pump)
Edited by SIDNEY FLEISCHER AND BECCA FLEISCHER

VOLUME 157. Biomembranes (Part Q: ATP-Driven Pumps and Related Transport: Calcium, Proton, and Potassium Pumps)
Edited by SIDNEY FLEISCHER AND BECCA FLEISCHER

VOLUME 158. Metalloproteins (Part A)
Edited by JAMES F. RIORDAN AND BERT L. VALLEE

VOLUME 159. Initiation and Termination of Cyclic Nucleotide Action
Edited by JACKIE D. CORBIN AND ROGER A. JOHNSON

VOLUME 160. Biomass (Part A: Cellulose and Hemicellulose)
Edited by WILLIS A. WOOD AND SCOTT T. KELLOGG

VOLUME 161. Biomass (Part B: Lignin, Pectin, and Chitin)
Edited by WILLIS A. WOOD AND SCOTT T. KELLOGG

VOLUME 162. Immunochemical Techniques (Part L: Chemotaxis and Inflammation)
Edited by GIOVANNI DI SABATO

VOLUME 163. Immunochemical Techniques (Part M: Chemotaxis and Inflammation)
Edited by GIOVANNI DI SABATO

VOLUME 164. Ribosomes
Edited by HARRY F. NOLLER, JR., AND KIVIE MOLDAVE

VOLUME 165. Microbial Toxins: Tools for Enzymology
Edited by SIDNEY HARSHMAN

VOLUME 166. Branched-Chain Amino Acids
Edited by ROBERT HARRIS AND JOHN R. SOKATCH

VOLUME 167. Cyanobacteria
Edited by LESTER PACKER AND ALEXANDER N. GLAZER

VOLUME 168. Hormone Action (Part K: Neuroendocrine Peptides)
Edited by P. MICHAEL CONN

VOLUME 169. Platelets: Receptors, Adhesion, Secretion (Part A)
Edited by JACEK HAWIGER

VOLUME 170. Nucleosomes
Edited by PAUL M. WASSARMAN AND ROGER D. KORNBERG

VOLUME 171. Biomembranes (Part R: Transport Theory: Cells and Model Membranes)
Edited by SIDNEY FLEISCHER AND BECCA FLEISCHER

VOLUME 172. Biomembranes (Part S: Transport: Membrane Isolation and Characterization)
Edited by SIDNEY FLEISCHER AND BECCA FLEISCHER

VOLUME 173. Biomembranes [Part T: Cellular and Subcellular Transport: Eukaryotic (Nonepithelial) Cells]
Edited by SIDNEY FLEISCHER AND BECCA FLEISCHER

VOLUME 174. Biomembranes [Part U: Cellular and Subcellular Transport: Eukaryotic (Nonepithelial) Cells]
Edited by SIDNEY FLEISCHER AND BECCA FLEISCHER

VOLUME 175. Cumulative Subject Index Volumes 135–139, 141–167

VOLUME 176. Nuclear Magnetic Resonance (Part A: Spectral Techniques and Dynamics)
Edited by NORMAN J. OPPENHEIMER AND THOMAS L. JAMES

VOLUME 177. Nuclear Magnetic Resonance (Part B: Structure and Mechanism)
Edited by NORMAN J. OPPENHEIMER AND THOMAS L. JAMES

VOLUME 178. Antibodies, Antigens, and Molecular Mimicry
Edited by JOHN J. LANGONE

VOLUME 179. Complex Carbohydrates (Part F)
Edited by VICTOR GINSBURG

VOLUME 180. RNA Processing (Part A: General Methods)
Edited by JAMES E. DAHLBERG AND JOHN N. ABELSON

VOLUME 181. RNA Processing (Part B: Specific Methods)
Edited by JAMES E. DAHLBERG AND JOHN N. ABELSON

VOLUME 182. Guide to Protein Purification
Edited by MURRAY P. DEUTSCHER

VOLUME 183. Molecular Evolution: Computer Analysis of Protein and Nucleic Acid Sequences
Edited by RUSSELL F. DOOLITTLE

VOLUME 184. Avidin-Biotin Technology
Edited by MEIR WILCHEK AND EDWARD A. BAYER

VOLUME 185. Gene Expression Technology
Edited by DAVID V. GOEDDEL

VOLUME 186. Oxygen Radicals in Biological Systems (Part B: Oxygen Radicals and Antioxidants)
Edited by LESTER PACKER AND ALEXANDER N. GLAZER

VOLUME 187. Arachidonate Related Lipid Mediators
Edited by ROBERT C. MURPHY AND FRANK A. FITZPATRICK

VOLUME 188. Hydrocarbons and Methylotrophy
Edited by MARY E. LIDSTROM

VOLUME 189. Retinoids (Part A: Molecular and Metabolic Aspects)
Edited by LESTER PACKER

VOLUME 190. Retinoids (Part B: Cell Differentiation and Clinical Applications)
Edited by LESTER PACKER

VOLUME 191. Biomembranes (Part V: Cellular and Subcellular Transport: Epithelial Cells)
Edited by SIDNEY FLEISCHER AND BECCA FLEISCHER

VOLUME 192. Biomembranes (Part W: Cellular and Subcellular Transport: Epithelial Cells)
Edited by SIDNEY FLEISCHER AND BECCA FLEISCHER

VOLUME 193. Mass Spectrometry
Edited by JAMES A. MCCLOSKEY

VOLUME 194. Guide to Yeast Genetics and Molecular Biology
Edited by CHRISTINE GUTHRIE AND GERALD R. FINK

VOLUME 195. Adenylyl Cyclase, G Proteins, and Guanylyl Cyclase
Edited by ROGER A. JOHNSON AND JACKIE D. CORBIN

VOLUME 196. Molecular Motors and the Cytoskeleton
Edited by RICHARD B. VALLEE

VOLUME 197. Phospholipases
Edited by EDWARD A. DENNIS

VOLUME 198. Peptide Growth Factors (Part C)
Edited by DAVID BARNES, J. P. MATHER, AND GORDON H. SATO

VOLUME 199. Cumulative Subject Index Volumes 168–174, 176–194

VOLUME 200. Protein Phosphorylation (Part A: Protein Kinases: Assays, Purification, Antibodies, Functional Analysis, Cloning, and Expression)
Edited by TONY HUNTER AND BARTHOLOMEW M. SEFTON

VOLUME 201. Protein Phosphorylation (Part B: Analysis of Protein Phosphorylation, Protein Kinase Inhibitors, and Protein Phosphatases)
Edited by TONY HUNTER AND BARTHOLOMEW M. SEFTON

VOLUME 202. Molecular Design and Modeling: Concepts and Applications (Part A: Proteins, Peptides, and Enzymes)
Edited by JOHN J. LANGONE

VOLUME 203. Molecular Design and Modeling: Concepts and Applications (Part B: Antibodies and Antigens, Nucleic Acids, Polysaccharides, and Drugs)
Edited by JOHN J. LANGONE

VOLUME 204. Bacterial Genetic Systems
Edited by JEFFREY H. MILLER

VOLUME 205. Metallobiochemistry (Part B: Metallothionein and Related Molecules)
Edited by JAMES F. RIORDAN AND BERT L. VALLEE

VOLUME 206. Cytochrome P450
Edited by MICHAEL R. WATERMAN AND ERIC F. JOHNSON

VOLUME 207. Ion Channels
Edited by BERNARDO RUDY AND LINDA E. IVERSON

VOLUME 208. Protein–DNA Interactions
Edited by ROBERT T. SAUER

VOLUME 209. Phospholipid Biosynthesis
Edited by EDWARD A. DENNIS AND DENNIS E. VANCE

VOLUME 210. Numerical Computer Methods
Edited by LUDWIG BRAND AND MICHAEL L. JOHNSON

VOLUME 211. DNA Structures (Part A: Synthesis and Physical Analysis of DNA)
Edited by DAVID M. J. LILLEY AND JAMES E. DAHLBERG

VOLUME 212. DNA Structures (Part B: Chemical and Electrophoretic Analysis of DNA)
Edited by DAVID M. J. LILLEY AND JAMES E. DAHLBERG

VOLUME 213. Carotenoids (Part A: Chemistry, Separation, Quantitation, and Antioxidation)
Edited by LESTER PACKER

VOLUME 214. Carotenoids (Part B: Metabolism, Genetics, and Biosynthesis)
Edited by LESTER PACKER

VOLUME 215. Platelets: Receptors, Adhesion, Secretion (Part B)
Edited by JACEK J. HAWIGER

VOLUME 216. Recombinant DNA (Part G)
Edited by RAY WU

VOLUME 217. Recombinant DNA (Part H)
Edited by RAY WU

VOLUME 218. Recombinant DNA (Part I)
Edited by RAY WU

VOLUME 219. Reconstitution of Intracellular Transport
Edited by JAMES E. ROTHMAN

VOLUME 220. Membrane Fusion Techniques (Part A)
Edited by NEJAT DÜZGUÜNES

VOLUME 221. Membrane Fusion Techniques (Part B)
Edited by NEJAT DÜZGÜNES

VOLUME 222. Proteolytic Enzymes in Coagulation, Fibrinolysis, and Complement Activation (Part A: Mammalian Blood Coagulation Factors and Inhibitors)
Edited by LASZLO LORAND AND KENNETH G. MANN

VOLUME 223. Proteolytic Enzymes in Coagulation, Fibrinolysis, and Complement Activation (Part B: Complement Activation, Fibrinolysis, and Nonmammalian Blood Coagulation Factors)
Edited by LASZLO LORAND AND KENNETH G. MANN

VOLUME 224. Molecular Evolution: Producing the Biochemical Data
Edited by ELIZABETH ANNE ZIMMER, THOMAS J. WHITE, REBECCA L. CANN, AND ALLAN C. WILSON

VOLUME 225. Guide to Techniques in Mouse Development
Edited by PAUL M. WASSARMAN AND MELVIN L. DEPAMPHILIS

VOLUME 226. Metallobiochemistry (Part C: Spectroscopic and Physical Methods for Probing Metal Ion Environments in Metalloenzymes and Metalloproteins)
Edited by JAMES F. RIORDAN AND BERT L. VALLEE

VOLUME 227. Metallobiochemistry (Part D: Physical and Spectroscopic Methods for Probing Metal Ion Environments in Metalloproteins)
Edited by JAMES F. RIORDAN AND BERT L. VALLEE

VOLUME 228. Aqueous Two-Phase Systems
Edited by HARRY WALTER AND GÖTE JOHANSSON

VOLUME 229. Cumulative Subject Index Volumes 195–198, 200–227

VOLUME 230. Guide to Techniques in Glycobiology
Edited by WILLIAM J. LENNARZ AND GERALD W. HART

VOLUME 231. Hemoglobins (Part B: Biochemical and Analytical Methods)
Edited by JOHANNES EVERSE, KIM D. VANDEGRIFF, AND ROBERT M. WINSLOW

VOLUME 232. Hemoglobins (Part C: Biophysical Methods)
Edited by JOHANNES EVERSE, KIM D. VANDEGRIFF, AND ROBERT M. WINSLOW

VOLUME 233. Oxygen Radicals in Biological Systems (Part C)
Edited by LESTER PACKER

VOLUME 234. Oxygen Radicals in Biological Systems (Part D)
Edited by LESTER PACKER

VOLUME 235. Bacterial Pathogenesis (Part A: Identification and Regulation of Virulence Factors)
Edited by VIRGINIA L. CLARK AND PATRIK M. BAVOIL

VOLUME 236. Bacterial Pathogenesis (Part B: Integration of Pathogenic Bacteria with Host Cells)
Edited by VIRGINIA L. CLARK AND PATRIK M. BAVOIL

VOLUME 237. Heterotrimeric G Proteins
Edited by RAVI IYENGAR

VOLUME 238. Heterotrimeric G-Protein Effectors
Edited by RAVI IYENGAR

VOLUME 239. Nuclear Magnetic Resonance (Part C)
Edited by THOMAS L. JAMES AND NORMAN J. OPPENHEIMER

VOLUME 240. Numerical Computer Methods (Part B)
Edited by MICHAEL L. JOHNSON AND LUDWIG BRAND

VOLUME 241. Retroviral Proteases
Edited by LAWRENCE C. KUO AND JULES A. SHAFER

VOLUME 242. Neoglycoconjugates (Part A)
Edited by Y. C. LEE AND REIKO T. LEE

VOLUME 243. Inorganic Microbial Sulfur Metabolism
Edited by HARRY D. PECK, JR., AND JEAN LEGALL

VOLUME 244. Proteolytic Enzymes: Serine and Cysteine Peptidases
Edited by ALAN J. BARRETT

VOLUME 245. Extracellular Matrix Components
Edited by E. RUOSLAHTI AND E. ENGVALL

VOLUME 246. Biochemical Spectroscopy
Edited by KENNETH SAUER

VOLUME 247. Neoglycoconjugates (Part B: Biomedical Applications)
Edited by Y. C. LEE AND REIKO T. LEE

VOLUME 248. Proteolytic Enzymes: Aspartic and Metallo Peptidases
Edited by ALAN J. BARRETT

VOLUME 249. Enzyme Kinetics and Mechanism (Part D: Developments in Enzyme Dynamics)
Edited by DANIEL L. PURICH

VOLUME 250. Lipid Modifications of Proteins
Edited by PATRICK J. CASEY AND JANICE E. BUSS

VOLUME 251. Biothiols (Part A: Monothiols and Dithiols, Protein Thiols, and Thiyl Radicals)
Edited by LESTER PACKER

VOLUME 252. Biothiols (Part B: Glutathione and Thioredoxin; Thiols in Signal Transduction and Gene Regulation)
Edited by LESTER PACKER

VOLUME 253. Adhesion of Microbial Pathogens
Edited by RON J. DOYLE AND ITZHAK OFEK

VOLUME 254. Oncogene Techniques
Edited by PETER K. VOGT AND INDER M. VERMA

VOLUME 255. Small GTPases and Their Regulators (Part A: Ras Family)
Edited by W. E. BALCH, CHANNING J. DER, AND ALAN HALL

VOLUME 256. Small GTPases and Their Regulators (Part B: Rho Family)
Edited by W. E. BALCH, CHANNING J. DER, AND ALAN HALL

VOLUME 257. Small GTPases and Their Regulators (Part C: Proteins Involved in Transport)
Edited by W. E. BALCH, CHANNING J. DER, AND ALAN HALL

VOLUME 258. Redox-Active Amino Acids in Biology
Edited by JUDITH P. KLINMAN

VOLUME 259. Energetics of Biological Macromolecules
Edited by MICHAEL L. JOHNSON AND GARY K. ACKERS

VOLUME 260. Mitochondrial Biogenesis and Genetics (Part A)
Edited by GIUSEPPE M. ATTARDI AND ANNE CHOMYN

VOLUME 261. Nuclear Magnetic Resonance and Nucleic Acids
Edited by THOMAS L. JAMES

VOLUME 262. DNA Replication
Edited by JUDITH L. CAMPBELL

VOLUME 263. Plasma Lipoproteins (Part C: Quantitation)
Edited by WILLIAM A. BRADLEY, SANDRA H. GIANTURCO, AND JERE P. SEGREST

VOLUME 264. Mitochondrial Biogenesis and Genetics (Part B)
Edited by GIUSEPPE M. ATTARDI AND ANNE CHOMYN

VOLUME 265. Cumulative Subject Index Volumes 228, 230–262

VOLUME 266. Computer Methods for Macromolecular Sequence Analysis
Edited by RUSSELL F. DOOLITTLE

VOLUME 267. Combinatorial Chemistry
Edited by JOHN N. ABELSON

VOLUME 268. Nitric Oxide (Part A: Sources and Detection of NO; NO Synthase)
Edited by LESTER PACKER

VOLUME 269. Nitric Oxide (Part B: Physiological and Pathological Processes)
Edited by LESTER PACKER

VOLUME 270. High Resolution Separation and Analysis of Biological Macromolecules (Part A: Fundamentals)
Edited by BARRY L. KARGER AND WILLIAM S. HANCOCK

VOLUME 271. High Resolution Separation and Analysis of Biological Macromolecules (Part B: Applications)
Edited by BARRY L. KARGER AND WILLIAM S. HANCOCK

VOLUME 272. Cytochrome P450 (Part B)
Edited by ERIC F. JOHNSON AND MICHAEL R. WATERMAN

VOLUME 273. RNA Polymerase and Associated Factors (Part A)
Edited by SANKAR ADHYA

VOLUME 274. RNA Polymerase and Associated Factors (Part B)
Edited by SANKAR ADHYA

VOLUME 275. Viral Polymerases and Related Proteins
Edited by LAWRENCE C. KUO, DAVID B. OLSEN, AND STEVEN S. CARROLL

VOLUME 276. Macromolecular Crystallography (Part A)
Edited by CHARLES W. CARTER, JR., AND ROBERT M. SWEET

VOLUME 277. Macromolecular Crystallography (Part B)
Edited by CHARLES W. CARTER, JR., AND ROBERT M. SWEET

VOLUME 278. Fluorescence Spectroscopy
Edited by LUDWIG BRAND AND MICHAEL L. JOHNSON

VOLUME 279. Vitamins and Coenzymes (Part I)
Edited by DONALD B. MCCORMICK, JOHN W. SUTTIE, AND CONRAD WAGNER

VOLUME 280. Vitamins and Coenzymes (Part J)
Edited by DONALD B. MCCORMICK, JOHN W. SUTTIE, AND CONRAD WAGNER

VOLUME 281. Vitamins and Coenzymes (Part K)
Edited by DONALD B. MCCORMICK, JOHN W. SUTTIE, AND CONRAD WAGNER

VOLUME 282. Vitamins and Coenzymes (Part L)
Edited by DONALD B. MCCORMICK, JOHN W. SUTTIE, AND CONRAD WAGNER

VOLUME 283. Cell Cycle Control
Edited by WILLIAM G. DUNPHY

VOLUME 284. Lipases (Part A: Biotechnology)
Edited by BYRON RUBIN AND EDWARD A. DENNIS

VOLUME 285. Cumulative Subject Index Volumes 263, 264, 266–284, 286–289

VOLUME 286. Lipases (Part B: Enzyme Characterization and Utilization)
Edited by BYRON RUBIN AND EDWARD A. DENNIS

VOLUME 287. Chemokines
Edited by RICHARD HORUK

VOLUME 288. Chemokine Receptors
Edited by RICHARD HORUK

VOLUME 289. Solid Phase Peptide Synthesis
Edited by GREGG B. FIELDS

VOLUME 290. Molecular Chaperones
Edited by GEORGE H. LORIMER AND THOMAS BALDWIN

VOLUME 291. Caged Compounds
Edited by GERARD MARRIOTT

VOLUME 292. ABC Transporters: Biochemical, Cellular, and Molecular Aspects
Edited by SURESH V. AMBUDKAR AND MICHAEL M. GOTTESMAN

VOLUME 293. Ion Channels (Part B)
Edited by P. MICHAEL CONN

VOLUME 294. Ion Channels (Part C)
Edited by P. MICHAEL CONN

VOLUME 295. Energetics of Biological Macromolecules (Part B)
Edited by GARY K. ACKERS AND MICHAEL L. JOHNSON

VOLUME 296. Neurotransmitter Transporters
Edited by SUSAN G. AMARA

VOLUME 297. Photosynthesis: Molecular Biology of Energy Capture
Edited by LEE MCINTOSH

VOLUME 298. Molecular Motors and the Cytoskeleton (Part B)
Edited by RICHARD B. VALLEE

VOLUME 299. Oxidants and Antioxidants (Part A)
Edited by LESTER PACKER

VOLUME 300. Oxidants and Antioxidants (Part B)
Edited by LESTER PACKER

VOLUME 301. Nitric Oxide: Biological and Antioxidant Activities (Part C)
Edited by LESTER PACKER

VOLUME 302. Green Fluorescent Protein
Edited by P. MICHAEL CONN

VOLUME 303. cDNA Preparation and Display
Edited by SHERMAN M. WEISSMAN

VOLUME 304. Chromatin
Edited by PAUL M. WASSARMAN AND ALAN P. WOLFFE

VOLUME 305. Bioluminescence and Chemiluminescence (Part C)
Edited by THOMAS O. BALDWIN AND MIRIAM M. ZIEGLER

VOLUME 306. Expression of Recombinant Genes in Eukaryotic Systems
Edited by JOSEPH C. GLORIOSO AND MARTIN C. SCHMIDT

VOLUME 307. Confocal Microscopy
Edited by P. MICHAEL CONN

VOLUME 308. Enzyme Kinetics and Mechanism (Part E: Energetics of Enzyme Catalysis)
Edited by DANIEL L. PURICH AND VERN L. SCHRAMM

VOLUME 309. Amyloid, Prions, and Other Protein Aggregates
Edited by RONALD WETZEL

VOLUME 310. Biofilms
Edited by RON J. DOYLE

VOLUME 311. Sphingolipid Metabolism and Cell Signaling (Part A)
Edited by ALFRED H. MERRILL, JR., AND YUSUF A. HANNUN

VOLUME 312. Sphingolipid Metabolism and Cell Signaling (Part B)
Edited by ALFRED H. MERRILL, JR., AND YUSUF A. HANNUN

VOLUME 313. Antisense Technology (Part A: General Methods, Methods of Delivery, and RNA Studies)
Edited by M. IAN PHILLIPS

VOLUME 314. Antisense Technology (Part B: Applications)
Edited by M. IAN PHILLIPS

VOLUME 315. Vertebrate Phototransduction and the Visual Cycle (Part A)
Edited by KRZYSZTOF PALCZEWSKI

VOLUME 316. Vertebrate Phototransduction and the Visual Cycle (Part B)
Edited by KRZYSZTOF PALCZEWSKI

VOLUME 317. RNA–Ligand Interactions (Part A: Structural Biology Methods)
Edited by DANIEL W. CELANDER AND JOHN N. ABELSON

VOLUME 318. RNA–Ligand Interactions (Part B: Molecular Biology Methods)
Edited by DANIEL W. CELANDER AND JOHN N. ABELSON

VOLUME 319. Singlet Oxygen, UV-A, and Ozone
Edited by LESTER PACKER AND HELMUT SIES

VOLUME 320. Cumulative Subject Index Volumes 290–319

VOLUME 321. Numerical Computer Methods (Part C)
Edited by MICHAEL L. JOHNSON AND LUDWIG BRAND

VOLUME 322. Apoptosis
Edited by JOHN C. REED

VOLUME 323. Energetics of Biological Macromolecules (Part C)
Edited by MICHAEL L. JOHNSON AND GARY K. ACKERS

VOLUME 324. Branched-Chain Amino Acids (Part B)
Edited by ROBERT A. HARRIS AND JOHN R. SOKATCH

VOLUME 325. Regulators and Effectors of Small GTPases (Part D: Rho Family)
Edited by W. E. BALCH, CHANNING J. DER, AND ALAN HALL

VOLUME 326. Applications of Chimeric Genes and Hybrid Proteins (Part A: Gene Expression and Protein Purification)
Edited by JEREMY THORNER, SCOTT D. EMR, AND JOHN N. ABELSON

VOLUME 327. Applications of Chimeric Genes and Hybrid Proteins (Part B: Cell Biology and Physiology)
Edited by JEREMY THORNER, SCOTT D. EMR, AND JOHN N. ABELSON

VOLUME 328. Applications of Chimeric Genes and Hybrid Proteins (Part C: Protein-Protein Interactions and Genomics)
Edited by JEREMY THORNER, SCOTT D. EMR, AND JOHN N. ABELSON

VOLUME 329. Regulators and Effectors of Small GTPases (Part E: GTPases Involved in Vesicular Traffic)
Edited by W. E. BALCH, CHANNING J. DER, AND ALAN HALL

VOLUME 330. Hyperthermophilic Enzymes (Part A)
Edited by MICHAEL W. W. ADAMS AND ROBERT M. KELLY

VOLUME 331. Hyperthermophilic Enzymes (Part B)
Edited by MICHAEL W. W. ADAMS AND ROBERT M. KELLY

VOLUME 332. Regulators and Effectors of Small GTPases (Part F: Ras Family I)
Edited by W. E. BALCH, CHANNING J. DER, AND ALAN HALL

VOLUME 333. Regulators and Effectors of Small GTPases (Part G: Ras Family II)
Edited by W. E. BALCH, CHANNING J. DER, AND ALAN HALL

VOLUME 334. Hyperthermophilic Enzymes (Part C)
Edited by MICHAEL W. W. ADAMS AND ROBERT M. KELLY

VOLUME 335. Flavonoids and Other Polyphenols (in preparation)
Edited by LESTER PACKER

VOLUME 336. Microbial Growth in Biofilms (Part A: Developmental and Molecular Biological Aspects) (in preparation)
Edited by RON J. DOYLE

VOLUME 337. Microbial Growth in Biofilms (Part B: Special Environments and Physicochemical Aspects) (in preparation)
Edited by RON J. DOYLE

VOLUME 338. Nuclear Magnetic Resonance of Biological Macromolecules (Part A) (in preparation)
Edited by THOMAS L. JAMES, VOLKER DÖTSCH, AND ULI SCHMITZ

VOLUME 339. Nuclear Magnetic Resonance of Biological Macromolecules (Part B) (in preparation)
Edited by THOMAS L. JAMES, VOLKER DÖTSCH, AND ULI SCHMITZ

VOLUME 340. Drug-Nucleic Acid Interactions (in preparation)
Edited by JONATHAN B. CHAIRES AND MICHAEL J. WARING

VOLUME 341. Ribonucleases (Part A) (in preparation)
Edited by ALLEN W. NICHOLSON

VOLUME 342. Ribonucleases (Part B) (in preparation)
Edited by ALLEN W. NICHOLSON

VOLUME 343. G Protein Pathways (Part A: Receptors) (in preparation)
Edited by RAVI SYENGAR AND JOHN D. HILDEBRANDT

VOLUME 344. G Protein Pathways (Part B: Effector Mechanisms) (in preparation)
Edited by RAVI SYENGAR AND JOHN D. HILDEBRANDT

Section I

Redox and Thiol-Dependent Proteins

[1] Ferredoxin and Related Enzymes from *Sulfolobus*

By TOSHIO IWASAKI and TAIRO OSHIMA

Introduction

The majority of thermophilic archaea are anaerobic organisms, because oxygen is often scarce in their environments.[1–3] One of the characteristic features in the central metabolic pathways of both anaerobic and aerobic archaea is the involvement in electron transport of ferredoxins.[4–9] Ferredoxins are simple iron–sulfur proteins with prosthetic groups composed of iron and sulfur atoms and function as intracellular electron carriers. The physiological significance of bacterial-type ferredoxins in several aerobic and thermoacidophilic archaea, such as *Sulfolobus* and *Thermoplasma*, was first recognized by Kerscher et al., when it was demonstrated that ferredoxins are an effective electron acceptor of a coenzyme A-acylating 2-oxoacid:ferredoxin oxidoreductase.[5] 2-Oxoacid:ferredoxin oxidoreductase is a key enzyme of the tricarboxylic acid cycle and of coenzyme A-dependent pyruvate oxidation in aerobic archaea[4,5,9–13] and takes the place of NAD^+-dependent 2-oxoacid dehydrogenase multienzyme complex in bacteria and eukarya.[4,6,9,12]

Ferredoxins from aerobic and thermoacidophilic archaea are characterized by relatively higher molecular masses for bacterial-type ferredoxins (~12–16 kDa) because of an N-terminal extension region and central loop attached to the ferredoxin core fold. They have low midpoint redox potential $[3Fe-4S]^{1+,0}$ and $[4Fe-4S]^{2+,1+}$ clusters, characteristic electronic spectra with Fe–S charge transfer bands at around 300 nm and 400 nm, and typical electron paramagnetic resonance (EPR) signals in both oxidized and reduced states. The most unusual feature of these ferredoxins is the presence of an isolated zinc center,[14–17] and hence they

---

[1] C. R. Woese, O. Kandler, and M. L. Wheelis, *Proc. Natl. Acad. Sci. U.S.A.* **87,** 4576 (1990).
[2] G. J. Olsen, C. R. Woese, and R. Overbeek, *J. Bacteriol.* **176,** 1 (1994).
[3] K. O. Stetter, *ASM News* **61,** 285 (1995).
[4] L. Kerscher and D. Oesterhelt, *FEBS Lett.* **83,** 197 (1977).
[5] L. Kerscher, S. Nowitzki, and D. Oesterhelt, *Eur. J. Biochem.* **128,** 223 (1982).
[6] L. Kerscher and D. Oesterhelt, *Trends Biochem. Sci.* **7,** 371 (1982).
[7] M. W. W. Adams, *Annu. Rev. Microbiol.* **47,** 627 (1993).
[8] M. W. W. Adams, *FEMS Microbiol. Rev.* **15,** 261 (1994).
[9] T. Iwasaki, T. Wakagi, Y. Isogai, K. Tanaka, T. Iizuka, and T. Oshima, *J. Biol. Chem.* **269,** 29444 (1994).
[10] L. Kerscher and D. Oesterhelt, *Eur. J. Biochem.* **116,** 587 (1981).
[11] L. Kerscher and D. Oesterhelt, *Eur. J. Biochem.* **116,** 595 (1981).
[12] T. Iwasaki, T. Wakagi, and T. Oshima, *J. Biol. Chem.* **270,** 17878 (1995).
[13] Q. Zhang, T. Iwasaki, T. Wakagi, and T. Oshima, *J. Biochem.* **120,** 587 (1996).
[14] T. Fujii, Y. Hata, T. Wakagi, N. Tanaka, and T. Oshima, *Nat. Struct. Biol.* **3,** 834 (1996).

are called the "zinc-containing ferredoxins." This chapter focuses on purification and some structural and functional properties of archaeal zinc-containing ferredoxins and several related metalloproteins: 2-oxoacid : ferredoxin oxidoreductase, red iron–sulfur flavoprotein, and sulredoxin.

Archaeal Zinc-Containing Ferredoxins

Purification

Organism and Cell Culture. Sulfolobus sp. strain 7 (originally named *S. acidocaldarius* strain 7 isolated from Beppu hot springs, Kyushu, Japan[18]) is a strictly aerobic and thermoacidophilic crenarchaeote that grows optimally at 75–80° and at pH 2.5–3.0.[9] The 16S rRNA sequence and biochemical analysis of the isolate suggested that it is a novel crenarchaeote, belonging to the genus *Sulfolobus*.[19]

Sulfolobus sp. strain 7 (JCM 10545) is cultivated aerobically and chemoheterotrophically in the shaking culture at pH 2.5–3 and 75–80°, and harvested in the late exponential phase of the growth as previously described.[9,20]

Preparation and Fractionation of Cytosol Fraction. Frozen cells are thawed and suspended in 100 mM Tris-HCl buffer, pH 7.3, containing 15 mM EDTA and 0.5 mM phenylmethylsulfonyl fluoride (PMSF) at a final concentration of 4 ml/g (wet weight) of the cells. The pH of the suspension is adjusted with NaOH because of the acidity of the growth media for thermoacidophilic archaea.[21] The suspension is passed twice through a French pressure cell (Otake Works, Tokyo) at 1500 kg/cm^2, and membranes are precipitated by ultracentrifugation with a Beckman 45Ti rotor at 130,000 g for 100 min at 15°. The precipitates are composed of two layers. The upper soft layer (membrane fraction) is carefully collected and suspended in 50 mM Tris-HCl buffer, pH 7.3, containing 10 mM EDTA and 0.5 mM PMSF. This mixture is ultracentrifuged at 130,000 g for 70 min at 15°. All supernatant fractions thus obtained are combined and used as the cytosol fraction.[9]

The archaeal zinc-containing ferredoxin and several related metalloproteins can be purified from the same batch of the cells, as described below.[9,12,13,22] For purification of the cognate 2-oxoacid:ferredoxin oxidoreductase[13] and red

---

[15] T. Iwasaki, T. Suzuki, T. Kon, T. Imai, A. Urushiyama, D. Ohmori, and T. Oshima, *J. Biol. Chem.* **272**, 3453 (1997).

[16] T. Fujii, Y. Hata, M. Oozeki, H. Moriyama, T. Wakagi, N. Tanaka, and T. Oshima, *Biochemistry* **36**, 1505 (1997).

[17] N. J. Cosper, C. M. V. Stålhandske, H. Iwasaki, T. Oshima, R. A. Scott, and T. Iwasaki, *J. Biol. Chem.* **274**, 23160 (1999).

[18] K. Inatomi, M. Ohba, and T. Oshima, *Chem. Lett.*, 1191 (1983).

[19] T. Suzuki, T. Iwasaki, T. Uzawa, K. Hara, N. Nemoto, T. Kon, T. Ueki, A. Yamagishi, and T. Oshima, submitted for publication (2000).

[20] T. Iwasaki, K. Matsuura, and T. Oshima, *J. Biol. Chem.* **270**, 30881 (1995).

[21] T. D. Brock, K. M. Brock, R. T. Belley, and R. L. Weiss, *Arch. Mikrobiol.* **84**, 54 (1972).

[22] T. Iwasaki, T. Isogai, T. Iizuka, and T. Oshima, *J. Bacteriol.* **177**, 2576 (1995).

iron–sulfur flavoprotein of unknown function,[12] the cytosol fraction is dialyzed against 30 mM Tris-HCl buffer, pH 8.5, containing 5 mM EDTA and 1 mM PMSF, to remove salts (contaminated in the growth media) and to enhance the recovery of these enzymes from a DEAE-Sephacel column chromatography step (see below). This treatment is not required when only zinc-containing ferredoxin[9,23] and sulredoxin[22] are the subjects of interest. Unless otherwise specified, the following purification steps are carried out at room temperature.

Purification of Zinc-Containing Ferredoxin from Sulfolobus sp. Strain 7. The isolation of ferredoxin from the aerobic thermoacidophilic crenarchaeote, *Sulfolobus acidocaldarius,* was described by Kerscher *et al.*[5] This procedure was adapted to purify ferredoxin[9,23] and related metalloproteins from the *Sulfolobus* sp. strain 7 cells.

The starting material is a frozen cell paste [typically 50–150 g (wet weight) of the cells]. The cytosol fraction prepared as described above is applied to a DEAE-Sephacel column (3.0 × 28 cm; Amersham Pharmacia Biotech, Piscataway, NJ) equilibrated with 30 mM Tris-HCl buffer, pH 7.3, containing 1 mM EDTA and 1 mM PMSF. The column is washed with 500 ml of the equilibration buffer, and the redox proteins bound to the column are eluted with a 1600-ml linear gradient of NaCl (0–400 mM) in the equilibration buffer, in the following order: red iron–sulfur flavoprotein[12], 2-oxoacid:ferredoxin oxidoreductase,[13] and zinc-containing ferredoxin (Fd-A).[9,23,24] Sulredoxin[22] is found in the flow-through fraction. In large-scale preparations, a small amount of a minor ferredoxin species (Fd-B) is also eluted from the column shortly after Fd-A.[24] Biochemical and spectroscopic characterization suggested that Fd-B is a 6Fe-containing form of zinc-containing ferredoxin, which can also be prepared from the 7Fe form (Fd-A) *in vitro.*[23,23a]

The zinc-containing ferredoxin fraction, recognized by the brown color, is diluted twofold with distilled water. This is applied onto a Bio-Rad (Hercules, CA) hydroxylapatite HTP column (equilibrated with Milli-Q water), is equilibrated with Milli-Q-purified distilled water, and is eluted by 2 mM potassium phosphate buffer, pH 6.8 (alternatively, a Toyopearl HW-55C hydrophobic column chromatography (Tosoh Corp, Japan) may be used at this step[9,25]). Finally, the pooled fraction is concentrated by pressure filtration through an Amicon (Danvers, MA) YM3 membrane, and passed through a preparative Sephadex G-50 gel filtration column (2.5 × 75 cm; Amersham Pharmacia Biotech) equilibrated with 80 mM potassium phosphate buffer, pH 6.8.

---

[23] T. Iwasaki and T. Oshima, *FEBS Lett.* **417,** 223 (1997).
[23a] T. Iwasaki, E. Watanabe, D. Ohmori, T. Imai, A. Urushiyama, M. Akiyama, Y. Hayashi-Iwasaki, N. J. Cosper, and R. A. Scott, *J. Biol. Chem.* **275,** 25391 (2000).
[24] T. Iwasaki, T. Fujii, T. Wakagi, and T. Oshima, *Biochem. Biophys. Res. Commun.* **206,** 563 (1995).
[25] T. Fujii, H. Moriyama, A. Takenaka, N. Tanaka, T. Wakagi, and T. Oshima, *J. Biochem.* **110,** 472 (1991).

Zinc-containing ferredoxin thus obtained gives a single band on 20% analytical polyacrylamide gel electrophoresis (PAGE) and has a purity index (A_{408}/A_{280}) of 0.70.[9] Approximately 40 mg of purified ferredoxin is routinely obtained from about 150 g (wet weight) of the cells, and this is stored in aliquots at $-80°$.

Crystallization. The crystallization procedure is taken from that described by Fujii et al.[16,25] For crystallization experiments, the *Sulfolobus* sp. ferredoxin solution obtained from a preparative Sephadex G-50 gel filtration column (Amersham Pharmacia Biotech) is concentrated by pressure filtration through an Amicon YM3 or YM10 membrane at $4°$ and made to 5 mg/ml in 0.5 M Tris–maleate–NaOH buffer, pH 5.0, containing 1% 2-methyl-2,4-pentanediol. Crystals suitable for X-ray diffraction analysis are obtained by a batch method performed under aerobic conditions.[25] Fine-powdered ammonium sulfate is slowly added to 300 μl of 5 mg/ml protein solution until the turbidity is observed to persist (1.9–2.1 M). The crystallization solution is stored at $37°$ in an incubator. Dark brown crystals with appropriate dimensions of $0.3 \times 0.3 \times 0.5$ mm are obtained in 3–5 weeks. Fujii et al.[16] reported that reproducibility of the crystallization is enhanced by seeding a drop of the mother liquor containing microcrystals into the crystallization solution just before the crystallization begins.

The dark brown crystals thus obtained have been reported to belong to the tetragonal space group $P4_32_12$, with the cell dimensions of $a = b = 50.12$ Å and $c = 69.52$ Å, and contain one zinc-containing ferredoxin molecule per asymmetric unit ($V_m = 1.88$ Å^3/Da).[25] The diffraction pattern extends beyond 2.0-Å resolution. The crystal structure (see below) has been determined by combining multiple isomorphous replacement and multiple anomalous dispersion (MAD) data collected at the Fe absorption edge, and refined at 2.0-Å resolution to a crystallographic R value of 0.173.[14,16]

Purification of Zinc-Containing Ferredoxin from Thermoplasma acidophilum. The isolation of ferredoxin from the aerobic thermoacidophilic euryarchaeote, *Thermoplasma acidophilum*, was described by Kerscher et al.[5] Similar results were obtained with *T. acidophilum* strain HO-62,[15] which was isolated from hot sulfur springs at Ohwakudani solfataric field in Hakone, Japan, and cultivated at pH 1.8 and at $56°$ as described by Yasuda et al.[26] The ferredoxin is separated from other redox proteins on a DEAE-Sepharose Fast Flow column (Amersham Pharmacia Biotech) connected to a Pharmacia Fast Protein Liquid Chromatography (FPLC) system. Chromatography gives well-resolved peaks, and subsequent preparative Sephadex G-50 column chromatography (Amersham Pharmacia Biotech) is sufficient to complete the purification. Purified ferredoxin, which shows a single band on 20% analytical polyacrylamide gel electrophoresis

---

[26] M. Yasuda, H. Oyaizu, A. Yamagishi, and T. Oshima, *Appl. Environ. Microbiol.* **61**, 3482 (1995).

in the absence of sodium dodecyl sulfate, has a purity index (A_{402}/A_{280}) of 0.53[5] and has been shown to be a zinc-containing ferredoxin.[15,17]

General Properties of Archaeal Zinc-Containing Ferredoxins

Zinc-containing ferredoxin is the most abundant ferredoxin in chemoheterotrophically grown *Sulfolobus* sp. strain 7[9,24] and *T. acidophilum*.[5,15] The types and spectroscopic properties of the iron–sulfur clusters and the isolated zinc center are very similar in archaeal zinc-containing ferredoxins (Table I). Zinc-containing ferredoxin from *Sulfolobus* sp. strain 7 (103 amino acids, 7 cysteines)[27] contains one $[3Fe-4S]^{1+,0}$ cluster (cluster I) with a midpoint redox potential of −280 mV, one $[4Fe-4S]^{2+,1+}$ cluster (cluster II) with a midpoint redox potential of −530 mV, and a tetragonal zinc center.[9,14,17,23,23a] *T. acidophilum* zinc-containing ferredoxin (142 amino acids, 9 cysteines)[17,28] is slightly larger than the ferredoxins from *Sulfolobus* and *Acidianus* species, but also contains one $[3Fe-4S]^{1+,0}$ cluster, one $[4Fe-4S]^{2+,1+}$ cluster, and a tetragonal zinc center.[15,17]

Sequence Features. The overall protein fold of archaeal zinc-containing ferredoxins is largely asymmetric because of the presence of a long N-terminal extension and the insertion of central loop region, as compared with those of regular bacterial-type ferredoxins (Figs. 1 and 2). However, the ferredoxin core fold shows the strict conservation of a pseudo-twofold symmetry with respect to the local two iron–sulfur cluster binding sites.[9,29,30] The deduced amino acid sequence of *Sulfolobus* sp. ferredoxin[27] contains seven cysteine residues. In spite of the presence of one $[3Fe-4S]^{1+,0}$ cluster and one $[4Fe-4S]^{2+,1+}$ cluster in purified proteins,[9,17,23] the distribution of the conserved cysteine ligand residues in archaeal zinc-containing ferredoxins is similar to those of regular 8Fe-containing dicluster ferredoxins, except for the presence of an aspartate residue (Asp-48 in *Sulfolobus* sp. ferredoxin and Asp-71 in *T. acidophilum* ferredoxin) in place of cysteine (Fig. 1).[9,17] It has been reported that the pattern of the hyperfine-shifted resonances of cluster II of *Acidianus ambivalens* zinc-containing ferredoxin resembles those of 4Fe- and 8Fe-containing ferredoxins, rather than those of other 7Fe-containing ferredoxins,[31,32] which may be due to the unique cysteine-containing sequence motif of archaeal zinc-containing ferredoxins.

---

[27] T. Wakagi, T. Fujii, and T. Oshima, *Biochem. Biophys. Res. Commun.* **225**, 489 (1996).
[28] S. Wakabayashi, N. Fujimoto, K. Wada, H. Matsubara, L. Kerscher, and D. Oesterhelt, *FEBS Lett.* **162**, 21 (1983).
[29] M. Bruschi and F. Guerlesquin, *FEMS Microbiol. Rev.* **54**, 155 (1988).
[30] H. Matsubara and K. Saeki, *Adv. Inorg. Chem.* **38**, 223 (1992).
[31] D. Bentrop, I. Bertini, C. Luchinat, J. Mendes, M. Piccioli, and M. Teixeira, *Eur. J. Biochem.* **236**, 92 (1996).
[32] I. Bertini, A. Dikiy, C. Luchinat, R. Macinai, M. S. Viezzoli, and M. Vincenzini, *Biochemistry* **36**, 3570 (1997).

TABLE I
ZINC-CONTAINING FERREDOXINS FROM THERMOACIDOPHILIC ARCHAEA

Zinc-containing ferredoxin	Number of amino acids in mature protein	Cys	Cluster type	Redox potential (mV)	Coordination environment of isolated zinc site
Sulfolobus sp. strain 7[9,14,16,17,23]	103	7	[3Fe–4S]$^{1+,0}$ [4Fe–4S]$^{2+,1+}$[b]	−280 −530[b]	X-ray structure[a] Zn–K edge EXAFS[b]
Sulfolobus acidocaldarius strain DSM 639[5,34,37]	103	7	[3Fe–4S]$^{1+,0}$ [4Fe–4S]$^{2+,1+}$	−275 −529	nd[d]
Acidianus ambivalens strain DSM 3772[31,33]	103	7[c]	[3Fe–4S]$^{1+,0}$ [4Fe–4S]$^{2+,1+}$	−270 −540	nd
Thermoplasma acidophilum strain HO-62[15,17]	142	9	[3Fe–4S]$^{1+,0}$ [4Fe–4S]$^{2+,1+}$	nd nd	Zn–K edge EXAFS

[a] The three-dimensional structure of *Sulfolobus* sp. ferredoxin contains two [3Fe–4S] clusters[14,16] and corresponds to the 6Fe form.[23]
[b] Determined with the as-isolated zinc-containing ferredoxin which contains one [3Fe–4S] cluster, one [4Fe–4S] cluster, and one isolated zinc center.[9,17,23]
[c] ^{1}H NMR evidence for the complete cysteinyl ligation to two iron–sulfur clusters has been reported.[31]
[d] nd, Not determined.

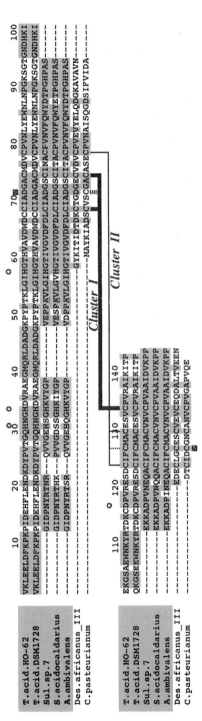

FIG. 1. Multiple amino acid sequence alignments of archaeal zinc-containing ferredoxins (*shaded*: T. acid.HO-62, *Thermoplasma acidophilum* strain HO-62 [N. J. Cosper, C. M. V. Stålhandske, H. Iwasaki, T. Oshima, R. A. Scott, and T. Iwasaki, *J. Biol. Chem.* **274**, 23160 (1999)]; T.acid.DSM1728, *Thermoplasma acidophilum* strain DSM1728 [S. Wakabayashi, N. Fujimoto, K. Wada, H. Matsubara, L. Kerscher, and D. Oesterheit, *FEBS Lett.* **162**, 21 (1983)]; Sul.sp.7, *Sulfolobus* sp. strain 7 [T. Wakagi, T. Fujii, and T. Oshima, *Biochem. Biophys. Res. Commun.* **225**, 489 (1996)]; S.acidocaldarius, *Sulfolobus acidocaldarius* strain DSM639 [Y. Minami, S. Wakabayashi, K. Wada, H. Matsubara, L. Kerscher, and D. Oesterhelt, *J. Biochem.* **97**, 745 (1985); J. L. Breton, J. L. C. Duff, J. N. Butt, F. A. Armstrong, S. J. George, Y. Pétillot, E. Forest, G. Schäfer, and A. J. Thomson, *Eur. J. Biochem.* **233**, 937 (1995)]; A.ambivalens, *Acidianus ambivalens* (f. *Desulfurolobus ambivalens*) strain DSM3772 [C. M. Gomes, A. Faria, J. C. Carita, J. Mendes, M. Regalla, P. Chicau, H. Huber, K. O. Stetter, and M. Teixeira, *J. Biol. Inorg. Chem.* **3**, 499 (1998)]) and bacterial dicluster-type ferredoxins (Des.africanus_III, *Desulfovibrio africanus* ferredoxin III [G. E. Bovier-Lapierre, M. Bruschi, J. J. Bonicel, and E. C. Hatchikian, *Biochim. Biophys. Acta* **913**, 20 (1987)]; C.pasteurianum, *Clostridium pasteurianum* [M. C. Graves, G. T. Mullenbach, and J. C. Rabinowitz, *Proc. Natl. Acad. Sci. U.S.A.* **82**, 1653 (1985)]). Conserved ligands to an isolated zinc site and two iron–sulfur clusters are marked, and a N^ϵ-methyl-L-lysine residue (at position 29 in *Sulfolobus* sp.–*S. acidocaldarius*; and *A. ambivalens* ferredoxins) is underlined. It should be noted that Asp-48 and Cys-86 do not serve as ligands to the [3Fe–4S] clusters in the X-ray crystal structure of *Sulfolobus* sp. zinc-containing ferredoxin [T. Fujii, Y. Hata, M. Oozeki, H. Moriyama, T. Wakagi, N. Tanaka, and T. Oshima, *Biochemistry* **36**, 1505 (1997)].

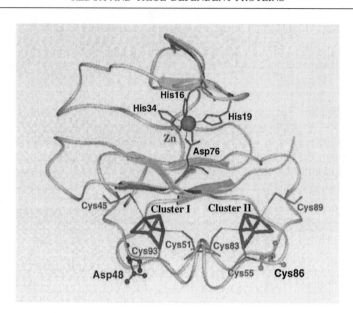

FIG. 2. The 2.0-Å resolution crystal structure of the 6Fe-containing form of zinc-containing ferredoxin from *Sulfolobus* sp. strain (PDB entry, 1XER [T. Fujii, Y. Hata, T. Wakagi, N. Tanaka, and T. Oshima, *Nat. Struct. Biol.* **3**, 834 (1996); T. Fujii, Y. Hata, M. Oozeki, H. Moriyama, T. Wakagi, N. Tanaka, and T. Oshima, *Biochemistry* **36**, 1505 (1997)]). Important residues are labeled. The model was drawn using the Insight II software (Molecular Simulations Inc.).

Interestingly, zinc-containing ferredoxins have been found specifically from the aerobic and thermoacidophilic archaea.[5,9,15,33,34] The N-terminal extension region with the consensus histidine-rich motif is rarely found in ferredoxins from euryarchaeotes, except for the thermoacidophilic *Thermoplasma*[15,17] an unexpected result based on the universal 16S rRNA-based sequence tree.[2,3] An analogous observation has been reported for functionally equivalent ferredoxins of aerobic and extremely halophilic euryarchaeotes,[4,35] which contain a single plant-type [2Fe-2S] cluster and exhibit the amino acid sequence similarity to those of the extremely halophilic cyanobacteria.[36] Based on these observations, we postulated that early zinc-containing ferredoxins might have evolved as an 8Fe-containing low-potential two-electron carrier, to which the N-terminal extension and central loop regions were attached in the later stage of molecular evolution, presumably shortly after divergence of the archaeal domain.[17]

[33] M. Teixeira, R. Batista, A. P. Campos, C. Gomes, J. Mendes, I. Pacheco, S. Anemüller, and W. R. Hagen, *Eur. J. Biochem.* **227**, 322 (1995).
[34] J. L. Breton, J. L. C. Duff, J. N. Butt, F. A. Armstrong, S. J. George, Y. Pétillot, E. Forest, G. Schäfer, and A. J. Thomson, *Eur. J. Biochem.* **233**, 937 (1995).
[35] L. Kerscher, D. Oesterhelt, R. Cammack, and D. O. Hall, *Eur. J. Biochem.* **71**, 101 (1976).
[36] F. Pfeifer, J. Griffig, and D. Oesterhelt, *Mol. Gen. Genet.* **239**, 66 (1993).

Iron–Sulfur Clusters. The spectroscopic data suggest that the as-isolated zinc-containing ferredoxin from *Sulfolobus* sp. strain 7 contains one $[3Fe–4S]^{1+,0}$ cluster (cluster I; $E_0 = -280$ mV) and one $[4Fe-4S]^{2+,1+}$ cluster (cluster II; $E_0 = -530$ mV) (Table I).[9,23] Very similar results have also been reported for other zinc-containing ferredoxins of Sulfolobales.[33,34] Cluster I is selectively reduced by the cognate 2-oxoacid:ferredoxin oxidoreductase during the steady-state turnover in the presence of 2-oxoglutarate and coenzyme A, whereas the bulk of cluster II remains in the oxidized state.[9] This suggests that the cluster I of *Sulfolobus* sp. ferredoxin plays a redox role in the physiological electron transfer. The X-ray structure analysis indicates that cluster I is bound to the polypeptide chain by three cysteinyl residues, Cys-45, Cys-51 and Cys-93 (Fig. 2).[16] Residue Asp-48 (a potential ligand for a fourth site, if the cluster I were a [4Fe–4S] cluster) is not bound and its carboxyl Oδ1 connects to the side chain Oγ and the main chain amide NH of Ser-50 by hydrogen bonds, away from the [3Fe-4S] cluster I, in the crystal structure. It has been reported that a one-electron reduced $[3Fe-4S]^0$ cluster I of the *Sulfolobales* ferredoxins undergoes a one-proton uptake reaction,[33,34] and that further two-electron hyperreduction, which also involves uptake of protons, reversibly produces a stable, hyperreduced $[3Fe-4S]^{2-}$ species containing the formal equivalent of three ferrous ions.[37]

An unexpected result from X-ray structural analysis is that two [3Fe-4S] clusters and one isolated zinc center are observed in the three-dimensional structure determined at 2.0-Å resolution; cluster II was found as a cubane [3Fe-4S] cluster and was connected to the polypeptide chain by three cysteinyl residues, Cys-55, Cys-83, and Cys-89 (Fig. 2).[16] The presence of two [3Fe-4S] clusters is very unusual in the bacterial-type dicluster ferredoxins.[23,38] Biochemical and spectroscopic analyses of *Sulfolobus* sp. ferredoxin have shown that this 6Fe-containing species is an artifact of the crystallization procedure after protein isolation.[23,23a] Thus, the crystal structure provides the first structural evidence for a [4Fe-4S] → [3Fe-4S] cluster conversion at the cluster II site in the dicluster-type ferredoxins. Interestingly, a potential ligand for a fourth site, Cys-86, is not bound in the crystal structure at 2.0-Å resolution. In addition, the polypeptide conformation in vicinity of the cubane [3Fe-4S] cluster II and Cys-86 is reportedly an intermediate form between the [3Fe-4S] and [4Fe-4S] cluster conformations.[16]

Although the electron density for Cys-86 is much lower than those of other cysteinyl ligand residues, it should be noted that the side chain of Cys-86 is apparently rotated toward the solvent, away from the cluster,[16] as observed for the nonligating cysteinyl residue (Cys^{II}) of *D. gigas* ferredoxin II with a single cubane

---

[37] J. L. C. Duff, J. L. J. Breton, J. N. Butt, F. A. Armstrong, and A. J. Thomson, *J. Am. Chem. Soc.* **118,** 8593 (1996).
[38] T. Imai, A. Urushiyama, H. Saito, Y. Sakamoto, K. Ota, and D. Ohmori, *FEBS Lett.* **368,** 23 (1995).

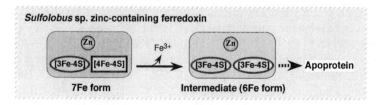

FIG. 3. Schematic illustration of the oxidative conversion of the [4Fe-4S] cluster II in *Sulfolobus* sp. zinc-containing ferredoxin, with a concomitant formation of a stable 6Fe-containing intermediate.

[3Fe-4S] cluster.[39] Given the pseudo-twofold symmetry of a ferredoxin core fold of bacterial-type ferredoxins, it appears that whenever a [3Fe-4S] cluster is present, the missing corner (Fe) of the cube is associated with either replacement (e.g., $Cys^{II} \rightarrow$ Asp, as observed for archaeal zinc-containing ferredoxins) or tilting away to the solvent of the second cysteine residue (Cys^{II}) in the $-Cys^{I}$–XaaXaa–Cys^{II}–XaaXaa–Cys^{III}–XaaXaaXaa–Cys^{IV}–(Pro)– motif (see Figs. 1 and 2). The oxidative cluster conversion of *Sulfolobus* sp. ferredoxin is schematically illustrated in Fig. 3, in which the 6Fe-containing species is a stable intermediate produced by oxidative degradation.[23,23a]

In the crystal structure of *Sulfolobus* sp. ferredoxin, the center-to-center distance of the two [3Fe-4S] clusters is 12.0 Å.[14,16] The crystallographic average Fe–S bond length of the [3Fe-4S] cluster II is 2.32 Å, slightly longer that of the [3Fe-4S] cluster I (2.29 Å). Temperature factors for the polypeptide chain around Cys-86 and the [3Fe-4S] cluster II are significantly higher than the average of temperature factors for all atoms, which is normal. The rms (root mean square) deviation in corresponding Cα atoms between the ferredoxin core fold of *Sulfolobus* sp. zinc-containing ferredoxin and *Clostridium acidiurici* 8Fe-containing ferredoxin is 1.8 Å,[16] indicating structural conservation of the ferredoxin core fold.

Isolated Zinc Site. Beside conventional iron–sulfur clusters, archaeal zinc-containing ferredoxin uniquely contains a stoichiometric amount of tightly bound zinc atom,[14,15,23] which cannot be removed by overnight dialysis against 20 m*M* potassium phosphate buffer, pH 6.8, containing 5 mM EDTA. The X-ray crystal structure of *Sulfolobus* sp. ferredoxin has shown that the zinc site is buried within the molecule (about 5 Å deep from the protein surface), in the boundary between the N-terminal extension and the cluster-binding ferredoxin core fold, connecting these together (Fig. 2).[14] This center is tetrahedrally coordinated with three histidine imidazole groups (contributed by His-16, His-19, and His-34 in the N-terminal extension region) and one carboxylate group (contributed by Asp-76 in the ferredoxin core fold). The crystallographic distances between the isolated

---

[39] C. R. Kissinger, L. C. Sieker, E. T. Adman, and L. H. Jensen, *J. Mol. Biol.* **219,** 693 (1991).

zinc site and cluster I or cluster II are 13.3 Å and 15.4 Å, respectively.[14,16] The N-terminal extension region of *Sulfolobus* sp. ferredoxin consists of three β strands and one α helix[14] and contains two *cis*-proline residues at the positions 24 and 28[16] and one monomethylated lysine residue at the position 29.[24,27] This region is slightly shorter than that of *T. acidophilum* ferredoxin, which also contains three consensus histidine residues (His-30, His-33, and His-57) but no monomethylated lysine residue.[15,17,28]

The X-ray absorption analysis of archaeal zinc-containing ferredoxins has unambiguously shown the presence of an isolated and structurally conserved zinc center, which is tetrahedrally coordinated with (most likely) three histidine imidazoles and one carboxylate, with the average Zn−N bond distance of 2.01 Å and the Zn−O bond distance in the range 1.89–1.94 Å.[17] These values are very similar to the average crystallographic Zn−N and Zn−O bond distance of 1.96 Å and 1.90 Å, respectively, in the three-dimensional structure of *Sulfolobus* sp. ferredoxin.[14] The sequence comparisons suggest that three histidine residues in the N-terminal extension region and one conserved aspartate in the ferredoxin core fold are strictly conserved (Fig. 1).[14,15,17] They probably serve as ligands to the isolated zinc center in archaeal zinc-containing ferredoxins. The evolutionary origin of this consensus N-terminal stretch remains unknown. A similarity search for zinc-containing ferredoxin homologs with the consensus sequence motif against nucleotide and amino acid sequence databases indicated a limited distribution only among hyperthermophilic organisms, and then only within the archaeal domain.[17] This implies that early zinc-containing ferredoxins might have appeared shortly after divergence of the early archaea, as also suggested by the phylogenetic analysis.[15]

Related Metalloproteins

Several iron–sulfur proteins can be purified from the same batch of the *Sulfolobus* sp. strain 7 cells together with the cognate zinc-containing ferredoxin. Purification and some properties of these proteins are described below.

2-Oxoacid:Ferredoxin Oxidoreductase

Purification. Purification of the 2-oxoacid:ferredoxin oxidoreductase of *Sulfolobus* sp. strain 7[12,13] is carried out by following the 2-oxoacid:ferredoxin oxidoreductase activity (as described below) and the absorption bands at 280, 408, and 450 nm of each fraction at different steps.

The cytosol fraction, prepared and dialyzed as described above, is loaded onto the DEAE Sephacel column (3.0 × 28 cm) equilibrated with 30 mM Tris-HCl buffer, pH 8.5, containing 1 mM EDTA and 1 mM PMSF. The column is washed with 500 ml of the equilibration buffer, and the material bound to the column is eluted with a 1600-ml linear gradient of NaCl (0–400 mM) in the equilibration buffer. The combined peak fractions are made to 1 M ammonium sulfate solution by adding fine solid ammonium sulfate at 4°. The suspension is gently degassed with an aspirator, and then directly adsorbed onto a Phenyl-Toyopearl 650M column (1.2 × 25 cm; Tosoh Corp, Japan) equilibrated with 20 mM potassium phosphate buffer, pH 6.8, containing 1 M ammonium sulfate. The column is washed with 100 ml of the equilibration buffer, and the proteins are eluted with a 200-ml linear gradient between 1 M and 0 M ammonium sulfate in 20 mM potassium phosphate buffer, pH 6.8.

The 2-oxoacid:ferredoxin oxidoreductase-containing fraction is dialyzed against 20 mM potassium phosphate buffer, pH 6.8, and applied to a Bio-Rad hydroxylapatite HTP column (1.0 × 23 cm) equilibrated with 10 mM potassium phosphate buffer, pH 6.8. The enzyme adsorbed onto the column is eluted with a 90-ml linear gradient of potassium phosphate buffer, pH 6.8 (10–300 mM), and the peak fraction is collected and concentrated by pressure filtration through an Amicon YM10 membrane at 4°.

The concentrated fraction (~2 ml per tube) is loaded onto a 34-ml glycerol density gradient (5–20%, v/v) in 20 mM potassium phosphate buffer, pH 6.8, which is placed on 0.5 ml of a 25% glycerol (v/v) cushion (containing 20 mM potassium phosphate buffer, pH 6.8). It is then ultracentrifuged in a Beckman SW 28 rotor at 28,000 rpm for 20 hr at 10°. The combined fractions collected by tubing are made to 2 M ammonium sulfate in 20 mM potassium phosphate buffer, pH 6.8, and gently degassed with an aspirator. This is applied to a phenyl-Toyopearl 650M column (1.0 × 25 cm; Tosoh Corp) equilibrated with 20 mM potassium phosphate buffer, pH 6.8, containing 2 M ammonium sulfate. The column is first washed with 40 ml of the equilibration buffer. It is eluted with a 160-ml linear gradient of 1–0 M ammonium sulfate in 20 mM potassium phosphate buffer, pH 6.8, and further with 10 mM potassium phosphate buffer, pH 6.8.

Although 2-oxoacid:ferredoxin oxidoreductase activity in the cytosol fraction is not very stable upon storage at room temperature, at 4° or −80°, the purified enzyme is quite resistant to oxygen and can be stored aerobically at −80° after dialysis against 20 mM potassium phosphate buffer, pH 6.8. Approximately 6–8 mg of purified enzyme is obtained from about 100 g (wet weight) of cells.[13]

Assay Methods. 2-Oxoacid:ferredoxin oxidoreductase activity is determined by following the absorbance at 550 nm, due to the ferredoxin-dependent reduction of horse heart cytochrome c (Sigma Chemicals, St. Louis, MO) in the presence of 2-oxoacid substrates, essentially as described by Kersher *et al.*[4,5] The assay is conducted at 50°, in 10 mM potassium phosphate buffer, pH 6.8, in the presence of 2–4 mM 2-oxoacids (2-oxoglutarate purchased from Nacalai Tesque, Japan, was mainly used), 50–100 μM coenzyme A (Kohjin, Japan), 17 μg of the *Sulfolobus* zinc-containing ferredoxin (purified as described above), 50 μM horse heart cytochrome c (Sigma Chemicals), and an appropriate amount of enzyme, in a total volume of 1 ml.[13] The reaction is initiated by addition of the enzyme, and nonenzymatic reduction of cytochrome c by coenzyme A at this temperature is corrected. Alternatively, the enzymatic activity is determined by using methyl viologen (1 mM; Sigma Chemicals) as an artificial electron acceptor of the enzyme,[40,41] instead of using ferredoxin and cytochrome c, by following absorbance changes at 600 nm.

Molecular Properties. Zinc-containing ferredoxin from *Sulfolobus* sp. strain 7 serves as an effective electron acceptor of a coenzyme A-acylating 2-oxoacid: ferredoxin oxidoreductase.[9] This enzyme has been purified, cloned, sequenced, and partially characterized.[12,13] The purified enzyme is an αβ-type heterodimer composed of two subunits, α and β. The purified enzyme contains one thiamin pyrophosphate (TPP), one [4Fe-4S]$^{2+,1+}$ cluster, and two magnesium atoms per αβ structure. The 70-kDa α subunit consists of 632 amino acid residues and contains no cysteine residues. This subunit is a structural fusion of two subunits, γ and α, of the hyperthermophilic αβγδ-type enzymes.[13,42] The 37-kDa β subunit consists of 306 amino acid residues and is probably involved in binding of the prosthetic groups, as indicated by the presence of a typical TPP-binding motif and four consensus cysteine residues (arranged in an atypical cysteine-containing sequence motif) that probably serve as ligands to the [4Fe-4S] cluster.[13] The coenzyme A and ferredoxin binding sites of the *Sulfolobus* sp. enzyme remain to be identified.

The purified enzyme shows a broad substrate specificity toward 2-oxoacids. The specific activity at 50° in the presence of the cognate zinc-containing ferredoxin (25 μM) and 4 mM substrate is in the order of 2-oxoglutarate ($K_m = 0.87$ mM) > 2-oxobutyrate > pyruvate ($K_m = 0.25$ mM). No ferredoxin reduction can be observed with glyoxylate.[13] The V_{max}/K_m for pyruvate is approximately 1.9-fold higher than that for 2-oxoglutarate. The binding of 2-oxoacid to *Sulfolobus* sp. enzyme results in the formation of an activated hydroxyalkyl-TPP radical intermediate,[13] as previously observed for *Halobacterium salinarium* enzymes.[6,10,11,43] The reduced [4Fe-4S] cluster in the β subunit has been postulated to serve as electron donor to the cognate ferredoxin in these enzymes.[13] The "reverse" 2-oxoacid synthase reaction for reoxidation of reduced ferredoxin and CO_2 fixation is thermodynamically unlikely to occur in *Halobacterium* and *Sulfolobus*. This is also in line with the operation of oxidative tricarboxylic acid cycle under chemoheterotrophic growth conditions, which is directly coupled to the membrane-bound aerobic respiratory chain at the level of succinate:quinone oxidoreductase (respiratory complex II).[20,44]

The molecular biological studies have suggested that 2-oxoacid:ferredoxin oxidoreductases of archaea, bacteria, and anaerobic amitochondrial eukarya with different sizes and subunit compositions are a phylogenetically closely related

---

[40] J. M. Blamey and M. W. W. Adams, *Biochim. Biophys. Acta* **1161**, 19 (1993).
[41] J. M. Blamey and M. W. W. Adams, *Biochemistry* **33**, 1000 (1994).
[42] A. Kletzin and M. W. W. Adams, *J. Bacteriol.* **178**, 248 (1996).
[43] R. Cammack, L. Kerscher, and D. Oesterhelt, *FEBS Lett.* **118**, 271 (1980).
[44] T. Iwasaki, T. Wakagi, and T. Oshima, *J. Biol. Chem.* **270**, 30902 (1995).

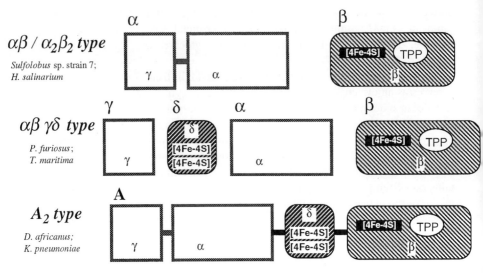

FIG. 4. Comparison of domains and subunits in various 2-oxoacid:ferredoxin oxidoreductases with different sizes and subunit compositions [Q. Zhang, T. Iwasaki, T. Wakagi, and T. Oshima, *J. Biochem.* **120,** 587 (1996); A. Kletzin and M. W. W. Adams, *J. Bacteriol.* **178,** 248 (1996)]. The arrangements of domains and subunits in these enzymes exhibit mosaic structures, implying that they might have evolved through multiple gene duplication, fusion, and reorganization events of primordial smaller fragments [Q. Zhang, T. Iwasaki, T. Wakagi, and T. Oshima, *J. Biochem.* **120,** 587 (1996); A. Kletzin and M. W. W. Adams, *J. Bacteriol.* **178,** 248 (1996)]. The X-ray crystal structure of the A_2-type pyruvate:ferredoxin oxidoreductase from *D. africanus* has been determined by Chabrière *et al.* [E. Chabrière, M.-H. Charon, A. Volbeda, L. Pieulle, E. C. Hatchikian, and J.-C. Fontecilla-Camps, *Nat. Struct. Biol.* **6,** 182 (1999)], and shown to contain one TPP cofactor, one Mg center, and three [4Fe-4S] clusters as prosthetic groups. The $\alpha\beta$-/$\alpha_2\beta_2$-type enzymes from aerobic archaea inherently lack the δ subunit/domain carrying two [4Fe-4S] clusters [Q. Zhang, T. Iwasaki, T. Wakagi, and T. Oshima, *J. Biochem.* **120,** 587 (1996); A. Kletzin and M. W. W. Adams; *J. Bacteriol.* **178,** 248 (1996)].

family (Fig. 4).[13,42] The arrangements of domains and subunits in these enzymes exhibit mosaic structures, implying that they might have evolved through multiple gene duplication, fusion, and reorganization events of primordial smaller fragments.[13,42]

Interestingly, the oxygen-resistant $\alpha\beta$-/$\alpha_2\beta_2$-type enzymes isolated from aerobic archaea, *Halobacterium* and *Sulfolobus* are different from other types of the enzymes from anaerobic organisms in that they inherently lack the δ subunit/domain (Fig. 4),[13,42] which is a low-potential dicluster-type ferredoxin carrying two [4Fe-4S] clusters.[45,46] Chabrière *et al.*[46] reported the X-ray structural determination of

---

[45] A. L. Menon, H. Hendrix, A. Hutchins, M. F. Verhagen, and M. W. W. Adams, *Biochemistry* **37,** 12838 (1998).

the A_2-type pyruvate:ferredoxin oxidoreductase from *D. africanus*, which contains one TPP cofactor, one Mg center, and three [4Fe-4S] clusters as prosthetic groups. In the three-dimensional structure of *D. africanus* enzyme, the pyrophosphate group of the TPP cofactor is located in the close vicinity of the proximal [4Fe-4S] cluster that is bound to atypical cysteine-containing sequence motif, at a C2 atom-to-center distance of 13.2 Å.[46] Magnesium ion is located in close vicinity of the TPP cofactor,[46] as previously assumed for *Sulfolobus* sp. enzyme.[13] The median [4Fe-4S] cluster is located away from the proximal cluster at a center-to-center distance of 15.1 Å, and the distal [4Fe-4S] cluster is farther away from the median cluster at a center-to-center distance of 12.3 Å.[46] Thus, these centers are suitably arranged to facilitate a plausible electron transfer pathway. Interestingly, the center-to-center distance of the median and distal [4Fe-4S] clusters in the δ domain (domain V) of *D. africanus* enzyme[46] is remarkably similar to that of the two iron–sulfur clusters in *Sulfolobus* sp. zinc-containing ferredoxin (12.0 Å),[16] implying that the physiological electron acceptor of the *Sulfolobus* enzyme may functionally take place of the δ subunit/domain of the A_2-/αβγδ-type enzymes of anaerobic organisms.

The types of ferredoxins utilized as a physiological electron acceptor by 2-oxoacid:ferredoxin oxidoreductases are very different. For instance, the physiological electron acceptor of *H. salinarium* enzyme is a plant-type ferredoxin carrying a single [2Fe-2S] cluster,[4,10,11,35] whereas that of the αβγδ-type enzymes from anaerobic hyperthermophiles is a monocluster ferredoxin.[40,41,47,48] *Sulfolobus* sp. enzyme utilizes zinc-containing dicluster ferredoxin as a physiological electron acceptor, although the enzymatic reduction of *Sulfolobus* sp. and *T. acidophilum* zinc-containing ferredoxins has suggested that only the [3Fe-4S] cluster (cluster I) is reduced during the steady-state turnover of the enzyme in the presence of 2-oxoacid and coenzyme A.[9,15] These observations suggest that archaeal 2-oxoacid:ferredoxin oxidoreductases preferably utilize a single electron carrier as a physiological electron acceptor.

Based on the biochemical and sequence data, we have proposed that 2-oxoacid : ferredoxin oxidoreductases with different subunit compositions have an essentially similar catalytic mechanism with one TPP and at least one [4Fe-4S] cluster as the minimal set of redox centers, which will include the intermediacy of a hydroxyalkyl-TPP radical, and that the δ subunit/domain carrying two [4Fe-4S] clusters may serve as (accessory) intramolecular electron transmitter from the [4Fe-4S] cluster in the β subunit/domain to the physiological electron acceptor

---

[46] E. Chabrière, M.-H. Charon, A. Volbeda, L. Pieulle, E. C. Hatchikian, and J.-C. Fontecilla-Camps, *Nat. Struct. Biol.* **6**, 182 (1999).

[47] R. C. Conover, A. T. Kowal, W. Fu, J.-B. Park, S. Aono, M. W. W. Adams, and M. K. Johnson, *J. Biol. Chem.* **265**, 8533 (1990).

[48] Z. H. Zhou and M. W. W. Adams, *Biochemistry* **36**, 10892 (1997).

such as ferredoxin/flavodoxin.[13] The kinetic studies of an A_2-type enzyme from *Clostridium thermoaceticum* by using rapid freeze-quench EPR and stopped-flow spectroscopy suggested a transient formation of a short-lived hydroxyalkyl-TPP radical intermediate in the first reductive half-reaction.[49] This seems to indicate that all 2-oxoacid:ferredoxin oxidoreductases with different subunit compositions and different cluster compositions should have essentially the same radical-based mechanism.

Other Metalloproteins

Red Iron–Sulfur Flavoprotein

In anaerobic archaea, ferredoxin functions as an intermediate electron acceptor of a variety of key steps in the central metabolic pathways involved in saccharolytic and peptide fermentation, and reduced ferredoxin thus formed donates its reducing equivalent to ferredoxin:NADP$^+$ oxidoreductase and hydrogenase.[7,8] In aerobic and thermoacidophilic archaea, the reoxidation steps of reduced zinc-containing ferredoxin are poorly characterized. The soluble fraction of *Sulfolobus* sp. strain 7 also contains an NADPH:ferredoxin oxidoreductase activity,[12] but this enzyme has not been purified and characterized. The following section describes the purification and partial characterization of a red iron–sulfur flavoprotein with a weak ferredoxin-reoxidizing activity from *Sulfolobus* sp. strain 7.[12]

Purification of Red Iron-Sulfur Flavoprotein. The red iron–sulfur flavoprotein[12] is eluted from the DEAE-Sephacel column shortly after (or sometimes together with) the cognate 2-oxoacid:ferredoxin oxidoreductase activity, as described above. The red-colored fractions are combined and made to 1 M ammonium sulfate solution by addition of solid ammonium sulfate at 4°, with stirring.

The suspension is gently degassed by aspirator at room temperature, and directly adsorbed onto a Phenyl-Toyopearl 650M column (1.2 × 25 cm; Tosoh Corp) equilibrated with 20 mM potassium phosphate buffer, pH 6.8, containing 1 M ammonium sulfate. The column is washed with 100 ml of the equilibration buffer, and a 200-ml linear gradient is run between 1 M and 0 M ammonium sulfate in 20 mM potassium phosphate buffer, pH 6.8. At this step, the red-colored fractions eluted at the end of the linear gradient are clearly separated from the 2-oxoacid:ferredoxin oxidoreductase and diaphorase [NAD(P)H:dye oxidoreductase] activities.[12] When required, the column is further washed by 20 mM potassium phosphate buffer, pH 6.8, to complete the elution of the red protein.

The red protein fraction thus obtained is diluted fivefold by adding 20 mM potassium phosphate buffer, pH 6.8, and adsorbed onto a Bio-Rad hydroxylapatite HTP column (1.2 × 23 cm) equilibrated with distilled water (Milli-Q). The column

---

[49] S. Menon and S. W. Ragsdale, *Biochemistry* **36**, 8484 (1997).

is washed with 100 ml of 10 mM potassium phosphate buffer, pH 6.8, and the red protein still bound to the column is eluted by a 250-ml linear gradient of potassium phosphate buffer, pH 6.8 (10–200 mM). The red protein elutes at around 110 mM potassium phosphate and is collected and concentrated by pressure filtration on an Amicon YM10 membrane under N_2 gas at 4°.

The concentrated material (3 ml or less) is carefully loaded onto a 34-ml glycerol density gradient (10–24%, v/v) in 20 mM potassium phosphate buffer, pH 6.8, which has been placed on 0.5 ml of 25% glycerol (v/v) cushion containing 20 mM potassium phosphate buffer, pH 6.8, and is ultracentrifuged with a Beckman SW28 rotor at 28,000 rpm for 20 hr at 10°. The red protein appears as a single band at ~19% glycerol (v/v) and is collected by tubing and analyzed by SDS–PAGE for purity. The purified protein thus obtained has the purity indices A_{341}/A_{278} and A_{428}/A_{278} of 0.22 and 0.17, respectively, and can be stored at $-80°$ until use. The extinction coefficient of oxidized protein, which refers to the concentration of a protomer ($\alpha\beta\gamma$ structure) (calculated and calibrated on the basis of Mo and Fe contents; see below), is $\varepsilon = 45$ mM^{-1} cm^{-1} at 428 nm.[50] Approximately 30 mg of pure protein is reproducibly obtained from ~60 g wet weight of the late-log-phase cells grown aerobically and chemoheterotrophically.[12]

Molecular Properties of Red Iron–Sulfur Protein. The purified enzyme is a thermostable, red-colored protein and consists of three nonidentical subunits with apparent molecular masses of 87, 32, and 19 kDa, respectively. The previous cofactor composition determined based on protein concentrations estimated by a colorimetric method[12] appears to be an underestimate, and more recent chemical and spectroscopic analyses suggest the presence of 2 Mo atoms (in the form of a molybdopterin type cofactor of unidentified chemical structure), 8 nonheme Fe atoms (in the form of [2Fe-2S] clusters), and about 3 noncovalently bound flavin, per $\alpha_2\beta_2\gamma_2$ structure.[50] The purified enzyme does not contain cobalt, nickel, tungsten, or zinc in significant amounts. Although the physiological role of the red iron–sulfur flavoprotein remains to be elucidated, it is likely to belong to an archaeal member of the mononuclear molybdenum-containing hydroxylase superfamily, which involves a variety of hydroxylases such as xanthine oxidase/dehydrogenase, aldehyde oxidoreductase, quinoline oxidoreductase, and CO dehydrogenase.[51–54] The iron–sulfur clusters and flavin centers of this enzyme were slowly reduced by the enzymatically reduced zinc-containing ferredoxin in the presence of the cognate 2-oxoacid:ferredoxin oxidoreductase and its substrate, but not by NAD(P)H.[12]

[50] T. Iwasaki and T. Oshima, unpublished results (1996); T. Iwasaki, Y. Kawarabayashi, N. J. Cosper, T. Oshima, and R. A. Scott, manuscript in preparation.
[51] R. C. Bray, *Q. Rev. Biophys.* **21,** 299 (1988).
[52] R. Hille and T. Nishino, *FASEB J.* **9,** 995 (1995).
[53] M. J. Romão, M. Archer, I. Moura, J. J. G. Moura, J. LeGall, M. Engh, M. Schneider, P. Hof, and R. Huber, *Science* **270,** 1170 (1995).
[54] R. Hille, *Chem. Rev.* **96,** 2757 (1996).

Sulredoxin

2-Oxoacid:ferredoxin oxidoreductase of *H. salinarium* has been reported to utilize a plant-type [2Fe-2S] ferredoxin as a physiological electron acceptor[4,35] and is highly homologous to the *Sulfolobus* sp. enzyme, which utilizes a zinc-containing ferredoxin as an acceptor.[9,13] This prompted us to search for other small and water-soluble iron–sulfur proteins in *Sulfolobus* sp. strain 7. Beside the 7Fe and 6Fe forms of zinc-containing ferredoxin,[9,23,24] a novel 12-kDa iron–sulfur protein has been purified from the soluble fraction of *Sulfolobus* sp. strain 7 and characterized.[22,55,56] Biochemical and spectroscopic analyses suggest that the purified protein is a novel Rieske-type iron–sulfur protein containing a single Rieske-type [2Fe-2S] cluster as described below and is called "sulredoxin."[22]

Purification of Sulredoxin. Sulredoxin[22] is monitored during its purification by measuring three different absorption bands at 280, 408, and 450 nm. Detergents are not included at any step of the purification procedure. The flow-through fraction obtained from the DEAE-Sephacel column chromatography, as described above, is combined and used as a starting material for purification of sulredoxin. Solid ammonium sulfate (0.2425 g/ml) is added to this solution, and the mixture is gently degassed by aspirator with gentle stirring. The suspension is directly adsorbed onto a butyl-Toyopearl 650M column (2.2 × 31.5 cm; Tosoh Corp) equilibrated with 20 mM potassium phosphate degassed buffer, pH 6.0, containing 234 g ammonium sulfate per liter (40% saturation). The column is washed with 250 ml of the equilibration buffer, and an 800 ml linear gradient is run between 40% and 0% saturation of ammonium sulfate in 20 mM potassium phosphate degassed buffer, pH 6.0. Sulredoxin can be recognized by its pink/purple color and is eluted around 18% ammonium sulfate saturation.

The combined sulredoxin fraction is concentrated by pressure filtration through an Amicon YM10 membrane under N_2 gas at 4° to ∼3 ml, and is applied onto a preparative Sephadex G-50 gel filtration column (2.5 × 75 cm; Amersham Pharmacia Biotech) equilibrated with 80 mM potassium phosphate buffer (pH 6.8) at 4°. The pink fractions, which show a single band on 20% SDS–PAGE, are combined. Purified sulredoxin can be stored at −80° until use.

Purified sulredoxin has purity indices A_{342}/A_{278} and A_{443}/A_{278} of 0.45 and 0.22, respectively. The extinction coefficient at 443 nm of oxidized protein, which refers to the concentration of [2Fe-2S] cluster (calculated on the basis of iron), is $\varepsilon([2Fe-2S]) = 5.8$ mM^{-1} cm^{-1}.[22] This is similar to the reported value at 458 nm of *T. thermophilus* respiratory Rieske protein [$\varepsilon([2Fe-2S]) = 6.0$ mM^{-1} cm^{-1} [57]]. Approximately 5 mg of purified sulredoxin is obtained from ∼60 g wet weight of the late-log-phase cells grown aerobically and chemoheterotrophically.

---

[55] T. Iwasaki, T. Imai, A. Urushiyama, and T. Oshima, *J. Biol. Chem.* **271**, 27659 (1996).

[56] T. Iwasaki and S. A. Dikanov, in "Abstract of 9th International Conference on Biological Inorganic Chemistry (ICBIC 9)", Minneapolis, MN (1999).

Molecular Properties of Sulredoxin. Although the plant-type ferredoxins contain a [2Fe-2S] cluster with four cysteinyl ligand residues and show an EPR signal with an average g-factor (g_{av}) of around 1.96 in the reduced state, Rieske-type [2Fe-2S] proteins elicit a characteristic EPR signal with unusually low g_{av} values of ~1.91 in the reduced state, as well as red-shifted visible spectra in the oxidized state.[58–63] The unique spectral properties of a Rieske-type [2Fe-2S] cluster have been interpreted in terms of the asymmetrical ligand environments, such that one iron atom of the [2Fe-2S] cluster is coordinated to the protein by two sulfide ligands contributed by two cysteine residues, while the other is coordinated by two δ-nitrogens of the imidazole rings contributed by two histidine residues.[57,63–73] The detailed coordination environments of several Rieske-type [2Fe-2S] centers have been determined by the X-ray structural analyses of mitochondrial and chloroplast respiratory Rieske protein fragments[74,75] and bacterial naphthalene 1,2-dioxygenase.[76]

Purified sulredoxin contains a single Rieske-type [2Fe-2S] cluster (per 12-kDa polypeptide) and shows a characteristic EPR signal with a g_{av} value of 1.90

---

[57] J. A. Fee, K. L. Findling, T. Yoshida, R. Hille, G. E. Tarr, D. O. Hearshen, W. R. Dunham, E. P. Day, T. A. Kent, and E. Münck, *J. Biol. Chem.* **259,** 124 (1984).

[58] B. L. Trumpower, *Biochim. Biophys. Acta* **639,** 129 (1981).

[59] J. A. Fee, D. Kuila, M. W. Mather, and T. Yoshida, *Biochim. Biophys. Acta* **853,** 153 (1986).

[60] B. L. Trumpower, *Microbiol. Rev.* **54,** 101 (1990).

[61] J. R. Mason and R. Cammack, *Ann. Rev. Microbiol.* **46,** 277 (1992).

[62] R. Malkin, *Photosynth. Res.* **33,** 121 (1992).

[63] T. A. Link and S. Iwata, *Biochim. Biophys. Acta* **1275,** 54 (1996).

[64] D. Kuila, J. A. Fee, J. R. Schoonover, W. H. Woodruff, C. J. Batie, and D. P. Ballou, *J. Am. Chem. Soc.* **109,** 1559 (1987).

[65] P. Bertrand, B. Guigliarelli, J.-P. Gayda, P. Beardwood, and J. F. Gibson, *Biochim. Biophys. Acta* **831,** 261 (1985).

[66] J. F. Cline, B. M. Hoffman, W. B. Mims, E. LaHaie, D. P. Ballou, and J. A. Fee, *J. Biol. Chem.* **260,** 3251 (1985).

[67] R. J. Gurbiel, C. J. Batie, M. Sivaraja, A. E. True, J. A. Fee, B. M. Hoffman, and D. P. Ballou, *Biochemistry* **28,** 4861 (1989).

[68] R. J. Gurbiel, T. Ohnishi, D. E. Robertson, F. Daldal, and B. M. Hoffman, *Biochemistry* **30,** 11579 (1991).

[69] R. D. Britt, K. Sauer, M. P. Klein, D. B. Knaff, A. Kriauciunas, C.-A. Yu, L. Yu, and R. Malkin, *Biochemistry* **30,** 1892 (1991).

[70] E. Davidson, T. Ohnishi, E. Atta-asafo-Adjei, and F. Daldal, *Biochemistry* **31,** 3342 (1992).

[71] J. K. Shergill, C. L. Joannou, J. R. Mason, and R. Cammack, *Biochemistry* **34,** 16533 (1995).

[72] R. J. Gurbiel, P. E. Doan, G. T. Gassner, T. J. Macke, D. A. Case, T. Ohnishi, J. A. Fee, D. P. Ballou, and B. M. Hoffman, *Biochemistry* **35,** 7834 (1996).

[73] S. A. Dikanov, L. Xun, A. B. Karpiel, A. M. Tyryshkin, and M. K. Bowman, *J. Am. Chem.Soc.* **118,** 8408 (1996).

[74] S. Iwata, M. Saynovits, T. A. Link, and H. Michel, *Structure* **4,** 567 (1996).

[75] C. J. Carrell, H. Zhang, W. A. Cramer, and J. L. Smith, *Structure* **5,** 1613 (1997).

[76] B. Kauppi, K. Lee, E. Carredano, R. E. Parales, D. T. Gibson, H. Eklund, and S. Ramaswamy, *Structure* **6,** 571 (1998).

in the reduced state.[22] Potentiometric and spectroscopic analyses[55] have shown that the Rieske-type [2Fe-2S] cluster in sulredoxin exhibits a redox-linked ionization behavior, as observed for regular respiratory Rieske proteins found as a part of cytochrome bc_1/b_6f complex and its analog[77–80]; the midpoint redox potential of sulredoxin [E_m(low pH), + 188 mV] is influenced by two ionization equilibria with p$K_{a,ox1}$ of 6.2 and p$K_{a,ox2}$ of 8.6, although redox-linked deprotonation of the reduced protein does not occur at least up to pH 10.[55]

In the 230 to 300 cm^{-1} region, low-temperature resonance Raman spectra of oxidized sulredoxin show a single band at 270 cm^{-1} at pH 5.5–7.4, and at higher pH values a 278-cm^{-1} band appears so that at pH 9.4 both are present in the spectrum.[55] These two bands at 270 and 278 cm^{-1} can be primarily attributable to Fe(III)–N(His)^t stretching motions of sulredoxin, and the appearance of the new 278 cm^{-1} band at alkaline pH may be due to the strengthening of the Fe–N bond on deprotonation. The ionizable group of oxidized sulredoxin with a p$K_{a,ox2}$ of ~8 apparently corresponds to those of the mitochondrial Rieske fragment (p$K_{a,ox}$ ~ 7.6)[79] and *T. thermophilus* respiratory Rieske protein (p$K_{a,ox}$ ~ 8)[78] found as a part of aerobic respiratory chain.

It should be noted that the redox-linked ionization has not been reported for other water-soluble Rieske-type proteins (Rieske-type ferredoxins) that are involved in some bacterial dioxygenase system.[61] Thus, although sulredoxin is a small water-soluble Rieske-type [2Fe-2S] protein[22] and is not a part of the cognate terminal oxidase complex that contains at least one *b*- and two *a*-type cytochromes and a Rieske-type cluster,[20] it provides a tractable model of the membrane-bound Rieske protein found as part of the respiratory chain. Genetic studies have indicated that sulredoxin has a consensus active site motif as observed for regular respiratory Rieske proteins.[56] Purified sulredoxin can be reduced by ascorbate, but not by the enzymatically reduced zinc-containing ferredoxin at 50° in the presence of the cognate 2-oxoacid:ferredoxin oxidoreductase, 2-oxoglutarate, and coenzyme A.[22] Hence, it is most likely involved in some unidentified sequence of the archaeal electron transfer chain.

Acknowledgments

The preparation of this chapter and part of the research work were supported in part by Grants-in-Aid from the Ministry of Education, Science and Culture of Japan. We thank numerous colleagues and collaborators whose names appear in the references.

---

[77] R. C. Prince and P. L. Dutton, *FEBS Lett.* **65,** 117 (1976).

[78] D. Kuila and J. A. Fee, *J. Biol. Chem.* **261,** 2768 (1986).

[79] T. A. Link, W. R. Hagen, A. J. Pierik, C. Assmann, and G. von Jagow, *Eur. J. Biochem.* **208,** 685 (1992).

[80] S. Anemüller, C. L. Schmidt, G. Schäfer, E. Bill, A. X. Trautwein, and M. Teixeira, *Biochem. Biophys. Res. Commun.* **202,** 252 (1994).

[2] Ferredoxin from *Thermotoga maritima*

By REINHARD STERNER

Introduction

Ferredoxins are small electron carrier proteins that contain clusters either of the 2Fe-2S type (mostly in plants and animals) or the 4Fe-4S (3Fe-4S) type (mostly in prokaryotes). Because of their ubiquitous occurrence and their involvement in a wide variety of metabolic processes, both structural and functional properties of ferredoxins have been studied extensively.[1-3] *Thermotoga maritima* is a hyperthermophilic bacterium with a maximum growth temperature of 90°.[4] It is a strictly anaerobic heterotroph that can gain energy by fermentation, using a wide range of substrates including monosaccharides, maltose, starch, and glycogen. Degradation of glucose is performed predominantly via a conventional Embden–Meyerhof pathway, and end products of fermentation include H_2, CO_2, acetate, and lactate.[5-7] The ferredoxin of *T. maritima* (tFdx) is involved in this fermentation process.

Native tFdx was purified from *T. maritima* and shown to contain one 4Fe-4S cluster[8] that can reversibly accept a single electron with a redox potential $E^{\circ\prime}$ of -388 mV and -453 mV at 25° and 80°, respectively.[9] The oxidation of a pyruvate:ferredoxin oxidoreductase by tFdx was demonstrated,[10] and the existence of a NADH:ferredoxin oxidoreductase to be oxidized by tFdx has been proposed.[6] However, a physiological electron acceptor for reduced tFdx remains to be identified.

Here the heterologous expression of the cloned t*fdx* gene in *Escherichia coli* as well as purification and characterization of the recombinant tFdx are described. Recombinant tFdx turns out to be a good model protein for the investigation of the structural basis of high protein thermostability, since it contains only 60 amino acid residues, is highly soluble, and is easy to crystallize.

---

[1] H. Matsubara and K. Saeki, *Adv. Inorg. Chem.* **38**, 223 (1992).
[2] J. J. G. Moura, A. L. Macedo, and P. N. Palma, *Methods Enzymol.* **234**, 165 (1994).
[3] H. Sticht and P. Rösch, *Prog. Biophys. Mol. Biol.* **70**, 95 (1998).
[4] R. Huber, T. A. Langworthy, H. König, M. Thomm, C. R. Woese, U. B. Sleytr, and K. O. Stetter, *Arch. Microbiol.* **144**, 324 (1986).
[5] M. W. W. Adams, *FEMS Microbiol. Rev.* **15**, 261 (1994).
[6] C. Schröder, M. Selig, and P. Schönheit, *Arch. Microbiol.* **161**, 460 (1994).
[7] M. Selig, K. B. Xavier, H. Santos, and P. Schönheit, *Arch. Microbiol.* **167**, 217 (1997).
[8] J. M. Blamey, S. Mukund, and M. W. W. Adams, *FEMS Microbiol. Lett.* **121**, 165 (1994).
[9] E. T. Smith, J. M. Blamey, Z. H. Zhou, and M. W. W. Adams, *Biochemistry* **34**, 7161 (1995).
[10] J. M. Blamey and M. W. W. Adams, *Biochemistry* **33**, 1000 (1994).

Heterologous Expression in *Escherichia coli*

Ferredoxin from *T. maritima* is encoded by the *tfdx* gene, which was cloned into the plasmid pDS56/RBSII for heterologous expression in *E. coli*.[11] In the pDS system, genes are under the control of the *lac* promoter/operator and are transcribed after induction with isopropylthiogalactoside (IPTG), which inactivates the *lac* repressor encoded by the repressor–plasmid pDM.[12] Expression is carried out in SG 200-50 cells,[13] which is a *lonA*$^-$(Tn5) derivative of strain Mc 4110 (F, $\Delta lacU$ 169, *araD*139, *rpsL*, *relA*, *thiA*, *flbB*).

Test Expression

It is advisable first to test about three individual colonies of transformants by small-scale expression. To this end, single colonies of freshly transformed cells are used to inoculate 20 ml LB-medium containing 0.1 mg/ml ampicillin (for maintenance of pDS) and 0.025 mg/ml kanamycin (for maintenance of pDM), grown at 37° to an OD_{600} of 0.5–0.7, induced with 1 mM IPTG, and grown for another 5 hr. The cells are centrifuged (Eppendorf, 5000 rpm, 5 min), resuspended in 1 ml 100 mM potassium phosphate buffer (pH 7.8), and lysed by sonification (Branson sonifier 250, small tip, 1 min, level 2, 50% pulse, 0°). Of the resulting suspension, 100 µl is removed and centrifuged, and the supernatant is discarded. The pellet is resuspended in 100 µl 100 mM potassium phosphate buffer (pH 7.8) and stored (fraction "pellet"). The residual 900 µl of lysed cell suspension is also centrifuged. Of the resulting supernatant, 100 µl is stored (fraction "crude extract"), and the remaining 800 µl is heated to 80° for 10 min, chilled on ice, and centrifuged. The supernatant should consist mainly of recombinant tFdx (fraction "heat step"). "Pellet," "crude extract," and "heat step" are electrophoresed on a 12.5% SDS–polyacrylamide gel. Since tFdx bands stain far more weakly with Coomassie Blue than conventional proteins, it is important to remove the iron–sulfur cluster before electrophoresis. To this end, samples are mixed with one volume of 3 M trichloroacetic acid to decompose the iron–sulfur cluster and agitated at room temperature for 60 min. The resulting suspension is centrifuged (Eppendorf, 14,000 rpm, 10 min), and the supernatant discarded. The pellet containing precipitated apo-tFdx is resuspended in 0.5 M Tris-HCl buffer (pH 8.5), containing 1 mM dithiothreitol (DTT), and agitated for 15 min. Samples are then mixed in a 1 : 1 ratio with 2× SDS sample buffer and SDS–PAGE is performed. Clones that provide the largest amounts of recombinant tFdx in the "crude extract" and "heat step" fractions are used for large-scale expression.

---

[11] B. Darimont and R. Sterner, *EMBO. J.* **13,** 1772 (1994).
[12] U. Certa, W. Bannwarth, D. Stüber, R. Gentz, M. Lanzer, S. LeGrice, F. Guillot, I. Wendler, G. Hunsmann, H. Bujard, and J. Mous, *EMBO J.* **5,** 3051 (1986).
[13] M. Y. Casabadan. *J. Mol. Biol.* **104,** 541 (1976).

Large-Scale Expression

Validated cells are grown overnight at 37° in 100 ml (5 × 20 ml) LB medium supplemented with 0.1 mg/ml ampicillin and 0.025 mg/ml kanamycin. Ten liters (5 × 2 liter) of the same medium are inoculated with these cultures and incubated at 37° until OD_{600} reaches a value of 0.5. The expression of tFdx is then induced by adding IPTG to a final concentration of 1 mM, and cells are grown for another 5 hr until OD_{600} attains 1.5. Cells are harvested by centrifugation (Sorvall H6000A, 4500 rpm, 30 min, 4°), washed with 50 mM Tris/HCl buffer (pH 7.4), and centrifuged again (Sorvall GSA, 12,000 rpm, 20 min, 4°). About 2.5 g cells (wet weight) are obtained per liter culture.

Purification

Because bacterial ferredoxins absorb in the visible wavelength region with a characteristic maximum at 390 nm, purification cannot only be monitored by SDS–PAGE, but also by the increase of the absorption ratio A_{390}/A_{280}. All purification steps of tFdx are performed under aerobic conditions and between 0° and 4°, unless otherwise stated. The cells are resuspended (2 ml buffer per 1 g wet mass) in 50 mM Tris/HCl buffer (pH 7.4) and lysed by sonification (Branson Sonifier 250, large tip, 3 × 1 min, 50% pulse). The resulting homogenate is centrifuged (Sorvall SS34, 18,000 rpm, 10 min) and the supernatant is heated to 80–90° for 10 min, chilled on ice, and again centrifuged (Sorvall SS34, 18,000 rpm, 20 min). This heat treatment results in the precipitation of most *E. coli* host proteins. From the resulting supernatant, nucleic acids are precipitated by adding protamine sulfate to a mass that corresponds to half of the mass of the nucleic acids. The suspension is stirred on ice for 15 min and centrifuged twice (Sorvall SS34, 18,000 rpm, 20 min). This protamine sulfate step is optional since it results in the coprecipitation of a significant amount of tFdx. The supernatant is loaded onto a DEAE-Sephacel column (3.6 × 25 cm; Pharmacia, Piscataway, NJ) being equilibrated with 50 mM Tris/HCl buffer (pH 7.4). The column is washed with 600 ml equilibration buffer, and bound proteins are eluted with 1.6 liter of a linear gradient of 0 to 450 mM NaCl in equilibration buffer. TFdx migrates as a brown zone on the column and is eluted at about 170 mM NaCl. Fractions containing the highest concentration of tFdx as determined by absorption measurement at 390 nm are pooled and concentrated by ultrafiltration, using YM3 membranes (Amicon, Danvers, MA). The retentate is loaded onto an hydroxylapatite column (2.4 × 20 cm) being equilibrated with 50 mM Tris/HCl buffer (pH 7.4). The column is washed with 200 ml of the equilibration buffer and eluted with 400 ml of a linear gradient of 0 to 50 mM Na_2HPO_4/NaH_2PO_4 in equilibration buffer. A single ferredoxin peak is obtained and concentrated to 5–10 mg/ml, again using YM3 membranes. Concentrated and pure tFdx typically shows an absorption ratio A_{390}/A_{280}

TABLE I
PURIFICATION OF *T. MARITIMA* 4 FE-4S FERREDOXIN FROM TRANSFORMED *E. COLI* CELLS[a]

Fraction	Volume (ml)	Protein[b] (mg)	Ferredoxin[c] (mg)	A_{390}/A_{280}
Crude extract	31	1470	<60	0.019
90° supernatant	38	179	<43	0.023
Protamine sulfate supernatant	56	157	<39	0.110
DEAE-Sephacel	21	25	16	0.820
Hydroxyl apatite	10	14	14	0.875

[a] Representative flow sheet based on a 10 liter culture yielding 30 g of wet cells. Adapted, with permission, from B. Darimont and R. Sterner, *EMBO J.* **13,** 1772 (1994). Copyright Oxford University Press.

[b] Determined according to O. H. Lowry, N. J. Rosenbrough, A. L. Farr, and R. J. Randall, *J. Biol. Chem.* **193,** 265 (1951) after extraction of the iron–sulfur cluster.

[c] Determined with $\varepsilon_{390} = 17{,}400\ M^{-1}\mathrm{cm}^{-1}$. Since other components of the cell extract might also absorb at 390 nm, for the first three purification steps only an upper limit of the amount of ferredoxin can be given.

of 0.84–0.88. The buffer is changed to 50 m*M* potassium phosphate (typically at pH 7.0), using small Sephadex G-25 columns (NAP, Pharmacia, Piscataway, NJ). The protein solution is dripped into liquid nitrogen and stored at $-70°$. The purification typically yields about 14 mg of tFdx per 30 g of wet cells, corresponding to 10 liter cell culture (Table I).

Association State

The molecular weight and thus the association state of purified tFdx is measured by sedimentation equilibrium runs in the analytical ultracentrifuge (Beckman, model Optima XL A). Assuming a partial specific volume of 0.67 ml/g, which was measured for the 4Fe-4S ferredoxin from *Clostridium thermoaceticum*,[14] a value of M_r of 6600 is obtained for tFdx. This value is in excellent agreement with the value of 6565 that is calculated for a monomeric holoferredoxin from the amino acid sequence, assuming the presence of a single 4Fe-4S cluster.[11] The determined molecular weight is independent of the ferredoxin concentration between 0.07 and 10 mg/ml. It is also independent of the NaCl concentration from 0.04 to 0.3 *M* and of the rotor speed between 28,000 and 56,000 rpm, as well as of the detection wavelength (280 or 390 nm). Native tFdx migrates with an apparent molecular mass of about 13–14 kDa on a calibrated Superose 12 column. A similar result is obtained with denatured apoferredoxin on SDS–PAGE. The reasons for these artificially high apparent molecular masses are not clear.[11]

[14] S.-S. Yang, L. G. Ljungdahl, and J. LeGall, *J. Bacteriol.* **130,** 1084 (1977).

Determination of the Iron Content and of the Extinction Coefficient

The iron content of tFdx is determined by atomic absorption spectroscopy at 248 nm using a Nanolab AG aa/ae spectrophotometer (Video 12 E). The spectrophotometer is calibrated with standard solutions of iron(II) sulfate hexahydrate. Correlation of the iron content with the amount of protein determined by an amino acid analysis of the same probe[15] reveals the presence of 4.2 mole of iron per mole of tFdx. This result and ^{1}H NMR measurements[16] show that recombinant tFdx purified under aerobic conditions contains a single and intact 4Fe-4S cluster. In contrast, the native 4Fe-4S cluster of the ferredoxin from the hyperthermophilic archaeon *Pyrococcus furiosus,* the amino acid sequence of which is 50% identical to that of tFdx, decays rapidly to the 3Fe-4S form in the presence of oxygen.[17] This difference is probably due to the replacement of one of the cysteine residues of tFdx coordinating the cluster by an aspartate in *P. furiosus* ferredoxin.

The concentration of tFdx is determined by measuring the absorbance at 390 nm. The corresponding molar extinction coefficient ε_{390} is 17,400 M^{-1}cm^{-1}, as determined by correlating the absorption of tFdx with a quantitative amino acid analysis of the same probe. With a molecular mass of 6565 Da, this corresponds to a specific extinction coefficient $A_{390}^{0.1\%} = 2.65$ cm^{2}mg^{-1}.

Preparation and Characterization of apo-tFdx

Native tFdx contains a 4Fe-4S cluster,[8] whereas the ferredoxin of *Escherichia coli* belongs to the 2Fe-2S cluster type.[18] Nevertheless, recombinant tFdx is correctly assembled within *E. coli,* proving that the apoprotein specifies the type of cluster that is incorporated. Similar results are obtained for the expression in *E. coli* of other ferredoxins containing single 4Fe-4S or 3Fe-4S clusters.[19,20] Along these lines, it has also been shown that *in vitro* 4Fe-4S clusters can be incorporated into the corresponding apoferredoxins without any additional cellular factors, suggesting a self-assembly process.[21] To produce and characterize apo-tFdx, the iron–sulfur cluster is removed by acid treatment as described under "Test Expression."

---

[15] R. Knecht and J.-Y. Yang, *Anal. Chem.* **58,** 2375 (1986).

[16] G. Wildegger, D. Bentrop, A. Ejchart, M. Alber, A. Hage, R. Sterner, and P. Rösch, *Eur. J. Biochem.* **229,** 658–668.

[17] R. C. Conover, A. T. Kowal, W. Fu, J.-B. Park, S. Aono, M. W. W. Adams, and M. K. Johnson, *J. Biol. Chem.* **265,** 8533 (1990).

[18] T. T. Ta and L. E. Vickery, *J. Biol. Chem.* **267,** 11120 (1992).

[19] A. Heltzel, E. T. Smith. Z. H. Zhou, J. M. Blamey, and M. W. W. Adams, *J. Bacteriol.* **176,** 4790 (1994).

[20] B. Chen, N. K. Menon. L. Dervertarnian, J. J. G. Moura, and A. E. Przybyla, *FEBS Lett.* **351,** 401 (1994).

[21] F. Bonomi, S. Pagani, and D. M. Kurtz, Jr., *Eur. J. Biochem.* **148,** 67 (1985).

The homogeneity of apo-tFdx is tested by reversed-phase HPLC. Protein (5 μg) is applied onto a C_4 Aquapore butyl column (2.1 × 30 mm), and elution is performed at a flow rate of 150 μl/min with a gradient of solution A (0.05% v/v trifluoroacetic acid) and solution B (0.035% trifluoroacetic acid in 80% v/v acetonitrile): 0 min: 2% B, 5 min: 2% B, 65 min: 85% B, 75 min: 85% B. The eluate is monitored at 214 nm, since apo-tFdx contains neither tryptophan nor tyrosine residues. Apo-tFdx elutes as a single peak after 30 min, indicating a homogeneous population of molecules. Equilibrium runs in the analytical ultracentrifuge (Beckman, model Optima XL A) are performed with apo-tFdx at 10° in 50 mM Tris/HCl buffer (pH 7.4), containing 10 mM DTE. The absorption is followed at 220 nm. Assuming a partial specific volume of 0.72 ml/g, which is calculated from the amino acid composition,[22] the determined molecular weight is 7150 Da, showing that apo-tFdx is monomeric. Far-UV CD spectroscopy is performed under identical buffer conditions at 22°C. The absence of minima at 208 and 222 nm in the apoferredoxin and the presence of a minimum at 200 nm indicate that almost no secondary structure is retained after the iron–sulfur cluster is removed.[11] The most plausible explanation to reconcile these results with the observed specificity of cluster insertion into 4Fe-4S ferredoxins is to assume an equilibrium between structured and unstructured conformations of the apoferredoxin, where the equilibrium is far on the side of the unstructured form. Cluster insertion would then lead to a stabilization of the structured protein and consequently shift the conformational equilibrium of the protein toward this form. Following insertion of the 4Fe-4S cluster into the apo form of the high-potential iron protein from *Chromatium vinosum,* an intermediate with native-like tertiary structure and probably 4 Fe^{2+} ions bound to the polypeptide chain was detected by a combination of NMR and mass spectrometry.[23] A similar intermediate may be populated on cluster assembly in tFdx.

Thermal Stability

The thermal stability of tFdx is measured by differential scanning calorimetry (DSC) and absorption spectroscopy at 390 nm, recording heat capacity changes and the release of the iron–sulfur cluster on unfolding, respectively.[24] In physiological buffers, thermal unfolding occurs beyond temperatures accessible to both techniques. Therefore, in order to labilize tFdx, the chaotropic reagents guanidine hydrochloride (GuHCl) or guanidine thiocyanate (GuHSCN) have to be added. There is a linear relationship between the observed transition temperature (T_m, temperature at which 50% of the tFdx molecules are denatured) and the concentration of denaturant, both for GuHCl and GuHSCN. Extrapolations of T_m to zero denaturant concentrations yield the same value of 124.5°. Unfolding of tFdx is

---

[22] H. K. Schachman, *Methods Enzymol.* **4,** 32 (1957).
[23] K. Natarajan and J. A. Cowan, *J. Am. Chem. Soc.* **119,** 4082 (1997).
[24] W. Pfeil, U. Gesierich, G. R. Kleemann, and R. Sterner, *J. Mol. Biol.* **272,** 591 (1997).

irreversible, probably because of the decomposition of the released iron–sulfur cluster. Nevertheless, since denaturation is not followed by further detectable enthalpy changes and because the calorimetric and van't Hoff enthalpies as measured by DSC and absorption spectroscopy are identical, tFdx unfolding can be analyzed thermodynamically using the "two-state" model. This analysis yields values for the free energy of stabilization (ΔG) of 39 kJ/mol at 45° and of 36 kJ/mol at 25°. Thus, at room temperature tFdx is thermodynamically not more stable than an average mesophilic protein,[25] despite its extreme thermostability. This apparent discrepancy is explained by the small heat capacity change of tFdx on unfolding [$\Delta C_p = 3.6$ kJ/mol per degree kelvin (K)], which renders the ΔG versus temperature profile very shallow, causing a broad stability range from about $-20°$ to $125°$.[24]

Crystallization and X-Ray Structure

The investigation of the structural basis of high protein thermostability is much simpler for small, monomeric proteins compared to large proteins and oligomers. TFdx is ideally suited for this task since it contains only 60 amino acid residues and is extremely thermostable (see above). Furthermore, X-ray structures are available for comparison to a number of thermolabile ferredoxins containing a single 4Fe-4S or 3Fe-4S cluster.[26–28] TFdx is crystallized using the hanging drop vapor diffusion method. Five microliters of 8 mg/ml protein in water is mixed with 10 μl of 3.2 M ammonium sulfate in 0.1 M Tris/HCl (pH 7.5) and equilibrated by diffusion against 1 ml of the latter solution, at 20°. Crystals grow within 1 week and are suitable for X-ray structure analysis. A comparison of the X-ray structure of tFdx to mesophilic ferredoxins shows that the large differences in thermostability are not reflected in large differences of the overall protein structure,[29,30] a result also suggested by an NMR-based model of the solution structure of tFdx.[31] The most striking differences include the formation of additional hydrogen bonding networks in tFdx, involving both side-chain and main-chain atoms. These networks connect mainly turns and fix the N terminus more strongly to the central core of the protein. Other potentially stabilizing features are the shortening of a solvent-exposed surface loop, the increased content

---

[25] W. Pfeil, "Protein Stability and Folding: A Collection of Thermodynamic Data." Springer, Berlin, 1998.
[26] K. Fukuyama, Y. Nagahara, T. Tsukihara, Y. Katsube, T. Hase, and H. Matsubara, *J. Mol. Biol.* **199**, 183 (1988).
[27] C. R. Kissinger, L. C. Sieker, E. T. Adman, and L. H. Jensen, *J. Mol. Biol.* **219**, 693 (1991).
[28] A. Séry, D. Housset, L. Serre, J. Bonicel, C. Hatchikian, M. Frey, and M. Roth, *Biochemistry* **33**, 15408 (1994).
[29] S. Macedo-Ribeiro, B. Darimont, R. Sterner, and R. Huber, *Structure* **4**, 1291 (1996).
[30] S. Macedo-Ribeiro, B. Darimont, and R. Sterner, *Biol. Chem.* **378**, 331 (1997).
[31] H. Sticht, G. Wildegger, D. Bentrop, B. Darimont, R. Sterner, and P. Rösch, *Eur. J. Biochem.* **237**, 726 (1996).

of alanines in one α helix, and the replacement by glycines of three residues close to the iron–sulfur cluster, which are in energetically unfavorable conformations in other ferredoxins.[29,30]

Acknowledgments

I thank Prof. Kasper Kirschner for continous support, stimulating discussions, and critical comments on the manuscript, and Dr. Beatrice Darimont for initiating the work on tFdx and close cooperation in the first steps of the project, as well as Andrea Löschmann and Ariel Lustig for protein purification and analytical ultracentrifugation. Analyses of the three-dimensional structure and stability of tFdx were performed in pleasant and fruitful cooperations with Drs. Sandra Macedo-Ribeiro and Robert Huber (Max Planck Institut für Biochemie, Martinsried), Detlef Bentrop, Heinrich Sticht, Gudrun Wildegger, and Paul Rösch (Universität Bayreuth), and Wolfgang Pfeil and Ulrike Gesierich (Universität Potsdam). *Thermotoga maritima* cells were a gift from Drs. Robert Huber and Karl Otto Stetter (Universität Regensburg). I thank Drs. Detlef Bentrop, Astrid Merz and Ralf Thoma for critical reading of the manuscript. Financial support was received by grants No. 31-32369.91 and 31-45855.95 from the Swiss National Science Foundation to K. Kirschner.

[3] Ferredoxin from *Pyrococcus furiosus*

By CHULHWAN KIM, PHILLIP S. BRERETON, MARC F. J. M. VERHAGEN, and MICHAEL W. W. ADAMS

Introduction

Ferredoxins (Fd) are small proteins that contain iron–sulfur clusters as a redox active group. They are ubiquitous in biological systems and play integral roles in a wide variety of electron transfer processes, including respiration, photosynthesis, and fermentation. There are two main varieties: the 4Fe-type, which contain one and sometimes two cubane-type [4Fe-4S] clusters, and the 2Fe-type, which contain a [2Fe-2S] cluster (the S represents inorganic sulfide). Both types are covalently attached to their proteins via Fe–S bonds between the Fe atoms of the cluster and the sulfur atoms of four cysteine residues. Ferredoxins have been purified from a variety of microbial and eukaryotic sources and have been extensively studied.[1–3] The first to be characterized from a hyperthermophile was from *Pyrococcus furiosus*; in fact, this was one of the first proteins to be obtained from such an organism.[4]

[1] R. Cammack, *Adv. Inorg. Chem.* **38,** 281 (1992).
[2] H. Matsubara and K. Saeki, *Advs. Inorg. Chem.* **38,** 223 (1992).
[3] H. Beinert, R. Holm, and E. Münck, *Science* **277,** 653–659.
[4] S. Aono, F. O. Bryant, and M. W. W. Adams, *J. Bacteriol.* **171,** 3433 (1989).

The ease of purification and remarkable stability of *P. furiosus* ferredoxin, together with the availability of the recombinant protein,[5] has enabled it to become one of the best studied of all ferredoxins, with numerous investigations into the properties of both the protein and of its single [4Fe-4S] cluster. These include the use of spectroscopic techniques, such as electron paramagnetic resonance (EPR),[6-12] electron nuclear double resonance (ENDOR),[7-12] Mössbauer,[13] mass spectrometry,[14,15] and nuclear magnetic resonance (NMR).[16-25] Such studies have utilized several mutants of the protein and of alternate forms of the [4Fe-4S] cluster. The latter include (a) the [3Fe-4S] form, where a unique Fe atom of the

---

[5] A. Heltzel, E. T. Smith, Z. H. Zhou, J. M. Blamey, and M. W. W. Adams, *J. Bacteriol.* **176,** 4790 (1994).
[6] R. C. Conover, A. T. Kowal, W. Fu, J.-B. Park, S. Aono, M. W. W. Adams, and M. K. Johnson, *J. Biol. Chem.* **265,** 8533 (1990).
[7] J. Telser, H.-I. Lee, E. T. Smith, H. Huang, P. S. Brereton, M. W. W. Adams, R. C. Conover, M. K. Johnson, and B. M. Hoffman, *Appl. Magn. Reson.* **14,** 305 (1998).
[8] J.-B. Park, C. Fan, B. M. Hoffman, and M. W. W. Adams, *J. Biol. Chem.* **266,** 19351 (1991).
[9] J. Telser, E. T. Smith, M. W. W. Adams, R. C. Conover, M. K. Johnson, and B. M. Hoffman, *J. Am. Chem. Soc.* **117,** 5133 (1995).
[10] J. Telser, H. Huang, H.-I. Lee, M. W. W. Adams, and B. M. Hoffman, *J. Am. Chem. Soc.* **120,** 861 (1998).
[11] J. Telser, R. Davydov, C.-H. Kim, M. W. W. Adams, and B. M. Hoffman, *Inorg. Chem.* **15,** 3550 (1999).
[12] W. Fu, J. Telser, B. M. Hoffman, E. T. Smith, M. W. W. Adams, and M. K. Johnson, *J. Am. Chem. Soc.* **116,** 5722 (1994).
[13] K. K. P. Srivastava, K. K. Surerus, R. C. Conover, M. K. Johnson, J. B. Park, M. W. W. Adams, and E. Münck, *Inorg. Chem.* **32,** 927 (1993).
[14] E. T. Smith, D. S. Cornett, I. J. Amster, and M. W. W. Adams, *Anal. Biochem.* **209,** 379 (1993).
[15] K. A. Johnson, M. F. J. M. Verhagen, P. S. Brereton, M. W. W. Adams, and I. J. Amster. *Anal. Chem.* **72,** 1410 (2000).
[16] S. A. Busse, G. N. La Mar, L. P. Yu, J. B. Howard, E. T. Smith, Z. H. Zhou, and M. W. W. Adams, *Biochemistry* **31,** 11952 (1992).
[17] Q. Teng, Z. H. Zhou, E. T. Smith, S. C. Busse, J. B. Howard, M. W. W. Adams, and G. N. La Mar, *Biochemistry* **33,** 6316 (1994).
[18] C. M. Gorst, Y.-H. Yeh, Q. Teng, L. Calzolai, Z. H. Zhou, M. W. W. Adams, and G. N. La Mar, *Biochemistry* **34,** 600 (1995).
[19] C. M. Gorst, Z. H. Zhou, K. Ma, Q. Teng, J. B. Howard, M. W. W. Adams, and G. N. La Mar, *Biochemistry* **34,** 8788 (1995).
[20] C. Calzolai, C. M. Gorst, Z. H. Zhou, Q. Teng, M. W. W. Adams, and G. N. La Mar, *Biochemistry* **34,** 11373 (1995).
[21] A. Donaire, Z. H. Zhou, M. W. W. Adams, and G. N. La Mar, *J. Biomol. NMR* **7,** 35 (1996).
[22] L. Calzolai, Z. H. Zhou, M. W. W. Adams, and G. N. La Mar, *J. Am. Chem. Soc.* **118,** 2513 (1996).
[23] P.-L. Wang, A. Donaire, Z. H. Zhou, M. W. W. Adams, and G. N. La Mar, *Biochemistry* **35,** 11319 (1996).
[24] L. Calzolai, C. M. Gorst, K. L. Bren, Z. H. Zhou, M. W. W. Adams, and G. N. La Mar, *J. Am. Chem. Soc.* **119,** 9341 (1997).
[25] P. L. Wang, L. Calzolai, K. L. Bren, Q. Teng, F. E. Jenney, Jr., P. S. Brereton, J. B. Howard, M. W. W. Adams, and G. N. La Mar, *Biochemistry* **38,** 8167 (1999).

cluster has been removed,[6,16,17,18] (b) the 4Fe form, where the unique Fe atom has bound ligands,[9,26] and (c) [M3Fe-4S] forms, where the Fe atom has been replaced by other metal ions (M), such as Zn,[13,27] Ni,[13,28] Co,[27] Ti,[12] Cs,[12] Cu,[29] Mn,[27] Cr,[29] and Cd,[29] which in some cases bind ligands such as cyanide, thiols, and carbon monoxide. The mutant forms of the ferredoxin that have been studied have focused mainly on the residues that bind the iron–sulfur cluster.[30–33] *P. furiosus* ferredoxin is also the primary electron acceptor for a variety of oxidoreductase-type enzymes in this organism,[34] and the structure of an oxidoreductase–Fd complex has been reported.[35] In this chapter we describe the purification of ferredoxin from *P. furiosus,* and the purification of the recombinant protein from *Escherichia coli,* together with several mutant forms. Some of the methods that have been developed in characterizing this protein are also described. The use of NMR spectroscopy to study *P. furiosus* ferredoxin is described elsewhere in this volume.[36]

Properties of Ferredoxin from *Pyrococcus furiosus*

P. furiosus ferredoxin is a small protein of 66 amino acids (7500 Da) and is distinguished from most other ferredoxins by the coordination of its single [4Fe-4S] cluster by three, rather than four, cysteinyl residues. The fourth ligand is provided by an aspartyl residue.[20] This is suggested by sequence comparisons with other 4Fe-type ferredoxins (Fig. 1). Their clusters are coordinated by a consensus sequence, Cys-X-X-Cys-X-X-Cys-$(X)_n$-Cys-Pro (where $n \sim 35$ residues). The *P. furiosus* protein is the notable exception as there is an Asp residue in place of the expected second Cys. Hence, the cluster in this ferredoxin is coordinated by

---

[26] R. C. Conover, J.-B. Park, M. W. W. Adams, and M. K. Johnson, *J. Am. Chem. Soc.* **113,** 2799 (1991).
[27] M. G. Finnegan, R. C. Conover, J.-B. Park, Z. H. Zhou, M. W. W. Adams, and M. K. Johnson, *Inorg. Chem.* **34,** 5358 (1995).
[28] R. C. Conover, J.-B. Park, M. W. W. Adams, and M. K. Johnson, *J. Am. Chem. Soc.* **112,** 4562 (1990).
[29] C. R. Staples, I. K. Dhawan, M. G. Finnegan, D. A. Dwinell, Z. H. Zhou, H. Huang, M. F. J. M. Verhagen, M. W. W. Adams, and M. K. Johnson, *Inorg. Chem.* **36,** 5740 (1997).
[30] Z. H. Zhou and M. W. W. Adams, *Biochemistry* **36,** 10892 (1997).
[31] P. S. Brereton, M. F. J. M. Verhagen, Z. H. Zhou, and M. W. W. Adams, *Biochemistry* **37,** 7351 (1998).
[32] R. E. Duderstadt, C. R. Staples, P. S. Brereton, M. W. W. Adams, and M. K. Johnson, *Biochemistry* **38,** 10585 (1999).
[33] P. S. Brereton, R. E. Duderstadt, C. R. Staples, M. K. Johnson, and M. W. W. Adams, *Biochemistry* **38,** 10594 (1999).
[34] M. W. W. Adams, and A. Kletzin, *Advs. Prot. Chem.* **48,** 101 (1996).
[35] Y. Hu, S. Faham, R. Roy, M. W. W. Adams, and D. C. Rees, *J. Mol. Biol.* **286,** 899 (1999).
[36] G. N. La Mar, *Methods in Enzymology* **334** [30] (2001) (this volume).

```
Pf (4Fe)   AWKVSVDQDT CIGDAICASL CPDVFEMNDE G-KAQPKVEVI EDEELYNCAK EAMEACPVSA ITIEEA
Tl (4Fe)   KVSVDKDA   CIGCGVCASI CPDVFEMDDD G-KAKALVAET D----LECAK EAAESCPTGA ITVE
Tm (4Fe)   KVRVDADA   CIGCGVCENL CPDVFQLGDD G-KAKVLQPET D----LPCAK DAADSCPTGA ISVEE
Da (4Fe)   ARKFYVDQDE CIACESCVEI APGAFAMDPE IEKAYVKDVEG AS---QEEVE EAMDTCPVQC IHWEDE
Dg (3Fe)          PIEVNDD CMACEACVEI CPDVFEMNEE GDKAVVINPDS D----LDCVE EAIDSCPAEA IVRS
Mb (3Fe)   PATVNADE   CSGCGTCVDE CPNDAITLDE E-KGIAVVDND E---CVECG- ACEEACPNQA IKVEE
Ct (4Fe)   KVTVDQDL   CIACGTCIDL CPSVFDWDDE G-LSHVIVDEV PEGAEDSCAR ESVNECPTEA IKEV
```

FIG. 1. Amino acid sequence alignments of some 4Fe-type ferredoxins to show consensus sequence of cysteinyl residues. The cluster-binding ligands are indicated in bold and with an asterisk. The vertical lines indicate the two additional Cys residues found in many of these proteins. The abbreviations are Pf, *Pyrococcus furiosus;* Tl, *Thermococcus litoralis;* Tm, *Thermotoga maritima;* Da, *Desulfovibrio africanus;* Dg, *Desulfovibrio gigas;* Mb, *Methanosarcina barkeri;* Ct, *Clostridium thermoaceticum.* The sequences were taken from P. S. Brereton, R. E. Duderstadt, C. R. Staples, M. K. Johnson, and M. W. W. Adams, Biochemistry **38**, 10594 (1999). Whether these proteins contain a [4Fe-4S] or a [3Fe-4S] cluster when they are purified is indicated by (4Fe) and (3Fe), respectively.

Cys-11, Asp-14, Cys-17, and Cys-56. One consequence of the partial noncysteinyl ligation to the 4Fe cluster is that the Fe atom that would be normally coordinated by a cysteinyl sulfur is more easily removed when coordinated by a carboxyl group of Asp. This means that the ferredoxin can be quantitatively converted to the 3Fe form by chemical oxidation that specifically removes the Asp-coordinated Fe atom. The *P. furiosus* protein also contains two other Cys residues, at positions 21 and 48 (Fig. 1), and these are redox active and can form a disulfide bond. The 4Fe center is also redox active, and like most other clusters of this type, it undergoes one-electron redox chemistry with a low midpoint potential (-370 mV, $23°$). Hence, *P. furiosus* ferredoxin can exist in four formal redox states, where the cluster is either reduced (Fd_{red}) or oxidized (Fd_{ox}), and the two Cys residues 21 and 48 are either in the form of a disulfide (form A) or exist as free thiols (form B). These four distinct redox states of the protein are stable and can be prepared at room temperature. The lack of rapid equilibration between the different states appears to be due to the high stability of the protein. For example, the wild-type ferredoxin exhibits limited degradation even when incubated at $95°$ for many days.

Purification of Ferredoxin from *Pyrococcus furiosus*

P. furiosus (DSM 3638) is obtained from the Deutsche Sammlung von Mikroorganismen, Germany. It is routinely grown at $90°$ in a 600-liter fermentor with maltose as the carbon source as described previously.[37,38] The ferredoxin is purified from cell-free extracts that are also used to purify a variety of other proteins from this organism. In fact, ferredoxin can be obtained as a side product from the purification of many *P. furiosus* proteins since it is a very acidic protein and is usually well

---

[37] M. F. J. M. Verhagen, A. L. Menon, G. J. Schut, and M. W. W. Adams, *Methods in Enzymology* **330** [3] (2000).
[38] F. O. Bryant and M. W. W. Adams, *J. Biol. Chem.* **264**, 5070 (1989).

separated from other proteins of interest during an initial anion-exchange step.[37] Ferredoxin is brown because of its iron–sulfur chromophore and can be easily followed during purification procedures by visible inspection of chromatography columns. Purity is assessed by SDS–PAGE using 20% (w/v) acrylamide and by measuring the absorbance ratio at 390 and 280 nm of the air-oxidized protein. The purified protein is completely stable in air, but it is recommended that the purification procedure be carried out under strictly anaerobic conditions. If the procedure is performed aerobically, there is some modification and degradation of the FeS cluster.[6] In our laboratory the protein is also purified using buffers containing reductants such as sodium dithionite and/or dithiothreitol, which are added to protect some O_2-sensitive enzymes that are purified during the same purification procedure.[37]

Frozen cells (400 g, wet weight) of *P. furiosus* are thawed and suspended under argon in 1.1 liters of buffer A [50 mM Tris/HCl, pH 8.0, 10% (v/v) glycerol, 2 mM dithiothreitol (DTT), and 2 mM sodium dithionite] containing lysozyme (1 mg/ml) and DNase I (10 µg/ml) and are lysed by incubation at 35° for 2 hr. A cell-free extract is obtained by centrifugation at 50,000g for 80 min. The supernatant (1.3 liters) is loaded onto a column (8 × 21 cm) of DEAE-Sepharose Fast Flow (Pharmacia LKB, Piscataway, NJ) equilibrated with buffer A. After washing with one column volume of equilibration buffer, the column is eluted with a linear gradient (9.0 liters) of 0 to 0.5 M NaCl in buffer A and 90-ml fractions are collected. The ferredoxin elutes at approximately 0.34 M NaCl. The fractions (350 ml) containing ferredoxin are combined ($A_{390}/A_{280} \geq 0.15$, 2000 mg) and applied directly at 3 ml/min to a column (5 × 10 cm) of hydroxyapatite (high resolution, Calbiochem, La Jolla, CA) equilibrated with 50 mM Tris/HCl, pH 8.0. After washing with one column volume of equilibration buffer, the ferredoxin is eluted with a 2.5 liter linear gradient from 0 to 0.2 M potassium phosphate. The ferredoxin elutes as 0.02 M phosphate is applied. Fractions with a A_{390}/A_{280} ratio of $\geq$0.44 are combined (800 mg) and concentrated to approximately 20 ml using an ultrafiltration concentrator (Amicon, Beverly, MA) fitted with a YM3 membrane. The concentrated sample is applied to a column (6 × 60 cm) of Superdex S-75 (Pharmacia LKB) equilibrated with 50 mM Tris/HCl, pH 8.0, containing 0.1 M NaCl at 3 ml/min. Ferredoxin-containing fractions with a A_{390}/A_{280} ratio of $\geq$0.56 are combined (220 ml, 200 mg), concentrated to 5 ml by ultrafiltration, and stored in liquid nitrogen. The protein concentration of the pure protein is determined by the visible absorption of air-oxidized sample, using a molar absorption coefficient of $\varepsilon_{390} = 17,000$ M^{-1}cm^{-1}.

Purification of Recombinant Ferredoxin from *Escherichia coli*

Plasmid pAH1993 containing the gene encoding the ferredoxin is used for the expression of the recombinant protein[5] and as a template to construct mutants.[30–33]

For mutagenesis, the gene is subcloned into the NcoI and PstI site of the plasmid pUC118N (Worthington, Freehold, NJ), a vector that can replicate in the single-stranded form, and this is transformed into *E. coli* MV1190 competent cells (Bio-Rad, Richmond, CA). Transformants are selected and successful insertion of the gene encoding the ferredoxin is confirmed by gel analysis of a NcoI/PstI restriction digest. The plasmids containing the required mutations are obtained using either the Mutagene kit (Bio-Rad) or by PCR (polymerase chain reaction) amplification. Mutations are confirmed by Sanger dideoxy sequencing (Sequenase kit, USB, Cleveland, OH). The mutant gene is subcloned into the NcoI and PstI sites of the expression vector pTrc99A (Pharmacia Biotech, Piscataway, NJ) and the construct is used to transform *E. coli* JM105 (Stratagene, La Jolla, CA). Expression of the ferredoxin gene is then under the control of the isopropylthiogalactoside (IPTG)-inducible *trc* promoter.

E. coli JM105 containing the ferredoxin gene within the p*Trc*99A vector is grown aerobically at 37° in Luria–Bertani broth containing ampicillin (100 μg/ml) and $FeCl_3$ (25 μM). Since relatively large amounts of ferredoxin are required for in-depth spectroscopic analyses, *E. coli* cells containing the wild-type gene and many of the mutant forms are routinely grown in 80 liter cultures in a 100 liter fermentor (New Brunswick, NJ). Cultures are grown until $OD_{600} \sim 0.5$ and IPTG (1.5 mM) is added to induce the *trc* promoter. The cells are harvested after a 12 hr induction period, immediately frozen in liquid N_2, and stored at −80° until required. Cell yields are approximately 6 g (wet weight)/liter of LB medium.

The purification scheme for the recombinant form of the ferredoxin, and for mutants thereof, is carried out under anaerobic conditions, whereby all solutions are degassed and flushed with argon and contain sodium dithionite (2 mM) to remove trace O_2 contamination. The buffer used throughout the purification is 50 mM Tris-HCl, pH 8.0. Frozen *E. coli* cells (500 g, wet weight) are suspended (2 ml/g) in buffer containing EDTA (1 mM), NaCl (0.1 M), phenylmethylsulfonyl fluoride (0.14 mM), and lysozyme (0.25 mg/ml) and are incubated at 25° for 1 hr. DNase I (6.7 μg/ml) and RNase (1.7 μg/ml) are then added, and the mixture is incubated for a further 30 min to lyse the cells. The suspension is then incubated at 70° for 1 hr and stored at 4° overnight to denature *E. coli* proteins. After centrifugation for 45 min (23,000 g at 4°), the supernatant is diluted threefold with buffer and applied to a column (5 × 27 cm) of Q-Sepharose Fast Flow (Pharmacia-LKB). After washing with one column volume of buffer, the adsorbed proteins are eluted with a gradient (3 liters) from 0 to 0.5 M NaCl. The ferredoxin elutes as 0.39–0.46 M NaCl is applied to the column (the precise salt concentration varies with the mutant form). The brown fractions, typically with $A_{390}/A_{280} \geq 0.25$, are combined and concentrated by diluting them threefold with buffer and applying them to a column (1.6 × 13 cm) of Q-Sepharose. The ferredoxin is eluted with a gradient (20 ml) from 0 to 1.0 M NaCl. The concentrated pool (20 ml) is applied to a column (6 × 60 cm) of Superdex 75 and eluted with buffer. The purity of the fractions is determined by electrophoretic analysis using Tris–Tricine polyacrylamide (16%, w/v; see below)[39] and by the maximum value of the absorbance ratio (A_{390}/A_{280}). Note that for electrophoresis, approximately 30 μg of Fd must be loaded into each lane because of the poor staining of the protein with the Coomassie dye. Those fractions judged to contain pure ferredoxin are combined and concentrated by ultrafiltration (YM3 membrane, Amicon), prior to storage under argon at $-80°$.

Characterization of Recombinant Ferredoxins

The purification scheme described above has been used to purify the wild-type recombinant form of the ferredoxin and the mutants A1K,[30] D14X, where X = S, C,[30] H, N, V, and Y,[31] and the double mutants D14C/C11S, D14C/C17S, and D14C/C56S.[33] For the wild-type protein, approximately 300 mg of pure recombinant protein is obtained from 500 g (wet weight) of *E. coli* cells. Similar yields are obtained for the mutants, with the exception of those containing two changes in sequence. Specifically, the D14C/C11S and D14C/C17S mutants are produced in amounts at about 4–5 times lower than that of the wild-type protein, with yields of approximately 0.15 mg/g (wet weight) of *E. coli* cell paste. The D14C/C56S protein is purified at significantly lower levels, approximately 0.01 mg/g (wet weight) of *E. coli* cells.

The gene encoding the ferredoxin in *P. furiosus* encodes an N-terminal methionine residue, but this is removed *in vivo* when the gene is expressed in *E. coli* to give a recombinant protein with an N-terminal Ala, like the native protein.[5] In *E. coli,* whether the N-terminal methionine of a protein is removed depends upon the nature of the second translated amino acid (Ala in the case of ferredoxin). Hence, as expected, when a mutant form of the ferredoxin was designed in which the second codon was changed from Ala to Lys (A1K), analysis of the resulting recombinant protein showed that its N terminus was *N*-Met-Lys- instead of *N*-Ala. However, the A1K mutant was expressed by *E. coli* in amounts similar to that of the wild-type protein, showing that N-terminal processing is unlikely to limit the extent of expression of any of the mutant ferredoxins. In fact, all of the mutant forms of the protein examined also lacked the N-terminal Met residue, with the exception of A1K.

All forms of the recombinant ferredoxin examined, with the exception of the D14C/C56S mutant, contain an FeS cluster, and the purified ferredoxins contain little or no apoprotein. The amount of apoprotein present is ascertained using native Tris–Tricine polyacrylamide gels, which can distinguish between intact holoferredoxin containing an FeS cluster and apoferredoxin lacking an FeS cluster. Under native, nondenaturing conditions the holo and apo forms run very differently

---

[39] H. Schägger and G. von Jagow, *Anal. Biochem.* **166**, 368 (1987).

during electrophoresis. The native folded form runs with the dye front while the apo or unfolded form migrates significantly more slowly and the resulting band is more diffuse. In the case of the D14C/C56S protein, further purification is required to remove the apoprotein. The sample is loaded (in 1 ml aliquots) onto a Mono Q column (1 ml bed volume) equilibrated in 50 mM Tris/HCl, pH 8.0, containing NaCl (0.1 M). A 30 ml linear gradient from 0.1 to 0.45 M NaCl in the same buffer is applied to the column. The D14C/C56S mutant elutes as 0.23–0.32 M NaCl is added. The ferredoxin-containing fractions are collected, concentrated by ultrafiltration, and loaded onto a column (2.6 × 60 cm) of Sephadex G-75 equilibrated with 50 mM sodium phosphate, pH 7.7, containing 5 mM dithionite at 0.5 ml/min. Electrophoretic analysis of the ferredoxin-containing fractions shows that they are virtually homogeneous with a very small amount of remaining apoprotein.

While all of the mutant ferredoxins, as well as the wild-type, contain an FeS cluster, the nature of the cluster depends on the type of mutation. The type of cluster present is most easily ascertained by electron paramagnetic resonance (EPR) spectroscopy.[6,33] This technique shows that the D14V, D14H, D14Y, and D14N proteins all contain a [3Fe-4S] cluster, rather than the [4Fe-4S] cluster found in the wild-type protein and in all of the other mutants examined (A1K, D14S, D14C, D14C/C11S, D14C/C17S, and D14C/C56S). Nevertheless, 4Fe and 3Fe clusters have similar visible absorption properties, so the 3Fe forms can also be monitored during purification by their brown color. Moreover, visible absorption is also a measure of the integrity of the cluster, which in turn indicates whether the protein is in the fully folded form. Therefore, changes in the visible absorption properties of the ferredoxin can be used to assess the stability of mutant forms, where loss of absorption indicates loss of cluster and protein denaturation. However, the wild-type ferredoxin and all of the mutants are extremely stable near neutral pH. For example, they show little change in visible absorption even after 24 hr at 95°.[30] It is therefore advisable to carry out stability analyses at low pH where denaturation occurs on a reasonable time scale. Thus, the time required for the wild-type protein containing a 4Fe cluster (0.1 mg/ml) to lose 50% of its visible absorption (at 390 nm) at pH 2.5 (in 50 mM glycine hydrochloride) at 80°C is about 15 min. Surprisingly, this increases to about 1 hr with the 3Fe form, but decreases to less than 5 min with the 3Fe form of the D14S mutant (the 3Fe form is generated by chemical oxidation).[31] The predominant factors that control the stability of this protein at high temperatures have yet to be elucidated, although cluster type and the nature of the residue at position 14 seem to play a role.[31,36]

Redox States and Biological Activity of Ferredoxin

When purified under anaerobic conditions in the presence of reductant (sodium dithionite), the ferredoxin is in the so-called Fd_B^{red} form, where the 4Fe cluster is reduced (red) and the two Cys residues not involved in coordinating the cluster (positions 21 and 48; see Fig. 1) exist as free thiols (form B). These two redox active groups, the cluster and the Cys residues, differ greatly in their ease of oxidation and reduction.[19] Hence, the cluster is oxidized by exposing the sample to air and with slow shaking until the sodium dithionite is oxidized. This is easily tested by taking a small sample (10 μl) with a syringe and injecting it slowly into a solution of methyl viologen (10 mM in 50 mM Tris/HCl, pH 8.0). The sample solution will reduce the methyl viologen and turn blue if any sodium dithionite remains. The protein is then in the Fd_B^{ox} form, where the cluster is oxidized but the Cys residues remain reduced. To generate the A form of the protein which contains the disulfide bridge, the ferredoxin must be exposed to 100% O_2 for at least 24 hr. Conversely, conversion of the Fd_A^{ox} form to the Fd_A^{red} is achieved by making the sample anaerobic and by the addition of excess sodium dithionite, which rapidly (within seconds) reduces the cluster but not the disulfide. Conversion to the B (free thiol) form is an extremely slow process if sodium dithionite is used as the reductant, with a half-time of about 10 hr (in 50 mM Tris/HCl, pH 8.0), but occurs rapidly if dithiothreitol is also added.[19]

The biological activity of wild-type ferredoxin and mutants thereof can be assessed in several different assays, as the protein is the physiological electron carrier for a variety of different oxidoreductases in *P. furiosus*[34] and both direct and coupled assays are available. The direct assay uses pyruvate ferredoxin oxidoreductase (POR),[40,41] which reduces ferredoxin according to Eq. (1). The coupled assay involved POR and ferredoxin: NADP oxidoreductase (FNOR).[42,43] FNOR accepts electrons from reduced ferredoxin and reduces NADP [Eq. (2)]. Hence with a combination of POR, FNOR, and Fd, pyruvate oxidation can be coupled to NADP reduction.

$$\text{Pyruvate} + \text{CoASH} + 2\text{Fd}_{ox} \rightarrow \text{acetyl-CoA} + \text{CO}_2 + \text{H}^+ + 2\text{Fd}_{red} \quad (1)$$

$$2\text{Fd}_{red} + \text{H}^+ + \text{NADP} \rightarrow \text{NADPH} + 2\text{Fd}_{ox} \quad (2)$$

For the direct assay, the reaction mixture (2.0 ml) contains 50 mM EPPS buffer, pH 8.0, pyruvate (10 mM), coenzyme A (0.2 mM), $MgCl_2$ (1 mM), and Fd (5–125 μM). The mixture is incubated at 80° and the reaction is initiated by the addition of POR (0.15 μg/ml; ~20 units/mg, see below). Reduction of ferredoxin is measured by the decrease in visible absorption at 390 nm. The activity of the POR is determined independently using the same assay conditions but with the artificial electron acceptor methyl viologen (1 mM) replacing ferredoxin. The reaction is followed by measuring the reduction of methyl viologen at 600 nm

---

[40] G. J. Schut, A. L. Menon and M. W. W. Adams, *Methods in Enzymology* **331** [12] (2001).
[41] J. M. Blamey and M. W. W. Adams, *Biochim. Biophys. Acta* **1161**, 19 (1993).
[42] K. Ma and M. W. W. Adams, *Methods in Enzymology* **334** [4] (2001) (this volume).
[43] K. Ma and M. W. W. Adams, *J. Bacteriol.* **176**, 6509 (1994).

($\varepsilon_{600} = 12{,}000 \ mM^{-1} \ cm^{-1}$). One unit of POR activity catalyzes the oxidation of 1 μmol of pyruvate per min. It should be noted that the specific activity of POR, measured using ferredoxin or methyl viologen as the electron acceptor, decreases as the concentration of the protein in the assay medium increases (over the range 0.15–10 μg/ml of reaction mixture). Hence, in comparing activities of different mutants of the ferredoxin, or of different samples of POR, care should be taken to make sure that similar amounts of POR are used in the assay.[31]

A variation of the direct assay shown in Eq. (1) involves the use of metronidazole. This artificial electron carrier is spontaneously and irreversibly reduced by various ferredoxins,[44,45] including that of *P. furiosus*.[30] Hence, the concentration of oxidized ferredoxin in the POR assay mixture effectively remains constant over the assay period and the kinetic results tend to be much more reproducible than those obtained while monitoring ferredoxin reduction directly.[30] In general, the activities measured with *P. furiosus* ferredoxin in the metronidazole system are about half of those calculated in the direct assay, but the K_m value(s) for ferredoxin (and mutants thereof) are an order of magnitude lower. The reason for this is presumably related to the spontaneous regeneration of oxidized ferredoxin. In these assays, the reaction conditions are the same as for the direct assay except that metronidazole (100 μM; [1-(2-hydroxyethyl)-2-methyl-5-nitroimidazole, obtained from Sigma Chemical, St. Louis, MO) is added and the reaction is measured by following metronidazole reduction at 320 nm. A molar absorption coefficient of 9300 $M^{-1}cm^{-1}$ is used for oxidized metronidazole,[44] and its reduction is assumed to be a one-electron process.[45] Metronidazole reduction occurs at a significant rate in the POR assay in the absence of added ferredoxin, and this must be taken into account when calculating enzyme activity. The absorption at 320 nm of a 100 μM solution (in 50 mM EPPS, pH 8.0) of metronidazole remains unchanged after a 20 min incubation at 80°, indicating that this compound does not undergo thermal degradation under assay conditions.[30] Note that lower amounts of substrate (ferredoxin) can be used in the metronidazole-linked assay than in the direct assay, which allow the reaction to approach V_m, i.e., $[S] \gg 5 \times K_m$.[31]

The biological function of a redox protein such as ferredoxin is more appropriately measured in a true coupled assay in which electron transfer (rather than just reduction or oxidation of the ferredoxin) is measured. This also lessens the effects that may be caused by nonspecific interactions. The POR/FNOR coupled assay illustrated by Eq. (2) is designed to specifically measure the interaction of the ferredoxin with FNOR, wherein an excess of POR ensures that the rate-limiting step is not the reduction of ferredoxin by POR. This assay is also performed at 80° in 50 mM EPPS buffer, pH 8.0, and the 2 ml reaction mixture contains pyruvate (10 mM),

---

[44] J.-S. Chen and D. K. Blanchard, *Anal. Biochem.* **93**, 216 (1979).
[45] S. M. J. Moreno, R. P. Mason, R. P. A. Muniz, F. S. Cruz, and R. Docampo, *J. Biol. Chem.* **258**, 4051 (1983).

coenzyme A (0.2 mM), MgCl$_2$ (1 mM), POR (~20 μg/ml; 10 units/mg), and ferredoxin (0.1–20 μM). The reaction is started by the addition of FNOR (5 μg/ml; ~50 units/mg, see below) and activity is measured by the reduction of NADP at 365 nm. Calculations use a molar absorbance of 3400 M^{-1}cm^{-1} for NADPH, and one unit of activity in this assay is 1 μmol pyruvate oxidized per min (equivalent to 2 μmol Fd oxidized per min). The activity of the FNOR is determined using benzyl viologen (1 mM) as the electron acceptor with NADH (0.3 mM) as the electron donor. One unit of activity of FNOR catalyzes the reduction of 2 μmol of benzyl viologen per min in 50 mM CAPS buffer, pH 10.3.[42]

All of the ferredoxin mutants described above have been analyzed in both the direct and coupled assay systems [Eqs. (1) and (2)]. Surprisingly, it is the reduction potential of the mutant that appears to be the predominant factor in determining the efficiency of electron transfer in both systems. The cluster type (4Fe or 3Fe), the nature of the residue at position 14 (Asp or Cys, or a variety of other residues), or the coordination environment of the 4Fe cluster (whether by four Cys, or by three Cys and one Ser in four different orientations) seems to have little effect on the interaction between, and transfer of electrons from, ferredoxin to FNOR.[30,31–33]

Acknowledgment

This research was supported by grants from the Department of Energy, the National Institutes of Health, and the National Science Foundation.

[4] Ferredoxin:NADP Oxidoreductase from *Pyrococcus furiosus*

By KESEN MA[1] and MICHAEL W. W. ADAMS

Introduction

Ferredoxin:NADP$^+$ oxidoreductase (FNOR, EC 1.18.1.2) is a flavoenzyme that catalyzes electron transfer between the redox protein, ferredoxin, and the pyridine nucleotide coenzymes, NADP(H) and/or NAD(H). Enzymes of this type have been characterized from many organisms, including from both the bacterial and eukaryotic domains.[1–3] However, only one such enzyme has been purified from

---

[1] A. K. Arakaki, E. A. Ceccarelli, and N. Carrillo, *FASEB J.* **11**, 133 (1997).
[2] J. K. Hurley, A. M. Weber-Main, A. E. Hodges, M. T. Stankovich, M. M. Benning, H. M. Holden, H. Cheng, B. Xia, J. L. Markley, C. Genzor, C. Gomez-Moreno, R. Hafezi, and G. Tollin, *Biochemistry* **36**, 15109 (1997).
[3] Y. S. Jung, V. A. Roberts, C. D. Stout, and B. K. Burgess, *J. Biol. Chem.* **274**, 2978 (1999).

the hyperthermophilic archaea, that from *Pyrococcus furiosus*.[4] The *P. furiosus* enzyme not only functions as a very efficient FNOR, but also catalyzes a variety of reactions, including the reduction of polysulfide using NADPH as electron donor.[4] The assay methods, purification procedure, and properties of *P. furiosus* FNOR are described in this chapter.

Assays for FNOR

Ferredoxin-Dependent Reduction of NADP

FNOR activity can be determined by measuring the formation of NADPH from NADP using reduced *P. furiosus* ferredoxin as the electron donor. Reduced ferredoxin[5] is generated by *P. furiosus* pyruvate ferredoxin oxidoreductase (POR), which oxidatively decarboxylates pyruvate to acetyl-CoA.[6] The reaction is carried out anaerobically under argon in a serum-stoppered cuvette (3 ml), since the reduced ferredoxin is rapidly oxidized in air.[5] The 2 ml reaction mixture contains 100 mM EPPS [N-(2-hydroxyethyl)piperazine-N'-(3-propanesulfonic acid)] buffer, pH 8.0, 10 mM pyruvate, 0.2 mM coenzyme A, 80 μg POR purified from *P. furiosus*,[7] 12.5 μg ferredoxin purified from *P. furiosus*, 0.3 mM NADP, and 10 μg FNOR.[8] The assay temperature is 80°. The reaction can be initiated by adding FNOR, ferredoxin, NADP, or POR as the final component. The production of NADPH is measured at 365 nm and the molar absorbance of 3400 M^{-1} cm^{-1} is used to calculate the concentration of NADPH. One unit of enzyme activity is equal to 1 μmol NADPH produced per min.

NADPH-Dependent Reduction of Elemental Sulfur and Polysulfide

Either NADPH or reduced ferredoxin can be used as the electron donor for the reduction of elemental sulfur (S°) and polysulfide to sulfide. This represents the so-called sulfide dehydrogenase activity of FNOR.[4] A solution of polysulfide (0.5 M) is prepared by mixing 12 g of Na$_2$S with 1.6 g of sublimed elemental sulfur in 100 ml of anoxic water.[9] The concentration of polysulfide is measured by cold cyanolysis.[10] The S°- and polysulfide-dependent oxidation of NADPH is measured at 340 nm (molar absorptivity of 6200 M^{-1} cm^{-1}) under anaerobic conditions. The assay mixture (2 ml) contains 100 mM EPPS, pH 8.0, 0.3 mM NADPH, 1.5 mM

---

[4] K. Ma and M. W. W. Adams, *J. Bacteriol.* **176**, 6509 (1994).
[5] S. Aono, F. O. Bryant, and M. W. W. Adams, *J. Bacteriol.* **171**, 3433 (1989).
[6] J. M. Blamey and M. W. W. Adams, *Biochim. Biophys. Acta* **1161**, 19 (1993).
[7] G. J. Schut, A. L. Menon, and M. W. W. Adams, Methods in *Enzymology* **331** [12] (2001).
[8] K. Ma, Z.-H. Zhou, and M. W. W. Adams. *FEMS Microbiol. Lett.* **122**, 245 (1994).
[9] S. H. Ikeda, T. Satake, T. Hisano, and T. Terazawa, *Talanta* **19**, 1650 (1972).
[10] T. Then and H. G. Trüper, *Arch. Microbiol.* **135**, 254 (1983).

polysulfide or 0.05% (w/v) colloidal sulfur or 5% (w/v) sublimed elemental sulfur. The assay temperature is 80° and the reaction is initiated with the addition of the enzyme. One unit of activity is equal to 1 μmol NADPH oxidized per min.

Ferredoxin-Dependent Reduction of Elemental Sulfur and Polysulfide

When reduced ferredoxin is used as the electron donor for polysulfide reduction, the reaction is monitored by the production of sulfide. The assays are performed in 8-ml serum-stoppered vials under Ar that are shaken at 150 rpm at 80°. The reaction mixture (2 ml) contains 100 mM EPPS buffer, pH 8.0, 10 mM pyruvate, 2 mM coenzyme A, 150 μg POR from *P. furiosus*, 25 μM ferredoxin, 40 μg FNOR, and a source of elemental sulfur. This is polysulfide (1.5 mM), sublimed elemental sulfur (0.5%, w/v; J. T. Baker, Marietta, GA), or colloidal sulfur (0.05%, w/v; Fluka, Ronkonkoma, NY). Note that sublimed elemental sulfur has a very low solubility, whereas colloidal sulfur generates a fine suspension. At 20 min intervals over 2 hr, aliquots of the reaction are removed with a syringe and the amount of sulfide produced is measured by methylene blue formation.[11] One unit of activity is defined as 1 μmol sulfide produced per min. When reactions are monitored by sulfide production, it is important that control assays be carried out without the addition of enzyme. This is particularly important when polysulfide is used as the source of elemental sulfur.

NADH-Dependent Reduction of Benzyl Viologen

FNOR also functions as a NADH:benzyl viologen oxidoreductase (NBVOR) *in vitro* and catalyzes the NADH-dependent reduction of the redox dye benzyl viologen. The assay is convenient and is routinely used for detecting FNOR during the purification procedures. The assay mixture (2 ml) contains 50 mM CAPS[3-cyclohexylamino)-1-propanesulfonic acid] buffer, pH 10.3, 0.3 mM NADH, and 1 mM benzyl viologen. The reaction is carried out at 80° and is initiated by the addition of FNOR. A molar absorbance of 7800 M^{-1} cm^{-1} is used to calculate the concentration of reduced benzyl viologen that is generated. One unit of activity equals the reduction of 2 μmol benzyl viologen per min.

A variety of other electron acceptors can be used to measure FNOR activity. These include (where their molar absorbances at the indicated wavelength are given in parentheses): methyl viologen (9700 M^{-1} cm^{-1} at 580 nm), 2,6-dichlorophenol indophenol (19,100 M^{-1} cm^{-1} at 600 nm), potassium ferricyanide (1020 M^{-1} cm^{-1} at 420 nm), cytochrome *c* (from horse heart, 19,520 M^{-1} cm^{-1} at 550 nm), methylene blue (30,500 M^{-1} cm^{-1} at 670 nm), flavin mononucleotide, and flavin adenine dinucleotide (11,300 M^{-1} cm^{-1} at 450 nm).

---

[11] J.-S. Chen and L. E. Mortenson, *Anal. Biochem.* **79**, 157 (1977).

Purification of FNOR

P. furiosus (DSM 3638) is obtained from the Deutsche Sammlung von Mikroorganismen, Germany. It is routinely grown at 90° in a 600-liter fermentor with maltose as the carbon source as described previously.[12,13] FNOR is purified from 400 g (wet weight) of cells under strictly anaerobic conditions at 23°.[4] Frozen cells are thawed in 1.2 liters of buffer A that contains 50 mM Tris/HCl, pH 8.0, 10% (v/v) glycerol, 2 mM dithiothreitol, 2 mM sodium dithionite, lysozyme (1 mg/ml), and DNase I (10 μg/ml). The cells are lysed by incubation at 35° for 2 hr. The cell-free extract is obtained by centrifugation at 50,000 g for 80 min.

DEAE-Sepharose Chromatography

The cell-free extract (1.3 liter) is loaded onto a column (10 × 20 cm) of DEAE-Sepharose Fast Flow (Pharmacia Biotech, Piscataway, NJ) equilibrated with buffer B (buffer A without lysozyme and DNase I). A 15 liter linear gradient of 0 to 0.5 M NaCl in buffer B is used and 90-ml fractions are collected. FNOR activity starts to elute as 0.18 M NaCl is applied to the column. Those fractions containing FNOR activity are combined (∼810 ml), concentrated by ultrafiltration (type PM30 membrane), and washed with buffer C (buffer B without sodium dithionite).

Blue Sepharose Chromatography

The concentrated sample from the previous column is loaded onto a column (5 × 12 cm) of Blue Sepharose (Pharmacia Biotech) equilibrated with buffer C. The column is eluted with a 1.4 liter linear gradient of 0 to 2 M NaCl in buffer C and 50-ml fractions are collected. FNOR activity starts to elute as 1.5 M NaCl is applied. It should be noted that three peaks of the NBVOR assay are separated by this chromatography step. The first peak to elute contains ∼20% of the total activity and the enzyme responsible has not been characterized. The third peak to elute represents ∼40% or the NBVOR activity and in this case BV reduction is catalyzed by NADPH rubredoxin oxidoreductase (NROR).[14] It is the second peak of NBVOR activity, which also contains about 40% of the total, that corresponds to FNOR, and this starts to elute as 1.7 M NaCl is applied to the column. Fractions containing FNOR are combined (600 ml) and concentrated to 30 ml by ultrafiltration (PM30 membrane).

Superdex 200 Gel Filtration Chromatography

The concentrated sample (30 ml) is applied to a column (6 × 60 cm) of Superdex 200 (Pharmacia Biotech) equilibrated with buffer C containing 50 mM

---

[12] F. O. Bryant and M. W. W. Adams, *J. Biol. Chem.* **264**, 5070 (1989).
[13] M. F. J. M. Verhagen, A. L. Menon, G. J. Schut, and M. W. W. Adams, *Methods in Enzymology* **330** [3] (2000).
[14] K. Ma and M. W. W. Adams, *J. Bacteriol.* **181**, 5530 (1999).

KCl, and fractions of 25 ml are collected. Those containing FNOR activity are combined (50 ml) and used directly for the next step.

Q-Sepharose Chromatography

The combined fractions from the Superdex 200 column are applied to column (2.6 × 15 cm) of Q-Sepharose (Pharmacia Biotech) equilibrated with buffer C. The column is eluted with a 500 ml gradient from 0 to 1.0 M KCl in buffer C and fractions of 20 ml are collected. FNOR activity starts to elute as 0.24 M KCl is applied. Those fractions containing FNOR that are judged homogeneous using SDS–PAGE are combined (80 ml), concentrated by ultrafiltration (PM-30 membrane), and stored in liquid nitrogen. This procedure yields approximately 45 mg of pure FNOR with a specific activity in the NBVOR assay of 160 units/mg.

Properties of FNOR

Molecular Properties

The FNOR holoenzyme has a molecular mass of 90,000 ± 5000 Da as estimated by gel filtration on a calibrated column of Superdex 200. The purified protein gives rise after analysis by SDS–PAGE to two protein bands with masses of 52,000 (α) and 29,000 (β) Da, indicating that the holoenzyme is a heterodimer. The holoenzyme contains two flavin groups, as determined by visible absorption spectroscopy, and 11 iron and 7 acid-labile sulfur atoms, as measured by colorimetric assays.[4] Electron paramagnetic resonance (EPR) spectroscopy indicates that these are present in the form of at least three FeS clusters.[4] The flavin can be extracted by acid treatment and was identified as FAD using thin layer chromatography. Pure FNOR is very thermostable, with the time required to lose 50% of its activity at 95° being about 12 hr.

Catalytic Properties

Purified FNOR will use NADPH to reduce a variety of compounds with a wide range of reduction potentials. These include (E_m values are given in parentheses) methyl viologen (−440 mV), benzyl viologen (−350 mV), FAD (−220 mV), FMN (−190 mV), methylene blue (+11 mV), 2,6-dichlorophenol indophenol (+220 mV), cytochrome c (+250 mV), ferricyanide (+360 mV), and oxygen (+820 mV). However, the enzyme will not reduce fumarate, succinate, nitrate, nitrite, sulfate, sulfite, or protons. The measured K_m values (measured at 80°) for NADPH and NADH (in the reduction of benzyl viologen), and for polysulfide, benzyl viologen, rubredoxin, and oxygen (in the oxidation of NADPH) and for reduced ferredoxin (in the reduction of NADP) are 11, 71, 1250, 125, 1.6, 240 and 0.7 μM, respectively. The corresponding k_{cat}/K_m values (in mM^{-1}s^{-1}) are 3.2×10^4, 3.5×10^3, 1.5×10^1, 3×10^3, 8.4×10^2, 9.3×10^2, and 1.5×10^4, respectively. These results suggest that NADP(H) and ferredoxin are the likely physiological substrates for FNOR. The reduction potentials of NADP(H) and *P. furiosus* ferredoxin are comparable, at least at 25°, with values of −320 and −370 mV,[15] respectively. However, FNOR exhibits only low activity (<0.1 units/mg at 80°) in catalyzing the NADPH-dependent reduction of oxidized ferredoxin. It is much more active (18 units/mg) in catalyzing the reduced ferredoxin-dependent reduction of NADP. Moreover, reduced ferredoxin, generated by the POR reaction, has an apparent K_m value of less than 1 μM, suggesting that the function of FNOR *in vivo* is to generate NADPH.

FNOR exhibits high activity in catalyzing the NADPH-dependent reduction of polysulfide to H_2S. In fact, at 80° (pH 8.0), the activity (14 units/mg) is comparable to that measured in the ferredoxin-dependent reduction of NADP (18 units/mg). However, the apparent K_m value for polysulfide is 1.25 mM, more than three orders of magnitude greater than that for reduced ferredoxin. The intracellular concentration of polysulfide is not known; therefore, the role of this enzyme in catalyzing S° reduction *in vivo* is at present uncertain. FNOR also catalyzes the NADPH-dependent reduction of oxygen at an extremely high rate (166 units/mg), comparable to that measured for benzy viologen (263 units/mg). However, in this case the apparent K_m value for oxygen is 0.24 mM, which is equivalent to air at more than 1.4 atm. It therefore seems unlikely that oxygen reduction by FNOR has any physiological significance.

Acknowledgments

This research was supported by grants from the Department of Energy.

[15] P. S. Brereton, M. F. J. M. Verhagen, Z. H. Zhou, and M. W. W. Adams, *Biochemistry* **37,** 7351 (1998).

[5] Rubredoxin from *Pyrococcus furiosus*

By FRANCIS E. JENNEY, JR. and MICHAEL W. W. ADAMS

Introduction

Rubredoxin (Rd) was first discovered in clostridia in 1965 by Lovenberg and Sobel[1] and is the simplest of all iron-sulfur proteins. Ironically, it has become one

[1] W. Lovenberg and B. E. Sobel, *Proc. Natl. Acad. Sci. U.S.A.* **54,** 193 (1965).

of the best characterized proteins of any type in terms of structure, even though its true physiological function remains elusive. Rubredoxins are small, redox-active proteins that range in size from 45 to 55 amino acids (Fig. 1). They all contain two conserved cysteine motifs ($CxxCx_nCPxC$, where x denotes any amino acid and $n = 22$–29) that coordinate a single iron atom via the cysteinyl sulfurs. More than a dozen rubredoxins have been purified and characterized (see, for example, refs. 2 and 3) and the protein and complete genome databases contain 26 rubredoxin-like sequences (Fig. 1). So far rubredoxin has been found exclusively in strict anaerobes, which includes both bacteria and archaea, and some organisms that might be considered microaerophiles (see Fig. 1).

There are also a number of much larger proteins that contain rubredoxin-like sequences or domains that are found in facultative anaerobes and aerobes. Such proteins vary considerably in size of, as well as location of and distance between, the cysteine motifs. These rubredoxin-related proteins can be divided into several classes. One class, with 10 members, contains the same cysteine motifs as rubredoxin ($CxxC_nCPxC$, where $n = 22$–29) but are slightly larger (10–40 extra amino acids distributed at both the amino and carboxyl termini). An example is the *Hup*I protein (GenBank accession P28151) from the aerobe *Rhizobium leguminosarum*. Other classes containing the same motifs are larger and include one containing eight proteins that range in size from 105 to 160 residues, e.g., from *Pseudomonas putida* (GenBank accession CAB54051), a class of five proteins that range from 160 to 178 residues, e.g., from *Ps. putida* (GenBank accession CAB54052), and another class of six proteins ranging from 279 to 661 residues, e.g., from *Pyrococcus horikoshii* (GenBank accession BAA30645). There are also a number of proteins that resemble true rubredoxins, except that the distance between the two cysteines in the amino-terminal motif is either less than (e.g., *Methanococcus jannaschii* Rd II, GenBank accession AAB98734) or greater than (*Desulfovibrio vulgaris* Rd-like protein, GenBank accession AAB39992) two residues. Other than the *Pseudomonas oleovorans* proteins, which are thought to be involved in alkane oxidation,[4] little is known about these rubredoxin-related proteins.

The canonical rubredoxin has been purified from only one hyperthermophile, the archaeon *Pyrococcus furiosus*.[5] The other sequences of hyperthermophilic rubredoxins shown in Fig. 1 are putative proteins derived from genome sequences. Note that while all rubredoxins are highly similar, the hyperthermophilic

---

[2] K. A. Richie, Q. Teng, C. J. Elkin, and D. M. Kurtz Jr., *Prot. Sci.* **5,** 883 (1996).

[3] R. Bau, D. C. Rees, D. M. Kurtz, Jr., R. A. Scott, H. Huang, M. W. W. Adams, and M. K. Eidsness, *J. Biol. Inorg. Chem.* **3,** 484 (1998).

[4] M. Kok, R. Oldenhuis, M. P. van der Linden, C. H. C. Meulenberg, J. Kingma, and B. Witholt, *J. Biol. Chem.* **264,** 5442 (1989).

[5] P. R. Blake, J.-B. Park, F. O. Bryant, S. Aono, J. K. Magnuson, E. Eccleston, J. B. Howard, M. F. Summers, and M. W. W. Adams, *Biochemistry* **30,** 10885 (1991).

FIG. 1. Sequence alignment of rubredoxin sequences. They were aligned using the GCG Pileup program (Wisconsin Package Version 10.0, Genetics Computer Group (GCG), Madison, WI). Hyperthermophilic rubredoxins are indicated in bold-face type. GenBank accession numbers are listed below in parentheses. Pfu (*Pyrococcus furiosus*, P24297), AfI (*A. fulgidus* RdI, AAB90363), AfIIs (*Archaeoglobus fulgidus* RdII, AAB89910; 19 residues at the putative amino terminus are not shown, see below), MthIs (*Methanobacterium thermoautotrophicum*, AAB84661; 10 residues at the putative amino terminus are not shown, see below), MthII (*M. thermoautotrophicum*, AAB84662), Pab (*Pyrococcus abyssi*, CAB49806), Tma (*Thermotoga maritima*, AAD35743), Aca (*Acinetobacter calcoaceticus*, Genbank Accession number Z46863), Bme (*Butyribacterium methylotrophicum*, JU0127), Cbu (*Clostridium butyricum*, CAA72620), Cce (*Clostridium cellulolyticum*, CAB41597), Cli (*Chlorobium limicola*, A27537), AAB89910), Cpa (*Clostridium pasteurianum*, AAA23279), Cpe (*Clostridium perfringens*, JU0074), Cst (*Clostridium sticklandii*, A33182), Dba (*Desulfoarculus baarsii*, X99543), Dde (*Desulfovibrio desulfuricans*, RUDVD), Dgi (*Desulfovibrio gigas*, P00270), DvH (*Desulfovibrio vulgaris*, Hildenborough, P00269), DvM (*Desulfovibrio vulgaris*, Miyazaki F, P15412), Hmo (*Heliobacillus mobilis*, P56263), Mel (*Megasphaera elsdenii*, P00271), Mtu (*Mycobacterium tuberculosis*, CAB08322), Pas (*Peptostreptococcus asaccharolyticus*, RUPE), Tpa (*Treponema pallidum*, AAC65947), Tth (*Thermoanaerobacterium thermosaccharolyticum*, P19500). Note that the open reading frames encoding putative rubredoxins (AfIIs and MthIs) in the complete genome sequences of *A. fulgidus* and *M. thermoautotrophicum* are proposed to begin 57 and 30 base pairs, respectively, before the sequence shown. However, the start sites indicated here (by the N-terminal methionine) are equally likely and the actual start site must await sequence data from purified proteins. A consensus sequence of the residues totally conserved in all of these rubredoxin sequences is given at the bottom (Cons).

archaeal sequences do cluster together, although the hyperthermophilic bacterial rubredoxin sequence (*Thermotoga maritima*) does not (Fig. 1). The aerobic, hyperthermophilic bacterium *Aquifex aeolicus*[6] and facultatively anaerobic hyperthermophilic archaeon *Pyrobaculum aerophilum*[7] do not appear to possess a rubredoxin, consistent with a physiological role for rubredoxin only in strictly anaerobic species. Genome sequences are available, however, for three species of *Pyrococcus*, *P. horikoshii*,[8] *P. abysii*,[9] and *P. furiosus*,[10] and, despite the fact that all three are strictly anaerobic heterotrophs, *P. horikoshii* does not appear to have a gene encoding the canonical rubredoxin.

The rubredoxin from *P. furiosus*[5] is perhaps the best-studied protein of this type. Its structure has been determined by NMR spectroscopy[11,12] and by X-ray crystallography[13] to a resolution of 0.95 Å.[14] Indeed, it was the first hyperthermophilic protein for which a structure became available.[11,13] The protein also serves as a model system for a metal ion coordinated by four cysteinyl sulfur atoms within a protein environment. The cadmium,[15–17] mercury,[16,17] nickel,[18,19] and zinc,[11,17] as well as the iron-containing, forms[20–23] have been investigated by a

[6] G. Deckert, P. V. Warren, T. Gaasterland, W. G. Young, A. L. Lenox, D. E. Graham, R. Overbeek, M. A. Snead, M. Keller, M. Aujay, R. Huber, R. A. Feldman, J. M. Short, G. J. Olsen, and R. V. Swanson, *Nature* **392**, 353 (1998).

[7] S. Fitz-Gibbon, personal communication.

[8] Y. Kawarabayasi, M. Sawada, H. Horikawa, Y. Haikawa, Y. Hino, S. Yamamoto, M. Sekine, S. Baba, H. Kosugi, A. Hosoyama, Y. Nagai, M. Sakai, K. Ogura, R. Otsuka, H. Nakazawa, M. Takamiya, Y. Ohfuku, T. Funahashi, T. Tanaka, Y. Kudoh, J. Yamazaki, N. Kushida, A. Oguchi, K. Aoki, and H. Kikuchi, *DNA Res.* **5**, 55 (1998).

[9] D. Prieur, P. Forterre, J. C. Thiery, J. Quérellou, *et al.*, Genoscope, http://www.genoscope.cns.fr/ (1999).

[10] R. Weiss, F. T. Robb, *et al.*, http://comb5-156.umbi.umd.edu/ (1999).

[11] P. R. Blake, J.-B. Park, Z. H. Zhou, D. R. Hare, M. W. W. Adams, and M. F. Summers, *Prot. Sci.* **1**, 1508 (1992).

[12] P. R. Blake, M. W. Day, B. T. Hsu, L. Joshua-Tor, J.-B. Park, Z. H. Zhou, D. R. Hare, M. W. W. Adams, D. C. Rees, and M. F. Summers, *Prot. Sci.* **1**, 1522 (1992).

[13] M. W. Day, B. T. Hsu, L. Joshua-Tor, J.-B. Park, Z. H. Zhou, M. W. W. Adams, and D. C. Rees, *Prot. Sci.* **1**, 1494 (1992).

[14] R. Bau, D. C. Rees, D. M. Kurtz, Jr., R. A. Scott, H. Huang, M. W. W. Adams, and M. K. Eidsness, *J. Biol. Inorg. Chem.* **3**, 484 (1998).

[15] P. R. Blake, J.-B. Park, M. W. W. Adams, and M. F. Summers, *J. Am. Chem. Soc.* **114**, 4931 (1992).

[16] P. R. Blake, M. F. Summers, M. W. W. Adams, J.-B. Park, Z. H. Zhou, and A. Bax, *J. Biomol. NMR* **2**, 527 (1992).

[17] P. R. Blake, B. Lee, J.-B. Park, Z. H. Zhou, M. W. W. Adams, and M. F. Summers, *New. J. Chem.* **18**, 387 (1994).

[18] Y.-H. Huang, I. Moura, J. J. G. Moura, J. LeGall, J.-B. Park, M. W. W. Adams, and M. K. Johnson, *Inorg. Chem.* **32**, 406 (1993).

[19] Y.-H. Huang, J.-B. Park, Z. H. Zhou, M. W. W. Adams, and M. K. Johnson, *Inorg. Chem.* **32**, 375 (1993).

variety of techniques, including NMR,[11,15–17] resonance Raman,[18] electron paramagnetic resonance,[19] and X-ray absorption and magnetic circular dichroism (CD) spectrocopy.[20–23] The protein has also been extensively used to investigate mechanisms of protein stability at extreme temperatures, using molecular modeling,[24,25] kinetics,[26–28] hydrogen exchange,[29,30] and mutant proteins,[31,32] and to study the influences of protein structure on the reduction potential of the iron site.[33,34] In fact, this protein is one of the most thermostable known, with a melting temperature estimated to be close to 200°.[29] In this chapter we describe the purification of rubredoxin from *P. furiosus* and of the recombinant protein from *Escherichia coli*. Included are methods to obtain the ^{15}N-labeled form, which is very useful for detailed structural and dynamic analyses using NMR spectroscopy.[29] The use of hydrogen exchange to investigate the thermodynamics of this protein is described elsewhere in this volume.[35]

Purification of Rubredoxin from *Pyrococcus furiosus*

Rubredoxins in their oxidized (Fe^{3+}) state are bright red in color, and this facilitates identification of the protein during the later stages of purification. However,

---

[20] S. J. George, J. van Elp, J. Chen, C. T. Chen, Y. Ma, J.-B. Park, M. W. W. Adams, F. M. F. de Groot, J. C. Fuggle, B. G. Searle, and S. P. Cramer, *J. Am. Chem. Soc.* **114**, 4426 (1992).

[21] J. van Elp, S. J. George, J. Chen, G. Pang, C. T. Chen, L. H. Tjeng, G. Meiga, H.-J. Lin, Z. H. Zhou, M. W. W. Adams, B. G. Searle, and S. P. Cramer, *Proc. Natl. Acad. Sci. U.S.A.* **90**, 9664 (1993).

[22] G. N. George, I. J. Pickering, R. C. Prince, Z. H. Zhou, and M. W. W. Adams, *J. Bioinorg. Chem.* **1**, 226 (1996).

[23] K. Rose, S. E. Shadle, M. K. Eidsness, D. M. Kurtz, R. A. Scott, B. Hedman, K. O. Hodgson, and E. I. Solomon, *J. Am. Chem. Soc.* **120**, 10743 (1998).

[24] J. E. Wampler, E. A. Bradley, D. E. Stewart, and M. W. W. Adams, *Prot. Sci.* **2**, 640 (1993).

[25] E. A. Bradley, D. E. Stewart, M. W. W. Adams, and J. W. Wampler, *Prot. Sci.* **2**, 650 (1993).

[26] S. Cavagnero, Z. H. Zhou, M. W. W. Adams, and S. I. Chan, *Biochemistry* **34**, 9865 (1995).

[27] S. Cavagnero, Z. H. Zhou, M. W. W. Adams, and S. I. Chan, *Biochemistry* **37**, 3377 (1998).

[28] S. Cavagnero, D. Debe, Z. H. Zhou, M. W. W. Adams, and S. I. Chan, *Biochemistry* **37**, 3369 (1998).

[29] R. Hiller, Z. H. Zhou, M. W. W. Adams, and S. W. Englander, *Proc. Natl. Acad. Sci. U.S.A.* **94**, 11329 (1997).

[30] G. Hernandez, F. E. Jenney, Jr., M. W. W. Adams, and D. M. LeMaster, *Proc. Natl. Acad. Sci. U.S.A.* (2000).

[31] M. K. Eidsness, R. A. Scott, and D. M. Kurtz, *FASEB J.* **9**, A1246 (1995).

[32] M. K. Eidsness, K. A. Richie, A. E. Burden, D. M. Kurtz, and R. A. Scott, *Biochemistry* **36**, 10406 (1997).

[33] P. D. Swartz and T. Ichiye, *Biochemistry* **35**, 13772 (1996).

[34] M. K. Eidsness, A. E. Burden, K. A. Richie, D. M. Kurtz, R. A. Scott, E. T. Smith, T. Ichiye, B. Beard, T. P. Min, and C. H. Kang, *Biochemistry* **38**, 14803 (1999).

[35] S. W. Englander and R. Hiller, *Methods in Enzymology* **334** [29] (2001) (this volume).

the cellular concentrations of rubredoxins in mesophilic bacteria are generally quite low, and this is also the case for the *P. furiosus* protein. Hence, it cannot be followed during the early stages of purification by its color. Moreover, the protein is purified from the same cell-free extracts that are also used to purify a variety of enzymes from *P. furiosus*. Such enzymes are usually sensitive to inactivation by oxygen and must be purified under strictly anaerobic and reducing conditions.[5] Although reduced rubredoxin is colorless, the protein readily autooxidizes in air, so its presence in column eluants can be assessed by shaking samples of fractions in air and determining if they turn reddish. Although rubredoxins have so far been found only in anaerobes, the protein is stable in air and reversibly oxidized by oxygen. Thus, it can be purified under aerobic conditions if desired.

P. furiosus (DSM 3638) is obtained from the Deutsche Sammlung von Mikroorganismen, Germany. It is routinely grown at 90° in a 600-liter fermenter with maltose as the carbon source as described previously.[36,37] All procedures are carried out at 23° using strictly anaerobic conditions.[5,36] All solutions contain 2 mM sodium dithionite and 2 mM dithiothreitol (except where indicated), with all vessels extensively degassed and flushed with argon. Frozen cells (400 g, wet weight) of *P. furiosus* are thawed and suspended under argon in 1.1 liters of buffer A [50 mM Tris/HCl, pH 8.0, 10% (v/v) glycerol, 2 mM dithiothreitol (DTT), and 2 mM sodium dithionite] containing lysozyme (1 mg/ml) and DNase I (10 μg/ml) and are lysed by incubation at 35° for 2 hr. A cell-free extract is obtained by centrifugation at 50,000g for 80 min. The supernatant (1.3 liters) is loaded onto a column (8 × 21 cm) of DEAE-Sepharose Fast Flow (Pharmacia LKB, Piscataway, NJ) equilibrated with buffer A. The column is eluted with a linear gradient (9.0 liters) of 0 to 0.5 M NaCl in buffer A and 90-ml fractions are collected. The presence of rubredoxin is monitored by air-oxidizing a small sample from each fraction, and determining the ratio of the absorbance values of the oxidized form at 494 nm ($\varepsilon_{494nm} = 9.22$ mM^{-1} cm^{-1}) and 280 nm.[5] Rubredoxin elutes near 0.30 M NaCl (pH 8.0). Rubredoxin containing fractions are pooled (1574 mg protein in 450 ml, $A_{490}/A_{280} = 0.15$), diluted to 1 liter with 50 mM Tris, pH 8.0, and applied to a column (5 × 20 cm) of DEAE-Sepharose Fast Flow. The column is eluted with a 350 ml gradient from 0 to 150 mM NaCl, followed by a 6.3 liter gradient from 150 to 400 mM NaCl at 7 ml/min. Rubredoxin elutes between 238 and 247 mM NaCl.[5]

The combined fractions from the DEAE Sepharose column (450 mg, 300 ml, $A_{490}/A_{280} = 0.2$), are loaded directly onto a column (2.5 × 25 cm) of hydroxyapatite (high resolution, Behring Diagnostics). The adsorbed proteins are eluted with a 2100 ml gradient from 0 to 500 mM potassium phosphate in 50 mM Tris/HCl,

---

[36] M. F. J. M. Verhagen, A. L. Menon, G. J. Schut, and M. W. W. Adams, *Methods in Enzymology* **330** [3] (2000).

[37] F. O. Bryant and M. W. W. Adams. *J. Biol. Chem.* **264**, 5070 (1989).

pH 8.0, at 3 ml/min. Care must be taken here as the rubredoxin elutes as approximately 10 mM phosphate is applied. Fractions containing rubredoxin (60 mg, 100 ml, $A_{490}/A_{280} = 0.33$) are concentrated to 5 ml by loading them directly onto a column (1 × 5 cm) of Q-Sepharose and eluting them with 500 mM NaCl in the same buffer.[5]

The concentrated rubredoxin fractions are applied to a column (2.5 × 100 cm) of Sephacryl S-200 previously equilibrated at 1 ml/min with 50 mM Tris-HCl, pH 8.0, containing 0.2 M NaCl. The rubredoxin-containing fractions judged pure by SDS–PAGE analysis using 20% (w/v) acrylamide are concentrated to 5 ml by Q-Sepharose chromatography as described above. Note that the protein migrates with the dye front during electrophoretic analysis. This procedure yields approximately 50 mg of protein with a ratio A_{490}/A_{280} of 0.36. The protein is stored frozen as pellets in liquid N_2.[5]

Purification of Recombinant Rubredoxin

There are two factors that must be taken into account in producing recombinant rubredoxin in *E. coli*. First, *E. coli* does not contain its own rubredoxin and, although the mechanism by which iron is inserted into proteins such as rubredoxin is not known, *E. coli* might insert the wrong metal or none at all. For example, the recombinant form of rubredoxin from the mesophilic bacterium *Clostridium pasteurianum* has been produced in *E. coli*. However, as much as 70% of the recombinant protein contains zinc instead of iron.[2,38] Similarly, when a synthetic gene encoding the *P. furiosus* protein was expressed in *E. coli*, a mixture of Zn- and Fe- recombinant forms also was produced.[32] A second problem concerns N-terminal processing. The native gene that encodes the *P. furiosus* protein (see below) encodes an N-terminal methionine, and this is processed in *P. furiosus* as the native protein lacks this residue (see Fig. 1). The synthetic gene encoding rubredoxin also included the N-terminal Met start codon, and when expressed in *E. coli*, a mixture of three different forms of the protein is produced.[32] One form was processed like the native protein and contained an N-terminal Ala residue, but incompletely processed proteins with N-terminal Met-Ala- and *N*-formyl-Met-Ala- were also purified. Thus, six different forms of recombinant rubredoxin were produced, the Zn and Fe derivatives of the *N*-formyl-Met-, *N*-Met-, and wild-type *N*-Ala forms. These different forms can be separated by a combination of anion exchange and hydroxyapatite chromatography.[32] The situation can be simplified, however, by the method used to grow *E. coli*. All of the above studies were carried out using cells grown on the "rich" Luria–Bertani (LB) medium. In the following procedure, the minimal M9 medium is used, and under these conditions only the

---

[38] M. K. Eidsness, S. E. O'Dell, D. M. Kurtz, Jr., R. L. Robson, and R. A. Scott, *Prot. Engineer.* **5**, 367 (1992).

Fe form of the rubredoxin is produced. This medium also has an advantage in that
E. coli can be grown on isotopically enriched media containing ^{15}N-, ^{34}S-, or ^{13}C-labeled compounds so that the isotopically enriched rubredoxins can be obtained.

To clone the gene encoding P. furiosus rubredoxin, oligonucleotide primers (5′-GTG GTA <u>CCA TGG</u> CAA AGT GGG TTT GTA-3′, and 5′-AAT CAT <u>CTG CAG</u> CAC CTC AAT CTT C-3′) are utilized in a polymerase chain reaction (PCR) using P. furiosus chromosomal DNA as the template. The primers have NcoI and PstI restriction sites on the 5′ and 3′ ends, respectively (underlined). The PCR product is cloned into the NcoI/PstI sites of the E. coli expression vector pTrc99a (Amersham Pharmacia, Piscataway, NJ) using standard techniques to produce plasmid pPfRd1. E. coli strain NCM533, which efficiently produces recombinant proteins when grown in minimal media,[39] is used to produce the rubredoxin.

E. coli NCM533/pPfRd1 is grown in a 100 liter fermenter (New Brunswick Scientific, NJ) with either glycerol (0.4%, v/v) or glucose (0.4%, w/v) as the carbon source in M9 minimal growth medium supplemented with 100 μM FeSO$_4$ (or FeCl$_3$), thiamin hydrochloride (0.05%, w/v), and a vitamin mixture,[40] together with 100 μg/ml ampicillin or carbenicillin. A one-liter culture grown for 14 hr using the same medium is used as the inoculum. The cells are concentrated by centrifugation (5000g for 10 min) and resuspended in 100 ml of fresh medium before inoculation. The fermenter is maintained at 37° and stirred at 180 rpm until the OD$_{600}$ reaches approximately 0.75 (after about 8 hr). The inducer, 1 mM isopropyl-β-D-thiogalactopyranoside (IPTG), is then added, and the culture is incubated for a further 10–12 hr before harvesting. The cells are frozen in liquid nitrogen and stored at −80° until needed. The typical yield under these conditions is about 5 g of cells (wet weight) per liter. For ^{15}N-labeling of the recombinant protein, 0.8 g/liter ^{15}NH$_4$Cl (>98% ^{15}N, Isotec, Miamisburg, OH) is substituted for the standard 1 g/liter NH$_4$Cl.[41]

The purification of recombinant rubredoxin is carried out under aerobic conditions at 23°. Frozen cells are broken into small pieces with a hammer and allowed to thaw by slowly stirring them in 50 mM Tris HCl, pH 7.5, containing 1 mM EDTA, 100 mM NaCl, and 0.5 mg/ml lysozyme (3 ml per gram of cells, frozen weight). The viscous suspension is sonicated (1 min per 100 ml of extract) in an ice bath using a Sonifier 450 (Branson Ultrasonics, Danbury, CT) and the protease inhibitor phenylmethylsulfonyl fluoride (PMSF) is added to a final concentration of 1 mM. The extract is heated in an 80° water bath until it reaches 80° (approximately 15 min per 100 ml extract) and then centrifuged (25,000g for 45 min at 4°).

---

[39] R. F. Shand, L. J. W. Miercke, A. K. Mitra, S. K. Fong, R. M. Stroud, and M. C. Betlach, *Biochemistry* **30**, 3082 (1991).
[40] R. A. Venters, T. L. Calderone, L. D. Spicer, and C. A. Fierke, *Biochemistry* **30**, 4491 (1991).
[41] P. L. Wang, L. Calzolai, K. L. Bren, Q. Teng, F. E. Jenney Jr., P. S. Brereton, J. B. Howard, M. W. W. Adams, and G. N. La Mar, *Biochemistry* **38**, 8167 (1999).

The supernatant is a deep burgundy color because of the recombinant (oxidized) rubredoxin. It is stored on ice 3 hr (or 4° overnight), and then centrifuged once more under the same conditions to remove denatured proteins that precipitated at 4°.

For protein purification, all columns and column materials are from Pharmacia Biotech (Piscataway, NJ) and rubredoxin is followed by the red color (A_{490}) of the oxidized protein. In a typical procedure, the heat-treated extract from 300 g (wet weight) of E. coli cells (approximately 20 g protein, $A_{490}/A_{280} \approx 0.1$) is loaded at 20 ml/min on to a column (10 × 15 cm) of DEAE-Sepharose Fast Flow equilibrated with 50 mM Tris/HCl, pH 7.5. The extract is diluted fivefold with the Tris buffer as it is applied to the column. A 1.2 liter linear gradient from 0 to 150 mM NaCl is applied to the column at 15 ml/min, followed by a 10.8 liter linear gradient from 150 to 500 mM NaCl at the same flow rate. Fractions containing rubredoxin elute as 200–240 mM NaCl is applied. There is some separation of the N-formyl-Met form from the other two forms at this step, but all rubredoxin-containing fractions from this column are combined as the different forms are more readily separated as described below. The combined fractions (1.6 g protein, $A_{490}/A_{280} = 0.25$) are concentrated in a stirred Amicon cell using a YM3 membrane (Millipore, Bedford, MA) to a final volume of approximately 25 ml. The concentrated rubredoxin fractions are applied at 4 ml/min to a column (6 × 60 cm) of Superdex G-75 equilibrated with 50 mM Tris/HCl, pH 7.5, containing 200 mM NaCl. Those fractions containing "pure" rubredoxin as judged by SDS–PAGE electrophoresis are combined (1.4 g protein, $A_{490}/A_{280} = 0.38$). Note that electrophoresis does not separate the three N-terminal forms.

Hydroxyapatite and anion exchange chromatography are used to separate the three N-terminally processed forms of the recombinant protein. The rubredoxin-containing fractions from the Superdex 75 column are loaded onto a column (6 × 18 cm) of ceramic hydroxyapatite (American International Chemical, Natick, MA) equilibrated with 20 mM 2-[N-morpholino]ethanesulfonic acid (MES), pH 6.0. The fractions are diluted fivefold with the equilibration buffer as they are applied to the column. The proteins are eluted at 10 ml/min using a 3.6 liter linear gradient from 0 to 200 mM potassium phosphate, pH 6.0. Those fractions eluting as 60–75 mM phosphate is applied contain a mixture of the native N-Ala form and the N-formyl-Met form (1.2 g protein). The N-Met form elutes as 100–220 mM phosphate is applied and is well separated from the other two forms. The yield of the N-Met-rubredoxin is about 140 mg ($A_{490}/A_{280} = 0.38$).

The mixture of the N-Ala and N-formyl-Met forms is loaded onto a column (3.5 × 10 cm) of Q-Sepharose High Performance equilibrated with 50 mM Tris/HCl, pH 7.5. The mixture is diluted fivefold with the buffer as it is loaded. The protein is eluted with a 2 liter linear gradient from 0 to 500 mM NaCl at 5 ml/min. Fractions containing the native N-Ala form elute as 190–220 mM NaCl is applied (1.0 g, $A_{490}/A_{280} = 0.38$), while the N-formyl-Met form elutes as 220–250 mM

NaCl is applied (160 mg, $A_{490}/A_{280} = 0.38$). Each form is concentrated by ultrafiltration using a YM3 membrane, flash-frozen in liquid nitrogen, and stored at $-80°$.

The typical yield from this purification procedure is about 4.5 mg of purified rubredoxin per gram of *E. coli* cells (wet weight). The three forms, *N*-Ala, *N*-formyl-Met, and *N*-Met, are obtained in the ratio of 7.1 : 1.2 : 1. The visible absorption spectrum of each is indistinguishable from the native rubredoxin purified from *P. furiosus*. The Zn-containing form of the rubredoxin appears not to be produced in significant amounts under the growth conditions for *E. coli* described above.

Properties of Rubredoxin

P. furiosus contains 53 amino acids with a predicted molecular weight of 5948 (including a single Fe atom).[5] The pure protein elutes from a gel filtration column with an apparent molecular weight 6800 ± 1000 in $1\ M$ NaCl, indicating that it is a monomeric protein. Using electrospray ionization mass spectrometry, the Fe-containing recombinant form has the expected molecular weight of 5949.4.[32] The UV/visible absorption spectrum of the oxidized protein has maxima at 280, 390, and 494 nm, but the visible absorption is lost upon reduction with sodium dithionite. Both the purified native and recombinant proteins have an A_{494}/A_{280} ratio of 0.38.[5,32] The redox midpoint potential of the iron site is near 0 mV, with only a small temperature dependence (-1.53 mV/°) and no pH dependence.[42]

The backbone structure of *P. furiosus* rubredoxin is superimposable on those of several rubredoxins from mesophilic sources that have been structurally characterized. Yet, the *P. furiosus* protein is virtually unaffected after many days at 95° (as determined by loss of visible absorption), whereas that from the mesophile *C. pasteurianum* is completely denatured within a few hours at the same temperature.[32] However, the mechanisms that lead to the enhanced stability of the hyperthermophilic protein are unclear.[14] *P. furiosus* rubredoxin unfolds irreversibly as it denatures, and the unfolding pathway is complex, with hydrophobic core relaxation, not loss of the metal cofactor, being the rate-determining step.[27] There are minor structural differences between the hyperthermophilic and mesophilic rubredoxins at the amino termini and in the triple-stranded β sheets that dominate these proteins. Their roles in differentially stabilizing the proteins have been dissected by mutagenesis and generation of β-sheet chimeras.[14,32] The results indicate that although both the amino-terminal electrostatic interactions and the more extensive H-bonding network in the β sheets contribute to thermostability, it is more likely a global combination of these interactions that makes the hyperthermophilic protein so stable.[14,32]

---

[42] P. L. Hagedoorn, M. C. P. F. Driessen, M. van den Bosch, I. Landa, and W. R. Hagen, *FEBS Lett.* **440,** 311 (1998).

Rubredoxins have been proposed to play multiple roles in various anaerobic microorganisms, including in acetogenesis[43] and in the reduction of nitrate,[44] sulfate,[45,46] sulfur,[47] and oxygen.[48] The *P. furiosus* protein has been shown to function as an electron donor to a new enzyme, superoxide reductase (SOR),[49] using electrons supplied by NADPH via NADPH:rubredoxin oxidoreductase.[50] In *P. furiosus* the gene encoding rubredoxin is 12 bps upstream of that encoding SOR (GenBank Accession AF156097), and thus they may be cotranscribed. However, a role for rubredoxin in oxygen detoxification or any other metabolic process has yet to be unequivocally demonstrated.

Acknowledgments

This research was supported by grants from the Department of Energy, the National Institutes of Health, and the National Science Foundation.

---

[43] J. Hugenholtz and L. G. Ljungdahl, *FEMS Microbiol. Rev.* **7**, 383 (1990).
[44] Y. Seki, S. Seki, M. Satoh, A. Ikeda, and M. Ishimoto, *J. Biochem.* **106**, 336 (1989).
[45] F. Shimizu, M. Ogata, T. Yagi, S. Wakabayashi, and H. Matsubara, *Biochimie* **71**, 1171 (1989).
[46] D. E. Stewart, J. LeGall, I. Moura, J. J. G. Moura, H. D. Peck Jr., A. V. Xavier, P. K. Weiner, and J. E. Wampler, *Eur. J. Biochem.* **185**, 695 (1989).
[47] K. Ma, R. N. Schicho, R. M. Kelly, and M. W. W. Adams, *Proc. Natl. Acad. Sci. U.S.A.* **90**, 5341 (1993).
[48] C. M. Gomes, G. Silva, S. Oliveria, J. LeGall, M.-Y. Liu, A. V. Xavier, C. Rodrigues-Pousada, and M. Teixeira, *J. Biol. Chem.* **272**, 22502 (1997).
[49] F. E. Jenney Jr., M. F. J. M. Verhagen. X. Cui, and M. W. W. Adams, *Science* **286**, 306 (1999).
[50] K. Ma and M. W. W. Adams, *J. Bacteriol.* **181**, 5530 (1999).

[6] NAD(P)H:rubredoxin Oxidoreductase from *Pyrococcus furiosus*

By KESEN MA and MICHAEL W. W. ADAMS

Introduction

NAD(P)H:rubredoxin oxidoreductase (NROR) catalyzes the reduction of the redox protein rubredoxin with NAD(P)H as the electron donor.[1] Rubredoxin is a small protein (~5 kDa) that contains a single Fe atom coordinated by the sulfur

---

[1] K. Ma and M. W. W. Adams, *J. Bacteriol.* **181**, 5530 (1999).

atoms of four cysteine residues. It has been purified from many anaerobic microorganisms, including the hyperthermophilic archaeon *Pyrococcus furiosus*.[2,3] The physiological role of rubredoxin has not been firmly established, but it has been proposed to function as the electron donor to a new enzyme, superoxide reductase (SOR), which reduces superoxide to hydrogen peroxide.[4] Unlike superoxide dismutase, the enzyme that protects aerobes from the toxic effects of oxygen, SOR does not catalyze the production of oxygen from superoxide and, therefore, confers a selective advantage to anaerobes.[4] Hence, NROR may play a role in detoxifying reactive oxygen species by providing the reducing equivalents to SOR via rubredoxin.

Two NRORs have been characterized from anaerobic bacteria, from the mesophilic acetogen *Clostridium acetobutylicum*[5] and from the mesophilic sulfate reducer *Desulfovibrio gigas*.[6] Both of these enzymes contain flavin adenine dinucleotide (FAD) as a cofactor, but the *C. acetobutylicum* enzyme is a monomer of 41 kDa, whereas *D. gigas* NROR is a dimer of 27 and 32 kDa and also contains flavin mononucleotide (FMN). This chapter describes the methods used to purify, assay, and characterize *P. furiosus* NROR.

Assay of NROR

NROR activity is determined by the NAD(P)H-dependent reduction of *P. furiosus* rubredoxin.[1] Rubredoxin is red in its oxidized state, but colorless when reduced. Its reduction can therefore be measured by the decrease in visible absorption at 494 nm (molar absorbance of 9220 M^{-1} cm^{-1}).[2] The reaction mixture (2.0 ml) contains 100 mM EPPS [N-(2-hydroxyethyl)piperazine-N'-(3-propanesulfonic acid)] buffer (pH 8.0), NAD(P)H (0.3 mM), and rubredoxin (10 μM)[3]. Reduced rubredoxin autoxidizes in air, so the reaction must be carried out under anaerobic conditions.[7] In addition, because of the extremely high activity of NROR at 80°, the assay temperature typically used for enzymes from *P. furiosus*,[7] rubredoxin reduction catalyzed by NROR can be readily measured at 25° if desired.

Because of the limited supply of rubredoxin available and for convenience, *P. furiosus* NROR is routinely measured during its purification by the NADH-dependent reduction of benzyl viologen. This BVNOR activity is carried out

---

[2] P. R. Blake, J. B. Park, F. O. Bryant, S. Aono, J. K. Magnuson, E. Eccleston, E. J. B. Howard, M. F. Summers, and M. W. W. Adams, *Biochemistry* **30**, 10885 (1991).

[3] F. E. Jenney, Jr. and M. W. W. Adams, *Methods in Enzymology* **334** [5] (2001) (this volume).

[4] F. E. Jenney, Jr., M. F. J. M. Verhagen, X. Cui, and M. W. W. Adams, *Science* **286**, 306 (1999).

[5] H. Petitdemange, R. Marczak, H. Blusson, and R. Gay, *Biochem. Biophys. Res. Commun.* **91**, 1258 (1979).

[6] L. Chen, M.-Y. Liu, J. LeGall, P. Fareleira, H. Santos, and A. V. Xavier, *Eur. J. Biochem.* **216**, 443 (1993).

at 80°. The assay mixture (2.0 ml) contains 50 mM CAPS [3-(cyclohexylamino)-1-propanesulfonic acid] buffer (pH 10.2), NADH (0.3 mM), and benzyl viologen (1 mM). A molar absorbance of 7800 M^{-1} cm^{-1} at 580 nm is used to calculate the concentration of reduced benzyl viologen. One unit of NROR activity catalyzes the reduction of 2 μmol of benzyl viologen per min.

NROR will reduce a variety of electron acceptors in addition to rubredoxin and benzyl viologen.[1] These include (where the molar absorbance at the indicated wavelength is given in parentheses) methyl viologen (9700 M^{-1} cm^{-1} at 580 nm), 2,6-dichlorophenol indophenol (19,100 M^{-1} cm^{-1} at 600 nm), cytochrome c (from horse heart, 19,520 M^{-1} cm^{-1} at 550 nm), FAD and FMN (11,300 M^{-1} cm^{-1} at 450 nm), iron(III) citrate (28,000 M^{-1} cm^{-1} at 562 nm), and 5,5′-dithiobis(2-nitrobenzoic acid) (DTNB; 13,600 M^{-1} cm^{-1} at 412 nm).

Purification of NROR

P. furiosus (DSM 3638) is obtained from the Deutsche Sammlung von Mikroorganismen, Germany. It is routinely grown at 90° in a 600-liter fermenter with maltose as the carbon source as described previously.[7,8] Virtually all (>94%) of the NROR activity, as measured by the NADPH-dependent reduction of rubredoxin, is in the supernatant fraction after ultracentrifugation of a cell extract of *P. furiosus* (110,000g for 2 h) indicating that it is a cytoplasmic enzyme. The BVNOR assay carried out at 80° is used to monitor the enzyme during purification. The procedure is performed under strictly anaerobic conditions at 23°.[1,7]

Cell-Free Extract

Frozen cells (400 g, wet weight) are thawed in 1.1 liter of buffer A [50 mM Tris/HCl (pH 8.0), 10% (v/v) glycerol, 2 mM dithiothreitol (DTT), and 2 mM sodium dithionite] containing lysozyme (1 mg/ml) and DNase I (10 μg/ml) and are lysed by incubation at 35° for 2 hr. A cell-free extract is obtained by centrifugation at 50,000g for 80 min.

DEAE-Sepharose Fast Flow Chromatography

The supernatant (1.3 liter) is loaded onto a column (8 × 21 cm) of DEAE-Sepharose Fast Flow (Pharmacia LKB, Piscataway, NJ) equilibrated with buffer A. The column is eluted with a linear gradient (9.0 liter) of 0 to 0.5 M NaCl in buffer A and 90-ml fractions are collected. NROR activity starts to elute as 0.2 M NaCl is applied to the column. Those fractions containing NROR activity are combined

---

[7] M. F. J. M. Verhagen, A. L. Menon, G. J. Schut, and M. W. W. Adams, *Methods in Enzymology* **330** [3] (2000).

[8] F. O. Bryant and M. W. W. Adams, *J. Biol. Chem.* **264,** 5070 (1989).

(810 ml), concentrated by ultrafiltration (type PM30 membrane; Amicon, Beverly, MA), and washed with buffer B. Buffer B is the same as buffer A except that sodium dithionite is omitted.

Blue Sepharose Chromatography

The concentrated sample (150 ml) is applied to a column (5 × 12 cm) of Blue Sepharose (Pharmacia LKB) equilibrated with buffer B. The column is eluted with a linear gradient (1.4 liters) from 0 to 2.0 M NaCl in buffer B and 50 ml fractions are collected. Three peaks of BVNOR activity are separated by this step. The first peak contains ~20% of the total activity, and the enzyme responsible has not been characterized. The second peak represents ~40% of the activity and, in this case, the reduction of benzyl viologen is catalyzed by ferredoxin:NADP oxidoreductase (FNOR).[9,10] The third peak of activity contains about 40% of the total and corresponds to NROR. This starts to elute as 1.7 M NaCl is applied to the column. Those fractions containing NROR activity are combined (600 ml) and concentrated to 30 ml by ultrafiltration (PM30 membrane).

Superdex 200 Chromatography

The concentrated sample is applied to a column (6 × 60 cm) of Superdex 200 (Pharmacia LKB) equilibrated with buffer B containing 50 mM KCl and fractions of 25 ml are collected. A large peak of BVNOR activity is observed, and this corresponds to NROR. It is preceded by a much smaller peak that represents some FNOR that is not completely separated from NROR by the previous chromatography step. Those fractions containing NROR activity are combined (50 ml).

Q-Sepharose Chromatography

The combined NROR fractions are applied to a column (2.6 × 10 cm) of Q-Sepharose (High Performance, Pharmacia LKB) equilibrated with buffer B. The column is eluted with a 500 ml gradient from 0 to 1.0 M KCl in buffer B and 20-ml fractions are collected. NROR activity starts to elute as 0.3 M KCl is applied to the column. Those fractions judged homogeneous by SDS–PAGE are combined (80 ml), concentrated by ultrafiltration (PM30 membrane), and stored in liquid nitrogen.

This procedure yields about 10 mg of purified NROR with a specific activity in the BVNOR assay of approximately 500 units/mg. The enzyme is purified over 600-fold and the final recovery of activity is about 17%. This is an underestimate of recovery, however, since at least two other enzymes in *P. furiosus* have BVNOR

[9] K. Ma and M. W. W. Adams, *Methods in Enzymology* **334** [4] (2001) (this volume).
[10] K. Ma and M. W. W. Adams, *J. Bacteriol.* **176,** 6509 (1994).

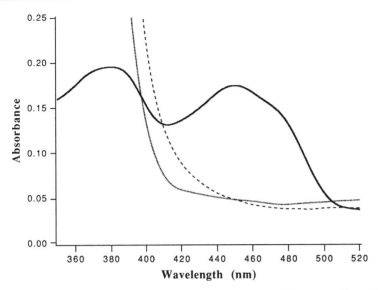

FIG. 1. UV–visible absorption spectra of NAD(P)H:rubredoxin oxidoreductase. The solid line (——) is the spectrum of the enzyme (0.5 mg/ml) as purified in 50 mM Tris/HCl (pH 8.0) containing 1 mM DTT. The dashed line (- - - -) is the spectrum after the addition of sodium dithionite (3 mM), and the dotted line (·····) is the spectrum after the addition of NADPH (0.2 mM).

activity. Thus, based on the assay involving the NADPH-dependent reduction of rubredoxin, NROR is purified 1830-fold by this procedure with a final recovery of activity of 47%.[1]

Properties of Purified NROR

Molecular Properties

Analysis of *P. furiosus* NROR by gel filtration (Superdex 200) and SDS–PAGE indicate that it is a monomeric protein of approximately 45 kDa. The presence of flavin in the enzyme is indicated by its UV–visible absorption spectra (Fig. 1). The air-oxidized enzyme exhibits peaks near 375 and 450 nm, and these decrease in intensity on the addition of a reductant such as sodium dithionite or NADPH, indicating the presence of flavin chromophore. Assuming that only flavin contributes to the absorption at 450 nm on reduction, and assuming a molar absorbance of 11,500 $M^{-1}cm^{-1}$ at 450 nm,[11] the change in absorbance on reduction

---

[11] K. Beaucamp, H. U. Bergmeyer, and H.-O. Beutler, *in* "Methods of Enzymatic Analysis" (H. U. Bergmeyer, ed.), p. 532. Verlag Chemie, Academic Press, New York, 1974.
[12] Holmgren and A. M. Björnstedt, *Methods Enzymol.* **252**, 199 (1995).

indicates that the holoenzyme contains one flavin moiety per mole. Acid extraction is used to determine the type of flavin present in NROR. To a 0.2 ml sample of the purified enzyme (1.2 mg/ml), 0.2 ml 10% (w/v) $HClO_4$ is added and the mixture is incubated in an ice bath for 10 min. After centrifugation ($5000g$ for 2 min), the supernatant is kept and the pellet is suspended in 0.1 ml 10% (w/v) $HClO_4$. A second supernatant is obtained by centrifugation ($5000g$ for 2 min) and the procedure is repeated. The three supernatant solutions are combined (0.6 ml) and neutralized by addition of saturated solution of $KHCO_3$ (0.25 ml), and the precipitate is removed by centrifugation ($5000g$ for 2 min). The supernatant (0.75 ml) is used for analysis by thin-layer chromatography. A sample (0.5 μl) is applied to a flexible TLC plate (5 × 8 cm, Selecto Scientific, Norcross, GA) and after air-drying, a further sample (0.5 μl) is added and this is repeated until 10 μl sample has been applied. The plate is developed for about 10 min using 5% (w/v) K_2HPO_4 as the solvent and then air dried. Riboflavin, FMN, and FAD are used as standards, and the spots are visible under a UV light. Such an analysis shows that the material that is acid-extracted from *P. furiosus* NROR corresponds to FAD (data not shown). This can be confirmed by mass spectrometry. The flavin in the acid-extracted material had a mass of 786 Da, which corresponds to FAD.[1]

Analysis of NROR by plasma emission spectroscopy reveals that the enzyme does not contain any metals in significant amounts (>0.1 g-atom/mol) and flavin appears to be its only prosthetic group. The enzyme is quite thermostable, with the time required for a 50% loss in catalytic activity when it is incubated at 80° and 95° of 12 and 1.6 hr, respectively. These values are determined by maintaining the purified enzyme (0.4 mg/ml) in 60 mM EPPS, pH 8.0) in serum-stoppered glass vilas at the desired temperature and periodically removing samples to determine residual BVNOR activity at 80°.

Catalytic Properties

NROR shows comparable activity with NADPH and NADH as electron donors, with apparent V_m values of about 500 units/mg at 80° using benzyl viologen (1 mM) as the electron acceptor. However, the apparent K_m value for NADPH (5 μM) is much lower than that for NADH (34 μM) and presumably NADPH is the preferred physiological electron donor. The enzyme will reduce various electron acceptors besides benzyl viologen, including methyl viologen, benzyl viologen, 2,6-dichlorophenol indophenol, cytochrome *c*, FAD, FMN, hydrogen peroxide, and iron(III) citrate,[1] with the highest activities observed with benzyl viologen and 2,6-dichlorophenol indophenol. Notably, NROR will not reduce *P. furiosus* ferredoxin with NADPH, which is the reaction catalyzed by FNOR.[9,10] The best substrate for NROR is *P. furiosus* rubredoxin. Remarkably, the enzyme exhibits an apparent V_m value for rubredoxin reduction at 80° of 20,000 units/mg, with an apparent K_m value of 50 μM, which corresponds to a k_{cat}/K_m of 300,000 mM^{-1}s^{-1}. The optimum pH for rubredoxin reduction is between pH 7.0 and 8.0.

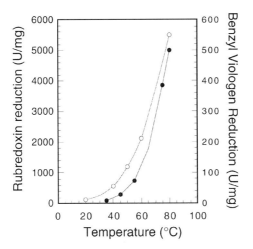

FIG. 2. Temperature dependence of the activity of *P. furiosus* NROR. Activities were measured at the indicated temperature using NADPH as the electron donor and either *P. furiosus* rubredoxin (○) or benzyl viologen (●) as the electron acceptor. For the rubredoxin assay, the 2 ml reaction mixture contained 100 mM EPPS pH 8.0, rubredoxin (9.5 μM) and NADPH (0.3 mM). For the benzyl viologen assay, the 2 ml mixture contained 50 mM CAPS buffer, pH 10.2, benzyl viologen (1 mM) and NADH (0.3 mM).

These kinetic data are consistent with rubredoxin being the physiological substrate of *P. furiosus* NROR.[1] As expected, the activity of the enzyme increases dramatically with temperature with a maximum well in excess of 80° (Fig. 2). However, the high catalytic activity with rubredoxin as the electron acceptor means that activity is readily measured even at 25°. Thus, the apparent V_m value for rubredoxin reduction at 25° is 290 units/mg, which is comparable to the activity measured with benzyl viologen at 80°.[1] Interestingly, the apparent K_m value for rubredoxin decreases fivefold (to 9.5 μM) with a decrease in temperature from 80 to 25°. This relatively high activity of NROR at low temperatures may have a physiological function. *P. furiosus* SOR is also very active at 25°, which is 75° below the organism's optimal growth temperature.[4] A combination of SOR, rubredoxin, and NROR would, therefore, rapidly reduce superoxide using NAD(P)H as the electron donor at temperatures far below the growth temperature of the organism. Hyperthermophiles such as *P. furiosus* inhabit geothermally heated marine environments, and it might be advantageous for them to have a mechanism to protect against exposure to cold, oxygen-saturated seawater.[4] Whether NROR is involved in such a physiological response remains to be established.

P. furiosus NROR is, therefore, an exceedingly active enzyme with a high specificity for rubredoxin as its electron acceptor. Its molecular properties are more similar to those of the monomeric NROR of *C. acetobutylicum*[5] than to those of the dimeric enzyme of *D. gigas*.[6] Interestingly, under comparable assay conditions at 25–30°, *P. furiosus* NROR is much more active (290 units/mg)[1] than either of these mesophilic enzymes (≤50 U/mg).[3,4] Such a comparison serves to emphasize the remarkable activity of the hyperthermophilic NROR, and perhaps adds some credence to its proposed physiological role at temperatures more than 70° below that which supports growth of *P. furiosus*.

Acknowledgment

This research was supported by grants from the Department of Energy.

[7] Protein Disulfide Oxidoreductase from *Pyrococcus furiosus*: Biochemical Properties

By SIMONETTA BARTOLUCCI, DONATELLA DE PASCALE, and MOSÈ ROSSI

Introduction

Protein disulfide oxidoreductases (PDI) are redox enzymes that catalyze dithiol–disulfide exchange reactions in eukaryotic and bacterial cells. These enzymes have a redox-active disulfide/dithiol group in the conserved motif CXXC. The two cysteine residues in the active site sequence can undergo reversible oxidation–reduction. Several proteins belonging to this superfamily have been identified and characterized; thioredoxin, glutaredoxin, protein disulfide-isomerase (PDI), disulfide bond forming (DsbA), and their homologs in prokaryotes are the most extensively studied members of this group.[1]

Thioredoxin and glutaredoxin, with a molecular mass of about 12 kDa, have at the N terminus the conserved active site sequence CGPC and CPYC, respectively.[2]

They typically transfer electrons from NADPH to the substrate in reactions coupled with other specific enzymes. Glutaredoxin derives its reducing equivalents from glutathione, which in turn is linked to the glutathione reductase/NADPH system, whereas thioredoxin utilizes the thioredoxin reductase/NADPH pathway for the same purpose. Thioredoxin is involved in maintaining the redox status of sulfhydryl groups inside the cell and participates in catalyzing and/or regulating a variety of cellular functions, including the ability to activate surface receptors and regulate microtubule assembly.[3] In other systems thioredoxin is

---

[1] H. Loferer and H. Hennecke, *Trends Biochem. Sci.* **19**, 169 (1994).
[2] A. Holmgren and F. Aslund, *Methods Enzymol.* **252**, 283 (1995).

reduced by a photoreduced ferredoxin via an iron–sulfur ferredoxin–thioredoxin reductase.[4]

The endoplasmic reticulum PDI, which catalyzes the reduction, oxidation, and reshuffling of protein disulfides in eukaryotes,[5,6] is a homodimer of two 57 kDa subunits. Each subunit contains two functional domains with significant sequence homology to *Escherichia coli* thioredoxin.[7] The functional equivalent of PDI in prokaryotes, DsbA, is a periplasmic, monomeric protein characterized by a low similarity to *E. coli* thioredoxin with the active site sequence motif CPHC.[8,9]

Little information is availble on protein disulfide oxidoreductases in Archaea. A glutaredoxin-like protein, isolated from *Methanobacterium thermoautotrophicum*, was shown to have a molecular mass of 9 kDa, and a low sequence identity (<20%) to known glutaredoxin, and not to function as glutaredoxin-dependent enzyme.[10]

A highly thermostable protein disulfide oxidoreductase was first isolated from *Sulfolobus solfataricus*.[11] From its ability to catalyze the reduction of insulin disulfides in the presence of dithiothreitol (DTT), the protein was considered a thioredoxin. The protein showed an unusually high molecular mass of 25 kDa and from amino acid composition analysis contained four cysteine residues. A homologous protein was subsequently purified from *Pyrococcus furiosus*.[12] From its amino acid sequence, which showed two distinct CXXC motifs, and from its thioltransferase activity the protein was considered to be a glutaredoxin-like protein.

More recently the protein was crystallized, and its three-dimensional (3-D) structure at high resolution revealed structural details suggesting it may be related to the multidomain eukaryotic PDI. For this reason the protein was more correctly named protein disulfide oxidoreductase from *P. furiosus* (*Pf* PDO).[13–15] In this paper we describe the purification of the *Pf* PDO and the cloning and its gene expression in *E. coli*.

---

[3] J. H. Wong, K. Kobrehel, and B. B. Buchanan, *Methods Enzymol.* **252**, 228 (1995).
[4] P. Schurmann, *Methods Enzymol.* **252**, 274 (1995).
[5] J. C. Edman, L. Ellis, R. Blacher, R. A. Roth, and W. J. Rutter, *Nature* **319**, 267 (1985).
[6] J. Kemmink, N. J. Darby, Dijkstra, M. Nilges, and T. E. Creighton, *Curr. Biol.* **7**, 239 (1997).
[7] J. Bardwell and J. Beckwith, *Cell* **74**, 769 (1993).
[8] J. L. Martin, J. C. A. Bardwell, and J. Kurigan, *Nature* **365**, 464 (1993).
[9] R. F. Golberg, C. J. Epstein, and C. B. Anfinsen, *J. Biol. Chem.* **238**, 628 (1968).
[10] S. McFarlan, C. A. Terrell, and H. P. C. Hogenkamp, *J. Biol. Chem.* **267**, 10561 (1992).
[11] A. Guagliardi, V. Nobile, S. Bartolucci, and M. Rossi, *Int. J. Biochem.* **26**, 375 (1994).
[12] A. Guagliardi, D. de Pascale, R. Cannio, V. Nobile, S. Bartolucci, and M. Rossi, *J. Biol. Chem.* **270**, 5748 (1995).
[13] B. Ren, G. Tibellin, D. de Pascale, M. Rossi, S. Bartolucci, and R. Ladenstein, *J. Struct. Biol.* **119**, (1997).
[14] B. Ren, G. Tibellin, D. de Pascale, M. Rossi, S. Bartolucci, and R. Ladenstein, *Nature Struct. Biol.* **5**, 602 (1998).
[15] B. Ren and R. Ladenstein, *Methods in Enzymology* **334** [8] (2001) (this volume).

Purification of Native Protein

Pf PDO is purified to homogeneity by two chromatography steps involving anion exchange and gel filtration. SDS–PAGE analysis performed according to Laemmli[16] of the active sample from the gel filtration column shows one protein band with molecular mass of about 26 kDa. The reductase activity was found to be associated with this protein throughout the purification steps. As it is difficult to quantify the assay of the activity, based on insulin precipitation, no yield and purification factors are calculated. In general 0.5 mg of homogeneous *Pf* PDO is obtained from 10 g of cells (wet weight).

Preparation of Crude Extract

P. furiosus is grown at 95° and pH 7.5 in a medium containing the following components (per liter): Bacto-peptone (5 g), yeast extract (1 g), NaCl (19.4 g), $MgSO_4$ (12.6 g), Na_2SO_4 (3.2 g), $CaCl_2$ (2.4 g), KCl (0.5 g), $NaHCO_3$ (0.16 g), KBr (0.08 g), $SrCl_2$ (0.057 g), H_3PO_3 (0.022 g), Na_2SO_3 (0.004 g), NaF (0.0024 g), KNO_3 (0.0016 g), Na_2HPO_4 (0.01 g), and sulfur (25 g). The culture is bubbled with N_2 (5 liter/min). The cells are harvested after 8 h, and immediately frozen in liquid N_2 and stored at $-80°$ until required.

For enzyme purification, 10 g (wet weight) of cells is freeze-thawed and resuspended in 40 ml of 20 mM Tris buffer, pH 8.4, 2 mM EDTA, 1 M NaCl, and homogenized in a French press for 20 min. The homogenate is centrifuged at 160,000g for 90 min at 4°; the supernatant (30 ml) represents the crude extract, and this catalyzes the reduction of insulin in the presence of DTT at 30° as described below.

Anion-Exchange Chromatography

The crude extract (1.0 g of protein in 30 ml) is extensively dialyzed against buffer A (20 mM Tris buffer, pH 8.4, 2 mM EDTA) and loaded onto a DEAE-Sepharose Fast Flow column (Pharmacia, Uppsala, Sweden, 2.2 × 18 cm) equilibrated in the same buffer. Bound proteins are eluted at a flow rate of 50 ml/hr by a linear gradient from 0 to 0.3 M NaCl in buffer A, and 5-ml fractions are collected. The active fractions are pooled, concentrated in a Savant centrifuge (Savant Speed Vac SC 110), and dialyzed against buffer A.

Gel-Filtration Chromatography

The active fractions from the DEAE-Sepharose Fast Flow column (0.1 g of protein in 6 ml) are loaded in three separate runs onto a HiLoad Superdex 75 column (Pharmacia, 2.6 × 60 cm) connected to an FPLC system (Pharmacia) eluted with buffer B (20 mM Tris buffer, pH 8.4, 2 mM EDTA, 1 M NaCl) at a

---

[16] U. K. Laemmli, *Nature* **227**, 680 (1970).

flow rate of 2 ml/min. The active fractions (3 ml each) are pooled and concentrated in a Savant centrifuge. The yield was 0.5 mg of homogeneous protein (95% of homogeneity). The protein concentration is determined by the Bio-Rad (Hercules, CA) dye-binding assay,[17] using bovine serum albumin (BSA) as the standard.

High-Performance Liquid Chromatographs

For analytical purposes *Pf* PDO can be further purified by high-performance liquid chromatography (HPLC) with a C_4 reversed-phase column (Vydac 0.46 × 25 cm, 0.5 μm size particles) with 0.1% trifluoroacetic acid TFA (v/v) in H_2O as buffer C, and 0.07% TFA in acetonitrile as buffer D. Starting from buffer C a linear gradient (0–100% D) was performed in 60 min.

N-Terminal Sequence Analysis

The N-terminal sequence of the homogeneous protein from *P. furiosus* can be determined by automatic Edman degradation using an Applied Biosystems 473A sequencer (Foster City, CA), according to the manufacturer's instructions. The following 29 residues were identified: GLISDADKKVIKEE FFSKMVNPVK-LIVFV.

Cloning, Overexpression, and Purification of *Pf* PDO

Since the purification procedure using *P. furiosus* required a large amount of cells, *Pf* PDO was produced as a soluble recombinant protein in *E. coli*. The recombinant *Pf* PDO is indistinguishable from the native protein isolated from *P. furiosus* in all activity assays tested, including insulin reduction.

Isolation of Chromosomal P. furiosus DNA

Ten grams of *P. furiosus* cells are suspended in 25 ml of 50 m*M* Tris-HCl buffer, pH 8.0, 10 m*M* EDTA, and centrifuged at 3000*g* for 10 min at 4°. Cells are resuspended in 20 ml in the same buffer containing 0.2% Triton X-100 and 1% SDS; the soluble fraction, cleared by centrifugation at 129,500*g* for 30 min at 4°, is heated for 10 min at 65° and centrifuged again in the same conditions. Cesium chloride (1 g/ml) and ethidium bromide (0.6 mg/ml) are added and the samples are ultracentrifuged at 352,000*g* for 16 hr at 16–18°C. Chromosomal DNA bands are revealed by irradiation using a 340-nm UV lamp and withdrawn as described by Sambrook *et al.*[18] Ethidium bromide and cesium chloride are removed, respectively, by

---

[17] M. M. Bradford, *Anal. Biochem.* **72,** 248 (1976).
[18] J. Sambrook, E. F. Fritsch, and T. Maniatis, "Molecular Cloning: A Laboratory Manual," 2nd ed. Cold Spring Harbor Laboratory, Cold Spring Harbor, NY, 1989.

extraction with isoamyl alcohol and extensive dialysis against 10 mM Tris-HCl buffer, pH 8.0, 1 mM EDTA. DNA concentration is determined spectrophotometrically at 260 nm, and the molecular weight is estimated by electrophoresis on 0.6% agarose gel in 90 mM Tris borate, 20 mM EDTA (TBE buffer) using suitable DNA molecular size markers (Roche) according to Sambrook et al.[18]

Construction of P. furiosus Gene Bank

A representative genomic bank in pGEM7Zf(+) (Promega, Madison, WI) was found to be highly efficient for the insertion of one single DNA fragment from *P. furiosus* genome per vector molecule.[19] To prepare the library, high molecular weight DNA of *P. furiosus* is partially digested with *Sau*3AI (1 hr of incubation with 0.15 units per μg of genomic DNA), and the fragments 2–4 kb in size are isolated by electroelution from 3.5% polyacrylamide gel electrophoresis. The fragments are subjected to partial filling in with *E. coli* DNA polymerase (Klenow fragment), dGTP, and dATP and inserted into the vector pGEM7Zf(+), which previously has been made end-compatible by linearization with *Xho*I and by partial filling in with the Klenow fragment, dCTP, and dTTP. *E. coli* BO3310-competent cells are transformed with the ligation mixture. Growth in Luria–Bertani medium (LB medium) (10 g Bacto-tryptone, 5 g Bacto-yeast, 5 g NaCl in 1 liter) containing 100 μg/ml ampicillin for 4 h[18] allows the propagation and amplification of the gene bank.

Cloning of Gene

The screening of the genomic bank of *P. furiosus* is performed using colony hybridization experiments carried out under different stringency conditions with 5× SSC, 0.5% SDS (two changes of 30 min each) at 45° and at 55°. The oligonucleotide mixture for the screening was designed on the N-terminal sequence of the protein from residue 8 to residue 20 and the codon usage in Archaea.[20] Ten pmol of 41-mer oligonucleotide mixture (5′-GATAAGAAGGT (G/T)AT(T/A)AAGGAGGA-GTTTTTT(T/A)(C/G)(T/G)AAGATGGT-3′) is end-labeled with [γ-^{32}P]ATP (Amersham International) by T4 polynucleotide kinase and utilized as probe. Replica filters are used as a control of signal specificity. Clones positive to a first selection are definitively isolated by a second colony hybridization screening performed according to the same protocol. After the sequential screenings, 5 clones out of 5×10^4 exhibiting positive hybridization signals are isolated; one clone, named pDR7, had a 2.3-kb region encompassing the complete gene sequence.

The whole coding sequence is labeled by incorporation of [α-^{32}P]dATP using the random priming method and the specific kit purchased from Roche Molecular Biochemicals (Mannheim, Germany). The hybridization of the whole coding

---

[19] R. Cannio, M. Rossi, and S. Bartolucci, *Eur. J. Biochem.* **222**, 345 (1994).
[20] J. Hain, W.-D. Reiter, U. Hudepohl, and W. Zilling, *Nucleic Acids Res.* **20**, 5423 (1992).

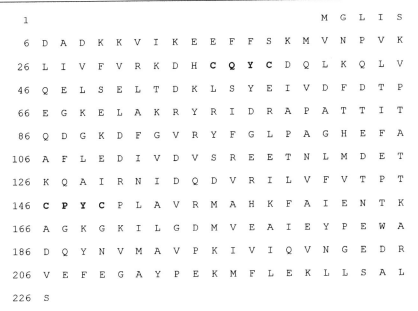

FIG. 1. The deduced amino acid sequence of *Pyrococcus furiosus* protein disulfide oxidoreductase (*Pf* PDO). The amino acids in bold-face type represent the two putative active sites.

sequence as a probe to the *P. furiosus* chromosomal DNA produces results consistent with the restriction maps of the isolated clone, thus confirming that the recombinant clone underwent no rearrangements during the cloning procedures.

The isolated clone is sequenced completely. Suitable restriction fragments are subcloned into pUC18 plasmid and sequenced using the universal M13 forward and reverse primers or specific synthetic oligonucleotides. DNA sequencing is carried out by the dideoxy chain termination method[21] with [^{35}S]dATP, using Sequenase version 2.0 sequencing kit (Amersham) on alkali-denatured double stranded templates.

A 675-bp open reading frame (ORF) starting from ATG and ending at TAA as well as extended flanking regions were found. Figure 1 shows the deduced primary structure of the *P. furiosus* protein. The 226 amino acid residues derived from the ORF matched the calculated size of the native protein whose N-terminal sequence lacks the initial methionine. The residues from Phe-141 to Cys-149 identify the FXXXXCXXC motif, which is a part of the active site of the enzymes that catalyze dithiol/disulfide interchange reactions. When the primary structure of the *P. furiosus* protein was compared with GenBank using the Fasta program, similarity with the active site sequence of the glutaredoxin family was detected,

[21] F. Sanger, S. Nicklen, and A. R. Coulson, *Proc. Natl. Acad. Sci. U.S.A.* **74,** 5463 (1977).

with the sequence CPYC at residues 146–149 typical of this family. The sequence CQYC at resides 35–38 is not conserved in the active site of any of the known protein disulfide oxidoreductases.

Construction of Expression Vector

The complete coding sequence of the gene is obtained by gene amplification performed according to previously described procedure[22] for 25 cycles at 55° annealing temperature, on a Perkin-Elmer (Norwalk, CT) cycler temp. The oligonucleotides used as primers were 5'-GgccATGGGATTGATTAGTGAC-3' (PfN) and 5'-gagAGTCGAGAGTCGACTAGATCTGCG-3' (PfC), where the underlined letters indicate the initiation and termination codons and small letters indicate point mutations inserted to generate the *Nco*I and *Xba*I sites. The pDR7 plasmid, containing the whole gene, is linearized with *Bam*HI and used as template for *Pfu* (Stratagene, La Jolla, CA, Cloning Systems) polymerase. The amplified fragment is digested with *Nco*I and *Xba*I and inserted into pTRC99A plasmid (Pharmacia), previously made end-compatible with the same restriction endonucleases. Restriction analysis and subsequent sequencing confirmed the correct sequence of the recombinant clone, which is designated as p*Pf* PDO. The newly constructed plasmid p*Pf*PDO is used as expression vector: in fact, the transcription of the recombinant gene is controlled by the strong hybrid regulatory sequence promoter *ptac,* derived from the fusion of the *lac* and *trp* promoters, and can still be inducible by isopropyl-β-D-thiogalactopyranoside (IPTG).

High Level Expression and Purification of Recombinant Protein

E. coli Rb791-competent cells are transformed with the newly constructed p*Pf* PDO expression vector and grown at 37° in 1 liter of Luria–Bertani medium. *E. coli* strain Rb791 cells transformed with pTRC99A (Pharmacia) represents a negative control. The expression of the *P. furiosus* protein is induced by the addition of 1 mM IPTG (final concentration) to the culture medium, when the growth was at A_{600} of 1. The induction time was 18 hr. Cells are harvested by centrifugation at 3000*g* for 10 min at 4°, washed with 30 ml of ice-cold 50 m*M* sodium phosphate buffer, pH 8.0, 0.1 *M* NaCl, 1 m*M* EDTA, and resuspended in 48 ml of the same buffer supplemented with 0.7 m*M* phenylmethanesulfonyl fluoride (PMSF). Cells are sonicated three times with a frequency of 20 kHz (Sonicator Ultrasonic liquid processor; Heat System Ultrasonic Inc., Farmingdale, NY) and ultracentrifuged at 160,000*g* for 90 min at 4°. The supernatant of the ultracentrifugation constitutes the crude extract.

---

[22] R. K. Saiki, "PCR Protocols: A Guide to Methods and Applications" (M. A. Innis, D. A. Gelfand, J. J. Sninski, and T. J. White, eds.). Academic Press, San Diego, 1990.

Heat Treatments

The crude extract undergoes two successive thermal precipitation steps. Most of the *E. coli* proteins become insoluble and are removed by centrifugation, whereas the recombinant protein is recovered in an active form in the supernatant.

The crude extract (85 mg of protein in 40 ml) is heated at 70° for 10 min and centrifuged at 5000g at 4° for 10 min; the supernatant is concentrated by ultrafiltration in an Amicon (Danvers, MA) cell (membrane cutoff 2000 Da), subjected to a second heating at 75° for 10 min, and centrifuged as above. Following the thermoprecipitation steps, about 80% of the *E. coli* proteins are removed.

Gel-Filtration Chromatography

The sample from heat treatment (28 mg of protein in 2 ml) is loaded onto a HiLoad Superdex 75 column (Pharmacia, 2.6 × 60 cm) connected to an FPLC (fast protein liquid chromatography) system (Pharmacia) eluted with buffer B (20 mM Tris buffer, pH 8.4, 2 mM EDTA, 1 M NaCl) at a flow rate of 2 ml/min. The active protein (18 mg) is stored at −20°. This protein is generally >90% homogeneous, as judged by (12%, w/v) SDS–PAGE.

The molecular mass of the recombinant protein is measured on a Bio-Q triple quadruple mass spectrometer (Micromass, Manchester, UK). Samples are dissolved in 1% (v/v) acetic acid/50% (v/v) acetonitrile and injected into the ion source at a flow rate of 10 ml/min using a Phoenix syringe pump. Spectra are collected and elaborated using the MASSLYNX software provided by the manufacturer.

Assay Methods

Insulin Activity Assays

Insulin reductase activity is assayed according to Holmgren[23] with a few modifications by measuring the catalytic reduction of insulin disulfide bonds at 30°. *Pf* PDO (approximately 20 μg) is added in 1 ml of 100 mM sodium phosphate buffer, pH 7.0, containing 2 mM EDTA and 1 mg of bovine insulin. A control cuvette contains only buffer and insulin. The reaction is started by addition of 1 mM dithiothreitol (DTT) to both cuvettes. Increasing turbidity from precipitation of insulin B chain is recorded at 650 nm. The stock solution of insulin (10 mg/ml) is prepared according to the Holmgren[23] protocol.

Thioltranferase Activity

Pf PDO has a glutathione (GSH)-dependent disulfide reductase activity (named the thioltransferase activity) that can be assayed by coupling to saturating

---

[23] A. Holmgren, *J. Biol. Chem.* **254,** 9627 (1979).

concentrations of glutathione reductase, which reduces glutathione, oxidized form GSSG in the presence of NADPH,[24] by using L-cystine as the disulfide substrate. The standard assay mixture consists of a 0.1 M sodium phosphate buffer, pH 8.0, 1 mM EDTA, 10 mM GSH, 2.5 mM L-cystine, 0.2 units of glutathione reductase (Sigma Aldrich, Milwaukee, WI), 0.3 mM NADPH in the absence or presence of 5 μg *Pf*PDO (final volume, 1 ml); enzyme activity is monitored at 340 nm and 30°. Activity dependence on pH is determined by the standard assay method except that 0.1 M sodium phosphate buffer is used in the pH range 6.0 to 8.0 and 0.1 M Tris-HCl buffer 8.0 to 9.0. The enzyme shows a sharp optimum at pH 8.0.

Oxidation Activity

The disulfide bond-forming activity of *Pf*PDO is monitored using the synthetic decapeptide NRCSQGSCWN containing two cysteine residues at position 3 and 8 designed by Ruddock *et al.*[25] The peptide contains a fluorescent group (tryptophan) on one side of one cysteine residue and a protonated group (arginine) on the other side of the second cysteine residue, and the two cysteine residues are separated by a flexible linker region. The linker is long enough to permit the formation of an unstrained disulfide bond, and the peptide is small and water soluble. Oxidation of this dithiol peptide to the disulfide state is accompanied by a change in tryptophan fluorescence emission intensity. In fact, on oxidation, the fluorescent group and the protonated group are brought close together and quenching of the fluorophore occurs where arginine is the charged quencher. Fluorescence quenching was used as the basis for monitoring the disulfide bond–forming activity of the eukaryotic enzyme protein disulfide-isomerase (PDI) and the bacterial enzyme DsbA at pH 7.0,[25] and with this assay it is possible to detect the disulfide bond activity of *Pf* PDO.

The catalyzed oxidation of the peptide was also observed with HPLC analysis[26] by incubating the peptide in the presence of GSH, GSSG, and *Pf* PDO.

Spectroscopic Determination of Oxidation of Substrate Peptide

The substrate peptide NRCSQGSCWN is synthesized from Primm s.r.l. Italia using a Shimadzu PSSM8 automated peptide synthesizer. The purified peptide is eluted in a single peak by HPLC analysis and stored at $-20°$ in the elution buffer (30% acetonitrile in 0.1% TFA). The peptide concentration is determined spectrophotometrically using an adsorption coefficient of 5600 M^{-1} cm^{-1} at 278 nm.

---

[24] K. Axelsson, S. Eriksson, and B. Mannervik, *Biochemistry* **17**, 2978 (1978).
[25] L. W. Ruddock, R. Hirst, and R. B. Freedman, *Biochem. J.* **315**, 1001 (1996).
[26] A. Tosco and L. Birolo, personal communication, 1999.

The assay is performed in McIlvaine buffer (0.2 M disodium hydrogen phosphate/0.1 M citric acid; pH 7.0) with 2mM GSH (stock solution 60.1 mg/ml), 0.5 mM GSSG (stock solution 30.7 mg/ml), and 5μM Pf PDO (stock solution, 1.0 mg/ml in Tris-Cl pH 8.4). It is placed in a fluorescence cuvette with a final assay volume of 1 ml.[25] After mixing, the cuvette is placed in a thermostatically controlled Perkin-Elmer LS50B spectrofluorimeter for 1 min to allow thermal equilibration of the solution to 50°C. Next, 5 μM substrate peptide (1.05 mM, in 30% acetonitrile/0.1% TFA) is added, mixed, and the change in fluorescence intensity (excitation 280 nm, emission 350 nm, slits 5/5 nm) is monitored over an appropriate time (15 min); 900 data points are collected. As a control, the same experiment is carried out in the absence of Pf PDO and the decrease in fluorescence intensity is not observed (Fig. 2). At pH 8.0 the spontaneous oxidation of the peptide substrate is observed, presumably due to air oxidation, but at pH 7.0 only the catalyzed oxidation of the substrate is measured.

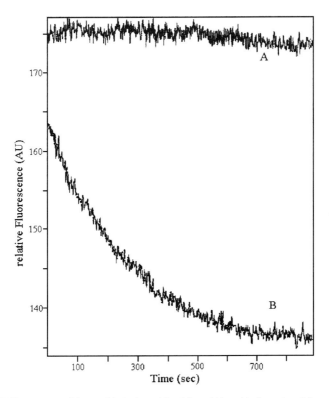

FIG. 2. Time course of the peptide in the oxidized form (A), and in the reduced form (B).

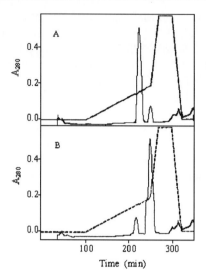

FIG. 3. HPLC analysis of the peptide in the oxidized form (A) and in the reduced form (B).

HPLC Analysis

High-performance liquid chromatography (Beckman Gold System, Palo Alto, CA) analysis is performed with a reversed-phase C_{18} column (Spherisorb S5ODS2) equilibrated in buffer E (0.1% TFA in H_2O). The assay mixture is prepared with 5 μM reduced peptide, 25 μM Pf PDO, 100 mM GSH, and 25 mM GSSG in McIlvaine buffer pH 7.0, and incubated for 30 min at 50°. After incubation the mixture is loaded onto the HPLC column. Chromatography is carried out with a linear gradient 0–100% buffer F (acetonitrile 95%, 0.1% TFA) in Buffer E at flow rate of 1 ml/min in 35 min. The reduced and oxidized form of the peptide have different retention times and are eluted separately, thus demonstrating the capability of Pf PDO to catalyze the formation of the disulfide bond. Indeed, when the same experiment was carried out in the absence of Pf PDO, only one peak of the reduced peptide was observed (Fig. 3). Using these assay conditions the disulfide bond-forming activity of Pf PDO shows a linear dependence on the protein concentration up to 50 μM.

Properties of Pf PDO

Homogeneous Pf PDO can be stored at $-20°$ for at least 3 months without any loss of activity; the presence of a reducing agent does not seem to be essential for its stability. Isoelectric focusing gel of the homogeneous protein reveals one band with isoelectric point of 4.9.

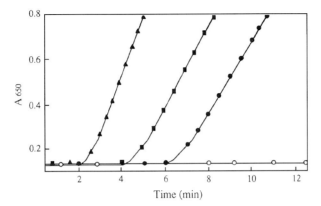

FIG. 4. The DTT-dependent reduction of bovine insulin disulfide as carried out in the absence (○) and in the presence at different concentration of pure Pf PDO: (▲) 1.2 μM, (■) 0.6 μM, (●) 0.2 μM.

The protein purified from *P. furiosus* is very thermostable and loses no activity after a 3-hr incubation at 90°. The molecular mass of the protein was determined by electrospray mass spectrometry; a value of 25,649 Da was calculated, which corresponds to the mass deduced from the amino acid sequence.

Pf PDO has the ability to act as a reductant at 30° and catalyzes the reduction of insulin disulfide in the presence of dithiothreitol. Figure 4 shows the DTT-dependent reduction of insulin disulfides at 30° in the presence of increasing concentrations of pure *P. furiosus* protein and in its absence (the spontaneous precipitation reaction). In this case the progress curve obtained for 5 μM of protein (not shown) was identical to that for 1.2 μM, indicating that the assay was saturated.

Because of substrate instability it is impossible to perform the assay at temperatures higher than 30°. The protein can also utilize glutathione, glutathione reductase, and NADPH to reduce disulfide substrates, and this demonstrates the thioltransferase activity of the enzyme. The maximal rate of activity with different concentrations of GSH and L-cystine is obtained with 10 mM GSH and 2.5 mM L-cystine; higher concentrations led to inhibition. The thioltransferase activity of the *P. furiosus* protein shows a linear dependence on the protein concentration up to 0.12 μM, and inhibition is detected at higher protein concentrations.

The functional data demonstrate that the archaeal protein is an oxidant and reductant; in fact, Pf PDO is able to catalyze the oxidation of dithiol as well as the reduction of disulfides. The studies of the disulfide bond rearrangement were not successful at the low temperature used (30°) testing as substrate scrambled RNAse.

Acknowledgment

This research was supported by grants from EU and MURT.

[8] Protein Disulfide Oxidoreductase from *Pyrococcus furiosus*: Structural Properties

By BIN REN and RUDOLF LADENSTEIN

Introduction

Proteins from hyperthermophilic archaea usually contain a decreased number of cysteine residues compared to their mesophilic counterparts. This is probably due to the easy oxidation and instability of sulfur-containing amino acid residues at high temperatures. Consequently, there have been arguments, although with little supportive evidence, that archaeal proteins contain hardly any disulfide bonds. However, a few protein disulfide oxidoreductases have been purified and characterized in archaea.[1-4] Their existence suggests that thiol–disulfide exchange reactions actually take place in archaea as in bacteria and eukarya. The determination of the first three-dimensional structure of a protein disulfide oxidoreductase from the archaeon *Pyrococcus furiosus* (*Pf* PDO) has provided the first view of active site disulfides in an archaeal protein.[5] It reveals a unique structural form compared to those of protein disulfide oxidoreductases in bacteria and eukarya.

Protein disulfide oxidoreductases have been well studied in bacteria and eukarya. They are ubiquitous redox enzymes consisting of the families of thioredoxin, glutaredoxin, protein disulfide-isomerase (PDI), DsbA, and their homologs. These enzymes catalyze the formation, breakage, and rearrangement of protein disulfide bonds. They share a CXXC sequence motif at their active sites, in which the two cysteines undergo reversible oxidation–reduction by shuttling between a dithiol and a disulfide form in the catalytic process. Thioredoxin is a general protein disulfide reductase that participates in a variety of biological activities.[6] It interacts with various proteins either for electron transport reactions or activity regulation via thiol-dependent control of the redox status. Glutaredoxin is also a disulfide reductase.[7,8] Unlike thioredoxin, glutaredoxin is a glutathione-dependent enzyme,

---

[1] F. Schlicht, G. Schimpff-Weiland, and H. Follmann, *Naturwissenschaften* **72,** 328 (1985).
[2] S. C. McFarlan, C. A. Terrell, and H. P. C. Hogenkamp, *J. Biol. Chem.* **267,** 10561 (1992).
[3] A. Guagliardi, V. Nobile, S. Bartolucci, and M. Rossi, *Int. J. Biochem.* **26,** 375 (1994).
[4] A. Guagliardi, D. de Pascale, R. Cannio, V. Nobile, S. Bartolucci, and M. Rossi, *J. Biol. Chem.* **270,** 5748 (1995); *J. Biol. Chem.* **272,** 20961 (1997); also see S. Bartolucci, D. de Pascale, and M. Rossi, *Methods Enzymol.* **334,** [7], 2001 (this volume).
[5] B. Ren, G. Tibbelin, D. de Pascale, M. Rossi, S. Bartolucci, and R. Ladenstein, *Nature Struct. Biol.* **5,** 602 (1998).
[6] A. Holmgren, *Ann. Rev. Biochem.* **54,** 237 (1985).
[7] W. W. Wells, Y. Yang, T. L. Deits, and Z. R. Gan, *Adv. Enzymol.* **66,** 149 (1993).
[8] A. Holmgren and F. Åslund, *Methods Enzymol.* **252,** 283 (1995).

which preferentially reacts with low molecular weight, especially GSH-mixed disulfides, rather than protein disulfide bonds. While thioredoxin and glutaredoxin catalyze the reduction of disulfides, DsbA and PDI catalyze the formation or rearrangement of disulfide bonds in protein folding processes. DsbA is a bacterial protein, which functions as an oxidase in the periplasmic space of gram-negative bacteria,[9] whereas PDI is an endoplasmic reticulum protein, which can catalyze both formation and rearrangement of protein disulfides in eukarya.[10]

These well-characterized protein disulfide oxidoreductases share a similarity in their three-dimensional structures: all of them use a common structural motif as the scaffold. It has been named the thioredoxin fold as it was first revealed in the crystal structure of *Escherichia coli* thioredoxin.[11] The thioredoxin fold consists of a four-stranded central β sheet and three flanking α helices, which are arranged in the order βI-αI-βII-αII-βIII-βIV-αIII.[12] The active site is located at the N terminus of αI. The thioredoxin fold does not include the first β strand and α helix present in the crystal structure of *E. coli* thioredoxin. Thioredoxin and glutaredoxin are small proteins with a molecular mass less than 12 kDa. They contain only one thioredoxin fold in their three-dimensional structures. DsbA is a 21 kDa monomeric protein. Its crystal structure consists of a thioredoxin-like domain and an additional α-helical domain.[13] PDI exists as a homodimer with two 57 kDa subunits. The three-dimensional structure of intact PDI is not yet determined, but its amino acid sequence and tertiary domain structures have revealed it to be a multidomain protein.[14-16] Unlike the other protein disulfide oxidoreductases, a PDI molecule possesses two thioredoxin-like domains with two active sites. In addition to the classical form of PDI, a growing number of proteins containing two or more thioredoxin-like regions in their amino acid sequences have been identified.[10] Although the exact functions of these proteins remain unclear, they have been suggested to form a family of PDI-like proteins.[10]

Protein Disulfide Oxidoreductases in Archaea

To date, only a few archaeal protein disulfide oxidoreductases have been isolated. The physiological functions of these enzymes, however, remain ambiguous. A small redox protein, with a molecular mass of 12 kDa, has been purified

---

[9] J. C. A. Bardwell, K. McGovern, and J. Beckwith, *Cell* **67**, 581 (1991).
[10] R. B. Freedman, T. R. Hirst, and M. F. Tuite, *Trends Biochem. Sci.* **19**, 331 (1994).
[11] A. Holmgren, B. O. Söderberg, H. Eklund, and C.-I. Bränden, *Proc. Natl. Acad. Sci. U.S.A.* **72**, 2305 (1975).
[12] J. L. Martin, *Structure* **3**, 245 (1995).
[13] J. L. Martin, J. C. A. Bardwell, and J. Kuriyan, *Nature* **365**, 464 (1993).
[14] J. C. Edman, L. Ellis, R. W. Blacher, R. A. Roth, and W. J. Rutter, *Nature* **317**, 267 (1985).
[15] J. Kemmink, N. J. Darby, K. Dijkstra, M. Nilges, and T. E. Creighton, *Biochemistry* **35**, 7684 (1996).
[16] J. Kemmink, N. J. Darby, K. Dijkstra, M. Nilges, and T. E. Creighton, *Curr. Biol.* **7**, 239 (1997).

from the archaeon *Methanobacterium thermoautotrophicum* by Schlicht *et al.*[1] and McFarlan *et al.*[2] This protein can catalyze the reduction of insulin disulfides and function as a hydrogen donor for *E. coli* ribonucleotide reductase. It was first described as a thioredoxin,[1] but the conservation of the active site motif CPYC in its amino acid sequence suggested that it was a glutaredoxin-like protein.[2] Surprisingly, however, this protein differs from other thioredoxins and glutaredoxins in that the reduced enzyme does not react with either thioredoxin reductase or glutathione.

Guagliardi *et al.* purified a protein disulfide oxidoreductase from the hyperthermophilic archaeon *Sulfolobus solfataricus*.[3] As it showed the ability to catalyze the reduction of insulin disulfides in the presence of dithiothreitol, the protein was named a thioredoxin. The monomeric form of the enzyme has an unusual molecular mass of 25 kDa. The protein is highly thermostable. Heat treatment at 90° for 3 hr did not cause decreased insulin reduction activity. Instead, an increase of activity was observed after exposure to heat. Another protein disulfide oxidoreductase was later purified from the hyperthermophilic archaeon *Pyrococcus furiosus*.[4] The *P. furiosus* protein showed close similarity to the *S. solfataricus* protein in molecular weight, dithiothreitol-dependent insulin reduction activity, and heat stability. In addition, both proteins displayed thioltransferase activity by catalyzing the reduction of disulfide bonds in L-cysteine in the presence of glutathione, glutathione reductase, and NADPH.[4] It was suggested that these two proteins are probably homologs. The gene of the *P. furiosus* protein has been cloned and the amino acid sequence has been deduced. The protein is about two times larger than normal thioredoxins and glutaredoxins. Its primary structure does not show overall similarity to the sequences of known protein disulfide oxidoreductases. The *P. furiosus* protein has the unusual feature of having two active site CXXC sequence motifs. A typical CPYC sequence motif is located at the C-terminal part of the protein, whereas this motif has been found at the N terminus in all glutaredoxins. In addition, a CQYC sequence exists at the N terminus of the protein, which has not been observed in any other protein disulfide oxidoreductase. Because of the existence of the CPYC motif and the observed thioltransferase activity, the protein was considered a glutaredoxin-like protein.[4] However, further exploration of the physiological function and the nature of the protein are needed. In fact, whether this protein functions as a glutathione-dependent enzyme *in vivo* is questionable. No NADPH-dependent glutathione reductase was detected in the crude extract of *P. furiosus* or *S. solfataricus*.[4] There is also no indication for the existence of glutathione or related peptides in archaeal cells. We have determined the crystal structure of the protein disulfide oxidoreductase from the archaeon *Pyrococcus furiosus* (*Pf* PDO). The structural analysis has dramatically changed our understanding of this enzyme. It provides the basis for further functional studies. Its unusual structural features suggest that this enzyme probably represents

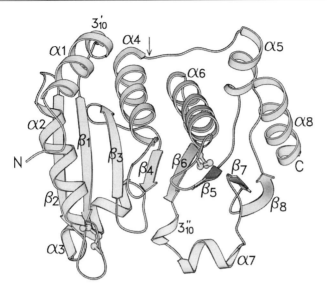

FIG. 1. Ribbon diagram of the structure of the *Pf* PDO monomer. The active site disulfides are shown in ball-and-stick representation. The conjunction point of the N and C units, which is between residue 117 and 118, is indicated by an arrow. This and the other ribbon diagrams were produced with Molscript [P. J. Kraulis, *J. Appl. Cryst.* **24**, 946 (1991)] and Raster3d [E. A. Merritt and D. J. Bacon, *Methods Enzymol.* **277**, 505 (1997)]. Adapted from B. Ren, G. Tibbelin, D. de Pascale, M. Rossi, S. Bartolucci, and R. Ladenstein, *Nature Struct. Biol.* **5**, 602 (1998).

a new member of the protein disulfide oxidoreductase superfamily rather than a thioredoxin or glutaredoxin as described previously.

Crystal Structure of *Pf* PDO and Its Implications

Pf PDO was crystallized by the sitting-drop vapor diffusion method.[17] The crystals belong to the enantiomorphic hexagonal space group $P6_522$ with cell dimensions of $a = b = 110.3$ Å and $c = 68.5$ Å and with one molecule in the asymmetric unit. The structure was solved by multiple isomorphous replacement including anomalous scattering data.[5] The structure has been refined to 1.9 Å resolution with a R factor of 0.192 and a free R factor of 0.217.

Monomer Structure and Comparison

Pf PDO has an α/β structure composed of eight β strands and eight α helices (Fig. 1). The eight strands form a central β sheet, in which $\beta 3$ and $\beta 7$ are antiparallel

---

[17] B. Ren, G. Tibbelin, D. de Pascale, M. Rossi, S. Bartolucci, and R. Ladenstein, *J. Struct. Biol.* **119**, 1 (1997).

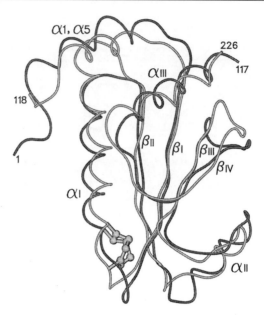

FIG. 2. Superposition of *Pf* PDO N and C units. The active site disulfides are shown in ball-and-stick representation. The secondary structure elements belonging to the thioredoxin fold are numbered with Roman numerals. Adapted from B. Ren, G. Tibbelin, D. de Pascale, M. Rossi, S. Bartolucci, and R. Ladenstein, *Nature Struct. Biol.* **5,** 602 (1998).

to the others. The β strands are organized in a rotational manner and the whole β sheet is twisted by about 114° when viewed from the side. The eight α helices are distributed asymmetrically on the two sides of the central β sheet. Two of them, α3 and α7, are located on one side and six on the other side. The two CXXC active site motifs form two intrachain disulfide bonds, Cys-35 Sγ–Cys-38 Sγ and Cys-146 Sγ–Cys-149 Sγ, respectively. The N-terminal cysteine in either disulfide, Cys-35 or Cys-146, is located in a loop which is between β1 and α2 (active site loop N) or between β5 and α6 (active site loop C), while the C-terminal cysteine, Cys-38 or Cys-149, is in the first turn of the helix α2 or α6.

It was found that the *Pf* PDO molecule can be almost equally divided into two structural units, the N-terminal one (residues 1 to 117) and the C-terminal one (residues 118 to 226). They are connected by the loop between α4 and α5 in the middle of the molecule. Either unit contains four β strands and four α helices and displays exactly the same fold (Fig. 2), despite the fact that their sequence identity is only about 20%. Their superposition reveals a root-mean-square (r.m.s.) deviation of 1.23 Å for 66 Cα atoms. Interestingly, each of the *Pf* PDO units is basically a thioredoxin fold motif but with an additional α helix, α1 or α5, inserted at the N terminus. As a result, the two units share a similar fold with other protein

TABLE I
ROOT-MEAN-SQUARE DIFFERENCES FOR Cα ATOMS ON SUPERPOSITIONS AND SEQUENCE IDENTITY
CALCULATIONS BETWEEN Pf PDO UNITS AND OTHER PROTEIN DISULFIDE OXIDOREDUCTASES[a]

Parameter	E. coli_Trx	T4_Grx	Human_PDI-a	E. coli_DsbA[b]
A. Superposition (r.m.s.d., Å/Cα atoms)				
Pf PDO_N_unit	1.56/65	1.57/39	1.39/40	1.66/42
Pf PDO_C_unit	1.38/76	1.85/17	0.99/43	1.48/47
B. Sequence identity (%)				
Pf PDO_N_unit	14.5	10.3	11.1	7.7
Pf PDO_C_unit	18.3	9.2	20.2	9.2

[a] The structural superpositions were performed with the program TOP (G. Lu, personal communication). The structures used in the comparison are *E. coli* thioredoxin, *E. coli*_Trx [S. K. Katti, D. M. LeMaster, and H. Eklund, *J. Mol. Biol.* **212**, 167 (1990)]; T4 glutaredoxin, T4_Grx [H. Eklund, M. Ingelman, B.-O. Söderberg, T. Uhlin, P. Nordlund, M. Nikkola, U. Sonnerstam, T. Joelson, and K. Petratos, *J. Mol. Biol.* **228**, 596 (1992)]; human PDI-*a* domain, Human_PDI-*a* [J. Kemmink, N. J. Darby, K. Dijkstra, M. Nilges, and T. E. Creighton, *Biochemistry* **35**, 7684 (1996)], and *E. coli* DsbA, *E. coli*_DsbA [J. L. Martin, J. C. A. Bardwell, and J. Kuriyan, *Nature* **365**, 464 (1993)].
[b] In DsbA, 75 residues, including mainly the whole α-helical domain, are excluded in structural superposition and sequence identity calculation.

disulfide oxidoreductases, including thioredoxin, glutaredoxin, DsbA, and PDI-*a* domain, although their sequence identities are extremely low (Table I). Structural comparison indicated that the two disulfides in *Pf* PDO units are located at exactly the same positions with respect to the active site disulfides in other protein disulfide oxidoreductases. Like in the other proteins, one *cis*-proline, *cis*-Pro-80 or *cis*-Pro-194, exists in the vicinity of the active site disulfide in each *Pf* PDO unit. The conserved *cis*-proline in thioredoxin has been shown to play an important role in the binding of substrates.[18,19]

Unlike thioredoxin, glutaredoxin, and DsbA, which possess only one thioredoxin-like motif, *Pf* PDO is the first protein disulfide oxidoreductase whose three-dimensional structure has been shown to contain two thioredoxin fold motifs with two active sites. Thus, *Pf* PDO shows structural resemblance to PDI and PDI-like proteins. This structural feature suggests that *Pf* PDO is probably not just a simple protein disulfide reductant like thioredoxin as described previously.[4] It may belong to the growing family of PDI-like proteins. From a structural point of view, *Pf* PDO may represent the simplest form of PDI.

Based on the structural findings, biochemical experiments have been carried out to investigate the activities of disulfide formation and rearrangement for *Pf* PDO.

[18] J. Qin, G. M. Clore, W. M. P. Kennedy, J. R. Huth, and A. M. Gronenborn, *Structure* **3**, 289 (1995).
[19] J. Qin, G. M. Clore, W. P. Kennedy, J. Kuszewski, and A. M. Gronenborn, *Structure* **4**, 613 (1996).

As hyperthermostable enzymes are usually less active at room temperatures, the major difficulty in these experiments is to develop novel methods to measure these activities at relatively high temperatures. The disulfide bond-forming activity was investigated by monitoring the fluorescence change of a decapeptide containing two cysteines, a method adapted from Ruddock et al.[20] At 50° and in the presence of glutathione, *Pf* PDO catalyzes the formation of a disulfide bond between the two cysteines of the peptide with a time-dependent fluorescence profile similar to that observed for DsbA at 25°.[21] As a result, *Pf* PDO is able to catalyze the oxidation of dithiols in addition to its ability to act as a protein disulfide reductant. Although little isomerase activity was observed at 30° for *Pf* PDO using scrambled RNase as the substrate, methods are being developed to measure such activity at appropriate temperatures.

Two Active Sites

Like PDI, *Pf* PDO possesses two active sites. Although the functional significance of the existence of two active sites in PDI has not yet been elucidated, comparison of the conformations of the two active site disulfides in *Pf*PDO enables us to speculate on their functional differences. In fact, the two active site disulfides in *Pf* PDO represent two extreme examples of disulfide conformations found in protein disulfide oxidoreductases. The C-terminal disulfide has the most relaxed conformation with a torsional angle ($\chi 3$) of 80° for the disulfide bond and a Cα–Cα distance of 5.5 Å between the two cysteines, whereas the N-terminal difsulfide has the most strained conformation with corresponding values of 47 ° and 5.0 Å, respectively. The calculated theoretical dihedral energy for the N-terminal disulfide is more than four times larger than that for the C-terminal disulfide (13.80 kcal mol^{-1} versus 3.04 kcal mol^{-1}). The strong conformational strain at the N-terminal active site may thus severely destabilize the disulfide bond. The highest density values around the N- and C-terminal disulfides are about 4σ and 8σ, respectively. Accordingly, the average *B* factor of the 14 atoms forming the N-terminal disulfide ring is about two times larger than that of those forming the C-terminal disulfide ring.

Generally, the atoms in the N unit have higher *B* factors than in the C unit (Fig. 3). The averaged *B* factors for the Cα atoms in the N and C units are 27.8 Å^2 and 22.8 Å^2, respectively. These differences suggest that the *Pf* PDO molecule may be considered as a modular structure consisting of two structural units with different conformational stabilities, even though it is not yet clear to what extent these two units are independent. They may display differences in their functional properties during disulfide shuffling. As the N-terminal disulfide is probably much

---

[20] L. W. Ruddock, T. R. Hirst, and R. B. Freedman, *Biochem. J.* **315**, 1001 (1996).
[21] S. Bartolucci, unpublished data (1998).

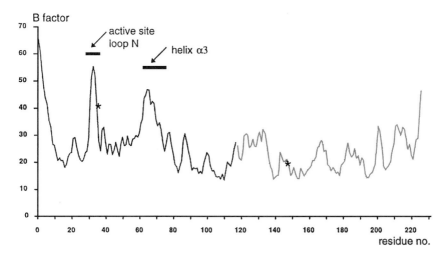

FIG. 3. B-factor plot of the Cα atoms in Pf PDO N and C units, represented by dark and light curves, respectively. The positions corresponding to the Cα atoms of Cys-35 and Cys-146 are marked by asterisks (∗). The regions with large B factors are indicated.

less stable than the C-terminal one, the nature of the N unit may be more oxidative and that of the C unit more reductive. Different conformational stabilities for the two thioredoxin domains and their active site disulfides were also suggested for the PDI molecule.[22]

Local structural differences have also been observed at the two active sites of Pf PDO. The active site loop N adopts a different conformation compared to the active site loop C, as if they have flipped close to or away from the active sites. As in the other protein disulfide oxidoreductases, the active site loop C has an "open" conformation so that the Sγ atom of the C-terminal disulfide is exposed to the solvent. In contrast, the active site loop N adopts a "closed" conformation, which exerts additional strain on the N-terminal disulfide by pushing it toward the interior of the molecule. As a result, the N-terminal disulfide becomes inaccessible. In accordance with the "closed" conformation of active site loop N, the orientation of the neighboring helix α3 that participates in the formation of the N-terminal active site groove deviates by 22°, compared to its counterpart (α7) of the C-terminal active site groove. To access the N-terminal active site and bind substrate would require considerable conformational adjustments of the active site loop N and helix α3. Inspection of the B-factor plot (Fig. 3) indicates that these two regions, in fact, possess the highest B-factors in the whole Pf PDO molecule besides the two termini, reflecting the intrinsic flexibility of these regions. Whereas

[22] N. J. Darby and T. E. Creighton, *Biochemistry* **34**, 11725 (1995).

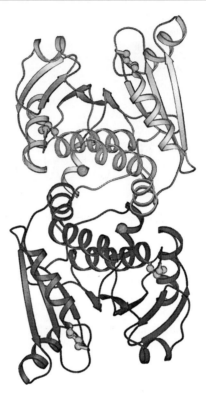

Fig. 4. Ribbon diagram of the *Pf* PDO dimer. The two zinc ions that are related by the crystallographic twofold axis are shown by spheres. The active site disulfides are shown in ball-and-stick representation. Adapted from B. Ren, G. Tibbelin, D. de Pascale, M. Rossi, S. Bartolucci, and R. Ladenstein, *Nature Struct. Biol.* **5,** 602 (1998).

the active site loop C is composed of hydrophobic or polar residues, all the residues in active site loop N are charged. They are populated around the N-terminal active site. Extensive electrostatic interactions may occur between these residues and substrates to induce necessary conformational movements in the catalytic process.

Zinc Binding Site and Pf PDO Dimer

It was found that in the crystal unit cell two neighboring *Pf* PDO molecules are associated by two zinc ions to form a dimer in the crystal (Fig. 4). The two zinc ions are related by a crystallographic twofold axis. Each ion is tetrahedrally coordinated to four ligands, including the Oε2 atom of Glu-15 in one molecule, the Nδ1 atom of His-102[*] from a symmetry-related molecule, and two water

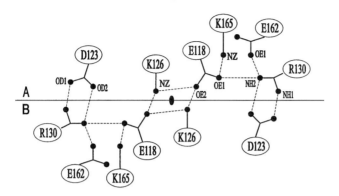

FIG. 5. Schematic representation of the 12-residue ion pair network at the dimer interface. Ion pairs were identified with a distance less than or equal to 4 Å.

molecules. The buried accessible surface area for each molecule is 783.3 Å^2, which accounts for 7.2% of the surface area of the molecule and is well within the range of buried areas observed for protein dimers.[23] The intersubunit contacts at the dimer interface consist of charged and hydrophobic interactions. The ratio of charged:polar:apolar for the interface area is 44.8%:13.7%:41.5%. There exist 10 intersubunit ion pairs. Six of them belong to a 12-residue ion pair network identified at the dimer interface (Fig. 5). This ion pair network is one of the largest ion pair networks observed at a protein subunit interface so far. These charged interactions are probably essential to stabilize the protein and subunit association at extreme temperatures. Similar features have been observed in other hyperthermophilic proteins, but not in mesophilic proteins.[24]

Substrate Binding Suggested by Crystal Packing Contacts

Thioredoxin is a general protein disulfide reductase. The hydrophobic nature of a protein surface around its active site has been suggested to be important for its interactions with a broad range of substrates.[25] The main residues forming such a hydrophobic surface patch include the highly conserved *cis*-proline (corresponding to *cis*-Pro-80 and *cis*-Pro-194 in *Pf* PDO units) and a glycine (corresponding to Gly-97 and Gly-210 in *Pf* PDO units) as well as their neighboring residues close to the active site. It has been shown that peptides (part of the transcription factor NF-κB or redox protein Ref-1) can bind to the active site of thioredoxin in two

[23] S. Jones and J. M. Thornton, *Prog. Biophys. Mol. Biol.* **63**, 31 (1995).
[24] R. Ladenstein and G. Antranikian, *Adv. Biochem. Eng. Biotechnol.* **61**, 37 (1998).
[25] H. Eklund, C. Cambillau, B.-M. Sjöberg, A. Holmgren, H. Jörnvall, J.-O. Höög, and C.-I. Brändén, *EMBO J.* **3**, 1443 (1984).

opposite orientations.[18,19] They can be arranged in either parallel or antiparallel fashion to the *cis*-proline and its preceding residue, in accordance with the versatility of substrate binding by this enzyme. Backbone–backbone hydrogen bonds form between the main-chain nitrogen and oxygen atoms of the residue preceding the *cis*-proline and the peptide substrates. With closely related structural characteristics, *Pf* PDO probably shares similar features with thioredoxin in substrate binding in order to act as a general protein disulfide oxidoreductase.

In addition to the dimeric interactions, the *Pf* PDO molecule makes a number of crystal packing contacts with symmetry-related molecules in the crystal. Although most of these contacts bury a surface area below 400 Å^2 per molecule, an unusually large interface is observed, which has a buried area of 755.9 Å^2 per molecule, representing 6.9% of the total surface area of the molecule. In contrast to that at the dimer interface, the interactions between these two *Pf* PDO molecules are predominantly hydrophobic with four possible hydrogen bonds. The ratio of charged:polar:apolar for this interface area is 4.8%:34.4%:60.8%. The crystal contacts at this interface mainly occur between the C-terminal active site areas of the two molecules, which are facing each other. The active site loop C in one molecule approaches the C-terminal active site of the other molecule and forms interactions with the loop between α7 and β7, preceding *cis*-Pro-194. These two loops are arranged in a parallel mode and the Oγ atom of Thr-145* can form a hydrogen bond with the main chain oxygen or nitrogen atom of Val-193. This interaction is very similar to the observed hydrogen bonding interactions in the human thioredoxin–peptide complexes. As the structures of the *Pf* PDO units and thioredoxin are highly superimposable, potential peptide binding to the *Pf* PDO molecule can be simulated by aligning the structure of a *Pf* PDO unit with that of human thioredoxin complexed with peptides.[18,19] With an r.m.s. deviation of 1.43 Å for 65 Cα atoms between the *Pf* PDO C-unit and human thioredoxin, the two peptides can be aligned into the *Pf* PDO C-terminal active site perfectly without collision. The positions of the aligned peptides superimpose very well with that of active site loop C from the symmetry related molecule (Fig. 6), which strongly suggests that the observed crystal contacts mimick protein–substrate interactions. Whereas the alignment of the human thioredoxin–peptide complexes with the *Pf* PDO C-unit indicates an easy access to the C-terminal active site, the alignment with the N-unit results in serious collisions between the peptides and local structures at the N-terminal active site, including active site loop N. Substrate binding to the N-terminal active site would require conformational adjustments of these local structures, such as the flip of active site loop N, in order to expose the active site.

Possible Common Ancestor

In addition to *Pf* PDO and other protein disulfide oxidoreductases, the thioredoxin fold has also been found in other proteins, including glutathione

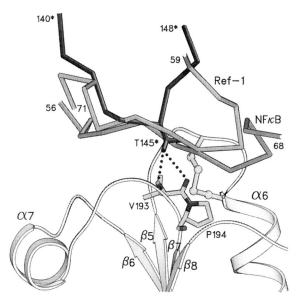

FIG. 6. Crystal packing contacts mimic substrate binding interactions. See text for description.

peroxidase,[26–28] glutathione S-transferase,[29] peroxiredoxin,[30] and phosducin.[31] Except phosducin, in which the thioredoxin fold motif plays a purely structural role, all the other proteins are involved in various redox reactions. Despite their functional differences, all the redox enzymes interact with substrates having a thiol or a disulfide group. Interestingly, on alignment of these structures, the active site of each protein is located in the same relative position in space, which is at or close to the N terminus of helix αI in the thioredoxin fold. Except glutathione peroxidase, every other enzyme possesses a *cis*-proline, which is positioned adjacent to the N-terminal end of strand βIII close to the active site.

The thioredoxin fold motif serves as the scaffold for the structures of these redox enzymes (Fig. 7). Various structural insertions into the thioredoxin fold lead to the diversity of these enzymes. Variations may occur at certain regions of the thioredoxin fold, including both termini and the loops between βII and αII, αII and βIII, and βIV and αIII.[12] Among the structures of these redox enzymes, the

---

[26] R. Ladenstein, O. Epp, K. Bartels, A. Jones, R. Huber, and A. Wendel, *J. Mol. Biol.* **134,** 199 (1979).
[27] O. Epp, R. Ladenstein, and A. Wendel, *Eur. J. Biochem.* **133,** 51 (1983).
[28] B. Ren, W. Huang, B. Åkesson, and R. Ladenstein, *J. Mol. Biol.* **268,** 869 (1997).
[29] P. Reinemer, H. W. Dirr, R. Ladenstein, J. Schäffer, O. Gallay, and R. Huber, *EMBO J.* **10,** 1997 (1991).
[30] H.-J. Choi, S. W. Kang, C.-H. Yang, S. G. Rhee, and S.-E. Ryu, *Nature Struct. Biol.* **5,** 400 (1998).
[31] R. Gaudet, A. Bohm, and P. B. Sigler, *Cell* **87,** 577 (1996).

thioredoxin

glutaredoxin

DsbA

PDI-*a*

glutathione S-transferase

peroxiredoxin

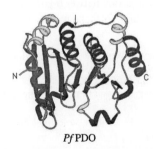

Pf PDO

glutathione peroxidase

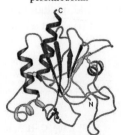

Pf PDO structure is the only one that has been shown to contain two thioredoxin fold motifs with two active sites so far. Each *Pf* PDO unit possesses an extra α helix, α1 or α5, at the N terminus of the thioredoxin fold. In the crystal structure, these two helices interact with a symmetry-related molecule and contribute to the formation of the dimer and zinc binding sites.

Based on the structural similarities among these redox enzymes, it is tempting to speculate that these proteins might have evolved from a common ancestor even though convergent evolution cannot be ruled out completely.[12,32] Such an ancestor might have a CXXC active site and a neighboring *cis*-proline.[12] It is generally believed that the existence of multiple thioredoxin-like domains in eukaryotic PDI and PDI-like proteins results from gene duplication events during evolution.[16,33] The two thioredoxin-like domains in PDI possess the same active site motif and display obvious sequence identities to each other (50%) and to thioredoxin (30%).[14] However, in *Pf* PDO the similarities among the two structural units and thioredoxin are only apparent upon the comparison of their three-dimensional structures. As the two *Pf* PDO units have distinct active site motifs and show rather low sequence identities (<20%) to each other and to thioredoxin, one may assume that the two *Pf* PDO units might have evolved more divergently than the thioredoxin-like domains in eukaryotic PDI. This assumption, however, is not in line with the hypothesis that hyperthermophiles are the least divergent organisms, which may represent the closest living descendants of ancestral life forms.[34–36] As the origin

---

[32] I. Sinning, G. J. Kleywegt, S. W. Cowan, P. Reinemer, H. W. Dirr, R. Huber, G. L. Gilliland, R. N. Armstrong, X. Ji, P. G. Board, B. Olin, B. Mannervik, and T. A. Jones, *J. Mol. Biol.* **232**, 192 (1993).

[33] S. Kanai, H. Toh, T. Hayano, and M. Kikuchi, *J. Mol. Evol.* **47**, 200 (1998).

[34] C. R. Woese, O. Kandler, and M. L. Wheelis, *Proc. Natl. Acad. Sci. U.S.A.* **87**, 4576 (1990).

[35] N. R. Pace, *Cell* **65**, 531 (1991).

[36] K. O. Stetter, *in* "Early Life on Earth" (S. Bengtson, ed.), p. 143. Columbia University Press, New York, 1994.

FIG. 7. Ribbon representations of the structures of *E. coli* thioredoxin [S. K. Katti, D. M. LeMaster, and H. Eklund, *J. Mol. Biol.* **212**, 167 (1990)], bacteriophage T4 glutaredoxin [H. Eklund, M. Ingelman, B.-O. Söderberg, T. Uhlin, P. Nordlund, M. Nikkola, U. Sonnerstam, T. Joelson, and K. Petratos, *J. Mol. Biol.* **228**, 596 (1992)], *E. coli* DsbA [J. L. Martin, J. C. A. Bardwell, and J. Kuriyan, *Nature* **365**, 464 (1993)], human PDI-*a* domain [J. Kemmink, N. J. Darby, K. Dijkstra, M. Nilges, and T. E. Creighton, *Biochemistry* **35**, 7684 (1996)], rat liver μ class glutathione *S*-transferase [X. Ji, P. Zhang, R. N. Armstrong, and G. L. Gilliland, *Biochemistry* **31**, 10169 (1992)], human peroxiredoxin [H.-J. Choi, S. W. Kang, C.-H. Yang, S. G. Rhee, and S.-E. Ryu, *Nature Struct. Biol.* **5**, 400 (1998)], *Pf* PDO [B. Ren, G. Tibbelin, D. de Pascale, M. Rossi, S. Bartolucci, and R. Ladenstein, *Nature Struct. Biol.* **5**, 602 (1998)], and human plasma glutathione peroxidase [B. Ren, W. Huang, B. Åkesson, and R. Ladenstein, *J. Mol. Biol.* **268**, 869 (1997)]. The active site residues and the secondary structure elements belonging to the thioredoxin fold are highlighted in dark.

of hyperthermophiles is still hotly debated,[37] the evolutionary implications of this observation remain puzzling.

Acknowledgments

We thank Gudrun Tibbelin for excellent technical assistance, Dr. Donatella de Pascale, Prof. Mosè Rossi, and Prof. Simonetta Bartolucci, University of Naples, Italy, for a fruitful collaboration. The financial support of the European Community (Project "Extremophiles as Cell Factories") is greatly acknowledged.

---

[37] J. Wiegel and M. W. W. Adams, "Thermophiles: The Keys to Molecular Evolution and the Origin of Life?" Taylor & Francis, London and Philadelphia, 1998.

Section II

Nucleic Acid Modifying Enzymes

[9] DNA Polymerases from Hyperthermophiles

By HOLLY H. HOGREFE, JANICE CLINE, AMY E. LOVEJOY, and KIRK B. NIELSON

With the emergence of the polymerase chain reaction (PCR), considerable attention has been focused on isolating and characterizing the DNA polymerases of thermophiles. Numerous laboratory techniques, including PCR, RT-PCR (reverse transcription-PCR), isothermal amplification, DNA sequencing, mutagenesis, and cloning, rely on thermostable DNA polymerases for carrying out DNA synthesis *in vitro* at high temperatures.

DNA polymerases perform essential roles in the replication and repair of genetic material. The DNA polymerases of several mesophiles have been extensively characterized, and the structural, biochemical, and kinetic properties of *Escherichia coli,* yeast, mammalian, and bacteriophage DNA polymerases have been the subject of numerous reviews.[1-6] Certain enzymes complement DNA polymerase activity with $3' \rightarrow 5'$-exonuclease (proofreading) activity and/or $5' \rightarrow 3'$-exonuclease activity, which typically reside in separate structural domains on the same polypeptide. To orchestrate DNA replication and repair, both eubacteria and eukaryotes possess multiple DNA polymerases, each with distinct properties, subunit compositions, and physiological roles. *In vivo,* DNA polymerases require the assistance of numerous accessory proteins to replicate and repair the genome (reviewed[7]). Based on amino acid sequence homology to *E. coli* enzymes, DNA polymerases have been classified into at least three distinct families: Family A (Pol I-like), Family B (Pol II-like), and Family C (Pol III-like).[8]

To date, more than 50 DNA polymerases have been cloned, sequenced, and characterized from thermophilic and hyperthermophilic eubacteria and archaea (for a comprehensive review, see Perler *et al.*[9]). The first thermostable DNA polymerases characterized, and hence the ones now commonly used in laboratory procedures, are those that are the most abundant and/or those that function and purify as monomers. These monomeric polymerases are generally 90–100 kDa in size.

[1] J. L. Campbell (ed.) *Methods Enzymol.* **262** (1995).
[2] C. M. Joyce and T. A. Steitz, *Annu. Rev. Biochem.* **63**, 777 (1994).
[3] T. S. Wang, *Annu. Rev. Biochem.* **60**, 513 (1991).
[4] A. Kornberg and T. A. Baker, "DNA Replication." Freeman, New York, (1992).
[5] H. Echols and M. F. Goodman, *Annu. Rev. Biochem.* **60**, 477 (1994).
[6] S. Linn, *Cell* **66**, 185 (1991).
[7] T. A. Baker and S. P. Bell, *Cell* **92**, 295 (1998).
[8] D. K. Braithwaite and J. Ito, *Nucleic Acids Res.* **21**, 787 (1993).
[9] F. B. Perler, S. Kumar, and H. Kong, *Adv. Protein Chem.* **48**, 377 (1996).

In (hyper) thermophilic eubacteria, the predominant DNA polymerase is structurally related to *E. coli* Pol I (Family A), and by analogy may play a role in repair. This class of polymerases is typified by *Taq* DNA polymerase (from *Thermus aquaticus*),[10] the first thermostable DNA polymerase developed for PCR.[11] Related Family A DNA polymerases have been described in other thermophilic eubacteria, including *Thermus* species [*T. thermophilus* (*Tth*),[12,13] *T. flavus* (*Tfl*),[14] and *T. brockianus* (*Tbr*)[15]], *Bacillus* species [*B. stearothermophilus* (*Bst*)[13,16] and *B. caldotenex* (*Bca*)[13,17]], and *Thermatoga maritima* (*Tma;* trade name UlTma[18]). The thermal stability and temperature optima of *Thermus* and *Thermatoga* DNA polymerases render them suitable for PCR, but only *Thermatoga maritima* is classified as a true hyperthermophile (growth T_{opt} 80°).

All eubacterial Family A DNA polymerases exhibit an associated $5' \rightarrow 3'$ structure-specific endonuclease (exonuclease) activity,[19] whereas only certain enzymes exhibit $3' \rightarrow 5'$-exonuclease-dependent proofreading activity. The N-terminal $5' \rightarrow 3'$-exonuclease domain of *E. coli* Pol I can be removed proteolytically to produce the Klenow fragment (KF),[20] and similar N-truncated enzymes have been generated from thermostable eubacterial DNA polymerases (Stoffel fragment from *Taq*[21]; UlTma from *Tma* Pol I[18]; BF from *Bst* strain Pol I[16]). Amino acid sequence alignments reveal that homologies among thermostable Family A DNA polymerases are most pronounced at the N terminus ($5' \rightarrow 3'$ exo domain) and C terminus (polymerase domain), whereas the absence or presence of associated proofreading activity has been attributed to structural variation in the internal $3' \rightarrow 5'$-exonuclease domain.[9] The crystal structures of thermostable Family A DNA polymerases (*Taq*[22,23]; BF[16]) reveal substantial homology to *E. coli* pol I, but offer few clues as to the structural features that provide DNA synthesis at high temperatures.[16] A second DNA polymerase has been described in *Thermus*

[10] F. C. Lawyer, S. Stoffel, R. K. Saiki, K. Myambo, R. Drummond, and D. H. Gelfand, *J. Biol. Chem.* **264,** 6427 (1989).
[11] R. K. Saiki, D. H. Gelfand, S. Stoffel, S. J. Scharf, R. Higuchi, G. T. Horn, K. B. Mullis, and H. A. Erlich, *Science* **239,** 487 (1988).
[12] N. Carballeira, M. Nazabal, J. Brito, and O. Garcia, *BioTechniques* **9,** 276 (1990).
[13] E. Sellman, K. L. Schroder, I. M. Knoblich, and P. Westermann, *J. Bacteriol.* **174,** 4350 (1992).
[14] A. A. Akhmetzjanov and V. A. Vakhitov, *Nucl. Acids Res.* **20,** 5839 (1992).
[15] Finnzymes Oy (Finland), unpublished, 1995.
[16] J. R. Kiefer, C. Mao, C. J. Hansen, S. L. Basehore, H. H. Hogrefe, J. C. Braman, and L. S. Beese, *Structure* **5,** 95 (1997).
[17] T. Uemori, Y. Ishino, K. Fujita, K. Asada, and I. Kato, *J. Biochem.* **113,** 401 (1993).
[18] "Guide to PCR Enzymes," Perkin-Elmer. (1993).
[19] V. Lyamichev, M. A. Brow, and J. E. Dahlberg, *Science* **260,** 778 (1993).
[20] H. Klenow and I. Henningsen, *Proc. Natl. Acad. Sci. U.S.A.* **65,** 168 (1970).
[21] F. C. Lawyer, S. Stoffel, R. K. Saiki, S. Y. Chang, P. A. Landre, R. D. Abramson, and D. H. Gelfand, *PCR Methods Appl.* **2,** 275 (1993).
[22] Y. Kim, S. H. Eom, J. Wang, D. S. Lee, S. W. Suh, and T. A. Steitz, *Nature* **376,** 612 (1995).
[23] Y. Li, S. Korolev, and G. Waksman, *EMBO J.* **17,** 7514 (1998).

thermophilus[24] that exhibits homology to *E. coli* Pol III (Family C), the multisubunit replicative DNA polymerase. The potential utility of this class of thermostable DNA polymerases in industrial applications is presently unknown.

In addition to thermostable eubacterial enzymes, a number of DNA polymerases from hyperthermophilic archaea have been isolated and characterized (reviewed[9]). The most abundant DNA polymerase(s) in archaea are homologs of Family B DNA polymerases, a group that includes eukaryotic Pol α, δ, and ε DNA polymerases and *E. coli* pol II. Within this group, archaeal DNA polymerases more closely resemble eukaryotic DNA polymerases, which *in vivo* are thought to carry out Okazaki fragment synthesis (Pol α) and leading and lagging strand DNA synthesis (Pol δ and Pol ε).[3,4,6] Archaeal homologs of Family B DNA polymerases are typified by Vent (from *Thermococcus litoralis*)[25] and *Pfu* (from *Pyrococcus furiosus*),[26,27] the first commercially available archaeal DNA polymerases. Other enzymes derived from hyperthermophilic archaea include Deep Vent (from *Pyrococcus* sp. *GB-D*),[9] Pwo (from *Pyrococcus woesei*[28]; identical to *Pfu* DNA polymerase[29]), KOD (from *Pyrococcus* strain KOD1),[30] and 9°N-7 (from *Thermococcus* species 9°N-7).[31] Sequence comparisons show that *Pyrococcus* and *Thermococcus* Family B DNA polymerases are highly related, exhibiting 75–90% identity.[9] Comparisons presented here, and elsewhere,[9] indicate that the *Pyrococcus* and *Thermococcus* Family B DNA polymerases exhibit similar yet distinct biochemical properties, which dictate their utility in particular laboratory applications.

The majority of archaeal Family B DNA polymerase homologs function as monomers, lack $5' \rightarrow 3'$-exonuclease activity, and possess an associated $3' \rightarrow 5'$-exonuclease (proofreading) activity. The editing capability of archaeal Family B DNA polymerases has led to their use in high-fidelity PCR applications. Sequence comparisons have identified three conserved motifs (exo I, II, III) in the $3' \rightarrow 5'$-exonuclease domain of DNA polymerases (reviewed[32]). Replacement of any of the conserved aspartic or glutamic acid residues with alanine has been shown to abolish the exonuclease activity of numerous DNA polymerases, including

---

[24] C. S. McHenry, M. Seville, and M. G. Cull, *J. Mol. Biol.* **272**, 178 (1997).
[25] H. Kong, R. B. Kucera, and W. E. Jack, *J. Biol. Chem.* **268**, 1965 (1993).
[26] K. S. Lundberg, D. D. Shoemaker, M. W. Adams, J. M. Short, J. A. Sorge, and E. J. Mathur, *Gene* **108**, 1 (1991).
[27] T. Uemori, Y. Ishino, H. Toh, K. Asada, and I. Kato, *Nucleic Acids Res.* **21**, 259 (1993).
[28] Boehringer Mannheim, unpublished.
[29] S. Dabrowski and J. Kur, *Protein Expr. Purif.* **14**, 131 (1998).
[30] M. Takagi, M. Nishioka, H. Kakihara, M. Kitabayashi, H. Inoue, B. Kawakami, M. Oka, and T. Imanaka, *Appl. Environ. Microbiol.* **63**, 4504 (1997).
[31] M. W. Southworth, H. Kong, R. B. Kucera, J. Ware, H. Jannasch, and F. B. Perler, *Proc. Natl. Acad. Sci.* **93**, 5281 (1996).
[32] V. Derbyshire, J. K. Pinsonneault, and C. M. Joyce, *Methods Enzymol.* **262**, 363 (1995).

archaeal DNA polymerases (Vent[25], Pfu^{33}), while conservative substitutions lead to reduced exonuclease activity ($9°N_m$ mutant of 9°N-7 DNA polymerase).[31] The crystal structure of an archaeal Family B DNA polymerase from *Themococcus gorgonarius* has been determined[34] and shown to exhibit the characteristic polymerase structure, resembling a right hand composed of palm, fingers, and thumb domains. Although fingers and thumb domains are highly diverse among different polymerase families, the exonuclease and palm domains of Family A and B DNA polymerases exhibit similar topologies and active site residues, implying common metal-assisted mechanisms for exonuclease and polymerase activities.[34,35]

Details on the mechanism of DNA replication and repair in archaea are largely unknown. Like eukaryotes and eubacteria, archaea are expected to possess multiple DNA polymerases that carry out specialized functions in conjunction with accessory proteins.[36,37] Indeed, analyses of archaeal genome sequences have revealed that, whereas certain archaea possess a single gene encoding a Family B DNA polymerase homolog (e.g., *Methanococcus jannaschii*[38]), two or three distinct homologs have been identified in other archaea (*Sulfolobus* species, *Pyrodictium occultum;* reviewed[39] and references therein). Moreover, a second class of DNA polymerases has been isolated from *Pyrococcus furiosus*.[40] This second DNA polymerase (pol II) is composed of two subunits, a catalytic subunit that appears to be structurally unrelated to any previously described polymerase family, and a noncatalytic/structural subunit that exhibits homology to the 50 kDa noncatalytic subunit of human pol δ.[37,40] To date, the physiological roles of the archaeal Family B and pol II DNA polymerases have not been elucidated. The archaeal pol II DNA polymerases are the subject of a chapter in this *Methods of Enzymology* volume.[41]

In this chapter, we present a compilation of methods used to assay various activities associated with thermostable DNA polymerases, including DNA polymerase activity (nucleotide incorporation), processivity, strand displacement, and exonuclease activity. We describe modifications to the nucleotide incorporation assay that allow measurements of thermostability, steady-state kinetic parameters, nucleotide analog incorporation, and reverse transcriptase activity. In addition, we provide a brief discussion regarding polymerase error rate determinations. These

---

[33] Stratagene, unpublished.
[34] K.-P. Hopfner, A. Eichinger, R. A. Engh, F. Laue, W. Ankenbauer, R. Huber, and B. Angerer, *Proc. Natl. Acad. Sci.* **96**, 3600 (1999).
[35] C. A. Brautigam and T. A. Steitz, *Curr. Opin. Struct. Biol.* **8**, 54 (1998).
[36] D. R. Edgell and W. F. Doolittle, *Cell* **89**, 995 (1997).
[37] I. K. Cann and Y. Ishino, *Genetics* **152**, 1249 (1999).
[38] C. J. Bult, O. White, G. J. Olsen, L. Zhou, R. D. Fleischmann, G. G. Sutton, J. A. Blake, L. M. Fitzgerald, R. A. Clayton, J. D. Gocayne, A. R. Kerlavage, B. A. Dougherty, J. F. Tomb, M. D. Adams, C. I. Reich, R. Overbeek, E. F. Kirkness, K. G. Weinstock, J. M. Merrick, A. Glodek, J. L. Scott, N. S. M. Geoghagen, and J. C. Venter, *Science* **273**, 1058 (1996).
[39] D. R. Edgell, H. P. Klenk, and W. F. Doolittle, *J. Bacteriol.* **179**, 2632 (1997).
[40] T. Uemori, Y. Sato, I. Kato, H. Doi, and Y. Ishino, *Genes Cells* **2**, 499 (1997).
[41] Y. Ishino and S. Dohino, *Methods in Enzymology* **334**, [21] (2001).

methods have been applied to purified thermostable DNA polymerases, including both eubacterial Family A and archaeal Family B DNA polymerases, and the properties of various DNA polymerases are presented here. The reader is referred to Perler et al.[9] for additional references describing the cloning, purification, and characteristics of thermostable DNA polymerases.

General Information for DNA Polymerase Assays

The assay conditions provided below should be considered as starting points for further optimization. As the optimal conditions for each activity may differ, researchers are encouraged to optimize salt concentration, pH, template concentration, and temperature for each DNA polymerase in each assay. A thorough review of assay methods for mesophilic DNA polymerases can be found in a *Methods of Enzymology* volume devoted entirely to DNA replication proteins.[1]

Table I summarizes the origin, source, assay temperature, and half-life ($t_{1/2}$ at 95°) of the purified DNA polymerases used in these studies. Except where noted, all measurements were performed at the assay temperature listed in Table I. For comparative purposes, measurements obtained by other investigators have been

TABLE I
Source, Thermostability, and Assay Temperature of DNA Polymerases

			DNA polymerase properties		
Polymerase name	Source (enzyme modifications)	Native organism (optimal growth temperature)[a]	Family	Assay temperature (°C)	Half-life at 95°C (hours)
Archaeal					
Pfu	Stratagene	*Pyrococcus furiosus* (100°)	B	72	19
Exo⁻ Pfu	Stratagene (3'-5'exo⁻)				
Deep Vent	NEB	*Pyrococcus* sp. GB-D (95°)	B	72	23[c]
Vent	NEB	*Thermococcus litoralis* (77°)	B	72	5, 6.7[c]
9°N$_m$	NEB (3'-5' exo reduced mutant)	*Thermococcus* sp. 9°N-7 (88–90°)	B	72	6.7[c]
Eubacterial					
Taq	Stratagene	*Thermus aquaticus* (70–75°)	A	72	≤1, 1.6[c]
UlTma	Perkin-Elmer[b]	*Thermatoga maritima* (80°)	A	72	
BF	Stratagene[b]	*Bacillus stearothermophilus* strain (72°)	A	65	<0.01
KF	Stratagene[b]	*E. coli* (37°)	A	37	

[a] F. B. Perler, S. Kumar, and H. Kong, *Adv. Protein Chem.* **48,** 377 (1996).
[b] N-truncated fragment (5'-3' exo⁻).
[c] New England BioLabs (NEB) catalog.

included, although somewhat different reaction conditions may have been used (e.g., DNA template, reaction buffer).

In general, purified DNA polymerases from (hyper) thermophilic eubacteria and archaea require the inclusion of nonionic detergents to both storage buffers and dilution buffers. The dilution/storage buffers used in the experiments described here consisted of 20–50 mM Tris-HCl (pH 7.5–8.2), 0.1–1 mM EDTA, 1 mM dithiothreitol (DTT) or 2-mercaptoethanol, 0.1–0.5% (v/v) Igepal [or Nonidet P-40 (NP-40)], 0.1–0.5% (v/v) Tween 20, 50% (v/v) glycerol, and 100 mM KCl (for *Taq* and BF buffers only). Nonionic detergents appear to be essential for long-term stability, as well as for stabilizing diluted enzymes in high-temperature reactions.

In addition to purified DNA polymerases, we have also been successful in directly assaying DNA polymerase activities in lysates prepared from *E. coli* expressing recombinant thermostable DNA polymerases. Typically, crude extracts are heated at 70° for 15–30 min and then centrifuged to obtain a cleared lysate.

Nucleotide Incorporation Assay

Introduction

All DNA polymerases catalyze the template-directed incorporation of deoxyribonucleotides (dNTPs) into DNA by addition at the 3′ OH termini of primer strands. Primer extension proceeds exclusively in the 5′ to 3′ direction, according to the reaction:

$$DNA_n + dNTP \rightarrow DNA_{n+1} + PP_i$$

Polymerase activity is assayed by measuring the incorporation of radiolabeled dNTPs into high molecular weight DNA. "Activated" calf thymus DNA, prepared by nicking (DNase I) genomic DNA, or primed single-stranded M13 DNA can be used as the DNA substrate. Poly(dA)/oligo(dT), a commonly used DNA substrate for mesophilic DNA polymerases, is unsuitable for assaying thermostable DNA polymerases at elevated temperatures (>50°).

Activated DNA is an uncharacterized mixture of single- and double-stranded ends, nicks, and gaps. In addition to being commercially available, activated DNA also has the advantage of providing a mixture of DNA substrates that, because of the length of the primers, are more resistant to thermal denaturation at temperatures where thermostable DNA polymerases are active. However, measurements carried out using activated DNA may be influenced by an enzyme's inherent strand displacement or 5′→3′-exonuclease activity. Enzymes that displace or degrade primers in their path exhibit greater incorporation (higher cpms) on activated DNA, compared to enzymes that stop synthesis on reaching downstream primers. As strand displacement and exonuclease activities can be significantly influenced by temperature, polymerization measurements should be carried out at the temperature at which the DNA polymerases will be ultimately used.

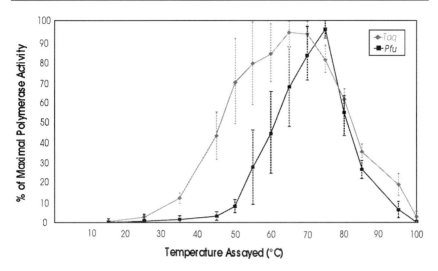

FIG. 1. Polymerase activity as a function of assay temperature. The polymerase activity of *Taq* (diamonds) and *Pfu* (squares) DNA polymerases was measured at the indicated temperatures using primed M13 DNA as template. Enzymes were compared at identical unit concentrations (as determined at 72°). Results are expressed as a percentage of maximal activity and represent the means (± standard deviations) from four measurements (duplicates in each of two assays). Additional controls were performed to verify that measurements were carried out at each temperature in the linear portion of the activity vs polymerase plot.

In contrast to activated DNA substrates, single-stranded templates annealed to oligonucleotide primers provide an unobstructed path for DNA polymerases, allowing one to monitor polymerization in the absence of strand displacement or $5' \rightarrow 3'$-exonuclease activity. DNA polymerases that exhibit high processivity, or the tendency to produce longer extensions, generally show higher activity using primed single-stranded M13 DNA. Determining the ratio of DNA polymerase activity on primed single-stranded DNA versus activated DNA is a common and quick assay that is used to identify enzymes that may be more processive. However, care must be taken to use oligonucleotide primers that anneal at the desired assay temperature. In practice, it is generally not possible to perform *in vitro* assays at the optimal temperature of many thermostable DNA polymerases (80–100°) due to denaturation of duplex DNA substrates (e.g., see Fig. 1).

Preparation of Primed M13mp18 Template

We routinely use a gel-purified 40-base oligonucleotide (5'GGT-TTT-CCC-AGT-CAC-GAC-GTT-GTA-AAA-CGA-CGG-CCA-GTG-C), which is suitable for priming extension reactions on single-stranded M13 at temperatures up to 75° (Fig. 1). To prepare primed M13 DNA substrate, we typically combine 76.2 nM (182 μg/ml) M13mp18 (+) strand (Pharmacia, Piscataway, NJ) and 762 nM

(10 μg/ml) of the 40-base primer. The substrate is heated at 90° for 3 min and then annealed at room temperature for 20 min.

Nucleotide Incorporation Assay Conditions

DNA polymerase activity is measured using activated calf thymus DNA or primed M13 DNA. DNA and dNTP substrates are used at concentrations that are at least 10× K_m and in large excess relative to the enzyme concentration. A typical DNA polymerase reaction cocktail contains:

50 mM Tris-HCl, pH 8.0
5 mM MgCl$_2$
1 mM DTT
50 μg/ml Bovine serum albumin (BSA)
4% (v/v) Glycerol
200 μM each dATP, dCTP, dGTP
195 μM TTP
5 μM [^{3}H] TTP (NEN 20.5Ci/mmol; partially evaporated to remove ethanol)
250 μg/ml of activated calf thymus DNA (Pharmacia) or 16.5 nM (39 μg/ml) primed M13 (see "Preparation of the Primed M13mp18 Template" above)

Polymerases are serially diluted in appropriate storage buffers and 1 μl of each enzyme dilution is added to 10-μl aliquots of polymerase cocktail. Polymerization reactions are conducted in duplicate or triplicate for 30 min at optimal temperature. The extension reactions are quenched on ice, and then 5-μl aliquots are spotted immediately onto DE81 ion-exchange filters (2.3 cm; Whatman, Clifton, NJ). Unincorporated [^{3}H]TTP is removed by six washes with 2× SCC (0.3 M NaCl, 30 mM sodium citrate, pH 7.0), followed by a brief wash with 100% ethanol. Incorporated radioactivity is measured by scintillation counting.

Reactions that lack enzyme are also set up along with sample incubations to determine "total counts per minute (cpms)" (omit filter wash steps) and "minimum cpms" (wash filters as above). Cpms bound is proportional to amount of dNTPs incorporated and can be converted into units of DNA polymerase activity. One unit of polymerase activity is defined as the amount of enzyme that catalyzes the incorporation of 10 nmol of total dNTP into polymeric form (binds to DE-81 paper) in 30 min at optimal temperature. To determine units, background (average "minimum cpms" value) is first subtracted from the average sample cpms. Units of polymerase activity can then be calculated using the following equation:

$$\text{Activity} = \frac{\text{(corrected sample cpms)}}{\text{total cpms}} \times \frac{\text{(8 nmol dNTPs)}}{\text{reaction}} \times \frac{\text{(1 unit)}}{\text{(10 nmol dNTPs incorporated)}}$$

Polymerase concentrations (U/ml) can be extrapolated from the slope of the linear portion of units versus enzyme volume plots (linear range is generally 0.03–0.003 U). Specific activity calculations (U/mg) can be carried out using protein concentrations determined relative to a BSA standard (Pierce, Rockford, IL) in a colorimetric assay (e.g., Coomassie Plus Protein Assay, Pierce). For polymerase activity comparisons (or when protein amounts are limiting), relative protein concentrations can be verified by SDS–PAGE analysis. Several aliquots of each polymerase preparation, ranging from 1 to 20 ng of total protein, are subject to SDS–PAGE electrophoresis and the intensities of silver- and/or Sypro orange (Molecular Probes, Eugene, OR)-stained bands are compared to standards.

The specific activity measurements obtained for purified archaeal and eubacterial DNA polymerases are summarized in Tables II and III, respectively. These experiments were conducted in a common assay buffer (50 mM Tris-HCl, pH 8.0, 5 mM MgCl$_2$, 1 mM DTT, 50 µg/ml BSA, and 4% glycerol) at the optimal temperature of each enzyme.

Nucleotide Incorporation Assay Modification: Measuring Thermostability

The nucleotide incorporation assay described above has been adapted to provide estimates of thermostability. Thermostability can be expressed in terms of half-life ($t_{1/2}$), as the length of time an enzyme can be incubated at a given temperature and still retain 50% of initial activity. We have determined the $t_{1/2}$ of various thermostable DNA polymerases by incubating diluted DNA polymerase solutions at 95° for varying periods of time, and then assaying the remaining DNA polymerase activity. Parameters to consider in determining relative thermostability include polymerase concentration and incubation buffer. As shown in Table I, thermostability ($t_{1/2}$) parallels the optimal growth temperature of the organism from which the DNA polymerase was derived.

Nucleotide Incorporation Assay Modification: Measuring Nucleotide Analog Incorporation

Nucleotide analog incorporation has been used to classify DNA polymerases and to assess the utility of particular thermostable DNA polymerases in laboratory applications (e.g., DNA sequencing). Unlike mesophilic Family B DNA polymerases, which are sensitive to aphidocolin, archaeal DNA polymerases have been shown to exhibit variation in sensitivity to this analog (reviewed,[9] and references therein). In addition, archaeal DNA polymerases can incorporate ddNTPs, 7-deaza-dGTP, dITP, and dUTP to varying degrees, but these enzymes cannot utilize templates containing dIMP[42] or dUMP.[43] Lasken *et al.* have shown that uracil-containing DNA inhibits nucleotide incorporation because archaeal DNA

---

[42] T. Knittel, and D. Picard, *PCR Methods Appl.* **2**, 346 (1993).
[43] G. Slupphaug, I. Alseth, I. Eftedal, G. Volden, and H. E. Krokan, *Anal. Biochem.* **211**, 164 (1993).

TABLE II
POLYMERASE ACTIVITIES OF ARCHAEAL FAMILY B DNA POLYMERASES

Polymerase	Specific activity (U/mg) × 10^4		K_m		k_{cat} (min^{-1})	Processivity bases	Strand displacement	Reverse transcriptase (U/mg) × 10^4	
	Activated DNA	Primed M13	DNA (nM)	dNTP (μM each)				Poly(rA)/ (dT)	Poly(rC)/ (dG)
Pfu	2.6 ± 0.07	2.0 ± 0.02	0.7	16 ± 2	560 ± 100	10	≥70°	<0.001	<0.001
exo$^-$ Pfu	4.1 ± 0.07	2.3	0.5	12	550 ± 60	11			
Vent	1.8^a	0.7^a	0.1^a	57^a	1000^a	11, 7^a	≥63°^{a}, ≥55°^{ad}		
Deep Vent			0.01^c	50^c			>60°^{cd}		
9°N-7	0.4^b	2.5^b	0.05^b	75^b			≥72°^{b}		

^{a} H. Kong, R. B. Kucera, and W. E. Jack, *J. Biol. Chem.* **268**, 1965 (1993).
^{b} M. W. Southworth, H. Kong, R. B. Kucera, J. Ware, H. W. Jannasch, and F. B. Perler, *Proc. Natl. Acad. Sci. U.S.A.* **93**, 5281 (1996).
^{c} F. B. Perler, S. Kumar, and H. Kong, *Adv. Protein Chem.* **48**, 377 (1996).
^{d} Measurement for exo minus version of the DNA polymerase.

TABLE III
POLYMERASE ACTIVITIES OF EUBACTERIAL FAMILY A DNA POLYMERASES

| Polymerase | Specific activity (U/mg) × 10⁴ | | K_m | | k_{cat} (min^{-1}) | Processivity bases | Strand displacement | Reverse transcriptase (U/mg) × 10⁴ | |
	Activated DNA	Primed M13	DNA (nM)	dNTP (M each)				Poly(rA)/ (dT)	Poly(rC)/ (dG)
Thermophilic									
Taq	2.2 ± 0.03	4.9	4.0 ± 0.7, 1.4[b]	24 ± 2, 16[b]	2800 ± 980, 3600[c]	10, 42[b]	+	0.06	0.001
BF	15	49	4.2 ± 0.3	13 ± 5	11,500 ± 2400	111	++	14.3	0.03
Mesophilic									
KF	0.9[d]	0.6[d]	1.8, 5[e]	2.3[e]	310, 168[f]	7, <10[g]	+[h], +[d,h]	9.1	0.001

[a] MMLV RT exhibited specific activities (U/mg) of 4 × 10⁵ poly(rA)/oligo(dT) and 2.1 × 10⁵ poly(C)/oligo(dG)
[b] H. Kong, R. B. Kucera, and W. E. Jack, *J. Biol. Chem.* **268**, 1965 (1993).
[c] M. A. Innis, K. B. Myambo, D. H. Gelfand, and M. A. D. Brow, *Proc. Natl. Acad. Sci. U.S.A.* **85**, 9436 (1988).
[d] Measured using the exo⁻ version of the DNA polymerase.
[e] *E. coli* Pol I; W. R. McClure and T. M. Jovin, *J. Biol. Chem.* **250**, 4073 (1975).
[f] A. H. Polesky, T. A. Steitz, N. D. F. Grindley, and C. M. Joyce, *J. Biol. Chem.* **265**, 14579 (1990).
[g] S. Tabor, H. E. Huber, and C. C. Richardson, *J. Biol. Chem.* **262**, 16212 (1987).
[h] G. T. Walker, M. C. Little, J. G. Nadeau, and D. D. Shank, *Proc. Natl. Acad. Sci.* **89**, 392 (1992).

polymerases exhibit a unique tendency to bind dU-DNA more tightly (>1000-fold) than normal DNA.[44] In contrast to archaeal DNA polymerases, eubacterial DNA polymerases such as *Taq* are resistant to aphidocolin and can utilize dUTP and dITP efficiently in PCR reactions.[42,43]

The nucleotide incorporation assay described above can be readily modified to measure the inhibition or incorporation of nucleotide analogs. A variety of analogs can be screened using this approach, which also has the advantage of not requiring labeled dNTP analogs. When assessing analog incorporation, researchers should consider using polymerases or polymerase mutants lacking $3' \rightarrow 5'$-exonuclease activity, as the editing function can excise incorporable dNMP analogs from primer termini.

To determine if an analog acts as a competitive inhibitor with one or more dNTPs, assays can be carried out using the "complete" nucleotide incorporation assay cocktail described above (all four dNTPs). If nucleotide incorporation (cpms) is unaffected by the addition of increasing amounts of dNTP analog, either the analog was not bound or incorporated by the polymerase, or it was incorporated and extended from as efficiently as the analogous dNTP. In contrast, dNTP analog inhibitors produce a decrease in cpms incorporated with increasing amounts of nucleotide analog. Inhibitory dNTP analogs can be classified as nonincorporable (e.g., aphidocolin) or incorporable, with the latter group providing either extensible primer termini (dUTP, rNTPs) or nonextensible primer termini (ddNTPs) (reviewed[45]). The nucleotide analog concentration that brings about 50% inhibition of DNA polymerase activity (I_{50}) can be used to compare the tendency of different DNA polymerases to incorporate dNTP analogs.

Conditions for "Complete" Assay. The nucleotide incorporation assay (activated DNA template) is carried out as described above, except that reaction mixtures contain varying concentrations of the nucleotide analog to be tested. For example, inhibition of *Pfu* DNA polymerase requires the addition of $\geq 500\,\mu M$ each ddNTP (data not shown). If desired, dNTP analog concentrations can be reduced by decreasing dNTP concentrations to 100 μM each dNTP. If the analog is expected to compete selectively with dTTP (e.g., dUTP, ddTTP), a radiolabeled dNTP other than [^{3}H]TTP should be used to monitor nucleotide incorporation. This avoids "pseudo" inhibition (reduction in cpms) that arises when an unlabeled analog simply substitues for a radiolabeled dNTP without significantly inhibiting the rate of polymerization.

Conditions for "Incomplete" Assay. To confirm that a nucleotide analog is incorporated and extensible primer termini are generated, an "incomplete" assay can be performed in which the dNTP known to be competitive with the nucleotide analog is omitted (e.g., TTP is replaced with dUTP, dATP is replaced with rATP). The

---

[44] R. S. Lasken, D. M. Schuster, and A. Rashtchian, *J. Biol. Chem.* **271,** 17692 (1996).
[45] N. C. Brown and G. E. Wright, *Methods Enzymol.* **262,** 202 (1995).

omission of one dNTP reduces the incorporation of labeled dNTPs significantly (but not entirely), depending on the enzyme and reaction conditions. The residual incorporation (assuming no interference from $3' \rightarrow 5'$-exonuclease activity) reflects extension of primer termini to the point at which the template dictates the binding and incorporation of the missing dNTP. The addition of incorporable analogs with extensible termini will bring about an increase in cpms incorporated relative to the level of residual incorporation in the absence of one dNTP.

As an alternative to this approach, one can directly measure dNTP analog incorporation and extension using a gel-based assay and a DNA template consisting of a $5'$-^{32}P-labeled primer, annealed to a single-stranded template (as an example, see ref. 44). Primer extension in the presence of three dNTPs produces a block that can be overcome with the addition of incorporable analogs with extensible $3'$-OH termini.

Nucleotide Incorporation Assay Modification: Measuring Reverse Transcriptase Activity

Certain thermostable DNA polymerases have been used to carry out cDNA synthesis at high temperatures *in vitro*.[46] To measure the reverse transcriptase activity associated with thermostable DNA polymerases, we have modified the nucleotide incorporation assay by substituting poly(rA)/oligo(dT)$_{12\text{-}18}$ or poly(rC)/oligo(dG)$_{12\text{-}18}$ for the DNA substrate. The poly(rC)/oligo(dG)$_{12\text{-}18}$ template has the advantage of stability at the high temperatures used to assay (hyper) thermophilic DNA polymerases (up to 72°). Although MMLV reverse transcriptase readily utilizes this artificial template, many DNA polymerases do not (see Tables II and Table III).

To prepare 1 ml of reaction cocktail containing the poly(rC)/oligo(dG)$_{12\text{-}18}$ template, we typically combine 530 μg poly(rC) (Pharmacia), 24 μg oligo(dG)$_{12\text{-}18}$ (Pharmacia), 100 μl 10× reaction buffer (see below), and water to 884 μl. The reactants are heated at 65°C for 5 min and allowed to cool to room temperature. Next, 15.9 μl of 15 mM dGTP and 100 μl of [^{3}H]dGTP (8.3 Ci/mmol, 1 mCi/ml; NEN, Boston, MA) are added, which brings the final concentrations of DNA template and nucleotides to ~5 μM and 250 μM, respectively. A typical reverse transcriptase reaction buffer consists of 50 mM Tris-HCl (pH 8.5), 30 mM KCl, and 3 mM MgCl$_2$. Reverse transcriptase reactions are then performed by adding 1 μl of DNA polymerase (0.003–2U/μl) to 10 μl of reaction cocktail. Reactions are incubated at appropriate temperatures up to 72°. Cpms incorporated are determined as described for the nucleotide incorporation assay (DNA template) and converted into units of reverse transcriptase (RT) activity. A unit of RT activity is defined as the amount of enzyme that catalyzes the incorporation of 1 nmol of dNTP into polymeric form (bound to DE81) in 10 min.

[46] T. W. Myers and D. H. Gelfand, *Biochem.* **30**, 7661 (1991).

To prepare reaction cocktails containing poly(rA)/oligo(dT)$_{12-18}$, poly(rA) (Pharmacia) is substituted for poly(rC) and oligo(dT)$_{18}$ is substituted for oligo (dG)$_{12-18}$. The reduced stability of poly(rA/dT) templates limits the maximum assay temperature to 48°.

The reverse transcriptase activities of commercial archaeal and eubacterial DNA polymerases are summarized in Tables II and III, respectively. Measurements were carried out in the presence of MgCl$_2$ using poly(rA)/oligo(dT) at 48° (50 mM Tris-HCl (pH 8.5), 30 mM KCl, 3 mM MgCl$_2$) and poly(rC)/oligo(dG) at optimal temperature [50 mM Tris-HCl (pH 7.8), 50 mM KCl, 6 mM MgCl$_2$, 1 mM DTT]. In these assays, Moloney murine leukemia virus reverse transcriptase (MMLV-RT) exhibited a specific activity of 4.0×10^5 U/mg using poly(rA)/oligo (dT) and 2.1×10^5 U/mg using poly (rC)/oligo(dG) (assayed at 37°). Certain eubacterial DNA polymerases were found to efficiently utilize poly(rA)/oligo(dT), with specific activities that were 23% (KF) and 36% (BF) of the specific activity of MMLV-RT (Table III). The lower RT activity observed for *Taq*, compared to KF and *Bacillus* fragment (BF) DNA polymerases, may partly reflect the use of a suboptimal reaction temperature (48° assay vs 70–80° optima for *Taq*). In contrast to the relatively high activity observed using poly (rA)/oligo (dT), KF and BF DNA polymerases utilized poly (rC)/oligo (dG) poorly, exhibiting specific activities that were only 0.005% and 0.14%, respectively, of the specific activity of MMLV RT. Archaeal DNA polymerases, on the other hand, appear to exhibit greater selectivity for DNA/DNA templates, as no detectable utilization of either poly (rA)/oligo (dT) or poly (rC)/oligo (dG) was observed (<0.003% the specific activity of MMLV RT) (Table II).

Although not tested in these experiments, the reverse transcriptase activity of certain DNA polymerases has been shown to increase with the addition of Mn^{2+}, or in some cases, the replacement of MgCl$_2$ by MnCl$_2$.[46]

Nucleotide Incorporation Assay Modification: Determining Steady-State Kinetic Parameters

Apparent K_m and k_{cat} values are determined by modifying the nucleotide incorporation assay described above such that variable concentrations of dNTPs and the primed M13 template are employed. To determine the DNA parameters, concentrations of dCTP, dGTP, and dATP are fixed at 200 μM and TTP is typically present at 100 μM (10 μM [^{3}H]TTP, 90 μM TTP). Polymerization rates are measured in triplicate in the presence of seven or eight different concentrations of the primed M13 template, chosen to bracket the expected K_m (see Tables II and III). For the dNTP parameters, the final M13 concentration is typically fixed at 15 nM and polymerization rates are measured at dNTP concentrations ranging from 1 to 200 μM each dNTP (equimolar concentration of all four dNTPs). Reactions are initiated with the addition of enzyme and allowed to proceed at optimal temperature for up to 15 min. Aliquots are withdrawn at various times and spotted onto DE-81 filters, and the amount of each dNTP incorporated into polymeric product is determined as described above. Initial rates are typically determined under conditions in which <5% of the dNTP pool is incorporated into product and are expressed in terms of micromolar each dNTP incorporated/minute. Apparent K_m and V_{max} values can be derived from Hanes–Woolf plots ($[S]/v$ vs $[S]$) of the data.

The kinetic parameters determined for purified archaeal and eubacterial DNA polymerases are summarized in Tables II and III, respectively. These experiments were conducted at each enzyme's optimal temperature in the common polymerase assay buffer used for specific activity measurements [50 mM Tris-HCl (pH 8.0), 5 mM MgCl$_2$, 1 mM DTT, 50 μg/ml BSA, and 4% glycerol].

Alternative Nucleotide Incorporation Assays

In situ gel assays with DNA substrates incorporated into the gel have also been described.[47] The advantage of these assays is that proteins with DNA polymerase activity can be clearly identified in heterogeneous samples. In addition, we have also successfully employed filter-based nucleotide incorporation assays.[48]

Processivity Assay

Introduction

Processivity is a measure of the number of nucleotides added by a DNA polymerase at the 3' end of a primer before the enzyme dissociates, and it reflects both the enzyme's polymerization rate and its dissociation constant (K_d). Each DNA polymerase has an intrinsic level of processivity, which is often increased by addition of accessory proteins[7] and is dependent on the type of substrate (e.g., poly [d(AT)] vs primed M13) and reaction conditions (salt, temperature) employed. Processivity measurements conducted *in vitro* have revealed that DNA polymerases incorporate a varying number of nucleotides, with the average number of additions ranging from a few to several hundred bases. DNA polymerases used for PCR, DNA sequencing, and primer extension reactions exhibit different degrees of processivity, but the effect of processivity on these applications is not well understood. Preliminary results have shown that increased processivity may contribute to replication fidelity of the Klenow fragment[49] and the ability of T7 DNA polymerase to perform uniform synthesis through regions of DNA secondary structure.[50]

To accurately measure processivity, primer-extension reactions are carried out using a large molar excess (>10-fold) of primer–template over DNA polymerase

---

[47] E. Karawya, J. A. Swack, and S. H. Wilson, *Anal. Biochem.* **135**, 318 (1983).
[48] G. Sagner, R. Ruger, and C. Kessler, *Gene* **97**, 119 (1991).
[49] K. A. Eckert and T. A. Kunkel, *J. Biol. Chem.* **268**, 13462 (1993).
[50] S. Tabor and C. C. Richardson, *J. Biol. Chem.* **264**, 6447 (1989).

to ensure that the extension products produced represent the products of a single round of processive synthesis (binding, extension, dissociation). In the absence of a polymerase "trap" (see below), one should analyze the distribution of products obtained at various primer–template : enzyme ratios. As the ratio is increased, the distribution of product sizes will decrease (as large products produced by two or more rounds of synthesis are eliminated) until a constant range of product sizes is achieved, reflecting a single round of processive synthesis.

Processive DNA synthesis is measured on either homopolymeric or heteropolymeric (natural) DNA templates. We have used primed M13 DNA below, although it tends to produce a less uniform distribution of products that reflects sequence-dependent pause sites unique to each DNA polymerase. In the protocol described below, we have also used a DNA trap (unlabeled activated DNA) to "trap" dissociated polymerase molecules, thereby ensuring that DNA substrates virtually never react with more than one enzyme during the reaction. To use a trap, the polymerase and DNA substrate are mixed in the absence of a necessary reaction component, such as dNTPs. To start the reaction, the dNTPs are added, along with a large excess of the trap. The polymerase molecules already bound to the DNA substrate carry out one round of processive synthesis and then dissociate and bind the trap.

To quantify processivity, primer-extension products are separated by gel electrophoresis and the distribution of extended primer lengths is visualized by autoradiography. At early reaction time points, extension products generated from one round of processive synthesis predominate and appear as a consistent series of bands that increase in intensity as the polymerization reaction proceeds. The different bands reflect variation in the number of nucleotides added onto each 5'-labeled primer molecule before the polymerase dissociates. At later reaction time points, higher molecular weight bands begin to appear as reinitiation of previously extended templates occurs. Processivity is expressed as both the range and average length of products synthesized. Average product length has been estimated by visual inspection of autoradiographs or, more recently, by quantitative methods such as scintillation counting of excised product bands and densitometry (or phosphorimaging) of autoradiographic exposures of extension reactions. Using these quantitative techniques, processivity is calculated as the ratio between the total number of nucleotides incorporated and the total number of product molecules synthesized.

Processivity Assay

Processivity measurements can be performed using the primed M13 template described above with some modifications. Typically, the 40-base primer is phosphorylated at the 5' end with [γ-^{32}P]ATP (>5000 Ci/mmol) and T4 polynucleotide kinase (e.g., KinAce-It kit, Stratagene, La Jolla, CA). The labeled oligo is purified using a suitable probe purification column (e.g., NucTrap column, Stratagene) and then annealed with single-stranded M13 at equimolar concentrations (100 nM). Polymerase reactions are routinely performed in the presence of a DNA trap using a molar ratio of 10–100 : 1 primed M13 : DNA polymerase. Labeled M13 template (0.2 pmol) is combined with 2–20 fmol of DNA polymerase in 35 μl of the enzyme's optimal PCR buffer (Tube A). (For BF and KF, extension reactions are conducted in 50 mM Tris-HCl, pH 7.5, 7 mM MgCl$_2$, and 1 mM DTT). In tube B, 1.25–2.5 μg of activated calf thymus DNA and 333 μM each dNTP are added to 15 μl of optimal PCR buffer. Tubes A and B are incubated at the appropriate temperature for 3 min, and the extension reactions are initiated by adding the contents of tube B to tube A. Ten-μl aliquots are removed after 1, 5, 15, and 30 min and the reactions are terminated by adding 3.3 μl of stop dye (95% formamide/20 mM EDTA/0.05% bromphenol blue/0.05% xylene cyanol). The reaction mixtures are then subjected to denaturing gel electrophoresis using 6–8% polyacrylamide/7 M urea gels, and the gels are dried down and exposed to autoradiographic film. The sizes of the extension products can be determined relative to DNA markers generated by Sanger dideoxy sequencing[51] using the same template.

To test the effectiveness of the DNA trap, control reactions are also performed in which the calf thymus DNA is omitted, or added prior to the binding of polymerase to the labeled primed M13 template (no detectable synthesis).

We routinely quantify the amount of each extension product by densitometry and integration using Stratagene's Eagle Eye II Still Video System and RFLPscan software. The average number of elongation steps (processivity) has been calculated as the ratio between the total number of nucleotides incorporated and the total number of product molecules synthesized, using the formula[52]:

$$\text{Processivity} = [[(1)(I_1)] + [(2)(I_2)] + \cdots + [(n)(I_n)]]/[I_1 + I_2 + \cdots + I_n]$$

where I equals the intensity of each product band generated by the addition of 1 to n nucleotides (determined relative to sizing markers). Densitometry analyses of different exposures of the same gel or of independent extension reactions typically give rise to very similar measurements of enzyme processivity.

A typical measurement of processivity is shown in Fig. 2. The products resulting from one round of synthesis were identified by the appearance of a consistent range of product lengths after 5–15 min of extension in the presence of competitor DNA (Fig. 2, lanes b and c) and after 1 min in the absence of competitor DNA (lane e). In the absence of competitor DNA, reinitiation of previously extended templates is evident by the longer extension products seen at 5- to 30-min time points (Fig. 2, lanes f–h).

---

[51] F. Sanger, S. Nicklen, and A. R. Coulson, *Proc. Natl. Acad. Sci. U.S.A.* **74**, 5463 (1977).
[52] M. Ricchetti and H. Buc, *EMBO J.* **12**, 387 (1993).

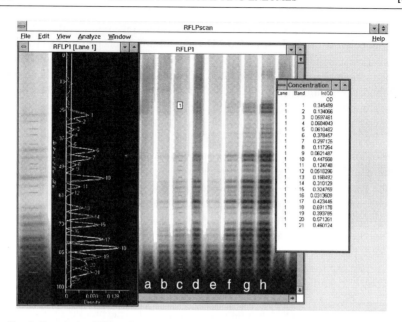

FIG. 2. Quantitating DNA polymerase processivity. Polymerization reactions were performed as described in the text in the presence (lanes a–d) or absence (lanes e–h) of calf thymus DNA. Extensions were stopped after 1 (lanes a and e), 5 (lanes b and f), 15 (lanes c and g), and 30 (lanes d and h) min and electrophoresed alongside molecular size markers (not shown). To quantitate extension products, the autoradiograph (in this example, lane c) was subject to RFLPscan software analysis using Stratagene's Eagle Eye II Still Video System [A. Lovejoy, T. Gackstetter, and H. Hogrefe, *Strategies in Molecular Biology* **8**, 63 (1995)]. Processivity was calculated as described in the text.

Processivity measurements carried out using commercial archaeal and eubacterial DNA polymerases are summarized in Tables II and III, respectively. Determinations were performed at optimal temperature in each enzyme's optimal PCR or reaction buffer (described above). The majority of DNA polymerases, including PCR enzymes (archaeal Family B DNA polymerases and *Taq*) and KF, exhibit limited processivities *in vitro*, incorporating ≤20 bases before dissociating from the growing primer–template. In contrast, BF DNA polymerase exhibited significantly higher processivity, extending over 300 nucleotides (mean 111 bases) before dissociation. High processivity is generally observed in the presence of accessory factors,[7] but in the case of the monomeric BF DNA polymerase, unique structural differences in the thumb and finger subdomains may contribute to the enzyme's relatively high processivity.[16]

In addition to the processivity assay described here, another common assay for processivity determinations has been developed by Bambara *et al.*[53] This method

---

[53] R. A. Bambara, D. Uyemura, and T. Choi, *J. Biol. Chem.* **253**, 413 (1978).

involves comparing rates of polymerization in the presence of limited vs a complete complement of all four dNTPs.

Strand Displacement Activity

Introduction

When DNA polymerases encounter a blocking downstream primer, some will displace the nontemplate strand of the DNA duplex, in a process referred to as strand displacement (SD). *In vitro* isothermal amplification strategies employ DNA polymerases with SD activity,[54] and these procedures have been improved with the use of thermostable DNA polymerases.[55] A common method for assaying SD activity utilizes a single-stranded DNA template to which is annealed a 5′-labeled upstream primer, followed by a second unlabeled "blocking" primer. The 5′ end of the blocking primer is typically positioned 20–70 bases downstream of the 3′ end of the labeled primer. In the absence of SD, the upstream labeled primer will extend to the position of the blocking primer (a known distance) and stop. In comparison, DNA polymerases with SD activity will produce labeled products that are greater in length than the distance between the 3′ of the upstream primer and the 5′ end of the blocking primer.

DNA polymerases with $5' \rightarrow 3'$-exonuclease activity (e.g., *Taq*) have the capacity to remove downstream primers through exonucleolytic degradation, although the preferred DNA substrate is a 5′ flap structure produced by SD activity.[19] To monitor both $5' \rightarrow 3'$-exonuclease and SD activity, an alternative strategy has been described that uses a single-stranded DNA template annealed to two labeled primers (reviewed[9] and citations therein). The primers are positioned near the 3′ end of a linear template such that two labeled runoff products of different sizes are produced. If $5' \rightarrow 3'$-exonuclease activity is present, both primers are extended, but the 5′ end of the blocking primer is also degraded.

As for polymerization and processivity measurements, SD activity is also influenced by the reaction conditions employed. In particular, archaeal Family B DNA polymerases exhibit SD in a temperature dependent manner. For example, primer extension by Vent DNA polymerase (*T. litoralis*) is completely blocked by downstream primers at 55°, but increasingly larger SD products are observed as the reaction temperature is raised from 63° to 72°.[25] Greater SD activity has also been noted for exo⁻ versions of Vent[25] and 9°N-7[31] DNA polymerases, possibly because the substrate spends less time in the exonuclease active site.[9]

---

[54] G. T. Walker, *PCR Methods Appl.* **3**, 1 (1993).
[55] C. A. Spargo, M. S. Frasier, M. Van Cleve, D. J. Wright, C. M. Nycz, P. A. Spears, and G. T. Walker, *Mol. Cell Probes* **10**, 247 (1996).

Strand Displacement Assay

Strand displacement measurements can be performed using the same radiolabeled primed M13 template described above for processivity measurements, with an additional downstream blocking primer. One suitable blocking oligonucleotide that we have used is the 47-base sequence 5′-CCCCGGGTACCGAGCTCGAATTCGTAATCATGGTCATAGCTGTTTCC, which anneals to a site 35 bases downstream from the radiolabeled 40-base primer. Using this template, DNA polymerases lacking SD activity will produce only a 75-base product (40-base oligonucleotide + 35-base extension), whereas DNA polymerases possessing SD activity will synthesize >75-base extension products.

To prepare the substrate, the gel-purified 40-base primer (∼500 ng) is phosphorylated at the 5′ end using 1 µl [γ-^{33}P]ATP (2000–4000 Ci/mmol; NEN #NEG-602H) and T4 polynucleotide kinase (e.g., KinAce-It kit, Stratagene) and then purified using a suitable probe purification column (e.g., NucTrap column, Stratagene). A typical reaction cocktail (200 µl total volume, ∼19 samples) consists of 4 nM M13mp18(+) strand (∼2 µg), ∼100 nM radiolabeled 40-base oligo (2 × 10^6 cpms, ∼300 ng), ∼500 nM of the gel-purified blocking oligo (∼1.7 µg), 1× PCR reaction buffer (20 µl), and water to a final volume of 200 µl. The reaction cocktail is heated to 95° for 5 min and then cooled to room temperature. dNTPs (1.6 µl, 25 mM each stock) are then added to give a final concentration of 200 µM each. SD reactions are carried out by combining 10 µl of reaction cocktail with ∼0.5 units of DNA polymerase. The reactions are incubated at the appropriate temperature(s) for 15 min.

Extension products are denatured by adding 20 µl of loading dye (Novex) and heating at 75° for 2 min. Samples (1.5 µl; ∼5000 cpm) are electrophoresed on denaturing 6% polyacrylamide/7 M urea gels (e.g., CastAway gels, Stratagene). The gels are dried down and exposed to film. Product sizes are determined relative to radiolabeled markers, which can be prepared by labeling suitable molecular weight size markers (e.g., 20 bp ladder from FMC #50320).

A typical assay for SD activity is shown in Fig. 3. "No DNA polymerase" controls are performed to visualize migration of the unextended primer (in this case, the 40-base oligonucleotide). Additional bands indicate the presence of contaminating DNA sequences produced during oligonucleotide synthesis, which can be significantly reduced by gel purifying the oligonucleotide primer. To verify annealing of the blocking oligonucleotide, we carry out controls lacking the blocking oligonucleotide using DNA polymerases and/or reaction conditions that produce low SD activity (e.g., *Pfu* at <68°). In the example shown (68°), *Pfu* DNA polymerase extended the primer 35 bases (75-base product) in the presence of the blocking primer (Fig. 3, lanes 4 and 5) and >160 bases in the absence of the blocking primer. In contrast, *Taq* DNA polymerase extended the primer >160 bases in both the absence (data not shown) and presence of the blocking primer (Fig. 3, lane 6).

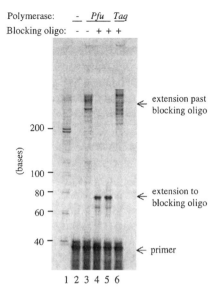

FIG. 3. Strand displacement assay. Polymerization reactions were performed at 68° as described in the text using *Pfu*Turbo (lanes 3, 4), *Pfu* (lane 5), or *Taq* (lane 6) DNA polymerase. (Stratagene's *PfuTurbo* DNA polymerase is a blend of cloned *Pfu* DNA polymerase and a proprietary PCR enhancing factor.) The blocking oligo was added to reactions shown in lanes 4–6. Lane 1 shows the radiolabeled 20 bp size marker (two bands are resolved on the denaturing gel for each size marker), while lane 2 is a "no enzyme" (template only) control.

As discussed above, an assay using two radiolabeled primers would allow one to monitor both the SD and $5' \rightarrow 3'$-exonuclease activity of *Taq* DNA polymerase.

Exonuclease Activity

Introduction

As discussed above, the presence of associated $3' \rightarrow 5'$ and/or $5' \rightarrow 3'$ exonuclease activities dictates the utility of thermostable DNA polymerases in certain laboratory applications (e.g., high-fidelity PCR, SD). The presence of $3' \rightarrow 5'$-exonuclease activity is routinely assayed by measuring the hydrolysis of single-stranded DNA:

$$\text{Single-stranded DNA}_n \rightarrow \text{DNA}_{n-1} + \text{dNMP}$$

Most $3' \rightarrow 5'$-exonucleases will hydrolyze the unpaired 3'-OH end of a DNA duplex or the 3'-OH terminus of a single-stranded chain. In the presence of dNTPs, $3' \rightarrow 5'$-exonuclease activity is suppressed in favor of polymerization on double-stranded DNA templates, whereas on single-stranded DNA substrates hydrolysis is unaffected by dNTPs. In comparison, the $5' \rightarrow 3'$ structure-specific endonuclease (exonuclease) activity of eubacterial DNA polymerases (e.g., *Taq*) is stimulated on activated DNA templates by dNTPs, as degradation of downstream blocking DNA primers proceeds in the path of the growing DNA strand. Eubacterial DNA polymerases preferentially cleave the junction between a 5' overhang and a duplex region.[19] The products generated by $5' \rightarrow 3'$ exonuclease activity are mono- and oligonucleotides, whereas only mononucleotides (dNMPs) are produced by $3' \rightarrow 5'$ exonuclease activity.

One assay we use to monitor exonuclease activity was adapted from a published method[25] that employs uniformly labeled DNA. Using double-stranded labeled DNA templates, we can determine specificity by measuring whether cpms increase ($5' \rightarrow 3'$ exonuclease) or decrease ($3' \rightarrow 5'$ exonuclease) upon the addition of dNTPs (10–100 μM). For specifically monitoring DNA polymerases with $5' \rightarrow 3'$ structure-specific endonuclease activity, duplex DNA templates which contain displaced 5' ends are preferred.[19] As is the case for polymerase activity measurements, exonuclease assays are significantly influenced by reaction conditions, and salt concentration, incubation temperature, and DNA concentration should be specifically optimized for each DNA polymerase.

Exonuclease Assay

Exonuclease activity can be measured using double-stranded λ DNA, which has been uniformly labeled with ^{3}H labeled *S*-adenosylmethionine (NEN) and *Sss*I methylase (NEB), and then restriction digested with *Pal*I. Exonuclease specific activities shown in Table IV were determined using this substrate, and a detailed procedure for its preparation can be obtained in the original reference.[25]

A typical exonuclease reaction cocktail consists of 20 μg/ml ^{3}H-labeled digested double-stranded λ DNA ($\sim 10^6$ cpm/ml) in 1× reaction buffer. In these studies, exonuclease activity was measured in each enzyme's optimal PCR buffer and in a common assay buffer consisting of 70 mM Tris-HCl (pH 8.8), 2 mM MgCl$_2$, 0.1% Triton-X, and 100 μg/ml BSA. Exonuclease reactions are performed (in triplicate) by adding 4 μl aliquots of serially diluted DNA polymerases (0.05–2.5 U *Pfu*; 2.5–25 U *Taq*) to 46 μl of reaction cocktail. Reactions are incubated for 1 hr at the appropriate temperature. Reactions lacking DNA polymerase are also set up along with sample incubations to determine "total cpms" (no TCA precipitation) and "minimum cpms" (TCA precipitation: see below).

Exonuclease reactions are stopped by transferring the tubes to ice. Sonicated salmon sperm DNA (150 μl; 2.5 mg/ml stock) and TCA (200 μl; 10% stock) are added to all but the "total cpms" tubes. The precipitation reactions are incubated for $\geq$15 min on ice, and then spun in a microcentrifuge at 14,000 rpm for 10 min. Then, 200 μl of the supernatant is removed, with care so as not to disturb the pellet, and transferred to scintillation fluid (Bio-Safe II, Research Products International

TABLE IV
EXONUCLEASE ACTIVITIES OF DNA POLYMERASES

Polymerase	Exonuclease activity 3'→5'	Exonuclease activity 5'→3'	Exonuclease specific activity (U/mg) Common buffer[a]	Exonuclease specific activity (U/mg) PCR buffer	Error rate (MF/bp/d)[c] (×10⁻⁶)	Terminal extendase activity
Archaeal						
Pfu	+	−	935	2500	1.3 ± 0.2	None[d]
Exo⁻ Pfu	−[f]	−	<0.9		47 ± 3	
Vent	+	−	1644, 1500[b]	7600	2.8 ± 0.9	Weak[d]
Deep Vent	+	−	1642	17,000	2.7 ± 0.2	
9°Nₘ	+[f]	−	26		57	
Eubacterial						
Taq	−	+	2		8.0 ± 3.9	Strong[d]
UlTma	+	−[e]	8	5	55 ± 2	
BF	−	−[e]	<0.05			
KF	+	−[e]	28			Weak[d]

[a] 70 mM Tris-HCl (pH 8.8), 2 mM MgCl₂, 0.1% Triton X, and 100 μg/ml BSA.
[b] H. Kong, R. B. Kucera, and W. E. Jack, *J. Biol. Chem.* **268**, 1965 (1993).
[c] MF is mutation frequency, bp is the number of detectable sites in *lac*I, and d is the number of template doublings; assay described in J. Cline, J. C. Braman, and H. H. Hogrefe, *Nucl. Acids Res.* **24**, 3546 (1996).
[d] G. Hu, *DNA and Cell Biology* **12**, 763 (1993).
[e] 5'→3' exonuclease domain removed in N-truncated polymerase fragment.
[f] Mutation in 3'→5' exonuclease domain.

Corp., Mount Prospect, Illinois). The samples are thoroughly mixed by inversion and then counted in a scintillation counter.

Cpms released are proportional to the amount of exonuclease activity present and can be converted into units of exonuclease activity. One unit of exonuclease activity is defined as the amount of enzyme that catalyzes the acid solubilization of 10 nmol of total dNMPs in 30 min at a defined temperature. To determine units, background (average "minimum cpms" value) is first subtracted from the average sample cpms. Nanomoles of dNMPs released is calculated using the following equation:

$$\text{dNMPs(nmol)} = \frac{\text{(corrected sample cpms)}}{\text{total cpms}} \times \frac{\text{(920 ng DNA)}}{\text{reaction}} \times \frac{\text{(1 nmol dNMP)}}{\text{(330 ng dNMP)}}$$

Units of exonuclease activity (in 30 minutes) can then be determined as:

$$\text{Activity} = \text{(nmol dNMPs released per hr)} \times \frac{\text{(1 unit)}}{\text{(10 nmol dNMPs released)}}$$

Exonuclease concentrations (U/ml) can be extrapolated from the slope of the linear portion of units versus enzyme volume plots. Specific activity (U/mg) calculations are carried out using protein concentrations determined as described above.

Table IV summarizes the results of exonuclease activity measurements (expressed as specific activity (U/mg)) conducted at optimal temperature in the absence of dNTPs. For certain DNA polymerases, assays were carried out in both the common assay buffer and in each enzyme's optimal PCR reaction buffer. Inclusion of ammonium sulfate in PCR buffers was found to significantly increase the exonuclease activity of archaeal DNA polymerases (data not shown). In assays employing uniformly labeled double-stranded DNA substrates, release of acid-soluble nucleotides can be attributed to the $3'\rightarrow 5'$-exonuclease activity associated with archaeal Family B DNA polymerases, UlTma, and KF, as these enzymes are reportedly devoid of detectable $5'\rightarrow 3'$-exonuclease activity (reviewed[9]). Release of acid-soluble nucleotides by nonproofreading DNA polymerases, such as *Taq*, can be attributed to $5'\rightarrow 3'$ exonuclease activity, and solubilized radioactivity increases in the presence of added dNTPs (data not shown). No detectable exonuclease activity was observed for the exo$^-$ *Pfu* DNA polymerase, whereas the 9°N mutant exhibited ~ 1–3% of the exonuclease activity of other archaeal DNA polymerases, consistent with the reduced exonuclease activity noted by the enzyme's manufacturer.

In addition to the assay described earlier, we have also performed routine qualitative measurements of exonuclease activity using [^{3}H]-*E. coli* genomic DNA (5.8 μCi/μg) (NEN), which has the advantage of being commercially available. A typical exonuclease reaction cocktail consists of 0.4 μg/ml ^{3}H-labeled *E. coli* genomic DNA in 1× reaction buffer, and assays are carried out as described above.

Exonuclease specificity can also be monitored using 3'- or 5'-radiolabeled DNA substrates. Oligonucleotides can be 3'-labeled with [α-^{32}P]dATP and terminal transferase using standard procedures. $5'\rightarrow 3'$-Exonuclease activity is monitored by incubating DNA polymerases with 3'-labeled oligos and looking for product-size reduction in denaturing gel electrophoresis. 5'-Radiolabeled oligonucleotides, prepared using [γ-^{32}P]dATP and T4 polynucleotide kinase (see kinasing and purification procedures above), can be used in a similar fashion to monitor $3'\rightarrow 5'$ exonuclease activity. 3'-Labeled double-stranded DNA substrates can also be prepared by filling-in a restriction enzyme cleavage site (5' overhang) with Klenow DNA polymerase. For example, we have used a heterogeneous mixture of *Taq*I-restricted λ DNA fragments, labeled at the 3' termini, by filling the 5' overhang with [^{3}H]dCTP and [^{3}H]dGTP.[26] $3'\rightarrow 5'$-Exonuclease activity can be monitored by quantifying the amount of radioactivity solubilized as described above.

Error Rate

Fidelity is an intrinsic property of DNA polymerases that reflects kinetic constraints favoring incorporation of correct nucleotides and editing of mismatched primer termini (reviewed[5,56,57]). As discussed above, error rate is one parameter that determines the utility of thermostable DNA polymerases in PCR applications. Various methods have been described for assaying the error rates of thermostable DNA polymerases, including single primer extension measurements,[58] DGGE analysis of PCR products,[59] and PCR-based forward mutation assays.[60,61] Error rates are influenced by many factors, including DNA polymerase fidelity (proofreading activity) and reaction conditions (e.g., [dNTP], [Mg^{2+}], pH). Conditions have been optimized for "high fidelity" PCR amplification with *Taq*,[59] Vent,[59] and *Pfu*[60] DNA polymerases.

Error rate measurements shown in Table IV were obtained by scoring phenotypic changes in *lacI*, a well-characterized mutational target gene. A complete description of the *lacI* (*lacIOZα*) fidelity assay has been published previously.[26,60] DNA polymerase fidelity is expressed in terms of error rate: that is, mutation frequency per base pair per duplication. *Pfu* DNA polymerase has the lowest error rate (1.3×10^{-6}) of any thermostable DNA polymerase described to date.[60] With an error rate of 1.3×10^{-6}, the probability of a base being mutated in a single round of replication is 1 per 769,000 bases. On the basis of this rate, after 17 duplications (2^{17}; 131,000-fold amplification), 2% of the products amplified from a 1-kb fragment will contain mutations. In comparison, DNA polymerases lacking $3' \rightarrow 5'$ exonuclease activity exhibit significantly higher error rates than enzymes possessing proofreading activity (e.g., *Taq*'s error rate is 8.0×10^{-6}). Interestingly, truncated or mutated DNA polymerases with diminished exonuclease activities (e.g., UlTma, exo$^-$ *Pfu*) are more error prone than DNA polymerases that naturally lack proofreading activity (*Taq*) (Table IV).

Terminal Extendase

Various PCR cloning strategies have been developed based on PCR enzymes that exhibit or lack terminal extendase activity. Terminal extendase activity refers to the tendency of a DNA polymerase to add non-template-directed nucleotides at the 3'-OH end of a DNA strand. Single-base extensions are often observed with DNA polymerases that lack proofreading exonuclease activity, such as *Taq*.[62] One can add a single-base extension to the 3' end of a product produced by a proofreading enzyme (*Pfu*) by treating with *Taq*, and one can blunt a single-base extension

---

[56] K. A. Johnson, *Annu. Rev. Biochem.* **62**, 685 (1993).
[57] T. A. Kunkel, *J. Biol. Chem.* **267**, 18251 (1992).
[58] M.-M. Huang, N. Arnheim, and M. R. Goodman, *Nucl. Acids Res.* **20**, 4567 (1992).
[59] L. L. Ling, P. Keohavong, C. Dias, and W. G. Thilly, *PCR Methods Appl.* **1**, 63 (1991).
[60] J. Cline, J. C. Braman, and H. H. Hogrefe, *Nucl. Acids Res.* **24**, 3546 (1996).
[61] J. M. Flaman, T. Frebourg, V. Moreau, F. Charbonnier, C. Martin, C. Ishioka, S. H. Friend, and R. Iggo, *Nucl. Acids Res.* **22**, 3259 (1994).
[62] G. Hu, *DNA Cell Biol.* **12**, 763 (1993).

product by *Taq* by treating with *Pfu*. Readers are referred to the published method of Hu[62] for carrying out terminal extendase measurements. In Hu's primer extension studies, summarized in Table IV, *Pfu* produced blunt ends exclusively, while Vent DNA polymerase exhibited weak terminal extendase activity and produced blunt ends 90% of the time. In contrast, DNA polymerases lacking proofreading activity (*Taq*) exhibited a strong tendency to add non-template-directed dNTPs at 3'-OH termini.

Conclusion

In general, the properties of commercial thermostable DNA polymerases are quite similar to monomeric DNA polymerases from mesophiles (e.g., KF comparisons in Tables III and IV) (reviewed[1–6,9]). As noted above, the (hyper) thermophilic archaeal and eubacterial DNA polymerases that have been described and commercialized represent the predominant DNA polymerase in the cell and/or the polymerase that is readily isolated in monomeric form. The presence of additional DNA polymerases and accessory factors in these (hyper) thermophilic organisms has been the subject of several publications.[24,36,37,40,41] Presumably, the DNA polymerases described here function with accessory factors *in vivo* in DNA replication and repair processes. Although commercial DNA polymerases exhibit limited capacity to carry out highly processive DNA synthesis, these limitations have not significantly affected their utility in a variety of laboratory applications, including PCR, DNA sequencing, mutagenesis, isothermal amplification, and cloning. Whether additional factors prove beneficial to such *in vitro* applications remains to be determined.

[10] Archaeal Histones and Nucleosomes

By KATHLEEN SANDMAN, KATHRYN A. BAILEY, SUZETTE L. PEREIRA, DIVYA SOARES, WEN-TYNG LI, and JOHN N. REEVE

Introduction

All organisms require architectural chromosomal proteins to compact and organize their genome, to provide a three-dimensional framework for chromosome activity,[1] and, in hyperthermophiles, to help stabilize double-stranded DNA against

---

[1] K. Sandman, S. L. Pereira, and J. N. Reeve, *Cell. Mol. Life Sci.* **54,** 1350 (1998).

thermal denaturation.[2] Almost all eukarya employ histones, small basic proteins that organize DNA into nucleosomes, which are further assembled into chromatin and chromosomes. Prokaryotic homologs of the nucleosome have been documented in the Euryarchaeota, one branch of the Domain Archaea.[1] In archaeal nucleosomes ~85 bp of DNA are wrapped around a histone tetramer,[3,4] forming a structure homologous to the structure at the center of the eukaryal nucleosome formed by the (H3·H4)$_2$ histone tetramer.[5] Archaeal and eukaryal histones share a common evolutionary ancestry: they exhibit both amino acid sequence similarity[2] and a conserved three-dimensional histone-fold structure.[6,7] Numerous techniques have been developed for the experimental manipulation of eukaryal histones and nucleosomes, but most are not applicable to their prokaryotic counterparts. Here, we provide protocols for the purification of archaeal histones and nucleosomes, and procedures for assays of DNA binding and archaeal nucleosome positioning.

Purification of Archaeal Histones

Histones are abundant proteins constituting ~4% of the soluble protein in the hyperthermophile *Methanothermus fervidus*.[3] In solution, histones are dimers, and the two histone polypeptides from *M. fervidus* (HMfA and HMfB) exist *in vivo* and *in vitro* as both homodimers and heterodimers.[8] Euryarchaeota with as few as two and as many as five histone-encoding genes have been identified. However, the conservation in the amino acid residues that form the monomer–monomer interface suggests that most homo- and heterodimer combinations of archaeal histones can form,[9] and therefore, the histone repertoire of an organism is probably all possible homo- and heterodimers. Growth phase changes in the abundance of HMfA and HMfB have been reported in *M. fervidus*,[8] and such regulation is likely to occur in other species. Histone preparations from these microorganisms will, therefore, be variable mixtures of different dimers, depending on the growth state at the time of harvest.

To obtain purified preparations of single histone homodimers, a system to synthesize and purify recombinant archaeal histones from *Escherichia coli* was

---

[2] K. Sandman, J. A. Krzycki, B. Dobrinski, R. Lurz, and J. N. Reeve, *Proc. Natl. Acad. Sci. U.S.A.* **87**, 5788 (1990).
[3] S. L. Pereira, R. A. Grayling, R. Lurz, and J. N. Reeve, *Proc. Natl. Acad. Sci. U.S.A.* **94**, 12633 (1997).
[4] K. A. Bailey, C. Chow, and J. N. Reeve, *Nucl. Acids Res.* **27**, 532 (1999).
[5] S. L. Pereira and J. N. Reeve, *Extremophiles* **2**, 141 (1998).
[6] M. R. Starich, K. Sandman, J. N. Reeve, and M. F. Summers, *J. Mol. Biol.* **255**, 187 (1996).
[7] W. Zhu, K. Sandman, G. E. Lee, J. N. Reeve, and M. F. Summers, *Biochemistry* **37**, 10573 (1998).
[8] K. Sandman, R. A. Grayling, B. Dobrinski, R. Lurz, and J. N. Reeve, *Proc. Natl. Acad. Sci. U.S.A.* **91**, 12624 (1994).
[9] K. Sandman, W. Zhu, M. F. Summers, and J. N. Reeve, *In* "Thermophiles: The Keys to Molecular Evolution and the Origin of Life" (M. W. W. Adams and J. Weigel, eds.), pp. 243–253.

developed.[8,10] Cloned archaeal histone genes are expressed in *E. coli* with no apparent toxicity, although some native archaeal histones have the N-terminal methionyl residue removed, and this reaction is often inefficient during heterologous expression in *E. coli*. To overcome this problem, the *E. coli* methionine aminopeptidase gene and the archaeal histone-encoding gene can be coexpressed in *E. coli*, which results in the synthesis of recombinant archaeal histones with N termini identical to those of their native counterparts.[11]

The following are detailed protocols used to purify native histones from methanogens and recombinant histones from *E. coli*, plus the modifications to the *E. coli* culture conditions needed for complete N-terminal processing of recombinant histones are described. Histones also exist in mesophiles, and the modifications used to purify such thermolabile histones from *Methanobacterium formicicum* are detailed elsewhere.[12]

Native Histones from M. fervidus (HMf) or M. thermoautotrophicum (HMt)

This procedure yields ~0.5 mg of histones per gram (wet weight) of starting cell paste. All steps are carried out aerobically at room temperature.

Preparation of Crude Extract. Cell paste is resuspended (2 ml buffer per gram of cell paste) in a buffer containing 0.1 M NaCl, 50 mM Tris, and 2 mM Na_2HPO_4 (pH 8.0), and the cells are lysed by two passages through a French pressure cell at 20,000 psi. The histone-containing soluble protein is separated from cellular debris by centrifugation, first at low speed (30,000g for 30 min) and then at high speed (240,000g for 90 min). The resulting supernatant is brought to 5 mM $MgCl_2$ and incubated with DNase I (20 μg/ml) for 2 hr at 37° to hydrolyze the fragmented chromosome and release histones into solution. Protease inhibitors are not included as the endogenous proteases are not active in the crude extract at this temperature.

Heat Precipitation of Bulk Protein. A heat precipitation step is included that takes advantage of the ability of HMf and HMt to refold fully after heat denaturation.[10] The DNase-treated crude extract is transferred to a Pyrex flask, solid NaCl is added to a final concentration of 3 M, and the mixture is heated until a massive precipitation of protein occurs (70–75°). This step is most easily performed on a heated magnetic stir plate. The mixture is allowed to cool to room temperature without agitation, and the refolded histones that remain in solution are separated from the precipitated material by filtration. The bulk of the precipitated protein is first removed by gravity filtration through cheesecloth using a funnel and collecting flask. The filtrate from this step is cloudy and is cleared by centrifugation

---

[10] W.-T. Li, R. A. Grayling, K. Sandman, S. Edmondson, J. W. Shriver, and J. N. Reeve, *Biochemistry* **37**, 10563 (1998).

[11] K. Sandman, R. A. Grayling, and J. N. Reeve, *BioTechnology* **13**, 504 (1995).

[12] T. J. Darcy, K. Sandman, and J. N. Reeve, *J. Bacteriol.* **177**, 858 (1995).

at 12,000g for 10 min. Passage through a syringe-mounted 0.45 μm filter then removes any remaining precipitated protein.

Heparin Affinity Chromatography. Histones are separated from the remaining contaminating protein by affinity chromatography using a HiTrap heparin column (Amersham Pharmacia Biotech, Inc., Piscataway, NJ). The protein is first dialyzed overnight at 4° against the binding buffer (30 mM potassium citrate, 50 mM Tris, pH 8.0) using dialysis tubing that retains proteins smaller than 15 kDa, the mass of an archaeal histone dimer. The dialyzed solution is loaded on the column and five column volumes of buffer are sufficient to elute all nonbinding proteins. The histones are then eluted using a 30 to 200 mM gradient of potassium citrate dissolved in 50 mM Tris (pH 8.0). Small samples of the material applied to the column and that flows through the column are retained for comparison by SDS–PAGE. Fractions are collected as a fixed number of drops, and the histone-containing fractions can be initially predicted as fractions having a decreased volume. Aliquots of these fractions are evaluated for protein content by Tricine–SDS–PAGE after microdialysis to exchange the potassium-based buffer for the sodium-based buffer used in the cell lysis step. Histone monomers migrate during SDS–PAGE consistent with a molecular mass of ~7 kDa. The histone-containing column fractions are then pooled, and a centrifugal filtration device (e.g., Ultrafree filters, Millipore Corp., Bedford, MA) is used to concentrate and exchange the buffer as needed.

Quantitation. Archaeal histones are enriched in arginine and lysine residues, and the methods of protein quantitation based on Coomassie Brilliant Blue G250 dye binding are inaccurate because the dye preferentially binds to these basic residues.[13] Most archaeal histones also lack tyrosine or tryptophan residues, precluding A_{280} measurement. Therefore, reliable quantitation of archaeal histone solutions requires a determination of the total amino acid composition after acid hydrolysis.[10]

Storage. Histones are stable in 1 M NaCl at 4° for several months, and indefinitely when stored frozen at −70°.

Purification of Recombinant Archaeal Histones

Several *E. coli* heterologous gene expression systems based on the *tac*, T7, and T7*lac* promoters have been used successfully to obtain recombinant archaeal histones, by following the commercial protocols specified for each expression system. The purification protocol detailed above for native archaeal histones also works well for recombinant archaeal histones, although the protease inhibitor phenylmethylsulfonyl fluoride (PMSF) must be added to a final concentration of 0.1 mM during the DNase I digestion. Almost 100% purity can be achieved from *E. coli* crude extracts.

---

[13] T. E. Creighton, "Proteins: Structures and Molecular Properties." W. H. Freeman and Co., New York, 1993.

Modifications to Purification Protocol. Most archaeal histones are positively charged at pH 8.0 in the Tris buffer recommended above; however, some histones have more neutral isoelectric points, when calculated using the Molecular Weight/Isoelectric Point program found at www.expasy.ch/tools/pi_tools.html. The buffer used should be at least one pH unit lower than the calculated isoelectric point for efficient binding to the heparin column.

The heat precipitation step cannot be used in the purification of recombinant HFo histones derived from a mesophile. These histones do not survive heating to 70–75°, and an ammonium sulfate precipitation step is therefore substituted.[10] Saturated ammonium sulfate solution is added dropwise to the DNase I-treated crude extract until the solution reaches 70% saturation. Precipitated protein is removed by centrifugation at 12,000g for 30 min and the histone-containing supernatant dialyzed overnight against the binding buffer used for the heparin column chromatography.

Growth Conditions used to Enhance N-Terminal Processing of Recombinant Histones Synthesized in E. coli. The enzymes responsible for N-terminal maturation of proteins in *E. coli* are deformylase, which catalyzes the removal of the formyl group from the N-terminal methionyl residue,[14] and methionine aminopeptidase, which catalyzes the hydrolysis of the peptide bond between a deformylated N-terminal methionyl residue and the second residue of the polypeptide.[15] The need for deformylase can be eliminated by growing the *E. coli* culture with trimethoprim.[16] This inhibits dihydrofolate reductase (DHFR), which catalyzes a step in the addition of the formyl group to the initiator methionine tRNA, and therefore protein synthesis proceeds with unformylated initiator methionine tRNA. Growth is slower, and to compensate, the culture is grown in SuperBroth[17] with thymidine added to compensate for this deficiency caused by the inhibition of DHFR.[16] The intracellular concentration of methionine aminopeptidase is increased by incorporating the encoding *map* gene and its promoter[18] into the histone expression plasmid.[11] The *map* gene can be obtained from the *E. coli* Genetic Stock Center (www.cgsc.biology.yale.edu).

Assays of Histone Binding

Four assays are described for archaeal histone binding to DNA. Two are electrophoretic mobility shift assays: one using agarose gels to detect compaction of

[14] J. M. Adams, *J. Mol. Biol.* **33**, 571 (1968).
[15] F. Sherman, J. W. Stewart, and S. Tsunasawa, *BioEssays* **3**, 27 (1985).
[16] D. Mazel, S. Pochet, and P. Marliére. *EMBO J.* **13**, 914 (1994).
[17] F. M. Ausubel, R. Brent, R. E. Kingston, D. D. Moore, J. G. Seidman, J. A. Smith, and K. Struhl (eds.), "Short Protocols in Molecular Biology." John Wiley and Sons, New York, 1992.
[18] A. Ben-Bassat, K. Bauer, S.-Y. Chang, K. Myambo, A. Boosman, and S. Chang, *J. Bacteriol.* **169**, 751 (1987).

long DNA molecules,[2] and the second using acrylamide gels to monitor archaeal nucleosome formation.[4] The third assay uses T4 DNA ligase to detect DNA circularization resulting from histone binding,[4] and the fourth assay follows a decrease in the circular dichroism of DNA at 275 nm (θ_{275}) that results from histone–DNA interaction.[19]

Gel Mobility Acceleration in Agarose. Histone wrapping physically compacts the DNA molecule, and this can be assayed with linear DNA molecules longer than ~2.0 kbp as increased mobility during agarose gel electrophoresis[2] (Fig. 1A). This effect is independent of the DNA source, and common plasmid DNA, such as pBR322 (4.5 kb), is routinely used. The plasmid DNA is linearized by restriction endonuclease (RE) digestion and stored in TE buffer (10 mM Tris, 1 mM EDTA, pH 8.0). Fifty ng of DNA is sufficient for detection, and a range of histone : DNA weight ratios from 0.1 to 1.5 : 1 (5 to 75 ng histones) should give a good assay of binding. The components are mixed in water, and after 5 min incubation at room temperature, sample buffer is added [40% w/v sucrose, 0.1% w/v bromphenol blue (BPB)]. A 0.8% agarose gel is prepared in TAE buffer (40 mM Tris–acetate, 1 mM EDTA, pH 8.0); not all agaroses perform equally well, but consistent results are obtained with biotechnology grade agarose I from Amresco (Solon, OH). To preserve the structure of the histone–DNA complexes, electrophoresis is undertaken at low voltage, typically 1 V/cm overnight.

Different archaeal histones have different "saturation" points in this assay, both in terms of the histone : DNA ratio that results in maximal enhancement of mobility, and in terms of the maximum extent of the gel shift[8] (Fig. 1A).

Gel Mobility Retardation in Polyacrylamide. This procedure assays the association of archaeal histones with linear DNA molecules 60–500 bp in length[4,20] and is used to monitor archaeal nucleosome assembly under different experimental conditions and to determine the dissociation constants (K_d) of different histone–DNA complexes.[21]

DNA substrates, prepared either by PCR amplification or as restriction fragments, are ^{32}P end-labeled[22,23] and purified by electrophoresis and excision from 8% nondenaturing polyacrylamide gels using an autoradiogram of the gel as a location guide. The DNA molecules are eluted from the excised gel fragment using a crush and soak procedure,[23] and two precipitations from 70% (v/v) ethanol are required to eliminate impurities. The ^{32}P-labeled DNA is stored in TE buffer.

The quantity of histone required to observe a gel shift depends on the length and concentration of the DNA molecule, and the affinity of the histone for the

[19] G. Oohara and A. Wada, *J. Mol. Biol.* **196**, 389 (1987).
[20] R. A. Grayling, K. A. Bailey, and J. N. Reeve, *Extremophiles* **1**, 79 (1997).
[21] J. Carey, *Methods Enzymol.* **208**, 103 (1991).
[22] G. Mukhopadhyay, J. A. Dibbens, and D. K. Chattoraj, *Methods Mol. Genet.* **6**, 400 (1995).
[23] J. Sambrook, E. F. Fritsch, and T. Maniatis, "Molecular Cloning: A Laboratory Manual." Cold Spring Harbor Laboratory Press, Cold Spring Harbor, NY, 1989.

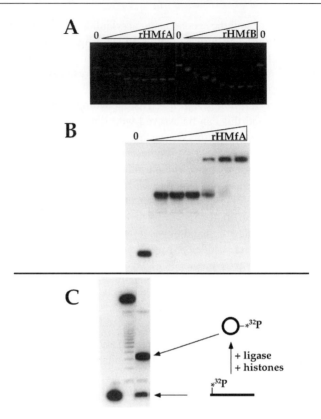

FIG. 1. Assays of archaeal histone–DNA interactions. (A) *Agarose gel-shift acceleration assay:* 50 ng of EcoRI-linearized pBR322 DNA was mixed with increasing amounts of rHMfA or rHMfB, recombinant histones from *M. fervidus,* at mass ratios of 0.1, 0.2, 0.3, 0.4, 0.6, 0.8, 1.0, and 1.2. Both histones wrap the DNA into complexes that migrate faster through agarose relative to free DNA. However, the histones have different saturation points beyond which histone–DNA mobility does not increase, and maximum magnitudes of mobility shift are different. Control lanes, labeled 0, had no histone added. (B) *Polyacrylamide gel-shift retardation assay:* Increasing amounts of rHMfA were mixed with a 180 bp DNA fragment. Retardation occurs in discrete steps as complexes with increasing integral numbers of nucleosomes form on the DNA. The control lane, labeled 0, had no histone added. (C) *Histone-mediated DNA circularization reaction:* As diagrammed on the right-hand side, the addition of T4 DNA ligase to a short linear double-stranded DNA molecule will result in circularization in the presence of histones. Left lane, DNA alone; center lane, DNA and ligase, illustrating DNA multimerization in the absence of histones; right lane, DNA, histones, and ligase.

DNA. Each DNA fragment is therefore titrated over a range of several orders of magnitude of histone concentration to determine the working range for the assay. DNA and histone are mixed in a siliconized tube in 100 mM KCl, 50 mM Tris (pH 8.0), incubated for 20 min at room temperature, 1/10th vol of sample buffer

[25% w/v Ficoll, 0.4% w/v BPB, 0.4% w/v xylene cyanol FF (XC)] added, and the mixture loaded into the well of a polyacrylamide gel prepared in 1× TBE buffer (90 mM Tris–borate, 2 mM EDTA, pH 8.0). Acrylamide concentrations ranging from 8 to 12% can be used, depending on the length of DNA fragment,[23] with an acrylamide : bisacrylamide ratio of 74 : 1. Electrophoresis is carried out at 8 V/cm and the gel dried under vacuum prior to complex detection by autoradiography. Complexes (archaeal nucleosomes) have a reduced mobility relative to free DNA, and complexes formed with DNA molecules >150 bp form several bands, reflecting different numbers of nucleosomes present on one DNA molecule (Fig. 1B).

DNA Circularization Promoted by Histones. In this assay, archaeal nucleosomes are exposed to T4 DNA ligase to form a DNA circle, which is especially useful in detecting unstable histone–DNA complexes that may dissociate during the mobility shift assays. DNA molecules of ∼100 bp are used for this assay, as molecules <88 bp are not long enough to encircle the histone tetramer[4] and molecules >150 bp circularize in the absence of histones.[24] The molecule must have one or more RE recognition sites for analysis of the ligation products, and blunt or cohesive ends as substrates for T4 DNA ligase. Appropriate DNA fragments can be obtained from plasmid multiple cloning sites such as the 98 bp *Xba*I/*Spe*I fragment from pLITMUS28 (New England Biolabs, Beverly, MA). Following digestion of the plasmid by the first RE and ^{32}P end-labeling, digestion with the second RE generates a fragment of the desired size with a single end labeled, which is then gel-purified.

The ^{32}P-labeled DNA fragment and archaeal histone at different mass ratios are assembled in a 9 μl reaction in ligase buffer in a siliconized tube and after a 20 min incubation at room temperature, 1 μl containing 40–80 U of freshly diluted T4 DNA ligase (New England Biolabs) is added and incubation continued for 1 hr. The DNA is prepared for electrophoresis by phenol–CHCl$_3$ extraction, precipitation from ethanol, and storage in TE buffer. The different topological forms of the DNA fragment are resolved by electrophoresis at 8 V/cm through 8% nondenaturing polyacrylamide gels in 1× TBE and visualized by autoradiography (Fig. 1C). Positive assays from the histone-containing reaction mixture will contain a predominant product that is not present in control reactions lacking histones. This product is excised from the gel, purified,[23] and treated with REs and exonuclease III to confirm its structure. Covalently closed circular DNAs are resistant to exonuclease III digestion, whereas nicked circles and linear DNA molecules are not; digestion of a monomer circle with an RE that has a single site will produce a linear fragment, the length of the initial substrate, whereas RE digestion of multimeric circles produces molecules of several different sizes resulting from head-to-tail and head-to-head ligations.[4]

[24] D. Shore, J. Langowski, and R. Baldwin, *Proc. Natl. Acad. Sci. U.S.A.* **78**, 4833 (1981).

DNA Circular Dichroism. There is a peak at 275 nm in circular dichroism spectra of DNA solutions that is depressed by interaction with histones (Fig. 2A). The presence of control proteins, such as bovine serum albumin (BSA), does not affect this signal (Fig. 2C), and neither free archaeal histones nor BSA generate a CD signal at this wavelength (Fig. 2B). Linearized plasmid DNA (~100 μg) and increasing amounts of histone are mixed in 100 mM KCl, 25 mM potassium phosphate (pH 7.0) and CD spectra recorded at room temperature from 200–320 nm using a 1 cm pathlength cylindrical quartz cuvette, with an averaging time of 5 sec between scans. The raw data, measured in millidegrees, are converted to $\Delta\varepsilon$ by using the equation:

$$\Delta\varepsilon = [(\theta \text{ millidegree})(660 \text{ Da/bp})/33000](1 \text{ cm path length/DNA concentration in mg/ml})$$

Purification of Archaeal Nucleosomes

Archaeal histones package most of the genome *in vivo* into archaeal nucleosomes,[3] and archaeal nucleosomes fixed *in situ* can be purified following exposure of the cells to formaldehyde. Following formaldehyde cross-linking and cell lysis, micrococcal nuclease (MN), an endo–exo-nuclease,[25] is used to digest protein-free DNA generating archaeal nucleosomes that contain a histone tetramer, and ~60 bp genomic DNA fragments.[3]

Cell paste, washed twice with cross-linking buffer (0.1 M NaCl, 50 mM HEPES pH 7.5, 1 mM EDTA, 0.5 mM EGTA),[26] is resuspended at 10 ml/g, the formaldehyde is added to a 1% concentration, and the reaction is allowed to proceed at room temperature. The time required for optimal cross-link fixation must be determined empirically, with 1 hr incubation used as a starting point. The reaction is terminated by adding an equimolar amount of ammonium acetate and continuing incubation at room temperature for 15 min. The cells are then washed twice, once in cross-linking buffer and once in MN buffer (1 mM CaCl$_2$, 50 mM Tris–acetate, pH 8.8), resuspended in MN buffer (3 ml/g of starting material), and lysed by passage through a French pressure cell (SLM instruments, Urbana, IL) at 10,000 psi. DNA in the resulting lysate is digested at 37° with MN, freshly diluted from a 0.5 U/μl stock to 1.5 to 10 U/g of starting material. Reactions are stopped by adding SDS to 0.2% (w/v) and EDTA pH 8.0 to 20 mM; the products are concentrated by ultrafiltration using a mini-stirred cell fitted with a YM1 membrane (Amicon, Beverly, MA) and loaded onto a Sephacryl S-200 column (Amersham Pharmacia Biotech, Inc. Piscataway, NJ) equilibrated in 150 mM NaCl, 25 mM TrisCl (pH 8.0). Fractions, eluted using the equilibration buffer, are collected after

---

[25] N. C. Mishra, "Molecular Biology of Nucleases." CRC Press, Boca Raton, FL, (1995).
[26] M. J. Solomon, P. L. Larsen, and A. Varshavsky, *Cell* **53**, 937 (1988).

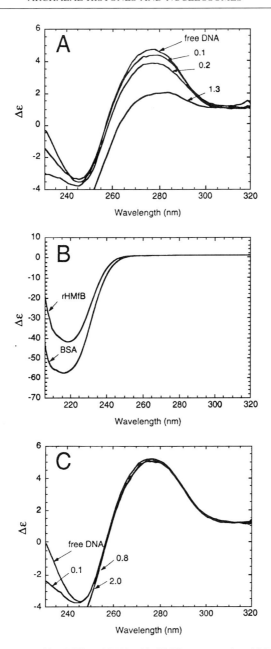

FIG. 2. (A) CD spectra of free DNA and DNA with rHMfB at mass ratios of 0.1, 0.2, and 1.3. The peak at 275 nm is depressed on histone binding. (B) CD spectra of rHMfB and BSA, showing no peaks at 275 nm. (C) CD spectra of DNA alone and mixed with BSA, showing no change in the 275 nm DNA peak.

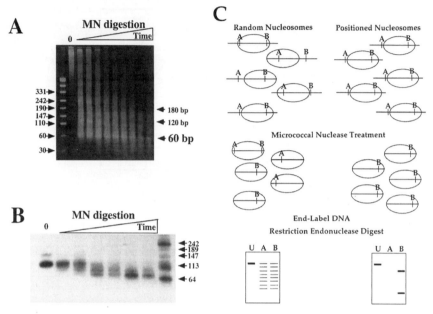

FIG. 3. (A) Time course of MN digestion of genomic DNA containing archaeal nucleosomes formaldehyde-cross-linked *in vivo*. Bands at multiples of 60 bp are visible with increasing time of digestions, resulting from the release of individual and adjacent archaeal nucleosomes. (B) Time course of MN digestion of nucleosomes assembled *in vitro* from rHMfB and a DNA template containing six repeats of the trinucleotide CTG.[31] Size standards are indicated on the right-hand side. (C) Diagram of the concept used to map the positions of nucleosome assembly on a DNA molecule.[32] The ovals indicate histones and horizontal lines the DNA fragment that is long enough to accommodate only one nucleosome. A and B are recognition sites for restriction endonucleases. On the left-hand side, the DNA fragment has no positioning elements, and nucleosomes assemble at random locations, whereas on the right-hand side, a strong nucleosome positioning signal exists. After MN digestion of DNA not protected by incorporation into a nucleosome, the DNA remaining is characterized by restriction endonuclease digestion. The lower panel shows the restriction digestion results predicted for the two assembly situations. Lane U contains the control full-length MN-resistant DNA, and lanes A and B show the DNA fragments resulting from digestion with the respective enzymes A and B.

the void volume and subjected to electrophoresis through a 4% (w/v) NuSieve GTG low melting temperature agarose gel (FMC Bioproducts, Rockville, ME) and ethidium bromide staining. Archaeal nucleosomes have an electrophoretic mobility similar to ~60 bp protein-free double-stranded DNA molecules (Fig. 3A).

Nucleosomal Proteins

Archaeal histones can be isolated from purified archaeal nucleosomes following excision from the agarose gel and agarase digestion of the gel slice. The gel is equilibrated in 10 vol β-agarase buffer (40 mM Bis–Tris-HCl, 40 mM NaCl, 1 mM EDTA, pH 6.0) (FMC Bioproducts) for 1 h; the buffer is then discarded, the gel melted by incubation at 70° for 45 min, and the agarose digested by incubation with β-agarase (25 U/ml) for several hours (or overnight) at 45°. The tube is centrifuged briefly to pellet any undigested agarose, and the nucleosomes precipitated after an overnight incubation at room temperature from 2.5 M ammonium acetate and 75% ethanol. The DNA present is digested by exposure to 100 μg DNase I (Sigma, St. Louis, MO)/ml for 30 min at 37°C, and the histones that remain can be visualized by silver staining following SDS–PAGE.

Nucleosomal DNA

To purify the genomic DNA fragments cross-linked into archaeal nucleosomes, the MN digestion is stopped as described above, and the lysate then treated with proteinase K (300 μg/ml) for 3–4 hr at 37°. The formaldehyde cross-links are reversed by incubation at 65° for 6 h,[27] and the nucleosomal DNA purified by phenol/chloroform extraction, chloroform extraction, and ethanol precipitation. Exposure to RNase A (40 μg/ml) for 30 min at 37°C yields a solution containing only ~60 bp nucleosomal DNA fragments.

Mapping Nucleosome Assembly Sites on DNA

Histones bind to DNA regardless of sequence, but some sequence-conferred DNA structures are preferentially wrapped into nucleosomes. Sequences that permit the alternate compression of the major and minor grooves are favored, as they readily adopt the distortions imposed by histone wrapping during nucleosome assembly.[28] A hierarchy of eukaryal sequences has been established that are preferentially incorporated into nucleosomes.[29,30] One of these, tandem repeats of the CTG trinucleotide, also positions archaeal nucleosome assembly,[31] consistent with archaeal and eukaryal histones responding similarly to DNA sequences that function as nucleosome positioning signals.

The procedures described below identify DNA regions protected from MN digestion by their incorporation into archaeal nucleosomes, and determine which sequences are incorporated preferentially within a nucleosome.[32]

---

[27] M. J. Solomon and A. Varshavsky, *Proc. Natl. Acad. Sci. U.S.A.* **82,** 6470 (1985).
[28] T. E. Shrader and D. M. Crothers, *Proc. Natl. Acad. Sci. U.S.A.* **86,** 7418 (1989).
[29] J. S. Godde and A. P. Wolffe, *J. Biol. Chem.* **271,** 15222 (1996).
[30] P. T. Lowary and J. Widom, *J. Mol. Biol.* **276,** 19 (1998).
[31] K. Sandman and J. N. Reeve, *J. Bacteriol.* **181,** 1035 (1999).
[32] S. L. Pereira and J. N. Reeve, *J. Mol. Biol.* **289,** 675 (1999).

Synthesis of Mapping Template

DNA fragments used as substrates for nucleosome positioning should be so short that only one nucleosome can form (<120 bp) but sufficiently long that several adjacent positions are available for assembly, in practice 90–115 bp (see Fig. 3). The substrate DNA is generated by PCR amplification in the presence of a ^{32}P-labeled nucleotide, after conditions are optimized for the yield and specificity of the PCR, and must contain at least one RE recognition site to map the ends of the protected fragments.

MN Digestion

MN (0.5 U/µl) remains stable, as a frozen stock solution, for ~6 months and is diluted immediately before use. The dilution must be determined empirically for each template–histone combination, but generally ~1000-fold dilution is appropriate, with the progress of MN digestion assayed by sampling at increasing times (Fig. 3B). Archaeal nucleosomes are assembled by mixing 500 ng of labeled template DNA with a 5- to 7-fold molar excess of archaeal histone tetramers in a total volume of 100 µl MN buffer at room temperature. A 10-µl (50 ng) aliquot is removed before MN is added to provide a zero-time control; 10-µl aliquots are then removed at time intervals after MN addition. A parallel control reaction lacking histones must be performed. Each aliquot is placed in a tube containing 1 µl of 0.1 mM EDTA (pH 8.0) to chelate the Ca^{2+} required for MN activity, an equal volume of phenol–$CHCl_3$ is added and mixed, and the tubes are stored on ice until the final sample is taken. The reaction mixtures are centrifuged briefly to separate the phases, the aqueous phase $CHCl_3$ extracted, and the DNA precipitated from 0.3 M sodium acetate, 10 mM $MgCl_2$, and 75% ethanol at $-70°$.

Depending on the resolution desired, gel electrophoresis using a small 8% nondenaturing polyacrylamide gel in 1× TBE (for resolution to 5–10 bp) (Fig. 3B) or a full-size DNA sequencing 10% polyacrylamide gel containing 8.3 M urea in 0.5× TBE buffer (for single base pair resolution)[31] can be used. The DNA is resuspended in sample buffer (95% formamide, 20 mM EDTA, 0.05% BPB, 0.05% XC) and applied to the gel and electrophoresis continued until the BPB is near the lower edge of the gel. The results are visualized by autoradiography.

As illustrated in Fig. 3B, MN digestion of archaeal nucleosomes results in a population of MN-protected ~70 bp molecules; however, the lengths of individual fragments vary with each DNA positioning sequence, ranging from 64 to 72 bp.

Mapping Ends of MN-Protected Fragments

Nucleosomes are prepared with unlabeled substrate DNA, and using the MN digestion conditions established as described above, the MN digestion is allowed to proceed until most of the DNA has been digested to the 60–70 bp single nucleosome form. The reaction is then stopped as described above, MN-protected DNA fragments purified from an ethidium bromide-stained 8% nondenaturing polyacrylamide gel,[23] and the protected fragments ^{32}P end-labeled.[23] Following exposure to REs that have cleavage sites within the substrate DNA, electrophoresis is used to separate the resulting DNA fragments to single nucleotide resolution.

If nucleosome assembly occurred at random locations, then all regions of the substrate DNA molecule will be present in the MN-protected 65–70 bp population of fragments. An autoradiogram of the restriction digests will show a continuum of bands that extend from the ends of the MN-protected fragments to the cleavage site (Fig. 3C). However, fewer bands will be observed if archaeal nucleosomes assembled at only a few locations on the substrate DNA, and the presence of only two bands would indicate the presence of a single nucleosome assembly site. In this case, the ends of the protected fragment can be identified based on their distance from the cleavage site, plus or minus any single-stranded nucleotide overhangs generated by the RE. After the RE digestion and electrophoresis, four single-stranded DNA fragments are present in the gel, but only two are labeled and therefore visible by autoradiography. The position of assembly should be confirmed, if possible, by using a second RE; however, some REs exhibit limited cleavage if their recognition sequence is located very close to the end of the MN-protected DNA fragment.

[11] DNA-Binding Proteins Sac7d and Sso7d from *Sulfolobus*

By STEPHEN P. EDMONDSON and JOHN W. SHRIVER

Introduction

The chromatin of the archaea *Sulfolobus acidocaldarius* and *Sulfolobus solfataricus* contains a number of small basic proteins ranging in molecular weight from 7000 to 10,000.[1] The 7-kDa proteins have been the most studied and are believed to be important in DNA compaction and stabilization of duplex DNA at the growth temperatures of 75° for *S. acidocaldarius* and 80° for *S. solfataricus*.[2–9] They

---

[1] J. Dijk and R. Reinhardt, in "Bacterial Chromatin" (C. Gualerzi and C. Pon, eds.). Springer-Verlag, Berlin, 1986.
[2] M. Kimura, J. Kimura, P. Davie, R. Reinhardt, and J. Dijk, *FEBS Lett.* **176,** 176 (1984).
[3] M. Grote, J. Dijk, and R. Reinhardt, *Biochim. Biophys. Acta* **873,** 405 (1986).
[4] R. Lurz, M. Grote, J. Dijk, R. Reinhardt, and B. Dobrinski, *EMBO J.* **5,** 3715 (1986).
[5] J. McAfee, S. Edmondson, P. Datta, J. Shriver, and R. Gupta, *Biochemistry* **34,** 10063 (1995).

have been shown to relieve positive supercoiling in DNA.[10] Other enzyme activities associated with these proteins include RNase,[11,12] "disulfide bond forming,"[13] ATPase,[14] and chaperone activity.[14]

The 7-kDa chromatin proteins of *S. acidocaldarius* consist of five species, designated Sac7a, b, c, d, and e in order of increasing basicity.[2,3,7] Sac7d (molecular weight 7477) and Sac7e (molecular weight 7338) differ by six amino acid residues and are coded by distinct genes[5] (Fig. 1). Sac7a and b are carboxy-terminal truncated forms of the Sac7d. Only one form of Sso7 has been characterized; it is referred to as Sso7d (MW 7019) based on homology to Sac7d.[6]

The 7-kDa *Sulfolobus* chromatin proteins are especially amenable to studies of both protein stability and DNA-binding.[5,15–17] They unfold reversibly over a wide range of temperature, pH, and salt concentration, thus permitting an extensive characterization of thermophile protein stability.[17] In addition, they bind to DNA with a relatively small site size, which facilitates a detailed study of the energetics of DNA-binding by site-directed mutagenesis.

The structures of recombinant Sac7d[18] (Fig. 2) and native Sso7d[19] have been determined by NMR and are similar with a 3.2 Å root mean square deviation of main-chain atoms. Although the proteins are small, they contain significant secondary structural elements, including an N-terminal two-stranded β ribbon, a three-stranded β sheet, and a C-terminal α helix with intervening single turns of 3_{10} helix. The interface between the elements forms a hydrophobic core of 10 residues. They have been shown to be homologous to a large group of proteins in eukaryotes known as chromo domains.[20] The NMR structure of Sso7d complexed with DNA shows significant unwinding of the DNA resulting from binding of the three stranded β sheet against the minor groove.[21] Unfortunately, the protein

---

[6] T. Choli, P. Henning, B. Wittmann-Liebold, and R. Reinhardt, *Biochim. Biophys. Acta* **950**, 193 (1988).
[7] T. Choli, B. Wittmann-Liebold, and R. Reinhardt, *J. Biol. Chem.* **263**, 7087 (1988).
[8] T. Reddy and T. Suryanarayana, *Biochim. Biophys. Acta* **949**, 87 (1988).
[9] T. R. Reddy and T. Suryanarayana, *J. Biol. Chem.* **264**, 17298 (1989).
[10] P. Lopez-Garcia, S. Knapp, R. Ladenstein, and P. Forterre, *Nucleic Acids Res.* **26**, 2322 (1998).
[11] P. Fusi, G. Tedeschi, A. Aliverti, S. Ronchi, P. Tortora, and A. Guerritore, *Eur. J. Biochem.* **211**, 305 (1993).
[12] P. Fusi, M. Grisa, E. Mombelli, R. Consonni, P. Tortora, and M. Vanoni, *Gene* **154**, 99 (1995).
[13] A. Guagliardi, L. Cerchia, M. De Rosa, M. Rossi, and S. Bartolucci, *FEBS Lett.* **303**, 27 (1992).
[14] A. Guagliardi, L. Cerchia, L. Camardella, M. Rossi, and S. Bartolucci, *Biocatalysis* **11**, 181 (1994).
[15] B. S. McCrary, S. P. Edmondson, J. W. Shriver, *J. Mol. Biol.* **264**, 784 (1996).
[16] J. G. McAfee, S. Edmondson, I. Zegar, and J. W. Shriver, *Biochemistry* **35**, 4034 (1996).
[17] B. S. McCrary, J. Bedell, S. P. Edmondson, and J. W. Shriver, *J. Mol. Biol.* **276**, 203 (1998).
[18] S. P. Edmondson, L. Qiu, and J. W. Shriver, *Biochemistry* **34**, 13289 (1995).
[19] H. Baumann, S. Knapp, T. Lundbach, R. Ladenstein, and T. Hard, *Nature Struct. Biol.* **1**, 808 (1994).
[20] L. J. Ball, N. V. Murzina, R. W. Broadhurst, A. R. Raine, S. J. Archer, F. J. Stott, A. G. Murzin, P. B. Singh, P. J. Domaille, and E. D. Laue, *EMBO J.* **16**, 2473 (1997).
[21] P. Agback, H. Baumann, S. Knapp, R. Ladenstein, and T. Härd, *Nature Struct. Biol.* **5**, 579 (1998).

```
            1              5              10             15
Sac7d   Val-Lys-Val-Lys*-Phe-Lys*-Tyr-Lys-Gly-Glu-Glu-Lys-Glu-Val-Asp-
Sso7d   Ala-Thr-Val-Lys*-Phe-Lys*-Tyr-Lys-Gly-Glu-Glu-Lys-Glu-Val-Asp-


            16             20             25             30
Sac7d   Thr-Ser-Lys-Ile-Lys-Lys-Val-|Trp|-Arg-|Val|-Gly-Lys-|Met|-Val-Ser-
Sso7d   Ile-Ser-Lys-Ile-Lys-Lys-Val-|Trp|-Arg-|Val|-Gly-Lys-|Met|-Ile-Ser-


            31             35             40             45
Sac7d   Phe-Thr-Tyr-Asp-Asp-Asn-Gly-    -Lys-Thr-Gly-Arg-Gly-Ala-Val-
Sso7d   Phe-Thr-Tyr-Asp-Glu-Gly-Gly-Gly-Lys-Thr-Gly-Arg-Gly-Ala-Val-


            46             50             55             60
Sac7d   Ser-Glu-Lys-Asp-Ala-Pro-Lys-Glu-Leu-Leu-Asp-Met-Leu-Ala-Arg-
Sso7d   Ser-Glu-Lys-Asp-Ala-Pro-Lys-Glu-Leu-Leu-Gln-Met-Leu-   -   -


            61             65
Sac7d   Ala-Glu-Arg -Glu-Lys -Lys
Sso7d       -Glu-Lys*-Gln-Lys*-Lys*
```

FIG. 1. Amino acid sequences of Sac7d and Sso7d from *Sulfolobus*. The lysines that are monomethylated in the native proteins are marked with asterisks. The single tryptophan (W23) and the two residues (V25 and M28) that intercalate into DNA are outlined. Bold type indicates differences.

affinity for short DNA oligomers is not strong enough to permit observation of a unique 1 : 1 complex in solution, so that data were collected under conditions of fast exchange with excess DNA leading to a mixture of species. The structures of both Sac7d and Sso7d DNA–protein complexes have also been determined by X-ray crystallography[22,23] (Fig. 3). These differ significantly from the NMR structure in that both proteins bend the double helix into the major groove by approximately 70° because of the intercalation of valine-25 and methionine-28. The induced bend is the largest bend observed to date at a single protein–DNA site. In contrast, the protein structure is only slightly modified on binding.

Names other than Sac7 and Sso7 have been used to refer to the 7-kDa *Sulfolobus* DNA-binding proteins. A search for ribonucleases in *Sulfolobus* led to the isolation of proteins referred to as P2 in *S. solfataricus*[11,12] and SaRD in

[22] H. Robinson, Y.-G. Gao, B. S. McCrary, S. P. Edmondson, J. W. Shriver, and A. H.-J. Wang, *Nature* **392,** 202 (1998).
[23] Y. Gao, S. Su, H. Robinson, S. Padmanabhan, L. Lim, B. McCrary, S. Edmondson, J. Shriver, and A. Wang, *Nature Struct. Biol.* **5,** 782 (1998).

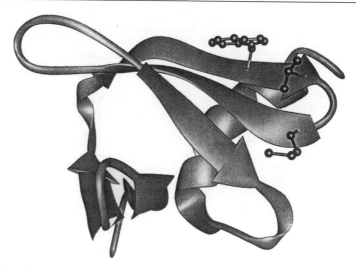

FIG. 2. A ribbon representation of the solution structure of recombinant Sac7d obtained from a full relaxation matrix refinement of 2-D ^{1}H NMR data [S. P. Edmondson, L. Qiu, and J. W. Shriver, *Biochemistry* **34,** 13289 (1995)] (PDB accession code 1sap). The side chains of W23, V25, and M28 are depicted on the surface of the β sheet. Prepared with MIDAS [T. Ferrin, C. Huang, L. Jarvis, and R. Langridge, *Mol. Graphics* **6,** 13 (1988)].

S. acidocaldarius.[24,25] Sequencing data indicate that these are Sso7d and Sac7d, respectively. In addition, the sequence of DBF ("disulfide bond forming") enzyme[13] is identical to Sso7d.

This chapter describes the purification of native Sac7 and Sso7 from *Sulfolobus* and recombinant proteins from *E. coli,* along with a brief description of various physical properties of the purified proteins. Assays for DNA binding and RNase activities are described and evaluated.

Purification

Growth of Sulfolobus

An initial cell culture of *Sulfolobus* is obtained by inoculating 10 ml Brock's medium,[26] supplemented with 0.2% sucrose with 1.0 ml of frozen cells. The culture is aerated with slow orbital oscillation in a temperature-regulated oil bath at 75° (*S. acidocaldarius*) or 80° (*S. solfataricus*). After 1 day, 200 ml of Brock's medium with 0.2% sucrose is inoculated with this culture and grown overnight

---

[24] D. Kulms, G. Schäfer, and U. Hahn, *Biol. Chem.* **378,** 545 (1997).
[25] D. Kulms, G. Schäfer, and U. Hahn, *Biochem. Biophys. Res. Comm.* **214,** 646 (1995).
[26] T. Brock, K. Brock, R. Belly, and R. Weiss *Arch. Microbiol.* **84,** 54 (1972).

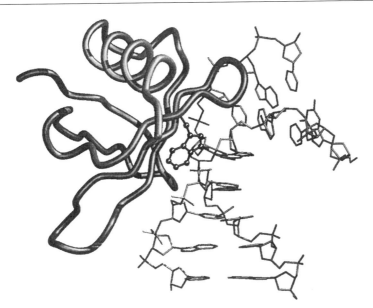

FIG. 3. The X-ray crystal structure of Sac7d–d (GTAATTAC)$_2$ complex determined to 1.9 Å resolution. The W23 side chain is shown at the protein–DNA interface. The intercalating valine and methionine are not shown for clarity, but lie in the pocket above W23 between unstacked base pairs 3 and 4 (PDB accession codes 1azq and 1azp). Prepared with MIDAS [T. Ferrin, C. Huang, L. Jarvis, and R. Langridge, *Mol. Graphics* **6**, 13 (1988)].

with oscillation in the same bath. The next day the 200 ml culture is used to inoculate 16 liters of DeRosa's medium,[27] supplemented with 0.1% glucose and 0.1% glutamic acid in a VirTis glass fermentor preheated to 75° (*S. acidocaldarius*) or 80° (*S. solfataricus*). Temperature is regulated with a water-circulating bath flowing through a submersed stainless steel coil. The 16-liter culture is aerated with a slow flow of air through a bubbling stone. Cells are grown to an A_{600} of 1.0 in about 5 days and harvested by centrifugation at 4500g in 1 liter bottles for 15 min in a Beckman J6B centrifuge at 4°. The pelleted cells can be stored at −70°.

Overexpression in E. coli

A pET-3b expression vector containing the Sac7d or Sso7d gene has been constructed[5] and transformed into *E. coli* strain BL21 (DE3) pLysS.[28] The pLysS plasmid is essential due to the toxicity of Sac7d in *E. coli*. The bacteria are grown in Luria–Bertani (LB) media (1% w/v Bacto-tryptone/1% w/v NaCl/0.5% w/v

[27] M. DeRosa and A. Gambacorta, *J. Gen. Microbiol.* **86**, 156 (1975).
[28] W. Studier, A. Rosenberg, J. Dunn, and J. Dubendorff, *Methods Enzymol.* **185**, 66 (1990).

yeast extract) by standard methods.[29] For protein preparation, a 20 ml culture of the transformant is grown overnight in LB broth containing 50 μg/ml ampicillin at 37°. From this, 4.0 ml is used to inoculate 200 ml of fresh medium. At an A_{600} of 0.3 to 0.6, 25-ml aliquots are used to inoculate 6 flasks each containing 1 liter of new medium; the flasks are then incubated at 37°. Sac7d expression is induced when the A_{600} reaches a value of 0.8 to 0.95 by adding isopropylthiogalactoside (IPTG) to a final concentration of 0.4 mM. After 1 hr the cells are harvested by centrifugation in 1 liter bottles at 4500g for 15 min in a Beckman J6B centrifuge at 4°.

Protein Isolation and Purification

Cells are suspended in lysis buffer [10 mM Tris-HCl buffer (pH 7.5), 0.5 mM phenylmethylsulfonyl fluoride, 0.1% Triton X-100, 5 mM EDTA] using 100 ml per 6-liter *E. coli* culture and 200 ml per 16-liter *Sulfolobus* culture. Frozen cells are slowly thawed with cold lysis buffer, and the suspended cells are lysed with brief sonication on ice. DNase I (5 mg/100 ml) is added to lysed *Sulfolobus* cells, and the suspension is incubated at 37° for 5 min. Cellular debris is removed by centrifugation at 280,000g for 60 min at 4°. The supernatant is cooled on ice and dialyzed in SpectraPor CE (1000 molecular weight cutoff) tubing against 0.2 M H_2SO_4 overnight at 4°. Precipitated protein is removed by centrifugation at 180,000g for 30 min at 4°, and the supernatant is dialyzed with three changes against 20 mM Tris-Cl (pH 7.4) with 1 mM EDTA. Any additional precipitate is removed by centrifugation at 180,000g for 30 min at 4°, and the protein is purified by high-performance liquid chromatography (HPLC) using a Pharmacia Hi-Trap SQ column with a linear 0.0 to 0.5 M NaCl gradient. Sac7 and Sso7 elute at approximately 0.3 M NaCl. Further purification can be accomplished by gel exclusion chromatography on Sephacryl S-100-HR but is generally unnecessary. Typical yields are 50 mg of protein from either 16 liters of *Sulfolobus* culture or 6 liters of *E. coli* culture.

Properties of Purified Proteins

Size

Nonreducing SDS gel electrophoresis[30] can be used to monitor the identity and purity of the 7-kDa proteins. Recombinant Sac7d shows a single band consistent with a molecular weight of 7600 that comigrates with the mixture of Sac7 native proteins. The band is absent in preparations from *E. coli* cells lacking the recombinant plasmid. Recombinant Sso7d runs slightly ahead of Sac7d, consistent with

---

[29] J. Sambrook, E. F. Fritsch, and T. Maniatis, Molecular Cloning: A Laboratory Manual. Cold Spring Harbor Laboratory, Cold Spring Harbor, N.Y, 1989.

[30] H. Schägger and G. von Jagow, *Anal. Biochem.* **166**, 368 (1987).

a molecular weight of 7150 (calculated from the sequence). Schägger–von Jagow gels do not resolve the individual native Sac7 and Sso7 species. The identity of the proteins can be confirmed by amino terminal sequencing, mass spectroscopy, and the various properties described below.

Sac7d and Sso7d have been shown to be monomeric in solution by small angle X-ray scattering. Data were measured using a 2-D phase-sensitive detector over a Q range of 0.01 to 0.22 Å^{-1}. The net scattering of the protein was determined by subtracting a normalized buffer spectrum measured in the same quartz capillary. Protein concentrations ranged from 5 to 20 mg/ml, and the data were extrapolated to zero concentration. The solution scattering data indicate that both Sac7d and Sso7d are globular in 0.01 M K_2HPO_4 (pH 7) with, or without, 0.3 M KCl. A mass of 7460 Da was obtained for Sac7d from the intensity at zero scattering angle, and a radius of gyration (R_g) of 12.4 Å was determined by Guinier analysis[31] of the scattering data. The R_g for Sso7d is slightly smaller (11.6 Å), as expected from the differences in the amino acid sequences. Both R_g values are large and indicate significant hydration of these highly basic proteins.

UV Absorbance

The UV extinction coefficient of both the native and recombinant Sac7d is 1.09 ± 0.04 ml/(mg · cm) at the wavelength of maximal absorption (278 nm).[5] The extinction coefficient for Sso7d is calculated to be 1.17 ml/mg · cm based on the difference in molecular weights. Use of the extinction coefficients to measure protein concentration requires an accurately calibrated spectrophotometer. The wavelength of the central most intense line of benzene vapor should be 252.88 nm. Absorbance accuracy is determined using 0.0400 gm/liter potassium chromate in 0.05 M KOH,[32] which should have an absorbance of 0.757 at 275 nm with a 1 cm cell at 25°.

Fluorescence

The intrinsic tryptophan fluorescence emission spectra of Sac7 and Sso7 (5 μM protein in 10 mM KH_2PO_4, pH 6.8) are obtained with excitation at 295 nm using 4 nm excitation and emission slit widths. Excitation at 295 nm prevents contributions from the two tyrosine and two phenylalanine residues. An emission maximum at 350 nm is similar to that of free tryptophan and indicates significant solvent exposure, consistent with the NMR solution structure. Addition of double-stranded DNA [e.g., duplex poly[d(GC)]] leads to quenching of tryptophan fluorescence by nearly 90%. A blue shift of the emission maximum to 340 nm is also observed,

---

[31] A. Guinier, *Ann. Phys. (Paris)* **12**, 161 (1939).
[32] A. Gordon and R. Ford, "The Chemist's Companion: A Handbook of Practical Data, Techniques, and References." John Wiley, New York, 1972.

indicating decreased solvent exposure of the tryptophan. This behavior is consistent with the X-ray crystal structure of a Sac7d–octanucleotide complex that shows the tryptophan at the DNA–protein interface (Fig. 3).

Circular Dichroism

CD spectra of Sac7d and Sso7d are measured in the far-UV spectral region (below 260 nm) with sample concentrations ranging from 0.01 to 1.0 mg/ml using cells of 1.0 to 0.01 cm path length, respectively. Shorter path lengths (and higher concentrations) permit CD measurements at shorter wavelengths, with 1 mm cuvettes working well for far-UV CD spectra. The choice of solvent can limit accessibility to the far-UV, and dilute simple buffers (e.g., 10 mM Na_2HPO_4) are the most transparent. The CD of the proteins in the near-UV (320–250 nm) is about 1% of that of the far-UV and is measured at concentrations of 1 mg/ml in a 1-cm path length cuvette. Spectra are collected at 1 nm intervals using averaging times of 8 to 15 s/nm, depending on the signal intensity, with a spectral bandwidth of 1.5 nm. The spectra are smoothed with a sliding 9- or 11-point convoluting quadratic function, as described by Savitsky and Golay.[33] Baselines are measured using water and subtracted from the sample spectrum. The molar CD per peptide bond ($\Delta \varepsilon = \varepsilon_L - \varepsilon_R$) is calculated using the UV extinction coefficient. The CD instrument is calibrated at 290.5 nm with d-camphor-10-sulfonic acid using $\Delta \varepsilon^{0.1\%} = 2.36$ and a ratio $\Delta \varepsilon_{192.5}/\Delta \varepsilon_{290.5}$ of -2.05[34]. Sample temperature is regulated with a computer-controlled thermoelectric cuvette holder to within $\pm 0.2°$ over a range of 10 to 80°. At 95° the error in the sample temperature is about $\pm 0.5°$.

No differences are detected in the far-UV or near-UV CD spectra of recombinant and native Sac7d. However, the CD of Sso7d differs significantly from that of Sac7d (Fig. 4), presumably because of a truncated and more disordered C-terminal α helix. The near-UV CD of protein is an induced optical activity of the aromatic side chains due to the asymmetric environment of the folded protein. Therefore, the near-UV CD is often used as an indicator of the protein tertiary structure. The far-UV CD is dominated by contributions from the peptide bond and, thus, is a measure of the protein secondary structure.

Stability

Both guanidine hydrochloride and urea promote unfolding of Sac7d, with guanidine hydrochloride much more effective than urea. Protocols for measuring free energies of unfolding by chemical denaturation have been described by Pace.[35]

---

[33] A. Savitsky and M. J. E. Golay, *Anal. Chem.* **36**, 1627 (1964).
[34] G. C. Chen and J. T. Yang, *Anal. Lett.* **10**, 1195 (1977).
[35] C. N. Pace, *Methods Enzymol.* **131**, 266 (1986).

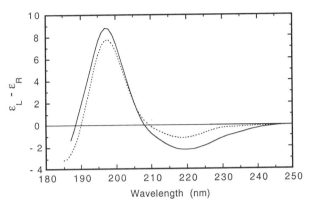

FIG. 4. The CD spectra of recombinant Sac7d and Sso7d (0.1 mg/ml) collected in 0.01 M KH$_2$PO$_4$ (pH 7) with a 1 mm path length cell.

At pH 7 and 20°, increasing guanidine hydrochloride leads to a characteristic sigmoidal change in CD with a midpoint concentration of 2.8 M denaturant. Linear extrapolation of the data to zero denaturant gives a free energy of unfolding of 5.3 (±0.3) kcal/mol and an m value of 1896 (±97) cal mol^{-1} M^{-1}.[15] Urea concentrations as high as 8 M cannot unfold Sac7d at pH 7. However, reversible unfolding is observed in urea at pH 4 with a midpoint at 5.3 M denaturant giving a free energy of unfolding of 3.4 (±0.3) kcal/mol and an m value of 633 (±64) cal mol^{-1} M^{-1} at this pH.

Thermal unfolding of Sac7d leads to significant changes in both the near- and far-UV CD spectra. The thermally unfolded protein has a CD spectrum similar to that observed in the presence of denaturing guanidine hydrochloride or urea at high temperature.[15] Thermal unfolding is performed under computer control in 1° increments at 205 nm, a wavelength that provides a strong CD signal in both the folded and unfolded states and displays a cooperative melting profile. After temperature equilibration, the CD signal is averaged for 20 sec before proceeding to the next temperature increment. The baseline for the melt is independent of temperature and is derived from the smoothed CD spectrum of water at 20°. Thermal melts of Sac7d monitored by CD at 205 nm, CD$_{obs}$, can be converted into fraction folded, f,

$$f = \frac{CD_{obs} - CD_u}{CD_n - CD_u} \qquad (1)$$

after defining the baseplanes due to temperature dependent changes in the CD of the folded and unfolded protein (CD$_n$ and CD$_u$, respectively).

The midpoint of the thermal unfolding transition at high pH (greater than 5) is 91° as indicated both by the temperature dependence by the CD and also by

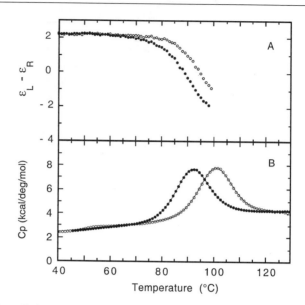

FIG. 5. Thermally induced unfolding of recombinant Sac7d (●) and Sso7d (○) followed by CD (upper, A) and DSC (lower, B). The CD intensity at 205 nm was collected in 1 cm path length cuvettes with 0.01 mg/ml protein in 0.001 M potassium acetate (pH 5.5). The DSC (partial molar heat capacity) of Sac7d and Sso7d (1–2 mg/ml) was collected in 0.01 M potassium acetate (pH 5.7) and 0.3 M KCl. Fitting of the DSC data gives a T_m for Sac7d of 90.4° and 100.2° for Sso7d.

differential scanning calorimetry (DSC)[36] (Fig. 5). Both techniques also demonstrate that unfolding is reversible. Thermally unfolded protein also regains full DNA binding activity on refolding (see below). The thermal stability of Sso7d is greater than that of Sac7d, as might be expected from the higher growth temperature of *S. solfataricus* compared to *S. acidocaldarius*.

Lysine Monomethylation

The ε-amino groups of a few specific lysines in native Sac7 and Sso7 are monomethylated.[6,7] The extent of monomethylation at each of the sites is not complete, but has been shown to increase with heat shock in Sso7d.[19] Lysine-4 and lysine-6 are monomethylated in both Sac7 and Sso7 (Fig. 1). In addition, C-terminal lysines are monomethylated in Sso7d (and also Sac7e). The T_m of native Sac7d and Sso7d is greater than that of the recombinant forms by about 7°. This may be due to the monomethylation of the native forms, but definitive evidence for this has not been provided.

[36] J. W. Shriver, W. B. Peters, N. Szary, A. T. Clark, and S. P. Edmondson, *Methods Enzymol.* **334**, [31], 2001 (this volume).

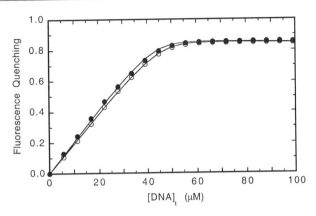

FIG. 6. Binding of Sac7d to DNA as indicated by the quenching of protein tryptophan fluorescence as a function of the concentration of duplex poly[d(GC)] in 0.001 M KH$_2$PO$_4$ (pH 6.8) and 0.025 M KCl at 25°C. The reversibility of thermal unfolding is shown by data (○) obtained on protein that had been unfolded by heating to 100° compared to protein that had not been unfolded (●). Nonlinear regression using the McGhee–von Hipple model gives a binding constant $K = 1.06 \times 10^7 M^{-1}$, a site size $n = 3.8$ base pairs, and $Q_{max} = 0.88$ for the protein not unfolded, and $K = 1.01 \times 10^7 M^{-1}$, $n = 4.0$ base pairs, and $Q_{max} = 0.88$ for the refolded protein.

DNA Binding Assays

The DNA binding of the 7-kDa proteins has been investigated using electron microscopy, filter binding assays, fluorescence and CD titrations, and isothermal titration calorimetry.[1–4,6,16,37] The fluorescence of the single tryptophan in Sac7d and Sso7d is quenched by nearly 90% on DNA binding, making this an optimal method for measuring the "activity" of the protein. Alternatively, structural transitions in DNA can be monitored by CD. Sac7 and Sso7 induce an unwinding and kink in duplex DNA[22] that leads to a large change in near-UV CD bands of DNA.[16] ITC is especially useful for binding studies at physiological growth temperatures of *Sulfolobus*.[36]

DNA Binding Followed by Fluorescence

Fluorescence DNA binding titrations are performed in the "reverse" mode by adding 5.0 μl aliquots of a 1.0 mM DNA stock to 3.0 ml of 5 μM protein in a 4 ml fluorescence quartz cell (Fig. 6). A carefully calibrated Unimetrics syringe with a utility stop permits reproducible, repetitive deliveries of a constant volume. The solution is mixed by constant stirring using a magnetic "flea" at the bottom of the cell. Nucleic acid and protein stock concentrations are determined by UV

---

[37] T. Lundbäck, H. Hansson, S. Knapp, R. Ladenstein, and T. Härd, *J. Mol. Biol.* **276**, 775 (1998).

absorption. Titrations are performed at constant temperature using a temperature-controlled cell holder. The fluorescence intensity reaches a plateau at saturating DNA concentrations, indicating that no correction is needed for an inner filter effect. Any decrease in fluorescence due to the inner filter effect is apparently balanced by scattering of the DNA–protein complexes.

DNA Binding Affinity and Site Size

DNA binding constants and site sizes are obtained from fluorescence titration data by nonlinear regression.[16] The observed intrinsic tryptophan fluorescence quenching, Q_{obs}, is defined by

$$Q_{obs} = (F_i - F_{obs})/F_i \tag{2}$$

where F_i and F_{obs} are the fluorescence intensity of the protein in the absence and presence of DNA, respectively. It is generally assumed that the fractional change in fluorescence quenching is equal to the fraction of protein bound, giving

$$L_b = (Q_{obs}/Q_{max})L_t \tag{3}$$

where L_b is the bound protein concentration, L_t is the total protein concentration, and Q_{max} is the maximal observed quenching at saturating DNA concentrations. This relation must be true if meaningful parameters (binding constant, site size, cooperativity) are to be extracted from binding isotherms.[5,16,37] It has been shown that this relation is valid for Sac7d binding to DNA by using the model-independent ligand-binding density function analysis method of Bujalowski and Lohman.[38] From Eq. (2), the free protein concentration is given by

$$L_f = L_t - L_b = L_t - \left(\frac{Q_{obs}}{Q_{max}}\right) L_t \tag{4}$$

Noncooperative binding of protein nonspecifically to an infinite lattice is described by the McGhee–von Hippel model[39] where

$$L_f = \frac{\nu}{K(1-n\nu)\left(\frac{1-n\nu}{1-(n-1)\nu}\right)^{n-1}} \tag{5}$$

K is the binding affinity, n is the binding site size (in nucleotide bases), and ν is the binding density given by

$$\nu = L_b/D_t \tag{6}$$

D_t is the total DNA concentration. Equation (5) cannot be arranged to give L_f (and therefore Q_{obs}) as a function of L_t and D_t, so Q_{obs} must be obtained using numerical methods for given values of n, K, Q_{max}. We have found that the Loehle adaptive

---

[38] W. Bujalowski and T. Lohman, *Biochemistry* **26**, 3099 (1987).
[39] J. McGhee and P. von Hippel, *J. Mol. Biol.* **86**, 469 (1974).

grid refinement Global Optimization routine (Loehle Enterprises, Naperville, IL) is especially robust for finding solutions to the McGhee–von Hippel equation. Nonlinear regression with this routine is used to obtain values for n, K, and Q_{max} given a data set of Q_{obs} values for a series of L_t and D_t concentrations.

Errors in the binding parameters derived from nonlinear regression are obtained using the Monte Carlo method for confidence interval estimation.[40–42] Random data sets (e.g., 50) are generated using a Gaussian random number generator[43] with the mean equal to the "true" Q_{obs} obtained from the optimized values for the parameters, and the standard deviation equal to the standard deviation of the fit. The distribution of the binding parameters determined from fitting the simulated data sets provides the confidence limits. In general, the Monte Carlo errors are much less than the experimental errors, i.e., ± 10, ± 5, and $\pm 0.3\%$ for K, n, and Q_{obs}.

The DNA-binding parameters for native and recombinant Sac7d are identical within experimental error.[5,16] The dissociation constant for duplex poly[d(GC)] is $9.3 \times 10^{-7} M$ in 10 mM KH$_2$PO$_4$ (pH 6.8) and 50 mM KCl at 25°. Double-stranded poly[d(AT)] binds slightly more weakly with a dissociation constant of $5 \times 10^{-6} M$ under the same conditions. Most other DNA sequences, including natural random sequence DNAs, are intermediate with dissociation constants of about $1.5 \times 10^{-6} M$ in 50 mM KCl. The binding affinity is salt dependent and decreases by about two orders of magnitude upon increasing the salt from 10 mM to 100 mM.[16] Native and recombinant Sso7d bind about 2-3-fold weaker than Sac7d (Shriver, 1999, unpublished data). The binding site sizes of Sac7d and Sso7d are rather small at about four base pairs per protein. There is no evidence of cooperativity in DNA binding from the fluorescence titration data.

Cooperative Structural Transitions Induced in DNA

Forward titrations of protein into DNA can be followed by CD since the spectrum of protein–DNA mixtures is dominated by contributions from the DNA at wavelengths greater than 250 nm (Fig. 7). Spectra are measured at 20° from 200 to 320 nm in 1-cm path length cuvettes starting with an initial DNA solution with an absorbance at 260 nm of 0.5 in 0.01 M K$_2$HPO$_4$ (pH 7). Aliquots of a concentrated protein solution are added gravimetrically, and the spectra are scaled by the $\Delta\varepsilon$ of the DNA. As expected, there is an increase in the negative short wavelength CD band at 220 nm due to increasing contributions of protein. However, the increase in the positive CD band at 285 nm is much greater than expected from

---

[40] U. Kamath and J. Shriver, *J. Biol. Chem.* **264**, 5586 (1989).
[41] W. Press, B. Flannery, S. Teukolsky, and W. Vetterling "Numerical Recipes: The Art of Scientific Computing (Fortran Version)." Cambridge University Press, Cambridge, U.K. 1989.
[42] M. Straume and M. Johnson, *Methods Enzymol.* **210**, 117 (1992).
[43] A. Miller, "Turbo BASIC Programs for Scientists and Engineers" SYBEX, San Francisco, 1987.

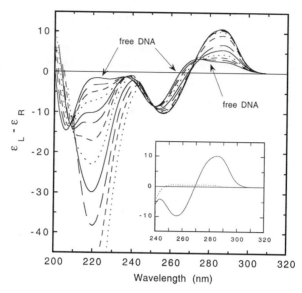

FIG. 7. The CD spectrum of duplex poly[d(GC)] with increasing amounts of Sac7d in 0.01 M KH$_2$PO$_4$ (pH 7.0, 25°). The initial concentration of DNA was 85 μM. Spectra are shown for protein/DNA ratios of 0.0, 0.01, 0.02, 0.03, 0.05, 0.07, 0.09, 0.12, 0.15, 0.19, 0.25, and 0.32 and are scaled to the Δε of the DNA. Inset: Comparison of the CD in the near-UV of free protein (dashed line) relative to the CD of the complex (solid line) at the highest protein/DNA ratio. Reproduced with permission from J. G. McAfee, S. Edmondson, I. Zegar, and J. W. Shriver, *Biochemistry* **35**, 4034 (1996).

addition of the protein alone (see insert in Fig. 7) and is attributed to changes in the DNA structure. The long wavelength CD band of duplex DNA is known to be very sensitive to DNA conformation,[44,45] and the CD changes measured upon binding Sac7d probably reflect unwinding and kinking of the DNA as observed in the X-ray crystal structure (Fig. 3).

The change in the near-UV CD as a function of the fraction of the DNA complexed with protein is sigmoidal (Fig. 8), indicating a cooperative conformational change in the DNA, even though Sac7d binding is noncooperative. Such behavior is a unique feature of Sac7d–DNA interactions and has not been observed in any other DNA-binding protein. The CD titration data can be fit using the McGhee–von Hippel model, assuming that the structural transition in DNA requires the binding of two protein monomers within a minimal distance. The maximum spacing (G_m, in base pairs) between bound proteins that can induce the conformational change

---

[44] W. A. Baase and W. C. Johnson, *J. Nucleic Acid Res.* **6**, 797 (1979).
[45] B. B. Johnson, K. S. Dahl, I. J. Tinoco, V. I. Ivanov, and V. B. Zhurkin, *Biochem.* **20**, 73 (1981).

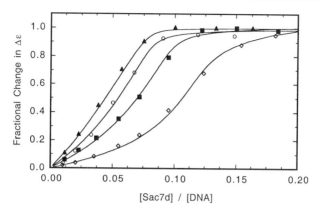

FIG. 8. The fractional change in the near-UV CD of DNA induced by binding Sac7d in 0.01 M KH$_2$PO$_4$ (pH 7.0, 25°). Results are shown for four DNA sequences: poly[d(AG)]-poly[d(CT)] (▲), poly[d(AC)]-poly[d(GT)] (○), poly[d(GC)]-poly[d(GC)] (■), and poly[d(IC)]-poly[d(IC)] (◇). The solid lines are simulations using the McGhee–von Hippel model with minimal separation required to induce a CD change of 8, 5, 3, and 0 base pairs. Reproduced with permission from J. G. McAfee, S. Edmondson, J. Zegar, and J. W. Shriver, *Biochemistry* **35**, 4034 (1996).

depends on the DNA sequence. Sac7d molecules must be bound within 8 base pairs ($G_m = 8$) to affect the structure of poly[d(AG)] · poly[d(CT)], while they must be immediately adjacent on poly[d(IC)] · poly[d(IC)] ($G_m = 0$). The susceptibility of different DNAs to the protein-induced conformational change reflects sequence dependent conformational preferences of DNA.

Stabilization of DNA Duplex

Thermal denaturation studies of DNA and DNA–protein complexes are performed on a UV spectrophotometer equipped with temperature controlled cell holders. Melting of the DNA duplex is indicated by a cooperative hyperchromism at 260 nm. Stabilization by Sac7d is most readily observed with poly[d(AT)]·poly[d(AT)] because of its intrinsic low stability, especially in low salt. The UV melting curve of poly[d(AT)] · poly[d(AT)] in 0.01 M K$_2$HPO$_4$ is sharp with a T_m of 43.5° (Fig. 9). In the presence of Sac7d, the melting profile of poly[d(AT)]·poly[d(AT)] broadens and the T_m increases by as much as 33°C for solutions with an excess of Sac7d. The observed T_m for the complex depends on the concentration of protein.

RNase Activity Assay

Fusi et al.[11,12] and Kulms et al.[25] have shown that the RNase activity associated with P2 (Sso7d) and SaRD (Sac7d) can be measured by two methods: either

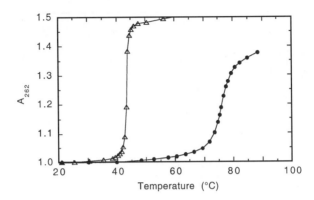

FIG. 9. Thermal denaturation of poly[d(AT)]-poly[d(AT)] (70 μM bases) alone (△) and saturated with Sac7d (350 μM) (●) monitored by the change in UV absorbance at 262 nm in 0.01 M KH$_2$PO$_4$ (pH 7). Reproduced with permission from J. McAfee, S. Edmondson, P. Datta, J. Shriver, and R. Gupta, *Biochemistry* **34**, 10063 (1995).

hydrolysis of yeast RNA measured by release of acid-soluble RNA fragments, or hydrolysis of tRNA or rRNA monitored by gel electrophoresis. We describe the former since the measurement of a specific activity permits a quantitative analysis. The method is derived from that of Anfinsen[46] and is based on measurement of the absorbance at 260 nm that remains following acid precipitation of polymeric RNA and protein. The measured activity is extremely low, and the background level of soluble RNA is typically high because of soluble ribonucleotides in commercial RNA samples, spontaneous hydrolysis of substrate at 60°, and contaminating RNases.

Purified Sac7 and Sso7 are chromatographed on Sephacryl S-100-HR (30 cm × 2.5 cm) in 0.02 M Tris (pH 7.4) to distinguish low levels of activity from the background. Column fractions are collected in glass tubes that have been heated to over 140° for 12 hr to eliminate contaminating RNase. Activities of column fractions are assayed quantitatively using essentially the procedure described by Fusi *et al.*[11] and Kulms *et al.*,[25] except for addition of a disulfide reducing agent. Neither Sac7 nor Sso7 contain cysteine, and therefore any activity due to these proteins should be unaffected by reducing agents. 2-Mercaptoethanol is chosen since dithiothreitol (DTT) and dithioerythritol (DTE) interfere with the absorbance measurement at 260 nm for RNA. Aliquots (200 μl) of each column fraction are preincubated at 60° in polypropylene microcentrifuge tubes for 5 min in 0.02 M Tris (pH 7.4), with or without 0.1 M 2-mercaptoethanol. RNA (140 μg of yeast RNA, Sigma, Type III) is added and the sample incubated at 60° for 30 min. The reaction is quenched with an equal volume of 1.2 M perchloric acid with 22 mM

---

[46] C. B. Anfinsen, R. R. Redfield, W. L. Choate, J. Page, and W. Carroll, **207**, 201 (1954).

lanthanum nitrate and left on ice for 30 min. Acid-induced precipitate is removed by centrifugation at 13,000 rpm for 10 min at 4°. A portion of the supernatant (200 µl) is diluted with 1.0 ml deionized water and the absorbance at 260 nm measured. A unit of activity produces an increase of 1.0 absorbance unit per minute.

Sso7d elutes from the Sephacryl column with a molecular weight consistent with a monomer. The RNase activity elutes slightly before the major protein peak. The specific activity associated with the peak is on the order of 1 to 10 units/mg Sso7d. The activity is low (namely, the activity of bovine RNase A is on the order of 5000 units/mg) and is largely eliminated with mercaptoethanol. The level of apparent activity remaining (0.15 ± 0.15 units/mg) is identical to that observed in the absence of RNA (0.15 ± 0.4 units/mg) and is attributed to the presence of acid soluble protein in the assay that contributes to the absorbance at 260 nm (i.e., the perchloric acid does not precipitate Sac7 and Sso7). All activities are reported relative to the background "activity" measured before and after the protein peak (i.e., in the absence of protein). These observations have been confirmed with recombinant Sso7d, and similar results are obtained with native and recombinant Sac7d.

The ability to largely eliminate the RNase activity associated with these proteins is consistent with a report that RNase A is a contaminant in Sso7d preparations.[47] The apparent activity that remains is due to the acid solubility of the proteins, which is expected given that acid solubility is used as an initial step in purifying these proteins.

Concluding Remarks

Sac7d and Sso7d are small monomeric chromatin proteins that bind nonspecifically to DNA with moderate affinity and stabilize the double helix. Binding occurs at a site of four base pairs and is associated with significant unwinding and bending of DNA due to intercalation of a methionine and valine. Although binding is noncooperative, the distortions induced in DNA require cooperative interactions between adjacent protein monomers.

Acknowledgment

This work was supported by the National Institutes of Health (GM 49686).

---

[47] U. C. Oppermann, S. Knapp, V. Bonetto, R. Ladenstein, and H. Jornvall, *FEBS Lett.* **432**, 141 (1998).

[12] Reverse Gyrases from Bacteria and Archaea

By ANNE-CÉCILE DÉCLAIS, CLAIRE BOUTHIER DE LA TOUR, and MICHEL DUGUET

Introduction

Among the enzymes that are responsible for shaping DNA, reverse gyrases are particularly interesting because of their structure and the unique reaction they catalyze: the formation of positive supercoils in circular DNA.[1,2] Reverse gyrases are classified as type I topoisomerases because they change the DNA linking number by increments of one.[3,4a] However, they are a unique form of ATP-dependent topoisomerase I, made of two domains, each representing roughly half of the sequence.[5] The C-terminal domain is related to bacterial topoisomerases I (type I-5' topoisomerases[4b]), whereas the N-terminal half of the enzyme, which contains the ATP binding site, has no similarity to any topoisomerase. It does contain motifs shared by RNA/DNA helicases,[5] although no such activity has been demonstrated so far. This complex structure provides some clues to understanding the mechanism of positive supercoiling, which is still not clear.[5,2,6]

Reverse gyrase was first discovered in *Sulfolobus acidocaldarius*,[1] a sulfur-metabolizing archaeon that grows at 75°. Remarkably, this peculiar enzyme appears widely distributed in both the archaeal[7,8] and bacterial[9-11] kingdoms, but seems restricted to hyperthermophilic strains. Its presence in eukaryotes has been postulated, although without strong evidence so far.[12,13] Table I shows that putative reverse gyrase genes are present in all the recently sequenced genomes of

---

[1] A. Kikuchi and K. Asai, *Nature* **309**, 677 (1984).
[2] M. Duguet, *in* "Nucleic Acids and Moleular Biology" (F. Eckstein and D. M. J. Lilley, eds.), p. 84. Springer-Verlag, Berlin, 1995.
[3] P. Forterre, G. Mirambeau, C. Jaxel, M. Nadal, and M. Duguet, *EMBO I.* **4**, 2123 (1985).
[4a] S. Nakasu and A. Kikuchi, *EMBO J.* **4**, 2705 (1985).
[4b] J. Roca, *Trends Biochem. Sci.* **20**, 156 (1995).
[5] F. Confalonieri, C. Elie, M. Nadal, C. Bouthier de La Tour, P. Forterre, and M. Duguet, *Proc. Natl. Acad. Sci. U.S.A.* **90**, 4753 (1993).
[6] C. Jaxel, C. Bouthier de la Tour, M. Duguet, and M. Nadal, *Nucleic Acids Res.* **24**, 4668 (1996).
[7] C. Bouthier de la Tour, C. Portemer, M. Nadal, K. O. Stetter, P. Forterre, and M. Duguet, *J. Bacteriol.* **172**, 6803 (1990).
[8] A. I. Slesarev, *Eur. J. Biochem.* **173**, 395 (1988).
[9] C. Bouthier de la Tour, C. Portemer, R. Huber, P. Forterre, and M. Duguet, *J. Bacteriol.* **173**, 3921 (1991).
[10] C. Bouthier de La Tour, C. Portemer, H. Kaltoum, and M. Duguet, *J. Bacteriol.* **180**, 274 (1998).
[11] L. Andera, K. Mikulik, and N. D. Savelyeva, *FEMS Microbiol. Lett.* **110**, 107 (1993).
[12] H. S. Koo, K. Lau, H. Y. Wu, and L. F. Liu, *Nucleic Acid Res.* **20**, 5067 (1992).

TABLE I
DISTRIBUTION OF REVERSE GYRASES IN PROKARYOTES

Organisms	Type[e]	Optimal growth temperatures (°C)	Refs.[f]
Archaebacteria			
Sulfolobales			
Sulfolobus acidocaldarius[a,c,d]	Cre	80	1, 2, 3
Sulfolobus shibatae B12[a,c,d]	Cre	85	3, 3a, 4
Acidianus infernus[a]	Cre	88	4
Thermoproteales			
Thermoproteus neutrophilus[a]	Cre	88	4
Pyrobaculum islandicum[a]	Cre	100	4
Staphilothermus marinus[a]	Cre	92	4
Pyrodictiales			
Pyrodictium occultum[a]	Cre	105	4
Desulfurococcus saccharovorans[a]	Cre	92	4
Desulfurococcus amylolyticus[a,d]	Cre	92	4, 5
Thermococcales			
Thermococcus celer[a]	Eur	88	4
Pyrococcus furiosus[b]	Eur	100	4, 6
Pyrococcus horikoshii OT3[b]	Eur	100	14
Archeoglobales			
Archeoglobus fulgidus[a,b]	Eur	80	4, 7
Methanopyrales			
Methanopyrus kandleri[a,c,d]	Eur	100	8, 9
Methanobacteriales			
Methanothermus fervidus[a]	Eur	85	4
Methanococcales			
Methanococcus janaschii[b]	Eur	85	10
Eubacteria			
Thermotogales			
Thermotoga maritima[a,c,d]		80	4, 11
Thermotoga thermarum[a]		70	4
Thermosipho africanus[a]		75	4
Fervidobacterium islandicum[a]		65	4
Aquificales			
Aquifex aeolicus[b]		90	12
Calderobacterium hydrogenophilum[d]		75	13

[a] Presence deduced from activity.
[b] Presence deduced from genome sequencing.
[c] Gene has been cloned and sequenced.
[d] Enzyme has been purified to homogeneity.
[e] Cre, Crenarchaeota; Eur, euryarchaeota.
[f] Key to References: (1) M. Nadal, C. Jaxel, C. Portemer, P. Forterre, G. Mirambeau, and M. Duguet, Biochemistry 27, 9102 (1998); (2) F. Confalonieri, C. Elie, M. Nadal, C. Bouthier de la Tour, P. Forterre, and M. Duguet, Proc. Natl. Acad. Sci. U.S.A. 90, 4753 (1993); (3) C. Jaxel, C. Bouthier de la Tour, M. Duguet, and M. Nadal, Nucleic Acids Res. 24, 4668 (1996); (3a) M. Nadal, E. Couderc, M. Duguet, and C. Jaxel, J. Biol. Chem. 269, 5255 (1994); (4) C. Bouthier de la Tour, C. Portemer, M. Nadal, K. O. Stetter, P. Forterre, and M. Duguet, J. Bacteriol. 172, 6803 (1990); (5) A. I. Slesarev, Eur. J. Biochem. 173, 395 (1988); (6) K. M. Borges, A. Bergerat, M. Bogert, J. DiRuggiero, P. Forterre, and F. Robb, J. Bacteriol. 179, 1721 (1997); (7) H. P. Klenk, R. A. Clayton, J. F. Tomb, O. When, E. K. Nelson, et al., Nature 390, 364 (1997); (8) S. A. Kozyavkin, R. Krah, M. Gellet, K. O. Stelter, J. A. Lake, and A. I. Slesarev, J. Biol. Chem. 269, 11081 (1994); (9) R. Krah, A. Kozyavkim, I. A. Slesarev, and M. Gellert, Proc. Natl. Acad. Sci. U.S.A. 93, 106 (1996); (10) C. J. Bult, O. White, J. G. Olsen, L. Zhou, D. R. Fleischmann, et al. Science 273, 1058 (1996); (11) C. Bouthier de la Tour, C. Portemer, H. Kaltoum, and M. Duguet, J. Bacteriol. 180, 274 (1998); (12) G. Deckert, P. V. Warren, T. Gaasterland, G. W. Young, A. L. Lenox, D. E. Graham, et al., Nature 392, 353 (1998); (13) L. Andera, K. Mikulik, and N. D. Savelyeva, FEMS Microbiol. Lett. 110, 107 (1993); (14) Y. Kawarabayasi, M. Sawada, et al. DNA Res. 5, 55 (1998).

hyperthermophiles, sometimes as two copies.[14] Furthermore, reverse gyrase activity is detectable in extracts of all hyperthermophilic species so far tested, provided that it is not masked by another prominent topoisomerase.[9] This distribution makes the enzyme a good marker of thermophily.

The biological function of reverse gyrases is still a matter of conjecture. However, since reverse gyrases seem essential in extremely thermophilic organisms, it was postulated that these enzymes participate in the stabilization of the DNA double helix.[15,16]

This chapter describes the purification of reverse gyrases from their natural hosts and from recombinant *Escherichia coli* strains, in addition to the various assays developed to test their biochemical properties. Finally, the main structural features of these enzymes and their putative mechanism are discussed, together with their possible *in vivo* functions.

Purification Protocols

Reverse gyrases have been purified from a number of organisms from both the Bacteria and Archaea Domains. More recently, recombinant versions have been purified from *E. coli*.

Purification from Hyperthermophilic Strains

To date, two bacterial[10,11] and four archaeal[17-20] reverse gyrases have been purified to homogeneity, essentially by using the protocol described for the *S. acidocaldarius* enzyme.[17] Frozen cells (40 g) are disrupted by thawing in 240 ml of buffer A [50 mM NaH$_2$PO4/Na$_2$HPO4, pH 7, 0.1 mM dithiothreitol (DTT), 1 mM EDTA] containing 1.2 M NH$_4$Cl, 1 mM EGTA, 1 mM sodium bisulfite, 1 mM Phenylmethylsulfonyl fluoride (PMSF), 1 mM leupeptin, and 1 mM pepstatin A. Cell lysis is achieved by homogenization with an Ultraturax (PT 18/2) apparatus. The resulting solution is centrifuged at 2400g for 10 min at 4°. The pellet is resuspended in the same buffer (100 ml), homogenized, and centrifuged as

---

[13] S. Gangloff, J. P. McDonald, C. Bendixen, L. Arthur, and R. Rothstein, *Mol. Cell. Biol.* **14,** 8391 (1994).

[14] G. Deckert, P. V. Warren, T. Gasterland, *et al., Nature* **392,** 353 (1998).

[15] A. Kikuchi, T. Shibata, and S. Kakasu, *Syst. Appl. Microbiol.* **20,** 5067 (1986).

[16] A. Kikuchi, in "DNA Topology and Its Biological Effects" (N. Cozzarelli and J. Wang, eds.), p. 285. Cold Spring Harbor Laboratory Press, Cold Spring Harbor, NY, (1990).

[17] M. Nadal, C. Jaxel, C. Portemer, P. Forterre, G. Mirambeau, and M. Duguet, *Biochemistry* **27,** 9102 (1988).

[18] K. M. Borges, A. Bergerat, A. M. Bogert, J. DiRuggiero, P. Forterre, and F. T. Robb, *J. Bacteriol.* **179,** 1721 (1997).

[19] S. A. Kosyavkin, R. Krah, M. Gellert, K. O. Stetter, J. A. Lake, and A. I. Slesarev, *J. Biol. Chem.* **169,** 11081 (1994).

[20] M. Nadal, E. Couderc, M. Duguet, and C. Jaxel, *J. Biol. Chem.* **269,** 5255 (1994).

indicated for lysis. The two supernatants are pooled (fraction I, 340 ml, 4360 mg of proteins).

Polymin P is added to fraction I to a final concentration of 0.36%. After gentle mixing during 15 min at 4°, the solution is centrifuged at 2400g for 30 min. The supernatant is further clarified by ultracentrifugation at 90,000g for 1 hr at 4° (fraction II, 405 ml, 3850 mg of proteins).

Ammonium sulfate is added to fraction II to a final concentration of 70% saturation, and the fraction is centrifuged at 24,000g for 20 min at 4°. The supernatant is diluted with 2.1 M NaCl in buffer A to give a final concentration of 0.8 M ammonium sulfate and 1.2 M NaCl. This solution is again clarified by centrifugation at 24,000g for 20 min and saved as fraction III at 4° (270 ml, 1040 mg of proteins).

Fraction III is loaded onto a phenyl-Sepharose column (2.6 × 18.5 cm) equilibrated with 0.8 M ammonium sulfate and 1.2 M NaCl in buffer A. The column is first washed with 1 liter of the same buffer at a flow rate of 40 ml/hr and then with 600 ml of 0.25 M NaCl in buffer A. It is further washed with 1 liter of 30% ethylene glycol and 0.25 M NaCl in buffer A and developed with a linear gradient of 30–60% (w/v) ethylene glycol (2 × 500 ml) in buffer A containing 0.25 M NaCl. Finally, the phenyl-Sepharose column is washed with 1% Triton X-100 in buffer A (450 ml). Active fractions are pooled, dialyzed against buffer A containing 10% ethylene glycol (buffer B), and referred to as fraction IV (480 ml, 86.5 mg of proteins).

Fraction IV is loaded onto a phosphocellulose column (2.6 × 15 cm) equilibrated with buffer B, and washed with 1 liter of the same buffer at a flow rate of 40 ml/hr. The bound proteins are eluted with 0.8 M NaCl in buffer B. Active fractions are pooled and dialyzed against buffer B containing 0.2 M NaCl (fraction V, 85 ml, 9.1 mg of proteins).

Fraction V is applied onto a heparin-Sepharose column (1.6 × 5 cm) equilibrated with 0.2 M NaCl in buffer B at a flow rate of 10 ml/hr. The column is washed with 50 ml of the same buffer and developed with a 0.2–1 M NaCl (2 × 60 ml) linear gradient. Active fractions are pooled and saved as fraction VI (31 ml, 0.8 mg of proteins).

Fraction VI is concentrated and equilibrated with buffer A containing 0.6 M NaCl on an Amicon (Danvers, MA) microconcentrator (Centricon 30). This fraction is loaded on a 5–20% sucrose gradient in buffer A containing 0.6 M NaCl and centrifuged at 175,000g for 40 hr at 4° in a SW 41 Beckman rotor. Fractions of 350 µl are collected. Active fractions are pooled, concentrated, and finally equilibrated with 25 mM NaH$_2$PO4/Na$_2$HPO4, pH 7, 0.5 mM DTT, 0.5 mM EDTA, and 100 mM NaCl (fraction VII, 0.36 ml, 0.08 mg of proteins).

The purification scheme is summarized in Table II. In the case of *Methanopyrus kandleri* reverse gyrase, the phosphocellulose step is carried out before the phenyl-Sepharose chromatography. The heparin-Sepharose step is followed with a Mono Q column and achieved with a gel filtration (Superdex 200).[19] For *Thermotoga maritima* reverse gyrase, the purification scheme includes, in order, phosphocellulose,

TABLE II
PURIFICATION OF REVERSE GYRASE FROM S. acidocaldarius[a]

Fraction	Step	Volume (ml)	Total protein (mg)	Total activity ($\times 10^{-3}$ units)	Specific activity ($\times 10^{-3}$ units/mg)
I	Crude extract	340	4360	nd[d]	nd
II	Polymin P	405	3850	nd	nd
III	Ammonium sulfate	270	1040	13000–18000	12.5–17.3
IV	Phenyl-Sepharose	480	86.5	12200	140
V	Phosphocellulose	85	9.1	4250	460
VI	Heparin-Sepharose	31	0.8[b]	4030	5000
VII	Sucrose gradient	0.36	0.08[c]	700	8750

[a] One unit of enzyme is defined as the amount of protein required to release 50% of the negatively-supercoiled pBR322 input under the standard assay conditions. From M. Nadal, C. Jaxel, C. Portemer, P. Forterre, G. Mirambeau, and M. Duguet, *Biochemistry* **27**, 9102 (1998).
[b] Amount of proteins estimated after silver staining of polyacrylamide gel.
[c] Amount of proteins determined by amino acid compositions.
[d] nd, Not determined.

single-stranded DNA agarose, phenyl-Sepharose, heparin-Sepharose chromatographies and a sucrose gradient ultracentrifugation.[10] The additional single-stranded DNA agarose chromatography is introduced to remove the ATP-independent topoisomerase activity present in the extracts.[10]

Expression and Purification from Recombinant E. coli

The entire coding sequence for reverse gyrase from *S. acidocaldarius* has been expressed in *E. coli* as a glutathione *S*-transferase (GST) fusion protein[20b], with the pGEX expression system described by Smith and Johnson.[21] Cells (DH5α) containing the construction are grown to an A_{600} of 0.8 and induced by isopropylthiogalactoside (IPTG) addition. After lysis, the fusion proteins are purified on glutathione agarose, and the GST tag is removed by thrombin cleavage. This procedure gives low levels of expression and a number of fragments produced by proteolysis or incomplete translation. Further purification of the recombinant enzyme is obtained by an additional step on phenyl-Sepharose, but the final yield is very low.

So far, the most successful expression of recombinant reverse gyrase has been from *M. kandleri,* a unique type of reverse gyrase, made of two subunits.[19] These

[20b] A. C. Déclais, J. Marsault, F. Confalonieri, C. Bouthier de La Tour, and M. Duguet, *J. Biol. Chem.* **275**, 19498 (2000).
[21] D. B. Smith and K. S. Johnson, *Gene* **67**, 31 (1988).

can be separately overexpressed in *E. coli* with good yields, and their mixing reconstitutes an active reverse gyrase.[22]

Physicochemical Properties

For an extensive discussion of the physical properties of reverse gyrase, see Nadal *et al.*[17] Independent of the species from which the enzyme was isolated, bacterial or archaeal reverse gyrases have similar physical properties. They are composed of a single polypeptide of 115–135 kDa, as determined by SDS–PAGE. These values are somewhat low in comparison with the data deduced from the sequence (for instance, 128 kDa instead of 143 kDa for *S. acidocaldarius* reverse gyrase[5]), possibly because of the hydrophobicity of these proteins. Hydrodynamic and electron microscopy measurements suggest a monomeric structure in native conditions.[17] A noticeable and to date unique exception is the reverse gyrase from *Methanopyrus kandleri*, which is made of two subunits, 138 kDa and 43 kDa, encoded by two different genes.[22,23]

Assays and Biochemical Properties

Reverse gyrases, and other topoisomerases, are not typical enzymes, for several reasons: (i) the substrate itself, the double helix of DNA, is a macromolecule and has limited diffusion and mobility; (ii) as the reaction proceeds, the structure of the substrate, and consequently its affinity for the enzyme, changes (see below); (iii) reverse gyrases not only catalyze chemical reactions, such as transesterifications (Fig. 1) and ATP hydrolysis, but also mechanical actions, such as DNA unwinding and DNA strand passage.

ATP-Dependent Linking Number Increase

In circular DNA, or generally in topologically constrained DNA, the number of topological links between the two strands, Lk, is related to the geometrical parameters Tw (twist) and Wr (writhe) by the equation[24]

$$Lk = Tw + Wr \qquad (1)$$

As a consequence, any change in the twist between the two strands (Tw) is compensated for a change in the supercoiling state of the molecule (Wr) and vice versa. The role of DNA topoisomerases in the cell is precisely to change the value of Lk, an operation that absolutely requires a transient strand cleavage (Fig. 1).

---

[22] R. Krah, M. H. O'Dea, and M. Gellert, *J. Biol. Chem.* **272,** 13986 (1997).
[23] R. Krah, A. Kozyavkim, I. A. Slesarev, and M. Gellert, *Proc. Natl. Acad. Sci. U.S.A.* **93,** 106 (1996).
[24] W. R. Bauer, F. H. C. Crick, and J. H. White, *Sci. Amer.* **243,** 100 (1980).

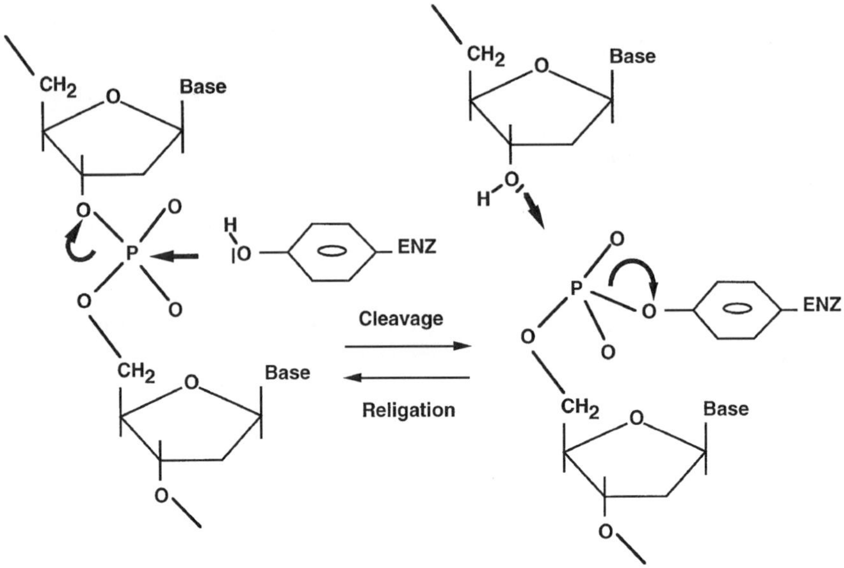

FIG. 1. Chemical reactions catalyzed by DNA topoisomerases. An initial transesterification involves the attack of the nucleotide phosphorus by the tyrosine of the enzyme active site. This reaction produces cleavage of the phosphodiester bond and formation of a covalent link between the tyrosine and the DNA end (in the example, the 5' end). A second transesterification, the attack of the phosphorus by the free hydroxyl DNA end, reverses the reaction, resealing the phosphodiester bond and liberating the tyrosine.

In the case of reverse gyrases, the best characterized activity of these enzymes is the production of a continuous increase in the DNA linking number. As a consequence, this introduces positive supercoils ($Wr > 0$) or produces an increase in the twist (Tw) of the double helix. Starting from a negatively supercoiled substrate, an increase in Lk by reverse gyrase first relaxes DNA, then produces positive supercoils (Fig. 2). Since the enzyme has a better affinity for negative supercoils,[8] the first part of the reaction (relaxation) is very fast, whereas the second part (positive supercoiling) is less efficient and requires higher amounts of enzyme. As expected, the reaction is inhibited by single-stranded DNA which competes for reverse gyrase binding.[4,8] Finally, positively supercoiled DNA is not a substrate for the enzyme.

The standard assay for reverse gyrases is the ATP-dependent positive supercoiling of circular DNA.[3,4,17] The assay described below is applicable to all reverse gyrases, with slight modifications. For simplicity, the substrate is usually a negatively supercoiled plasmid, but it is possible to start from relaxed DNA. The reaction mixture (usually 20 µl final volume) contains 50 mM Tris-HCl, pH 8.0, at 20° (or MOPS, pH 6.6), 1 mM ATP, 0.5 to 1 mM DTT, 0.5 mM EDTA, 30 µg/ml

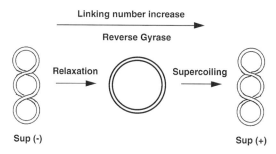

FIG. 2. Main reaction catalyzed by reverse gyrases: ATP-dependent linking number increase. Sup⁻: negatively supercoiled DNA (right-handed superhelix), equivalent to an underwinding of the double helix; Sup⁺: positively supercoiled DNA (left-handed superhelix), equivalent to a overwinding of the double helix.

bovine serum albumin, 3 to 10 mM MgCl$_2$, 30 to 200 mM NaCl (usually 120 mM), 300 ng of negatively supercoiled plasmid DNA (pTZ 18), and 1 μl of the fraction to be assayed. Spermidine (1 mM) or polyethylene glycol (7–8%) is sometimes added to increase the efficiency of the reaction.[3,4] Finally, paraffin oil (4 μl) is added on top of the reaction mixture to avoid evaporation.

After 10 to 30 min incubation at 75°C (in some cases 80 to 90°, depending on the source of reverse gyrase), the reaction is stopped by addition of 1% SDS, 10 mM EDTA, 0.01% bromphenol blue, and 15% sucrose or 10% glycerol (final concentrations). The reaction products are analyzed by two-dimensional 1% agarose gel electrophoresis.[25] The first dimension is a standard, long migration (4 hr at 3.5 V/cm) in TEP buffer (36 mM Tris, 30 mM NaH$_2$PO$_4$, 1 mM EDTA, pH 7.8), whereas the second dimension, in a perpendicular direction, contains 3 to 4 μg/ml chloroquine and is run at 0.9 V/cm for 10 hr. After chloroquine elimination and staining with ethidium bromide, followed by destaining with 1 mM MgSO$_4$, the image is recorded on a CCD camera.

Figure 3 shows the disappearance of the form I DNA substrate (Fig. 3a) and the appearance of positively supercoiled DNA that forms the right part of the arch (Fig. 3c). In processive conditions, i.e., low ionic strength (30 mM NaCl), and in the presence of higher enzyme concentrations or of condensing agents, efficient positive supercoiling (superhelical density $\sigma > +0.05$) is obtained (Fig. 3d).[3] By contrast, in the absence of ATP, an inefficient DNA relaxation occurs, but there is no positive supercoiling (Fig. 3b).

Determination of precise kinetic parameters is difficult, since the affinity for the DNA substrate changes at each cycle of topoisomerization. A global quantitation of the activity is usually made by measuring the disappearance of the input DNA band,

---

[25] L. J. Peck and J. C. Wang, *Proc. Natl. Acad. Sci. U.S.A.* **80**, 6206 (1983).

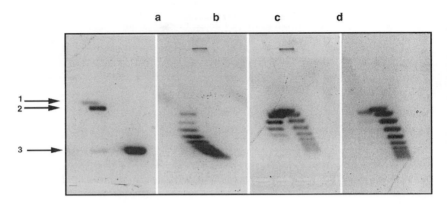

FIG. 3. Analysis of the reverse gyration by two-dimensional gel electrophoresis. (a) Negatively supercoiled pTZ18 control; (b) incubation with reverse gyrase from *S. acidocaldarius* (4 ng) in the absence of ATP; (c and d) incubations in the presence of ATP with 1 and 4 ng reverse gyrase, respectively. In this analysis, negatively supercoiled topoisomers form the left branch of the arch (b and c), and positively supercoiled topoisomers the right branch (c and d). The arrows indicate respectively Form II (1), Form I dimer (2), and Form I monomer (3).

independent of the extent of positive supercoiling obtained. One unit of enzyme is defined as the amount of protein required to relax 50% of form I substrate in the conditions of the assay.[17] Another difficulty comes from the temperature difference between the incubation mixture (75°) and the electrophoresis (25°), which tends to reduce the apparent positive supercoiling in the gels. With a 3 kbp plasmid, the change in the twist is about 3 to 4 turns.[26] Thus, the DNA relaxed at 75° appears with 3–4 negative supercoils in the agarose gel at 25°, such that the actual extent of positive supercoiling is higher than estimated from the gel.

ATPase Activity

An extensive analysis of the ATPase activity of reverse gyrase has been performed by Shibata *et al.* on the enzyme from *Sulfolobus*.[27] Classical assays for ATPase activity are applicable to reverse gyrases, with the same reaction mixture as that used for the assay of positive supercoiling and an incubation of 10 min at 75°. Quantitation of the product (ADP) and substrate (ATP) is obtained by high-performance liquid chromatography (HPLC) or by thin-layer chromatography (TLC).

As ATP-dependent topoisomerases, reverse gyrases hydrolyze ATP to ADP, but not the three other ribonucleoside triphosphates[27]; this is a property consistent

---

[26] M. Duguet, *Nucleic Acids Res.* **21**, 463 (1993).
[27] T. Shibata, S. Nakasu, K. Yasui, and A. Kikuchi, *J. Biol. Chem.* **262**, 10419 (1987).

with their inability to drive the positive supercoiling reaction. The ATP hydrolysis is strictly dependent on DNA, with a preference for single-stranded over double-stranded DNA consistent with a higher affinity of the enzyme for single-stranded regions. This property, and the unusually low ATP concentration required for positive supercoiling (1–10 μM), are atypical of a topoisomerase and reminiscent of ATP hydrolysis by helicases.[28] This is supported by the observation that the putative ATP binding site resides in the helicase-like domain of reverse gyrases,[5] although no helicase activity has been demonstrated. Furthermore, Krah et al.[22] showed that in the case of the two subunit reverse gyrase, the large subunit (RgyB) is responsible for the ATPase activity.

DNA Cleavage Specificity

Reverse gyrases are classified as type I topoisomerases because they perform transient single-strand breaks in a DNA duplex. In order to analyze the sequence specificity of a topoisomerase, one has to stabilize the cleaved intermediate, which is usually achieved by using drugs or "killer substrates." However, no such drug has been described so far for reverse gyrases, and killer substrates are usually short DNA duplexes, which would not be stable enough at the high temperature required for these enzymes. Nevertheless, a cleavage reaction can be observed with stoichiometric amounts of reverse gyrase incubated at 75–85° with DNA.[29] Analysis of the cleavage products revealed that reverse gyrase remains bound to the 5′ DNA end. This cleavage is enhanced in the absence of ATP or in the presence of nonhydrolyzable analogs (ATPγS or ADPNP), presumably because reverse gyrase remains bound to DNA for a longer time under these conditions.[29]

The cleavage reaction is performed in a mixture nearly identical to that used for positive supercoiling, except that ATP is omitted and the substrate is a linear DNA labeled either at the 3′ or 5′ end on only one strand.[29] A molar ratio of about 20 protein molecules per DNA fragment is used, and the mixture is incubated for 5 min at 75°. The reaction is stopped by addition of SDS to 1% and proteinase K to 400 μg/ml. After incubation for 30 min at 60°C, the samples are extracted twice with chloroform/isoamyl alcohol and precipitated by ethanol. The pellets are resuspended, denatured by heating at 85°, and analyzed by electrophoresis in agarose or polyacrylamide gels. Details of the cleavage protocoles are given in Jaxel et al.[30]

Analysis of 31 cleavage sites of reverse gyrase from *Desulfurococcus amylolyticus* on pBR322 has not revealed a precise consensus, but a strong preference for a C in position −4 from the cleavage point.[31] This preference has also been

---

[28] T. M. Lohman and K. P. Bjornson, *Annu. Rev. Biochem.* **65,** 169 (1996).
[29] C. Jaxel, M. Nadal, G. Mirambeau, P. Forterre, M. Takahashi, and M. Duguet, *EMBO J.* **8,** 3135 (1989).
[30] C. Jaxel, M. Duguet, and M. Nadal, *Eur. J. Biochem.* **260,** 103 (1999).
[31] O. I. Kovalsky, S. A. Kozyavkin, and A. I. Slesarev, *Nucleic Acids Res.* **19,** 2801 (1990).

described for *Methanopyrus kandleri*[23] and seems to be a common feature of type I-5′ topoisomerases.[32] Analysis has been done on the cleavage specificity of reverse gyrase from *Sulfolobus shibatae* toward one of its natural substrates, the genomic DNA of SSV1 virus. This DNA is highly positively supercoiled in the infected *Sulfolobus* cells, presumably to allow its packaging into the viral particles.[33] The strongest sites could be selected and their analysis again revealed no consensus sequence, except the C(-4) position. However, these sites are highly specific, since their export into a foreign sequence creates a strong reverse gyrase site.[30] Moreover, the prefered sequences are AT-rich, consistent with a more energetically favorable distortion of the helical structure of DNA by reverse gyrase binding to these regions (see below).

Noncovalent Stoichoimetric Binding

In order to monitor the effect of reverse gyrase DNA binding on the helical parameters, circular DNA containing one single-strand break per circle is incubated at 75° with stoichiometric amounts of reverse gyrase for 10 min in the absence of ATP. The details of the protocol are given in Jaxel *et al.*[29] Briefly, after this first incubation, the DNA is quickly sealed by addition of a thermophilic ligase, deproteinated by proteinase K digestion and chloroform extraction, and analyzed by two-dimensional electrophoresis on agarose. The result of this experiment is a progressive appearance of negatively supercoiled DNA as the ratio of reverse gyrase to DNA increases. This indicates that reverse gyrase binding produces a distortion of the double helix, so that its linking number is reduced after covalent closure by ligase.[29] A minimum of -0.5 turn per reverse gyrase molecule is calculated for ΔLk. The nature of the distorsion, base pair opening, unwinding, left-handed wrapping, or bending, is not presently known. Whatever its precise nature, this distorsion may provide a favorable substrate for the subsequent steps of the supercoiling mechanism. Similar results were obtained by Kikuchi,[16] who showed that ATP or ADP supressed the distortion and suggested that ATP hydrolysis is used for enzyme dissociation from DNA.

Structural and Mechanistic Features

Sequence Data

The first gene encoding a reverse gyrase, TopR, was cloned by screening a λgt11 expression library[34] of *Sulfolobus acidocaldarius* genomic DNA, with polyclonal

---

[32] Y. C. Tse, K. Kirkegaard, and J. C. Wang, *J. Biol. Chem.* **255,** 5560 (1980).
[33] M. Nadal, G. Mirambeau, P. Forterre, W. D. Reiter, and M. Duguet, *Nature* **321,** 256 (1986).
[34] T. V. Huynh, R. A. Young, and R. W. Davis, *in* "DNA Cloning" (D. M. Glover, Ed.), p. 49. IRL, Oxford (1985).

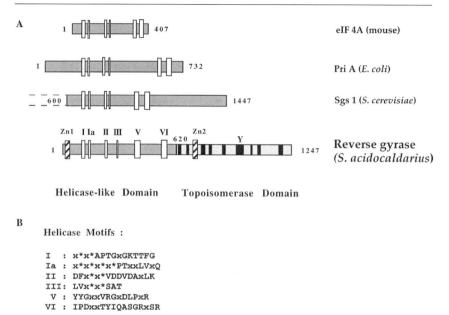

FIG. 4. Primary structure of reverse gyrases. (A) Schematic representation of reverse gyrase sequence with its two domains: comparison of the N-terminal domain with various helicases. eIF4A is an eukaryotic initiation factor for translation [J. P. Nielson and H. Trachsel, *EMBO J.* **7,** 2097 (1988)]. PriA is a component of the primosome in bacteria (P. Nurse, R. J. DiGate, K. H. Zavitz, and K. J. Marians, *Proc. Natl. Acad. Sci. U.S.A.* **87,** 4615 (1990)). Sgs1 is a helicase interacting with topoisomerases II and III in yeast [S. Gangloff, J. P. McDonald, C. Bendixen, L. Arthur, and R. Rothstein, *Mol. Cell. Biol.* **14,** 8391 (1994)]. (B). Amino acid sequences of the helicase motifs in reverse gyrases. x* indicates a hydrophobic residue. The sequences shown are a consensus where at least four positions out of six are conserved in reverse gyrases.

antibodies against purified reverse gyrase.[17,5] The sequence that was identified is represented schematically in Fig. 4. It reveals two clearly distinct domains in the coding sequence, each representing about half of the polypeptide length. The N-terminal domain contains the different characteristic motifs of DNA/RNA helicases, whereas the C-terminal domain is clearly related to the ATP-independent topoisomerases I found essentially in bacteria (type I-5′). This two-domain structure is remarkably conserved in all of the other archaeal and bacterial reverse gyrases.[10] The size of reverse gyrases is also approximately conserved, between 1100 and 1250 amino acids in a unique polypeptide. Two exceptions to this scheme were described: (i) *M. jannaschii* reverse gyrase contains an intein (494 amino acids) within the topoisomerase domain.[35] In this case, when the intein sequence is removed, the size (1119 amino acids) and the organization in two domains is

[35] C. J. Bult, O. White, *et al., Science* **273,** 1058 (1996).

conserved. (ii) *M. kandleri* reverse gyrase is made of two subunits. Surprisingly, the subunits of this enzyme do not correspond to the previously described domains. Nevertheless, it is possible to recognize in the large subunit (138 kDa) the helicase-like domain and the N-terminal part of the topoisomerase domain, whereas the small subunit contains the remaining of the topoisomerase.[22] Another peculiar feature of reverse gyrases is the presence of two putative zinc fingers. One is in the N terminus of the helicase-like domain. The other is in the topoisomerase domain, but contrary to type I-5' topoisomerase, the zinc motif is in the N-terminal region of the domain. Thus, reverse gyrases form a new, rather homogenous group within the type I-5' topoisomerase family, a feature that is confirmed by sequence comparisons.[10]

Mechanistic Models for Positive Supercoiling

The first model to explain the mechanism of reverse gyration was based on sequence data and postulated the cooperation of two enzymatic activities: helicase and topoisomerase I. The former was supposed to progress through the DNA duplex, producing two waves of supercoiling, positive ahead and negative behind, according to Liu and Wang's transcription theory.[36] The topoisomerase I activity was then supposed to specifically relax these negative supercoils, leading to a net positive supercoiling (Fig. 5A). This early model is, however, difficult to reconcile with the following arguments: (i) to produce efficient tortional stress, reverse gyrase should present a large enough momentum of inertia and must move rapidly through the double helix; (ii) the existence of sequence-specific cleavage sites suggests that, on the contrary, reverse gyrase may be static on the DNA for some time at these discrete positions; and (iii) the model does not explain the advantage for the domains to be part of the same polypeptide.

Based on the presently available data, other models that do not need far tracking along the DNA have been proposed[2,6] that suggest a mechanism close to that of the bacterial topoisomerases. In the case of these enzymes, the substrate, an unwound DNA, allows the topoisomerase to bind and provides the energy ($\Delta G°$ of supercoiling) necessary for strand passage. This explains why the reaction ends when the DNA substrate is relaxed. The first task of reverse gyrases is precisely to create the proper substrate for the activity of its topoisomerase I domain. The new models (Fig. 5B) postulate that the binding of reverse gyrases produces the partition of a relaxed circular DNA in two domains: an unwound domain, which is a good substrate for the topoisomerase I activity, and a positively supercoiled domain. Then, ATP binding would allow relaxation of the unwound domain by the topoisomerase activity. Finally, ATP hydrolysis would produce the dissociation of reverse gyrase from DNA, yielding positively supercoiled circles.

---

[36] L. F. Liu and J. C. Wang, *Proc. Natl. Acad. Sci. U.S.A.* **84**, 7024 (1987).

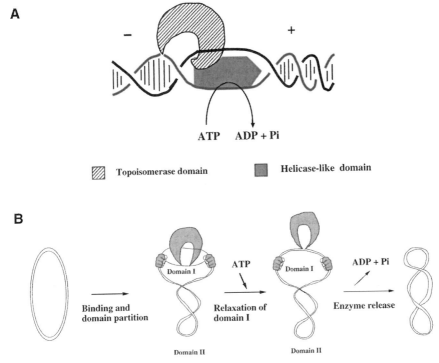

FIG. 5. Models for positive supercoiling by reverse gyrases. (A) Dynamic model based on the tracking of DNA by the helicase-like domain. (B) Static model: domains I and II refer to as DNA domains in a circular molecule.

Reverse Gyrase Functions

DNA Topology in Hyperthermophiles

The unique enzymatic activity of reverses gyrases, i.e., the efficient production of positively supercoiled DNA, was used by a number of authors to estimate the impact of such a DNA structure on various biological mechanisms. These include chromatin stability,[37] recombination,[38] and transcription.[39] However, the *in vivo* role of reverse gyrases in hyperthermophiles remains a matter of speculation. It was suggested long ago that one of the main functions of topoisomerases is to tightly

---

[37] A. Hamiche, V. Carot, M. Alilat, F. De Lucia, M. F. O'Donohue, B. Revet, and A. Prunell, *Proc. Natl. Acad. Sci. U.S.A.* **93**, 7588 (1996).
[38] H. M. Lim, H. J. Lee, C. Jaxel, and M. Nadal, *J. Biol. Chem.* **272**, 18434 (1997).
[39] S. D. Bell, C. Jaxel, M. Nadal, P. Kosa, and S. P. Jackson, *Proc. Natl. Acad. Sci. U.S.A.* **95**, 15218 (1998).

control the superhelical density (σ) of DNA.[40] In mesophilic bacteria, this function is fulfilled by a balance between DNA gyrase that produces negative supercoils and topoisomerase I that removes them, so that a σ value of about -0.05 is maintained.

In hyperthermophilic archaea, reverse gyrase is the main topoisomerase I activity. Moreover, some highly positively supercoiled DNA in the SSV1 viral particles has been discovered, in correlation with the presence of reverse gyrase in this virus's host, *Sulfolobus shibatae*.[33] However, the episomal form of SSV1 contains a large spectrum of topoisomers, from negatively supercoiled to relaxed and positively supercoiled. For technical reasons, it has not been possible to measure the superhelical state of chromosomal DNA. Analysis of the DNA topology in a series of plasmids discovered in Thermococcales and Sulfolobales shows that they exist in a relaxed to positively supercoiled state.[41] By comparison, plasmids from mesophiles (archaea or bacteria) are negatively supercoiled. The higher linking number found in archaeal thermophiles might be useful to prevent local openings of the double helix. Since reverse gyrase is the main topoisomerase I in these organisms, it was suggested that DNA topology was regulated by a balance between reverse gyrase and the type II topoisomerase (now referred to as topo VI[42]), which is also present[43] (see Fig. 6). However, these organisms also contain histone-like proteins, which play a part in the regulation of DNA supercoiling[44-46]: some of them stabilize negative supercoiling, whereas in other organisms, they stabilize positive supercoiling. Hence, no unique scheme of supercoiling regulation can be proposed for hyperthermophilic archaea.

In hyperthermophilic bacteria,[47,14] gyrase and reverse gyrase are both present,[48] but surprisingly, the plasmids they contain appear negatively supercoiled. This suggests that gyrase is the main determinant of supercoiling in these bacteria. Indeed, contrary to the case of archaea, the major topoisomerase I is not reverse gyrase but an ATP-independent topoisomerase equivalent to TOPA from *E. coli*.[49] It is possible that global regulation of supercoiling is, as in mesophilic bacteria, the result

[40] K. Drlica, *Mol. Microbiol.* **6,** 425 (1992).
[41] P. Lopez-Garcia and P. Forterre, *Mol. Microbiol.* **23,** 1267 (1997).
[42] A. Bergerat, B. de Massy, D. Gadelle, P. C. Varoutas, A. Nicolas, and P. Forterre, *Nature* **386,** 414 (1997).
[43] P. Forterre, A. Bergerat, and P. Lopez-Garcia, *FEMS Microbiol. Lett.* **18,** 237 (1996).
[44] D. R. Musgrave, K. M. Sandman, and J. N. Reeve, *Proc. Natl. Acad. Sci. U.S.A.* **88,** 10397 (1991).
[45] C. Teyssier, F. Toulme, J. P. Touzel, A. Gervais, J. C. Maurizot, and F. Culard, *Biochemistry* **35,** 7954 (1996).
[46] P. Lopez-Garcia, S. Knapp, R. Ladenstein, and P. Forterre, *Nucleic Acids Res.* **26,** 2322 (1998).
[47] R. Huber, T. A. Langworth, H. Koning, M. Thomm, C. R. Woese, U. B. Sleytr, and K. O. Stetter, *Arch. Microbiol.* **324,** (1986).
[48] O. Guipaud, E. Marguet, K. M. Noll, C. Bouthier de la Tour, and P. Forterre, *Proc. Natl. Acad. Sci. U.S.A.* **94,** 10606 (1997).
[49] H. Kaltoum, C. Portemer, F. Confalonieri, M. Duguet, and C. Bouthier de La Tour, *System. Appl. Microbiol.* **20,** 505 (1997).

A : Hyperthermophilic Archaea

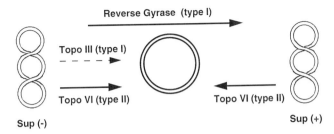

B : Hyperthermophilic Bacteria

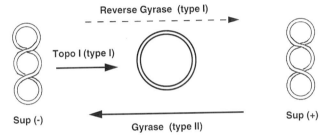

FIG. 6. DNA topology in hyperthermophiles. The dashed arrows for topo III (A) and reverse gyrase (B) indicate that this activity is presumably not a major determinant of the supercoiling density.

of a balance between gyrase and this topoisomerase I, and that reverse gyrase may play a more local role to maintain the structural parameters of the double helix (see below).

Ubiquitous Roles of Reverse Gyrases

The situation in thermophilic bacteria, where reverse gyrase is not the main determinant of DNA topology, suggests alternative roles for this enzyme.[2] For instance, local positive supercoiling would remove structural accidents in the DNA double helix, such as cruciforms, triple helices, or left-handed or unpaired regions. These structures are naturally favored at temperatures of 75–85° where the kinetics of melting and reannealing are fast, but also exist in mesophilic organisms. Efficient progression of replication forks and transcription complexes requires that these structural accidents be removed. However, results indicate that DNA topology has little influence on transcription levels, at least *in vitro*.[39]

Another local role suggested for reverse gyrase is the possibility to rewind the double helix after the passage of a transcription complex, preventing extensive strand separation. Kikuchi *et al.*[15] proposed the term *renaturase* for this potential role of reverse gyrase. This may be extended to many other mechanisms (e.g., replication, repair, and recombination) where the DNA double helix is partly or totally unwound, and it was suggested that reverse gyrases are general *repressors* of DNA metabolism.[2]

The organization of reverse gyrases in two distinct domains, a helicase-like and a topoisomerase I, suggests that these enzymes are involved in recombination, with possible equivalents in eukaryotes.[13] Indeed, the activity of reverse gyrase may repress recombination in two ways: (i) by acting on the initiation of recombination, which is unfavored by positive supercoiling[38]; and (ii) by disrupting illegitimate recombination intermediates.[50] It was proposed that reverse gyrases in thermophiles, and the association of a helicase with a topoisomerase in mesophiles, including eukaryotes, may play this role.[51]

Finally, a potential role of reverse gyrase in the remodeling of chromatin-like structures has been suggested, since a local stress on the double helix by positive supercoiling may remove these structures[51] and open chromatin for replication or transcription machineries.

Acknowledgments

We thank J. Marsault and C. Portemer for technical assistance. This work was supported by grants from Ministère de la recherche (Action Microbiologie n°99N60/0221) and CNRS (UMR 8621). Our laboratory (Enzymologie des Acides Nucléiques) is supported by the Association pour la Recherche contre le Cancer (ARC).

[50] P. Morel, J. A. Heijna, S. D. Ehrlich, and E. Cassuto, *Nucleic Acids Res.* **21**, 3205 (1993).
[51] M. Duguet, *J. Cell Sci.* **110**, 1345 (1997).

[13] DNA Gyrase from *Thermotoga maritima*

By OLIVIER GUIPAUD and PATRICK FORTERRE

Introduction

DNA gyrases are members of the Topo IIA family of type II DNA topoisomerases. They are unique among these enzymes in their ability to introduce negative supercoiling in a covalently closed circular DNA at the expense of ATP.[1] DNA gyrase is a heterotetramer composed of two subunits, GyrA and GyrB. GyrB is involved in the binding and hydrolysis of ATP, whereas GyrA is involved in

[1] J. C. Wang, *Annu. Rev. Biochem.* **65**, 635 (1996).

the transient breaking of the DNA double helix via its nicking–closing activity. During the reaction, GyrA is transiently linked to the DNA by a phosphotyrosine bond with the 5' end of the DNA break. The coordinated movement of GyrA and GyrB allows the transfer of a double helix through the break across the protein.[2] The complete DNA gyrase structure comprises an entry gate, through which the second DNA molecule enters into the protein, and an exit gate from which it leaves the complex once the transfer has occurred.[3] These characteristics are shared by all members of the Topo IIA family. What makes DNA gyrase special among these enzymes is its ability to wrap the DNA around itself to form a positive superturn.[4] The crossing of another DNA segment from the same covalently closed molecule transforms this positive superturn into a negative one (a mechanism that has been described as sign inversion[5]). Because positive wrapping has already produced a compensatory negative superturn in the covalently closed molecule, each reaction cycle introduces two negative superturns.

The main structural difference between DNA gyrase and the other Topo IIA enzymes resides in the C-terminal region of GyrA, which appears to be essential for the wrapping of the DNA around enzyme. Indeed, removal of this C-terminal region transforms DNA gyrase into a classical Topo II that can only relax either negatively or positively supercoiled DNA.[6]

The two genes encoding DNA gyrase subunits GyrA and GyrB are present in all completely sequenced bacterial genomes, including the hyperthermophilic bacterium *Thermotoga maritima* (with only one possible exception; see below). They are also found in two archaea, the halophile *Haloferax* sp. strain AA2.2 and the hyperthermophile *Archaeoglobus fulgidus*. All other thermophilic archaea whose genomes have been completely sequenced have neither DNA gyrase nor Topo IIA genes, but all possess genes encoding Topo VI, which is the prototype of another Topo II family (Topo IIB) (see Bocs *et al.*[6a] in this volume). DNA gyrase is also absent from eukaryotes that contain a Topo IIA enzyme devoid of DNA gyrase activity.[7,8]

The identification of DNA gyrase genes in completely sequenced genomes is not trivial because of the existence of another Topo IIA member, called Topo IV, in many bacteria. This enzyme is a heterotetramer composed of two subunits, ParC

[2] R. J. Reece and A. Maxwell, *Crit. Rev. Biochem. Mol. Biol.* **26,** 335 (1991).
[3] J. H. Morais Cabral, A. P. Jackson, C. V. Smith, N. Shikotra, A. Maxwell, and R. C. Liddington, *Nature* **388,** 903 (1997).
[4] R. J. Reece and A. Maxwell, *Nucleic Acids Res.* **19,** 1399 (1991).
[5] N. R. Cozzarelli, *Science* **207,** 953 (1980).
[6] S. C. Kampranis and A. Maxwell, *Proc. Natl. Acad. Sci. U.S.A.* **93,** 14416 (1996).
[6a] C. Bocs, C. Buhler, P. Forterre, and A. Bergerat, *Methods in Enzymology* **334** [14] (2001) (this volume).
[7] M. I. Baldi, P. Benedetti, E. Mattoccia, and G. P. Tocchini-Valentini, *Cell* **20,** 461 (1980).
[8] T. Hsieh and D. Brutlag, *Cell* **21,** 115 (1980).

and ParE, which are homologous to GyrA and GyrB, respectively.[9] However, Topo IV has no DNA gyrase activity but, as is the case with the eukaryotic enzyme, can relax both positively and negatively supercoiled DNA.[10] In the absence of ATP, DNA gyrase can also relax DNA, but with a lower efficiency.[11,12] Discriminating between DNA gyrase and Topo IV can be done tentatively *in silico* by examination of the C-terminal region of GyrA and ParC genes.[13] This region is well conserved in all GyrA polypeptides but highly variable in ParC. This could be related to the importance of this region for the specific DNA wrapping around DNA gyrase. Table I shows the distribution of DNA gyrase and Topo IV in all completely sequenced prokaryotic genomes based on this criterion. It appears that all bacteria contain a DNA gyrase, with the possible exception of *Aquifex aeolicus,* but that only some bacteria have both DNA gyrase and Topo IV. *A. aeolicus* contains a GyrB gene and a ParC gene only. Thus, it could have a unique Topo IIA, a hybrid of gyrase and Topo IV, and with no gyrase activity (see below). It is interesting that *A. aeolicus* also contains two reverse gyrase genes, suggesting that selection pressure at high temperature favors reverse gyrase over DNA gyrase. However, it should be noted that the predictions of Table I have been confirmed in only a few cases by the purification and characterization of the enzyme. Thus, it would be interesting to systematically look for Topo II activities in organisms whose genomes have been sequenced.

This report describes the purification of DNA gyrase from *T. maritima,* a hyperthermophilic anaerobic bacterium from the order of Thermotogales, with an optimal growth temperature of 80°.[14] This hyperthermophile contains both a DNA gyrase and a reverse gyrase. Also, the plasmid pRQ7, from another *Thermotoga* species, has been found to be negatively supercoiled.[13,15] This indicates that when the two enzymes are present in the same organism, DNA gyrase activity predominates over that of the reverse gyrase *in vivo.*

The purification protocol used to purify the *T. maritima* DNA gyrase was based on a modification of the affinity chromatography method of Staudenbauer and Orr.[16] This method exploits the strong interaction of DNA gyrase (via the

---

[9] J. Kato, Y. Nishimura, R. Imamura, H. Niki, S. Hiraga, and H. Suzuki, *Cell* **63**, 393 (1990).
[10] J. Kato, H. Suzuki, and H. Ikeda, *J. Biol. Chem.* **267**, 25676 (1992).
[11] M. Gellert, K. Mizuuchi, M. H. O'Dea, T. Itoh, and J. I. Tomizawa, *Proc. Natl. Acad. Sci. U.S.A.* **74**, 4772 (1977).
[12] A. Sugino, C. L. Peebles, K. N. Kreuzer, and N. R. Cozzarelli, *Proc. Natl. Acad. Sci. U.S.A.* **74**, 4767 (1977).
[13] O. Guipaud, E. Marguet, K. M. Noll, C. Bouthier de la Tour, and P. Forterre, *Proc. Natl. Acad. Sci. U.S.A.* **94**, 10606 (1997).
[14] R. Huber, T. A. Langworthy, H. König, M. Thomm, C. R. Woese, U. B. Sleytr, and K. O. Stetter, *Arch. Microbiol.* **144**, 324 (1986).
[15] C. Bouthier de la Tour, C. Portemer, H. Kaltoum and M. Duguet, *J. Bacteriol.* **180**, 274 (1998).
[16] W. L. Staudenbauer and E. Orr, *Nucleic Acids Res.* **9**, 3589 (1981).

TABLE I
Gyrases and Topoisomerases IV in Complete Sequenced Genomes[a]

Group	Species	DNA gyrase GyrA	DNA gyrase GyrB	Topo IV ParC	Topo IV ParE	DNA gyrase activity	Ref.[b]
Bacteria							
Thermotogales	*Thermotoga maritima*	+	+	−	−	+	1
Aquificales	*Aquifex aeolicus*	−	+	+	−	Not determined	
Deinococcaceae	*Deinococcus radiodurans*[c]	+	+	(−)	(−)	Not determined	
Cyanobacteria	*Synechocystis* sp.	+	+	+	−	Not determined	
Chlamydiales	*Chlamydia trachomatis*	+	+	+	+	Not determined	
	Chlamydia pneumoniae	+	+	+	+	Not determined	
Gram positives							
High GC content	*Mycobacterium tuberculosis*	+	+	−	−	+ In *M. smegmatis*	2
Low GC content	*Bacillus subtilis*	+	+	+	+	+	3
	Mycoplasma genitalium	+	+	+	+	Not determined	
	Mycoplasma pneumoniae	+	+	+	+	Not determined	
Spirochaetales	*Borrelia burgdorferi*	+	+	+	+	Not determined	
	Treponema pallidum	+	+	−	−	Not determined	
Proteobacteria							
α subdivision	*Rickettsia prowazekii*	+	+	+	+	Not determined	
β subdivision	*Neisseria gonorrhoeae*[d]	+	+	+	+	Not determined	
ε subdivision	*Helicobacter pylori*	+	+	−	−	Not determined	
γ subdivision	*Escherichia coli*	+	+	+	+	+	4
	Haemophilus influenzae	+	+	+	+	+	5
Archaea[e]							
Archaeoglobales	*Archaeoglobus fulgidus*	+	+	−	−	Not determined	
Methanococcales	*Methanococcus jannaschii*	−	−	−	−	Not determined	
Methanobacteriales	*Methanobacterium thermoautotrophicum*	−	−	−	−	Not determined	
Thermococcales	*Pyrococcus horikoshii*	−	−	−	−	Not determined	

[a] References and genome DNA sequences can be found on the Genomes On Line Database, 1.0 (http://geta.life.uiuc.edu/~nikos/genomes.html).
[b] References: (1) O. Guipaud, E. Marguet, K. M. Noll, C. Bouthier de la Tour, and P. Forterre, *Proc. Natl. Acad. Sci. U.S.A.* **94**, 10606 (1997); (2) V. Revel-Viravau, Q. C. Truong, N. Moreau, V. Jarlier, and W. Sougakoff, *Antimicrob. Agents Chemother.* **40**, 2054 (1996); (3) A. Sugino and K. F. Bott, *J. Bacteriol.* **141**, 1331 (1980); (4) M. Gellert, K. Mizuuchi, M. H. O'Dea, and H. A. Nash, *Proc. Natl. Acad. Sci. U.S.A.* **73**, 3872 (1976); (5) J. K. Setlow, D. Spikes and M. Ledbetter, *J. Bacteriol.* **158**, 872 (1984).
[c] Not published. Genome DNA sequence is complete but unavailable. Preliminary sequence data were obtained from The Institute for Genomic Research Web site at http://www.tigr.org.
[d] Not published. Genome DNA sequence is available on http://www.genome.ou.edu/gono.html.
[e] In all these cases, another type II topoisomerase is present, the Topo VI.

GyrB subunit) with novobiocin, a universal inhibitor of DNA gyrases that competes with ATP for the ATP-binding site located on GyrB. This method has been proven to be successful in the purification of native DNA gyrase from *Bacillus subtilis*,[17] *Mycobacterium smegmatis*,[18] *Citrobacter freundii*,[19] and *Campylobacter jejuni*,[20] and in the purification of the recombinant GyrB protein of *Myxococcus xanthus*.[21] Novobiocin-Sepharose affinity chromatography has also been used to look for DNA gyrase activity in the hyperthermophilic bacterium *Aquifex aeolicus*; however, no activity was detected. This is in agreement with the hypothesis that this hyperthermophilic bacterium has no DNA gyrase activity, as suggested by the absence of GyrA in the completely sequenced genome (Table I).

Purification

Preparation of Novobiocin-Sepharose

(See ref. 16.) Two g of epoxy-activated Sepharose 6B (Pharmacia) are resuspended in 250 ml distilled water for 1 hr at room temperature and washed on a sintered glass filter with 500 ml of 0.3 M sodium carbonate pH 9.5 (buffer I). The gel is then mixed with a solution of 200 mg novobiocin (Sigma, St. Louis, MO) in 12 ml buffer I and gently shaken for 20 hr at 37°. Excess epoxy groups are blocked by addition of ethanolamine (final concentration 1 M), and shaking is continued for another 4 hr at 37°. The product is then alternately washed in 250 ml each of 0.5 M NaCl in buffer I, distilled water, 0.5 M NaCl in 0.1 M sodium acetate pH 4.0, and distilled water.

Bacterial Strain

Frozen cells of *T. maritima* (strain MSB8) have been generously provided by Dr. Huber (Regensburg, Germany). *T. maritima* can be cultivated anaerobically at 80° in a modification of the VSM medium[22] (supplemented with glucose and with cysteine omitted): artificial seawater (0.34 mM NaCl, 3.3 mM KCl, 1.5 mM KH$_2$PO$_4$, 0.42 mM KBr, 0.32 mM H$_3$BO$_3$, 0.04 mM SrCl$_2$·6H$_2$O, 1.7 mM trisodium citrate dihydrate), 1 mg/liter resazurin, 1 g/liter yeast extract, 4 g/liter peptone, 9 mM PIPES, 0.5 g/liter glucose, 4 mM MgSO$_4$·7H$_2$O, 0.3 mM CaCl$_2$·2H$_2$O, 0.4 mM KH$_2$PO$_4$, pH 6.8.

[17] E. Orr and W. L. Staudenbauer, *J. Bacteriol.* **151,** 524 (1982).
[18] V. Revel-Viravau, Q. C. Truong, N. Moreau, V. Jarlier, and W. Sougakoff, *Antimicrob. Agents Chemother.* **40,** 2054 (1996).
[19] H. Aoyama, K. Sato, T. Fujii, K. Fujimaki, M. Inoue, and S. Mitsuhashi, *Antimicrob. Agents Chemother.* **32,** 104 (1988).
[20] T. D. Gootz and B. A. Martin, *Antimicrob. Agents Chemother.* **35,** 840 (1991).
[21] Y. Paitan, N. Boulton, E. Z. Ron, E. Rosenberg, and E. Orr, *Microbiology* **144,** 1641 (1998).
[22] E. Marguet, Y. Zivanovic, and P. Forterre, *FEMS Microbiol. Lett.* **142,** 31 (1996).

Cell Lysis

Ten g of *T. maritima* frozen cells stored at $-80°$ are thawed by gently stirring in ice-cold 50 mM Tris-HCl (pH 7.5) containing 2 mM dithiothreitol (DTT), 1 mM phenylmethylsulfonyl fluoride (PMSF), 1 μg/ml leupeptin, and 1 μg/ml pepstatin A. All subsequent steps are performed at 0–4°. Cells are then broken by sonication in a final volume of 26 ml with a Labsonic U sonicator (B. Braun) using a 12T probe, a power level of 40 W, and a repeating duty cycle of 0.7 sec for 1 min, four times.

KCl and Magnesium Acetate Precipitation

Proteins are separated from DNA by slow addition of KCl, with continuous stirring, to a final concentration of 0.66 M, following which the solution is gently shaken for 15 min. Nucleic acids and ribosomes are precipitated by slow addition of magnesium acetate, with continuous stirring, to a final concentration of 5 mM, and the solution is mixed for another 1.5 hr. The resulting solution is then centrifuged for 1.5 hr at 30,000 rpm in a Beckman 70 Ti rotor, and the supernatant is saved.

Ammonium Sulfate Precipitation

Solid ammonium sulfate is added to the supernatant to reach 35% saturation (0.194 mg/ml). After stirring for 1 hr at 4°, the precipitate is removed by 20 min of centrifugation at 10,000 rpm (8000g) in a Beckman JA20 rotor. The ammonium sulfate concentration in the supernatant is then adjusted to 55% by adding 0.118 g of the salt per ml. After 1 hr of stirring at 4°, the precipitate is collected by centrifugation and resuspended in 10 ml of 25 mM HEPES (pH 8.0), 1 mM EDTA, 1 mM DTT, 10% ethylene glycol, and 0.2 M KCl (buffer A). This 35 to 55% ammonium sulfate fraction is dialyzed overnight against buffer A.

Novobiocin-Sepharose Affinity Chromatography

The resulting effluent is applied to a 0.7 × 2 cm novobiocin-Sepharose column (see above), equilibrated with the same buffer. Proteins are first eluted with 10 ml of 20 mM ATP (adjusted to pH 8.0) in buffer A, then with 10 ml of 5 M urea in buffer A, and collected in 0.5 ml fractions. After dialysis overnight against buffer B [50 mM potassium phosphate (pH 7.6), 1 mM EDTA, 1 mM DTT], the protein content of each fraction is analyzed by SDS–PAGE. The fractions are finally concentrated by dialysis against buffer B containing 50% glycerol for 20 min, and 30 μl aliquots are rapidly frozen in a liquid nitrogen bath and stored at $-80°$. Fractions containing *T. maritima* DNA gyrase activity are identified by the supercoiling assay described below.

Conservation of Fractions

Fractions containing *T. maritima* DNA gyrase maintain their activity when conserved as 30-μl aliquots at −80° in buffer B. However, it was observed that DNA gyrase activity is completely abolished when samples are stored at −20° in 40% glycerol. Once thawed, aliquots can be stored at 4°, but DNA gyrase activity is lost in about 1 month under these conditions.

Comments about Purification

The present protocol for *T. maritima* DNA gyrase purification differs in some extent from the published method for the *E. coli* enzyme. The main difference is that only one chromatographic step is used in order to obtain essentially pure enzyme.

A 35–55% ammonium sulfate fraction, containing about 230 mg of protein, was obtained from crude extract of *T. maritima* and directly loaded onto a novobiocin-Sepharose column at low ionic strength (0.2 M KCl). This unique chromatographic step was sufficient to obtain a DNA gyrase with more than 90% purity (Fig. 1A), as described below. In contrast, in the case of *E. coli* DNA gyrase, heparin-Sepharose chromatography was used prior to the novobiocin-Sepharose step. The loading of the column with the 35–55% fraction led to the elimination of numerous proteins (about 200 mg of proteins was detected in the pass-through).

Elution of nonspecific proteins bound to novobiocin with a buffer containing 20 mM ATP (whose novobiocin is a competitor) eliminated another 10 mg of

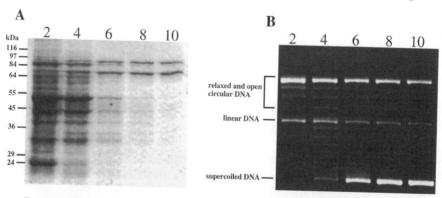

FIG. 1. Purification of the DNA gyrase from *T. maritima* using a novobiocin-Sepharose column. (A) SDS–PAGE of fractions containing novobiocin-binding proteins from *T. maritima*. Crude extracts were dialyzed against buffer A and applied to a novobiocin-Sepharose column (see Purification section). The column was successively eluted with buffer A containing 20 mM ATP and buffer A containing 5 M urea. Samples (10 μl) from several dialyzed fractions of the 5 M urea eluates were applied to SDS–PAGE (10%). (B) Supercoiling activity. Two μl of each fraction from the *T. maritima* 5 M urea eluates analyzed on SDS–PAGE was incubated with relaxed pBR322 DNA at 84° for 10 min. Samples were analyzed on a 1% agarose gel. Lane numbers refer to the numbers of fractions eluted from the novobiocin-Sepharose column by urea buffer.

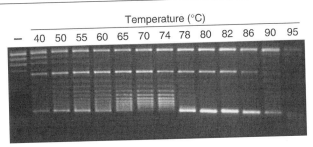

FIG. 2. Temperature dependence of the *T. maritima* DNA gyrase supercoiling activity. Relaxed pBR322 DNA was incubated with the fraction containing active DNA gyrase for 10 min at various temperatures. Samples were analyzed on a 1% agarose gel. Lane—: Assay without enzyme.

protein. It was not possible to establish if the fractions eluted by ATP contained supercoiling activity, since they had a strong restriction endonuclease activity that cut the circular DNA substrate (data not shown). Therefore, it was not possible to determine if *T. maritima* DNA gyrase was partially eluted in this step, as is the case for the *Mycobacterium smegmatis* DNA gyrase.[18]

Finally, a supercoiling activity was eluted with 5 M urea (Fig. 1B). After the elution with 6 column volumes (4.5 ml), almost all contaminating polypeptides were removed as shown in Fig. 1A (fraction 10 is the fraction harvested after 4.5 ml of urea buffer). Pure proteins were then regularly eluted with the rest of the urea buffer (5.5 ml).

Analysis of the protein content of active fractions revealed two bands in the fraction eluted by urea (Fig. 1A). The apparent molecular masses of the proteins, 90 and 70 kDa, corresponded well to the molecular masses calculated from the deduced amino acid sequences of GyrA (90,460 kDa) and GyrB (72,459 Da).[13,23] The concentration of DNA gyrase (GyrA and GyrB) in fractions was estimated to be 40 ng/μl, i.e., about 120 nM with a calculated molecular mass of 325,838 Da for the heterotetramer A_2B_2.

The main problem was whether urea elution followed by dialysis, in order to allow protein folding, is sufficient enough to restore all DNA gyrase activity. In particular, some linear DNA was generated during the supercoiling assay (see Fig. 1B and Fig. 2). This linear DNA could have been created by a DNA gyrase being incompletely refolded after dialysis (the closing activity could have been lost).

Supercoiling Assay

The topoisomerase II activity of the *T. maritima* DNA gyrase has been investigated by assessing the supercoiling activity of the purified enzyme. In order to

[23] O. Guipaud, B. Labedan, and P. Forterre, *Gene* **174**, 121 (1996).

characterize the catalytic activity of the DNA gyrase over time, preliminary supercoiling assays are carried out in a standard reaction mixture with 0.24 pmol of purified enzyme. The assay conditions are subsequently optimized.

Original Reaction

The standard reaction (20 μl) contains 35 mM Tris-HCl, pH 8.5 (at 25°C), 1 mM EDTA, 2 mM DTT, 4 mM MgCl$_2$, 5 mM spermidine, 1 mM ATP, 0.25 mg/ml BSA, 0.5 μg of relaxed pBR322 DNA from Lucent Ltd. (Leicester, England), and 2 μl (about 0.24 pmol of DNA gyrase) of each fraction eluted from the novobiocin-Sepharose column. The mixture is then incubated for 10 min at 84°, the optimal temperature of supercoiling activity, and the reaction is stopped by placing samples on ice. After addition of 2 μl of 10% SDS and 2 μl of 10 mg/ml proteinase K, samples are incubated at 50° for 15 min in order to remove proteins from DNA. A loading mixture (2.5 μl) (20% sucrose, 0.1% bromphenol blue) is then added, and samples are analyzed after electrophoresis on 1% ultrapure agarose (Indubiose 37NA; Biosepra, France) gels at room temperature for 17 hr at 55 V (2.75 V/cm) in TBE buffer (100 mM Tris–borate, 2 mM EDTA). The gels are stained with ethidium bromide and examined under UV illumination.

Temperature Dependence of Reaction

The temperature inside a control tube is measured with a HI 9063 thermometer (Hanna instruments) in order to verify the temperature of the reaction mixture. Supercoiling assays are carried out from 40 to 95°C for 3 (not shown) or 10 min (Fig. 2). When the reaction is performed for 3 min, *T. maritima* DNA gyrase is active from 50 to 95°, with an optimal temperature activity around 82–86°. With an incubation time of 10 min, supercoiling activity is detected at very low temperature (40°), but the reaction is only complete with temperatures in the range of 80 to 86°. At 95°, it is possible to detect supercoiling activity; however, the DNA is quickly degraded and denatured at this temperature in these conditions. With incubation times shorter than 3 min at 84°, a measurable transition of relaxed to supercoiled pBR322 DNA is detected, although the enzyme never carries out a complete conversion (not shown). An incubation temperature of 84° is, therefore, considered to be optimal for the *T. maritima* DNA gyrase. This temperature has been used for further investigations on DNA gyrase catalytic properties.

Optimization of Reaction

The initial reaction mixture is based on different topoisomerase II buffers cited in the literature. In order to study the catalytic properties of the *T. maritima* enzyme, several parameters, such as ATP, Mg^{2+}, pH, bovine serum albumin (BSA), and spermidine, have been tested. The DNA gyrase completely converts 200 ng

(0.07 pmol) of relaxed pBR322 DNA into supercoiled DNA in a pH range of pH 8.2 to 9.3 at 25° (35 mM HEPES), i.e., pH 7.4 to 8.5 at 84°C, with an ATP concentration of 800 μM and a MgCl$_2$ concentration of 3 mM. The reaction is enhanced by 0.25 mg/ml BSA, but partially inhibited by spermidine.

Definition of Unit

An optimal reaction mixture for the *T. maritima* DNA gyrase is defined that contains 35 mM HEPES, pH 8.2 (at 25°), 1 mM EDTA, 2 mM DTT, 3 mM MgCl$_2$, 0.8 mM ATP, and 0.25 mg/ml BSA. A unit of *T. maritima* DNA gyrase is defined as the amount of activity that supercoils 200 ng of relaxed pBR322 DNA in 5 min at 84° under these conditions.

Concluding Remarks

The finding of a regular DNA gyrase in a hyperthermophile was a surprise, since it had been previously thought that these microorganisms contain only a reverse gyrase. Purified reverse gyrase from *T. maritima* will be useful for studying the interaction of a DNA gyrase and a reverse gyrase *in vitro* and the behavior of negatively supercoiled DNA at high temperature. Also, it can be used to produce negatively supercoiled DNA substrate in tests performed at high temperature with other hyperthermophilic enzymes. For example, this enzyme could be used to maintain a negatively supercoiled state of the DNA together with the elimination of topological constraints during *in vitro* transcription performed with hyperthermophilic RNA polymerases on circular templates.

Acknowledgments

The authors are grateful to Robert Huber for the gift of *T. maritima* and *A. aeolicus* cells, and to Purificación López-García, Sarah Tite, and Emily Umpleby for critical reading of the manuscript. Work in our laboratory is supported by grants from the Association pour la Recherche sur le Cancer and the European Community program Cell factory (BIO4-CT 96-0488).

[14] DNA Topoisomerases VI from Hyperthermophilic Archaea

By CHANTAL BOCS, CYRIL BUHLER, PATRICK FORTERRE, and AGNÈS BERGERAT

Introduction

DNA topoisomerases are enzymes that can transiently break the DNA backbone and force the crossing of one or two DNA strands through the other, thereby interconverting DNA topological isomers.[1] They have been classified into two types according to their mechanistic properties. The type I DNA topoisomerases transiently break a single strand of the DNA helix, pass the other strand through this break, and religate the broken strand, thereby altering the linking number by an increment of one. Type II topoisomerases function by cleaving both strands of a DNA molecule, passing an intact double strand through this break, and religating the broken strands, thereby changing the linking number by an increment of two.

The DNA topoisomerases VI are type II DNA topoisomerases only found in archaea. These enzymes exhibit all the biochemical properties of type II DNA topoisomerases. However, although eukaryotic DNA topoisomerase II, DNA gyrase, or DNA topoisomerase IV are clearly homologous (Topo IIA family), archaeal DNA topoisomerases VI share no sequence similarity with these enzymes, except for a common ATP binding site. Thus, DNA topoisomerase VI is a member of a new family of type II DNA topoisomerase (Topo IIB).[2]

Like all type II DNA topoisomerases, the DNA topoisomerases VI are ATP-dependent enzymes that relax both positively and negatively supercoiled DNA and decatenate intertwined DNA. They are heterotetramers composed of two subunits. The A subunit carries the active tyrosine involved in the nicking–closing reaction and the B subunit binds and hydolyzes ATP. The DNA topoisomerase VI genes are widespread in the whole archaeal kingdom but absent in Bacteria. Homologs of the A subunit alone have been found in Eukarya, Spo11 (*Saccharomyces cerevisiae*), Rec12 (*Schizosaccharomyces pombe*), and Mei-W68 (*Drosophila melanogaster*), whose function has been clarified by the discovery of archaeal DNA topoisomerases VI.[2] Indeed, these homologs, which are involved in the initiation of meiotic recombination, are responsible for the breakage of the two DNA strands at the onset of this mechanism. After cleavage, Spo11 is covalently linked to the 5'

---

[1] J. C. Wang, *Annu. Rev. Biochem.* **65**, 635 (1996).
[2] A. Bergerat, B. de Massy, D. Gadelle, P. C. Varoutas, A. Nicolas, and P. Forterre, *Nature* **386**, 414 (1997).

end of the DNA break via a tyrosine, which is conserved with the archaeal DNA topoisomerases VI.[3]

Two DNA topoisomerases VI have been purified to homogeneity from archaeal strains. The first purification reported was that of the enzyme from *Sulfolobus shibatae*, an extremely thermophilic crenoarchaeota with an optimal growth temperature of 85°.[4] This article presents the purification of the DNA topoisomerase VI from *Pyrococcus furiosus*, an extremely thermophilic anaerobic euryarchaeota with an optimal growth temperature of 95°. These two enzymes share many properties and can be purified by a similar procedure. However, the purification of the DNA topoisomerase VI from *P. furiosus* has been simplified and some modifications, required by the extreme thermophily of this enzyme, have been introduced into DNA topoisomerase assays. Finally, physical and enzymatic properties of DNA topoisomerase VI are discussed.

Purification

Cell Culture

P. furiosus (DSM 3638) is grown at 95° in a 15 liter culture in synthetic seawater supplemented with a vitamin mixture, $FeCl_3$ (25 mM), elemental sulfur [5 g/liter (w/v)], tryptone (10 g/liter), and yeast extract (5 g/liter) as previously described.[5] Cells are stored at −80° until purification.

Cell Lysis and Polyethyleneimine Precipitation

P. furiosus frozen cells (3 g, wet weight) are disrupted by thawing and blending with an Ultraturax (TP18/2) in a total volume of 60 ml of buffer A [50 mM Tris-HCl pH 8, 1 mM EDTA, 2 mM dithiothreitol (DTT), 1.2 M NH_4Cl, 1 mM EGTA], 1 μg/ml leupeptin, and 1 μg/ml pepstatin A, and 1 mM phenylmethylsulfonyl fluoride (PMSF). All subsequent steps are performed at 4°. The resulting solution is centrifuged at 7500g for 10 min and the supernatant is saved. Nucleic acids are precipitated by slow addition of polyethyleneimine, with continuous stirring, to a final concentration of 0.33%.[6] After 15 min of mixing, the solution is centrifuged at 7500 g for 10 min. The supernatant is further clarified at 100,000g for 1 h (fraction I, Table I).

Ammonium Sulfate Precipitation

Proteins are precipitated by addition of ammonium sulfate to 70% saturation and centrifuged at 7500g for 45 min. The pellet is dissolved in 100 ml of buffer B

---

[3] S. Keeney, C. N. Giroux, and N. Kleckner, *Cell* **88**, 375 (1997).
[4] A. Bergerat, D. Gadelle, and P. Forterre, *J. Biol. Chem.* **269**, 27663 (1994).
[5] G. Fiala and K. O. Stetter, *Arch. Microbiol.* **145**, 56 (1986).
[6] M. Cull and C. S. McHenry, this series, Vol 182, p. 147.

TABLE I
PURIFICATION OF DNA TOPOISOMERASE VI FROM *Pyrococcus furiosus*

Fraction	Volume (ml)	Protein (mg)	Enzyme activity[a] (units 10^{-3})	Specific activity (units 10^{-3} mg^{-1})
Polyethyleneimine supernatant(I)	60	150	ND[b]	ND
70% (NH$_4$)$_2$SO$_4$ pellet (II)	42	115	ND	ND
Phenyl-Sepharose pool (III)	73	29	300	10
Heparin-Sepharose pool 1 (IVa)	32	3,5	100	28
Heparin-Sepharose pool 2 (IVb)	10	0,8	50	62
Sucrose gradient[c] (V)	0,04	0,03	30	2000

[a] Units of enzyme actvity determined by k-DNA decatenation assay.
[b] ND, Not determined.
[c] Sucrose gradient has been performed on fraction IVb.

(50 mM potassium phosphate, 1 mM EDTA, 1 mM DTT) with ammonium sulfate at 20% saturation (fraction II).

Phenyl-Sepharose Chromatography

Fraction II is loaded on a phenyl-Sepharose column, 2.5 × 10 (diameter × length in centimeters) equilibrated with 20% (w/v) ammonium sulfate in buffer B. The column is washed successively with 320 ml of 20% (w/v) ammonium sulfate in buffer B, 450 ml of 0.25 M NaCl in buffer B (buffer C), and 750 ml of 30% ethylene glycol in buffer C. The column is then developed with a 200 ml linear gradient of 30% to 45% ethylene glycol, a 140 ml step of 45% ethylene glycol, and a 200 ml step of 60% ethylene glycol in buffer C. The peak of type II DNA topoisomerase activity is eluted between 45 and 60% ethylene glycol. Active fractions from the phenyl-Sepharose chromatography are pooled and referred to as fraction III.

Heparin-Sepharose Chromatography

Fraction III is diluted twice in buffer B containing 80% ethylene glycol to adjust the ethylene glycol concentration to 60%. Triton X-100 is added to 0.01% final concentration. This fraction is loaded on a (1.5 × 9 cm) heparin-Sepharose column equilibrated with buffer D (buffer C with 60% ethylene glycol and 0.01% Triton X-100). The column is washed with the same buffer until the eluate's protein concentration drops down to 20 μg/ml. The column is then developed with a 150 ml linear gradient of 0.25 to 1 M NaCl in buffer D. The type II DNA topoisomerase active fractions are eluted between 0.3 and 0.7 M NaCl. Fractions eluted between 0.3 and 0.4 M and between 0.4 and 0.7 M are pooled and referred to as

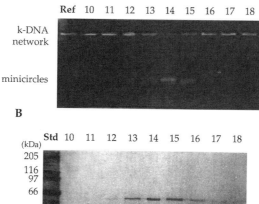

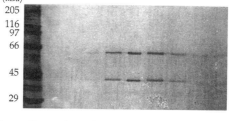

FIG. 1. Sucrose gradient sedimentation *P. furiosus* DNA topoisomerase VI. Fraction IV was sedimented through a linear 5–20% (v/v) sucrose gradient. Fractions were collected from the bottom of the gradient and were analyzed for decatenation activity. (A) Decatenation activity. Lane Ref, assay without enzyme. Other lane numbers refer to the sucrose gradient fraction numbers. k-DNA network does not migrate from the well. (B) Twenty μl of each fraction was analyzed on 10% SDS–PAGE and silver-stained. Lane Std, molecular mass standards. The other lane numbers refer to the sucrose gradient fraction numbers.

fractions IVa and IVb, respectively. Fraction IVa has higher levels of reverse gyrase activity.

Sucrose Gradient

The concentration of ethylene glycol in fraction IVb is decreased to 6% by successive dilutions and concentrations using Centricon C100 (Amicon, Inc., Beverly, MA). This aliquot is concentrated 70-fold before loading on a 5–20% sucrose gradient in purification buffer B, supplemented with 0.01% Triton X-100 and 0.25 M KCl, and centrifuged at 40,000 rpm for 15 hr in a SW41 rotor. Fractions of 0.33 ml are collected. These fractions are analyzed on sodium dodecyl sulfate–polyacrylamide gel electrophoresis (SDS–PAGE) stained by the silver method (Fig. 1B). DNA topoisomerase II activity is detected by decatenation activity in the presence of ATP (Fig. 1A). The active fractions are pooled, concentrated, and equilibrated in buffer E (Tris-HCl, pH 7.5, 50 mM, 1 mM EDTA, 1 mM DTT, 50%

ethylene glycol, 0.01% Triton X-100) by two concentration and dilution steps on Centricon C100 and stored at −70° as fraction V.

Comments about Purification

The present protocol for DNA topoisomerase VI purification differs to some extent from the previously published method for this enzyme from *S. shibatae*.[4] It gives similar yields and the same degree of purity, but the phosphocellulose, DNA cellulose, and hydroxyapatite chromatographic steps have been omitted.

The main problem encountered during the purification of hyperthermophilic DNA topoisomerases VI was separation of its activity from reverse gyrase activity. The reverse gyrase is a peculiar ATP-dependent type I DNA topoisomerase that introduces positive superturns in DNA, and it has been found in all hyperthermophilic organisms. To solve this problem we took advantage of the hydrophobicity difference between the DNA topoisomerases VI and the reverse gyrase. It appeared that this type I DNA topoisomerase was eluted from phenyl-Sepharose between 30 and 45% ethylene glycol as the DNA topoisomerase VI elution required slightly more than 45% ethylene glycol. A long washing step at 45% ethylene glycol eliminated a significant part of the reverse gyrase activity. DNA topoisomerase VI was eluted by a 60% ethylene glycol step, both to keep the samples as concentrated as possible and to elute the two subunits in the same fractions. Indeed, the purification of the recombinant subunits of the *S. shibatae* enzyme suggests that the two subunits of DNA topoisomerases VI do not exhibit the same hydrophobicity.[7] Also, they did not appear to be tightly associated (unpublished data). Part of the reverse gyrase activity can be removed by heparin-Sepharose chromatography, as the reverse gyrase appears less positively charged than the DNA topoisomerase VI.

However, the high hydrophobicity of the DNA topoisomerases VI, as well as their basicity (the calculated p*I* of the subunits is about 9) made it necessary to keep at least 0.01% of Triton X-100 in all the buffers and to introduce ethylene glycol (60%) in the heparin-Sepharose chromatography buffers to prevent aggregation and adsorption on tubes. Ethylene glycol appeared to be an excellent stabilizer for the purified enzyme, since activity was retained for 12 months when enzyme was stored at −70°. In contrast, it was observed that DNA topoisomerase VI activity was completely lost when stored in 50% (v/v) glycerol.

Decatenation Assay

The standard reaction (20 μl) contains 50 mM Tris-HCl, pH 8.8 (at 25°), 1 mM Na$_2$EDTA, 10 mM MgCl$_2$, 2 mM DTT, 0.5 M potassium glutamate, 1 mM ATP, 0.2 μg of kinetoplast DNA (k-DNA) from TopoGEN, Inc. (Columbus, OH), and

---

[7] C. Buhler, D. Gadelle, P. Forterre, J. C. Wang, and A. Bergerat, *Nucleic Acids Res.* **26**, 5157 (1998).

2 μl of the enzyme sample (eventually diluted in purification buffer D). Reactions are incubated for 5 minutes in a water bath heated at 90° and are stopped on ice. After addition of 1 μl of 10% SDS and of 1 μl of 10 mg/ml proteinase K, the sample is incubated at 50° for 15 min. Five μl of a loading mixture, containing 10% sucrose and 0.05% bromophenol, is added to the reaction. The sample is then subjected to electrophoresis on a 1% agarose gel in TBE running buffer (89 mM Tris base, 89 mM boric acid, and 2.5 mM Na$_3$EDTA).

Definition of Units

One unit of DNA topoisomerase VI is defined as the amount of enzyme required to decatenate 0.2 μg of k-DNA into open circular or covalently closed minicircles under standard reaction conditions.

Notes on Assay Method

The ATP-dependent topoisomerization reactions, used to follow DNA topoisomerase II activities, are relaxation of supercoiled DNA, catenation, decatenation, and unknotting of double-strand circular DNA. The ATP relaxation activity cannot be used to follow the DNA topoisomerase VI activity because of the presence of the contaminating reverse gyrase activity throughout the purification.

The k-DNA decatenation, which requires a double-strand cleavage and the transfer of a DNA segment, is a specific reaction for type II DNA topoisomerases. The kinetoplast DNA from trypanosomid is a large network of thousands of catenated minicircles. Under the action of type II DNA topoisomerases, these DNA minicircles are released from the network. On 1% agarose gel, the minicircles are very quickly visualized, as the network remains in the well. This assay requires careful control of the ionic strength of the reaction medium. When this decatenation reaction was performed in presence of 5 mM spermidine and 10 mM MgCl$_2$, a minimum concentration of 25 mM of monovalent cations had to be maintained in the reaction medium in order to prevent the reverse reaction (catenation). This cation concentration must not exceed 100 mM to avoid inhibition of the DNA topoisomerases VI activities. Because the k-DNA also appeared to be unstable and easily degraded after extensive incubation at high temperature, the holding times were limited to 5 min at 80° for the *S. shibatae* enzyme, and to 5 min at 90° for the *P. furiosus* enzyme. These temperatures correspond to the water bath temperatures, but the actual temperatures of the reaction media, determined by a microprobe, appeared to be 5° less.

In the contrast to the *S. shibatae* enzyme, the *P. furiosus* DNA topoisomerase VI needs the presence of a protein stabilizer in the reaction medium. In the absence of 0.5 M potassium glutamate, no activity was detectable. This strong stabilization effect of potassium glutamate has already been observed for other extreme

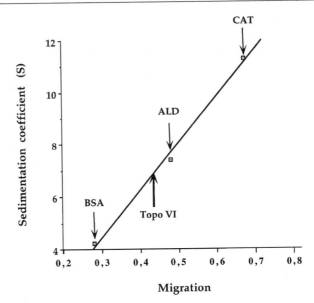

Fig. 2. The sedimentation for the decatenation activity was compared with markers centrifuged in separate tubes at the same time: catalase (CAT), 11.2 S; aldolase (ALD), 4.6 S; chymotrypsinogen A (CHY) 2.6 S. Relative migration is plotted against the sedimentation coefficient.

thermophilic enzymes such as the DNA topoisomerase V of *Methanopyrus kandleri*. However, the reverse gyrase of *P. furiosus*, purified in parallel to the DNA topoisomerase VI, did not require a stabilizer to be active at very high temperature.[8]

The P4 unknotting assay described by Liu *et al.* can be used as an alternative to the k-DNA decatenation assay.[9] However, the naturally knotted DNA isolated from the bacteriophage T4 capsids is nicked on both strands and needs to be ligated before the assay is performed at high temperature.

Physical Properties

An estimate of the native molecular mass of the *S. shibatae* DNA topoisomerase VI was calculated by combining the results of sedimentation studies and gel filtration chromatography, following the method of Siegel and Monty.[10] The enzyme had a sedimentation coefficient of 7.8 S in 5–20% sucrose gradient containing

---

[8] A. I. Slesarev, J. A. Lake, K. O. Stetter, M. Geller, and S. A. Kozyavkin, *J. Biol. Chem.* **269**, 3295 (1994).

[9] L. F. Liu, J. L. Davis, and R. Calendar, *Nucleic Acids Res.* **9**, 3979 (1981).

[10] L. M. Siegel, and K. J. Monty, *Biochim. Biophys. Acta.* **112**, 346 (1966).

200 mM KCl and a Stokes radius of 69 Å, yielding a calculated native molecular mass of approximately 230 kDa. On SDS–PAGE, this enzyme appeared to be composed of two polypeptides of 60 and 47 kDa, indicating that the native DNA topoisomerase II of *S. shibatae* is a heterotetramer.

The decatenation activity isolated from *P. furiosus* cosediments with two polypeptides on sucrose gradient (Fig. 1). The sizes of these two fragments, 65 and 40 kDa, are very similar to those of the *S. shibatae* enzyme subunits, suggesting that they also carry the DNA topoisomerase VI activity. The sedimentation coefficient of *P. furiosus* DNA topoisomerase VI was estimated to be 7 S (Fig. 2). These data indicate that this enzyme is also a heterotetramer.

Enzymatic Properties

The *S. shibatae* and *P. furiosus* DNA topoisomerase VI are both ATP-dependent enzymes. DNA topoisomerases VI, like type IIA DNA topoisomerases, can apparently promote the passage of one DNA duplex through another in the presence of a nonhydrolyzable ATP analog.[7] They relax both positively and negatively supercoiled plasmids and unknot phage P4 DNA. The *S. shibatae* enzyme alters the linking number of an unique topoisomer only in steps of two, confirming the type II nature of the DNA topoisomerases VI.[4]

Using the recombinant *S. shibatae* enzyme, catenation–decatenation reactions of the DNA topoisomerase VI have been investigated.[7] The maximum activity of the DNA topoisomerases VI is obtained in the presence of 5 mM spermidine and of 10 mM MgCl$_2$. In the absence of NaCl, the enzyme tends to catenate a negatively supercoiled plasmid DNA. In the presence of 25 mM NaCl or more, this catenation reaction is eliminated. It has also been shown that a covalent complex between the recombinant enzyme and DNA can be trapped by rapid addition of a denaturing agent.[7] As with all the type II DNA topoisomerases, the DNA topoisomerases VI thus introduce a transient double-strand break in DNA where the enzyme becomes covalently attached to the cleaved DNA.

[15] Topoisomerase V from *Methanopyrus kandleri*

By Alexei I. Slesarev, Galina I. Belova, James A. Lake, and Sergei A. Kozyavkin

Introduction

Topoisomerases are enzymes that alter the topological state (e.g., over- and underwinding, knotting, or tangling) of nucleic acids by generating transient breaks in the sugar–phosphate backbone of DNA.[1] In order to maintain the integrity of the genetic material during this process, topoisomerases form covalent bonds with DNA termini created by their actions. This covalent linkage is a hallmark of all topoisomerases.

There are two classes of topoisomerases, known as type I and type II enzymes. Type I DNA topoisomerases transiently break one DNA strand at a time for the enzyme-mediated passage of another strand, whereas the type II enzymes break both strands of a double helix. Both types use the same basic chemistry in their transient breakage of DNA—an enzyme tyrosyl group attacks and breaks a DNA backbone bond, then becomes linked to one side of the break. Reversal of this reaction disrupts the enzyme–DNA covalent link and rejoins the broken DNA strand.

Type I DNA topoisomerases can further be divided into two unrelated groups.[1,2] Group A comprises enzymes making a transient covalent complex with the 5' end of the broken DNA strand. These enzymes act on negatively, but usually not on positively, supercoiled DNA and are inhibited by single-stranded competitor DNA. A divalent cation is required for the catalytic activity of these enzymes. Analysis of the DNA cleavage site specificity and gene sequences indicates that type IA topoisomerases form two major evolutionarily distinct subfamilies.[3] The first subfamily, consisting of enzymes designated prokaryotic topoisomerase I, is represented by reverse gyrase in hyperthermophiles and by DNA-relaxing enzymes in other prokaryotes.[4–7] The other subfamily comprises DNA-relaxing enzymes, named topoisomerase III, which are found in mesophililc and thermophilic prokaryotes and in eukaryotes.[8–11]

Type IB topoisomerases differ from group A in that they form a covalent intermediate with the 3' end of DNA, in their ability to relax both positive and negative supercoils, and in their activity in the absence of divalent cations. Initially, all type IB topoisomerases were thought to form a single subfamily and were found only in eukaryotes and poxviruses.[12] Topoisomerase I versions

---

[1] J. C. Wang, *Annu. Rev. Biochem.* **65**, 635 (1996).
[2] A. I. Slesarev, J. A. Lake, K. O. Stetter, M. Gellert, and S. A. Kozyavkin, *J. Biol. Chem.* **269**, 3295 (1994).
[3] C. Bouthier de la Tour, C. Portemer, H. Kaltoum, and M. Duguet, *J. Bacteriol.* **180**, 274 (1998).
[4] Y. C. Tse, K. Kirkegaard, and J. C. Wang, *J. Biol. Chem.* **255**, 5560 (1980).
[5] F. Dean, M. A. Krasnow, R. Otter, M. M. Matzuk, S. J. Spengler, and N. R. Cozzarelli, *Cold Spring Harb. Symp. Quant. Biol.* **47**, 769 (1983).
[6] F. B. Dean and N. R. Cozzarelli, *J. Biol. Chem.* **260**, 4984 (1985).
[7] O. I. Kovalsky, S. A. Kozyavkin, and A. I. Slesarev, *Nucleic Acids Res.* **18**, 2801 (1990).
[8] R. J. DiGate and K. J. Marians, *J. Biol. Chem.* **263**, 13366 (1988).
[9] A. I. Slesarev, D. A. Zaitzev, V. M. Kopylov, K. O. Stetter, and S. A. Kozyavkin, *J. Biol. Chem.* **266**, 12321 (1991).
[10] J. W. Wallis, G. Chrebet, G. Brodsky, M. Rolfe, and R. Rothstein, *Cell* **58**, 409 (1989).
[11] R. Hanai, P. R. Caron, and J. C. Wang, *Proc. Natl. Acad. Sci. U.S.A.* **93**, 3653 (1996).
[12] J. J. Champoux, in *"DNA Topology and Its Biological Effects"* (N. R. Cozzarelli and J. C. Wang, eds.) pp. 217–242. Cold Spring Harbor Laboratory, Cold Spring Harbor, NY, 1990.

purified from different eukaryotic organisms share a consensus sequence for DNA cleavage sites, are inhibited by camptothecin, and are immunologically cross-reactive. The first, and so far the only, prokaryotic type IB topoisomerase, named topoisomerase V,[2,13,14] was discovered in *Methanopyrus kandleri*, a hyperthermophilic methanogen, which grows in temperatures up to 110°[15] and represents a separate lineage distinct from other methanogens.[16,17] It has been shown that topoisomerase V is a single polypeptide with a molecular mass of about 110 kDa. The enzyme relaxes positive and negative supercoils with a rate similar to or even higher then other topoisomerases, does not require divalent cations for activity, and binds to the 3' end of the cleaved DNA strand.

In this chapter the purification and characterization of DNA topoisomerase V from *Methanopyrus kandleri* cells is described. Furthermore, current applications of topoisomerase V in DNA technology are also presented, including discussion of those features of the enzyme that are of interest for DNA sequencing.

Assay

The assay for topoisomerase V activity is based on its ability to relax positively and negatively supercoiled DNA without magnesium ions. Typically, the relaxation of 3-4 kb supercoiled plasmid DNA (e.g., pBR322 DNA) in the presence of EDTA is used to monitor the activity. Briefly, 1 μl of topoisomerase V preparation is incubated with 0.1 μg each of positively and negatively supercoiled DNA in "standard" buffer: 30 mM Tris/HCl, pH 8.0 at 25°, 1 M potassium glutamate, and 5 mM Na$_2$EDTA in 10 μl reaction mixture at 88° for 15 min. The reaction is terminated by rapidly cooling to 0° and adding sodium dodecyl sulfate (SDS) to 1%. For the topoisomerase assay in crude extracts, the reactions terminated by SDS is treated with proteinase K (400 mg/ml) at 37° for 1 hr and then heated at 80° for 2 min. The topoisomerization products are analyzed by 1.5% agarose gel electrophoresis with 1.6 mg/ml chloroquine at 3 V/cm for 10 hr. One unit of activity is defined as the amount of enzyme required to relax 50% of form 1 pBR322 DNA (0.2 μg) under standard conditions.

DNA degradation by nucleases, a common problem in the assays of Mg-dependent topoisomerases, is not observed for these standard assay conditions. However, a substantial change in the electrophoretic mobility of supercoiled or open circular DNA is noted after incubation with early fractions, unless the samples are heated to 80° in the presence of SDS or treated with proteinase K. The aberrant migration of DNA is apparently caused by DNA-binding proteins.

---

[13] A. I. Slesarev, K. O. Stetter, J. A. Lake, M. Gellert, R. Krah, and S. A. Kozyavkin, *Nature* **364**, 735 (1993).

[14] S. A. Kozyavkin, A. V. Pushkin, F. A. Eiserling, K. O. Stetter, J. A. Lake, and A. I. Slesarev, *J. Biol. Chem.* **270**, 13593 (1995).

[15] R. Huber, M. Kurr, H. W. Jannash, and K. O. Stetter, *Nature* **342**, 833 (1989).

[16] S. Burggraf, K. O. Stetter, P. Rouviere, and C. R. Woese, *System. Appl. Microbiol.* **14**, 346 (1991).

[17] M. C. Rivera and J. A. Lake, *Int. J. Syst. Bacteriol.* **46**, 348 (1996).

Purification

DNA topoisomerase V from *M. kandleri* can be purified by the following procedure: (1) lysis of the cells; (2) precipitation with polyethyleneimine and separation of the supernatant; (3) precipitation with ammonium sulfate; (4) phosphocellulose chromatography; (5) chromatography on heparin; and (6) gel filtration.

Cells

M. kandleri strain AV-19, DSM 24, Braunschweig, Germany, is used as source of topoisomerase V. Batch cultures are grown in "BSM" medium[15] at 100° in a 300-liter enamel-protected fermentor (HTE, Bioengineering, Wald, Switzerland), with stirring and continuous gasing (H_2/CO_2, 80:20, v/v). Exponentially growing cells are rapidly cooled, harvested with a separator (Westfalia, Germany), and stored at −70°.

Lysis

Lysis is carried out in a disrupting device such as a French pressure cell. The preferred lysis buffer contains 100 mM Tris-HCl, pH 8.0 at 25°C, 0.5 M NaCl, 10 mM 2-mercaptoethanol, 50 μg/ml of phenylmethylsulfonyl fluoride (PMSF), 50 μg/ml of *N*-tosyl-L-phenylalanine chloromethyl ketone (TPCK), 50 μg/ml of *N*-α-tosyl-L-lysine chloromethyl ketone (TLCK), 50 μg/ml of pepstatin A, 50 μg/ml of leupeptin, and 1 mM benzamidine. Typically, the cell sample is thawed in a water bath at room temperature in 1 ml of lysis buffer per 2 g of cells, and passed through a French pressure cell at 16,000 psi to produce a lysate.

Polyethyleneimine Precipitation

The next step is the precipitation of the lysate with polyethyleneimine (Polymin P) and the collection of the supernatant. Typically, the lysate is diluted with lysis buffer to a final volume five times the original volume of lysis buffer, and centrifuged for 2 hr at high speed [40,000 rpm in a Beckman Instruments (Fullerton, California) Ti-50 rotor (144,600g)]. To the recovered solution, a 5% (v/v) solution of polyethyleneimine (pH 7.0) is added to a final polyethyleneimine concentration of 0.3%. After mixing for about 30 min at 0°, the solution is centrifuged at 12,000 rpm (23,750g) for 20 min and the supernatant recovered. The supernatant includes topoisomerase V.

Precipitation with Ammonium Sulfate

In the next purification step, the supernatant from the polyethyleneimine precipitation step is precipitated with saturated ammonium sulfate. Briefly, a volume of 4 M ammonium sulfate equal to about 0.9 times the volume of the polyethyleneimine supernatant is added while stirring, and then solid ammonium sulfate is added to 90% saturation. The precipitate is collected by centrifugation after decantation of the supernatant. About half of the protein content of *M. kandleri* is in the precipitate. Essentially all of the topoisomerase V activity is found in the pellet.

Phosphocellulose Chromatography

The next step is phosphocellulose chromatography. The pellet from ammonium sulfate precipitation is redissolved in a suitable starting buffer for phosphocellulose chromatography. The preferred starting buffer is 0.2 M NaCl, 30 mM sodium phosphate, pH 7.4 at 25°, 10 mM 2-mercaptoethanol, 10% glycerol, 25 μg/ml each of PMSF, TPCK, and TLCK, 5 μg/ml of pepstatin A, 1 μg/ml of leupeptin, and 1 mM benzamidine, referred to as Buffer A + 0.2 M NaCl. The redissolved pellet is dialyzed against starting buffer and loaded onto the column. The column is washed with the starting buffer after loading. Topoisomerase V is eluted with a linear gradient running from 0.2 to 1.0 M NaCl of a buffer containing the other components of starting buffer, followed by a linear gradient from 1.0 to 2.0 M NaCl in the same buffer. Active topoisomerase V elutes between 0.55 and 0.73 M NaCl (fraction I) and also between 0.73 and 1.45 M NaCl (fraction II). The resulting fractions are dialyzed against buffer B (10 mM sodium phosphate, pH 7.4 at 25°, 10% glycerol, 2 mM 2-mercaptoethanol) + 0.5 M NaCl, if necessary, after concentration.

Heparin Chromatography

The next step is heparin chromatography of the active fractions from the phosphocellulose chromatographic step. Preferably, fractions I and II are processed separately on heparin. The dialyzed fractions are loaded onto a heparin column equilibrated with buffer B + 0.5M NaCl. After washing with the starting buffer, the enzyme is eluted with a linear gradient from 0.5 to 1.5 M NaCl in buffer B. Active topoisomerase V elutes from about 1.0 to about 1.25 M NaCl for fraction I, and from 0.95 to 1.25 M NaCl for fraction II. At this stage, a protein band of 110 kDa should be visible and should be the major band in the most active fractions, as revealed by SDS–polyacrylamide gel electrophoresis. It is recommended to concentrate the eluted enzyme by processing it again on a smaller heparin column, after decreasing the sodium chloride concentration to 0.5 M by dilution with buffer B without sodium chloride. The enzyme is then reeluted from the smaller column.

Gel Filtration

The final step in the purification process is gel filtration on a column capable of resolving proteins in the molecular mass range from approximately 50,000 to

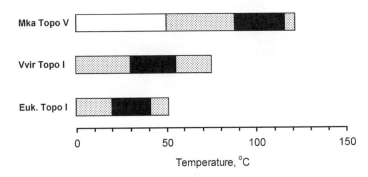

FIG. 1. Temperature range of activities of type IB topoisomerases. Temperature ranges for optimal activities are shown in black. Dotted areas show temperatures where topoisomerase activities can still be detected. White regions represents temperatures where topoisomerase V is not active.

200,000 Da. A preferable chromatography column is a HiLoad 16/60 Superdex 200 PG column (Pharmacia Biotech, Piscataway, NJ). The concentrated topoisomerase from the heparin chromatographic step is passed through the column, which has been equilibrated with 10 mM sodium phosphate, pH 7.4 at 25°, 1 M NaCl, 5% (v/v) glycerol, 2 mM 2-mercaptoethanol. The gel filtration product represents >3000-fold purified topoisomerase V. A typical yield is approximately 1 mg of topoisomerase V, having a specific activity of approximately 2.0 × 10^6 units per milligram from 100 g $M.$ $kandleri$ cells (wet weight).

Comparison of Topoisomerase V with Other Type 1B Topoisomerases

Topoisomerase V has a molecular mass of 110 kDa, which is within the 62–165 kDa size range of topoisomerase I purified from different eukaryotic sources. Many eukaryotic topoisomerases are immunologically cross-reactive. For example, topoisomerase V is recognized by antibody against human topoisomerase I.[13] A major biochemical difference between these enzymes is the conditions in which they are enzymatically active. Eukaryotic topoisomerases are active below 41° and repidly lose activity at 60°. Poxviral topoisomerase is slightly more active at 55° than at 30° and retains some activity at 75° (Fig. 1). $Methanopyrus$ topoisomerase V is active up to at least 122°. It is active in an unusually wide range of ionic strengths (from 10 mM to 3.1 M) (Table I). The highest activity is at 0.35 M NaCl, 0.5 M KCl, or 1.5 M potassium glutamate. This result is not unexpected because of the high osmolarity inside $M.$ $kandleri$ cells. Cyclic 2,3-diphosphoglycerate, a trivalent anion, is present at 1.1 M.[18] Figure 2 compares salt concentration ranges for activity of type IB topoisomerases.

[18] M. Kurr, R. Huber, H. König, H. W. Jannasch, M. Fricke, A. Trincone, J. K. Kristjansson, K. O. Stetter, $Arch.$ $Microbiol.$ **156**, 239 (1991).

TABLE I
EFFECT OF DIFFERENT IONS ON DNA RELAXATION ACTIVITY OF TOPOISOMERASE V[a]

Conditions	Concentration (M) at which activity is minimal			Highest concentration (M) at which activity was detected		
	K-Glu	KCl	NaCl	K-Glu	KCl	NaCl
80°, EDTA	1.5	0.45	0.30	2.1	0.55	0.45
80°, MgCl$_2$	1.5	0.45	0.30	2.1	0.45	0.45
88°, EDTA	1.5	0.5	0.35	3.1	0.65	0.65

[a] K-Glu, Potassium glutamate.

DNA Unlinking Activity of Topoisomerase V

A remarkable feature of DNA topoisomerase V is its ability to both relax and unlink covalently closed circular DNA. At high temperatures, DNA melting drives the catalysis of DNA unlinking by thermostable DNA-relaxing topoisomerases (Fig. 3). To explain this effect, one should take into consideration the DNA

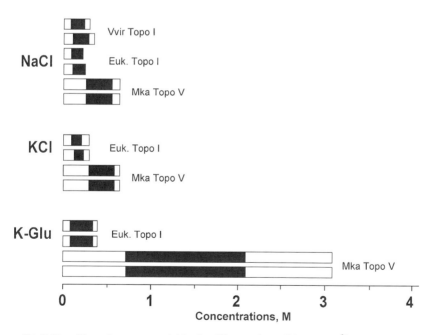

FIG. 2. Type IB topoisomerases activities for different salt conditions ($\pm$Mg^{2+}). Top bar in each pair represents activity in the presence of Mg^{2+}. Black regions show salt ranges for optimal activities.

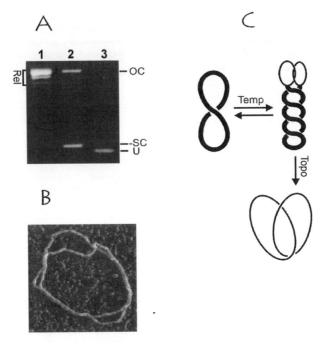

Fig. 3. Unlinking and melting of covalently closed DNA by topoisomerase V. (A) Electrophoresis of control relaxed plasmid DNA (1), negatively supercoiled (2), and the same DNA as in (1) but after incubation at 95°C with topo V (3). "−SC" and "U" indicate negatively supercoiled and unlinked DNA, respectively. (B) Unlinked DNA strands of pBR322 DNA by electron microscopy. The molecule shown here is magnified approximately 120,000. ssDNA circles are linked once. (C) An illustration of incomplete melting of circular DNA at high temperature and its complete melting and unlinking by topoisomerase V. Duplex regions and melted strands are shown by thick and thin lines.

helix–coil transition. In the melting range of linear DNA, the covalently relaxed topoisomer should be partly melted. If its degree of denaturation is $1 - \theta_0$, then its linking number, Lk_0, is less than that of the completely relaxed duplex DNA, Lk_0^d, and is equal to $\theta_0 \cdot Lk_0^d$. The higher the temperature, the less θ_0 is and the smaller Lk_0 is, compared with Lk_0^d. Thus, whatever the initial LK of a topoisomer is, at some temperature it becomes equal to $Lk_0 = \theta_0 \cdot Lk_0^d$. Below this temperature, a topoisomerase relaxes DNA by increasing Lk in a negatively supercoiled topoisomer, and by decreasing Lk in a positively supercoiled one. Above the melting range of linear DNA, complete relaxation means complete separation of DNA strands, i.e., $Lk_0 = 0$. At these conditions, thermostable relaxing topoisomerases always reduce Lk. Thus, topoisomerase V can be used for generation of topoisomers with required Lk up to $Lk = 0$. DNA unlinking at high temperature was first observed using *Desulfurococcus* topoisomerase III.[9]

Electron Microscopy of Unlinked DNA

The composition of the unlinked DNA complexes, assayed by electron microscopy, is shown in Fig. 3B. To prepare unlinked DNA, 0.1 μg of relaxed pBR322 DNA was incubated in 30 mM Tris-HCl (pH 8.0), 1.5 M potassium glutamate, 10 mM magnesium acetate, 1.1 M betaine with 10 units of topoisomerase V at 106° for 5 min. To prevent boiling, the reaction mixture of 4 μl was placed inside a 25 μl fast protein liquid chromatography (FPLC) sample loop and closed. The recovered DNA was then treated by glyoxal at 62° for 30 min, desalted, and treated with exonuclease VII to degrade single-stranded DNA. The products were then mixed with RecA protein (~1 : 40 mass ratio) in 50 μl containing 25 mM Tris-HCl (pH 7.6), 2.5 mM MgCl$_2$, and 0.5 mM ATPγS. After 30 min incubation at 37°, the samples were prepared for electron microscopy using the single carbon method with some modifications.[14,19]

The principal product in the sample consists of pairs of concatenated single-stranded (ss) DNA circles linked once (Fig. 3B). From these data, the rate of DNA unlinking at 106° was established to be about four cycles/sec/enzyme monomer, or about 16 times faster than the rate of DNA relaxation at 90°.[13] Thus, topoisomerase V can indeed reduce the linking between the complementary strands down to a single link.

Topoisomerase V in DNA Technology

The ability of topoisomerase V to bind and unlink DNA at high temperature makes it extremely useful for thermal cycling processes. In PCR or DNA sequencing topological constraints and strong secondary structures in DNA may inhibit normal enzymatic processes.

It is generally accepted, in spite of some technical problems, that the use of double-stranded (ds) DNA for sequencing is preferable to the construction and use of single-stranded DNA. Moreover, it becomes increasingly important to be able to sequence robustly long double-stranded multikilobase and megabase-long DNA templates, as it reduces the time it takes to close gaps in large-scale projects.[20] The major problems in this case are associated with DNA denaturation and reassociation. The reassociation of denatured DNA strands is pertinent to all DNA samples and results in fewer readable bases. Sequencing at higher temperatures may solve this problem, but then one must deal with the thermodegradation of nucleic acids. The latter decreases the yield and quality of sequencing fragments, particularly long ones, and results in false fragments accumulating from the sequence-dependent apurination. An enzymatic preparation based on

---

[19] R. C. Valentine and M. Green, *J. Mol. Biol.* **27**, 615 (1967).
[20] S. Kozyavkin, A. Slesarev, and A. Malykh, *Microb. Compar. Genom.* **3**, C-76 (1998).

topoisomerase V (ThermoFidelase, Fidelity Systems, Inc.) can in many cases solve these and other problems associated with DNA sequencing.

Topoisomerase V to Increase Signal Intensity in Radioactive Sequencing

A 5.5 kb pHMK plasmid that contains a G + C-rich insertion in pET21bt vector is manually sequenced using a *fmol Taq* cycle sequencing kit (Promega, Madison, WI) and ^{33}P label. The reactions are run with and without topoisomerase V and analyzed by denaturing gel electrophoresis. The autoradiogram in Fig. 4 shows that the intensity of bands is stronger in the reactions with topoisomerase V compared to the control. Densitometer quantitation of the bands shows that the intensity is increased 2 to 10 times; the effect is more pronounced for weak bands. In this experiment the following protocol is used:

For each set of four sequencing reactions, mix the following reagents in a microcentrifuge tube:

Template DNA	0.5 μg
Topoisomerase V	50 units (1 μl)
Primer (20-mer)	3.0 pmol
[^{33}P]dATP (10μCi/μl)	0.5 μl
fmol Sequencing 57× Buffer	5 μl
Sterile H$_2$O to final volume	16 μl

Add 1.0 μl of *Taq* DNA polymerase (5 U/μl) to the mix. Add 4 μl of the resulting mix to each tube containing 2 μl of d/ddNTP (deoxy-dideoxynucleotide triphosphate) Mix. Place the reaction tubes in a thermal cycler that has been preheated to 95° and start the cycling program as follows:

95° for 2 min, then:
95° for 30 sec (denaturation)
42° for 30 sec (annealing)
70° for 1 min (extension)
30 cycles total, then 4°

ThermoFidelase to Increase Accuracy of Base Calling on ABI Machines

pGEM3 plasmid was sequenced using the dye-primer *Taq* cycle sequencing kit (Perkin-Elmer, Norwalk, CT) with and without ThermoFidelase and analyzed by an ABI-373A sequencer. The accuracy of base calling is increased dramatically by the addition of ThermoFidelase. The error rate of cycle sequencing with dye–primers and *Taq* polymerase is presented in Table II. The total accuracy of sequencing has been significantly improved in all cases.

Another measure of DNA sequencing accuracy is the maximal correct read length of contiguous sequence that does not contain frame shifts and wrong

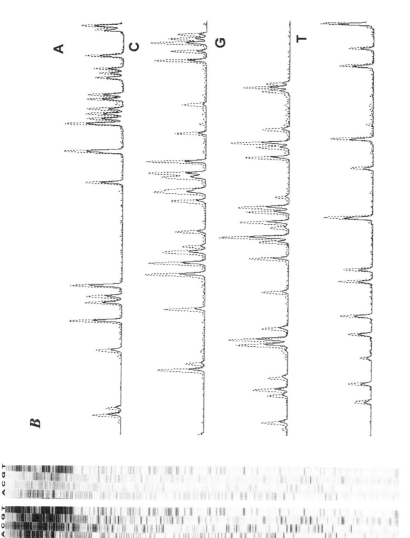

FIG. 4. Manual sequencing of plasmids with topoisomerase V. Autoradiogram of the gel (A) and densitometer scans (B) of the lanes (+Topo V, dotted lines, −Topo V, solid lines). *fmol* DNA sequencing kit from Promega was used.

TABLE II
ACCURACY OF AUTOMATED DNA CYCLE SEQUENCING

Plasmid (sequencer): ThermoFidelase:	pGEM (ABI) −	pGEM (ABI) +	pBSC (ABI) −	pBSC (ABI) +	PSPC (ABI) −	PSPC (ABI) +	pUC (ALF) −	pUC (ALF) +
Read length	546	542	544	549	422	421	476	479
Insertion/Deletion								
Total	2	2	4	1	1	0	8	4
%	0.4	0.4	0.7	0.2	0.2	0.0	1.7	0.8
Wrong base calls								
Total	8	2	2	2	0	0	3	1
%	1.5	0.4	0.4	0.4	0.0	0.0	0.6	0.2
Ambiguous base calls								
Total	11	0	7	4	10	3	5	4
%	2.0	0.0	1.3	0.7	2.4	0.7	1.0	0.8
Correct base calls								
Total	527.3	540	536.3	543	412.5	417.8	469.8	475
%	97.0	99.3	97.9	99.1	98.2	99.5	97.3	98.3

base calls. This parameter is directly related to the throughput of large-scale sequencing projects, in which raw data from multiple clones are compared and any conflicts in the base calling have to be resolved. Table III shows that ThermoFidelase dramatically improves this parameter for different automated sequencers.

Data indicate that topoisomerase V version of ThermoFidelase indeed improves sequencing reactions. It increases the yield of sequencing fragments, makes sequencing profiles more even and with less background, and dramatically increases the accuracy and correct read length.

Other DNA Sequencing Protocols with ThermoFidelase

The following approach has been adopted at the Human Genome Sequencing Center of Washington University, St. Louis, and has proven successful in sequencing through stop regions of M13 or pUC subclones, and PCR products.

TABLE III
CORRECT READ LENGTH FOR DNA SEQUENCING ACCURACY

Plasmid (sequencer): ThermoFidelase:	PGEM (ABI) −	PGEM (ABI) +	pBSC (ABI) −	pBSC (ABI) +	PSPC (ABI) −	PSPC (ABI) +	pUC (ALF) −	pUC (ALF) +
Maximal length	308	462	347	515	398	421	345	392
%increase		50.0		48.4		5.8		13.6

1. ThermoFidelase Binding

   Combine:

DNA	1 μl (about 0.2 μg)
Primer	1 μl (3 μM)
ThermoFidelase	1 μl
H_2O	6 μl
	9 μl

   Place into a thermal cycler. Heat at 95° for 5 min.

2. Extension/Termination

   To each reaction, add:

Prism premix	8 μl
Dimethyl sulfoxide (DMSO)	2 μl

   Quick spin to mix.
   Place into a thermal cycler and cycle on a two-step *Taq* terminator program:

   95° for 15 sec
   58° for 2 min
   ×15 cycles
   Hold at 4°.

Dr. Cheryl Heiner at PE- Applied Biosystems Advanced Center for Genetic Technology has developed a ThermoFidelase-based protocol for direct sequencing from genomic DNA. In the Heiner protocol, the following reaction setup is used.

REACTION SETUP

Component	Volume
Big Dye terminator premix	16 μl
ThermoFidelase:	1 μl
Primer:	13 pmol
DNA template:	2–3 μg
Distilled H_2O:	sufficient quantity for a total volume of 40 μl

CYCLING CONDITIONS

95°, 5 min initial hold
95°, 30 sec
55°, 20 sec
60°, 4 min
For 45 cycles
Hold at 4°

With this protocol the following templates were successfully sequenced: *Ureaplasma urealyticum*, 0.75 Mb genome; *Mycoplasma fermentans*, 1.2 Mb genome; *Streptococcus pneumoniae*, 2.3 Mb genome; *Escherichia coli*, 4.6 Mb genome.

Conclusion

The discovery of DNA topoisomerase V resolved a long-standing evolutionary enigma: why the ubiquitous topoisomerase I enzymes are so different in the eukaryotic and prokaryotic kingdoms. The discovery of yeast DNA topoisomerase III, which is homologous to prokaryotic topoisomerase I, provided one clue to answer this puzzle, and now topoisomerase V provides another.

A remaining puzzle is how *M. kandleri* DNA is sustained in a duplex conformation in the presence of the powerful DNA unlinking activity of Topo V. Topo V would denature *M. kandleri* DNA around 100° in a matter of minutes unless its unlinking activity is inhibited by unknown mechanism(s). Our current working hypothesis is that *M. kandleri* contains endogenous factors capable of blocking the unlinking activity of topoisomerase V while allowing its other activities such as DNA binding and relaxation.

It is important to stress that the ability of topoisomerase V to unlink DNA at high temperature made it, and potentially other thermostable topoisomerases, extremely useful for thermal cycling processes in which topological constraint inhibits normal enzymatic processes.

Acknowledgments

We thank Karl Stetter and Martin Gellert for their initial support of this work. The work was supported by an International Research Scholar Award 75195544202 of the Howard Hughes Medical Institute, by the U.S. Civilian Research and Development Foundation under Award No. RB1-248 (to A.I.S.), and by grants from the National Institutes of Health 5R21HG01640-02 (to J.A.L.) and 1R43GM55485 (to S.A.K).

[16] pGT5 Replication Initiator Protein Rep75 from *Pyrococcus abyssi*

By STÉPHANIE MARSIN and PATRICK FORTERRE

Introduction

Plasmids are extrachromosomal elements that can replicate autonomously in host cells. They are well known and studied in bacteria and have allowed the development of essential genetic tools in many bacterial species. Plasmids have also been very useful in studying fundamental cellular mechanisms, such as DNA replication. The plasmid pGT5 (3.4 kb), isolated from the euryarchaeon *Pyrococcus abyssi*, was the first plasmid studied in hyperthermophilic archaea.[1] It was completely sequenced and shown to replicate via the rolling circle (RC) mechanism (see below). This asymmetrical replication mechanism, which was first described for ΦX174 and m13 coliphages,[2] is also used by many bacterial plasmids and eukaryotic viruses.[3–8] RC replicons encode a replication initiator protein (usually named Rep) that exhibits a site-specific endonuclease/ligase activity. Rep proteins recognize the plasmid double-stranded origin (*dso*) and cleave the plus strand, leaving a free 3′-hydroxyl end that will be used as primer for DNA chain elongation by the host replicase. During the replication round, Rep proteins usually remain covalently linked to the 5′ end of the displaced DNA chain. At the end of one replication round, Rep proteins cleave the *dso* again and religate the new synthesized strand, generating a circular single-stranded form that is converted into a double-stranded plasmid by host proteins.

Several Rep proteins encoded by bacterial or eukaryal RC replicons have been studied *in vivo* and *in vitro*, and they all share common characteristics.[9–14] *In vitro*,

---

[1] G. Erauso, S. Marsin, N. Benbouzid-Rollet, M. F. Baucher, T. Barbeyron, Y. Zivanovic, D. Prieur, and P. Forterre, *J. Bacteriol.* **178**, 3232 (1996).
[2] A. Kornberg and T. Baker, "DNA Replication," 2nd ed. W. H. Freeman & Co. 1992.
[3] G. del Solar, R. Giraldo, M. J. Ruiz-Echevarria, M. Espinosa, and R. Diaz Orejas, *Microbiol. Mol. Biol. Rev.* **62**, 434 (1998).
[4] M. Espinosa, G. del Solar, F. Rojo, and J. C. Alonso, *FEMS Microbiol. Lett.* **130**, 111 (1995).
[5] A. Gruss and S. D. Ehrlich, *Microbiol. Rev.* **53**, 231 (1989).
[6] S. A. Khan, *Microbiol. Mol. Biol. Rev.* **61**, 442 (1997).
[7] E. V. Koonin and T. V. Ilyina, *J. Gen. Virol.* **73**, 2763 (1992).
[8] R. P. Novick, *Ann. Rev. Microbiol.* **43**, 537 (1989).
[9] R. Hanai and J. C. Wang, *J. Biol. Chem.* **268**, 23830 (1993).
[10] M. F. Noirot-Gros, V. Bidnenko, and S. D. Ehrlich, *EMBO J.* **13**, 4412 (1994).
[11] J. Laufs, S. Schumacher, N. Geisler, I. Jupin, and B. Gronenborn, *FEBS Lett.* **377**, 258 (1995).
[12] C. D. Thomas, D. F. Balson, and W. V. Shaw, *J. Biol. Chem.* **265**, 5519 (1990).
[13] M. Moscoso, R. Eritja, and M. Espinosa, *J. Mol. Biol.* **268**, 840 (1997).
[14] R. A. Hoogstraten, S. F. Hanson, and D. P. Maxwell, *MPMI* **9**, 594 (1996).

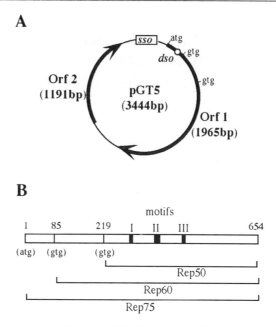

FIG. 1. pGT5 organization. (A) Map of pGT5. The two major ORFs (1 and 2) and their orientation are indicated by thick arrows. The three possible initiation codons for Orf1 and the putative *dso* and *sso* are noted. (B) Map of Rep75 encoded by ORF1. The three motifs (1–3) common to Rep proteins from the ΦX174/pC194 superfamily of RC replicons are indicated by black boxes. The positions of the three possible initiation codons are indicated.

they can cleave and religate an oligonucleotide harboring the *dso* sequence; the cleavage is always done by a transesterification with a tyrosine at the active site. Several superfamilies of RC replicons have been identified, based on sequence similarities among their Rep proteins.[15] One of them, the ΦX174/pC194 superfamily, includes coliphages, bacterial plasmids, and eukaryal geminiviruses. In particular, the plasmid pRQ7[16] from the hyperthermophilic bacterium *Thermotoga* sp. RQ7 and pGRB[17] from the archaeon *Halobacterium* sp. GRB belong to this superfamily. Rep proteins of the ΦX174/pC194 superfamily are characterized by three conserved motifs, 1, 2, and 3, which are present in the same arrangement from the N- to C-terminal.[15] The motif 3 contains either one or two active tyrosines involved in the cleavage and reactions of transphosphorylation. The two putative proteins encoded by pGT5 (orf1 and orf2, Fig. 1A) have no detectable similarity

[15] T. V. Ilyina and E. V. Koonin, *Nucleic Acids Res.* **20,** 3279 (1992).
[16] J. S. Yu, and K. M. Noll, *J. Bacteriol.* **179,** 7161 (1997).
[17] N. R. Hackett, M. P. Krebs, S. DasSarma, W. Goebel, U. L. RajBhandary, and H. G. Khorana, *Nucleic Acids Res.* **18,** 3408 (1990).

with other proteins in databases using search programs such as FASTA or BLAST. However, the three motifs characteristic of the ΦX174/pC194 Rep superfamily can be detected by visual inspection in the putative protein of 75 kDa encoded by orf1.[1] Moreover, an 11 nucleotide sequence, identical to the pC194 *dso,* is present in the pGT5 sequence.[1] These two observations suggested that pGT5 replicated via the RC mechanism and, as a result, could be another archaeal member of the ΦX174/pC194 superfamily.[15] This prediction has been confirmed by biochemical analyses. It was shown that: (1) cells of *P. abyssi* contain a single-stranded form of pGT5 that has the characteristic of a RC replication intermediate[1]; (2) the purified Rep protein of pGT5 (Rep75) exhibits a highly thermophilic and specific nicking-closing (NC) activity on *single-stranded* oligonucleotides containing the pGT5 *dso* sequence[18]; and (3) replacement of the only tyrosine located in motif 3 by a phenylalanine abolishes this activity.[19] In addition to its expected activities (NC), Rep75 exhibits an unusual site-specific nucleotidyl-terminal transferase (NTT) activity, never described before for a Rep protein.[18] This transfer occurs only at the 3' end of the nicking site with an adenine (or a deoxyadenine) nucleotide monophosphate, and it was proposed that this NTT activity plays a role in the regulation of pGT5 replication.[19]

The protein Rep75 can be overproduced in *Escherichia coli* and purified to near homogeneity. The method used, as well as the activity tests, is described in detail here and could be relevant to the study of other hyperthermophilic Rep proteins. It is still not known if Rep75 corresponds to the physiological form of the pGT5 Rep protein, since the orf1 of pGT5 contains three in-frame initiation codons (Fig. 1), and thus could encode for three putatives Rep proteins of 75, 66, and 50 kDa, respectively (Rep75, Rep60, and Rep50). In particular, since all Rep proteins from bacterial plasmids and viruses, as well as Rep proteins from geminiviruses, are smaller than Rep75 (in the range of 20 to 40 kDa), the physiological form of pGT5 Rep might have been Rep50 or Rep60. This chapter also describes the purification of the Rep50 protein expressed in *E. coli* and its biochemical characterization. Results have shown that Rep50 cannot be the physiological form of pGT5 Rep and provide novel information about the mechanism of Rep75 activity. Unfortunately, the *rep60* gene has yet to be cloned in *E. coli,* because of the instability of this DNA fragment in pET3b vector.

Purification of Recombinant pGT5 Rep75 Protein

Cloning of rep75 Gene

The Rep proteins of RC plasmids are usually present in very low copy number in the host cell. In bacteria, these proteins have been successfully purified from

---

[18] S. Marsin and P. Forterre, *Mol. Microbiol.* **27,** 1183 (1998).
[19] S. Marsin and P. Forterre, *Mol. Microbiol.* **33,** 537 (1999).

host cells only when it has been possible to take advantage of mutations that upregulate Rep expression *in vivo*. In the absence of such system, Rep proteins should be overproduced using expression vectors. The bacteriophage T7 expression system (pET3b plasmid from Stratagene, La Jolla, CA) is very efficient to obtain high amount of the pGT5 Rep75 proteins in *E. coli*. In this plasmid, the gene of interest is under the control of the bacteriophage T7Φ10 promoter and located downstream of a strong ribosome binding site. T7 RNA polymerase is provided by isopropylthiogalactoside (IPTG) induction of the T7 RNA polymerase gene under control of the lac repressor in a defective integrated lambda bacteriophage. The *rep75* gene has been cloned into the pET3b plasmid from PCR products amplified from the pGT5 plasmid using Vent polymerase[18] (New England Biolabs, MA). The recombinant plasmid is called pETRep75.

Cell Culture and Induction

Cells of *E. coli* BL21(λDE3) pLysS, carrying the plasmid pETRep75, are grown at 37° in L-broth containing 100 μg/ml ampicillin and 25 μg/ml chloramphenicol with vigorous aeration up to an absorbance of 0.6 at 600 nm (A_{600}). The Rep75 protein is induced by addition of 0.5 mM IPTG followed by growth for 3 or 4 hr up to an A_{600} close to 1. After induction, Rep75 can represent up to 30% of *E. coli* proteins (compare T0 and T4 on Fig. 2). The cells are then harvested by centrifugation without previous chilling, to prevent lysis, and kept frozen at −20°.

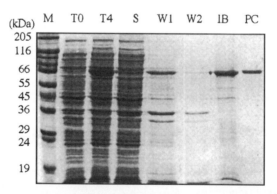

FIG. 2. SDS–PAGE analysis of Rep75 purification. The proteins were analyzed by electrophoresis using a 10% gel stained with Coomassie Brilliant Blue and photographed. M, molecular size marker; T0, *E. coli* crude extract of uninduced culture; T4, *E. coli* crude extract after 4 hr of induction; S, soluble proteins after sonication; W1 and W2, first and second wash, respectively, in buffer C of insoluble proteins; IB, aggregated Rep75 resolubilized in buffer D; PC, Rep75 elution from the phosphocellulose column.

Purification of Inclusion Bodies

Cells from 100 ml culture are resuspended at an A_{600} of 30 in 3 ml of ice-cold buffer A300 (25 mM Tris-HCl, pH 8.8, 300 mM NaCl, 2 mM dithiothreitol, and 1 mM EDTA) and broken by sonication. Three ml of ice-cold TE buffer is then added, and the viscosity due to DNA is reduced by incubation with 10 μg/ml DNase I in the presence of 8 mM MgCl$_2$ for 1 hr at 4°. Three ml of ice-cold buffer B [10 mM Tris-HCl, pH 8.8, 100 mM NaCl, 1% Nonidet P-40 (NP-40), 0.5% deoxycholate, and 1% Triton X-100) is added and mixed to solubilize membranes. Inclusion bodies containing Rep75 are collected by centrifugation at 10,000g for 10 min. The supernatant contains very little detectable soluble Rep75 protein (compare fractions T4 and S in Fig. 2). The pellet is washed twice in ice-cold buffer C (50 mM Tris-HCl, pH 8.8, 500 mM NaCl, and 1 mM EDTA). Part of Rep75 is lost in the first wash (W1), but this step eliminates some contaminants (compare fractions W1 and IB, Fig. 2). The insoluble material present in the pellet is solubilized in 1 ml of buffer D (10 mM Tris-HCl, pH 8.8, and 6 M guanidinium hydrochloride) (fraction IB, Fig. 2). Rep75 is completely solubilized by this procedure, whereas it is only partly solubilized in 8 M urea.

Renaturation and Purification of Solubilized Rep75

Fraction IB is diluted in buffer D down to a protein concentration of 0.1 mg/ml. Solubilized proteins present are renatured by continuous dialysis overnight at 4° against buffer A200 (buffer A300 with 200 mM NaCl instead of 300 mM). A large amount of protein precipitates during this step (see comments) and the aggregate is eliminated by centrifugation for 10 min at 15,000g. The supernatant is loaded on a 2 ml P11 cellulose phosphate (Whatman, Clifton, NJ) column equilibrated in buffer A200 and then washed with five volumes of buffer A300. Rep75 is eluted with 5 ml of buffer A450 (buffer A300 with 450 mM NaCl instead of 300 mM) (fraction PC, Fig. 2). Rep75 can be stored in this buffer with 0.08% sodium azide for 1 month at 4°, and for at least 1 year at $-80°$ without loss of activity.

Comments about Rep75 Purification

Soluble recombinant Rep75 protein could not be obtained by changing the conditions of induction, such as growth temperature or IPTG concentration. A purification procedure was therefore designed that takes advantage of Rep75 insolubility, using the isolation of inclusion bodies as an enrichment step. There were significant amounts of Rep75 in inclusion bodies, corresponding to about 5 mg per 100 ml of culture. Rep75 was recovered after solubilization of the inclusion bodies followed by renaturation of the protein. During the renaturation step, 70 to 80% of the protein aggregated. This percentage increases if the initial Rep75 concentration is more than 0.1 mg/ml. Aggregation also occurs in the phosphocellulose column if the column volume is less than 1 ml for 1 mg of soluble protein. Moreover, after elution, less than 0.25 mg/ml of Rep75 protein was obtained, and it was not possible to concentrate Rep75 further. Nevertheless, between 0.5 mg and 1 mg of pure recombinant Rep75 could be obtained routinely per 100 ml of culture. Rep75 is >95% homogeneous as determined by Coomassie Blue stained SDS–PAGE.

Rep75 preparations obtained after denaturation/renaturation could *a priori* contain both inactive (incorrectly folded) and active proteins. This is likely, in view of the amount of protein required to get *in vitro* activities (see below). It is not possible to perform quantitative enzymatic analysis with recombinant Rep75 since it was not possible to determine the proportion of active protein in the final fraction (PC, Fig. 2).

Preparations sometimes contained low amount of polypeptides that have molecular weights, slightly lower than those of Rep75 and that cross-react with Rep75 antibodies (data not shown). These proteolytic products of Rep75 are as thermostable as full-size Rep75 protein (Fig. 3A).

Figure 3A shows that only 60% of recombinant Rep75 remained soluble when the purified protein is incubated for 1 hr at 100°. However, the protein preparation retains activity and its level remains constant (this is the case, for example, for the nicking activity illustrated in Fig. 3B). This suggests that the polypeptides that disappear during incubation at 100° are those that have not been correctly refolded during the renaturation step of the purification procedure. One can thus conclude that correctly folded and active Rep75 proteins are fully stable for at least 1 hour at 100°.

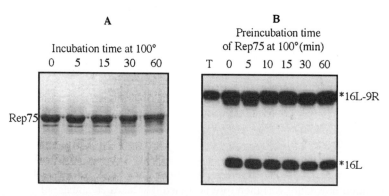

FIG. 3. Rep75 thermostability. (A) Rep75 remains soluble after incubation at 100°. Purified Rep75 was incubated at 100° during the indicated time and centrifuged for 10 min at 15,000g to eliminate insoluble material. Soluble Rep75 proteins were then loaded on a 10% SDS–PAGE. The gel shown was stained with Coomassie Brillant Blue. (B) Rep75 remains active after incubation at 100°. The Rep75 protein incubated at 100° during the indicated time was used in the nicking test. The labeled oligonucleotide substrate *16R-9L and its cleavage product *16L are indicated.

Recombinant pGT5 Rep50 Purification

The *rep50* gene has been cloned in the pET3 plasmid, using the same strategy as for the *rep75* gene. However, in contrast to the plasmid pETRep75, the plasmid harboring the *rep50* gene (pETRep50) is unstable in the BL21pLysS strain and is lost during the culture. To express Rep50, several colonies of transformed cells are thus selected on plates and used immediately as inoculum for the overproducing culture. Rep50 was then expressed, as described for Rep75. The protein Rep50 is also insoluble and can be recovered from purified inclusion bodies. Rep50 is solubilized in 1 ml of buffer D and renatured simply by a 10-fold dilution in buffer A300. Some aggregated material is then removed by centrifugation for 10 min at 15,000g. In contrast to Rep75, Rep50 does not bind to a phosphocellulose column; however, the protein Rep50 is at least 90% pure (similar to the degree of Rep75 purity in fraction IB). Approximately 0.5 mg of Rep50 protein is obtained from 100 ml of culture.

Assay Method

The activity of Rep proteins from RC plamids can be assayed on the basis of their ability to support *in vitro* replication initiation of RC plasmids. This is very tedious since it requires one to set up a complete *in vitro* system with host proteins. The basic function of Rep proteins is nicking–closing activity, which is usually specific for DNA-containing the *dso* plasmid sequence. This activity can be detected on supercoiled plasmid as a topoisomerase activity, or on single-stranded oligonucleotides containing the *dso* nicking site. The latter facilitates better analysis of the reaction, since it is possible to work with small oligonucleotides (less than 50 nucleotides) of defined sequence and to test either nicking or closing activity, as well as the novel terminal transferase activity discovered in Rep75. In tests for the nicking–closing activity of Rep75 and Rep50, it was noted that oligonucleotides as short as 15–25 nucleotides could be used. The oligonucleotides substrates

TABLE I
OLIGONUCLEOTIDES USED IN ACTIVITY TESTS[a]

Name	Sequence	Size
16L-9R	5′-CGTTGGG**TTTATCTTG**/**ATA**TATCCA-3′	25 nt
9L-16R	5′-**TTTATCTTG**/**ATA**TATCCACAACCAA-3′	25 nt
16L	5′-CGTTGGG**TTTATCTTG**	16 nt

[a] The characters in boldface type indicate the conserved nucleotides between the putative *dso* pGT5 sequence and the *dso* sequence from pC194. The slash indicates the cleavage site.

routinely used are presented in Table I. They are radiolabelled at their 5′ end using [γ-^{32} P]ATP. To study the affinity of Rep75 or Rep50 proteins to DNA, the gel shift technique was used, which turned out to be suitable for following binding of the protein on both single-stranded or double-stranded oligonucleotides.

Nicking–Closing

For nicking-closing reactions, 20 fmol of the indicated oligonucleotide is incubated at 105° with Rep75 in 10 μl of buffer R (50 mM HEPES–HCl, pH 8,200 mM Sodium glutamate, 1 mM DTT, 5 mM MnCl$_2$, 0.1% Triton X-100, 1 mM EDTA, and 50 μg/ml BSA). For ligation tests, the incubation is performed in two steps, first 5 min at 105° and then 5 min at 75°, because of the different temperature requirement between nicking and closing activities (see below). Reactions are stopped on ice by addition of EDTA 25 mM and incubation with 15 μg of proteinase K for 30 min at 55°. Five μl of loading buffer is then added (50 mM Tris–borate, pH 8.3, 80% deionized formamide, 1 mM EDTA, 0.1% xylene cyanol, and 0.1% bromphenol blue). The reaction products are separated on a 20% polyacrylamide (20 : 1) gel containing 8 M urea. Autoradiograms are performed and quantified by scanning the gel with a Phosphoimager ImageQuant.

Nucleotide Terminal Transferase

16L oligonucleotide (Table I), harboring only the left part of the nicking site, is radiolabeled and used in NTT tests. The assay is performed at 75° in the same buffer as described for the NC test but with 1 mM of ATP or dATP.

Gel Shift

Twenty fmol of 5′-radiolabeled single-stranded or double-stranded oligonucleotides is incubated with 1.3 pmol of the Rep protein (protein to DNA molar ratio of 60 : 1) for 5 min. The reaction is stopped by addition of 25 mM EDTA and 5 μl of loading buffer (30% glycerol, 0.25% xylene cyanol, and 0.25% bromphenol blue). The products are then loaded on a 8% polyacrylamide gel (30 : 1) in 5% glycerol, 0.25× TBE, and run in 0.25× TBE at room temperature. Incubation temperature is not important for migration under native conditions, and no differences are observed when protein and DNA are previously incubated together either at 105° and 4°.

Comments about Rep75 Activities

Rep75 Nicking–Closing Activity

Figure 4 illustrates Rep75 nicking–closing activities on single-stranded oligonucleotides harboring the cleavage sequence of pGT5 *dso*. The 5′ radiolabeled *16L-9R is cleaved by Rep75, so that only the *16L nicking product can be

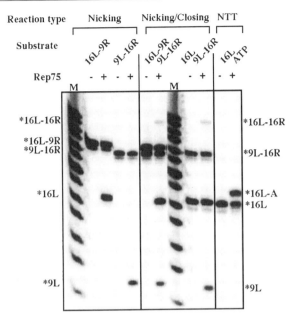

FIG. 4. *In vitro* Rep75 activities. Results of nicking, nicking–closing, and NTT tests were run on a polyacrylamide gel and autoradiographed. The radiolabeled oligonucleotides, used as substrates, are indicated on the upper part. The oligonucleotides detectable on the gel are indicated on the sides of the figure. M, markers.

detected. The same activity is obtained with the *9L–16R oligonucleotide, which is cleaved to give the *9L product. When the two substrates *16L–9R and *9L–16R are incubated together with Rep75, the two nicking products are obtained and the closing product *16L–16R can be detected (the ligation efficiency is too weak here to observe the *9L–9R closing product). A *16L–16R religation product is also obtained when *9L–16R is incubated with 16L and Rep75. No religation is observed when the test is performed with oligonucleotides containing only the right or the left part (16L and 16R oligonucleotides) of the nicking site (data not shown). This is because Rep75 should activate the DNA by nicking and covalent fixation before the religation step at the 3′ end.[18]

Rep75 NTT Activity

In the presence of ATP or dATP, Rep75 exhibits an unusual site-specific nucleotidyl transferase activity, i.e., it transfers one AMP or dAMP to the 3′OH extremity of an oligonucleotide containing the left part of the nicking site. Figure 4 illustrates an experiment in which the *16L oligonucleotide (Table II) is transformed into a 17-mer long oligonucleotide (*16L-A) by this activity. The

TABLE II
ACTIVITY COMPARISON BETWEEN REP75 AND REP50[a]

Source	Activities			DNA binding			
				ssDNA		dsDNA	
Protein	Nicking	Closing	NTT	Left part	Right part	Left part	Right part
Rep75	+	+	+	+	+	−	+
Rep50	+	−	−	−	−	nd	nd

[a] The activity assays were performed under the same conditions with the two proteins. DNA binding was performed by gel shift assay as described in the text, on oligonucleotide single-stranded (ss) DNA or double-stranded (ds) DNA, these substrates containing the left part or the right part of the nicking site. nd, Not determined.

transfer is slightly better with dATP than with ATP as substrate (Fig. 5C), and the activity is already maximal at 50 μM. The NTT activity is nucleotide specific, since Rep75 cannot use (d)CTP, (d)GTP, (d)TTP, AMP, ADP, or AMP-PNP. The NTT protein active site at least partly overlaps the NC active site, as an arginine located in the motif 3 of Rep75 is essential for both ligation and NTT activities. However, the nicking and NTT activities can be uncoupled *in vitro*, since the active tyrosine in motif 3 is dispensable for the NTT activity, and the arginine of motif 3 is dispensable for the nicking activity.[19]

The NC and NTT activities of Rep75 reach equilibrium after 15 min of incubation. Optimal results are obtained when Rep75 is added at a very high protein to DNA molar ratio (60 : 1),[18] indicating the low activity of the protein preparation. However, this is not unusual for this type of protein. In nicking assays, one never observes more than 50% of cleavage for the conditions studied, whereas it is possible to observe nearly complete transfer in NTT assays. Moreover, it is difficult to obtain the same specific activity for different protein preparations, probably due to the denaturation step in the purification process.

Comparison of Different Activities of Rep75

The different Rep75 activities are optimal at different assay conditions. The most important and surprising difference is their temperature optima. Figure 5A shows that NTT activity is optimal at 75°, whereas the nicking activity increases with temperature at least up to 105° (verified in a test tube using a thermoprobe). The closing activity is also known to be optimal around 85°.[19] The reason for these differences is presently unclear.

Another important observation is the difference in the divalent cation requirement between nicking and NTT activities (Fig. 5B). The two activities are optimal

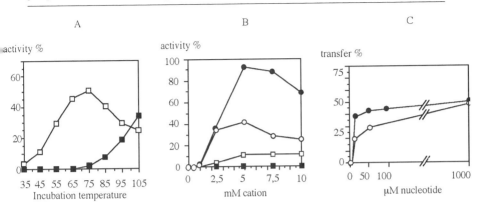

FIG. 5. Different characteristics of Rep75 activities. (A) Temperature dependence of Rep75 activities. Percentage activity corresponds to the percentage of modificated oligonucleotides, appearance of cleavage product in the case of nicking test (black squares), and apparition of transfer product in the case of NTT assay (white squares). (B) Effect of variation in Mg^{2+} and Mn^{2+} concentration on Nicking and NTT activities. The activities assays were performed in presence of Mn^{2+} (circles) or Mg^{2+} (squares). NTT activity (in black) and nicking activity (in white) were quantified as presented in A. (C) Effect of variation in ATP and dATP concentration on NTT activity. The NTT assays were performed for different concentrations of ATP (open circles) or dATP (black circles).

for 5 mM of $MnCl_2$. However, the NTT activity is more specific for this salt, since only 25% of the nicking activity is conserved when $MnCl_2$ is replaced by $MgCl_2$. On the other hand, the NTT activity cannot be observed at all in the presence of $MgCl_2$.

Comparison between Rep75 and Rep50

Partially purified Rep50 and Rep75 were compared (Table II). This makes sense since partially purified Rep75 (fraction IB, Fig. 2) exhibits the same activities as the completely pure protein (fraction S). Rep50 can perform site-specific DNA cleavage and its nicking activity is optimal at 105°, as in the case of Rep75. However, unlike Rep75, Rep50 does not have closing or NTT activities. This clearly indicates that Rep50 cannot be the native protein produced by pGT5 *in vivo*. The disappearance of the ligation and NTT activities in Rep50 can be explained by the inability of this truncated protein to recognize single-stranded DNA. Indeed, unlike Rep75,[18] Rep50 cannot bind single-stranded DNA fragment harboring the left part of the nicking site, as indicated by gel retardation experiment (Table II). Rep50 could have lost these properties because the single-stranded DNA binding site of Rep75 is located in its N-terminal region, which has been removed from Rep50, or because the absence of this region induces an incorrect protein folding. This suggests that the nicking activity (conserved in Rep50) does not involve the

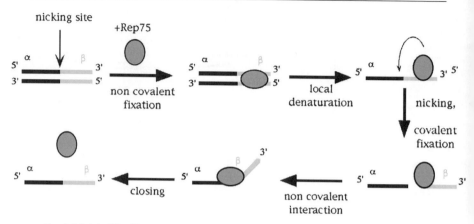

FIG. 6. Model of Rep75 origin recognition. α is the left part of the nicking site, β is the right part of the nicking site. The model is explained in the text.

strong interaction detected by gel shift between single-stranded DNA and Rep75. In contrast, this interaction should be essential for the closing and NTT activities.

Comparison between Rep75 and Rep50 suggests a mechanism for origin recognition by Rep75 (Fig. 6). Rep75 first interacts with the right part of the nicking site (double-stranded) and induces the local melting of DNA. This is similar to the model for recognition of ΦX174 *dso* by gpA, which has been proposed by Baas and Jans.[20] The single-stranded DNA region thus produced is then cleaved by Rep75, which remains covalently linked to the 5' end. The covalently linked protein could then interact with the left part (still single-stranded) of the nicking site and religates DNA. This last step cannot be performed by Rep50, since it is unable to interact with the left part of the nicking site.

The proteins Rep75 and Rep50 appear to be an interesting model system to study the mechanism of action of RC Rep proteins, in general. Overproduction of these proteins could pave the way for structural studies on an exciting new family of enzymes that can perform different types of DNA manipulations (cleavage, ligation, and terminal transfer) using a single active site.

Acknowledgments

This work was supported by grants from the Association pour la Recherche contre le Cancer and the European Community Programm Cell Factory (BIO4-CT 96-0488).

---

[20] P. D. Baas and H. S. Jansz, *Curr. Top. Microbiol. Immunol.* **136**, 31 (1988).

[17] Stability and Manipulation of DNA at Extreme Temperatures

By EVELYNE MARGUET and PATRICK FORTERRE

Introduction

The problem of DNA stability at temperatures near the boiling point of water rivaled that of protein stability to those scientists who first studied hyperthermophiles. It was even believed that the T_m of DNA might have determined the upper temperature limit for life on Earth,[1] and that a high G + C content of genomic DNA was a characteristic of hyperthermophiles. These two assumptions turned out to be wrong: life exists well above the T_m of linear DNA (up to 113°),[2] and there is no correlation between the G + C content of chromosomal DNA and the optimal growth temperature of hyperthermophiles.[3] This lack of correlation is not so surprising, since the T_m of a *topologically open* DNA molecule (such as a linear or a nicked DNA used in classical T_m experimental determination) is actually not relevant to the situation of intracellular DNA. The latter, either circular or linear, is *topologically closed* by cellular structures preventing the free rotation of the two strands of the helix. Unlike the situation with a linear or open circular DNA molecule, denaturation of a topologically closed DNA molecule does not result in two independent single-stranded molecules, but rather in a random-coiled structure in which the two single strands are still intertwined by the same number of topological links [the linking number (Lk)]. As a consequence, a topologically closed DNA molecule is highly resistant to thermodenaturation.[4] It has been shown that a bacterial plasmid remains double-stranded at least up to 107°.[5] This extraordinary resistance probably explains the absence of correlation between the genomic G + C content and the growth temperature of hyperthermophiles. Indeed, even in destabilizing buffer (7.3 M $NaClO_4$), the denaturation of topologically closed DNA does not depend appreciably on the G + C content.[6]

The *in vivo* characteristics of DNA are important not only to rationalize the existence of functional DNA at high temperature, but also for studying hyperthermophilic enzymes acting on DNA substrates *in vitro*. When using a topologically open DNA substrate, the upper temperature of the reaction with a

---

[1] E. S. Kemper, *Science* **142**, 1318 (1963).
[2] E. Blochl, R. Rachel, S. Burggraf, D. Hafenbradl, H. W. Jannasch, and K. O. Stetter, *Extremophiles* **1**, 14 (1997).
[3] K. O. Stetter, G. Fiala, G. Huber, R. Huber, and A. Segerer. *FEMS Microbiol. Rev.* **75**, 117 (1990).
[4] J. Vinograd, J. Lebowitz, and R. Watson, *J. Mol. Biol.* **33**, 173 (1968).
[5] E. Marguet and P. Forterre, *Nucleic Acids Res.* **22**, 1681 (1994).
[6] A. V. Gagua, B. N. Belintsev, and Y. Lyubchenko, *Nature* **294**, 662 (1981).

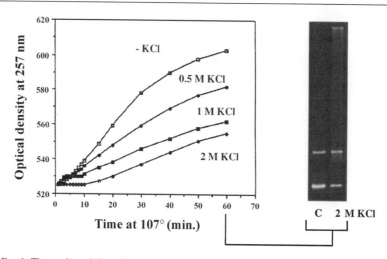

FIG. 1. Thermodegradation of plasmid DNA in the presence of various KCl concentrations. *Left:* Thermodegradation time course measured by spectrophotometry as in ref. 5. *Right:* DNA sample was taken from the above experiment after 1 hr of incubation at 107° in 2 M KCl and run on agarose neutral gel with plasmid control as described.

hyperthermophilic enzyme will be defined neither by the temperature dependence of the enzyme nor by its stability, but rather by denaturation of the substrate. In contrast, a topologically closed DNA molecule will not be denatured as long as the phosphate–sugar backbone remains intact. For example, the upper temperature limit for an *in vitro* transcription system in *Pyrococcus furiosus* is about 80° when linear DNA is used as template,[7] but it is 95°–100° when the template is topologically closed circular DNA.[8]

The major problem for topologically closed DNA at high temperature is degradation. This phenomenon can be followed by spectrophotometry, as illustrated in Fig. 1. When plasmid DNA is incubated at extreme temperatures (here, 107°), one observes a time-dependent increase in optical density at 257 nm that corresponds to the denaturation of the molecule. The denaturation kinetics reflect the introduction of nicks in the DNA backbone. Indeed, as soon as the topological constraints are relieved in one molecule by cleavage of one strand, this molecule is denatured at the temperature of the experiment, which is well above the T_m of nicked DNA. The rate of denaturation is thus proportional to the number of intact molecules remaining at any time in the solution. It is, therefore, very important to limit the rate of DNA cleavage as much as possible, both *in vivo* for hyperthermophiles

---

[7] C. Hethke, A. C. Geerling, W. Hausner, W. M. De Vos, and M. Thomm, *Nucleic Acids Res.* **24**, 2369 (1996).

[8] C. Hethke, A. Bergerat, W. Hausner, P. Forterre, and M. Thomm, *Genetics* **152**, 1325 (1999).

themselves, and *in vitro* when studying hyperthermophilic DNA-binding proteins or enzymes with DNA substrates. For example, it has been shown that DNA thermodegradation occurs in commonly recommended denaturation procedures[9] and is responsible for the low yield of large PCR (polymerase chain reaction) products compared to smaller ones.[10]

Mechanism of DNA Thermodegradation

The cleavage of DNA chains at high temperature occurs in two steps: first, depurination, and then hydrolysis of the phosphodiester bond adjacent to the apurinic site.[11] Depurination is the rate-limiting step, since both DNA cleavage and depurination increases at low pH and are inhibited by $MgCl_2$,[12] whereas hydrolysis of the phosphodiester bond at an apurinic site is pH-independent and stimulated by $MgCl_2$.[13] The protection of DNA against thermodegradation by $MgCl_2$ reflects the difference between RNA and DNA, since $MgCl_2$ catalyzes RNA hydrolysis.[14] For RNA, thermodegradation is not due to depurination, but to a direct attack of the phosphodiester bond by the $2'$-OH oxygen of ribose.[15] The susceptibility of DNA to depurination (compared to RNA) specifically arises from the lack of the oxygen in $2'$ position of the ribose that weakens the N-glycosyl bond (for review, see Lindahl[16]).

In addition to $MgCl_2$, KCl also protects double-stranded DNA against cleavage.[5] It was first supposed that this effect simply arises from stabilization of the double helix at high ionic strength, because depurination occurs more rapidly in single-stranded DNA than in double-stranded DNA.[17] However, this is not the case, since $MgCl_2$ and KCl also protect single-stranded DNA against depurination and cleavage.[12] $MgCl_2$ and KCl thus should directly interact with DNA to stabilize the N-glycosidic bond.

The effect of salt on DNA thermodegradation is especially relevant *in vivo* and for *in vitro* manipulation. Figure 1 shows that about 40% of topologically closed, circular DNA remains double-stranded after 1 hr of incubation at 107° in 2 M KCl. Because high concentrations of K^+ are present in some hyperthermophiles, this ion could play a key role in protecting DNA against thermodegradation.[12] However,

---

[9] D. Porter and B. M. Sanborn, *BioTechniques* **13**, 406 (1992).
[10] C. E. Gustafson, A. A. Alm, and T. J. Trust, *Gene* **123**, 241 (1993).
[11] T. Suzuki, S. Oshumi, and K. Makino, *Nucleic Acids Res.* **22**, 4997 (1994).
[12] E. Marguet and P. Forterre, *Extremophiles* **2**, 115 (1998).
[13] T. Lindahl and A. Andersson, *Biochemistry* **11**, 3618 (1972).
[14] T. Lindahl, *J. Biol. Chem.* **242**, 1970 (1966).
[15] W. Ginoza, C. J. Hoelle, K. B. Vessey, and C. Carmack, *Nature* **203**, 606 (1964).
[16] T. Lindahl, *Nature* **362**, 715 (1993).
[17] T. Lindahl and B. Nyberg, *Biochemistry* **11**, 3610 (1972).

salts have varying effects on DNA at high temperature, such that it is not always appropriate to extrapolate *in vitro* results to *in vivo* situations.

From the above discussion, it is clear that, when examining an enzyme *in vitro* with a DNA substrate at high temperature, a compromise needs to be made between salt and pH requirements of this enzyme and those for DNA stability. It is always essential to determine the sensitivity of the DNA substrate to the particular reaction conditions.

They are ways to easily check for DNA thermodegradation using agarose gels. Although these methods are rapid and accurate for thermodegradation analysis, one should proceed with caution when monitoring thermodenaturation, since the DNA can renature in the well at low temperature and high salt concentration (see below). A protocol for the detection of apurinic sites using alkaline agarose gels is also provided, as are methods for preparation of depurinated DNA. These methods can be useful for checking the actual amount of intact DNA after temperature incubation, since the presence of apurinic site does not change the migration of DNA in neutral gels (as long as the nearby phosphodiester bond has not been cleaved). It is also useful for checking the effect of depurination on various DNA-metabolizing enzymes. In the course of this work, it was also noticed that some combinations of salt, buffer, and temperature produce abnormal migrations in gels (neutral or alkaline) that can be troublesome when using this method to test enzymatic activities dealing with DNA at very high temperature. Some of these effects that have not been described previously are presented here, both as a warning and in the hope of stimulating additional work on this interesting aspect of DNA thermostability.

Agarose Gel Assays for DNA Thermodegradation

DNA thermodegradation is analyzed by incubating 0.5 µg of DNA in 20 µl incubation mixture covered with 150 µl of H_2O-saturated paraffin oil to prevent evaporation. The pH and temperature of the incubation mixtures are controlled through temperature and pH probes. The pH determined by probes can be slightly different from the pH calculated for pH–temperature compensation,[18] depending on salt present or buffer strength.

Plasmid pTZ18 isolated from *Escherichia coli* JM109 is routinely used as source of double-stranded supercoiled DNA, whereas for single-stranded DNA, M13 DNA isolated from bacteriophage M13mp19 is used. To prepare depurinated plasmid DNA, 500 ng of pTZ18 or 1 µg M13mp19 is incubated in 20 µl of 50 mM Na_2HCO_3, pH 5.5, from 10 to 30 min at 75°. Depurinated DNA is precipitated by addition of 20 µl 3 M sodium acetate and 600 µl of ethanol, and resuspended in 200 µl HEPES buffer at 50 mM, pH 7.5. The extent of depurination can be estimated by comparison between neutral and alkaline gels (see below).

---

[18] V. S. Stoll and J. S. Blanchard, *Methods Enzymol.* **182**, 24 (1990).

For neutral gel electrophoresis, 20 μl DNA samples are run at room temperature in 0.7% indubiose gel in TBE buffer (45 mM Tris–borate, 1 mM EDTA, pH 8) for 16 hr at 2 V cm^{-1}. Indubiose (Bio-Rad, Hercules, CA) is used instead of agarose for better resolution of both DNA topoisomers and single-stranded DNA forms.

For alkaline gel electrophoresis, DNA samples are first incubated before loading for at least 2 hr in 0.5 M NaOH to denature DNA and cleave DNA strands at apurinic sites (see below). Gels are prepared in 50 mM NaCl, 4 mM EDTA, and soaked for at least 1 hr in running buffer containing 30 mM NaOH and 2 mM EDTA. Alkaline-treated samples are run for 16 hr at 1 V cm^{-1} in a 0.7% agarose gel (with recirculation of the buffer).

For DNA detection, the gels are washed with water and stained with ethidium bromide. Polaroid photographs are taken under UV transillumination at 254 nm. Visual examination of the gels is sufficient to get useful information about DNA stability. Densitometric analysis can be used to determine approximately the relative amount of the different DNA forms present in the gel. However, such determination is not very precise at high DNA concentrations, since labeling with ethidium bromide does not increase linearly with DNA under these conditions. For accurate results, it is necessary to radioactively label DNA and use a PhosphoImager.

Comments on the Gel Assays

Thermodegradation of supercoiled double-stranded DNA (topologically closed, form I) first produces circular nicked molecules (topologically open, form II). Form II molecules are thus linearized (form III) when two nicks are located closed to each other, and finally smaller fragments are produced. At high temperature, form II is immediately denatured and gives two single-stranded forms: a linear one (ss1) and a circular one (ssc). These two forms are the only degradation products visible on agarose gels, since the other products have different sizes and either produce a smear or are not visible at all. A time course for thermodegradation of plasmid pTZ18 is shown in Fig. 2. To quantify the extent of thermodegradation, one should follow the decrease of Form I, since the amount of ssc and ss1 forms are in equilibrium balancing their production (e.g., ssc from FI) and destruction (e.g., ssc to ss1).

In neutral gels, the exact position of forms ssc and ss1, relative to FI and FII, depends on several parameters, such as the size of the plasmid, the percentage of agarose, or the salt concentration in running buffers. For example, whereas both single-stranded forms of pTZ18 migrate clearly between FI and FII in TBE 0.5× (Fig. 2), the form ss1 comigrates with FI in TBE 1× buffer (see Fig. 3, upper panel, and Fig. 4, neutral gel). To distinguish the ss1 band from FI, one could run neutral gels in the presence of an intercalating agent, which relaxes the negatively supercoiled FI but does not affect the migration of single-stranded DNA. One can use either ethidium bromide or chloroquine at the low concentrations which partly

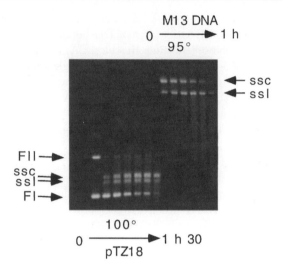

FIG. 2. Kinetics of thermodegradation of double-stranded DNA and single-stranded DNA visualized in a neutral gel. Plasmid pTZ18 was incubated at 100° in 50 mM Tris, pH 7.5 (room temperature) for 0, 15, 30, 45, 60, 75, and 90 min. M13 DNA was incubated at 95° in the same buffer for 0, 10, 20, 30, 60, 45, and 60 min.

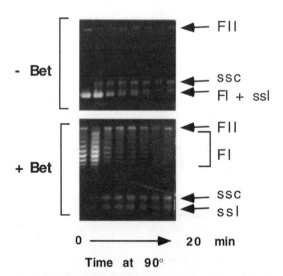

FIG. 3. Kinetics of thermodegradation of double-stranded DNA in neutral gels with or without ethidium bromide. Plasmid pTZ18 was incubated at 90° in 50 mM Tris buffer, pH 7.5 (room temperature) for 0, 5, 10, 12, 14, 16, 18, and 20 min. Samples were divided in two parts and run either in a normal neutral gel or in a neutral gel with 10 ng/ml of ethidium bromide both in the buffer used to prepare the gel itself and in the running buffer.

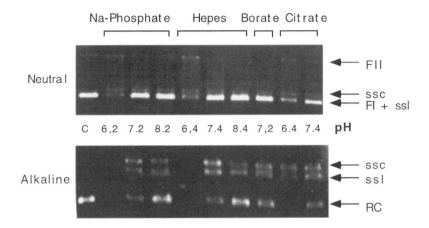

FIG. 4. Effect of various pH and buffers on double-stranded DNA thermodegradation analyzed using neutral and alkaline gels. Plasmid pTZ18 was incubated for 20 min at 100° in various buffers at the indicated pH (measured at the incubation temperature with a pH probe). After incubation, samples were divided into two parts and run in either neutral or alkaline gels. RC, Random-coiled molecules.

relax supercoiled plasmids. Figure 3 (lower panel) shows a gel run in the presence of 10 ng/ml of ethidium bromide. One can see the different topoisomers that were present in the FI and have been relaxed by the drug. These topoisomers are clearly separated from the form ss1, which comigrates with FI in the gel run in the absence of the drug.

In the case of M13, used to test the effect of thermodegradation on single-stranded DNA, the preparation initially contains both circular and linear forms (ssc and ss1) (Fig. 2). On incubation at high temperature, the circular form is converted into the linear one by the first nick. The linear form is then converted into smaller fragments. Thermodegradation should, thus, be followed by the disappearance of the form ssc.

The products of DNA thermodegradation have different patterns of migration in neutral and alkaline gels. In alkaline gels, supercoiled plasmids are denatured, but the two strands remain topologically closed, producing a random coiled DNA (RC). This form migrates faster than FI or ss DNA. In Fig. 4, the two upper bands visible in the alkaline gel correspond to the forms ss1 and ssc. The lower band corresponds to the RC form. Alkaline gels also allow one to discriminate between ssDNA and FI. In addition, if some FI molecules are partly depurinated, they are cleaved by the alkaline treatment. Indeed, the phosphodiester bond located nearby an apurinic site is labile in alkali. Accordingly, determination of the amount of residual RC form allows one to estimate the DNA that remains completely intact (without abasic sites) after incubation at very high temperature. Figure 4 shows an illustrative comparison between alkaline and neutral gels and illustrates the effect of pH on thermodegradation. In this experiment, the DNA has been analyzed after 20 min of incubation at 100° at various pH values (measured *in situ*) in different buffers. The pH decreases with increasing temperature; for example, the pH of Tris buffers decreases by about two units between 25 and 95°.[18] In this experiment, buffers were used in which pH is less sensitive than in Tris buffer to temperature variation (i.e., HEPES, phosphate, and borate buffers). One can see that the relative amount of DNA which migrates as a topologically closed molecule is higher in the neutral gel than in the alkaline gel. This indicates that the form FI detected in neutral gel after incubation at high temperature corresponds to a mixture of intact FI molecules and FI molecules containing apurinic sites. The intensity of form I in neutral gel is similar in sodium phosphate buffer at pH 7.2 and 8.2, whereas the intensity of the RC form in alkaline gels is lower at pH 7.2 than at pH 8.2. The DNA that has not been cleaved by thermal treatment at low pH thus contains more apurinic site than the DNA incubated under the same conditions at higher pH.

In the case of M13 DNA, only the ss1 form was observed in alkaline gels,[12] indicating that typical preparations of M13 DNA contain apurinic sites.

Effect of Salts on DNA at High Temperature and on Detection of Degradation Products

In addition to protecting DNA against thermodegradation, the presence of salt often alters the pattern of DNA migration in agarose gels. First, salts counteract the denaturation of double-stranded DNA at high temperature after cleavage, and promote renaturation of single-stranded products of thermodegradation as soon as the temperature of the incubation mixture is lowered. When thermodegradation of plasmid DNA is studied in the presence of salt, the reaction product in agarose gel will thus be restricted to form II in neutral gels (Fig. 1, right panel). High salt concentrations will also make plasmid DNA resistant to alkaline denaturation.[12] The FI form, which has not been denatured by the alkaline treatment, migrates more slowly than the RC form in alkaline gel, either slightly in front of the ss1 form or at the same position.[5,12]

All salts tested also induce abnormal migrations in agarose gels, such as aggregation in the wells, smearing, or retardation. These effects, which are detailed below for some commonly used salts, are not observed at low temperatures, indicating that salts induce irreversible modifications in the DNA structure at high temperature. These are concentration-dependent and differ from one salt to another, depending on the incubation buffer. Some of these effects are only visible at very high concentrations, but others are present at concentrations used for enzymatic assays.

Perturbation of DNA migration after incubation with magnesium at high temperature has already been reported.[5,9] In summary, incubation of double-stranded or single-stranded DNA at high temperature with concentrations of $MgCl_2$ above 5–10 mM produces DNA aggregates that do not enter into the gel. Migration of single-stranded and double-stranded DNA is also disturbed by $MgCl_2$, forming smears in alkaline gels.

In the case of plasmid duplex DNA, $MgCl_2$ probably induces the aggregation of the ssDNA produced by denaturation at high temperature. Indeed, such aggregation does not take place when KCl is also present (concentrations as low as 25 mM KCl are sufficient to prevent aggregation, probably because, in combination with $MgCl_2$, it prevents denaturation at high temperature).

$CaCl_2$ also protects DNA against thermodegradation at low concentrations (0.5–5 mM) and produces either aggregates or smears at high concentrations, depending on the buffer used (Fig. 5). In particular, this salt does not protect DNA at high concentration in sodium phosphate buffer. The effects of $MnCl_2$ are similar to those of $MgCl_2$ and $CaCl_2$ when the DNA is incubated at high temperatures in sodium phosphate buffer (Fig. 5), but this salt does not protect DNA in HEPES or Tris buffers (not shown).

In the gel shown in Fig. 5, one can see an additional band corresponding to the dimeric form I (ID). This form, as well as open circular dimers (IID), is produced by

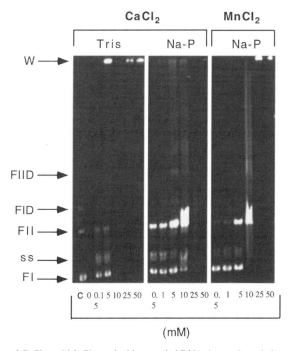

FIG. 5. Effect of $CaCl_2$ and $MnCl_2$ on double-stranded DNA thermodegradation. Plasmid pTZ18 was incubated for 30 min at 95° either in 25 mM Tris buffer, pH 7.5 (room temperature), or 25 mM sodium phosphate buffer and run in a neutral gel. c, Control without incubation at high temperature; W, well; FIID, dimer of form II; FID, dimer of form I.

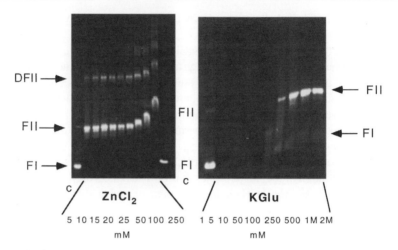

FIG. 6. Effect of ZnCl$_2$ and potassium glutamate (KGlu) on double-stranded DNA thermodegradation (neutral gel). To test the effect of ZnCl$_2$, plasmid pTZ18 was incubated for 30 min at 95° in 50 mM Tris buffer, pH 7.5 (room temperature). To test the effect of KGlu, plasmid pTZ18 was incubated for 1 hr at 95° in 25 mM Tris buffer, pH 7.5 (room temperature). c, Control without incubation at high temperature; DFII, dimer of form II.

in vivo plasmid recombination and is present in some control plasmid preparation (see also Fig. 6).

The effect of ZnCl$_2$ is quite different, since this salt protects DNA at relatively low concentrations (0.4 M) (not shown). This is another dramatic difference between DNA and RNA, since Zn^{2+} ions strongly stimulate DNA degradation at high temperature.[19] However, protection by ZnCl$_2$ is not complete since supercoiled plasmids are never recovered (Fig. 6). Furthermore, high concentrations of ZnCl$_2$ (50–250 mM) also disturb plasmid migration (Fig. 6). The absence of FI in protection experiments by ZnCl$_2$ suggests that this salt specifically cleaves double-stranded DNA at high temperature when it is supercoiled. Indeed, low concentration of ZnCl$_2$ completely protects linear double-stranded DNA against thermodegradation (not shown).

Finally, abnormal DNA migration was also observed in the presence of potassium or sodium glutamate. This is troublesome since these salts are now often used preferentially to test the activity of hyperthermophilic enzymes. Indeed, many enzymes apparently can support higher ionic strength in potassium or sodium glutamate than in KCl or NaCl. Glutamate salts somehow mimic the compatible solutes that are present in high concentration in many hyperthermophiles and stabilize proteins at very high temperatures.[20] As shown in Fig. 6, potassium

[19] J. J. Butzow and G. L. Eichorn, *Nature* **254**, 358 (1975).
[20] R. Hensel and H. König, *FEMS Microbiol. Rev.* **49**, 75 (1988).

glutamate (K Glu), like KCl, protects DNA against thermodegradation, but at high concentrations of KGlu the supercoiled form of plasmid DNA is not recovered, in contrast to what happens with KCl (Fig. 1), and the migration pattern is highly disturbed. In preliminary experiments, the potassium salt of a native compatible solute present in hyperthermophiles, mannosyl glycerate, had little protective effect on DNA at high temperature and induced even more perturbation of gel migration than glutamate salts, even at low temperature (not shown).

All these unusual salt effects should be taken into account in PCR experiments using multiple amplification rounds or long incubation time at high temperatures, or else when studying hyperthermophilic DNA enzymes *in vitro*. It appears that salt and DNA interact at very high temperatures in various ways that are not yet understood. It will be interesting to explore the chemistry of these unusual interactions in the future, and to determine if these interactions could also occur *in vivo*.

Acknowledgments

Our work was supported by grants from the Association pour la Recherche contre le Cancer and the European Community Program Cell Factory (BIO4-CT 96-0488).

[18] Ribonucleotide Reductase from *Pyrococcus furiosus*

By MARC FONTECAVE

Introduction: Three Classes of Ribonucleotide Reductases with Different Metal Cofactors and Free Radicals

DNA synthesis depends on a balanced supply of the four deoxyribonucleotides.[1] In all living organisms, with no exception to date, this is achieved by reduction of the corresponding ribonucleoside diphosphates, NDPs, or triphosphates, NTPs (Scheme 1). This reaction is catalyzed by a fascinating family of allosterically regulated metalloenzymes, named ribonucleotide reductases (RNRs).[2-5]

It is now generally accepted that life was first based on RNA and that the emergence of a ribonucleotide reductase was the key event that allowed the

---

[1] P. Reichard, *Ann. Rev. Biochem.* **57**, 349 (1988).
[2] P. Reichard and A. Jordan, *Ann. Rev. Biochem.* **67**, 71 (1998).
[3] B.-M. Sjöberg, *Structure and Bonding* **88**, 139 (1997).
[4] J. Stubbe, *Adv. Enzymol. Related Areas Mol. Biol.* **63**, 349 (1990).
[5] M. Fontecave, *Cell. Mol. Life Sci.* (1998).

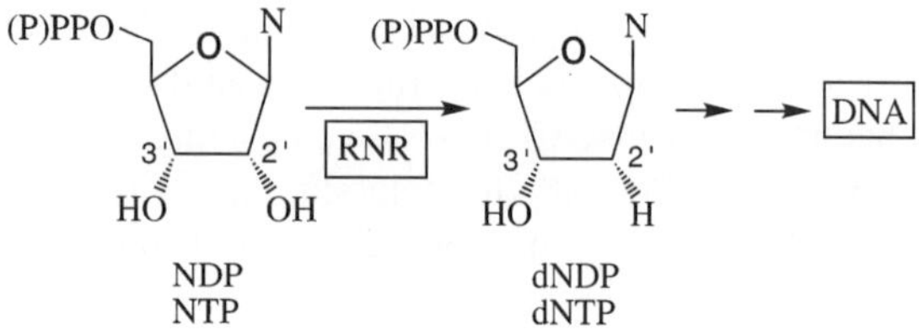

SCHEME 1. The reaction catalyzed by ribonucleotide reductases.

transition from the RNA to the DNA world.[2,6,7] According to that concept, one would expect to find only one type of enzyme with the same general structure in all organisms. Instead, in contemporary metabolism, at least three distinct classes of RNR are found, which probably are the products of divergent evolution from a common ancestor.[6] The evolutionary relationship between these three classes has been discussed in a number of excellent review articles.[2,6,8]

Class I RNRs are strictly aerobic $\alpha_2\beta_2$ enzymes, and their substrates are ribonucleoside diphosphates. They are divided into two subclasses, Ia and Ib. Class Ia RNRs are found in all types of eukaryotes and several viruses, as well as a few prokaryotes and bacteriophages. Both proteins R1 (α_2) and R2 (β_2) from *Escherichia coli* have been crystallized and their three-dimensional structure determined at high resolution.[9–12] Protein R1 contains the binding sites for both substrates and allosteric effectors. Complexes of protein R1 with substrates and effectors have been structurally characterized[13] showing that the substrate site contains the three conserved redox-active cysteines, which were previously suggested to participate in ribonucleotide reduction.[14,15] Protein R2 contains a tyrosyl radical essential for enzyme catalysis and a nonheme diiron center, in which the ferric ions are linked by an oxo and a bidentate glutamate bridge, on each polypeptide chain.[9,11]

---

[6] P. Reichard, *Trends Biochem. Sci.* **22**, 81 (1997).

[7] J. Riera, F. T. Robb, R. Weiss, and M. Fontecave, *Proc. Natl. Acad. Sci. U.S.A.* **94**, 475 (1997).

[8] P. Reichard, *Science* **260**, 1773 (1993).

[9] P. Nordlund and H. Eklund, *J. Mol. Biol.* **232**, 123 (1993).

[10] U. Uhlin and H. Eklund, *Nature* **370**, 533 (1994).

[11] M. Fontecave, P. Nordlund, H. Eklund, and P. Reichard, *Adv. Enzymol. Related Areas Mol. Biol.* **65**, 147 (1992).

[12] B.-M. Sjöberg, *in* "Nucleic Acids and Molecular Biology" (F. Eckstein and D. M. J. Lilley, eds.), Vol. 9, p. 192. Springer-Verlag, Berlin, 1995.

[13] M. Eriksson, U. Uhlin, S. Ramaswamy, M. Ekberg, K. Regnström, B.-M. Sjöberg, and H. Eklund, *Structure* **5**, 1077 (1997).

[14] A. Åberg, S. Hahne, M. Karlsson, A. Larsson, M. Ormö, A. Ahgren, and B.-M. Sjöberg, *J. Biol. Chem.* **264**, 12249 (1989).

[15] S. S. Mao, T. P. Holler, G. X. Yu, J. M. Bollinger, S. Booker, M. I. Johnston, and J. Stubbe, *Biochemistry* **31**, 9733 (1992).

Class Ib RNRs are also $\alpha_2\beta_2$ enzymes found in bacteria.[2] They are closely related to class Ia enzymes, with similar Fe-radical center and amino acid sequences, except for the lack of the first 50 N-terminal amino acid residues in the large α_2 (R1) protein. As the N terminus provides residues for binding the allosteric effectors, ATP and dATP, this lack results in differences in the allosteric regulation of ribonucleotide reduction.[16]

Class II RNRs are found in bacteria and archaea.[2,7,17,18] They are active both aerobically and anaerobically. A class II RNR is characterized by the requirement for adenosylcobalamin (AdoCbl). The enzyme facilitates homolysis of the Co–C bond of Adobl for generating an essential cysteinyl radical.[19] A three-dimensional structure is not available to date, but elegant studies by Stubbe and co-workers[20] have demonstrated that ribonucleotide reduction depends on the presence of three essential redox-active cysteines in the active site and proceeds much as in class I RNRs.

Class III RNRs are oxygen-sensitive enzymes found in some facultative anaerobes and bacteriophages.[2,21,22] On the basis of sequence comparisons, it seems likely that methanogens also use a class III enzyme for deoxyribonucleotide synthesis.[2,3] The prototype is the enzyme discovered in 1989 in anaerobically growing *Escherichia coli* cells.[21] This class has not been structurally characterized yet. However, biochemical and spectroscopic studies have shown that it is an $\alpha_2\beta_2$ enzyme.[23] The large component α_2 contains the substrate and the allosteric effector binding sites and, in its active form, a glycyl radical (Gly-681 in *E. coli*) absolutely required for catalysis.[24–26] The small component β_2 contains an iron-sulfur center which catalyzes the reduction of *S*-adenosylmethionine by flavodoxin to generate a putative 5′-deoxyadenosyl radical. The latter is supposed to be a precursor of the glycyl radical on protein α_2.[23,27–29]

[16] R. Eliasson, E. Pontis, A. Jordan, and P. Reichard, *J. Biol. Chem.* **271**, 26582 (1996).
[17] J. Harder, *FEMS Microbiol. Rev.* **12**, 273 (1993).
[18] A. Tauer and S. Benner, *Proc. Natl. Acad. Sci. U.S.A.* **94**, 53 (1997).
[19] S. Licht, G. J. Gerfen, and J. Stubbe, *Science* **271**, 477 (1996).
[20] S. Booker, S. Licht, J. Broderick, and J. Stubbe, *Biochemistry* **33**, 12676 (1994).
[21] M. Fontecave, R. Eliasson, and P. Reichard, *Proc. Natl. Acad. Sci. U.S.A.* **86**, 2147 (1989).
[22] P. Young, M. Öhman, and B.-M. Sjöberg, *J. Biol. Chem.* **269**, 27815 (1994).
[23] S. Ollagnier, E. Mulliez, J. Gaillard, R. Eliasson, M. Fontecave, and P. Reichard, *J. Biol. Chem.* **271**, 9410 (1996).
[24] X. Sun, J. Harder, M. Krook, H. Jörnvall, B.-M. Sjöberg, and P. Reichard, *Proc. Natl. Acad. Sci. U.S.A.* **90**, 577 (1993).
[25] X. Sun, S. Ollagnier, P. P. Schmidt, M. Atta, E. Mulliez, L. Lepape, R. Eliasson, A. Gräslund, M. Fontecave, P. Reichard, and B.-M. Sjöberg, *J. Biol. Chem.* **271**, 6827 (1996).
[26] P. Young, J. Andersson, M. Sahlin, and B.-M. Sjöberg, *J. Biol. Chem.* **271**, 20770 (1996).
[27] E. Mulliez, M. Fontecave, J. Gaillard, and P. Reichard, *J. Biol. Chem.* **268**, 2296 (1993).
[28] E. Mulliez and M. Fontecave, *Chem. Ber.* **130**, 317 (1997).
[29] S. Ollagnier, E. Mulliez, P. P. Schmidt, R. Eliasson, J. Gaillard, C. Deronzier, T. Bergman, A. Gräslund, P. Reichard, and M. Fontecave, *J. Biol. Chem.* **272**, 24216 (1997).

In 1996, when the ribonucleotide reductase in *Pyrococcus furiosus* and various *Sulfolobus* strains was first investigated, nothing was known about ribonucleotide reduction in thermophilic organisms. The assumption was that these organisms might have been a source of a new class of RNR, on the basis that the aerobic class I RNR could not be active in the anaerobic strains and that the cofactors AdoCbl (in class II) and AdoMet (in class III) could not be used at high temperature, because of their heat sensitivity. This assumption was wrong, as the enzymes from *P. furiosus, Sulfolobus shibatae,* and *S. acidocaldarius* required AdoCbl and could thus be classified as a class II RNR.[7] At the same time, Benner[18] isolated the enzyme from another archaea, *Thermoplasma acidophilum,* and Reichard[30] reported on the RNR from the hyperthermophilic eubacteria *Thermotoga maritima*. All these enzymes proved to be AdoCbl-dependent. Finally, the sequenced genomes of *Archaeoglobus fulgidus* and *Methanobacterium thermoautotrophicum* revealed the presence of a RNR highly homologous to that of *P. furiosus*, indicating that growth of these organisms also likely depends on a class II RNR. So far, all isolated thermophilic RNRs belong to the class II. The *M. thermoautotrophicum* genome has been shown to contain genes homologous to class III RNR genes as well. However, the corresponding enzyme has not been purified and it is not known whether it is expressed and active.

This article deals with the enzyme from *P. furiosus* as a prototype for thermophilic RNRs. Some aspects, such as allosteric regulation studied with other RNRs, will also be discussed.

Adenosylcobalamin-Dependent RNR from *P. furiosus:* Class II RNR

Purification of Enzyme

RNR activity is readily detectable in soluble cell-free extracts of *S. shibatae, S. solfataricus,* and *P. furiosus,* incubated with cytidine diphosphate (CDP), dithiothreitol (DTT), and adenosylcobalamin (AdoCbl), at high temperatures (70°–80°). Sodium acetate stimulates the activity and is included in the assay. Acetate probably acts as an allosteric effector. The activity is not air-sensitive and purification can be achieved aerobically at room temperature, reflecting the extreme stability of the enzyme. The activity can be purified 3000-fold in three steps (ammonium sulfate precipitation, phenyl-Sepharose and dATP-Sepharose chromatography) (Table I). The key purification step is affinity chromatography on dATP-Sepharose (400-fold purification), a method used successfully for the purification of several other RNRs.[31,32] Binding to this ligand is an indication that the thermophilic RNR is

---

[30] A. Jordan, E. Torrents, C. Jeanthon, R. Eliasson, U. Hellman, C. Wernstedt, J. Barbé, I. Gibert, and P. Reichard, *Proc. Natl. Acad. Sci. U.S.A.* **94,** 13487 (1997).
[31] R. Eliasson, E. Pontis, M. Fontecave, C. Gerez, J. Harder, H. Jörnvall, M. Krook, and P. Reichard, *J. Biol. Chem.* **267,** 25541 (1992).
[32] Y. Engström, S. Eriksson, L. Thelander, and M. Akerman, *Biochemistry* **18,** 2941 (1979).

TABLE I
PURIFICATION OF RIBONUCLEOTIDE REDUCTASE FROM *P. FURIOSUS*

Source	Protein (mg)	Specific activity	Total units	Purification factor
Extract	1020	0.17	173	1
Ammonium sulfate	380	0.35	133	2
Phenyl-Sepharose	58	1.3	77	7.6
dATP-Sepharose	0.1	510	51	3000

allosterically regulated by nucleoside triphosphates, a characteristic for all of the RNRs studied to date.[2] Chromatography on dATP-Sepharose is also used for the preparation of RNRs from *T. acidophilum* and *T. maritima*.[30]

In a typical experiment, 13 g of tightly packed anaerobically grown *P. furiosus* cells are sonicated in 20 ml Tris-HCl 50 mM, pH 8, EDTA 4 mM, DTT 5 mM (buffer A) in the presence of phenylmethylsulfonyl fluoride (PMSF) 1 mM and centrifuged at 180,000g for 90 min. Soluble extracts (1.02 g protein) are first treated with 3% (w/v) streptomycin sulfate, centrifuged at 40,000g for 15 min and then most of the activity precipitated with 50% saturated ammonium sulfate. After overnight dialysis against buffer A at 4°, the active solution, containing 0.38 g protein, is loaded onto a 25 ml phenyl-Sepharose (Pharmacia, Biotech) column, equilibrated with buffer A containing 0.5 M KCl. The column is eluted with buffer A plus 0.5 M KCl (0.1 ml/min), then 0 M KCl (0.5 ml/min). When the absorption at 280 nm is below 0.05, the activity is eluted with water (1 ml/min). A significant amount of enzyme is retained on the column and is eluted with 30% ethylene glycol (v/v). The two last fractions are combined and exchanged with buffer A.

The solution (58 mg) is then loaded onto a 2 ml dATP-Sepharose column, prepared according to ref. 33, and equilibrated with buffer A + 10% glycerol (buffer B) containing 0.5 M KCl. The column is washed with the equilibration buffer (0.2 ml/min). The reductase is then eluted with 2.5 mM ATP in buffer B. After overnight dialysis against buffer B, and concentration by centrifugation in Centricon 30 (Millipore Corporation, Bedford, MA) microconcentrators, aliquots of the enzyme solution are stored in liquid nitrogen. Gel filtration on Ultrogel AcA-34 shows that the enzyme is a monomer of about 90 kDa.

Assay for Enzyme Activity

In a standard assay, the protein sample is incubated at 80° for 30 min, in the presence of 5 μM adenosylcobalamin, 0.5 M sodium acetate, 40 mM dithiothreitol, and 1 mM [^{3}H] CDP in 50 mM Tris-HCl, pH 7.5, in a final volume of 50 μl. The

---

[33] O. Berglund and F. Eckstein, *Eur. J. Biochem.* **28**, 492 (1972).

TABLE II
ENZYMATIC PROPERTIES OF THERMOPHILIC RNRs

Source	Structure	K_m(AdoCbl)	Temperature (°C)	K_m(NDP) μM	Specific activity	Refs.
P. furiosus	α	1 μM	80	70	510	7
T. acidophilum	α_2	0.1 mM	55	64	300	18,30
T. maritima	nd[a]	nd	80	nd	1100	30

[a] nd, Not determined

reaction is stopped by addition of 0.5 ml of 1 M HClO$_4$. After heating for 15 min in boiling water, the amount of [^{3}H] dCMP formed is determined after separation from CMP on Dowex-50.[34] One unit of activity is 1 nmol of dCMP formed per minute. The specific activity of the purified enzyme is 500-600 units per mg protein.

Catalytic Properties

As shown in Fig. 1, the enzyme catalyzes the reduction of ribonucleoside diphosphates (Fig. 1A) by dithiothreitol (Fig. 1B). K_m values for CDP and DTT are 70 μM and 20 mM, respectively. The requirement for a dithiol suggests that, as for other class II RNRs, such as the extensively studied enzyme from *Lactobacillus leichmannii*, the hydrogen donor is very likely to be a dithiol protein such as thioredoxin or glutaredoxin.[2] However, there is still no experimental evidence that an archaeal thioredoxin operates as an electron source for RNRs. The enzyme also requires AdoCbl for which a K_m value of 1 μM has been obtained (Fig. 1C). Finally, the reaction has an optimal temperature of 80° (Fig. 1D), with very little activity at 30°. How AdoCbl resists such a high temperature and how the enzyme controls Co–C bond homolysis required for catalysis in thermophilic AdoCbl-dependent enzymes is an intriguing question. These properties are shared by other isolated thermophilic class II RNRs (Table II).

Sequence Analysis

The RNR gene was discovered during a genomic survey of the genome of *P. furiosus*. Surprisingly, the open reading frame, commencing with the 5' end corresponding exactly to the purified protein as shown from the identity of the first 15 amino acids of the N-terminal end, was 4.5 kb long. The deduced protein sequence contained 1740 amino acids with M_r of 200,000 whereas the native protein had M_r of 90,000. This discrepancy arises from the presence of two inteins (containing 454 and 382 residues, respectively) within the 200-kDa protein which

[34] L. Thelander, B.-M. Sjöberg, and S. Eriksson, *Methods Enzymol.* **51,** 227 (1978).

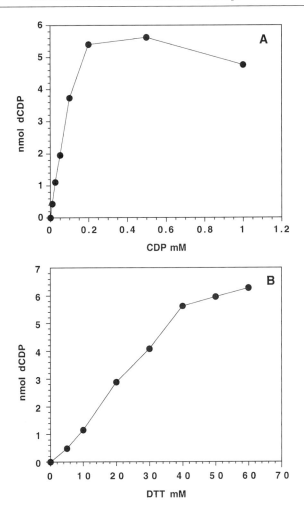

FIG. 1. Enzyme reduction of CDP (A) depends on the presence of DTT (B) and AdoCbl (C); (D) shows the temperature dependence. Reaction conditions: 0.4 μg enzyme, 0.5 M sodium acetate, 30 min reaction. DTT is 40 mM in (A), (C), and (D). AdoCbl is 5 μM in (A), (B), and (D). CDP is 1 mM in (B), (C), and (D). The temperature is 80° in (A), (B), and (C).

presumably are spliced out to generate the 90-kDa mature enzyme. The splicing sites, in particular the conserved short sequence His-Asn-Cys/Thr found at the intein-C extein border, and the characteristic LAGLI-DADG motif, are clearly identified by comparison with previously described inteins.[35] The class II RNR

[35] H. Fsihi, V. Vincent, and S. T. Cole, *Proc. Natl. Acad. Sci. U.S.A.* **93**, 3410 (1996).

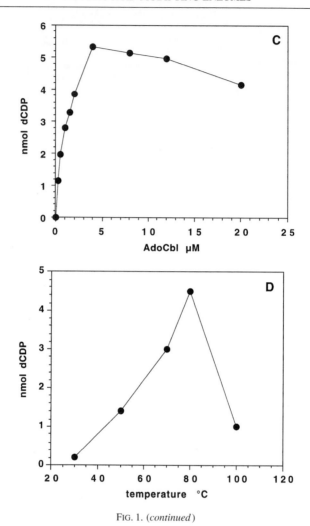

FIG. 1. (*continued*)

from *M. thermoautotrophicum* also contains one intein, with only 134 amino acid residues, integrated at the same position of one of the two *P. furiosus* RNR inteins, whereas other thermophilic enzymes from *T. acidophilum* and *T. maritima* do not.[31] Two inteins have been discovered within the *Thermococcus litoralis* DNA polymerase[10] and a single intein has been identified in the *P. furiosus* type III topoisomerase homolog. There is apparently a significant predisposition for inteins to insert within genes encoding DNA and nucleotide modifying enzymes of archaea. This is a weak trend in other organisms.

```
                       273
         P f    L S A C F V V P
         T a    L S A C F V L P
         A f    L S A C F V I P
         M t    L S A C F V L P
         T m    L S A C F V V P
         L l    L N N C W F V A


                  461 463              472
         P f    N P C G E E P L Y E Y E S C N L A S
         T a    N P C G E Q P L L P Y E S C N L G S
         A f    N P C G E Q P L L P Y E S C N G G S
         M t    N P C G E Q P L L T H E S C N L G S
         T m    N P C G E I G L S D Y E A C N L G S
         L l    N P C G E I S L A N G E P C N L F E
```

FIG. 2. The essential cysteines in the Class II *L. leichmannii* (Ll) RNR are present in *P. furiosus* (Pf), *T. acidophilum* (Ta), *A. fulgidus* (Af), *M. thermoautotrophicum* (Mt), and *T. maritima* (Tm) RNR. The numbers are from the Pf sequence. Cys-273 and Cys-472 designate the cysteines that directly reduce the ribonucleotide. Cys-461 is the cysteinyl radical. E463 is the conserved glutamate present in the active site.

The deduced archaeal protein sequence aligned well with that of the large R1 subunit of eukaryal RNR (for example, 28% identity and 48% similarity with murine RNR) but also, albeit at a slightly lower significance level, with that of the large R1 subunits of class I (22% identity and 47% similarity with *Escherichia coli*) and class III (23% identity and 46% similarity with *E. coli*) bacterial RNRs and with that of class II (22% identity and 48% similarity with *Lactobacillus leichmannii*) bacterial enzymes. In class I and class II RNRs, two essential cysteines are directly involved in nucleotide reduction.[14,15,20] A third cysteine is proposed to be the transient protein radical required for substrate activation.[19,36,37] It is interesting to note that the thermophilic RNRs contain these cysteines correctly placed within regions with strong similarity to both class I and class II RNRs, suggesting similar enzyme mechanisms (Fig. 2).

---

[36] S. S. Mao, G. X. Yu, D. Chalfoun, and J. Stubbe, *Biochemistry* **31**, 9752 (1992).
[37] G. J. Gerfen, S. S. Licht, J.-P. Willems, B. M. Hoffman, and J. Stubbe, *J. Am. Chem. Soc.* **118**, 8192 (1996).

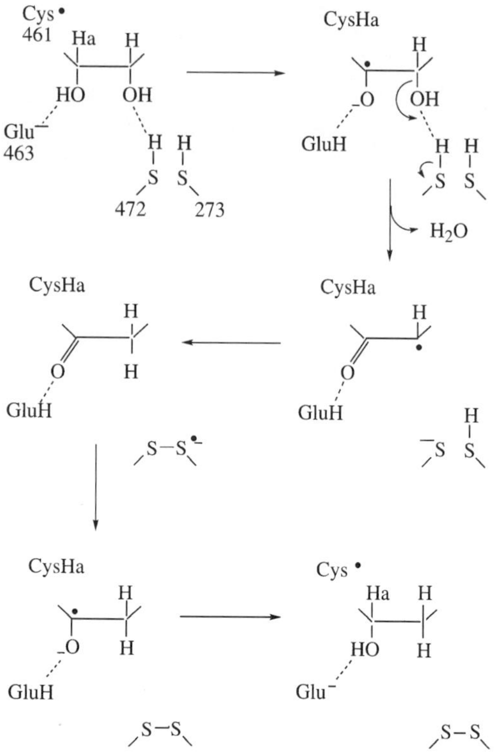

FIG. 3. Reaction mechanism of class II RNR.

Reaction Mechanism for Ribonucleotide Reduction

No mechanistic studies have been carried out on thermophilic class II RNRs to date. However, ribonucleotide reduction is likely to proceed by a radical mechanism similar to that common to class I *E. coli* and class II *L. leichmannii* enzymes.[38] In the following, this mechanism is described, with numbers of the key aminoacid residues from the *P. furiosus* RNR sequence (Fig. 3).

The reaction is absolutely dependent on the formation of a thiyl radical (located on Cys-461) on the β face of the ribonucleotide. The model postulates that the enzyme catalyzes the homolysis of the cobalt–carbon bond of AdoCbl to generate, in a concerted fashion, this thiyl radical together with 5′-deoxyadenosine and cob(II)alamin.[19,37,39] The radical initiates the reaction by abstraction of H_a the hydrogen atom at the 3′ position of the substrate ribose, to give a 3′-radical.[38] The

---

[38] J. Stubbe and W. Van der Donk, *Chem. Biol.* **2,** 793 (1995).
[39] S. S. Licht, S. Booker, and J. Stubbe, *Biochemistry* **38,** 1221 (1999).

presence of a base (Glu-463) facilitates H abstraction by deprotonating the 3′-OH group (40, 41). In *E. coli*, the corresponding Glu residue has been mutated resulting in an inactive enzyme.[40] Hydrogen bonding/protonation of the OH group at the 2′ position by a cysteine pair (Cys-273 and Cys-472), present on the α face of the nucleotide, follows. After loss of H_2O, a new intermediate carbonyl conjugated radical receives a hydrogen atom from the two cysteines, thus generating a 3′-ketodeoxyribonucleotide. Its reduction by the disulfide radical anion and return of H_a to the 3′ position completes the synthesis of the deoxyribonucleotide with regeneration of the initiating cysteinyl radical and formation of a protein disulfide. The latter needs to be reduced for another turnover. The reducing equivalents in both class I and class II enzymes are provided by NADPH with the help of an electron transfer chain, such as thioredoxin reductase–thioredoxin. The enzyme from *E. coli* (class I) or from *L. leichmannii* (class II) contains a second cysteine pair, residing on the C-terminal part of the protein, which participates in the transfer of the reducing equivalents from thioredoxin to the active site cysteine pair.[14,15] Whether this is also true in thermophilic RNRs remains to be shown. However, considering the requirement for DTT and the presence of a number of cysteines in the C terminus, this could very likely be the case.

Allosteric Regulation

Among the fascinating properties of ribonucleotide reductases is their allosteric regulation of substrate specificity, reviewed in an article by P. Reichard and A. Jordan.[2] Effectors are nucleoside triphosphates, ATP, and the four dNTPs. By binding to specific sites, they prepare the enzyme for the selective reduction of a given substrate so that they provide a mechanism for the balanced supply of the four deoxyribonucleotides to the cell for an efficient DNA synthesis. Allosteric regulation has been extensively studied in class I RNRs. It is now well established that they all contain an allosteric site, named the substrate specificity site, with the following rule: binding of ATP or dATP to that site induces activity toward pyrimidine ribonucleotides, binding of dTTP toward guanine ribonucleotide, and binding of dGTP toward adenine ribonucleotides. Class Ia, but not class Ib, RNRs have an additional allosteric site, named the activity site, that regulates their overall activity, with ATP promoting and dATP inhibiting enzyme activity. The second site has a much lower affinity for nucleotides and effects are seen at larger concentrations of nucleotides.

Until very recently, our knowledge of allosteric regulation of class II enzymes relied on early studies carried out on the *L. leichmannii* enzyme.[42] A reanalysis

---

[40] A. L. Persson, M. Eriksson, B. Katterle, S. Pötsch, M. Sahlin, and B.-M. Sjöberg, *J. Biol. Chem.* **272**, 31533 (1997).

[41] D. J. Silva, J. Stubbe, V. Samano, and M. J. Robins, *Biochemistry* **37**, 5528 (1998).

[42] S. Booker and J. Stubbe, *Proc. Natl. Acad. Sci. U.S.A.* **90**, 8352 (1993).

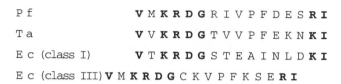

```
Pf             VMKRDGRIVPFDESRI
Ta             VVKRDGTVVPFEKNKI
Ec (class I)   VTKRDGSTEAINLDKI
Ec (class III) VMKRDGCKVPFKSERI
```

FIG. 4. Comparison of the N-terminal sequences of *P. furiosus* (Pf) and *T. acidophilum* (Ta) RNR, and the large subunits of *E. coli* (Ec) Class I and Class III RNR.

of allosteric regulation in class II RNR has included the same enzyme and the ribonucleoside diphosphate reductases from the thermophilic eubacteria *Thermotoga maritima* and the archaea *Thermoplasma acidophilum*.[30,43] Allosteric effects in the two last cases are more pronounced at high temperature even though binding occurs in the cold room. Apparently, the increased flexibility of the proteins at high temperature allows better communication between the effector and substrate sites.

This study led to the remarkable conclusion that the allosteric regulation of class II enzymes responds to the same rules as class I, described above. However, some of the class II RNRs, for example the one from *L. leichmannii* or that from *T. maritima*, contain only the specificity site, as in class Ib. This site has a very high affinity for nucleotides with a K_d value for dATP of 15 nM for the thermophilic RNR. The allosteric effectors seem to decrease the apparent K_m value for the appropriate substrate and the K_m value for AdoCbl. Many lines of evidence indicate that the binding of allosteric effectors induces a considerable conformational change, as shown from changes of the CD spectrum and of the sedimentation coefficients, quenching of protein fluorescence, and changes of the EPR spectrum of protein-bound cob(II) alamin.

The enzyme from *T. acidophilum* contains a second site that binds dATP and ATP. However, this site does not seem to respond as an "activity site" since the enzyme could not be inhibited by dATP even at very high concentrations, in contrast to what occur in class Ia RNRs.[30,43]

The elucidation of the structure of the class Ia *E. coli* reductase complexed with effectors now gives an explanation for these differences.[13] In the *E. coli* enzyme, the specificity site is at the interface between the two subunits of R1. It uses amino acids in regions between 200 and 300 amino acids from the N terminus. In this region, class II enzymes, from *T. maritima* and *P. furiosus*, seem to contain amino acids homologous to the effector binding residues of protein R1.

On the other hand, the activity site of R1 is constructed from the first 100 N-terminal amino acids, with a lot of key residues in the first 25 amino acids (Fig. 4).

For example, the sequence VXKRDG at the N terminus is one of the signatures for the activity site, with V, K and R contributing to effector binding.[13] The RNR

---

[43] R. Eliasson, E. Pontis, A. Jordan, and P. Reichard, *J. Biol. Chem.* **274**, 7182 (1999).

from *T. maritima* lacks these key residues, explaining why dATP does not bind to a second site and why it does not inhibit activity. More surprising is the fact that, while this sequence is present in the class II RNRs from *T. acidophilum* and *P. furiosus* in agreement with the observed binding of ATP and dATP, the second site seems to have lost functionality.[30,43] The chemical basis for the fact that binding to this site is not translated to the catalytic site is unclear and awaits structural characterization of these enzymes.

Conclusion

An unexpected finding during the work on the isolation and characterization of a thermophilic ribonucleotide reductase was that the enzyme was from the class II and absolutely required adenosylcobalamin. It remains to be understood how this rather fragile cofactor is protected from inactivation. Probably the enzyme itself plays a role in that protection, as shown from preliminary experiments in which the half-life of AdoCbl at 80° proved to be greatly increased in the presence of *P. furiosus* RNR. It became obvious when more thermophilic RNRs became available that the great majority of thermophilic organisms have preferred a class II RNR for growth, and not RNRs from other classes I or III, whether they are anaerobic or not, further showing the importance of AdoCbl for life at high temperature.

There is still no crystal structure of a class II RNR. The enzymes isolated from thermophilic bacteria and archaea may be good starting material for obtaining this important information. This is currently under investigation.

[19] Preparation of Components of Archaeal Transcription Preinitiation Complex

By YAKOV KORKHIN, OTIS LITTLEFIELD, PAMLEA J. NELSON, STEPHEN D. BELL, and PAUL B. SIGLER

Transcription initiation is a key point in the regulation of the expression of most genes. The centerpiece of the molecular mechanism of eukaryal gene transcription is the preinitiation complex (PIC), which assembles on the promoter. The eukaryal PIC consists of a multisubunit enzyme RNA polymerase (RNAP) and general transcription factors (GTFs), as well as a large (and ever-growing) number of other auxiliary proteins.[1] The archaeal transcription PIC is homologous to the eukaryal one; however, it is a much simpler system, requiring only the RNA

---

[1] A. J. Berk, *Curr. Opin. Cell. Biol.* **11**(3), 330 (1999).

polymerase and two GTFs: the TATA binding protein (TBP) and the transcription factor B (TFB) (the homolog of eukaryal TFIIB).[2] The comparable complex in Eukarya represents the minimum core for transcriptional initiation *in vitro* from Pol II promoters. Recent crystallographic studies have shown that the structures of the preinitiation complexes consisting of TBP, TFIIB, and a promoter fragment are essentially the same in Archaea and Eukarya.[3-5] Archaeal DNA-dependent RNAP consists of approximately 12 subunits, which have high degree of sequence identity with their analogs from eukaryal Pol I, Pol II, and Pol III.[6] Thus, the archaeal transcription PIC is amenable to structural investigations and represents a simplified model with which to study the underlying stereochemical principles of nonbacterial gene transcription. Presented below are preparation procedures for various components of transcription PIC from three species of Archaea: *Sulfolobus acidocaldarius, Sulfolobus shibatae,* and *Pyrococcus furiosus.*

RNA Polymerase Preparation

Sulfolobus Acidocaldarius

Media Formulation. Major ingredient solution A: per each liter of deionized water add 1.3 g $(NH_4)_2SO_4$, 0.28 g KH_2PO_4, 0.25 g $MgSO_4$, 0.07 g $CaCl_2$, 2 g Bactopeptone, pH 3.3 (adjusted with H_2SO_4). Minor ingredient solution B: dissolve 5 g $FeCl_3$, 0.45 g $MnCl_2$, 1.13 g $Na_2B_4O_7$, 0.055 g $ZnSO_4$, 0.013 g $CuCl_2$, 0.008 g Na_2MoO_4, 0.008 g $VOSO_4$, 0.003 g $CoSO_4$ in 100 ml of 1 M HCl. Add 0.4 ml of solution B per 1 liter of solution A. Adjust the temperature of the mixture to 75° and add 10 g of sucrose per each liter of media. This media formulation is adapted from Robb and Pace.[7]

Growth of Archaea. Ten ml of media in a 25-ml flask are inoculated with 0.1–0.3 g of frozen cell pellet or with a revived ATCC frozen cell stock (ATCC Rockville, MD number 33909). Place the flask into a water bath filled with 50% ethylene glycol at 75°. After 4 days of incubation, transfer the 10-ml growth into a 250-ml flask filled with 100 ml of media and incubate for another 4 days (or until OD_{600} reaches 0.4–0.5) in a water bath at 75°. Transfer the 100-ml growth into 3 liter of media in a 4-liter glass jar equilibrated at 75° on a hot plate with mild aeration and stirring (Fig. 1a). Grow cells for 4 days ($OD_{600} \approx 3.0$). This serves as an inoculum for further cell growth of 115 liter conducted at 75° in a modified

---

[2] J. Soppa, *Mol. Microbiol.* **31**(5), 1295 (1999).
[3] D. B. Nikolov, *et al., Nature* **377**(6545), 119 (1995).
[4] P. F. Kosa, *et al., Proc. Natl. Acad. Sci. U.S.A.* **94**(12), 6042 (1997).
[5] O. Littlefield, Y. Korkhin, and P. B. Sigler, *Proc. Natl. Acad. Sci. U.S.A.* **96**(24), 13668 (1999).
[6] W. Zillig, *et al., in* "The Biochemistry of Archaea (Archaebacteria)" (M. Kates, D. J. Kushner, and A. T. Matheson, eds.), pp. 367–391. Elsevier, Amsterdam, 1993.
[7] F. T. Robb and A. R. Place, *in* "Archaea: A Laboratory Manual" (F. T. Robb, ed.), p. 172. Cold Spring Harbor Laboratory Press, Cold Spring Harbor, NY, 1995.

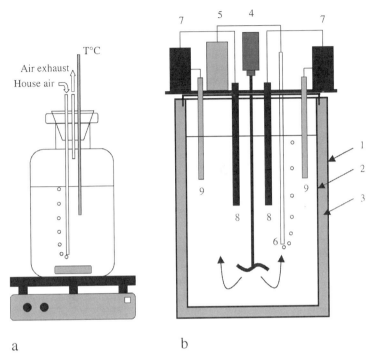

FIG. 1. Schematic representation of *Sulfolobus acidocaldarius* and *Sulfolobus shibatae* fermentation: (a) medium growth (3 Liter); (b) large-scale growth (115 liter), a 30-gal of Nalgene polypropylene tank (2) is placed inside a 44-gal Rubbermaid garbage container (1) separated with a layer of glass-fiber insulation (3). The medium is agitated with a 4 inch impeller placed on a long shaft controlled by a motor (4). Air is supplied through four glass tubes (6) connected to an air pump (5). The medium is heated with two 1000 W glass immersion heaters (8) each connected to a separate temperature controller (7) equipped with a temperature regulator (9). All equipment is mounted on a thick polypropylene board attached to the cover of the 30-gal tank.

version of the fermenter described in ref. 8 (Fig. 1b). Typically after 4 days of fermentation ($OD_{600} \approx 4.0$), 500 g of wet cell mass can be harvested by continuous flow centrifugation from a 115-liter growth. Flash freeze cells in liquid nitrogen and store at $-70°$ in 200-g pellets.

Purification Procedure. All buffers are prepared at room temperature; all procedures are carried out at $4°$ and are adapted with modifications from ref. 9. The procedure listed below typically yields 0.2 mg of pure enzyme per 1 liter of archaeal culture growth.

---

[8] D. Searcy, in "Archaea: A Laboratory Manual" (F. T. Robb, ed.), pp. 51–66. Cold Spring Harbor Laboratory Press, Cold Spring Harbor, NY, 1995.

[9] S. A. Qureshi, S. D. Bell, and S. P. Jackson, *EMBO J.* **16**(10), 2927 (1997).

Buffer A': 50 mM Tris/HCl, pH 7.8, 22 mM NH$_4$Cl, 10 mM 2-mercaptoethanol (2-ME)

Buffer A: Buffer A' + 10% (v/v) glycerol

Lysis buffer: Buffer A + 0.5 mM EDTA + 7 μg/ml pepstatin A + 0.5 μg/ml leupeptin + 0.1 μg/ml aprotinin + 35 μg/ml phenylmethylsulfonyl fluoride (PMSF)

1. Remove a 200-g cell pellet from storage and homogenize it with a hand-held blender in 600 ml of lysis buffer.
2. Disrupt cells by liquid extrusion in four passes through liquid fluidizer (Microfluidic Corp.) at 70 psi, bring the lysate volume to 800 ml.
3. While stirring, slowly add 64 ml of 5% neutralized polyethyleneimine (PEI) hydrochloride solution.
4. Sediment precipitate in Sorval GSA rotor at 11,000 rpm for 30 min.
5. Resuspend precipitate in 1 liter of buffer A' + 200 mM NH$_4$Cl, sediment as above, decant, and discard the supernatant. Repeat this step two more times.
6. Resuspend precipitate in 1 liter of buffer A' + 1.2 M NH$_4$Cl, sediment as above, collect the supernatants. Repeat the step two more times.
7. Combine three supernatants from step 6 and gradually add dry (NH$_4$)$_2$SO$_4$ to 65% (w/v). Sediment in Sorval GSA rotor at 11,000 rpm for 60 min. Discard supernatant and solubilize precipitate in 100 ml of buffer A.
8. Dialyze the sample against 2 liter buffer A with three buffer exchanges.
9. Clarify the sample by centrifugation in Beckman TI-45 rotor at 40,000 rpm for 45 min.
10. Apply the sample at a flow rate of 3 ml/min to a Q Sepharose Fast Flow XK50/20 column (Pharmacia, Piscataway, NJ) that has been equilibrated with buffer A. Wash the column with seven column volumes of buffer A, develop the column with a linear gradient from buffer A to buffer A + 0.5 M KCl over five column volumes at a flow rate of 3 ml/min, and collect 25-ml fractions. The polymerase elutes at a KCl concentration around 270 mM (Fig. 2a). Fractions can be tested for the presence of polymerase by enzyme-linked immunosorbent assay (ELISA) with a primary antibody described below.
11. Pool polymerase containing fractions and dilute the pool twofold with buffer A. Apply the sample at a flow rate of 1 ml/min to a 2 × 5 ml Hi-Trap heparin column (Pharmacia) equilibrated with buffer A. Wash column with 100 ml of buffer A, develop the column with a 100-ml linear gradient from buffer A to buffer A + 1 M KCl at a flow rate of 1 ml/min, and collect 2.5-ml fractions. The polymerase elutes in the major peak centered at 470 mM KCl. Fractions can be tested for the presence of polymerase by SDS–PAGE.

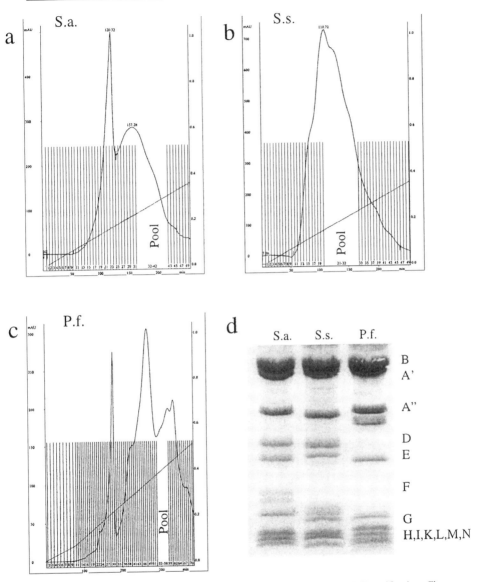

FIG. 2. (a)–(c) Q Sepharose chromatography elution profiles during RNAP purification. Chromatography runs are described in the text. (d) Coomassie stained SDS–PAGE analysis of RNAPs from *Sulfolobus acidocaldarius* (S.a.), *Sulfolobus shibatae* (S.s.), and *Pyrococcus furiosus* (P.f.).

12. Pool polymerase containing fractions and concentrate the pool by ultrafiltration in a Schleicher & Schuell (Keene, NH) concentrator (membrane molecular weight cutoff 75,000) to a final volume of 0.5 ml.
13. Apply the sample at a flow rate of 0.5 ml/min to a Superdex 200 16/60 column (Pharmacia) equilibrated with buffer A + 300 mM KCl. The polymerase elutes as the first major peak.
14. Pool polymerase containing fractions (SDS–PAGE analysis shown in Fig. 2d), concentrate as above, and store at −70°.

Sulfolobus Shibatae

Revive cells from ATCC frozen cell stock (ATCC number 51178). Cell growth and polymerase purification are performed according to procedures described above for *S. acidocaldarius* except for the following differences:

Buffer A′: 50 mM Tris/HCl, pH 7.6, 25 mM MgCl$_2$, 10 mM 2-mercaptoethanol (2-ME)

Purification procedure: In step 5 (above) the pellet is washed with buffer A′ without addition of NH$_4$Cl; in step 10 (above) the polymerase elutes at a KCl concentration around 170 mM (Fig. 2b); in step 11 (above) the polymerase elutes at a KCl concentration around 520 mM. SDS–PAGE analysis of purified polymerase is shown in Fig. 2d.

Pyrococcus Furiosus

Cells (ATCC number 43587) are grown according to procedure developed by M. Adams[10] and were kindly provided by the author. Polymerase (pfRNAP) purification is performed according to procedure described above for *S. acidocaldarius*, except for the following differences:

Buffer A′: 25 mM HEPES/KOH, pH 7.5, 25 mM KCl, 1 mM EDTA
Buffer A: buffer A + 20% (v/v) glycerol
Purification procedure: In step 5 (above) the pellet is washed with buffer A′ without addition of NH$_4$Cl; in step 10 (above) the pfRNAP elutes at KCl concentration around 390 mM (Fig. 1b); in step 11 (above) the pfRNAP elutes at KCl concentration around 470 mM. After size exclusion chromatography the sample is diluted threefold with buffer A and loaded at a flow rate of 2 ml/min onto a Mono Q HR 10/10 column (Pharmacia) equilibrated with buffer A. The column is washed with 100 ml of buffer A and developed with a 160-ml linear gradient from buffer A to buffer

---

[10] M. W. W. Adams, in "Archaea: A Laboratory Manual" (F. T. Robb, ed.), pp. 47–49. Cold Spring Harbor Laboratory Press, Cold Spring Harbor, NY, 1995.

A + 1 M KCl at a flow rate of 2 ml/min; 3-ml fractions are collected. The polymerase elutes in the major peak at 365 mM KCl. SDS–PAGE analysis of purified pfRNAP is shown in Fig. 2d.

Immunoaffinity Purification of Archaeal RNA Polymerases

Choice of Suitable Epitopes

Subunit A″ of archaeal RNA polymerases has substantial sequence identity with the C-terminal region of the large subunit of eukaryal RNAPs, suggesting a high degree of structural homology (Fig. 3a). In Eukarya the C-terminal domain (CTD) of the large subunit is composed of a series of seven polar amino acids repeats and is exposed to the kinase activity of auxiliary factors during various stages of transcription. The solvent exposed nature of CTD is confirmed by the fact that it can be easily removed by limited proteolysis. The CTD is not present in archaeal RNAPs, but the preceding sequence is nearly identical in Archaea and Eukarya, suggesting that the eukaryal CTD and the corresponding C-terminal residues in Archaea eminate from the same structure. Since the CTD is almost certainly solvent exposed, the corresponding residues in Archaea would also be exposed and, therefore, serve as a good epitope. To test this hypothesis, we generated antiserum against an 18-residue peptide from the C terminus of the A″ subunit of pfRNAP (Fig. 3b). Peptides for antibody generation against other archaeal polymerases can be selected in a similar way. We tested the antiserum against pfRNAP by ELISA and Western blot analysis. Results confirmed high selectivity of the antibody for pfRNAP in native and denatured forms, respectively. The antibody was subsequently used in ELISA assays to detect pfRNAP during early stages of the purification procedure described above. For immunoafinity polymerase purification, the anti-pfRNAP antibody was peptide-affinity purified and immobilized on Affi-Gel 10 resin (Bio-Rad, Hercules, CA). Peptide synthesis, antiserum production, and antipeptide antibody affinity purification were performed by Zymed Laboratories, Inc.

Purification Procedure

Buffer A′: 50 mM Tris/HCl, pH 7.8, 50 mM NH$_4$Cl, 10 mM 2-ME.
Buffer A: buffer A′ + 10% (v/v) glycerol.
Buffer A*: buffer A + 1.2 M NH$_4$Cl.

Resuspend 50 g of *Pyrococcus furiosus* cells in 200 ml of buffer A′. Disrupt cells by liquid extrusion in three passes through liquid fluidizer (Microfluidic Corp.) at 70 psig. Immediately after lysis add Triton X-100 to cell lysate (250 ml) to a final concentration of 0.1% (w/v). Clarify cell lysate by centrifugation in Sorval GSA rotor at 11,000 rpm for 1 hr. Add 20 ml of 5% PEI (titrated to pH 7.5 with

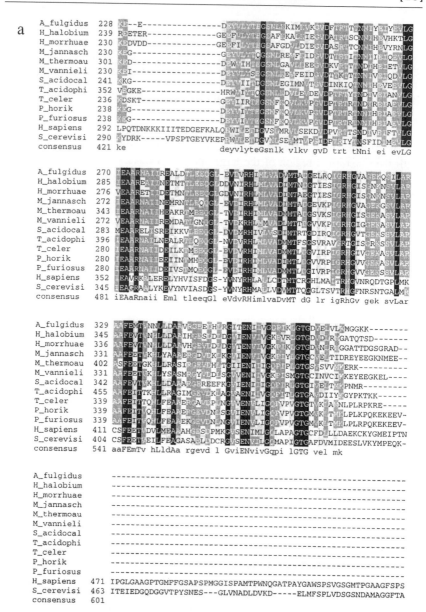

b

CKLTMKLPLRPQKEKEEV

FIG. 3. (a) Sequence alignment of the C termini of archaeal RNAP A″ subunit with subunit RPB1 of eukaryal Pol II. (b) Peptide sequence used for anti-pfRNAP antiserum generation.

HCl) slowly to the supernatant while stirring. Sediment the precipitate by centrifugation in a Sorval GSA rotor at 11,000 rpm for 30 min. Wash the pellet (about 45 g) in 135 ml of buffer A and resediment; repeat this procedure three times. Elute bound proteins from the PEI DNA pellet by washing the pellet in 135 ml of buffer A*. Repeat elution three times. While stirring, slowly add dry ammonium sulfate to 80% (w/v) to combined supernatant. Allow proteins to precipitate for 1 hr. Collect by centrifugation in Sorval GSA rotor at 11,000 rpm for 45 min. Solubilize precipitate in 50 ml phosphate-buffered saline (PBS) and dialyze against PBS with three buffer exchanges. Clarify the sample after dialysis by centrifugation in Beckman TI-45 rotor at 40,000 rpm for 45 min. Incubate the sample with immobilized antibody (5 ml Affi-Gel 10 resin equilibrated in PBS) for 1 hr while gently shaking. Pack the resin by gravity into a 1.5/10 column while collecting the flow-through. Elute bound pfRNAP by incubating the resin for 1 hr with 15 ml of 1 mM peptide (Fig. 3b) in PBS. Elute the bound peptide from the antibody with 15 ml of PBS + 3 M KSCN, followed immediately by equilibration with PBS to prevent antibody degradation. After regeneration the antibody column can be reused. Concentrate eluted polymerase in Schleicher & Schuell concentrator (molecular weight cutoff 75,000) to a final volume of 0.5 ml and apply to a Superdex 200 16/60 column equilibrated in buffer A' + 300 mM KCl. Develop the column at a flow rate of 0.5 ml/min. pfRNAP elutes as the major peak.

Preparation of General Transcription Factors

TATA Binding Protein

Sulfolobus Acidocaldarius. *Sulfolobus acidocaldarius* TBP (saTBP) is cloned as an insert in pET30a vector (Novagen) between *Nde*I and *Xho*I restriction sites (plasmid prepared by S. D. Bell) and overexpressed in *E. coli* strain BL21 (DE3) (Novagen). Cells are grown in three 6-liter shaker flasks, each containing 2 liter of 2 × YT media, with agitation at 200 rpm at 37° in an incubator–shaker in the presence of 20 μg/ml kanamycin. Protein expression is induced with 0.5 mM isopropylthiogalactoside (IPTG) at OD_{600} 0.5, and after 3 hr of induction, typically 15 g of cells are harvested by continuous flow centrifugation.

Purification procedure. All procedures are carried out at room temperature.

Buffer A: 200 mM Potassium acetate, 10 mM Tris/HCl, pH 8.3
Buffer B: 100 mM Potassium acetate, 100 mM K_2HPO_4, pH 7.5

1. Homogenize cell pellet in 150 ml of buffer A and disrupt cells by liquid extrusion in three passes (see procedure for *S. acidocaldarius* above).
2. Clarify cell lysate in Sorval SS34 rotor at 15,000 rpm for 20 min.
3. Incubate the supernatant at 70° for 10 min in a water bath.
4. Clarify the supernatant in Sorval SS34 rotor at 10,000 rpm for 20 min.
5. Apply the sample at a flow rate of 4 ml/min to a 30-ml SP Sepharose Fast

Flow column (Pharmacia) equilibrated with buffer A. Wash the column with 10 column volumes of buffer A; develop the column with a linear gradient from buffer A to buffer A + 1 M potassium acetate over seven column volumes at a flow rate of 4 ml/min. saTBP elutes as the major peak centered at 440 mM potassium acetate.

6. Pool saTBP containing fractions and dialyze the pool against 20 volumes of buffer B two times.
7. Apply the sample at 2 ml/min to a hydroxyapatite (Bio-Rad Macro-Prep Ceramic hydroxyapatite, Type I) HR 10/10 column (Pharmacia) equilibrated with buffer B. Wash the column with 80 ml of buffer B; develop the column with a linear gradient from buffer B to buffer B + 1 M K$_2$HPO$_4$/KH$_2$PO$_4$ pH 7.5 over 80 ml at a flow rate of 2 ml/min. saTBP elutes as the major peak centered at 440 mM K$_2$HPO$_4$/KH$_2$PO$_4$.
8. Pool saTBP containing fractions and dialyze the pool against 20 volumes of buffer A three times.
9. Apply the sample at 2 ml/min to a Mono S HR 10/10 column equilibrated with 20 column volumes of buffer A. Wash the column with 80 ml of buffer A, develop the column with a linear gradient from buffer A to buffer A + 0.5 M potassium acetate over 80 ml at a flow rate of 2 ml/min. saTBP elutes as the major peak at the beginning of the gradient.
10. Pool saTBP containing fractions and store at $-70°$.

Pyrococcus Furiosus. Pyrococcus Furiosus TBP (pfTBP) is cloned as an insert in pET11a vector (Novagen) (plasmid prepared by Brian DeDecker) and overexpressed in *E. coli* strain BL21 (DE3) (Novagen). Cells are grown in three 6-liter shaker flasks, each containing 2 liter of LB media, with agitation at 200 rpm at 37° in an incubator–shaker in the presence of 100 μg/ml ampicillin. Protein expression is induced with 1 mM IPTG at OD$_{600}$ 0.6, and after 3 hr of induction typically 10 g of cells are harvested by continuous flow centrifugation.

Purification procedure. All procedures are carried out at room temperature.

Buffer A: 10 mM K$_2$HPO$_4$/KH$_2$PO$_4$, pH 7.4

1. Homogenize cell pellet in 200 ml of buffer A + 400 mM NaCl and disrupt cells by liquid extrusion in three passes as described above.
2. Clarify cell lysate in Sorval SS34 rotor at 15,000 rpm for 20 min.
3. Incubate the supernatant at 90° for 10 min in a water bath.
4. Clarify the supernatant in Sorval SS34 rotor at 10,000 rpm for 20 min.
5. While stirring, slowly add 16 ml of 5% neutralized polyethyleneimine hydrochloride.
6. Clarify the supernatant in Sorval SS34 rotor at 15,000 rpm for 30 min.

7. Add dry $(NH_4)_2SO_4$ to the supernatant to 75% (w/v) final concentration. Sediment precipitate in Sorval SS34 rotor at 15,000 rpm for 30 min. Solubilize precipitate in 50 ml of buffer A.
8. Dialyze the sample in 3 liter of buffer A.
9. Apply the sample at 4 ml/min to a hydroxyapatite (Bio-Rad Macro-Prep Ceramic hydroxyapatite, Type I) HR 16/10 column (Pharmacia) equilibrated with buffer A. Wash the column with 200 ml of buffer B; develop the column with a linear gradient from buffer A to buffer A + 300 mM K_2HPO_4/KH_2PO_4 over 200 ml at a flow rate of 4 ml/min. pfTBP elutes as the major peak centered at 200 mM K_2HPO_4/KH_2PO_4.
10. Pool pfTBP containing fractions and concentrate the pool in a Centriprep-10 cencentrator to a final volume of 1 ml.
11. Apply the sample at a flow rate of 0.5 ml/min to a Superdex 75 16/60 column (Pharmacia) equilibrated with 100 mM K_2HPO_4/KH_2PO_4, pH 7.4. pfTBP elutes as the major peak.
12. Pool pfTBP-containing fractions and store at $-70°$.

TFB

Sulfolobus Acidocaldarius. *Sulfolobus acidocaldarius* TFB (saTFB) is cloned as an insert in pET30a vector (Novagen) between *Nco*I and *Xho*I restriction sites and overexpressed in *E. coli* strain BL21-CodonPlus (DE3)-RIL (Stratagene, La Jolla, CA) with a hexahistidine (His_6) tag at the N terminus. Cells are grown in six 6-liter shaker flasks, each containing 2 liter of LB medium, with agitation at 200 rpm at 25° in an incubator–shaker in the presence of 20 μg/ml kanamycin and 37 μg/ml chloramphenicol. Protein expression is induced with 0.5 mM IPTG at OD_{600} 0.5, and after 8 hr of induction typically 15 g of cells are harvested by continuous flow centrifugation.

Purification procedure. All procedures are carried out at room temperature.

Buffer A: 50 mM Tris/HCl, pH 7.8, 500 mM KCl, 10% (v/v) glycerol, 10 mM 2-ME
Buffer B: Buffer A + 20 mM imidazole

1. Homogenize cell pellet in 150 ml of buffer B and disrupt cells by liquid extrusion in three passes.
2. Clarify cell lysate in Sorval SS34 rotor at 15,000 rpm for 20 min.
3. Incubate the supernatant at 70° for 15 min in a water bath.
4. Clarify the supernatant in Sorval SS34 rotor at 10,000 rpm for 20 min.
5. Apply the sample by gravity to a 30-ml Ni-NTA agarose column (Qiagen, Chatsworth, CA) equilibrated with buffer B. Wash the column with 10 column volumes of buffer B; develop the column with a linear gradient from buffer B to buffer B + 0.5 M imidazole over 7 column volumes at a

flow rate of 4 ml/min. saTFB elutes as the major peak centered at 150 mM imidazole.
6. Pool saTFB containing fractions and concentrate the pool in a stirred cell (Amicon) to a final volume of 2–3 ml.
7. Apply the sample at a flow rate of 0.5 ml/min to a Superdex-75 26/60 column (Pharmacia) equilibrated with buffer A. saTFB elutes as the major peak.
8. Pool saTFB-containing fractions and store at −70°.

Pyrococcus Furiosus. *Pyrococcus furiosus* TFB (pfTFB) is cloned as an insert in pET11a vector (Novagen) (plasmid prepared by Gouri Ghosh) and overexpressed in *E. coli* strain BL21-CodonPlus (DE3)-RIL (Stratagene). Cells are grown in six 6-liter shaker flasks, each containing 2 liter of LB medium, with agitation at 200 rpm at 37° in an incubator–shaker in the presence of 100 μg/ml ampicillin and 37 μg/ml chloramphenicol. Protein expression is induced with 0.5 mM IPTG at OD_{600} 0.5, and after 3 hr of induction typically 15 g of cells are harvested by continuous flow centrifugation.

Purification procedure. All procedures are carried out at room temperature.

<u>Buffer A</u>: 25 mM Tris/HCl pH 8.3, 100 mM potassium acetate

1. Homogenize cell pellet in 200 ml of buffer A + 100 mM potassium acetate and disrupt cells by liquid extrusion in three passes.
2. Clarify cell lysate in Sorval SS34 rotor at 15,000 rpm for 20 min.
3. Incubate the supernatant at 90° for 10 min in a water bath.
4. Clarify the supernatant in Sorval SS34 rotor at 10,000 rpm for 20 min.
5. Apply the sample at a flow rate of 4 ml/min to a 30-ml SP Sepharose Fast Flow column (Pharmacia) equilibrated with buffer A. Wash the column with 10 column volumes of buffer A; develop the column with a linear gradient from buffer A to buffer A + 1M potassium acetate over seven column volumes at a flow rate of 4 ml/min. pfTFB elutes as the major peak centered at 850 mM potassium acetate.
6. Pool pfTFB-containing fractions and concentrate the pool in a stirred cell (Amicon) to a final volume of 2–3 ml (protein concentration should be less than 2 mg/ml to avoid precipitation).
7. Apply the sample at a flow rate of 1 ml/min to a Superdex-75 26/60 column (Pharmacia) equilibrated with buffer A + 1 M potassium acetate. pfTFB elutes as the major peak.
8. Pool pfTFB containing fractions and store at −70°.

General Notes

Purification of general transcription factors, especially TBP, has to be carried out at the maximum possible rate to prevent protein aggregation over time. Concentration of GTFs can be performed only in a stirred cell to prevent formation of protein concentration gradients that lead to precipitation of proteins. GTFs are best stored at low concentration. Once mixed with DNA and other components of transcription PIC, the resulting complexes can be further concentrated without causing precipitation. All the components of the archaeal transcription PIC can be stored at $-70°$ without loss of activity. Once removed from storage, proteins are best thawed at room temperature to avoid precipitation.

[20] Methylguanine Methyltransferase from *Thermococcus kodakaraensis* KOD1

By MASAHIRO TAKAGI, YASUSHI KAI, and TADAYUKI IMANAKA

Introduction

The alkylation of DNA after administration of alkylating agents, such as N-methyl-N'-nitro-N-nitrosoguanidine (MNNG) and N-methyl-N-nitrosourea (MNU), has been implicated in the production of mutations and cancer through covalent modification of the cellular genome to generate miscoding alkylated base derivatives. Among the many methylated bases formed, O^6-methylguanine (O^6-meG) appears to be a major premutagenic lesion in organisms.[1–3] Conversion of guanine–cytosine to an adenine–thymine pair in DNAs occurs when O^6-meG is formed and paired with thymine during DNA replication.[4,5] O^6-meG is efficiently repaired by specialized DNA repair enzymes, O^6-meG DNA methyltransferases (MGMT) transfer methyl groups from O^6-meG [as well as a minor methylated base, O^4-methylthymine (O^4-meG)] of the DNA to the cysteine residue of its own molecule, repairing DNA lesions in a single-step reaction.[6–10] Because this reaction irreversibly inactivates the enzyme, the capacity for repair of O^6-meG depends on the number of active enzyme molecules per cell. This DNA repair enzyme is present in various organisms from bacteria to human cells, and their characteristic

[1] A. Loveless, *Nature* **223**, 206 (1969).
[2] B. Strauss, D. Scudiero, and E. Henderson, "Molecular Mechanisms for Repair of DNA," Part A, p. 13. Plenum, New York, 1975.
[3] S. Sukumar, V. Natario, D. Martin-Zanca, and M. Barbacid, *Nature* **306**, 658 (1983).
[4] C. Coulondre and J. H. Miller, *J. Mol. Biol.* **117**, 577 (1977).
[5] T. Ito, T. Nakamura, H. Maki, and M. Sekiguchi, *Mutant Res.* **314**, 273 (1994).
[6] P. Karran and T. Lindahl, *Nature* **280**, 76 (1979).
[7] H. Kawate, K. Ihara, K. Kohda, K. Sakumi, and M. Sekiguchi, *Carcinogenesis* **16**, 1595 (1995).
[8] G. Koike, H. Maki, H. Takeya, and M. Sekiguchi, *J. Biol. Chem.* **265**, 14754 (1990).
[9] M. Olsson and T. Lindahl, *J. Biol. Chem.* **255**, 10569 (1980).
[10] P. Zak, K. Kleibl, and F. Laval, *J. Biol. Chem.* **269**, 730 (1994).

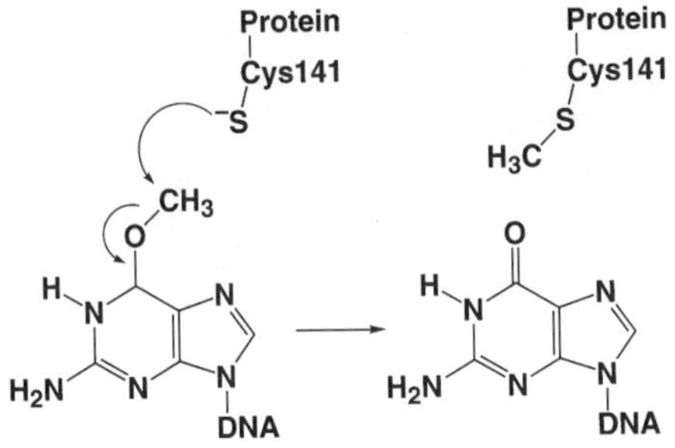

FIG. 1. Reaction catalyzed by the O^6-methylguanine-DNA methyltransferase (MGMT) transferase from *T. kodakaraensis* (*Tk* MGMT).

suicidal reaction mechanism is conserved in all cells.[11] Enzymes derived from the hyperthermophilic archaeon *Thermococcus kodakaraensis* KOD1 (previously reported as *Pyrococcus kodakaraensis* KOD1) have been shown to be extremely thermostable, and amino acid sequences of those enzymes are closely related to eukaryal homologs.[12] Here the purification and biochemical characterization of the MGMT from *T. kodakaraensis* KOD1 (*Tk*-MGMT) is described, the amino acid sequence of which exhibits homology with similar enzymes from mammals, yeast, and bacteria.[13]

Assay Procedures

Principle

O^6-Methylguanine-DNA methyltansferase (MGMT) transfers an alkyl group, substituted at the guanine O^6, to one of its own cysteine residues. From nucleotide sequence analysis (see below) of MGMT from *T. kodakaraensis*, Cys-141 functions as an active cysteine residue (Fig. 1). Methylated MGMT is inactive in further O^6-methylguanine–DNA repair, because the active-site cysteine is blocked by covalent methylation. Indeed, MGMT is often designated as a suicidal DNA repair protein. All previously reported methods for assay of MGMT activity are based on quantification of the transferred methyl group from the alkylated DNA to the cysteine residue at the active site of the MGMT protein. From nucleotide sequence analysis, it was shown that Cys-141 is the putative active site of the *Tk*-MGMT (Fig. 2).[13]

---

[11] B. Demple, B. Sedgwick, P. Robins, N. Totty, M. D. Waterfield, and T. Lindahl, *Proc. Natl. Acad. Sci. U.S.A.* **82**, 2688 (1985).

[12] S. Fujiwara, S. Okuyama, and T. Imanaka, *Gene* **179**, 165 (1996).

[13] M. M. Leclere, M. Nishioka, T. Yuasa, S. Fujiwara, M. Takagi, and T. Imanaka, *Mol. Gen. Genet.* **258**, 69 (1998).

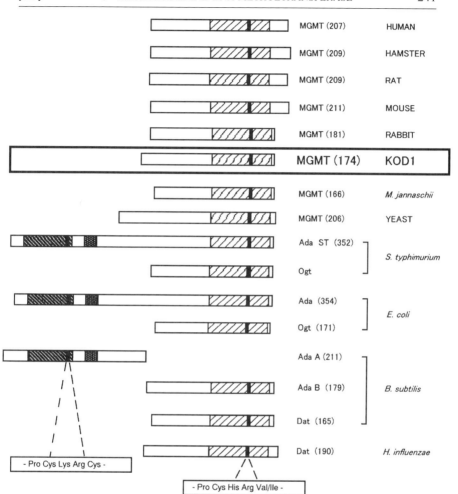

FIG. 2. Schematic comparison of amino acid sequence from known MGMT proteins. Black bars indicate active site regions, including cysteine residues that accept methyl group from methyl guanine. Black and hatched regions indicate conserved regions.

Preparation of Substrate

DNA substrate containing O^6-meG residues is prepared by alkylation of calf thymus DNA with N-[^{3}H]methyl-N-nitrosourea ([^{3}H]MNU).[14] A 1-ml aliquot of calf thymus DNA (2 mg/ml in 50 mM Tris-HCl, pH 9.0) is incubated with 40 ml of [^{3}H]MNU (0.2 mCi) at 37° for 4 hr. The DNA is precipitated by the addition of 0.1 volume of 3 M sodium acetate, and two volumes of cold ethanol. After

[14] J. M. Bogden, A. Eastman, and E. Bresnick, *Nucleic Acids Res.* **13**, 3089 (1981).

centrifugation, the DNA pellet is washed several times with cold 95% (v/v) ethanol, dried by desiccation, and redissolved overnight at 4° in 0.15 N sodium chloride containing 0.015 N sodium citrate pH 7.0. The DNA is reprecipitated, washed, dried as above, and stored at −20°. The methylated substrate prepared in this manner has a specific activity of about 3000 cpm/mg.

Assay Method

The enzymatic activity of MGMT can be measured by detection of [^{3}H]methyl transfer from the substrate DNA to the cysteine residue at the active site of the Tk MGMT.[8] In the standard assay, the reaction mixture [100 μl containing 70 mM HEPES–KOH (pH 7.5), 1 mM dithiothreitol (DTT), 5 mM EDTA, 10 μg of bovine serum albumin (BSA), 13.5 μg [^{3}H]MNU-treated calf thymus DNA, and the enzyme] is incubated at 50° for 1 hr, and the reaction is terminated by adding 500 μl of 5% trichloroacetic acid and placing the sample on ice for 5 min. In control samples, enzyme is added after the incubation period and immediately before ethanol precipitation. The mixture is heated at 90° for 30 min to hydrolyze the DNA and placed on ice for 5 min; then, 100 μl of BSA solution (1mg/ml) is added. The hydrolyzed DNA is transferred onto a circular Whatman (Clifton, NJ) GF/C filter (25 mm in diameter) under suction, and the filter is washed several times with 5% trichloroacetic acid and once with ethanol. The radioactivity of enzymes trapped on the filter is measured in a liquid scintillation counter.

Alternative Method

The alternative method measures the disappearance of O^6-methylguanine from DNA. The enzyme sample is contained in the reaction mixture (30 μl) with 8 μg of [^{3}H]DNA in 50 mM 4-(2-hydroxyethyl)-1-piperazineethanesulfonic acid–KOH (pH 7.8), 10 mM dithiothreitol, 1 mM EDTA, 50 μM spermidine hydrochloride, and 5% (v/v) glycerol. Mixtures are incubated for 10 min at 50°, chilled, and precipitated with an equal volume of cold 0.8 M trichloroacetic acid. After centrifugation in Eppendorf microtubes, the supernatants (which contain less than 5% of the radioactive material) are carefully removed and the pellets are hydrolyzed in 30 μl of 0.1 M HCl at 70° for 30 min to release purines from DNA. Background interference in the assay from HCl-soluble material, probably resulting from alkylated pyrimidine oligonucleotide released on hydrolysis of DNA, can be reduced by dilution of each hydrolyzate with nine volumes of H$_2$O and passage through a small dispensable Dowex-1 column directly into scintillation vials.[15]

Definition of Enzyme Activity

One unit of methyltransferase activity is defined as the activity that accepts 1 pmol of methyl group.[15]

---

[15] B. Demple, A. Jacobsson, M. Olsson, P. Robins, and T. Lindahl, *J. Biol. Chem.* **257**, 13776 (1982).

Expression of Recombinant Tk-MGMT

Gene Cloning of Tk-MGMT

All the O^6-meG-DNA methyltransferases have significant homology within the core motif I (region I) and a very high homology within the putative active site (region II) containing the methyl acceptor cysteine. In order to obtain the MGMT gene from KOD1, PCR (polymerase chain reaction) is carried out using primers with sequences based on the conserved regions I and II:

Forward primer: 5'-AGGGCAGT(G/A/T/C)GG(G/A/T/C)G(G/C) (G/A/T/C) GC(G/A/T/C)-3'

Reverse primer: 5'-GACCCT(A/G)TG(A/G)CA(G/A/T/C)AC-3')

The DNA fragment can be amplified by using the genomic DNA (1μg) as a template in a 100 μl reaction containing 200 pmol of each primer, 10 μl of 10× Vent DNA polymerase buffer (New England Biolabs, Beverly, MA), 10 μl of 2 mM dNTP mixture (Takara Shuzo, Kyoto, Japan), and 1 U of Vent DNA polymerase. The mixture is subjected to 30 cycles of heat treatment in a thermal cycler (Gene Amp PCR System 2400, Perkin-Elmer Applied Biosystems, Foster City, CA) at 94°, 55°, and 72° for 1, 1, and 2 min, respectively. The PCR product is cloned into pUC 19 vector, and the sequencing analysis shows that the amplified DNA fragment is a part of the methyltransferase gene. The entire DNA region encoding Tk-MGMT is cloned using the PCR fragment labeled with digoxigenin-dUTP (DIG DNA labeling system, Boehringer Mannheim GmbH, Mannheim, Germany) as a probe for Southern hybridization. The plasmid library is constructed from 2- to 2.5-kbp *Sac*I fragments of KOD1 chromosomal DNA in the pUC 18 vector, and a positive clone harboring a 2.3-kbp insert should be obtained. DNA sequencing analysis is performed by the dideoxy-chain termination method with fluorescent primers and T7 DNA polymerase (A.L.F. DNA Sequencer, Amersham Pharmacia Biotech, Uppsala, Sweden).[13]

Amino Acid Sequence Homology

The deduced amino acid sequence of the Tk-MGMT from the DNA sequence has been compared with MGMTs from various sources (Fig. 3). Significant homology found in their C-terminal halves and their sequences are aligned in Fig. 3. *Escherichia coli* HM174(DE3)pLysS cells, harboring the recombinant plasmid, are grown overnight at 37° in NZCYM medium (per liter, 10g NZ-amine, 5g NaCl, 5g Bacto-yeast extract, 1 g casamino acids, 2 g $MgSO_4 \cdot 7H_2O$) containing ampicillin (100 μg/ml). The overnight culture is inoculated (1% inoculation) into fresh NZCYM medium containing ampicillin (100 μg/ml), and incubation at 37° is continued until the optical density at 660 nm reaches approximately 0.4. The

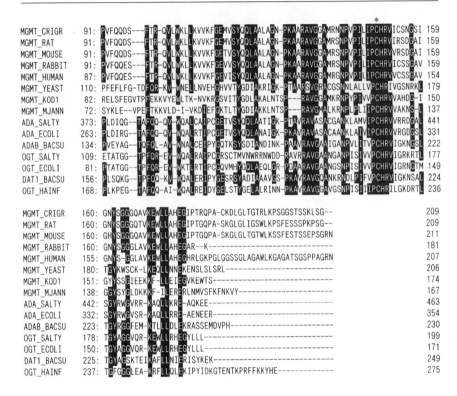

FIG. 3. Comparison of the amino acid sequences of the C-terminal halves of the O^6-meG-DNA methyltransferases from various sources. Spaces, shown by "-", were introduced to obtain maximum matches. Black and white letters indicate the amino acid residues conserved in more than 8 of 15 sources. The catalytic methyl acceptor cysteine is indicated with an asterisk. The abbreviations are as follows (the entry names or accession numbers in protein or DNA databanks are in parentheses): PIR, Protein information Resource; SP, Swiss-Prot; GB, GenBank. MGMT_CRIGR, *Cricetulus griseus* (Chinese hamster) (SP P26186); MGMT_RAT, rat (SP P24528); MGMT_MOUSE, mouse (PIR A41809); MGMT_RABBIT, rabbit (GB177047); MGMT_HUMAN, human (SP P16455); MGMT_YEAST, *Saccharomyces cerevisiae* (SP P26188); MGMT_KOD1, *Thermococcus kodakaraensis* KOD1 (GB D86335); MGMT_MJANN, *Methanococcus jannaschii* (PIR H64490); ADA_SALTY, Ada *Salmonella typhimurium* (PIR XYEBOT); ADA_ECOLI, Ada *Escherichia coli* (SP P06134); OGT_ECOLI, Ogt *Escherichia coli* (SP P09168); OGT_SALTY, Ogt *Salmonella typhimurium* (SP P37429); ADAB_BACSU, Ada-B *Bacillus subtilis* (SP P19220); DAT1_BACSU, Dat1 *Bacillus subtilis* (SPP11742); OGT_HAINF, Dat1 *Haemophilus influenzae* (SP P44687).

inducer isopropylthiogalactoside (IPTG, final concentration, 1 mM) is then added and culture is incubated for 6 hr at 37°. Cells are harvested by centrifugation (8,000g for 10 min) and then washed with buffer A (50 mM Tris-HCl pH 8.0, 0.1 mM EDTA).[13]

Enzyme Extraction

The cell pellet is resuspended into buffer A. The washed cells are disrupted by sonication and centrifuged (8000g for 60 min), and the supernatant is taken as a crude extract (Fraction I). Fraction I is incubated at 90° for 10 min, and then centrifuged (8000g for 60 min) to remove denatured proteins. The clear supernatant (Fraction II) is brought to 80% ammonium sulfate saturation and kept at 4° for 2 hr. The precipitate is collected by centrifugation (8,000g for 60 min), dissolved in buffer A, and dialyzed overnight against the same buffer (Fraction III).[13]

Ion-Exchange Chromatography

For initial purification, Fraction III is applied to a Sepharose Q column (1.6 cm × 2.5 cm, Hi-Trap Q) (Amersham Pharmacia Biotech, Piscataway, NJ), equilibrated with buffer A and eluted with a linear gradient of NaCl (0 to 1 M) at a flow rate of 1 ml/min. Flow-through fractions (Fraction IV) containing MGMT activity are pooled, dialyzed overnight against buffer B (50 mM potassium phosphate pH 6.5, 0.1 mM EDTA, 50 mM KCl), and applied to a Mono S column (Amersham Pharmacia Biotech) previously equilibrated with buffer B. The homogeneous enzyme is eluted with a linear gradient of KCl (50 mM to 1 M) at a flow rate of 1 ml/min. Eluted MGMT fractions (Fraction V) are pooled and dialyzed again against buffer A containing 0.15 M KCl.[13]

Gel Filtration

The eluted enzyme fractions are concentrated with Centriprep-3 (Amicon, Inc., Beverly, MA). Concentrated MGMT fractions are applied to a Superose 6 column (Amersham Pharmacia Biotech) equilibrated with buffer A containing 0.15 M KCl and eluted at a flow rate of 0.5 ml/min. The molecular mass of recombinant Tk-MGMT is estimated from gel-filtration chromatography.[13] Results of SDS–PAGE analysis of MGMT purification are shown in Fig. 4.

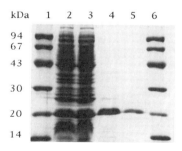

FIG. 4. SDS–PAGE analysis of Tk-MGMT purification. Lanes 1 and 6, molecular markers; lane 2; cell extract; lane 3, Fraction I (crude extract); lane 4, fraction II (heat treated); lane 5, purified MGMT.

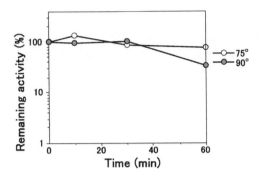

FIG. 5. Thermal denaturation curve of Tk-MGMT.

Thermostability of Tk-MGMT

The homogeneous enzyme is incubated at 75° and 90° in 50 mM Tris-HCl buffer (pH 8.0), for 10, 30, and 60 min, and the remaining activity is assayed at 50° in the standard assay condition described above (Fig. 5).[13]

In Vivo Assay of Tk-MGMT in E. coli Mutant Strain

The plasmid pKM-1m containing the Tk-MGMT gene is digested with XbaI and BamHI restriction enzymes. The resulting fragment containing the ribosomal binding site (RBS), the translational start and stop codons, is subcloned into pUC 19 previously digested with the same restriction enzymes. The new plasmid, named pKM-2m, is used to transform E. coli strain KT233, which lacks both ogt and ada genes.[16] A 3 ml aliquot of the culture is added to 30 ml of NZCYM broth containing 50 μg/ml of ampicillin. Exponentially growing cells (OD_{660} of 0.3–0.4) are induced with 1 mM IPTG at 37°. After 3 hr of incubation, the cells are harvested and suspended in the same volume of M9 salt medium. A 1-ml aliquot of the suspension is added to 1 ml of M9 salt medium containing different concentrations of MNNG. After incubation at 37° for 10 min, 2 ml of LB broth is added and, after appropriate dilutions, the cells are plated on LB (50 μg/ml ampicillin) plates. The number of viable cells (colonies) is counted after overnight incubation at 37° (Fig. 6).[13]

Crystallization Method

The purified Tk-MGMT is dialyzed against 50 mM Tris-HCl buffer at pH 8.0 containing 0.1 mM EDTA. The dialyzed protein is concentrated to 10 mg/ml.

---

[16] K. Takano, T. Nakamura, and M. Sekiguchi, *Mutat. Res.* **254,** 37 (1991).

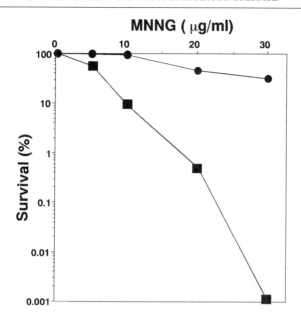

FIG. 6. Survival of cells after treatment with N-methyl-N'-nitro-N-nitrosoguanidine (MNNG). *E. coli* KT233 ($\Delta ada, \Delta ogt$) cells carrying one of the following plasmids were treated with various concentrations of MNNG at 37° for 10 min and the survival frequencies were determined. Closed box, pKM-2m: KOD1 MGMT clone; closed circle, pUC-19: control plasmid.

Crystallization procedures are carried out using the hanging-drop vapor diffusion technique. The droplet (typically 4 μl) is prepared by mixing equal volumes (2 μl) of the protein and reservoir solutions.[17]

Form 1 crystals are obtained by the following procedure. The reservoir solution is prepared by mixing 400 μl of 50% (w/w) polyethylene glycol (PEG) 8000, 200 μl of 1.0 M zinc acetate, 100 μl of 1.0 M sodium cacodylate pH 6.5, and 300 μl distilled water, yielding final concentrations of 20% PEG 8000, 200 mM zinc acetate, and 100 mM sodium cacodylate. Form 1 crystals are thin plates with a maximum size of 0.2 × 0.1 × 0.05 mm (Fig. 7a).[17]

Form 2 crystals are obtained as follows. First, needle-shaped microcrystals are grown against a reservoir solutions of 15% PEG 8000 and 200 mM ammonium sulfate. If 15% (w/w) PEG 20,000 [300 μl 50% (w/w) PEG 20,000] is used instead of PEG 8000, rod-shaped crystals appear. When the concentration of PEG 20,000 is decreased to 12% (w/w), crystals reach dimensions of 0.8 × 0.1 × 0.1 mm in a few weeks (Fig. 7b).[17]

---

[17] H. Hashimoto, M. Nishioka, T. Inoue, S. Fujiwara, M. Takagi, T. Imanaka, and Y. Kai, *Acta Crystallogr.* **D54**, 1395 (1998).

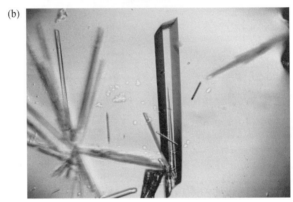

FIG. 7. Crystals of *Tk*-MGMT. (a) Form 1 crystals of *Tk*-MGMT with maximum dimensions 0.2 × 0.1 × 0.05 mm. (b) Form 2 crystals of *Tk*-MGMT with the maximum dimensions 0.8 × 0.1 × 0.01 mm.

Hyperthermophilic archaea live in an environment that is very similar to that of the primitive earth. Early organisms must have been exposed to deleterious, high-temperature environments. It was therefore essential to develop effective DNA repair systems. Hence, the study of archaeal MGMT is expected to provide interesting clues concerning primitive systems of DNA repair and their high thermostability. The detailed structural analysis is now in progress using the crystals mentioned above. Three-dimensional structure will allow us to elucidate the relationship between structure and function of their enzymes.

[21] DNA Polymerases from Euryarchaeota

By YOSHIZUMI ISHINO and SONOKO ISHINO

Introduction

Since the establishment of DNA cloning techniques, many genes coding for DNA polymerase have been cloned and sequenced, and their deduced amino acid sequences have been compared. Based on the sequence similarities, DNA polymerases are now classified into four families, family A, B, C, and X.[1] These families are represented by *Escherichia coli* DNA polymerase I (family A), DNA polymerase II (family B), DNA polymerase III α subunit (family C), and others including eukaryotic DNA polymerase β and terminal transferase (family X).

In isolating the entire set of DNA polymerase activities from one hyperthermophilic archaeon, a new DNA polymerase was discovered that does not belong to any family.[2,3] This enzyme isolated from *Pyrococcus furiosus* is composed of the two subunits, DP1 (69 kDa) and DP2 (143 kDa), with the larger protein containing the catalytic subunit. The two genes that code for the proteins homologous to these two subunits were found in the genomes of four euryarchaeotic strains, *Methanococcus jannaschii*, *Archaeoglobus fulgidus*, *Methanobacterium thermoautotrophicum*, and *Pyrococcus horikoshii*.[4,5] These DNA polymerases were designated Pol II as the second DNA polymerase to distinguish them from their typical family B DNA polymerase. Biochemical properties of *P. furiosus* Pol II suggest the euryarchaeal Pol II to be the replicational DNA polymerase.[3] Because there is no sequence similarity of the catalytic subunit of the euryarchaeal Pol II to any family of DNA polymerases described above, we have proposed the family D for the euryarchaeal Pol II and the designation of Pol BI, Pol BII, and Pol D, respectively, for archaeal DNA polymerases identified to date, to clarify which family the enzymes belong to.[6] Pol BI and Pol D are used hereafter for Pol I and Pol II, respectively.

---

[1] J. Ito and D. K. Braithwaite, *Nucleic Acids Res.* **19,** 4045 (1991).
[2] M. Imamura, T. Uemori, I. Kato, and Y. Ishino, *Biol. Pharm. Bull.* **18,** 1647 (1995).
[3] T. Uemori, Y. Sato, I. Kato, H. Doi, and Y. Ishino, *Genes Cells* **2,** 499 (1997).
[4] Y. Ishino, K. Komori, I. K. O. Cann, and Y. Koga, *J. Bacteriol.* **180,** 2232 (1998).
[5] I. K. O. Cann, K. Komori, H. Toh, S. Kanai, and Y. Ishino, *Proc. Natl. Acad. Sci. U.S.A.* **95,** 14250 (1998).
[6] I. K. O. Cann and Y. Ishino, *Genetics* **152,** 1249 (1999).

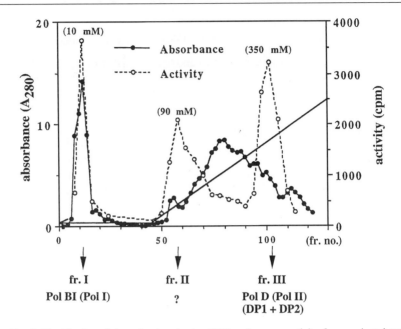

FIG. 1. Identification of three fractions having DNA polymerase activity from sonicated crude *P. furiosus* cell extract separated by anion-exchange chromatography (Fractogel EMD TMAE 650, Merck, Germany). DNA polymerase activity was measured by [^{3}H]TTP incorporation using calf-thymus activated DNA as template primers.

Identification of Pol D in *P. furiosus*

Starting from the crude cell extract of *P. furiosus*, DNA polymerase activity could be separated into three fractions (I, II, and III) by anion-exchange chromatography; these activities were eluted at 10, 90, and 350 m*M* NaCl concentration, respectively, during the development of this linear gradient (Fig. 1). The DNA polymerase activities from fraction I and II were aphidicolin-resistant, whereas the activity from fraction III was aphidicolin-sensitive. A DNA polymerase having a typical family B sequence was isolated from fraction I.[7] Fraction III was further separated by four additional steps, yielding a protein with a molecular mass of 130–135 kDa and having DNA polymerase activity by *in situ* activity gel analysis.[2] This activity was designated DNA polymerase II (Pol II → Pol D), while the enzyme having a family B-like sequence from fraction I was designated DNA polymerase I (Pol I → Pol BI).[2]

---

[7] T. Uemori, Y. Ishino, H. Toh, K. Asada, and I. Kato, *Nucleic Acids Res.* **21,** 259 (1993).

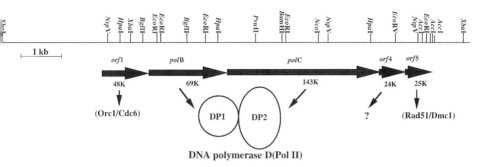

FIG. 2. The region containing the genes for Pol D in genomic DNA of *P. furiosus*. A restriction map and encoded open reading frames (ORFs) in the region containing Pol D gene are shown. Five ORFs are arranged in tandem in this region, and they are transcribed together to a single mRNA. The subunit proteins of Pol D, DP1 and DP2, are located in the second and third positions, respectively. The deduced sequences of ORF1 and ORF5 are similar to those of the eukaryotic Orc1/Cdc6-like proteins and Rad51/DMC1-like proteins, respectively.

Cloning of Genes for Pol D

A cosmid library containing *P. furiosus* genomic DNA fragments of 35–50 kbp was constructed. Five hundred *Escherichia coli* strains transformed from the cosmid library were independently cultured and the host proteins in the cell extracts were inactivated by heat treatment (100° for 10 min). The resulting heat-stable protein library was screened for DNA polymerase activity. The *P. furiosus* DNA fragment was isolated from the clone having the heat-stable DNA polymerase activity, and a physical map was constructed by restriction enzyme digestions. The subsequent deletion analyses and nucleotide sequencing showed that the genes coding for the DNA polymerase activity were the second and the third *orfs* in five continuous *orfs* found in a 10 kbp *Xba*I fragment (Fig. 2). The molecular masses of the deduced proteins from these two genes were 69,294 Da and 143,161 Da, respectively. The N-terminal amino acid sequence of the larger ORF corresponded to the sequence that was experimentally determined from the protein having activity by the *in situ* activity gel analysis described above. DNA polymerase activity was confirmed after independently cloning these two *orfs*, then expressing and purifying the proteins produced in *E. coli*. The distinct DNA polymerase activity was detected by the conventional incorporation assay of [^{3}H]TTP in solution only when the two proteins were mixed. Purified DP2 itself had a little activity (2% of the full activity from DP1 + DP2), which was consistent with the previous detection by *in situ* DNA incorporation assay. From these results, it was concluded that Pol D is composed of at least these two subunits, and the two proteins were named DP1 and DP2.[3] The genes for Pol D are arranged in tandem in one operon on the genome of the pyrococci

(*P. furiosus,*[3] *P. horikoshii,*[8] and *P. abyssi*[9]), but randomly located in the other euryarchaeotes.[10–12]

Recombinant Pol D

Bacterial Strain and Growth

The operon containing the genes for Pol D was inserted into an expression vector, pTV118N (Takara Shuzo, Kyoto, Japan), and the initiation codon, ATG, of the ORF1 was adjusted to *Nco*I (-CC<u>ATG</u>G-). The resulting plasmid was designated pFU1001.[3] Pol D can be produced in *E. coli* carrying pFU1001. The target genes are under the control of the *lac* promoter on the expression vector, and therefore, their expression is induced by isopropylthiogalactoside (IPTG). *E. coli* JM109 carrying pPF1001 is grown at 37° in a shaker bath in 1 liter of LB broth containing 100 µg/ml ampicillin. When the A_{600} of the culture reach 0.5–0.6, IPTG is added to 1 mM and the culture is incubated with aeration for an additional 5 hr.

Cell Disruption and Nucleic Acid Removal

The following procedure for the purification of Pol D was modified from that originally reported.[3] After centrifugation at 10,000g for 15 min at 4°, harvested cells (3.0 g of wet weight) are frozen and thawed, and then resuspended in 40 ml of buffer A (50 mM Tris-HCl, pH 8.5, 10% glycerol, 2 mM 2-mercaptoethanol, 0.1 M NaCl). The cells can be disrupted by sonication, and the cell debris is removed by centrifugation at 48,000g for 30 min at 4°. The supernatant is heated at 80° for 15 min, and then centrifuged again. Various heat treatments led to similar results; comparable levels of recovered activity and purification level were found with temperatures at 80°, 85°, and 90°. To the heated supernatant (40 ml), polyethyleneimine (Sigma, St. Louis, MO) is added to 0.15% (w/v). The solution is stirred on ice for 30 min, following which the precipitate is recovered by centrifugation at 22,000g for 10 min. The precipitate is then suspended in 30 ml of buffer A containing 0.3 M of ammonium sulfate, and stirred on ice for 30 min. To the supernatant recovered by centrifugation by 30,000g for 20 min, ammonium sulfate is added to 80% saturation and the solution is stirred on ice for 30 min. The precipitate recovered by centrifugation includes Pol D without DNA.

---

[8] Y. Kawarabayashi, M. Sawada, H. Horikawa, Y. Haikawa, Y. Hino *et al., DNA Res.* **5,** 55 (1998).
[9] J. Querellou, personal communication (1998).
[10] C. J. Bult, O. White, G. J. Olsen, L. Zhou, R. D. Fleischmann *et al., Science* **273,** 1058 (1996).
[11] H. P. Klenk, R. A. Clayton, J. Tomb, O. White, K. E. Nelson *et al., Nature* **390,** 364 (1997).
[12] D. R. Smith, L. A. Doucette-Stamm, C. Deloughery, H.-M. Lee, J. Dubois *et al., J. Bacteriol.* **179,** 7135 (1997).

TABLE I
PURIFICATION OF *P. FURIOSUS* DNA POLYMERASE D

Step	Total protein (mg)	Total activity (U)	Specific activity (U/mg)	Recovery (%)	Purification (-fold)
Heat	28	1000	36	100	
HiTrap Q	1.2	340	283	34	8
HiTrap Heparin	0.5	240	480	24	13
Superdex 200	0.3	210	700	21	20

Chromatography

The precipitate from the ammonium sulfate is resuspended with 8 ml of buffer A and dialyzed against the same buffer overnight. The dialyzate is applied to an anion-exchange column (HiTrap Q; 5 ml, Pharmacia, Piscataway, NJ), preequilibrated with buffer A using an HPLC system (AKTA explorer 10S, Pharmacia), and the separation is done with a linear gradient of 0.1–1 M NaCl. The Pol D–containing fractions are eluted at 0.45 M NaCl. Then, these fractions are applied to a heparin-Sepharose (HiTrap heparin, Pharmacia) after dialysis against buffer B containing 50 mM Tris-HCl, pH 8.0, 0.5 mM DTT, 0.1 mM EDTA, and 10% (v/v) glycerol. The column is again developed with 0.1 to 1 M NaCl gradient. The active fractions, eluted at 0.45 M NaCl, are loaded onto a gel filtration column (Superdex 200, Pharmacia) equilibrated with buffer B containing 0.15 M NaCl. An illustrative example of the recombinant Pol D preparation using the described procedure is summarized in Table I and Fig. 3. The specific activity in one nanomole of Pol D is usually 150–200 units, which is about twofold higher than the value obtained from the purification of Pol BI using a recombinant *E. coli* carrying its gene on the pFU3 expression vector.[13] (To compare the biochemical properties, Pol BI was prepared by using the recombinant *E. coli*, which is modified from the original report[6].) Pol D activity is eluted from a gel filtration column at the position corresponding at approximately 220 kDa, which suggests that Pol D exists as a heterodimer of DP1 and DP2 (1 : 1).

Comments

Two subunits can be purified independently from *E. coli* JM109/pPFDP1 and *E. coli* JM109/pPFDP2, respectively, as described previously.[3] DP1 is easy to purify to homogeneity. However, it is very difficult to purify DP2, because it is very unstable in *E. coli* when produced by itself.[3] An immunoprecipitation experiment confirmed that DP2 coprecipitated with DP1 by an anti-DP1 antibody

---

[13] K. Komori and Y. Ishino, *Protein Eng.* **13**, 41 (2000).

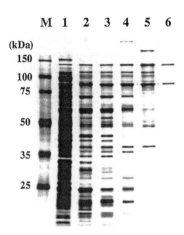

FIG. 3. SDS–PAGE of fractions from a purification of *P. furiosus* Pol D. Samples from each purification step were analyzed by electrophoresis through a 10% polyacrylamide gel. Lanes: 1, sonicated cell extract; 2, heat-treated supernatant; 3, polyethyleneimine and ammonium sulfate fraction; 4, Hi-Trap Q fraction; 5, HiTrap Heparin fraction; 6, Superdex 200 fraction. Lane M contains size marker proteins. The protein bands were detected by silver staining. The DP1 protein always moves on the gel more slowly than expected from its actual size estimated by the size-strandard markers.

from the crude extract of *P. furiosus* cells.[5] Moreover, DP1 and DP2 elute together in the same fraction (fraction III of Fig. 1) by anion-exchange chromatography.[14] In contrast to the case of production of these recombinant proteins in *E. coli*, the native DP2 produced in *P. furiosus* cells is stable and the native DP1 is rapidly degraded in *P. furiosus* cells, and therefore, it is somewhat difficult to purify distinct amount of the native Pol D from *P. furiosus* cell extract as we originally presented.[2]

Biochemical Properties

Pol D requires Mg^{2+} for activity with an optimum at 5 mM. Activity is stimulated up to 300% by 50–100 mM of KCl in comparison with its activity without KCl, but is inhibited by a KCl concentration higher than 200 mM. This is in contrast to Pol BI, which is sensitive to this salt; activity is inhibited to 55% and 6% by 50 mM and 100 mM KCl, respectively. Pol D is resistant to 4 mM aphidicolin, which completely inhibits Pol BI activity, as is the case for the first fraction from anion-exchange chromatography. Pol D is sensitive to N-ethylmaleimide (NEM), as activity decreases to 45% in 12 mM of NEM, although more than 90% of Pol BI activity is retained. Pol D activity is completely inhibited with 20 mM NEM.

[14] K. Komori, S. Ishino, and Y. Ishino, unpublished results (1998).

TABLE II
ACTIVITIES OF DNA POLYMERASES ON VARIOUS COMBINATIONS
OF TEMPLATE–PRIMERS

Template–primer	Relative activity (%)			
	Pfu Pol D	*Pfu* Pol BI	*Taq*	*Bca*
Calf thymus activated DNA	100	100	100	100
Heat-denatured calf thymus activated DNA	340	87	130	230
M13 ssDNA–45-mer primer	170	23	90	180
M13 ssDNA–RNA 18-mer primer	52	0.49	38	23
Poly(dA)–Oligo(dT) (20:1)	94	390	290	960
Poly(A)–Oligo(dT) (20:1)	0.085	—	0.063	7.8

Optimal reaction temperature and pH, using DNase I-activated calf thymus DNA as a substrate, are approximately 75° and pH 7.7 (at 75°) for both Pol BI and Pol D. The heat stability of Pol D is much lower than that of Pol BI; however, more than 50% of Pol D activity is retained after incubation at 94° for 20 min.

The primer–template preference of Pol D, as compared with other DNA polymerases including Pol BI, *Taq* polymerse,[15] and *Bacillus caldotenax (Bca)* polymerase,[16] is shown in Table II. Pol D prefers primer-extension type substrates to gap-filling type template–primers, such as the DNase I-activated DNA. This tendency is similar to the results from *Taq* polymerase and *Bca* polymerase, but in contrast to that of *Pfu* Pol BI. Pol D uses RNA primer more efficiently than other enzymes. Very little reverse transcriptase activity, comparable to the level of *Taq* polymerase, is detected. Figure 4 shows the excellent primer extension ability of Pol D. Here the *in vitro* primer extension rate, using the linearized M13 single-stranded DNA as a template, is about 4–5 bases/sec and 60 bases/sec for Pol BI and Pol D, respectively.

DNA polymerases, especially replicases, have to synthesize DNA strands having the sequences complementary to the template strands as correctly as possible to preserve the genetic information. To ensure accuracy, DNA polymerases have the $3' \rightarrow 5'$-exonuclease activity to remove nucleotides that have been incorporated incorrectly. *P. furiosus* Pol BI, also referred to as *Pfu* DNA polymerase, is known to be one of the most accurate DNA synthesis.[17–19] The $3' \rightarrow 5'$-exonuclease

---

[15] Y. Ishino, T. Ueno, M. Miyagi, T. Uemori, M. Imamura, S. Tsunasawa, and I. Kato, *J. Biochem.* (Tokyo) **116**, 1019 (1994).

[16] T. Uemori, Y. Ishino, K. Fujita, K. Asada, and I. Kato, *J. Biochem.* (Tokyo) **113**, 401 (1993).

[17] K. S. Lundberg, D. D. Shoemaker, M. W. W. Adams, J. M. Short, J. A., Sorge, and E. J. Mathur, *Gene* **108**, 1 (1991).

[18] K. Hollung, O. S. Gabrielsen, and K. S. Jakobsen, *Nucleic Acids Res.* **22**, 3261 (1994).

[19] J. Cline, J. C. Braman, and H. H. Hogrefe, *Nucleic Acids Res.* **24**, 3546 (1996).

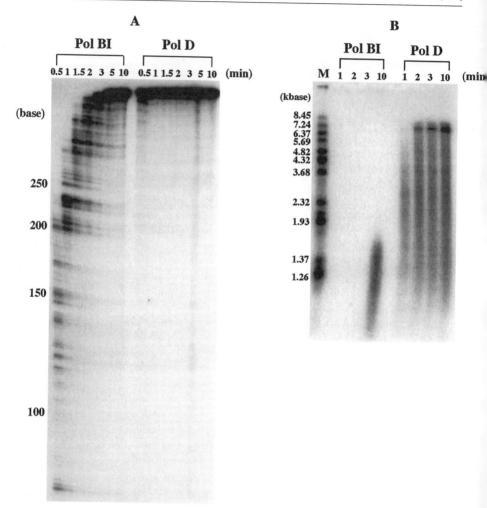

FIG. 4. Primer extension abilities of *P. furiosus* Pol BI and Pol D. A ^{32}P-labeled d30-mer annealed to the linearized M13 single-stranded DNA was used as the template–primer. To 1 μg of the substrate, 0.25 units of Pol BI or Pol D were added and the reactions in 20 μl were started at 75°. Equal amounts of the reaction mixture were sampled at 0.5, 1, 1.5, 2.0, 3.0, 5.0, and 10.0 min, and stop solution (98% deionized formamide, 1 mM EDTA, 0.1% xylene cyanol, 0.1% bromphenol blue) was added. Products were then analyzed by an 8% PAGE containing 8 M urea (A) and a 1.2% alkali agarose gel (B).

activities associated with *P. furiosus* Pol BI and Pol D were compared. Comparable activities were detected from Pol BI and Pol D under the same assay conditions for the same DNA polymerase activities described previously.[3] These activities are shown visually by autoradiography (Fig. 5). Both enzymes prefer single-stranded

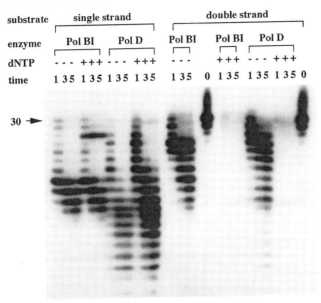

FIG. 5. Detection of the 3′ → 5′-exonuclease activities of *P. furiosus* Pol BI and Pol D. A ^{32}P-labeled d30-mer was used as a substrate with (for a double-stranded substrate) or without (for a single-stranded substrate) annealing to the M13 single-stranded DNA. To 0.25 pmol of the substrate, 0.25 units of Pol BI or Pol D was added and the reactions in 20 μl were started at 75°. Equal amounts of the reaction mixture were sampled at the indicated time and analyzed by an 8% PAGE containing 8 *M* urea. To investigate the effect of dNTPs, the mixture of dATP, dGTP, dCTP, and dTTP was added to the reaction to 125 μ*M*. In the cases of the double-stranded substrate, primers were extended in the presence of dNTPs in both Pol BI and Pol D reactions.

DNA to double-stranded DNA as substrates. Degradation of the single-stranded DNA by these enzymes is independent of added dNTPs, and the reaction of Pol D is slightly faster than for Pol BI. The primer annealed to the template (the double-stranded substrate) was degraded equally by Pol BI and Pol D. These results suggest that Pol D may also have the ability to synthesize the DNA strands as accurately as Pol BI. Preliminary experiment using a simple method, which counts the number of colonies converted from blue to white because of mutations in the *lacZ* gene on the plasmid, shows that mutation frequency from the synthesis by Pol D is 3.5-fold higher than that of Pol BI, but less than half of the value from *Taq* polymerase. Further experiments using more accurate methods are in progress.

We cloned the genes for the Pol D homolog from *M. jannaschii* and expressed them in *E. coli*.[4] The DP1 and DP2 proteins from *M. jannaschii* produced in *E. coli* react with anti-*P. furiosus* Pol D polyclonal antibody, and the subunits of *P. furiosus* Pol D can complement *M. jannaschii* Pol D to yield DNA polymerase activity and vice versa. Biochemical properties of *M. jannaschii* Pol D are basically the same

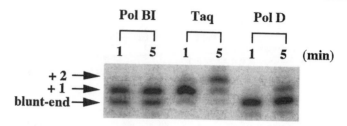

FIG. 6. Detection of +1 addition products synthesized by three DNA polymerases. M13 single-stranded DNA was linearized (cut the circular DNA at a *Pst*I site) and annealed with a ^{32}P-labeled d30mer at the position 55 bases upstream of the cutting site. To 0.25 μg of this substrate, 0.25 units of Pol BI, Pol D, or *Taq* polymerase was added (in 15 μl) and the equal amounts of the reaction mixture at 75° were sampled after 1 and 5 min. The reaction products were analyzed by an 8% PAGE containing 8 *M* urea.

as those of *P. furiosus* Pol D. The differences are sensitivity to salt (KCl) and heat stability. *M. jannaschii* Pol D is not stimulated with 50–100 m*M* KCl and is much less heat stable. Optimal reaction temperature of *M. jannaschii* Pol D, using the calf thymus activated DNA as template–primers, is around 65°. Preincubation at 94° for 2 min completely inactivates *M. jannaschii* Pol D. Intersubunit complementation between Pol D of *P. furiosus* and *M. jannaschii* showed that *M. jannaschii* DP2 is especially heat-sensitive.[20]

Because of its thermostability and excellent *in vitro* primer-extension ability, *P. furiosus* Pol D is useful for PCR. To clone PCR products into a plasmid, the efficiency of a particular DNA polymerase in catalyzing the nontemplated addition of nucleotides to the 3′ termini of blunt-ended DNA duplexes should be known.[21,22] An electrophoretic analysis of the nontemplated addition for three DNA polymerases is shown in Fig. 6. *P. furiosus* Pol D made blunt-ended product and +1 product at about a 1 : 1 ratio. On the other hand, Pol BI made only blunt-ended product after 1 min, but made some +1 product (about 20% of the total products) after 5 min. *Taq* polymerase, which is usually used for the TA cloning method, made +1 product with more than 80% efficiency after 1 min, and, moreover, +2 product after 5 min in this model experiment. This result shows that the PCR products produced by Pol D may be cloned efficiently into TA-cloning vectors.

Sequence Comparison

Amino acid sequences homologous to DP1 and DP2 proteins from *P. furiosus* and *M. jannaschii* are conserved in the genomes of other euryarchaeotes as

---

[20] Y. Ishino and I. K. O. Cann, *Genes Genet. Syst.* **73**, 323 (1998).
[21] J. M. Clark, *Nucleic Acids Res.* **16**, 9677 (1988).
[22] G. Hu, *DNA Cell Biol.* **12**, 763 (1993).

described above. No proteins or ORFs having obviously similar sequences to DP2 are found in the available databases. In contrast, from detailed sequence comparisons, it was found that euryarchaeotic DP1 proteins have considerable sequence similarity to the second subunit of eukaryotic DNA polymerase δ (Pol δ), one of the essential DNA polymerases for eukaryotic DNA replication. Even though the amino acid identities between the proteins ranged from 12.9 to 19.6%, the significance of these identities was statistically confirmed.[5] It has been shown that the second subunits of Pol δ from mammalian cells are required for efficient stimulation of the polymerase processivity by the proliferating cell nuclear antigen (PCNA).[23,24] Furthermore, it has been shown more recently that the second subunits of eukaryotic DNA polymerase α and ε (Pol α and Pol ε) also have some similarity with that of Pol δ and DP1, and that they consist of a family of DNA polymerase-associated B subunits.[25] These subunits might have originated from a single ancestor. The second subunit of Pol α has been implicated in cell-cycle control and regulation,[26,27] and a stable Pol ε complex essential for chromosomal replication requires the second subunit.[28] It is, therefore, interesting to consider what the functions of the euryarchaeotic DP1s are.

The amino acid sequences of euryarchaeotic DP2s do not have significant similarity to any protein or ORF in the databases, even though there is more than 50% amino acid identity among known DP2s. The crystal structures of nucleotide polymerases determined so far share a common folding pattern that resembles a right hand composed of fingers, thumb, and palm subdomains. Within the palm subdomain are two motifs containing two invariant carboxylates, which are thought to constitute part of the polymerase active site.[29–33] Because of the finding that *P. furiosus* DP2 by itself has some weak polymerase activity, the motifs containing invariant carboxylates, which can constitute part of the polymerase active site (palm subdomain), were sought by visual inspection of four DP2 proteins. Two regions, including the candidate motif A and C, have been proposed,[5] and experimental confirmation is in progress.

[23] J. Q. Zhou, H. He, C. K. Tan, K. M. Downey, and A. G. So, *Nucleic Acids Res.* **25**, 1094 (1997).
[24] Y. Sun, Y. Jiang, P. Zhang, S.-J. Zhang, Y. Zhou, B. Q. Li, N. L. Toomey, and M. Y. W. T. Lee, *J. Biol. Chem.* **272**, 13013 (1997).
[25] M. Makiniemi, H. Pospiech, S. Kilpelainen, M. Jokela, M. Vihinen, and J. E. Syvaoja, *Trends. Biochem. Sci.* **24**, 14 (1999).
[26] H. Araki, R. K. Hamatake, L. H. Johnston, and A. Sugino, *Proc. Natl. Acad. Sci. U.S.A.* **88**, 4601 (1991).
[27] H.-P. Nasheuer, A. Moore, A. F. Wahl, and T. S.-F. Wang, *J. Biol. Chem.* **266**, 7893 (1991).
[28] M. Foiani, F. Marini, D. Gamba, G. Lucchini, and P. Plevani, *Mol. Cell. Biol.* **14**, 923 (1994).
[29] D. L. Ollis, P. Brick, R. Hamlin, N. G. Xuong, and T. A. Steitz, *Nature* **313**, 762 (1985).
[30] L. A. Kohlstaedt, J. Wang, J. M. Friedman, P. A. Rice, and T. A. Steitz, *Science* **256**, 1783 (1992).
[31] R. Sousa, Y. J. Chung, J. P. Rose, and B.-C. Wang, *Nature* **364**, 593 (1993).
[32] Y. Kim, S. H. Eom, J. Wang, D.-S. Lee, S. W. Suh, and T. A. Steitz, *Nature* **376**, 612 (1995).
[33] J. L. Hansen, A. M. Long, and S. C. Schultz, *Structure* **5**, 1109 (1997).

From these results, a new family of DNA polymerases was proposed, which classify the euryarchaeotic Pol D (DP1+DP2) into family D.[6]

Future Aspects

A new DNA polymerase, composed of DP1 and DP2, has been found in Euryarchaeota. It seems to be the replicative enzyme, because it has very excellent primer extension ability and strong $3' \rightarrow 5'$-exonuclease activity, both of which were measured *in vitro* for the *P. furiosus* enzyme. Further studies are necessary to prove this, and it would be helpful to clone and analyze the homologs from the methanogenic and halophilic euryarchaeotes, whose host–vector systems are available for the genetic analyses. Both Pol BI and Pol D in *P. furiosus* can interact with a PCNA homolog found in *P. furiosus*,[34] which suggests that both enzymes may have roles in the DNA replication machinery. Family D DNA polymerases have been looked for in the Crenarchaeota, but the results have consistently been negative. Instead, the crenarchaeotes have two family B DNA polymerases (Pol BI and Pol BII).[35,36] Furthermore, it has been predicted from the nucleotide sequences that Sulfolobals may have the third family B DNA polymerase.[37] Therefore, it would be very interesting to know the difference between the DNA replication mechanisms of these two subdomains of Archaea.

The structure–function relationship of this DNA polymerase is also very interesting and would help in understanding more about the mechanism of DNA strand synthesis. Because DP2 proteins do not have sequence similarities to other known DNA polymerases, crystallographic analyses of Pol D will contribute to understanding how convergent and divergent the structure necessary for DNA polymerase activity is. Studies on the molecular mechanism of the DNA replication in the thermophilic archaea will contribute greatly to understanding eukaryotic DNA replication. The proteins cased for this process in Archaea are structurally similar to eukaryotic proteins, for example, ORC, PCNA, RFC, RPA, and MCM.[6,20,38,39] Proteins from hyperthermophilic organisms are very stable and suitable for structural and biochemical analyses investigating protein–protein or protein–DNA interactions. Studies examining the interactions of these proteins with DNA polymerases are also now in progress.

---

[34] I. K. O. Cann, S. Ishino, I. Hayashi, K. Komori, H. Toh, K. Morikawa, and Y. Ishino, *J. Bacteriol.* **181**, 6591 (1999).
[35] T. Uemori, Y. Ishino, H. Doi, and I. Kato, *J. Bacteriol.* **177**, 2164 (1995).
[36] I. K. O. Cann, S. Ishino, N. Nomura, Y. Sako, and Y. Ishino, *J. Bacteriol.* **181**, 5984 (1999).
[37] D. R. Edgell, H.-P. Klenk, and W. F. Doolittle, *J. Bacteriol.* **179**, 2632 (1997).
[38] D. R. Edgell and W. F. Doolittle, *Cell* **89**, 995 (1997).
[39] R. Bernander, *Mol. Microbiol.* **29**, 955 (1998).

[22] RecA/Rad51 Homolog from *Thermococcus kodakaraensis* KOD1

By NAEEM RASHID, MASAAKI MORIKAWA, SHIGENORI KANAYA, HARUYUKI ATOMI, and TADAYUKI IMANAKA

Introduction

Recombination and transposition generate new gene combinations. They are immensely important in evolution because they markedly enlarge the genetic repertoire from which natural selection chooses. General recombination is the process by which DNA strands are exchanged between homologous chromosomes, essentially at any site. In bacteria, the RecA protein is essential for homologous genetic recombination, recombinational DNA repair, and induction of the SOS response.[1] RecA catalyzes the strand exchange reactions between a single-stranded DNA (ssDNA) and a double-stranded DNA (dsDNA) or between two dsDNAs, which is believed to be the central mechanism of homologous genetic recombination. Many RecA protein homologs have been reported, not only from bacteria,[2] but also from eukaryotes.[3,4] In eukaryotic cells, the Rad51 protein corresponds to RecA in its binding properties to both ssDNA and dsDNA and its DNA-dependent ATPase activity, suggesting structural and functional similarity between these proteins.[4,5] Structural similarity has also been observed between Rad51 and DMC1, a meiosis-specific protein in yeast.[6] Genes encoding RecA/Rad51 homologs have been reported from 11 genera of archaea, including hyperthermophiles.[7] The genes can be divided into two subfamilies, the *radA* and *radB* subfamilies. The enzymatic properties of these archaeal RecA/Rad51 homologs are of high interest, as they may represent prototypes of the bacterial and eukaryotic enzymes. Here, the properties of a RecA/Rad51 homolog (*Tk*-REC) from *Thermococcus kodakaraensis* KOD1 (previously reported as *Pyrococcus kodakaraensis* KOD1) are described. *Tk*-REC is a member of the RadB subfamily and is the smallest RecA/Rad51 homolog that has been characterized.[8–10]

---

[1] M. M. Cox and I. R. Lehman, *Ann. Rev. Biochem.* **216,** 229 (1987).
[2] S. Karlin, G. M. Weinstock, and V. Brendel, *J. Bacteriol.* **177,** 6881 (1995).
[3] F. E. Benson, A. Stasiak, and S. C. West, *EMBO J.* **13,** 5764 (1994).
[4] A. Shinohara, H. Ogawa, and T. Ogawa, *Cell* **69,** 457 (1992).
[5] E. M. Zaitseve, E. N. Zaitsev, and S. C. Kowalczykowski, *J. Biol. Chem.* **274,** 2907 (1999).
[6] D. K. Bishop, D. Park, L. Xu, and N. Kleckner, *Cell* **69,** 439 (1992).
[7] S. J. Sandler, P. Hugenholtz, C. Schleper, E. F. Delong, N. R. Pace, and A. J. Clark, *J. Bacteriol.* **181,** 907 (1999).
[8] N. Rashid, M. Morikawa, and T. Imanaka, *Mol. Gen. Genet.* **253,** 397 (1996).
[9] N. Rashid, M. Morikawa, K. Nagahisa, S. Kanaya, and T. Imanaka, *Nucleic Acids Res.* **25,** 719 (1997).

Comparison of Tk-REC with Other RecA Protein Families

A gene encoding DNA recombination protein, *Tk*-REC, is located 3.2 kb upstream of the 16S rRNA gene. It encodes 220 amino acid residues (DDBJ/EMBL/GenBank accession No. D83176). The amino acid sequences of RecA family proteins from bacteria, eukaryotes, and archaea were compared, and a phylogenetic tree is shown in Fig. 1a. Most of the archaeal sequences were deduced from putative genes. Proteins that have been actually identified or characterized are the RadB from *Pyrococcus furiosus*,[11] RadA from *Haloferax volcanii*,[12] RadA from *Sulfolobus solfataricus*,[13] RadA from *Desulfococcus amylolyticus*,[14] RadA from *Pyrobaculum islandicum*,[15] and *Tk*-REC.[8–10] Amino acid sequences in the main central domain of the proteins (amino acid residues 42 to 195 in the case of RecA from *Escherichia coli*) were used for the analysis. Alignment of amino acid sequences and estimation of the number of amino acid substitutions per site or evolutionary distances between sequences was calculated with the CLUSTALW program provided by the DNA Data Bank Japan (DDBJ). Based on the evolutionary distance matrix, a phylogenetic tree was constructed by the neighbor joining method. Bootstrap resampling was repeated 1000 times. Both RecA proteins from bacteria and Rad51 proteins from eukaryotes form clear clusters in the tree. The proteins from archaea form two diverse clusters, clearly corresponding to the two subfamilies, RadA and RadB. *Tk*-REC is a member of the RadB subfamily, where other members have also been found from *Archaeoglobus fulgidus*, *Methanobacterium thermoautotrophicum*, *Methanococcus jannaschii*, *Pyrococcus furiosus*, and *P. abyssi*. Southern blot analysis has indicated the presence of a *radA* gene in strain KOD1 (data not shown), indicating that in all archaeal strains which possess a *radB* gene, a *radA* gene is also present.

Tk-REC is quite small (24,543 Da, p*I* 9.69) compared to its bacterial and eukaryotic counterparts, with the N- and C-terminal regions being obviously truncated (Fig. 1b). The crystal structure of *E. coli* RecA protein–ADP complex revealed a major domain that bound to ATP and single- and double-stranded DNAs, and two smaller subdomains at the N and C termini protruded from the major domain.[16] Deletion analysis of the N-terminal 33 amino acid residues of RecA

---

[10] N. Rashid, M. Morikawa, S. Kanaya, H. Atomi, and T. Imanaka, *FEBS Lett.* **445**, 111 (1999).
[11] I. Hayashi, K. Morikawa, and Y. Ishino, *Nucleic Acids Res.* **27**, 4695 (1999).
[12] W. G. Woods and M. L. Dyall-Smith, *Mol. Microbiol.* **23**, 791 (1997).
[13] E. M. Seitz, J. P. Brockman, S. J. Sandler, A. J. Clark, and S. C. Kowalczykowski, *Genes Dev.* **12**, 1248 (1998).
[14] Y. V. Kil, D. M. Baitin, R. Masui, E. A. Bonch-Osmolovskaya, S. Kuramitsu, and V. A. Lanzov, *J. Bacteriol.* **182**, 130 (2000).
[15] M. Spies, Y. Kil, R. Masui, R. Kato, C. Kujo, T. Ohshima, S. Kuramitsu, and V. Lanzov, *Eur. J. Biochem.* **267**, 1125 (2000).
[16] R. M. Story, I. T. Wever, and T. A. Steitz, *Nature (London)* **355**, 318 (1992).

a)

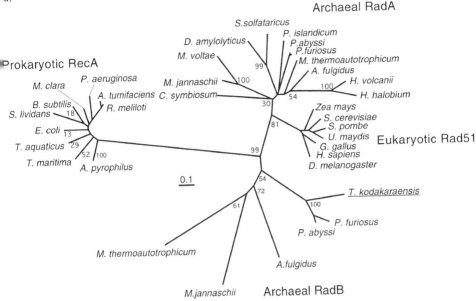

FIG. 1. (a) Phylogenetic tree for RecA family proteins. Calculations were performed by the CLUSTALW program provided by DDBJ. Segments corresponding to an evolutionary distance of 0.1 are shown. Each name at the terminus represents the species from which the protein originated. The DDBJ/EMBL/GenBank protein accession numbers are as follows. Archaeal RadA: *Archaeoglobus fulgidus* (O29269), *Cenarchaeum symbiosum* (AAD16063), *Desulfurococcus amylolyticus* (AAD33955), *Halobacterium halobium* (AAD16062), *Haloferax volcanii* (Q48328), *Methanobacterium thermoautotrophicum* (AAB85860), *Methanococcus jannaschii* (Q49593), *Methanococcus voltae* (AAC23499), *Pyrobaculum islandicum* (BAA88984), *Pyrococcus abyssi* (CAB49165), *Pyrococcus furiosus* (AAC34998), *Sulfolobus solfataricus* (Q55075). Archaeal RadB: *A. fulgidus* (AAB89159), *M. thermoautotrophicum* (AAB86165), *M. jannaschii* (Q57702), *P. abyssi* (CAB49042), *P. furiosus* (AAC34999), *Thermococcus kodakaraensis* KOD1 (BAA11830). Bacterial RecA: *Agrobacterium tumefaciens* (JC1377), *Aquifex pyrophilus* (AAA67702), *Bacillus subtilis* (P16971), *Escherichia coli* (AAC75741), *Methylomonas clara* (P24542), *Pseudomonas aeruginosa* (P08280), *Rhizobium meliloti* (P27865), *Streptomyces lividans* (P48294), *Thermotoga maritima* (AAA27417), *Thermus aquaticus* (JX0292). Eukaryotic Rad51: *Drosophila melanogaster* (BAA04580), *Gallus gallus* (P37383), *Homo sapiens* (Q06609), *Saccharomyces cerevisiae* (P25454), *Schizosaccharomyces pombe* (P36601), *Ustilago maydis* (AAC61878), *Zea mays* (AAD32030). (b) Domain structure of RecA family proteins. Domain I (N-terminal) is essential for self-association or interaction with other proteins. Domain II (central) is an important major domain that contains specific ATP binding motifs A (for phosphate binding) and B (for coordination of ATP-associated Mg^{2+} ion) and is responsible for recombination, UV resistance, and formation of active oligomer. Domain III (C-terminal) contributes for protein–protein interaction, binding to nucleotides, and LexA. The acidic C-terminal region is indicated by a black box.

(b)

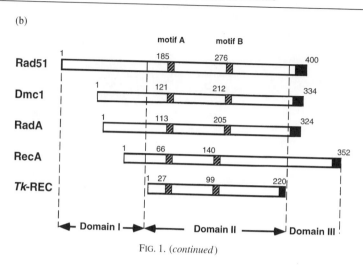

Fig. 1. (continued)

protein demonstrated that this region contributed to protein self assembly,[17,18] which suggested poor formation of normal nucleoprotein filaments by *Tk*-REC. The C-terminal truncated RecA protein bound at a higher rate to ssDNA (10 times) and dsDNA, and more effectively promoted the LexA repressor cleavage reaction and ATP hydrolysis.[19] These facts suggest the possibility that *Tk*-REC maintains high DNA repairing activity and binding activity to ssDNA and dsDNA even at neutral pH.[20]

Complementation of UV Sensitivity

In order to examine the DNA repairing activity of *Tk*-REC *in vivo*, the complementation ability of UV resistance was tested in *E. coli*. Survival of *E. coli recA* null mutant strain HMS174 (DE3)pLysS, harboring plasmids containing *Tk-rec* gene in pET-8c vector (Novagen, Madison, WI), was determined under various UV irradiation conditions. The method used is as follows: Single colonies are cultured to the mid-log phase in NZCYM medium containing ampicillin. Gene expression is induced by the addition of 1mM isopropylthiogalactoside (IPTG). After incubation for 4 hr, the cells are exposed to UV light for various periods of time, diluted, and plated onto LB plates containing ampicillin to determine the number of colony-forming units (cfu).[21] UV dose is measured with an ultraviolet

---

[17] M. Dutreix, B. Burnett, A. Bailone, C. M. Radding, and R. Devoret, *Mol. Gen. Genet.* **232**, 489 (1992).
[18] T. Mikawa, R. Masui, T. Ogawa, H. Ogawa, and S. Kuramitsu, *J. Mol. Biol.* **250**, 471 (1992).
[19] S. Tateishi, T. Horii, T. Ogawa, and H. Ogawa, *J. Mol. Biol.* **223**, 115 (1992).
[20] B. F. Pugh and M. M. Cox, *J. Biol. Chem.* **262**, 1326 (1987).
[21] R. Kato and S. Kuramitsu, *J. Biochem.* **114**, 926 (1993).

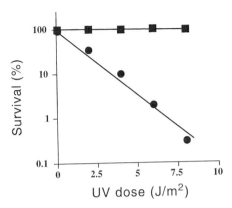

FIG. 2. UV survival curve for the *E. coli recA* mutant strain HMS174(DE3)pLysS harboring the *Tk-rec* gene, following UV irradiation. Data are average values of three independent experiments. Squares, *E. coli* HMS174(DE3)pLysS carrying *Tk-rec* gene; circles, *E. coli* HMS174(DE3)pLysS carrying pET-8c only.

intensity meter J225 (UVP, Inc., Upland, CA). As shown in Fig. 2, the cells that express *Tk-rec* gene show remarkable UV light resistance, whereas the cells with vector only are UV-sensitive. RecA mutation of *E. coli* HMS174 (DE3)pLysS is a null mutation (*recA1*), which produces an ATPase-deficient, inactive RecA protein but still can bind to DNA in normal fashion.[1] RecA1 would compete with *Tk*-REC in the DNA binding event. The results shown above indicate that the *Tk*-REC protein is functional in the *E. coli* cell, which is, to our knowledge, the first demonstration of a nonbacterial RecA homolog complementing an *E. coli recA* defect. This is interesting from two aspects. First, *Tk*-REC is a member of the RadB, and not RadA, subfamily. Members of the RadA subfamily have been reported to display ability to bind to ssDNA, DNA strand exchange activity, and DNA-dependent ATPase activity, which are typical of the RecA family proteins.[13–15] On the other hand, the RadB protein from *P. furiosus* was shown to bind to the second subunit of DNA polymerase II and has been suggested to be one of its components.[11] Further studies will be needed to determine the physiological role of the RadB subfamily proteins. Second, *Tk*-REC maintains only the major central domain of RecA proteins. Mutations in *recA* from *E. coli*, *recA441*, *recA730*, *recA803*, and *recA1211* that provide enhanced recombination activity locate in the region encoding either or both of the N-terminal or C-terminal domains that is lacking in *Tk*-Rec.[22] Moreover, *recA1202* contains the mutation from Gln-184 to Lys-184 and also increases the recombination activity of RecA. The amino acid

---

[22] S. C. Kowalczykowski, D. A. Dixon, A. K. Eggleston, L. D. Lauder, and W. M. Rehrauer, *Microbiol. Rev.* **58**, 401 (1994).

residue corresponding to Gln184 in *Tk*-Rec is Lys144. Each mutation is expected to increase the rate of association of the RecA protein with ssDNA by discouraging the formation of RecA protein aggregates that decrease the rate of RecA protein binding to ss-DNA.[22] The mechanism through which *Tk*-Rec renders UV resistance to the *E. coli recA* null mutant is of high interest.

Production and Purification of *Tk*-REC

E. coli strain HMS174(DE3)pLysS carrying the expression plasmid of *Tk-rec* gene is grown overnight at 37° in NZCYM medium containing ampicillin. The preculture is inoculated (1%) into fresh NZCYM medium, and the cultivation is continued until the optical density at 660 nm reaches 0.35 to 0.4 (mid-log phase). The culture is then induced with 1 mM (final concentration) of IPTG and incubated for another 4 hr at 37°. Cells are harvested by centrifugation at 6,000g for 10 min at 4° and washed with 50 mM sodium phosphate buffer (pH 7.0). The cell pellet is resuspended in the same buffer and cells are disrupted by sonication. Soluble and insoluble fractions are separated by centrifugation (15,000g for 30 min at 4°). Recombinant *Tk*-REC is recovered mainly in the soluble fraction. The soluble fraction is heat treated at 80° for 20 min, leading to the precipitation of most proteins from the host cells. The supernatant, after centrifugation at 15,000g for 20 min at 4°, is then applied to a HiTrap Q column (Amersham Pharmacia Biotech, Piscataway, NJ). When a 0 to 1.0 M NaCl gradient in 50 mM sodium phosphate buffer (pH 7.0) is applied, *Tk*-REC elutes from the column at a salt concentration of 0.25 to 0.35 M NaCl. Fractions containing *Tk*-REC are pooled and dialyzed (1:1000, v/v) against 50 mM sodium phosphate buffer (pH 7.0) and applied to a Mono Q column (Amersham Pharmacia Biotech). Again, *Tk*-REC elutes from the column at a salt concentration of 0.25 to 0.35 M NaCl. The eluate, after dialysis, is further applied to a HiTrap Heparin affinity column (Amersham Pharmacia Biotech), and *Tk*-REC is eluted as a single peak at 0.9–0.95 M NaCl (Fig. 3a). Purity of the protein for 10 consecutive fractions containing *Tk*-REC is examined by polyacrylamide gel electrophoresis in the presence of sodium dodecyl sulfate (SDS–PAGE) (Fig. 3b). A mixture of rabbit muscle phosphorylase *b* (94,000), bovine serum albumin (67,000), egg white ovalbumin (43,000), bovine erythrocyte carbonic anhydrase (30,000), soybean trypsin inhibitor (20,100), and bovine milk β-lactalbumin (14,400) is used for molecular weight standards. Protein concentration is determined with a bicinchoninic acid (BCA) protein assay kit (Pierce, Rockford, IL) according to the manufacturer's instructions using bovine serum albumin as a standard. We found a single protein band corresponding to the molecular weight of *Tk*-REC in 10 consecutive fractions (Fig. 3b).

The molecular weight of the native enzyme is determined by analytical gel filtration chromatography using a Superose 6 [HR 10/30] column (Amersham Pharmacia Biotech). Molecular weight standards [bovine thyroglobulin (670,000),

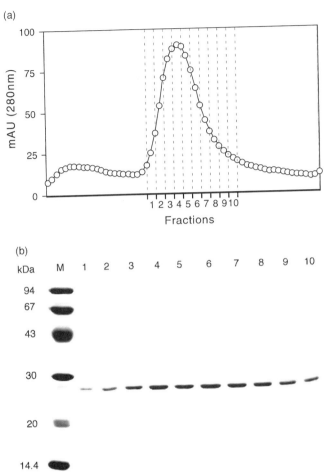

FIG. 3. (a) Elution profile of *Tk*-REC when applied to HiTrap Heparin affinity column. The fractions containing *Tk*-REC are indicated. (b) Coomassie Brilliant Blue stained SDS–PAGE of the 10 consecutive fractions containing purified recombinant *Tk*-REC eluted from the HiTrap Heparin affinity column. Lane M, molecular mass standards; lanes 1–10, purified recombinant *Tk*-REC from fractions 1 to 10. (c) DNase activity of purified recombinant *Tk*-REC from fractions 1 to 10. Lane C, pUC19 (2 μg) without addition of *Tk*-REC; lanes 1–10, pUC19 (2 μg) with *Tk*-REC (10 μl) of fractions 1 to 10. Arrows indicate nicked DNA with slow migration rates.

bovine β-globin (158,000), chicken ovalbumin (44,000), horse myoglobin (17,000), and vitamin B_{12} (1,500)] are purchased from Bio-Rad (Bio-Rad Labs., Hercules, CA). *Tk*-REC elutes at a position corresponding to molecular weight 49 kDa, indicating that *Tk*-REC exists as a dimer. Since *Tk*-REC lacks the N-terminal

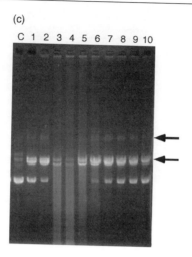

FIG. 3. (continued)

domain of RecA from *E. coli* (Fig. 1b), it is understandable that it does not aggregate in a multimeric form but instead forms a dimer during purification. It should be noted that *Tk*-REC also aggregates when the protein concentration is over 1 mg/ml in a low salt concentration buffer, <100 mM KCl.

DNase Activity

When the purified recombinant *Tk*-REC is tested for its DNA binding and strand exchange activity, a common and basic characteristic of RecA protein family, it has been previously noted that it has DNA degrading (DNase) activity.[9] The nuclease activity is detected both for ssDNA and dsDNA, but no activity is observed for RNA (tRNA). DNase activity is measured as follows: A reaction mixture (30 μl) containing double (φX174 vector plasmid)- or single (M13 mp19)-stranded circular DNA (5 μg) and purified recombinant *Tk*-REC (1 μg) is incubated at 60° for 10 min in 30 mM Tris-HCl (pH 8.0) and 10 mM MgCl$_2$. To each reaction sample, 5 μl of sample loading buffer [40 mM Tris–acetate (pH 8.0), 1 mM EDTA, 50% (v/v) glycerol, and a trace of bromphenol blue] is added and kept on ice to stop the reaction, and the reaction mixtures are analyzed by 1% agarose gel electrophoresis in Tris–acetate/EDTA buffer or 15% native polyacrylamide gel in Tris–borate/EDTA buffer. DNA is visualized under UV light by ethidium bromide staining. When the effect of metal ions is examined, metal ion-free enzyme is first prepared as follows. Metal ions in the enzyme solution are removed by chelating with 20 mM EDTA and subsequent dialysis for 36 hr at 4° against Milli-Q water. Mg^{2+} (MgCl$_2$), in the reaction mixture described above is replaced by

Ca^{2+} ($CaCl_2$), Cd^{2+} ($CdCl_2$), Co^{2+} ($CoCl_2$), Cu^{2+} ($CuSO_4$), Fe^{2+} ($FeSO_4$), Mn^{2+} ($MnCl_2$), Ni^{2+} ($NiCl_2$), Pb^{2+} ($PbCl_2$), Sr^{2+} ($SrCl_2$), or Zn^{2+} ($ZnCl_2$). It has been found that the enzyme utilizes several divalent ions other than Mg^{2+}, such as Ca^{2+}, Cd^{2+}, Cu^{2+}, Mn^{2+}, Ni^{2+}, Pb^{2+}, and Sr^{2+}, but DNase activity is reduced to some extent. When Mg^{2+} is substituted with Co^{2+} or Fe^{2+}, no detectable change is observed. It should be noted that the protein exhibits no DNase activity and even no DNA binding activity in the presence of Zn^{2+} ion. Because the protein exhibits a very weak DNase activity even in the absence of any metal ion, it seems likely that Zn^{2+} ion binds to the Mg^{2+} binding sites of the protein and inhibits the DNase activity.[9]

As there have been no other reports of a RecA family protein harboring DNase activity, we have performed further detailed analysis. The 10 consecutive fractions after elution from the HiTrap heparin affinity column, containing purified Tk-REC, are independently analyzed for DNA binding activity (Fig. 3c). We have found that fractions 3 to 6 display significant DNase activity, whereas the other fractions do not. As no other proteins besides Tk-REC are visible on the SDS–PAGE gel (Fig. 3b), there is a possibility that the recombinant Tk-REC may exist in two distinct forms: one with DNase activity, and the other without activity. There is also the possibility of a trace contamination of DNase from the host cell. However, in this case, the DNase must be quite thermostable. One reason is that significant heat treatment (80° for 20 min) is performed during purification of Tk-REC. Another is the fact that the optimal temperature of the DNase activity that we previously observed was 60°.[9] We are not aware of any reports of such DNase from E. coli.

ATPase Assay

Taking into account the differences in DNase activity among the purified fractions of Tk-REC, we have examined ATPase activity for each fraction independently. ATPase activity of Tk-REC is assayed in a reaction mixture containing 1 μg Tk-REC, 20 mM Tris-HCl (pH 8.0), 2 mM DTT, 0.1 mM ATP, 2 mM $MgCl_2$, 100 μg/ml BSA, and with or without 120 μM of ϕX-174 DNA to make a final volume of 25 μl. Each reaction contains 100 nCi [α-^{32}P]ATP. Reaction mixtures are set up on ice and incubations are made at 55° for 30 min. After chilling on ice 1 μl is then spotted directly onto Polygram CEL 300 polyethyleneimine (PEI) thin-layer chromatography plates (Macherey-Nagel, GmbH & Co., Duren, Germany). The substrate and products of the reaction are separated by one-dimensional chromatography using 1 M LiCl.

Interestingly, the reaction product of ATPase activity of Tk-REC is different between fractions that do or do not display DNase activity. Fractions displaying DNase activity show complete hydrolysis of ATP with AMP as the product.

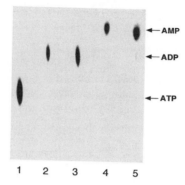

FIG. 4. Reaction product of ATPase activity of Tk-REC. Separation of the reaction product by thin-layer chromatography. Lane 1, negative control with no Tk-REC protein. Lane 2, Tk-REC from fraction 8 was added. Lane 3, Tk-REC from fraction 8 was added along with 1 μg pUC19. Lane 4, Tk-REC from fraction 4 was added. Lane 5, Tk-REC from fraction 4 was added along with 1 μg pUC19.

Fractions without DNase activity lead to ADP as the product. A representative result is shown in Fig. 4. At present, the reason for this correlation between DNase and ATPase activities is unknown. However, unlike other RecA family proteins, all fractions exhibit ATPase activity independent of both ssDNA and dsDNAs.

All of these results demonstrate that Tk-REC is highly functional both *in vivo* and *in vitro*. Archaea in general, and hyperthermophilic archaea in particular, are thought to evolve from and still live in environments very similar to that of primitive earth. Cells at that time might have been exposed to harsh conditions including fairly strong UV light. Under this stress it would be essential to maintain effective DNA repair systems.

[23] Hyperthermophilic Inteins

By FRANCINE B. PERLER

Introduction

Inteins[1] are intervening sequences that are posttranslationally excised from protein precursors. They are the protein equivalent of introns, which are intervening sequences that splice from precursor RNAs. The sequences flanking both sides of

---

[1] F. B. Perler, E. O. Davis, G. E. Dean, F. S. Gimble, W. E. Jack, N. Neff, C. J. Noren, J. Thorner, and M. Belfort, *Nucleic Acids Res.* **22**, 1125 (1994).

the intein are called exteins.[1] During protein splicing, the intein is excised from a precursor protein and the flanking exteins are joined by a peptide bond. This ligation of exteins differentiates protein splicing from other forms of proteolytic processing. The self-catalytic protein splicing reaction is mediated by the intein plus the first carboxy-extein amino acid (aa), which are capable of splicing in heterologous exteins. However, each intein has its own "substrate" specificity that dictates allowable proximal extein residues. As of December 31, 1999, there were 100 putative inteins listed in the Intein Registry,[1] representing all three domains of life (see InBase[2] at <http://www.neb.com/neb/inteins.html>); 74% of these inteins are found in thermophilic organisms, mainly Archaea.

Thermophilic inteins were among the first inteins discovered[3] and played a key role in establishing protein splicing as a fundamental method of protein biosynthesis. The proof that inteins were spliced from precursor proteins rather than from precursor RNAs[4] and the mechanism of protein splicing were initially demonstrated using archaeal inteins.[4-10] Since their discovery in 1990, inteins have been harnessed to perform numerous protein engineering processes.

Identifying Inteins

A combination of criteria should be used to identify inteins, since one must differentiate true inteins from in-frame insertions present because of sequence variability or other types of insertion elements. Experimentally, an intein may be indicated when the observed size of the protein is smaller than the predicted size of the gene product and a second protein is also produced; since there may be other explanations, such as aberrant electrophoretic mobility or protease processing, the gene should be examined for the presence of intein motifs. Most putative inteins have been identified by sequence comparison, rather than experimentally.[2,11-14] A large (>100 aa) in-frame insertion in a sequenced gene that is absent in other

---

[2] F. B. Perler, *Nucleic Acids Res.* **28**, 344 (2000).
[3] F. B. Perler, D. G. Comb, W. E. Jack, L. S. Moran, B. Qiang, R. B. Kucera, J. Benner, B. E. Slatko, D. O. Nwankwo, S. K. Hempstead, C. K. S. Carlow, and H. Jannasch, *Proc. Natl. Acad. Sci. U.S.A.* **89**, 5577 (1992).
[4] M. Xu, M. W. Southworth, F. B. Mersha, L. J. Hornstra, and F. B. Perler, *Cell* **75**, 1371 (1993).
[5] M. Xu and F. B. Perler, *EMBO J.* **15**, 5146 (1996).
[6] M. Xu, D. G. Comb, H. Paulus, C. J. Noren, Y. Shao, and F. B. Perler, *EMBO J.* **13**, 5517 (1994).
[7] Y. Shao, M. Q. Xu, and H. Paulus, *Biochemistry* **34**, 10844 (1995).
[8] Y. Shao, M.-Q. Xu, and H. Paulus, *Biochemistry* **35**, 3810 (1996).
[9] Y. Shao and H. Paulus, *J. Pept. Res.* **50**, 193 (1997).
[10] R. A. Hodges, F. B. Perler, C. J. Noren, and W. E. Jack, *Nucleic Acids Res.* **20**, 6153 (1992).
[11] F. B. Perler, G. J. Olsen, and E. Adam, *Nucleic Acids Res.* **25**, 1087 (1997).
[12] S. Pietrokovski, *Protein Sci.* **3**, 2340 (1994).
[13] S. Pietrokovski, *Protein Sci.* **7**, 64 (1998).
[14] J. Z. Dalgaard, M. J. Moser, R. Hughey, and I. S. Mian, *J. Comput. Biol.* **4**, 193 (1997).

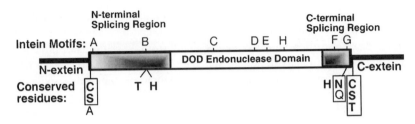

FIG. 1. The organization and conserved sequences of a typical protein splicing precursor with a DOD homing endonuclease. The amino-terminal and carboxy-terminal splicing regions from a single protein domain that carries out the splicing reaction. The splicing regions are separated by a homing endonuclease domain as depicted, or by a shorter linker region in mini-inteins. Intein Blocks A through G are shown above the precursor, and conserved residues in selected motifs are shown below the precursor using the single-letter amino acid code. Block G contains the (His)/(Asn or Gln)/(Ser, Thr or Cys) residues and includes the first residue of the carboxy-extein. Nucleophiles are boxed.

sequenced homologs and that contains one or more intein motifs suggests that this gene may contain an intein.

Inteins have several signature motifs (Fig. 1 and Table I). Blocks A, B, F, and G,[11,12] also known as Blocks N1, N3, C2, and C1,[13] respectively, are present in all inteins. Block A begins with the first residue of the intein and Block B is usually 60–100 amino acids from the start of the intein. Block F closely precedes Block G, which includes the end of the intein and the first residue of the carboxy-extein. Blocks C, D, E, and H (also known as Blocks EN1–4[13]) are sometimes present between Blocks B and F and include the signature motifs of one class of endonuclease, termed the dodecapeptide (DOD) or LAGLI-DADG family of homing endonucleases[15,16] (see below). Mini-inteins (<200 aa) have a small linker in place of the homing endonuclease domain.

The consensus sequence for each conserved intein motif is listed in Table I. No single position is invariant in all sequenced inteins. Any member of the amino acid group defined in Table I may be present at a given position, even when a specific predominant residue is indicated.[2,11–13] Most inteins begin with Cys or Ser and end in the dipeptide His-Asn. The carboxy-extein usually begins with Ser, Thr, or Cys. All but one intein have a conserved His in Block B. The Ser, Thr, Cys, and Asn residues are the nucleophiles that perform the chemical steps, whereas the conserved His residues facilitate these nucleophilic displacements. As more inteins have been identified and characterized, polymorphisms at these positions have become evident, including inteins beginning with Ala, ending with Gln, and with penultimate residues other than His.[2]

[15] M. Belfort and R. J. Roberts, *Nucleic Acids Res.* **25**, 3379 (1997).
[16] M. S. Jurica and B. L. Stoddard, *Cell. Mol. Life Sci.* **55**, 1304 (1999).

TABLE I
CONSERVED INTEIN MOTIFS[a]

Motif	Conserved sequence[b]
Block A (N1):	ChxxDpxhhhxxG
Block B (N3):	GxxhxhTxxHxhhh
Block C (EN1):	LhGxhhaG
Block D (EN2):	KxIPxxh
Block E (EN3):	LxGhFahDG
Block H (EN4):	pxSxxhhxxhxxLLxxhGI
Block F (C2):	rVYDLpV [1-3 residues]axx[H/E]NFh
Block G (C1):	NGhhhHNp

[a] Blocks C, D, E, and H are only present in DOD family homing endonuclease domains.

[b] An uppercase letter indicates that the amino acid is present in ≥50% of inteins; no single position is invariant in all inteins. Lowercase letters represent amino acid groups: x, any residue; h, hydrophobic residue (G, A, V, L, I, M); p, polar residue (S, C, T); a, acidic residue (D or E); r, aromatic residue (F, Y, W).

How one identifies new inteins depends on whether one is analyzing the sequence of a favorite gene or searching databases for new inteins. Although intein sequences are very divergent, even in the conserved motifs, inteins can usually be detected by running BLAST or Psi BLAST searches[17] using a few individual intein sequences, searching for intein motifs,[11–13] or by using an intein-trained Hidden Markov model.[18]

Protein Splicing Reaction

Protein splicing requires proteolytic cleavage of the precursor protein at both splice junctions and formation of a native peptide bond between the exteins. This complex pathway is mediated by a highly coordinated series of simple chemical reactions (four nucleophilic displacements). The elements directing these chemical reactions are contained within the intein plus the first downstream extein residue. However, splicing in foreign proteins is generally not as efficient as splicing in the native context, suggesting that the exteins may play a role in the splicing reaction by helping to align the splice junctions or by enabling proper folding of the intein catalytic core. The extein amino acid preceding the intein is the target of 3/4 of

---

[17] S. F. Altschul, T. L. Madden, A. A. Schaffer, J. Zhang, Z. Zhang, W. Miller, and D. J. Lipman, *Nucleic Acids Res.* **25**, 3389 (1997).
[18] J. Z. Dalgaard, *Trends Genet.* **10**, 306 (1994).

the nucleophilic displacements. Some inteins are more tolerant of proximal extein sequences than are other inteins.[19–22]

A review of the mechanism can be found in Noren et al.[23] Protein splicing is initiated by an acyl rearrangement of the Ser or Cys at the beginning of the intein to form a (thio)ester bond between the intein and the amino-extein (Fig. 2). The upstream splice junction is cleaved when the side chain of the Ser, Thr, or Cys at the beginning of the carboxy-extein attacks this (thio)ester bond, resulting in extein ligation and formation of a branched intermediate. The intein is released from the branched intermediate when the intein carboxy-terminal Asn or Gln cyclizes, cleaving the downstream splice junction. In the absence of the intein, the ligated exteins undergo a spontaneous acyl rearrangement to form a peptide bond. Structural and mutagenesis data indicate that the His in Block B assists in the nucleophilic displacements at the upstream splice junction and the intein penultimate His facilitates Asn cyclization.[5,24–26]

Intein Expression in Heterologous Proteins and Organisms

In all cases examined, protein splicing is rapid and efficient in the native organism, suggesting that inteins may have little effect on extein expression. However, that is not always the case when intein containing genes are expressed in heterologous organisms, such as *Escherichia coli,* or when inteins are cloned into model exteins. Most often, the rate of splicing is slower, some precursors fail to splice, dead-end single splice junction cleavage products accumulate, and precursors are proteolyzed, suggesting that the precursors are not optimally folded. If precursors are degraded by cellular proteases, yields may be improved by inducing expression at lower temperatures (12–20°)[4] or by using protease-deficient strains. Insertion of an intein into a foreign extein sometimes results in splicing at one temperature, but not at another temperature. For example, the *Pyrococcus* sp. GB-D DNA polymerase intein only spliced in a chimeric extein at temperatures above 25–30°,[4] whereas the *Mycobacterium xenopi* gyrase subunit A intein only spliced in a chimeric extein at temperatures below 20°.[21] Both inteins spliced at

---

[19] S. Chong, K. S. Williams, C. Wotkowicz, and M. Q. Xu, *J. Biol. Chem.* **273,** 10567 (1998).
[20] M. W. Southworth, K. Amaya, T. C. Evans, M. Q. Xu, and F. B. Perler, *BioTechniques* **27,** 110 (1999).
[21] A. Telenti, M. Southworth, F. Alcaide, S. Daugelat, W. R. Jacobs Jr., and F. B. Perler, *J. Bacteriol.* **179,** 6378 (1997).
[22] S. Nogami, Y. Satow, Y. Ohya, and Y. Anraku, *Genetics* **147,** 73 (1997).
[23] C. J. Noren, J. Wang, and F. B. Perler, *Angew. Chemi. Int. Ed. Engl.* **39,** 450 (2000).
[24] M. Kawasaki, S. Nogami, Y. Satow, Y. Ohya, and Y. Anraku, *J. Biol. Chem.* **272,** 15668 (1997).
[25] T. Klabunde, S. Sharma, A. Telenti, W. R. Jacobs, Jr., and J. C. Sacchettini, *Nat. Struct. Biol.* **5,** 31 (1998).
[26] X. Duan, F. S. Gimble, and F. A. Quiocho, *Cell* **89,** 555 (1997).

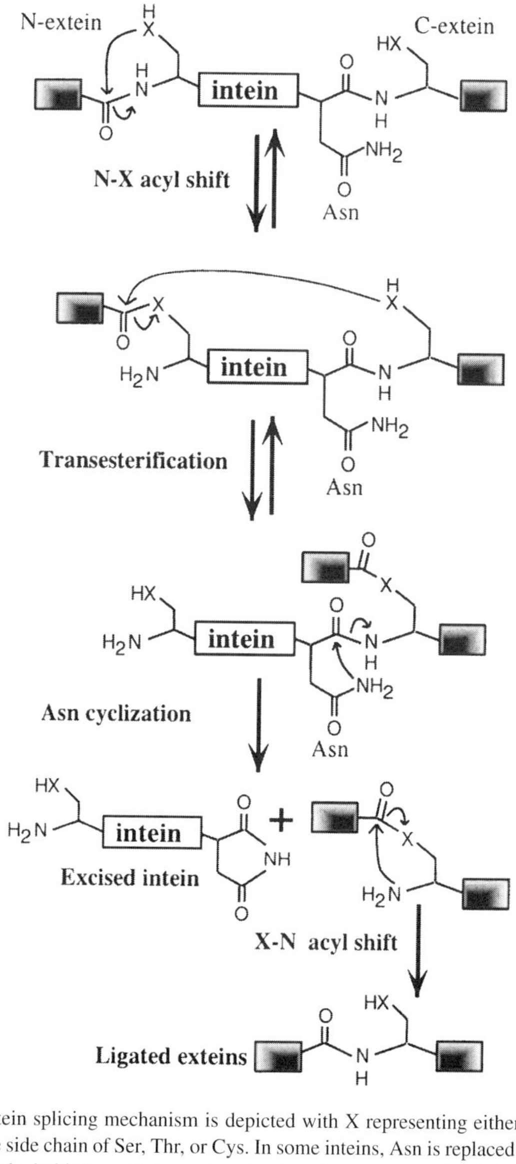

FIG. 2. The protein splicing mechanism is depicted with X representing either the oxygen or the sulfur present in the side chain of Ser, Thr, or Cys. In some inteins, Asn is replaced by Gln, which also can cyclize. All tetrahedral intermediates and proton transfer steps are omitted for clarity. The shaded rectangles represent the exteins.

the nonpermissive temperatures when in their native extein context. Although the reason for this temperature dependent splicing is unknown, it probably reflects a suboptimal extein context.

In attempting to study splicing of any intein, one may have to examine constructs that have partial extein sequences to ensure proper active site architecture. However, many inteins splice in a variety of contexts, while others do not, suggesting that each intein has its own degree of flexibility with respect to "substrate" (extein) specificity. When cloning an intein into a foreign extein, it is often better to choose an insertion site whose sequence is similar or identical to the native extein for 1–5 amino acids on each side of the intein. Analysis of native intein insertion sites failed to indicate a preferred type of secondary structure or sequence.[2,11–14] Inteins tend to be found in conserved extein motifs or active sites, which are likely to be surface accessible since they need to interact with substrates and cofactors. Therefore, inserting an intein into a surface location might improve splicing and result in less proteolysis due to precursor misfolding.

Inteins may also be cytotoxic in heterologous organisms, because of homing endonuclease activity.[15,16] The *Thermococcus litoralis* DNA polymerase gene (Vent pol) contains two inteins that are active endonucleases (PI-Tli I and PI-Tli II).[3,10] Each of these enzymes cleaves the *E. coli* genome several times (T. Davis and F. B. Perler, New England Biolabs, 1993, unpublished data), rendering the *T. litoralis* DNA polymerase unclonable on a multicopy plasmid in *E. coli* unless one of the two inteins is deleted.[3] Expression of intein containing genes may thus be improved by inactivating the endonuclease, which can be accomplished by mutating the conserved Asp residues in Blocks C and E.[10]

Intein Distribution

Are inteins an archaeal phenomenon, since 70% of inteins are found in archaea?[2] Six fully sequenced eubacterial genomes have no inteins,[2] whereas many fully sequenced archaeal genomes contain multiple inteins (Table II), including *Methanococcus jannaschii* (19 inteins), *Pyrococcus horikoshii OT3* (14 inteins), *Pyrococcus abyssi* (14 inteins), *Pyrococcus furiosus* (10 inteins), *Methanobacterium thermoautotrophicum* (delta H strain) (1 intein), *Aeropyrum pernix* (1 intein), and *Archaeoglobus fulgidus* (no identified inteins). Inteins have also been identified in three thermophilic eubacteria, Deinococcus radiodurans (2 inteins), Rhodothermus marinus (1 intein), and Aquifex aeolicus (1 intein). Archaeal genes often contain two or three inteins, whereas no eubacterial or eukaryal gene identified to date has more than one intein. In several cases, inteins represent greater than 50% of the coding region of the precursor. Thirty-four different intein insertion sites have been found in 27 genes from thermophilic organisms. Inteins present at the same insertion site in homologous genes from different organisms are considered intein homologs or alleles. However, the presence of an intein in a gene

TABLE II
INTEINS IN SEQUENCED ARCHAEAL GENOMES[a]

Intein[b] or (insert site)	Pab (14)[c]	Pho (14)	Pfu (10)	Mja (19)	Ape (1)	Mth (1)
(CDC21-a)	164	168	—	—	—	—
(CDC21-b)	268	260	367	—	—	—
GF-6P	—	—	—	499	—	—
Helicase	—	—	—	501	—	—
Hyp-1	—	—	—	392	—	—
Hyp	—	—	—	—	468	—
IF2	394	444	387	546	—	—
KlbA	196	520	522	168	—	—
LHR	—	475	—	—	—	—
Lon	333	474	401	—	—	—
Moaa	455	—	—	—	—	—
PEP	—	—	—	412	—	—
(pol-a)	—	—	—	369	—	—
(pol-b)	—	460	—	476	—	—
(pol-c)	—	—	—	—	—	—
Pol II (DP2)	185	166	—	—	—	—
RadA	—	172	—	—	—	—
(RFC-a)	499	525	525	548	—	—
(RFC-b)	—	—	—	436	—	—
(RFC-c)	608	—	—	543	—	—
(r-Gyr-a)	—	410	373	494	—	—
(RIR1-a)	399	—	454	—	—	—
(RIR1-b)	382	385	382	—	—	134
(RIR1-c)	438	—	—	—	—	—
RNR-1	—	—	—	453	—	—
RNR-2	—	—	—	533	—	—
RpolA″	—	—	—	471	—	—
RpolA′	—	—	—	452	—	—
RtcB-2	436	390	380	488	—	—
TFIIB	—	—	—	335	—	—
UDP GD	—	—	—	454	—	—
VMA (VMA-b)	429	376	425	—	—	—

[a] The number of amino acids in each intein is indicated, while the absence of an intein is indicated by a dash (—). Pab, Pyrococcus abyssi; Pho, Pyrococcus horikoshii OT3; Pfu, Pyrococcus furiosus; Mja, Methanococcus jannaschii; Mth, Methanobacterium thermoautotrophicum (delta H strain); Ape, Aeropyrum pernix.

[b] The intein name is listed unless more than one insertion site is present in a gene, in which case the intein insertion site is listed in parentheses. CDC21, Cell division control protein 21; GF-6P, glutamine–fructose-6-phosphate transaminase; Hyp, hypothetical protein; IF2, translation initiation factor IF2; KlbA, kilB operon orfA; LHR, large helicase-related protein; Lon, ATP-dependent protease LA; Moaa, molybdenum cofactor biosynthesis homolog; PEP, phosphoenolpyruvate synthase; pol, DNA polymerase; Pol II, DNA polymerase II, DP2 subunit; RadA, RadA DNA repair protein; RFC, replication factor C; r-gyr, reverse gyrase; RIR1, ribonucleoside-diphosphate reductase, α subunit; RpolA″, RNA polymerase subunit A″ RpolA′, RNA polymerase subunit A′; RtcB, RNA terminal phosphate cyclase operon orfB; TFIIB, transcription factor IIB; UDP GD, UDP-glucose dehydrogenase; VMA, vaculor ATPase, subunit A.

[c] The number in parentheses indicates the number of inteins per sequenced genome.

from one species does not ensure that a homolog from a closely related organism will also contain an intein (Table II). Some inteins are more widely dispersed than others. For example, RIR1 (ribonucleoside-diphosphate reductase α subunit) intein alleles are found in all three domains of life.[2] Several reports have commented upon the observation that inteins tend to be found in proteins involved in transcription, translation, replication, and DNA repair. In thermophilic organisms, 51% of inteins are present in these types of proteins. It is possible that this distribution of inteins reflects how they may be acquired.

Homing Endonuclease Domain and Intein Mobility

There are several classes of homing endonucleases, which are named for their signature motifs.[15,16] Homing endonucleases are a type of site-specific endonuclease that makes a double-stranded break, usually leaving a four-base overhang. They tend to have large, degenerate recognition sites (18–40 base pair) and fail to cleave DNA isolated from their native organism. Most inteins contain a centrally located sequence that has similarities to the DOD class of homing endonucleases (Fig. 1). One intein, the *Synechocystis* sp. PCC6803 DNA Gyrase B intein, contains a second class of homing endonuclease termed an HNH homing endonuclease. Endonuclease activity has only been demonstrated experimentally with a few inteins, including the *T. litoralis* and *Pyrococcus* sp. GB-D DNA polymerase inteins,[3,4,10] the two *Pyrococcus furiosus* ribonucleoside-diphosphate reductase α subunit inteins,[27,28] and the *Pyrococcus kodakaraensis* KOD DNA polymerase inteins.[29] Many inteins do not have essential residues in their endonuclease active sites,[2,10,15,16] such as the conserved Asp in Blocks C and E, and may thus be inactive as endonucleases. Some inteins contain intermediate sized inserts between Blocks B and F that appear to lack one or more homing endonuclease motifs[2]; these may represent inteins in the process of becoming mini-inteins, losing their endonuclease domains by gradual mutation and deletion.

Homing endonucleases are generally found within mobile intron or intein genes, although some, such as the yeast HO endonuclease involved in mating type switching, are present as free-standing genes.[15,16] Homing endonucleases cut at or near the site in the extein where the intein gene is usually inserted. This double-stranded break in the intein minus gene initiates an extremely efficient, unidirectional gene conversion event, resulting in transfer of the intein gene into the homologous extein gene that previously lacked the intein.[15,16,30] Once the

---

[27] K. Komori, K. Ichiyanagi, K. Morikawa, and Y. Ishino, *Nucleic Acids Res.* **27**, 4175 (1999).
[28] K. Komori, N. Fujita, K. Ichiyanagi, H. Shinagawa, K. Morikawa, and Y. Ishino, *Nucleic Acids Res.* **27**, 4167 (1999).
[29] M. Nishioka, S. Fujiwara, M. Takagi, and T. Imanaka, *Nucleic Acids Res.* **26**, 4409 (1998).
[30] M. Belfort and P. S. Perlman, *J. Biol. Chem.* **270**, 30237 (1995).

intein insertion site is cut, the only gene present to repair this DNA break is the intein-containing gene.

Lateral transfer of an intein gene may occur when a piece of DNA that encodes an intein containing gene enters the cell by any means (conjugation, transformation, infection, mating, etc.). Intein homing by homologous recombination is very efficient,[15,16,30,31] whereas insertion of the intein into a nonhomologous site would be very rare because of the inefficiency of illegitimate recombination. Lateral transmission of inteins is borne out in phylogenetic studies demonstrating that inteins present at the same insertion site in genes from vastly different organisms are more closely related to each other than multiple inteins present in the same gene.[11,14,32] Although gain of intein genes with homing endonuclease activity is simple and efficient, loss of inteins is not. Introns can be lost by retrotransposition of spliced mRNA back into the genome, but intein-containing genes cannot be lost in this way, since their RNA is never spliced. Instead, there must either be recombination with intein minus genes (which would only occur efficiently if the endonuclease was inactive) or a precise deletion of the intein coding sequence that would maintain the extein reading frame and not leave extra residues that would inhibit extein activity. The latter constraint may also explain why inteins are found in essential extein sequences.

Given the sporadic distribution of inteins and their ability to invade genes, it seems likely that present-day inteins are relatively recent acquisitions that arrived by lateral transfer. There are several instances when GC content or codon usage of the intein is different from the rest of the genome. Inteins may have accumulated in archaea because archaea are naturally competent to take up free DNA, have efficient conjugation systems, or have broad host range viruses, all of which would enhance the spread of inteins. The possibility that inteins are spread by viruses or plasmids may account for their increased presence in genes involved in transcription, translation, replication, and repair, since these types of genes are more likely to be present on episomal elements than genes for other types of biochemical pathways. Hyperthermophiles might also have very efficient recombination systems, since life at high temperatures may lead to increased DNA damage, requiring efficient DNA repair. Studies in mesophiles indicate that cellular repair and recombination systems are required for endonuclease-mediated gene conversion. The question of whether inteins are recent or ancient elements has yet to be answered. In ancient times, inteins may have been involved in evolution of better enzymes by domain shuffling, since splicing *in trans* can be very efficient.[23] The inteins that we see today may be remnants of these ancient inteins that spread to different sites as the splicing domains picked up different homing endonucleases.

[31] F. S. Gimble and J. Thorner, *Nature* **357**, 301 (1992).
[32] J. Z. Dalgaard, A. J. Klar, M. J. Moser, W. R. Holley, A. Chatterjee, and I. S. Mian, *Nucleic Acids Res.* **25**, 4626 (1997).

Intein genes are selfish DNA, since they can invade extein genes, but this event is not lethal because inteins splice efficiently to generate active extein protein. The unanswered question is whether inteins have any effect on their hosts. Many have suggested that inteins may have a role in biological regulation. As people begin to harness the genome and its products, we may be able to begin asking if there are any advantages or disadvantages of inteins under various physiological stresses, remembering that the rapid self-catalytic protein splicing mechanism leaves little room for environmental influences other than at the level of message stability, protein folding, extremes of pH, or *trans*-acting inhibitors.

Inteins in Protein Engineering

Understanding intein function has led to their use in a variety of applications.[23] Temperature-dependent inteins[5] can be inserted into proteins to control expression of the active extein *in vivo* or *in vitro,* generating conditional knockouts or allowing expression of inactive cytotoxic proteins that can be activated *in vitro* by protein splicing. Hyperthermophilic inteins have been split in their endonuclease domains and precursor fragments have been shown to reassemble and splice.[33,34] Protein splicing *in trans* allows segmental modification of a protein, which has been used to improve resolution for NMR spectroscopy,[33] but can also be used to express extremely cytotoxic proteins. Protein purification vectors have been developed, based on modified inteins that only cleave at a single splice junction. When the target protein is cloned at the amino-terminal splice junction, it is recovered with a carboxy-terminal α-thioester; this reactive group has been used to perform several different types of chemoselective condensations, including protein semisynthesis and incorporation of modified amino acids, biosensors, or tags. The list of applications involving inteins is exponentially expanding as more uses of autocleavage elements and reactive carboxy-terminal α-thioesters are being devised.

Acknowledgments

I thank Don Comb for support and encouragement and Maurice Southworth, Eric Adam, and other members of the New England Biolabs protein splicing groups.

---

[33] T. Yamazaki, T. Otomo, N. Oda, Y. Kyogoku, K. Uegaki, N. Ito, Y. Ishino, and H. Nakamura, *J. Am. Chem. Soc.* **120,** 5591 (1998).

[34] M. W. Southworth, E. Adam, D. Panne, R. Byer, R. Kautz, and F. B. Perler, *EMBO J.* **17,** 918 (1998).

Section III

Biophysical and Biochemical Aspects of Protein Stability

# [24] Assaying Activity and Assessing Thermostability of Hyperthermophilic Enzymes

By ROY M. DANIEL and MICHAEL J. DANSON

Introduction

There is now a wide variety of intra- and extracellular enzymes available from organisms growing above 75° and having sufficient stability to allow assay well above this temperature. For some of these enzymes, to assay below even 95° will involve measurement below the optimal growth temperature for the organism. The purpose of this chapter is to cover practical aspects of enzyme assay procedures that are specific to high temperatures. Since by far the commonest routine assessment of enzyme stability is activity loss, and because it is always unwise to measure enzyme activity without being confident of its stability during the assay, we include an outline of procedures for measuring enzyme activity loss/stability at high temperatures.

There are a number of useful reviews of the effects of temperature on enzyme activity,[1] and these apply as much to reactions at 100° as at 37°. However, enzymes stable at 100° have a number of advantages as research subjects. For example, they can be used to investigate enzymes that are particularly unstable when isolated from mesophiles, to slow reactions without the use of cryosolvents, and to probe the effect of temperature on enzyme and protein behavior over a very wide temperature range. They can also be used to study enzyme behavior under conditions that would denature most enzymes, since enzymes resistant to heat are also, at room temperature, resistant to organic solvents, chaotropic agents, and proteolysis. Clearly, such stable enzymes also have a variety of applications in biotechnology, where protein stability may be an important practical and economic factor.

Enzyme activity increases with temperature, usually by a factor of 1.4–2.0 per 10°, depending on the Arrhenius activation energy. As the temperature is raised, at some point the enzyme will begin to denature during the assay. The extent will depend on the stability of the enzyme and the temperature, but also on the duration of the assay, and on factors such as the buffer composition and the degree of stabilization of the enzyme by substrate/cofactor. This combination of activity acceleration and increasing denaturation with temperature has unfortunately led to the production of activity versus temperature graphs that exhibit a so-called "temperature-optimum" peak for the enzyme. The position of this peak can of course be shifted by tens of degrees by varying the assay duration, and these

---

[1] K. J. Laidler, *Methods Enzymol.* **63**, 234 (1979); M. Dixon, E. C. Webb, C. J. R. Thorne, and K. F. Tipton, in "Enzymes." Longman, London, 1979.

graphs give little biochemical information above the temperature at which denaturation becomes significant during the assay. As a tool to aid in the development of enzyme technology processes, with the assay duration linked to the process time, these graphs may be useful in comparing enzymes. Their use should otherwise be avoided, as should the term temperature optimum derived from them.

However, there is evidence[2] that some enzymes may have genuine temperature optima. That is, at some point in the temperature profile, an enzyme may actually become less active as the temperature is raised, but this is not caused by denaturation. Although it is not clear yet if this phenomenon is widespread among enzymes, where it is present this genuine temperature optimum will be an important and diagnostic characteristic of an individual enzyme, alongside pH optimum, stability, and kinetic properties.

The acceleration of enzyme rates with temperature obviously means that, in the absence of denaturation, the specific activity of an enzyme increases as the temperature is raised. However, as far as it is possible to judge, enzymes from hyperthermophiles growing optimally above 95° have similar specific activities to those from organisms growing optimally at 25° when assayed at their respective *in vivo* temperatures. Moreover, both sets of enzymes will have similar stabilities at their optimal or "evolved" growth temperature. This and much other evidence has led to the notion that enzyme activity and stability are connected via dynamics, molecular motions being clearly required for catalysis, whereas too much movement will lead to unfolding/denaturation. Although this is a gross generalization, which begs many questions regarding local or global dynamics and time scales, it explains most of the evidence so far.[3] However, it should be remembered that (relatively) small changes in activity and stability can be generated separately by *in vitro* directed evolution if the appropriate selection pressures are applied,[4] indicating that some changes in thermostability are not necessarily linked to changes in activity and that the observed inverse correlation of the two properties may reflect in part the process by which any enzyme has adapted.

Enzyme Assays above 75°

General Comments

As for all enzyme assays, continuous assays are to be preferred to discontinuous, although at high temperature a number of factors render continuous assays more difficult. The unavailability of "linker" enzymes stable above 75° will often

---

[2] T. M. Thomas and R. K. Scopes, *Biochem. J.* **330**, 1087 (1998); U. Gerike, N. J. Russell, M. J. Danson, and D. W. Hough, *Eur. J. Biochem.* **248**, 49 (1997); C. L. Kydd, H. Connaris, D. W. Hough, C. D. Reeve, and M. J. Danson, submitted (1999).

[3] R. M. Daniel, M. Dines, and H. H. Petach, *Biochem. J.* **317**, 1 (1996).

[4] S. Akanuma, A. Yamagishi, N. Tanaka, and T. Oshima, *Eur. J. Biochem.* **260**, 499 (1999); L. Giver, A. Gershenson, P.-O. Freskgard, and F. H. Arnold, *Proc. Natl. Acad. Sci. U.S.A.* **95**, 12809 (1998).

prevent the use of coupled enzyme assays, while only electrically heated spectrophotometer cuvette holders are practical above 80°, and pressurized cuvettes will be needed above 100°. Therefore, at high temperatures, discontinuous assays will be more frequently used, with the concomitant need for duplicate/triplicate assays and very careful controls. It is especially important that all controls be run at the same temperature as the assay, given the instability of many substrates, products, and cofactors at elevated temperatures. In addition, control of assay conditions, especially temperature and pH, poses particular problems at high temperatures.

In the light of these difficulties, there will be a strong temptation to do assays at lower (or as they are sometimes called, more realistic) temperatures than the growth temperature of the source organism. Where practical this should be resisted, but in the case of reactions involving heat-labile substrates or cofactors, for example, there may be no alternative but to run routine lower temperature assays, after seeking validation/comparison of these with those run at higher temperatures (possibly for very short periods). In our structure–function studies on citrate synthase,[5] for example, one of the substrates, oxaloacetate, has a half-life of approximately 20 sec at 100°. Therefore, routine assays of the enzyme from *Pyrococcus furiosus* (optimal growth temperature of 100°) are carried out at 55°, although, with a K_m for oxaloacetate of 10 μM, brief assays at 100° are possible with a starting concentration of substrate in excess of 1 mM.

Assessment of enzyme activity or stability at or above 100° needs considerable care. Spills of water/reaction mixture into oil baths can result in small explosions and sprays of hot oil, and the failure of any closed vessel incubating at these high temperatures can have similar results. Oil baths should be run in fume hoods, with lids raised only for manipulation. Operators should wear lab coats that include cover of the neck area, polycarbonate face shields, and long insulated gloves and should use tongs for handling vessels

Before embarking on work above 100°, a batch of the chosen reaction vessels should be tested, filled to the experimental level and at somewhat above the experimental temperature, to assess the likely frequency of failure. Small vessels tend to be safer than larger ones. Normal Eppendorf tubes tend to pop their lids around 100°; the screw-top version with O rings is satisfactory up to about 120°, but not at 130°. For fast temperature equilibration and integrity at high temperatures, sealed capillary tubes are ideal. We use 2 mm OD, 100 mm long, soda glass melting-point capillaries, containing up to 80 μl of liquid. With a > 10 mm air space at each end these can be sealed with a narrow flame without risk to the contents. Integrity is largely dependent on the end seal: we have used them up to 190° and had some failures at that temperature, but with careful sealing they can be used with some confidence up to at least 160°. If larger vessels are absolutely essential, we have had relatively few failures up to 135° with crimped Bellco aluminum seal culture tubes

---

[5] M. J. Danson and D. W. Hough, *Trends Microbiol.* **6**, 307 (1998).

(Bellco Glass, NJ) using new seals, but these failures can be violent and special precautions should be taken. Note that variation between batches and manufacturers is to be expected, and all incubations above 100° should be treated as hazardous.

Temperature Control

To equilibrate a cuvette in a spectrophotometer at 37° is a relatively quick procedure, although a plastic cuvette may still require more than 5 min.[6] At higher temperatures, not only is this slower temperature equilibration a more serious drawback, but the cuvettes may distort. Above 60°, not only is equilibration of any cuvette slow, but for liquid-jacketed cuvette holders heat losses during circulation may necessitate a significant offset between the (higher) water-bath temperature circulating the liquid and the temperature required in the water-jacketed cuvette holder. The water bath will need to be controlled by a temperature sensor actually in the cuvette for the most reliable results. Above 80° even this becomes impractical, and electrically heated cuvette holders are required. In all cases, the *only* reliable estimate of the temperature of the cuvette contents is a direct measurement. Temperature gradients within cuvettes can be significant at high temperatures despite lowered viscosity, and if stirred cuvettes are not used, temperature measurements at both top and bottom of the cuvette need to be checked under experimental conditions.

Above 80°, even equilibration of test tube/vial contents in a water bath needs substantial time. Above 110°, we have found that only containers that are totally immersed in the oil bath will achieve oil bath temperature, although equilibration is usually slow. In any event, above 100° containers should be completely immersed to prevent distillation within the container. Above 100°, the only reliable way to run short assays is to seal the contents in a capillary tube and start the reaction by immersion of the capillary. Temperature equilibration can be achieved in 2–20 sec in the temperature range 105–140°. Reactions can be stopped by plunging the capillary into iced water. In all these systems, there is no acceptable substitute for direct temperature measurement using a vessel and contents identical to those used for the reaction. A small, fast-response thermistor is essential.

Buffers

There are many factors to consider when choosing a buffer,[7] but at high temperatures the temperature coefficient will be particularly important. Although in theory any buffer can be used at high temperatures provided the pH is adjusted

---

[6] R. A. John, *in* "Enzyme Assays, a Practical Approach" (R. Eisenthal and M. J. Danson, eds.) p. 59. Oxford University Press, Oxford, UK, 1992.

[7] L. Stevens, *in* "Enzyme Assays, a Practical Approach" (R. Eisenthal, and M. J. Danson, eds.) p. 317. Oxford University Press, Oxford, UK, 1992; R. J. Beynon and J. S. Easterby, "Buffer Solutions: The Basics." BIOS Scientific Publishers, Oxford, UK, 1996.

TABLE I
BUFFERS WITH LOW $d(pK_a)/dt$ VALUES AND pK_a VALUES BETWEEN pH 6–8 AT 90°[a]

Buffer	$pK_{a,90}$[b]	$d(pK_a)/dT$	$pK_{a,25}$[b]	ΔpH 90°C→25°
PIPES	6.1	−0.0085	6.7	0.6
MOPS	6.5	−0.011	7.2	0.7
Phosphate($pK_{a,2}$)	6.6	−0.0028	6.8	0.2
HEPES	6.7	−0.014	7.6	0.9
HEPSO	6.9	−0.014	7.8	0.9
POPSO	7.0	−0.013	7.8	0.8
HEPPS/EPPS	7.3	−0.011	8.0	0.7

[a] Data are taken from L. Stevens, in "Enzyme Assays, a Practical Approach" (R. Eisenthal and M. J. Danson, eds.), p. 317, Oxford University Press, Oxford, UK, 1992, and R. J. Beynon and J. S. Easterby, " Buffer Solutions: The Basics," BIOS Scientific Publishers, Oxford, UK, 1996.

[b] The pK_a values given above are apparent rather than thermodynamic values, assuming a 100 mM buffer solution. The temperature at which the pK_a applies is denoted by subscript.

at the temperature of use, in practice there are great advantages in using buffers with low temperature coefficients [$d(K_a)/dt$]. This will minimize pH change upon cooling, which is quite likely to be necessary at some point, especially if the assay is discontinuous; it will also enable buffers to be made up at room temperature without too heavy a dependence on $d(pK_a)/dT$ values, few of which have been derived or tested above 60°. There are relatively few such buffers with "apparent" or "working" pK_a values near neutrality at 90°. (Loosely speaking, the apparent pK_a of a buffer is the actual pH value of the buffer solution with 50% of each species present, at the temperature and ionic strength of the solution: for the values quoted here we have assumed an ionic strength arising from a 100 mM buffer solution. See ref. 7 for a fuller discussion. Table I lists buffers with an apparent pK_a between pH 6 and 8 at 90° that change in pK_a by less than 1 pH unit when cooled from 90° to 25°. It is clear that effective buffering around neutrality with only moderate changes of pH with large temperature changes can be obtained only with phosphate, which will allow cooling from 90° to 20° with a pH change of <0.2 units; the corresponding change with Tris buffer would be 2.0! Phosphate buffer will thus allow activity measurements over a temperature range of 40° or so in the same buffer. For most of the buffers in the table, different buffer solutions will be needed to compare activities over any greater temperature range than 15°. There are few buffers available with $pK_{a,90}$ values between 7.3 and 8.7 (borate $pK_{a,90} = 8.7$), and none that will not change in pH by more than 1 pH unit when cooled from 90 to 25°. CHES has a convenient $pK_{a,90}$ of about 8.1, but a $d(pK_a)/dT$ of −0.018.

For further details, see ref. 7, or the Web site at http://www.bi.umist.ac.uk/buffers.html

Although thermal stability is less likely to be an issue with inorganic buffers, imidazole is known to be unstable at high temperatures,[7] and little work has been done on the high temperature stability of the commonly used Good buffers.

There are no reports to date of enzymes having markedly different pH optima at different temperatures, but given the high $d(pK_a)/dt$ of the side chains of the ionizable amino acids, especially histidine, lysine, and arginine, it will not be surprising if this is found to be the case.

Assay Component Stability

If the enzyme substrate(s) or cofactor(s) are significantly heat-labile during the assay period, results will obviously be affected. The same applies to the product if product measurement is being used as the assay. This is especially important in the case of NAD(P), where oxidation/reduction is very widely used to follow enzyme reaction progress, but which is quite unstable at high temperature.[8] An additional complication is the possibility that the product of thermal degradation may be inhibitory; little work has been done on this possibility, and indeed for many cases where enzyme reaction components are known to degrade at high temperature, the products of this degradation are unknown or poorly characterized. Overall, not enough is known of reactant temperature stability, and if a reactant is known to occur *in vivo* in a hyperthermophile, this does not mean it will be stable in an *in vitro* enzyme assay, since the organism may overcome reactant instability by a variety of means.[9] NADH and NADPH, for example, occur in hyperthermophiles (although many dehydrogenases in these organisms are linked to nonheme iron protein), but have a half-life of only about 2 min at neutral pH at 95°,[8] and an even shorter half-life at lower pH values. Glutamine degrades significantly even at 80°,[10] but is present in hyperthermophiles.

Table II gives approximate stability data for a variety of metabolites and coenzymes. It can be seen that for short (<10 min) assays at 95°, of the compounds listed only NAD (and including NADP, NADH, NADPH) and acetyl phosphate will be degraded at a rate that renders assay results useless, but losses of ATP and ADP may be as high as 20%. The situation is further complicated by the influence of reaction conditions on metabolite/coenzyme stability. For example, ATP stability is greatly affected by pH and metal ions,[11] and NADH stability is very pH sensitive.[12]

---

[8] R. M. Daniel and M. J. Danson, *J. Mol. Evol.* **40**, 559 (1995).
[9] R. M. Daniel, in "Thermophiles" (J. Weigel and M. W. W. Adams, eds.), p. 299. Taylor & Francis, London, 1998; R. M. Daniel and D. A. Cowan, *Cell. Mol. Life Sci.* **57**, 250 (2000).
[10] H. D. Ratcliffe and J. W. Drozd, *FEMS Micro. Lett.* **3**, 65 (1978).
[11] M. Tetas and J. M. Lowenstein, *Biochemistry* **2**, 350 (1963); F. Ramirez, J. F. Marecek, and J. Szamosi, *J. Org. Chem.* **45**, 4748 (1980).
[12] K. A. J. Walsh, R. M. Daniel, and H. W. Morgan, *Biochem. J.* **201**, 427 (1983).

TABLE II
SOME METABOLITE AND COENZYME STABILITIES[a]

Metabolite/coenzyme	Percentage remaining after 1 hr/95°
NAD	<5
FAD	100
FMN	75
Acetyl phosphate	<10
CoASH	100
Acetyl-CoA	100
ATP	40
ADP	50
AMP	95

[a] Data from R. M. Daniel, in "Thermophiles" (J. Weigel and M. W. W. Adams, eds.), p. 299. Taylor & Francis, London, 1998.

There are few satisfactory solutions to reactant instability. The best solution may be to run very short assays. In some cases it may be possible to modify reaction conditions (pH, metal concentrations) to enhance stability, but this is likely to imply assay under suboptimal conditions. The obvious alternative, lowering the temperature, may be preferable, providing the effects of changed K_m, V_{max}, etc., are acceptable.

An additional complication is that many metabolities/cofactors have temperature-dependent extinction coefficients; those for NADH and for potassium ferricyanide, for example, are about 10% lower at 80° than at 20°.[12] Fourage et al.[13] point out that the effect of temperature on the absorbance and λ_{max} of p-nitrophenol can lead to substantial errors in k_{cat} values measured by continuous release of p-nitrophenol, if the calibration curve is not determined at the same temperature as the assay.

Effects of Temperature on Kinetic Parameters

Variation of K_m with Temperature

A variety of enzymes are known to have different K_m values at different temperatures (Table III). For reasons that are not completely clear, in most cases a drop in temperature results in a drop in K_m, or little change. For some enzymes, the biggest changes in K_m occur close to the "evolved" or growth temperature of the organism. For 3-phosphoglycerate kinase from *Thermoanaerobacter*, for example, the K_m

---

[13] L. Fourage, M. Helbert, P. Nicholet, and B. Colas, *Anal. Biochem.* **270**, 184 (1999).

TABLE III
K_m Variation with Temperature

Enzyme	Substrate	K_m (mM)	Temperature (°C)
3-Phosphoglycerate kinase (*Thermoanaerobactor* sp.) [T. M. Thomas and R. K. Scopes, *Biochem. J.* **330**, 1087 (1998)]	3-Phosphoglycerate	0.8	40
		1.1	65
		1.9	76
	ATP	0.7	40
		0.7	65
		1.5	77
Glutamate dehydrogenase (*Thermococcus zilligii* strain AN1) [R. C. Hudson, L. D. Ruttersmith, and R. M. Daniel, *Biochim. Biophys. Acta* **1202**, 244 (1993)]	2-Oxoglutarate	0.2	14
		0.52	60
		1.7	80
	NH_4Cl	2.0	14
		5.7	60
		15.5	80
	Glutamate	1.39	40
		1.93	60
		9.12	80

values for both ATP and 3-phosphoglycerate are relatively unchanged over the range 40–65°, but rise sharply above this temperature.[2] A similar effect was found for the 3-phosphoglycerate kinases from two measophiles, *Zymomonas mobilis* and an unidentified soil bacterium, but with the increase in K_m values occurring at a lower temperature. Data from the glutamate dehydrogenase from *Thermococcus zilligii* are consistent with this pattern.[14] These findings tend to emphasize the need to carry out enzyme assays/determination of properties at the growth temperature of the source organism if the results are to have physiological significance.

Variation of Catalytic Activity with Temperature

In the absence of significant denaturation or degradation of an enzyme during the assay, the rate of an enzyme-catalyzed reaction will increase with a rise in temperature in a manner empirically described by the Arrhenius equation:

$$k = Ae^{-(E_a/RT)}$$

where k is the first-order rate constant for the conversion of substrate to products; A, Arrhenius constant; E_a, Arrhenius activation energy; R, universal gas constant; and T, temperature.

---

[14] R. C. Hudson, L. D. Ruttersmith, and R. M. Daniel, *Biochim. Biophys. Acta* **1202**, 244 (1993).

Using transition state theory, a similar equation can be derived:

$$k = \frac{k_B T}{h} e^{-(\Delta G^{\circ*}/RT)}$$

where k_B is Boltzmann's constant; h, Planck's constant; and $\Delta G^{\circ*}$, change in standard free energy between the ground and transition states of the substrate. From both equations, a plot of ln[Enzyme velocity] vs [$1/T$] will be linear in the absence of a temperature-induced loss of catalytic activity.

There is no evidence to date that Arrhenius activation energies are systematically different in enzymes from hyperthermophiles (i.e., in very stable enzymes) and, perhaps surprisingly, for at least some of these enzymes the Arrhenius activation energies are unchanged over very large temperature ranges. The glutamate dehydrogenase from *T. zilligii*, for example, gives a linear Arrhenius plot in buffer between 5° and 90°,[14] and in 70% (v/v) methanol between 0° and −85°.[15]

For a more detailed discussion, see ref. 1.

Assessing Enzyme Stability

An assessment of enzyme stability is an important exercise before assays can be designed and carried out.

Denaturation and Degradation

In principle, there are two ways in which enzyme activity can be lost at high temperatures: denaturation, which in principle consists of the reversible loss of the active conformation with no loss of primary structure, and degradation, which consists of irreversible inactivation involving covalent bond disruption. These are not always easy to separate. Deamidation of the amide side chain of Asn and Gln residues, succinimide formation at Glu and Asp, and oxidation of His, Met, Cys, Trp, and Tyr are the most facile and common amino acid degradations. The rates of protein degradation by these mechanisms are greatly accelerated at high temperatures and can thus play an important role in the thermoinactivation of enzymes (for a brief discussion, see ref. 3). Both denaturation and degradation can be assessed by measuring activity loss, but if the loss is not reversible, this cannot be taken as showing that degradation is the cause, as in practice denaturation is usually not readily reversible.

Although there is clear evidence that denaturation can occur in the absence of degradation, the converse is not as obviously the case. In proteins at neutral pH, degradation only occurs at conveniently measurable rates at high temperatures, and almost all studies of protein degradation have used proteins that were

---

[15] N. More, R. M. Daniel, and H. H. Petach, *Biochem. J.* **305**, 17 (1995).

of mesophilic origin and were therefore fully denatured at the temperatures used. In those few cases where very stable proteins have been investigated, there is evidence that at least in the temperature range 85–110°, denaturation still precedes degradation.[3,16] This fits the evidence that the chemical mechanisms for irreversible degradation in proteins require a certain local molecular flexibility. For example, at 37° the rate of deamidation has been shown to be higher both in small peptides with high flexibility and in denatured proteins than in folded proteins. A survey of environments around Asp and Asn residues in known three-dimensional protein structures suggests that the rigidity of the folded protein greatly decreases the intramolecular imide formation necessary for further degradation. In the numerous X-ray crystal structures studied, the peptide-bond nitrogen could not approach the side-chain carbonyl carbon closely enough to form the succinimide ring (see ref. 3).

A detailed assessment of protein degradation processes and rates is beyond the scope of this work. However, the commonest degradative processes are deamidation of the side chain of asparagine and glutamine residues and succinamide formation at glutamate and aspartate resides (in both cases often followed by hydrolysis of the resulting succinamide), and tests for ammonia evolution or peptide bond cleavage (such as gel electrophoresis or electrospray mass spectrometry) are probably the simplest to use if degradation is suspected.

Assessing Activity Loss at High Temperatures

The commonest method used for assessing enzyme stability is the determination of activity loss with time at a given temperature. Although this provides little information on the nature of the process or the end products, it has the virtue of ease and simplicity and it gives the information on enzyme stability that is essential for assay development.

The process mostly consists of incubating an enzyme at a given temperature, removing samples at various time intervals, rapidly cooling, and assessing activity at some lower temperature at which activity loss is known not to occur. This procedure is open to criticism because of the possibility that the enzyme may be denatured at high temperature and then renatured (thus regaining activity) rapidly enough to display activity in the cooled assayed sample. Although we know of no cases where such rapid renaturation of a very thermophilic enzyme has occurred, direct high-temperature measurements of activity are a necessary accompaniment of stability measurements made in this way. The activity determined is likely to be only an approximation because of denaturation losses occurring during the assay. This, together with the effects of the presence of substrate and product in the assay, will prevent a good correlation between stability assessed and remaining

[16] R. Hensel, L. Jacob, H. Scheer, and L. Lottspeich, *Biochem. Soc. Symp.* **58**, 127 (1992).

activity (as above), but will still greatly reduce the likelihood that an apparent high-temperature stability is actually the result of reversible denaturation.

Stabilities determined in this way are dramatically dependent upon conditions. Buffer type, pH, and ionic strength can have major effects, as well as agents such as substrate, product, cofactors, and metal ions (especially calcium). Stability comparisons made between enzymes on this basis must be regarded as approximations only, and the conditions carefully defined.

For extracellular enzymes, *in vitro* assessments of stability may have some physiological significance, although it is by no means clear that in nature such enzymes are free in solution rather than associated with a surface. However, the physiological significance of *in vitro* stabilities of intracellular enzymes is less obvious. Apart from the potential stabilizing factors mentioned above, the intracellular low water activity and the presence of high concentrations of protein are both powerful stabilizing agents. In addition, some, but not all, thermophiles are known to contain metabolites that strongly stabilize proteins, although it is not certain that this is their primary function. The most effective of these agents are cyclic 2,3-diphosphoglycerate and di-*myo*-inositol 1,1'-phosphate, found in *Methanothermus fervidus* and *Pyrococcus woesei*, respectively. In the presence of potassium, these agents increase the half-lives of some enzymes by up to 130-fold at 90°.[17]

Acknowledgments

We thank the Royal Society of New Zealand for the award of the James Cook Fellowship to RMD, and the Biotechnological and Biological Sciences Research Council, UK, for financial support to MJD. We also thank our graduate students for their contribution over the years toward solving the problems associated with high temperature enzyme assays.

[17] R. Hensel and L. Jacob, *Syst. Appl. Micobiol.* **16**, 742 (1994).

[25] Chaperonin from *Thermococcus kodakaraensis* KOD1

By SHINSUKE FUJIWARA, MASAHIRO TAKAGI, and TADAYUKI IMANAKA

Introduction

Chaperonins are a group of molecular chaperones that are classified into the GroEL/HSP60 (heat shock protein 60) family.[1,2] They are widely distributed from prokaryotes to eukaryotes and first came to light because of their specific induction during the cellular response of all organisms to heat shock. It is now clear that the majority of these proteins are expressed constitutively and abundantly in the absence of any stress, and genetic studies show that many of them are essential for cell viability under normal conditions of growth.[2] Many HSPs do not respond significantly to heat shock and are induced under a variety of other stress conditions, whose common denominator may be the accumulation of unfolded or malfolded proteins in cells.[2] They also mediate the correct assembly of polypeptides and the translocation of proteins across membranes, but are themselves not components of the final structures.[3]

In hyperthermophilic archaea, chaperonins play an essential role in hindering protein denaturation.[4-7] New heat shock proteins identified from mammals and archaea have been classified into a new group of the chaperonin family.[8,9] Even though primary structures of archaeal chaperonins are different from GroEL, they have similar quaternary structures, and they may be involved in *de novo* protein folding and assembly. The mammalian chaperonin protein is a t-complex polypeptide-1 (TCP-1), a ubiquitous eukaryotic protein, organized as a multisubunit toroid, which requires ATP for activity and has the ability to catalyze the refolding of denatured proteins.[8,10] The other protein is the thermophilic factor 55 (TF55), isolated from the thermophilic archaeon *Sulfolobus shibatae*.[4,11] TF55 is a hetero-oligomeric complex (α and β) having two stacked nine-membered rings, which binds unfolded polypeptides *in vitro* and has ATPase activity.[5,12,13] In the GroEL/GroES chaperonin system, the binding and releasing of target proteins are associated with ATP hydrolysis. TCP-1 and TF55 also require ATP for refolding of

---

[1] S. M. Hemmingsen, C. Woolford, S. M. Vies van der, K. Tilly, D. T. Dennis, C. P. Georgopoulos, R. W. Hendrix, and R. J. Ellis, *Nature* **333**, 330 (1988).
[2] R. J. Ellis, *Ann. Rev. Biochem.* **60**, 321 (1991).
[3] P. J. Kang, J. Ostermann, J. Shilling, W. Neupert, E. A. Craig, and N. Pfanner, *Nature* **345**, 137 (1990).
[4] J. D. Trent, J. Osipiuk, and T. J. Pinkau, *J. Bacteriol.* **172**, 1478 (1990).
[5] Q. Quaite-Randall, J. D. Trent, R. Josephs, and A. Joachimiak, *J. Biol. Chem.* **270**, 28818 (1995).
[6] Z. Yan, S. Fujiwara, K. Kohda, M. Takagi, and T. Imanaka, *Appl. Environ. Microbiol.* **63**, 785 (1997).
[7] M. Izumi, S. Fujiwara, M. Takagi, S. Kanaya, and T. Imanaka, *Appl. Environ. Microbiol.* **65**, 1801 (1999).
[8] V. A. Lewis, G. M. Hynes, D. Zheng, H. Saibil, and K. Willison, *Nature* **358**, 249 (1992).
[9] M. B. Yaffe, G. W. Farr, D. Miklos, A. L. Horwich, M. L. Sternlicht, and H. Sternlicht, *Nature* **358**, 245 (1992).
[10] Y. Gao, J. O. Thomas, R. L. Chow, G. H. Lee, and N. J. Cowan, *Cell* **69**, 1043 (1992).
[11] J. D. Trent, E. Nimmesgern, J. S. Wall, F. U. Hartl, and A. L. Horwich, *Nature* **354**, 490 (1991).
[12] B. M. Phipps, D. Typke, R. Hegerl, S. Volker, A. Hoffmann, K. O. Stetter, and W. Baumeister, *Nature* **361**, 475 (1993).
[13] H. K. Kagawa, J. Osipiuk, N. Maltsev, R. Overbeek, E. Quaite-Randall, A. Joachimiak, and J. D. Trent, *J. Mol. Biol.* **253**, 712 (1995).

denatured proteins. Several protein folding cycles have been proposed for bacterial chaperonin. It has been suggested that the refolding cycle of archaeal chaperonin includes a change in conformation and in oligomerization state.[5] According to this model, binding or hydrolysis of ATP acts as a switch between two conformational forms of chaperonin: open and closed complexes. Thermodynamic barriers, separating protein bound and free archaeal chaperonin states, are overcome by ATP hydrolysis. Previous studies, however, revealed that CpkB (β subunit of *Thermococcus kodakaraensis* chaperonin) functions as a chaperonin in the absence of ATP when it is present in an excess amount.[6]

Thermostable chaperonins are potentially useful for industrial application. In this article, methods to study archaeal chaperonins are discussed, which use as model systems recombinant forms of CpkA and CpkB from the hyperthermophilic archaeon *Thermococcus kodakaraensis* KOD1, previously reported as *Pyrococcus kodakaraensis* KOD1.[14,15] The enhancing effect of chaperonin on enzyme stability is introduced using yeast alcohol dehydrogenase (ADH) as a model protein. In addition, a unique approach to prevent insoluble complex formation in *Escherichia coli* by coexpressing CpkB with a target protein is mentioned.

Expression System for Recombinant Chaperonin

Chaperonins (HSP60 family members) have been expressed from heterogeneous promoter systems on multicopy plasmids. These recombinant constructs have enabled the overexpression of chaperonins. Genes encoding chaperonins are now cloned easily through PCR (polymerase chain reaction) techniques. While amplifying DNA, convenient restriction enzyme sites for facilitating cloning in a variety of expression vectors are created by designing primer sequences. Furthermore, some additional modifications, such as codon changes, are also possible. Recombinant plasmids, containing *cpkA* or *cpkB* under the control of the T7 promoter, have been constructed by introducing genes into the *Nco*I and *Bam*HI sites of pET-8c (Novagen, Madison, WI). Generally, rare codons located near the 5′ end of the gene are not suitable for efficient translation.[16,17] The 5′ terminus region of *cpkA* is occupied by rare codons that are not efficiently utilized in *Escherichia coli*. In order to achieve an efficient expression, such rare codons, CTT for Leu4, AGT for Ser-5 and GGA for Gly-6, were replaced by CTG, AGC, and GGC, respectively. Recombinant plasmids for *cpkA* and *cpkB* expression are designated as pCPAE[7] and pECPK,[6] respectively.

---

[14] M. Morikawa, Y. Izawa, N. Rashid, T. Hoaki, and T. Imanaka, *Appl. Environ. Microbiol.* **60**, 4559 (1994).

[15] S. Fujiwara, M. Takagi, and T. Imanaka, *Biotechnol. Annu. Rev.* **4**, 259 (1998).

[16] S. C. Makrides, *Microbiol. Rev.* **60**, 512 (1996).

[17] E. Goldman, A. H. Rosenberg, G. Zubay, and F. W. Studier, *J. Mol. Biol.* **245**, 467 (1995).

Overexpression and Purification

Expression of *cpkA* or *cpkB* from the T7 promoter has been achieved in the *E. coli* strain BL21 (DE3), which possesses a stable chromosomal copy of the T7 RNA polymerase gene under control of the *lacUV5* promoter. Both chaperonin subunits, CpkA and CpkB, have been purified in the same way. Details for CpkB purification are described below.

E. coli cells harboring pECPK are induced by 0.1 mM isopropyl-β-D-thiogalactopyranoside (IPTG) at mid-exponential phase and incubated for 6 hr at 37°. Cells (from 3 liter culture) are centrifuged, following which the pellet is washed with 50 mM phosphate buffer (pH 7.5). Cells are disrupted by sonication, and the supernatant fraction is recovered by centrifugation at 27,000g for 20 min at 4°. The supernatant is heat-treated at 90° for 20 min and centrifuged again at 27,000g for 15 min at 4°. This simple heat treatment results in an effective fractionation for thermostable chaperonin. The resulting supernatant is exposed to 75% ammonium sulfate saturation and kept at 4° overnight. The precipitate is collected by centrifugation at 27,000g for 30 min, dissolved in 20 ml of 50 mM phosphate buffer (pH 7.5), and dialyzed overnight against the same buffer. The dialyzate is applied to an anion-exchange column (HiTrap Q, Amersham Pharmacia Biotech, Uppsala, Sweden) and equilibrated in 50 mM phosphate buffer (pH 7.5), and CpkB is eluted by a linear gradient of NaCl using FPLC (fast protein liquid chromatography) system (Amersham Pharmacia Biotech). CpkB is further purified by repeating anion-exchange chromatography using a Resource Q column (Amersham Pharmacia Biotech).

Conventional Method for ATPase Assay

An ATPase assay is performed by monitoring ADP formation using [α-^{32}P] ATP (400 Ci/mmol, Amersham Pharmacia Biotech) on polyethyleneimine (PEI) cellulose thin-layer sheets (TLC sheet, Polygram cel 300 PEI, Macherey-Nagel, Germany), based on the procedure reported previously.[6,18,19]

1. Prepare reaction mixture A containing 40 mM HEPES, pH 7.2, 75 mM KCl, 4.5 mM MgCl$_2$, 1.5 mM CaCl$_2$ and 1.0 mM ATP.
2. Add 1 μl of [α−^{32}P] ATP (400 Ci/m mol, Amersham) to 1.0 ml of mixture A just before use.
3. Preincubate the prepared mixture (20 μl) in an Eppendorf tube at specified reaction temperature for 5 min.
4. Add the enzyme solution and incubate for 5 min at the specified temperature. After reaction, keep the samples on ice.

---

[18] K. Liberek, J. Marszalek, D. Ang, C. Georgopoulos, and M. Zylicz, *Proc. Natl. Acad. Sci. U.S.A.* **88**, 2874 (1991).

[19] M. E. Cheetham, A. P. Jackson, and B. H. Anderton, *Eur. J. Biochem.* **226**, 99 (1994).

5. Spot aliquots (2 μl) on PEI cellulose thin-layer sheets and dry. Develop with 1 M LiCl by one-dimensional chromatography.
6. Dry immediately and examine the sheet by autoradiography. Subsequently, quantify by densitometry. If corresponding spots are cut out, the radioactivity can be measured by liquid scintillation counting. Spontaneous ADP formation needs to be examined and subtracted prior to calculations for rates of ATP hydrolysis.

Aggregation Prevention of Denatured Rhodanese by Chaperonin

The ability to prevent aggregation of unfolded polypeptides diluted from denaturant can be used to examine the chaperonin activity toward various proteins. Bovine mitochondrial rhodanese, a monomeric protein of 33,000 Da, is known as an excellent model substrate of GroEL because of its pronounced tendency to aggregate during attempted *in vitro* refolding.[20,21] Rhodanese is also applicable to test the function of archaeal chaperonin. The refolding process of chemically denatured rhodanese can be monitored by light scattering.

1. *Denaturation of rhodanese:* Dissolve 10 mg of rhodanese (thiosulfate sulfurtransferase; EC 2.8.1.1; type II, Sigma, St. Louis, MO) in 100 mM Tris-HCl (pH 7.8), containing 5 mM dithiothreitol (DTT). Take 10 μl of rhodanese and mix with 60 μl of denaturation solution containing 6 M guanidine hydrochloride, 100 mM HEPES (pH 7.5), 100 mM KCl, 25 mM NaCl, and 10 mM DTT. Incubate the mixture for 1 hr at 25°.

2. *Dilution of the denatured rhodanese:* Prepare aggregation solution containing 200 mM HEPES (pH 7.5), 200 mM KCl, 50 mM NaCl, and 20 mM MGCl$_2$. Put 500 μl of the aggregation solution into an Eppendorf tube and add 4 μl of 100 mM ATP (if necessary), the specified amount of chaperonin, and distilled water to adjust total volume to 950 μl. Transfer the mixture to 1 cm cuvette and add 50 μl of the denatured rhodanese.

3. Measure the absorbance at 320 nm with UV-visible recording spectrophotometer. Aggregation in buffer alone at 10 min is set to 100% (1 of arbitrary units).

Typical refolding profiles by CpkA in the presence of ATP are shown in Fig. 1.

Protection of Enzyme from Irreversible Heat Denaturation by Chaperonin

Influence of archaeal chaperonins on thermal inactivation of enzymes can be examined by various targets. Experimental conditions depend on the features of the

---

[20] J. M. Martin, T. Langer, R. Boteva, A. Schramel, A. L. Horwich, and F.-U. Hartl, *Nature* **352**, 36 (1991).
[21] J. A. Mendoza, B. Demelar, and P. M. Horwitz, *J. Biol. Chem.* **269**, 2447 (1994).

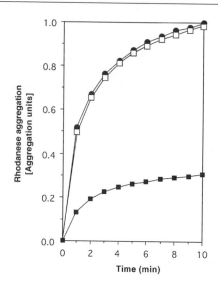

FIG. 1. Prevention of denatured rhodanese aggregation by CpkA as measured by light scattering. Denatured rhodanese was diluted 140-fold (2.16 μM) into buffer alone (□) or into buffer containing CpkA (■, 2.24 μM) or BSA (●, 3 μM). Aggregation of rhodanese in buffer alone at 10 min is set to 100% (1 of arbitrary units).

target protein of interest. Several parameters should be considered for chaperonin-dependent actions. From past experience, spontaneous folding of target proteins depends strongly on both temperature and protein concentration. The molar ratio of chaperonin to target protein needs to be optimized for efficient monitoring. As an example, experimental conditions for yeast alcohol dehydrogenase (ADH) are provided here.

Yeast alcohol dehydrogenase (ADH) has its highest activity at 30°, and this decreases sharply with increase of temperature above 50°. No activity is observed at 70°. This inactivation process can be monitored by measuring the remaining activity of the enzyme during the incubation at elevated temperatures. ADH from yeast is assayed by monitoring ethanol-dependent NAD^+ reduction at 340 nm[22,23]; ADH activity is expressed as micromoles of NADH produced per minute with a molar absorption coefficient of 6.22 mM^{-1} cm^{-1}. The influence of molecular chaperonin on ADH inactivation can be examined as follows.

Method 1: Measuring Residual Activity of ADH at Elevated Temperatures

1. Add specified amount of chaperonin to a reaction buffer, containing 30 mM Tris-HCl, pH 7.2, 30 mM NaCl, 50 mM KCl, 2 mM DTT, 10 mM ATP, and

[22] R. J. Lamed and J. D. Zeikus, *Biochem. J.* **195,** 183 (1981).
[23] H. Sakoda and T. Imanaka, *J. Bacteriol.* **174,** 1397 (1992).

125 mM magnesium acetate preincubated for 5 min. Then, add 0.025 μM yeast ADH (alcohol dehydrogenase, alcohol: NAD^+ oxidoreductase from baker's yeast; EC 1.1.1.1, Sigma) and incubate mixtures at 30°, 50°, and 70° for a specified time in a total volume of 200 μl.

2. Take 100 μl of the heat-treated solution from (1) and mix with 980 μl of 100 mM glycine buffer (pH 8.8) and 10 μl of 100 mM β-NAD^+. Incubate mixtures for 5 min at 25° and initiate the reaction by adding 10 μl of 10 M ethanol. Measure the absorbance at 340 nm with UV–visible recording spectrophotometer.

Here, yeast ADH (0.025 μM) is incubated for 6 min at temperatures between 30° and 70° in the absence or presence of purified recombinant KOD1 CpkB (0.25 μM), and residual ADH activity is assayed. The ADH activity decreases sharply with increase of temperature above 50°. In contrast, when CpkB is incubated with ADH, the rate of thermal inactivation decreases (Fig. 2).

Method 2: Measuring Remaining Activity after Heat Treatment at 50°

The time course of the inactivation process can be monitored by measuring the remaining activity of the enzyme during various periods of incubation at 50°.

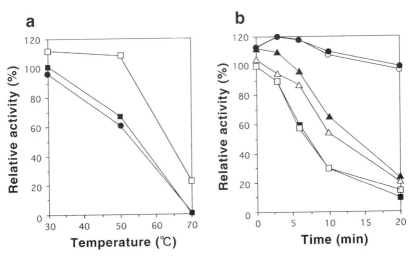

FIG. 2. Effect of KOD1 CpkB on heat stabilization of ADH. (a) Thermal inactivation of ADH and CpkB- mediated reactivation *in vitro*. The yeast ADH (0.025 μM) was incubated for 6 min at temperatures between 30° and 70° in the absence (■) or presence (□) of purified recombinant CpkB (0.25 μM) and remaining ADH activity was assayed. (●) indicates remaining ADH activity in the presence of bovine serum albumin (0.25 μM). (b) CpkB dependent stabilization of ADH. The ADH (0.025 μM) was incubated at 50° either without (■) or with 0.25 μM CpkB (●); 0.25 μM CpkB and 10 mM ATP (○); 0.025 μM CpkB (△); 0.025 μM CpkB and 10 mM ATP (▲); and 0.25 μM bovine serum albumin (□). ADH activity was measured at the times indicated.

In vitro chaperonin function for maintaining ADH in an enzymatically active state at 50° can be examined as follows.

1. Incubate the mixture containing 30 mM Tris-HCl, pH 7.2, 30 mM NaCl, 50 mM KCl, 2 mM DTT, 10 mM ATP, 125 mM magnesium acetate 0.025 μM yeast ADH, and chaperonin (0.025 μM) at 50° in a total volume of 600 μl.
2. Take 100 μl of the heat treated solution at various incubation times (e.g., every 5 min). Keep the samples on ice. Mix with 980 μl of 100 mM glycine buffer (pH 8.8) and 10 μl of 100 mM β-NAD$^+$. Incubate mixtures for 5 min at 25° and start the ADH reaction by adding 10 μl of 10 M ethanol. Measure the absorbance at 340 nm with UV–visible recording spectrophotometer.

In experiments with CpkB, the remaining activity of ADH is about 11% after 20 min heat treatment in the absence of CpkB, but approaches 100% in the presence of CpkB. The stabilization effect is more significant with a further increase in CpkB concentration.[6]

Effect of Chaperonin Coexpression on Decreasing Insoluble Form of Foreign Proteins

When recombinant proteins are overexpressed in *E. coli*, insoluble inclusion bodies are often formed. It has been observed that molecular chaperonin prevents insoluble body formation when it is coexpressed.[24] Chaperonin activity can be examined by monitoring the decreasing rate of insoluble formation. The *cobQ* gene of KOD1 strain, which encodes cobyric acid synthase, leads to an insoluble inclusion complex when it is overexpressed in *E. coli*.[6] The model experiment using CobQ as a target for CpkB is described below.

Plasmid Constructions for Coexpression

Plasmid pACYC184 is compatible with pET-8c in *E. coli* cells. The *cobQ* and *cpkB* genes are cloned into plasmid pET-8c and pACYC184 and the constructed plasmids are designated pCOB and pCPK, respectively.[6] Both *cobQ* and *cpkB* are inducible by T7 promoter system with IPTG addition.

Coexpression in E. coli

E. coli BL21 (DE3) is transformed by plasmid pCOB and/or pCPK and grown in NZCYM medium (100 ml) containing ampicillin (50 μg/ml) and chloramphenicol (34 μg/ml), and overexpression of proteins is induced by IPTG addition

---

[24] T. Yasukawa, C. Kanei-Ishii, T. Maekawa, J. Fujimoto, T. Yamamoto, and S. Ishii, *J. Biol. Chem.* **270**, 25328 (1995).

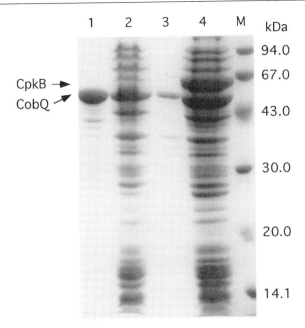

FIG. 3. Solubilization of CobQ by coexpression with CpkB in *E. coli*. Lanes: 1, insoluble fraction from *E. coli* (pCOB); 2, soluble fraction from *E. coli* (pCOB1); 3, insoluble fraction from *E. coli* (pCOB and pCPK); 4, soluble fraction from *E. coli* (pCOB1 and pCPK); M, molecular mass marker (phosphorylase *b*, 94,000; albumin, 67,000; ovalbumin, 43,000; carbonic anhydrase, 30,000; trypsin inhibitor, 20,100; α-lactalbumin, 14,400).

(0.1 mM). After induction for 3 hr, cells are harvested by centrifugation. The pellet is suspended into 1 ml of 30 mM Tris-HCl pH 8.0, 30 mM NaCl buffer. After cells are disrupted by sonication and centrifuged, supernatant is rescued as the soluble fraction. The pellet is washed by 1 ml of 30 mM Tris-HCl pH 8.0, 30 mM NaCl, and suspended in 1 ml of sample buffer [50 mM Tris-HCl pH 6.8, 100 mM dithiothreitol, 2% sodium dodicyl sulfate (SDS), 0.1% bromphenol blue, 10% (v/v) glycerol], boiled for 3 min, and centrifuged. The supernatant is rescued as the insoluble fraction. Each fraction (20 μl) is loaded on an SDS (0.1%)–PAGE (10%) followed by Coomassie Brilliant Blue R-250 staining.

Typical results are shown in Fig. 3. A significant reduction in the amount of insoluble CobQ is obtained by coexpressing CpkB.

Acknowledgments

This work was supported by the grant from CREST (Core Research for Evolutional Science and Technology, Japan).

[26] Organic Solutes from Thermophiles and Hyperthermophiles

By HELENA SANTOS and MILTON S. DA COSTA

Introduction

Research on thermophilic and hyperthermophilic bacteria and archaea has intensified for several years primarily because of the biotechnological importance of their thermostable enzymes. However, the diversity of these organisms, their physiology and biochemistry, and our perception that they represent ancient lines of evolution are sufficient to justify research on these fascinating organisms. Although there is no biological definition that distinguishes a thermophile from a hyperthermophile, we can, as a general rule, define a thermophile as an organism with optimum growth between 60° and 80° and a hyperthermophile as one that grows optimally at 80° or above.[1]

Many thermophilic and hyperthermophilic organisms originate from inland geothermal areas where the salinity of the water is very low, but increasingly, the isolation and description of novel microorganisms has shifted to marine geothermal areas where the levels of NaCl can reach those of seawater. Organisms originating from inland hot springs are generally sensitive to salt and do not grow in media containing more than 1 to 2% NaCl. However, some thermophilic bacteria, having higher growth rates without appreciable concentrations of NaCl are designated halotolerant, because they can grow at salt levels that would inhibit growth of related species. For example, the species *Thermus thermophilus* are frequently isolated from marine hot springs but do not require salt for growth.[2] These organisms, unlike other species of the same genus, are able to grow in media containing 3 to 4% NaCl. Other thermophilic bacteria, such as *Rhodothermus marinus*, are exclusively isolated from marine geothermal environments and require NaCl for growth.[3] These organisms are slightly halophilic with a NaCl requirement in the range of 0.5 to 6.0%. In contrast to the thermophilic bacteria known, most hyperthermophilic bacteria and archaea originate from marine geothermal areas and require NaCl for growth. For example, all the species of the genera *Pyrococcus* and *Thermococcus*, except for *Tc. zilligii*, were isolated from marine geothermal areas and are slightly halophilic.[1,4]

---

[1] E. Blöchl, S. Burggraf, G. Fiala, G. Lauerer, G. Huber, R. Huber, R. Rachel, A. Segerer, K. O. Stetter, and P. Völkl, *World J. Microbiol. Biotechnol.* **11**, 9 (1995).
[2] C. M. Manaia, B. Hoste, M. C. Gutierrez, M. Gillis, A. Ventosa, K. Kersters, and M. S. da Costa, *System. Appl. Microbiol.* **17**, 526 (1994).
[3] O. C. Nunes, M. M. Donato, and M. S. da Costa, *System. Appl. Microbiol.* **15**, 92 (1992).

All microorganisms must, within intrinsic limits, adapt to fluctuations in the external osmotic pressure, since a positive turgor is necessary for cell division. Osmotic downshock will lead to an influx of water, and the concomitant efflux or metabolism of low molecular weight solutes, or to cell lysis if the osmotic adjustment is not adequate. On the other hand, osmotic upshock will lead to a loss of cytoplasmic water. In order to adjust to higher osmolarities of the environment, microorganisms must accumulate intracellular osmolytes to reestablish the cell turgor pressure and/or cell volume, and to protect intracellular enzymes and other macromolecules from dehydration. Failure to adjust osmotically to higher salt or sugar concentrations in the environment will necessarily result in the cessation of growth or death. Comprehensive reviews on osmoadaptation are available in the literature.[5–8]

To convey the idea that the accumulation of intracellular solutes, sometimes to very high levels (in the molar range), in response to osmotic stress cannot interfere with cell function, Brown[9] coined the term "compatible solute." This term has come to define intracellular solutes that preserve metabolism and protect cell components under conditions of water stress and dehydration. The term compatible solute is equally applicable to solutes that appear to protect cell components against stress conditions, such as supraoptimal temperatures, which are not necessarily associated with water stress.[8]

There are two principal mechanisms of osmoadaptation in microorganisms that illustrate two distinct strategies to cope with water stress. One strategy for maintaining osmotic equilibrium across the membrane involves the selective influx of salts into the cytoplasm and has been called the salt-in-the-cytoplasm or the saline cytoplasm type of osmoadaptation.[6] The extremely halophilic archaea of the family Halobacteriaceae and the bacteria of the order Haloanaerobiales accumulate very large quantities of inorganic ions (K^+, Na^+, Cl^-) and for this reason have developed proteins and other macromolecules that are dependent on the high intracellular salt concentrations for activity. However, the most common type of osmoadaptative strategy involves the accumulation of specific organic osmolytes (compatible solutes) with the exclusion of NaCl and the majority of the organic solutes found in the environment. In these organisms, intracellular macromolecules have not undergone adequate modifications and are, therefore, sensitive to high concentrations of salts and many organic solutes.

---

[4] R. S. Ronimus, A.-L. Reysenbach, D. R. Musgrave, and H. W. Morgan, *Arch. Microbiol.* **168,** 245 (1997).

[5] L. N. Csonka and A. D. Hanson, *Annu. Rev. Microbiol.* **45,** 569 (1991).

[6] E. A. Galinski, *Adv. Microbial Physiol.* **37,** 272 (1995).

[7] A. D. Brown, "Microbial Water Stress Physiology. Principles and Perspectives." Wiley, Chichester, 1990.

[8] M. S. da Costa, H. Santos, and E. A. Galinski, *Adv. Biochem. Eng. Biotechnol.* **61,** 117 (1998).

[9] A. D. Brown, *Bacteriol. Rev.* **40,** 803 (1976).

Many archaea appear to possess an intermediate mechanism of osmoadaptation where K^+ accumulates to high levels and contributes to a very high turgor pressure.[10,11] These organisms also accumulate anionic compatible solutes that counterbalance the positive charge and contribute to the osmotic balance of the cells, as in the extremely halophilic *Natronococcus occultus* and *Natronobacterium* spp. that accumulate the novel negatively charged trehalose derivative sulfotrehalose.[12]

A large variety of compatible solutes have been encountered in microorganisms that are capable of osmotic adaptation. These include amino acids and derivatives, sugars, polyols, and their derivatives, betaines, ectoine, and hydroxyectoine. Some compatible solutes are widespread in the three domains of life, but others are restricted to a small number of microorganisms. Although polyols, primarily glycerol, arabitol, and mannitol, are the major compatible solutes of yeasts and fungi, only a very small number of bacteria accumulate polyols (mannitol and sorbitol) in response to salt stress. On the other hand, glutamate, glycine betaine, and the ectoines are widespread in mesophilic bacteria and archaea.[6,8]

Halophilic and halotolerant microorganisms generally prefer uptake of compatible solutes from the medium over *de novo* synthesis. For this reason organisms grown in complex media will normally accumulate compatible solutes, such as trehalose and glycine betaine, from yeast extract. Presumably, the same preference for solute uptake occurs in the environment where compatible solutes are available from cell death.

It is not surprising that most of our knowledge on osmoadaptation in microorganisms comes from research conducted in mesophiles, and that little is known about the response to salt stress in thermophiles and hyperthermophiles. However, the same interest that fuels research into so many aspects of life at extremely high temperatures also fuels research into osmotic adaptation in thermophilic and hyperthermophilic organisms, leading to the rapid increase in the knowledge of the compatible solutes themselves, their role in the protection of macromolecules, and the mechanisms of osmotic adaptation.

Compatible Solutes of Thermophilic and Hyperthermophilic Organisms

Many compatible solutes of thermophiles and hyperthermophiles occur in no other organisms examined (Fig. 1), although a few are also encountered in mesophiles. Trehalose, for example, is widespread in Eukarya, Bacteria, and Archaea. The hyperthermophilic archaea of the order Sulfolobales are not halophilic, but

---

[10] D. D. Martin, R. A. Ciulla, and M. F. Roberts, *Appl. Environ. Microbiol.* **65**, 1815 (1999).
[11] D. E. Robertson and M. F. Roberts, *Biofactors* **3**, 1 (1991).
[12] D. Desmarais, P. Jablonski, N. S. Fedarko, and M. F. Roberts, *J. Bacteriol.* **179**, 3146 (1997).

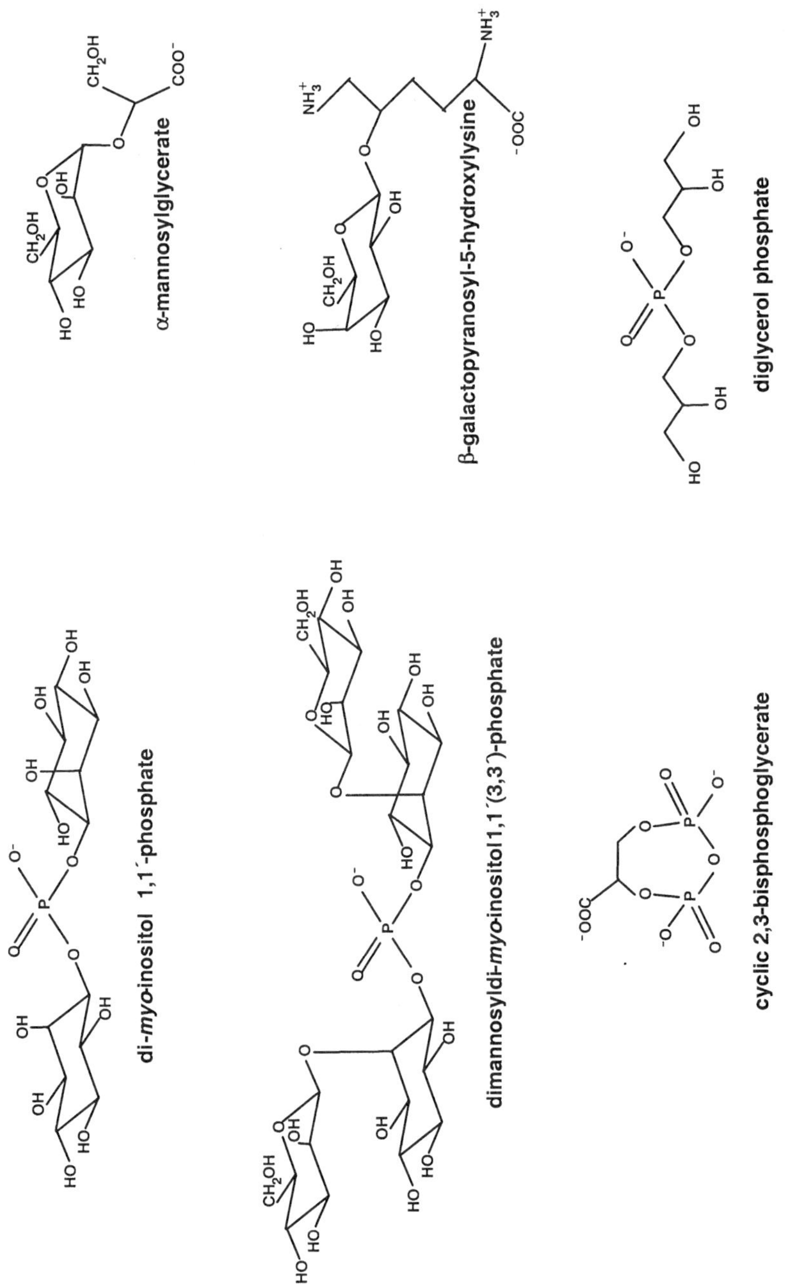

FIG. 1. Unusual compatible solutes restricted to hyperthermophiles (organisms with optimal growth temperature $\geq 80°$).

accumulate large amounts of trehalose, whose function remains unknown.[13] The halotolerant *Thermus thermophilus* and the halophilic *Rhodothermus marinus* also accumulate trehalose at high salinities. In the former trehalose is the major intracellular organic solute, where it may act as an osmolyte, but it is only a minor component in the latter.[14,15] Trehalose is also the only organic solute detected in the microaerophilic marine species *Pyrobaculum aerophilum* under optimum growth conditions; however, organic solutes are not detected in the closely related nonhalotolerant species *Pyrobaculum islandicum*.[16]

Glutamate is a common compatible solute in many mesophilic bacteria and archaea where it generally serves as a counterion for K^+ during low level osmotic adjustment. Almost all halotolerant or halophilic thermophiles and hyperthermophiles examined accumulate α-glutamate during growth in media containing low levels of NaCl.[16,17] The rare β-amino acids, namely β-glutamate, β-glutamine, and N^ε-acetyl-β-lysine, protect several mesophilic and thermophilic methanogens from salt stress.[10,18] The hyperthermophilic and slightly halophilic bacterium *Thermotoga neapolitana* also accumulates β-glutamate (and α-glutamate) during osmotic adaptation.[17] However, the major compatible solutes of this species, and the closely related species *Tt. maritima*, are derivatives of di-*myo*-inositol phosphate.[17,19] As a whole, dimannosyldi-*myo*-inositol phosphate, di-*myo*-inositol 1,1′-phosphate (DIP), and di-*myo*-inositol 1,3′-phosphate are the major organic solutes of these organisms. On the other hand, an increase in the growth temperature above the optimum led to a profound increase in the intracellular levels of dimannosyldi-*myo*-inositol phosphate and di-*myo*-inositol 1,3′-phosphate. Dimannosyldi-*myo*-inositol phosphate and di-*myo*-inositol 1,3′-phosphate have only been identified in these *Thermotoga* spp.,[17] whereas DIP has also been detected in the hyperthermophilic bacterium *Aquifex pyrophilus*[20] and is widespread in hyperthermophilic crenarchaeotes and euryarchaeotes.

---

[13] B. Nicolaus, A. Gambacorta, A. L. Basso, R. Riccio, M. de Rosa, and W. D. Grant, *System. Appl. Microbiol.* **10,** 215 (1988).

[14] O. C. Nunes, C. M. Manaia, M. S. Da Costa, and H. Santos, *Appl. Environ. Microbiol.* **61,** 2351 (1995).

[15] Z. Silva, N. Borges, L. O. Martins, R. Wait, M. S. Da Costa, and H. Santos, *Extremophiles* **3,** 163 (1999).

[16] L. O. Martins, R. Huber, H. Huber, K. O. Stetter, M. S. Da Costa, and H. Santos, *Appl. Environ. Microbiol.* **63,** 896 (1997).

[17] L. O. Martins, L. S. Carreto, M. S. Da Costa, and H. Santos, *J. Bacteriol.* **178,** 5644 (1996).

[18] D. E. Robertson, M. F. Roberts, N. Belay, K. O. Stetter, and D. R. Boone, *Appl. Environ. Microbiol.* **56,** 1504 (1990).

[19] V. Ramakrishnan, M. F. J. M. Verhagen, and M. W. W. Adams, *Appl. Environ. Microbiol.* **63,** 347 (1997).

[20] L. O. Martins, N. Raven, M. S. Da Costa, and H. Santos, unpublished results (1997).

DIP was initially identified in *Pyrococcus woesei*[21] and *Methanococcus igneus*.[22] Later, this solute was detected in other hyperthermophilic archaea, namely in *Pyrodictium occultum,* and in *Pyrococcus* and *Thermococcus* spp. In several species of the Thermococcales large increases in the levels of DIP are observed at growth temperatures above the optimum, leading to the view that this solute has a thermoprotective role in these organisms.[16,23,24]

The most commonly encountered compatible solute of thermophilic and hyperthermophilic organisms is α-mannosylglycerate (MG). This sugar derivative was initially detected in a few species of red algae, where it accumulates during the initial stages of osmoadaptation before protecting levels of mannitol have been reached.[25] Two forms of MG, originally identified as β-mannosylglycerate and α-mannosylglycerate, are found in the thermophilic halophilic bacterium *Rhodothermus marinus*.[14] However, a thorough investigation of both compounds by NMR, mass spectrometry, and elemental analysis, after purification, led to the conclusion that the two forms of MG were α-mannosylglycerate and α-mannosylglyceramide.[15] The negatively charged form has also been detected in several thermophilic bacteria such as *Thermus thermophilus,* the hyperthermophilic archaeal species of the genera *Pyrococcus* and *Thermococcus,* and the species *Methanothermus fervidus*.[14–16,23,24] In contrast to DIP, the concentration of MG generally increases concomitantly with the NaCl concentration of the medium and appears to serve as a compatible solute under salt stress. The neutral MG derivative mannosylglyceramide has only been found, in addition to MG, in several strains of *R. marinus,* where this solute accumulates preferentially during salt stress at lower growth temperatures, whereas MG accumulates during salt stress at supraoptimal growth temperatures.[15]

Mannosylglycerate is also detected, along with roughly similar amounts of trehalose and glycine betaine, in *Petrotoga miotherma,* a moderately thermophilic and slightly halophilic member of the bacterial order Thermotogales.[17] It also came as a surprise to identify mannosylglycerate in *Methanothermus fervidus*.[16] Moreover, MG was recently detected in one culture of *Archaeoglobus fulgidus*[26] and in the crenearchaeote *Aeropyrum pernix*.[27]

---

[21] S. Scholz, J. Sonnenbichler, W. Schäfer, and R. Hensel, *FEBS Lett.* **306,** 239 (1992).
[22] R. A. Ciulla, S. Burggraf, K. O. Stetter, and M. F. Roberts, *Appl. Environ. Microbiol.* **60,** 3660 (1994).
[23] L. O. Martins and H. Santos, *Appl. Environ. Microbiol.* **61,** 3299 (1995).
[24] P. Lamosa, L. O. Martins, M. S. da Costa, and H. Santos, *Appl. Environ. Microbiol.* **64,** 3591 (1998).
[25] U. Karsten, K. D. Barrow, A. S. Mostaert, R. J. King, and J. A. West, *Plant Physiol. Biochem.* **32,** 669 (1994).
[26] T. Q. Faria, R. Huber, and H. Santos, unpublished results (1999).
[27] T. Q. Faria, N. Raven, and H. Santos, unpublished results (1999).

Cyclic 2,3-bisphosphoglycerate (cBPG) has been detected in some methanogens with optimum growth temperatures that vary between 37° and 98°.[10,28–31] In most cases, there seems to be a positive correlation between the levels of intracellular cBPG, K^+, and higher growth temperatures of the organisms culminating in the accumulation of very large concentrations of cBPG in *Methanopyrum kandleri*.[31] Another solute found only in methanogens is 1,3,4,6-tetracarboxyhexane (TCH). This solute accumulates in *Methanobacterium thermoautotrophicum* and, with four negative charges, may be important in balancing the positive charge of potassium.[32] However, TCH is a structural unit of methanofuran and thus may have two roles in this methanogen.

Other unusual organic solutes also seem to play a role in the osmotic adaptation of thermophilic and hyperthermophilic organisms. For example, the glycerol derivative diglycerol phosphate has only been found in *Archaeoglobus fulgidus*, where it is by far the major compatible solute during salt stress. However, DIP, also present in this organism, becomes the major solute when the temperature is raised above the optimum for growth.[16]

Two unusual organic solutes found only in *Thermococcus litoralis* grown in medium containing peptone are hydroxyproline and β-galactopyranosyl-5-hydroxylysine, although asparate, MG, and DIP are the primary organic solutes in this organism.[24] Hydroxyproline, present in low levels, does not appear to serve as a compatible solute. Aspartate has been encountered before in *Methanococcus thermolithotrophicus*, where the slight increase in the intracellular levels of aspartate in response to salt stress indicates that this amino acid contributes to the compatible solute pool of the organisms.[18] It is interesting to note that the majority of the solutes of *Tc. litoralis* (aspartate, hydroxyproline, β-galactopyranosyl-5-hydroxylysine, and trehalose) are uptaken from peptone and yeast extract of the medium. In the absence of these supplements MG, DIP, and, in some cases, trehalose are the only solutes that accumulate. Table I gives an overview of the distribution of compatible solutes in thermophiles and hyperthermophiles.

Solute Uptake vs *de Novo* Synthesis in Hyperthermophiles

Unlike *Tc. litoralis,* hyperthermophiles do not appear to uptake many osmolytes from the medium, resorting, in general, to the synthesis of specific compatible solutes such as cBPG, MG, DIP, and diglycerol phosphate. For example, *P. furiosus* resorts exclusively to *de novo* synthesis of MG for osmotic adaptation, and DIP

---

[28] S. Kanodia and M. F. Roberts, *Proc. Natl. Acad. Sci. U.S.A.* **80,** 5217 (1983).
[29] R. Hensel and H. König, *FEMS Microbiol. Lett.* **49,** 75 (1988).
[30] R. J. Seely and D. E. Fahrney, *J. Biol. Chem.* **258,** 10835 (1983).
[31] R. Huber, M. Kurr, H. W. Jannasch, and K. O. Stetter, *Nature* **342,** 833 (1989).
[32] A. Gorkovenko, M. F. Roberts, and R. H. White, *Appl. Environ. Microbiol.* **60,** 1249 (1994).

TABLE 1
DISTRIBUTION OF COMPATIBLE SOLUTES OF THERMOPHILES AND HYPERTHERMOPHILES[a]

Organisms	Optimal Temperature (°C)	cBPG	Tre	MG	MGA	DIP	DIP'	DMDIP	DGP	TCH	α-Glu	β-Glu	Asp	GalHL	GB	Pro	NAL	Ref.
Archaea																		
Pyrodictium accultum	105										+							16
Pyrobaculum aerophilum	100		+++															16
Pyrobaculum islandicum	100																	16
Pyrococcus furiosus	100			+++		+++												23
Methanopyrus kandleri	98	+++																16
Aeropyrum pernix	90			+++		+					+							27
Methanococcus igneus	88			+++		+++					+++	++						18,22
Thermoproteus tenax	88		+++															16
Thermococcus stetteri	87			+++		+++					++		+++					24
Thermococcus celer	87			+++		+++					+		+++					24
Thermococcus litoralis	85		+++	+++		+++					+++		+++	+				24
Methanothermus fervidus	83	+++		++														16
Archaeoglobus fulgidus	83				+				+++		+							16
Acidianus ambivalens	80		+++															16
Thermococcus zilligii	75		+++	+++														24
Sulfolobus solfataricus	75		+															16
Metallosphaera sedula	75																	16
Methanobacterium thermoautotrophicum	70	+++								+	++							32
Methanococcus thermolithotrophicus	65										++	++	+				++	10,18
Bacteria																		
Thermotoga maritima	80					+++	++	++			++							17
Thermotoga neapolitana	80					+++	++	+			++							17
Thermosipho africanus	75											+				++		17
Thermotoga thermarum	70																	17
Fervidobacterium islandicum	70																	17
Thermus thermophilus	70		+++	++							+				+			14
Rhodothermus marinus	65		+	+++	+++						+							14
Petrotoga miotherma	55		+++	+++											+++			17

[a] The number of plus signs indicates the relative abundance of the solutes.
[b] cBPG, cyclic 2,3-bisphosphoglycerate; Tre, trehalose; MG, α-mannosylglycerate; MGA, α-mannosyldi-*myo*-inositol 1, 1'-phosphate; DIP, di-*myo*-inositol 1, 1'-phosphate; DIP″, di-*myo*-inositol 1,3'-phosphate; DMDIP, dimannosyldi-*myo*-inositol 1,1'-phosphate; DGP, diglycerol phosphate; TCH, 1,3,4,6-tetracarboxyhexane; α-Glu, α-glutamate; β-Glu, β-glutamate; Asp, aspartate; GalHL, β-galactopyranosyl-5-hydroxylysine; GB, glycine betaine; Pro, proline; NAL, N$^\epsilon$-acetyl-β-lysine.

for growth at temperatures above the optimum.[23] On the other hand, *Tc. litoralis* represents a dramatic example of the ability of a microorganism to scavenge compatible solutes from the environment.[24] Some of the most important osmolytes of mesophiles, namely ectoine and glycine betaine, have never been encountered in hyperthermophilic organisms, although glycine betaine has been detected in the moderate thermophilic bacterium *Petrotoga miotherma* and in low levels in some strains of *Thermus thermophilus*.[14,17] The absence of glycine betaine and ectoine in hyperthermophiles corroborates our view that some common compatible solutes of mesophiles cannot be used by organisms that grow at high temperatures. Perhaps they are unstable at higher temperatures or do not meet the requirements of the organisms for osmotic adaptation and thermoprotection of macromolecules.

Phylogenetic Distribution of Compatible Solutes in Thermophiles and Hyperthermophiles

As seen above, most thermophilic and hyperthermophilic organisms do not use common compatible solutes, namely glycine betaine or ectoine, for osmotic adjustment. However, the distribution of unusual compatible solutes among thermophiles and hyperthermophiles may not necessarily be related to their high growth temperatures, but rather constrained by phylogenetic relationships. If this prediction is correct, closely related organisms should have similar characteristics that could also include the type of compatible solute used in response to salt or temperature stress. In fact, the species of the euryarchaeotal genera *Pyrococcus* and *Thermococcus* are closely related and synthesize the same major compatible solutes, namely MG and DIP, and several other euryarchaeotes such as *Mb. fervidus* and *A. fulgidus* also accumulate MG (Fig. 2). However, MG is distributed among thermophilic bacteria, namely *Petrotoga miotherma, Thermus thermophilus,* and *Rhodothermus marinus,* that are clearly unrelated to each other or to the hyperthermophilic archaea. Surprisingly, MG was also detected for the first time in the Crenarchaeota (*Aeropyrum pernix*).[27] DIP is also widely distributed among hyperthermophilic Crenarchaeota and Euryarchaeota, as well as the hyperthermophilic bacteria *Aquifex pyrophilus, Thermotoga neapolitana,* and *Tt. maritima.*

There is growing evidence for the horizontal transfer of genetic material between the domains Archaea and Bacteria, so the synthesis of MG and DIP may reflect this process, especially if we take into account that these organisms inhabit similar environments. The ability of thermophilic and hyperthermophilic organisms to synthesize these solutes could also represent the conservation of ancient characteristics that have been selected for their role in osmo- or thermoprotection. The relationship between the ability of some red algae and thermophilic and hyperthermophilic microorganisms to synthesize MG is, however, more difficult to explain.

The accumulation of MG and DIP by a wide range of thermophiles and hyperthermophiles seems to indicate that the phylogenetic relationship of the organisms

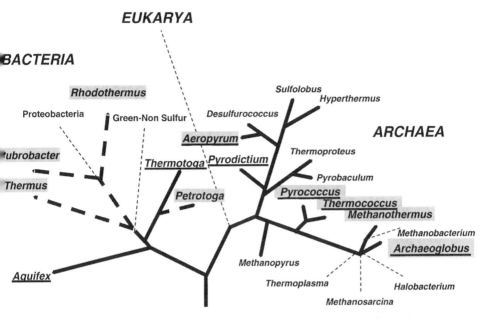

FIG. 2. Distribution of mannosylglycerate (shaded name) and di-*myo*-inositol 1,1'-phosphate (underlined name) among the Archaea and the Bacteria.

is not the sole criterion leading to the accumulation of these solutes. Therefore, growth temperature may be the most important constraint on the accumulation of unusual compatible solutes by halophilic bacteria and archaea.

Compatible Solutes as Stabilizers of Enzymes

There is a wealth of information showing that the majority of well-characterized enzymes from thermophiles and hyperthermophiles are intrinsically stable to heat.[33,34] It is unlikely that it could be otherwise, and that the thermostability of these enzymes would be exclusively due to extrinsic factors. However, some enzymes are less thermostable than expected from the growth temperature range of the organism, implying the existence of extrinsic stabilizers and/or very high turnover rates. There is now a growing amount of evidence showing that compatible solutes from thermophiles contribute to the thermostability of enzymes, at least *in vitro*. The accumulation of high levels of cBPG in thermophilic and hyperthermophilic methanogens (up to 1 M in *Methanopyrus kandleri*), in contrast to mesophilic species, led to the hypothesis that cBPG could have a role as an enzyme

[33] R. Hensel, *New Comp. Biochem.* **26,** 209 (1993).
[34] M. W. W. Adams, *Ann. Rev. Microbiol.* **47,** 627 (1993).

thermoprotectant. In accordance with this hypothesis, the potassium salt of cBPG (300 mM) had a thermostabilizing effect on glyceraldehyde-3-phosphate dehydrogenase (GAPDH) and malate dehydrogenase from *Mt. fervidus;* potassium phosphate also had a significant effect on enzyme stabilization. However, cBPG did not have an effect on the thermostabilization of rabbit GAPDH.[29] Recently, the stabilizing effect of cBPG was also demonstrated on two enzymes involved in methane formation in *Methanopyrus kandleri*.[35]

The role of DIP in the thermostabilization of enzymes remains questionable. This solute was shown to have a positive effect on the stabilization of glyceraldehyde-3-phosphate dehydrogenase,[21] but did not increase the stability of hydrogenase and pyruvate ferredoxin oxidoreductase of *Thermotoga maritima*[19] or rabbit muscle lactate dehydrogenase.[36]

Several thermophiles and hyperthermophiles synthesize mannosylglycerate, and the effect of this compatible solute has also been evaluated on thermoprotection and protection against desiccation for enzymes of mesophilic, thermophilic, and hyperthermophilic origin.[37] The thermostability of alcohol dehydrogenases from *P. furiosus* and *Bacillus stearothermophilus* and glutamate dehydrogenases from *Thermotoga maritima* and *Clostridium difficile* was significantly improved by the potassium salt of MG (0.5 M) at supraoptimal temperatures, but no effect on the thermostability of the extremely stable glutamate dehydrogenase from *P. furiosus* was observed. On the other hand, a remarkable effect of MG was found on the resistance to thermal inactivation of rabbit muscle lactate dehydrogenase (Fig. 3), baker's yeast alcohol dehydrogenase, and bovine liver glutamate dehydrogenase. In all cases MG was a better thermoprotectant than trehalose. A comparative study using MG, DIP, and diglycerol phosphate revealed the superiority of MG to protect rabbit muscle lactate dehydrogenase against heat inactivation.[36]

It should be mentioned that lyotropic salts, such as ammonium sulfate and potassium phosphate, have a thermoprotective effect on enzymes that rivals the effect of organic solutes *in vitro*. However, it is unlikely that these salts could play a role *in vivo,* since the large concentrations required for protection would interfere with enzyme activity and metabolic regulation, and have, in fact, never been found *in vivo*.

Pathways for Biosynthesis of Compatible Solutes in Thermophiles and Hyperthermophiles

The elucidation of metabolic pathways for the synthesis of compatible solutes is fundamental for the elucidation of the regulatory mechanisms during osmotic

---

[35] S. Shima, D. A. Herault, A. Berkessel, and R. K. Thauer, *Arch. Microbiol.* **170**, 469 (1998).

[36] N. Borges, A. Ramos, N. D. H. Raven, R. J. Sharp, and H. Santos, unpublished results (2000).

[37] A. Ramos, N. D. H. Raven, R. J. Sharp, S. Bartolucci, M. Rossi, R. Cannio, J. Lebbink, J. van der Oost, W. M. de Vos, and H. Santos, *Appl. Environ. Microbiol.* **63**, 4020 (1997).

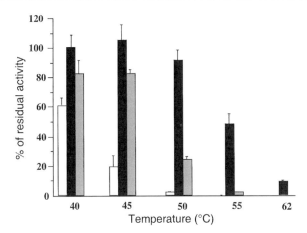

FIG. 3. Protecting effect of mannosylglycerate and trehalose against thermal inactivation of rabbit muscle lactate dehydrogenase. Aliquots containing the enzyme at a concentration of 50 μg ml^{-1} and the solutes (final concentration 0.5 M) were incubated for 10 min at the temperatures indicated, cooled in an ice bath, and assayed immediately for residual activity. Mannosylglycerate, black bars; trehalose, shaded bars; no additions, open bars. Data from A. Ramos, N. D. H. Raven, R. J. Sharp, S. Bartolucci, M. Rossi, R. Cannio, J. Lebbink, J. van der Oost, W. M. de Vos, and H. Santos, *Appl. Environ. Microbiol.* **63**, 4020 (1997).

adaptation and the ultimate determination of their physiological role. The biosynthetic pathway for cBPG has been studied in *Methanobacterium thermoautotrophicum* and *Methanothermus fervidus*. Initially, cBPG was reported to be formed from gluconeogenic intermediates, 2-phosphoglycerate, 3-phosphoglycerate, or 1,3-bisphosphoglycerate.[38] Later reports led to the conclusion that cBPG is formed from 2-phosphoglycerate via 2,3-bisphosphoglycerate using two ATP-dependent reactions catalyzed by 2-phosphoglycerate kinase and cBPG synthetase.[39,40] Both enzymes were purified and characterized from *Mt. fervidus*,[39] and the gene encoding cBPG synthetase has been cloned and expressed in *Escherichia coli*.[41] The cBPG synthetase from *Mb. thermoautotrophicum* has been characterized in detail.[42]

Two different pathways for the synthesis of DIP have been investigated in *Pyrococcus woesei* and *Methanococcus igneus*. In *P. woesei* the synthesis was reported to proceed via two steps: glucose 6-phosphate is converted into

[38] J. N. S. Evans, C. J. Tolman, S. Kanodia, and M. F. Roberts, *Biochemistry* **24**, 5693 (1985).
[39] A. Lechmacher, A.-B. Vogt, and R. Hensel, *FEBS Lett.* **272**, 94 (1990).
[40] G.-J. W. van Alebeek, C. Klaassen, J. T. Keltjens, C. van der Drift, and G. D. Vogels, *Arch. Microbiol.* **156**, 491 (1991).
[41] K. Matussek, P. Moritz, N. Brunner, C. Eckerskorn, and R. Hensel, *J. Bacteriol.* **180**, 5997 (1998).
[42] G.-J. W. van Alebeek, G. Tafazzul, M. J. J. Kreuwels, J. T. Keltjens, and G. D. Vogels, *Arch. Microbiol.* **162**, 193 (1994).

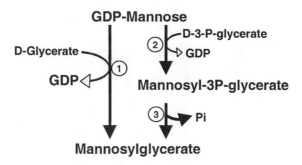

FIG. 4. Biosynthesis of mannosylglycerate in *Rhodothermus marinus*. The reaction scheme was proposed based on measurements of enzyme activities in crude cell extracts and ^{13}C-labeling experiments. The numbers refer to the following enzymes: 1, mannosylglycerate synthase; 2, mannosyl-3-phosphoglycerate synthase; 3, mannosyl-3-phosphoglycerate phosphatase. Data from L. O. Martins, N. Empadinhas, J. D. Marugg, C. Miguel, C. Ferreira, M. S. da Costa, and H. Santos, *J. Biol. Chem.* **274**, 35407 (1999).

L-*myo*-inositol 1-phosphate, two molecules of which are subsequently condensed to form di-*myo*-inositol 1,1′-phosphate.[43] In *Methanococcus igneus* the final step involves the condensation of CDP-inositol and *myo*-inositol to produce DIP and CMP. Inositol 1-phosphate is activated to CDP-inositol by the action of a CTP-inositol cytidylyltransferase. Glucose 6-phosphate is converted to inositol 1-phosphate, part of which is hydrolyzed to *myo*-inositol.[44]

The biosynthetic reaction scheme for the compatible solute mannosylglycerate in *Rhodothermus marinus* has been proposed based on enzymatic activities in cell-free extracts and *in vivo* ^{13}C-labeling experiments.[45] The synthesis proceeds via two alternative pathways (Fig. 4). In one, GDP-mannose is condensed with D-glycerate to produce MG in a single reaction catalyzed by mannosylglycerate synthase, whereas in the other pathway, mannosyl-3-phosphoglycerate synthase catalyzes the conversion of GDP-mannose and D-3-phosphoglycerate into a phosphorylated intermediate, which is subsequently converted to MG by the action of a phosphatase. Mannosylglycerate synthase was also purified and characterized, and the encoding gene (*mgs*) was sequenced, cloned, and overexpressed in *E. coli*. The enzyme has a molecular mass of 46 kDa and is highly specific for GDP-mannose and D-glycerate.

Concluding Remarks

Several questions have already been answered regarding osmotic adaptation in organisms that live at high temperatures, but many remain to be answered. It

---

[43] S. Scholz, S. Wolff, and R. Hensel, *FEMS Microbiol. Lett.* **168**, 37 (1998).
[44] L. Chen, E. T. Spiliotis, and M. F. Roberts, *J. Bacteriol.* **180**, 3785 (1998).
[45] L. O. Martins, N. Empadinhas, J. D. Marugg, C. Miguel, C. Ferreira, M. S. da Costa, and H. Santos, *J. Biol. Chem.* **274**, 35407 (1999).

seems well established that thermophilic and hyperthermophilic microorganisms have selected a few novel solutes, namely di-*myo*-inositol phosphate and derivatives, mannosylglycerate, diglycerol phosphate, and 2,3-bisphosphoglycerate, to cope with osmotic or heat stresses. Therefore, one could hypothesize that these solutes act as thermoprotectants of cell components *in vivo*, being part of the strategies for adaptation to high temperature. However, the accumulation of compatible solutes is not a general rule among thermophiles and hyperthermophiles, this characteristic being restricted to halophilic or halotolerant species. In fact, the nonhalophilic organisms *Thermotoga thermarum, Thermococcus zilligii, Fervidobacterium islandicum,* or *Pyrobaculum islandicum* do not accumulate compatible solutes, although high amounts of solutes are found in the halophilic members of the same genera. This indicates that these novel solutes are primarily related with osmoadaptation in halophilic thermophiles and hyperthermophiles rather than with intrinsic adaptations to high temperature; they may, however, combine roles in osmoprotection and thermoprotection.

An interesting feature of the unusual solutes restricted to hyperthermophiles is the predominance of a net electric charge. Some results indicate that the negatively charged compatible solutes serve to counterbalance the positive charge of K^+ that accumulates to high levels in methanogens and hyperthermophilic archaea, but the negatively charged compatible solutes could also enable more specific interactions with cell components that would enhance their protecting ability at high temperature.

The definite elucidation of the physiological role of compatible solutes in thermophiles and hyperthermophiles must await the development of suitable genetic tools. Compatible solutes in thermophiles and, particularly, in hyperthermophiles serve as excellent examples of evolutionary strategies developed to cope with adaptation to selective pressures, and their role in cell protection and their function in osmotic adaptation or thermal protection will, undoubtedly, lead to surprising new results.

Addendum Added in Proof

Our initial detection of mannosylglycerate in *A. fulgidus* was not confirmed in later attempts. We, therefore, stress that the presence of MG in *A. fulgidus* is, at present, uncertain.

Acknowledgments

We thank the many collaborators and coauthors cited in the reference list, who contributed decisively to our knowledge on solutes from thermophiles and hyperthermophiles. This work was supported by the European Community Biotech Programme (Extremophiles as Cell Factories, BIO4-CT96-0488), and by PRAXIS XXI and FEDER, Portugal (PRAXIS/2/2.1/BIO/1109/95).

[27] Pressure Effects on Activity and Stability of Hyperthermophilic Enzymes

By MICHAEL M. C. SUN and DOUGLAS S. CLARK

Introduction

Along with high temperatures, high pressures [ca. 200–300 atm (1 atm = 101.29 kPa)] are a salient feature of deep-sea hydrothermal vents, one of the most violent environments on earth and a rich source of hyperthermophilic organisms. Thermophilic bacteria have also been isolated from the cores of oil and sulfur wells,[1] and the existence of microbial life in other environments of high temperature and pressure appears likely. Although much work has been devoted to the investigation of high-temperature adaptation of enzymes from thermophiles and hyperthermophiles,[2–4] relatively few studies have focused on the adaptation of these enzymes to high pressure.[5–7] However, high pressure may play an important role in determining the functional properties of enzymes from organisms conditioned to high temperature–pressure environments. Moreover, high pressure can be a very useful tool for studying the structure and function of proteins, as evidenced by high-pressure studies of protein folding/unfolding pathways,[8,9] the kinetics of enzyme-catalyzed reactions,[10] and enzymatic reaction mechanisms in nonaqueous media.[11]

From a practical perspective, high-pressure food processing is an expanding technology,[12,13] and the prospect of new bioprocesses where pressure is used to regulate or improve biological reactions is of growing interest.[11,14,15] This chapter

---

[1] C. E. ZoBell, *Producers Monthly* **22,** 12 (1958).
[2] R. Jaenicke and G. Böhm, *Curr. Opin. Struct. Biol.* **8,** 738 (1998).
[3] M. J. Danson and D. W. Hough, *Trends Microb.* **6,** 307 (1998).
[4] F. T. Robb and D. L. Maeder, *Curr. Opin. Biotech.* **9,** 288 (1998).
[5] P. C. Michels, D. Hei, and D. S. Clark, *Adv. Protein Chem.* **48,** 341 (1996).
[6] M. Gross and R. Jaenicke, *Eur. J. Biochem.* **221,** 617 (1994).
[7] J. L. Silva and G. Weber, *Annu. Rev. Phys. Chem.* **44,** 89 (1993).
[8] G. J. A. Vidugiris, J. L. Markley, and C. A. Royer, *Biochemistry* **34,** 4909 (1995).
[9] G. J. A. Vidugiris and C. A. Royer, *Biophys. J.* **75,** 463 (1998).
[10] S. Kunugi, *Ann. N. Y. Acad. Sci.* **672,** 293 (1992).
[11] P. C. Michels, J. S. Dordick, and D. S. Clark, *J. Am. Chem. Soc.* **119,** 9331 (1997).
[12] H. Ludwig (ed.), "Advances in High Pressure Bioscience and Biotechnology." Springer-Verlag, Heidelberg, 1999.
[13] B. R. Thakur and P. E. Nelson, *Food Rev. Intl.* **14,** 427 (1998).
[14] S. V. Kamat, B. Iwaskewycz, E. J. Beckman, and A. J. Russell, *Proc. Natl. Acad. Sci. U.S.A.* **90,** 2940 (1993).
[15] V. V. Mozhaev, K. Heremans, J. Frank, P. Masson, and C. Balny, *Tibtech* **12,** 493 (1994).

describes some of the special techniques and equipment used for high-pressure activity and stability studies of hyperthermophilic enzymes in the authors' laboratory, along with results obtained from investigations of hyperthermophilic enzymes.

Buffers for High-Temperature, High-Pressure Experiments

An important consideration for high-pressure, high-temperature enzymatic studies is the effect of pressure and temperature on buffer pH. It is important to choose a buffer for which the changes in pH with respect to both pressure and temperature are known. Because data for the pressure dependence of buffer pH are scarce and not easily measurable, it is most convenient to choose one of the few buffers whose pH–pressure dependence is known or can be approximated. The pressure dependence of buffer pH has been found experimentally to be well represented by Eq. (1):[16,17]

$$\text{pH}(P) - \text{pH}(1 \text{ atm}) = \frac{\overline{\Delta V}^\circ P - 0.5\overline{\Delta \kappa}^\circ P^2}{2.303 RT} \quad (1)$$

where pH(P) is the pH at any pressure P (in atm), $\overline{\Delta V}^\circ$ (in ml/mol^{-1}) is the volume change for weak acid ionization at 1 atm, $\overline{\Delta \kappa}^\circ$ (in ml mol^{-1} atm^{-1}) is the compressibility term that reflects the pressure dependence of $\overline{\Delta V}^\circ$, R is the gas constant (82.1 atm ml mol^{-1} K^{-1}), and T (in kelvin) is the temperature at which the pH–pressure experiments were performed. Table I lists values of $\overline{\Delta V}^\circ$ and $\overline{\Delta \kappa}^\circ$ (where available) for the ionization of some common weak acids covering a pH-buffering range of 3.7–10.7. The reaction volumes for many more compounds, including other weak acids, can be found in the literature.[18,19] For pressures below ~2000 atm, the correction due to the $\overline{\Delta \kappa}^\circ$ term is small and ignoring the compressibility of $\overline{\Delta V}^\circ$ does not change the ΔpH [= pH (P) – pH (1 atm)] term substantially. For example, the calculated ΔpH for phosphoric acid (whose $\overline{\Delta \kappa}^\circ$ is

---

[16] R. C. Neuman, Jr. W. Kauzmann, and A. Zipp, *J. Phys. Chem.* **77**, 2687 (1973).

[17] The pressure-induced change in pH is related to the change in pK_a by

$$\partial(\text{pH})/\partial P = \partial(\text{p}K_a)/\partial P + \partial \log([A^-]/[HA])/\partial P$$

where K_a is the equilibrium constant for the dissociation of the buffer acid, HA, into its conjugate base, A$^-$, and a proton, H$^+$. Neuman *et al.*[16] noted that when buffer concentrations are higher than 0.05–0.1 M (as was the case in their work), the pressure-induced change in pH does not significantly alter the concentration ratio, and the change in pH is approximately given by the change in the pK_a. Thus, the following approximation was used throughout their work and in obtaining Eq. (1):

$$\partial(\text{pH})/\partial P \approx \partial(\text{p}K_a)/\partial P$$

[18] T. Asano and W. J. Le Noble, *Chem. Rev.* **78**, 407 (1978).

[19] N. S. Isaacs, "Liquid Phase High Pressure Chemistry." John Wiley & Sons, New York, 1981.

TABLE I
REACTION VOLUMES AND PRESSURE DEPENDENCE OF REACTION VOLUMES
FOR DISSOCIATION OF SOME COMMON WEAK ACIDS[a]

Buffer acid	pH buffering range	$\overline{\Delta V}^\circ$ (ml/mol)	$\overline{\Delta \kappa}^\circ \times 10^3$ (ml mol^{-1} atm^{-1})
Acetic[b]	3.7–5.6	−11.2	−1.49
H$_2$PO$_4{}^{-\,[b]}$	4.0–6.0	−24.5	−2.98
Cacodylic[b]	5.0–7.4	−13.2	−2.00
MES[c]	5.5–6.7	+5.0	ND[e]
ImidazoleH$^{+\,[d]}$	6.2–7.8	−2.4	ND[e]
HEPES[c]	6.8–8.2	+2.7	ND[e]
TrisH$^{+\,[b]}$	7.0–9.0	+1	∼0
TAPS[c]	7.7–9.1	+0.4	ND[e]
Glycine–NaOH[b]	8.6–10.6	∼+1	∼0

[a] Parameters for the dissociation of these buffer acids were all determined at 25°.
[b] R. C. Neuman, Jr., W. Kauzmann, and A. Zipp, *J. Phys. Chem.* **77**, 2687 (1973).
[c] S. Kunugi, *Ann. N. Y. Acad. Sci.* **672**, 293 (1992).
[d] M. Tsuda, I. Shirotani, S. Minomura, and Y. Terayama, *Bull. Chem. Soc. Jpn.* **49**, 2952 (1976).
[e] Not determined.

largest in magnitude in Table I) due to a pressure increase of 2000 atm is −0.76 pH units when the $\overline{\Delta \kappa}^\circ$ term is included, whereas a value of −0.87 pH units is obtained if $\overline{\Delta \kappa}^\circ$ is assumed to be zero. Unfortunately, $\overline{\Delta V}^\circ$ and $\overline{\Delta \kappa}^\circ$ values for the ionization of most weak acids (including the acids listed in Table I) were determined at a single temperature (usually near room temperature).

The effect of temperature alone on the pH of common buffers is tabulated in the literature.[20] However, it is still advisable to measure the pH as a function of temperature whenever possible, especially if high concentrations of salts or other solutes are added as part of the thermoinactivation solution.[21]

High-Pressure, High-Temperature Activity Assays

Hydrogenase Activity Measured with High-Pressure, High-Temperature Fiber Optic Probe

The methyl viologen-reducing activity of hydrogenase in crude extracts from the deep-sea thermophile *Methanococcus jannaschii* was assayed using the high-pressure, high-temperature reactor system shown in Fig. 1.[22] The reaction vessel

[20] R. M. C. Dawson, *et al.*, "Data for Biochemical Research." Clarendon Press, Oxford, 1986.
[21] Such measurements require a pH meter and probe with temperature compensator control and calibration with standard buffers at the desired temperature.
[22] J. F. Miller, C. M. Nelson, J. M. Ludlow, N. N. Shah, and D. S. Clark, *Biotech. Bioeng.* **34**, 1015 (1989).

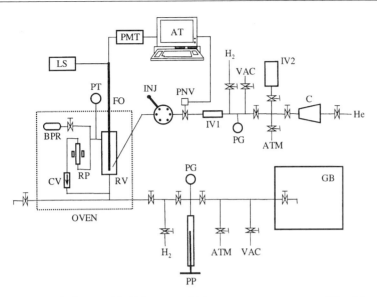

FIG. 1. Schematic diagram of the high-pressure, high-temperature reactor system: RV, stainless steel reaction vessel; RP, magnetically driven vapor recirculation pump; CV, check valve; BPR, back pressure regulator; PT, pressure transducer; FO, fiber optic probe; LS, light source; PMT, photomultiplier tube, water-cooled housing, and photometer; AT, computer; PNV, pneumatic valve; INJ, six-port valve; IV1, 1-ml injection vessel; H_2, hydrogen supply; PG, pressure gauge; IV2, 100-ml injection vessel; C, two-stage compressor; He, helium supply; GB, anaerobic glove box; PP, piston pump. The abbreviations ATM and VAC refer to atmosphere and vacuum, respectively. Adapted from Miller et al.[22] with permission; copyright 1989 John Wiley & Sons, Inc.

(Fig. 2), a back-pressure regulator (NUPRO, Willoughby, OH), and a magnetically driven pump (for a description, see Miller et al.[23]) for recirculating substrate gas through the liquid phase are housed in a forced-air oven (Blue M Corp., Blue Island, IL) that permits temperature control to within 0.1°. A homemade fiber optic probe (Fig. 2A) is inserted into the reaction vessel for spectrophotometric measurements (for a description, see Miller et al.[22]). When necessary, oxygen-sensitive enzyme and reactant solutions are prepared in the anaerobic glove box (Coy Laboratory Products, Ann Arbor, MI). The primary functions of the remainder of the apparatus are fluid transfer to and from the vessel and pressure maintenance. All high-pressure tubing, connections, and valves were obtained from High Pressure Equipment, Inc. (Erie, PA).

The optical fibers (SFS 200/240T, Radiant Communications Corp., East Hanover, NJ) of the probe function at temperatures up to 350° and operate over

[23] J. F. Miller, E. L. Almond, N. N. Shah, J. M. Ludlow, J. A. Zollweg, W. B. Streett, S. H. Zinder, and D. S. Clark, *Biotech. Bioeng.* **31**, 407 (1988).

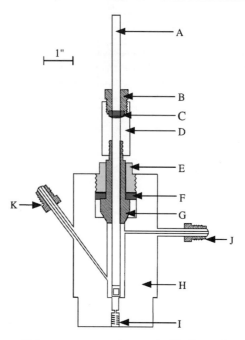

FIG. 2. High-pressure, high-temperature reaction vessel with fiber optic probe: (A) SS316 tube containing optical fibers, (B) 1/4-inch o.d. tube taper-seal gland, (C) 1/4-inch o.d. tube taper-seal sleeve, (D) modified 1/4-inch o.d. tube taper seal-to-1/4-inch NPT coupling, (E) custom 7/8-inch high-pressure gland, (F) washer, (G) 7/8-inch o.d. × 1/4-inch i.d. high-pressure plug with 1/4-inch NPT connector, (H) SS316 reaction vessel, (I) 1/8-inch high-pressure female opening, and (J,K) 1/4-inch high-pressure collar and gland. Adapted from Miller et al.[22] with permission; copyright 1989 John Wiley & Sons, Inc.

the range of 180 to 2400 nm. The transmitting fiber is butt-coupled to a light-emitting diode with a maximum emission of 590 nm and a bandwidth of 20 nm (Model LDY 5393, Siemens Components, Inc., Cupertino, CA). Light from the receiving fiber is measured by a photomultiplier tube (Model 3520, Hamamatsu Corp., Middlesex, NJ) and the output from the photomultiplier tube is received by a digital photometer (Model 124, Pacific Instruments). Alternatively, the fiber optic probe can be replaced by a capillary sampling line that runs from the headspace of the reactor chamber to a gas chromatograph, thereby providing a second method to analyze reactions involving gaseous substrates or products.

Hydrogenase-activity assays consist first of reactivating 15–60 μl of filtered *M. jannaschii* crude extract under an atmosphere of 98% N_2 and 2% H_2 (in the anaerobic glove box that is connected to the reaction vessel) by incubation in 500 μl of reactivation buffer (50 mM Tris, 10 mM 2-mercaptoethanol, 50 μM methyl viologen, pH 8.0). Reactivation of reversibly inactivated hydrogenase is necessary when cell extracts of *M. jannaschii* are prepared aerobically. Two hundred μl of the reactivated extract is then diluted into 39.8 ml of reducing buffer (50 m*M* Tris, 2 m*M* 2-mercaptoethanol, pH 8), and 4 ml of the resulting reaction solution are transferred to the reaction vessel, which already contains a mixture of about 7.5 atm hydrogen and enough helium to establish the final desired pressure. Transfer of the reaction solution is accomplished by drawing the solution from the anaerobic glove box into the screw-actuated piston pump (High Pressure Equipment, Inc., Erie, PA), and manually operating the pump to force the liquid into the reactor chamber. After a 15-min equilibration period, a small volume of concentrated methyl viologen solution is injected into the reactor chamber by high-pressure gas injection to obtain a final methyl viologen concentration of about 1.1 m*M*. High-pressure gas injection is accomplished by first transferring the methyl viologen solution from an anaerobic syringe to the helium-flushed sample loop of the six-port valve (Series 210A, Beckman, San Ramon, CA). A pneumatic valve is then opened, and a mixture with the same ratio of hydrogen and helium (an injection pressure of 340 atm is used for a vessel pressure of 250 atm) is released from injection vessel 1 (volume ≈ 1 ml) and/or injection vessel 2 (volume ≈ 100 ml) to push the liquid from the sample loop into the reactor chamber. Note that the amount of gas injected is small enough so that the gas injection process does not significantly alter the partial pressure of hydrogen inside the reaction vessel (∼7.5 atm). Six seconds after the start of injection, data are collected at a rate of 10 points per minute.

Figure 3 shows representative activity assays of methyl viologen reduction at 86° catalyzed by hydrogenase at total pressures of 8.5 and 250 atm. The initial rate

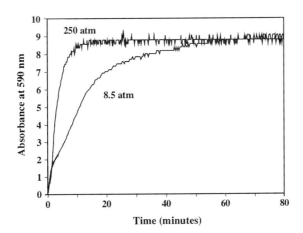

FIG. 3. Time course of methyl viologen reduction at 86° catalyzed by hydrogenase at 250 and 8.5 atm. Reprinted from Miller *et al.*[25] with permission; copyright 1989 John Wiley & Sons, Inc.

of hydrogenase activity is more than tripled by an increase in pressure from approximately 7.5 ± 1 to 260 ± 7 atm; the rates are 52.7 ± 7.2 and 171 ± 23 units/mg, respectively.[24] The rate enhancement of hydrogenase activity by pressure translates to an overall activation volume, ΔV^*, of -140 ± 32 ml/mol, as determined by the Johnson–Eyring equation:

$$\Delta V^* = RT \frac{\ln(\nu_1/\nu_2)}{(P_2 - P_1)} \qquad (2)$$

where ν_1 and ν_2 are the reaction rates at P_1 and P_2, respectively. This activation volume is more negative than those typically observed for enzymes that are activated by pressure (for example, activation volumes for enzymes from a variety of nonmethanogenic sources are compiled in the review by Morild[25] and range from -69 to 175 ml mol^{-1}). Because hydrogenase is a central enzyme in the methanogenesis pathway, it is interesting to note that growth and methanogenesis of *M. jannaschii* under similar conditions of temperature and pressure are also enhanced by pressure with an overall activation volume of $\Delta V^* = -80$ ml/mol.[26] The pressure activation of hydrogenase from *M. jannaschii* raises the question of whether reactions catalyzed by other enzymes from barophilic organisms also exhibit unusually large (i.e., negative) activation volumes.

Proteolytic Activity

The amidase activity of a putative protease[27] from *M. jannaschii* was studied at high pressures and temperatures using the substrate CBz-Ala-Ala-Leu-*p*-nitroanilide (Bachem Biosciences, Torrence, CA) and the reactor system shown in Fig. 1 with the following modifications.[28] The vapor recirculation pump (RP), check valve (CV), and back pressure regulator (BPR) are replaced by a 300-ml gas reservoir. The high-pressure, six-port valve (INJ) is replaced with a low-pressure injection port equipped with a high-pressure guard valve connected to the slanted 1/4-inch high-pressure opening of the reaction vessel (Fig. 2, K). The horizontal 1/4-inch high-pressure opening (Fig. 2, J) and the 1/8-in. high-pressure female opening (Fig. 2, part I) are connected to the 300-ml gas reservoir and to a high-pressure sampling valve, respectively. The tube containing optical fibers (Fig. 2, A) is removed, and parts B, C, and D are replaced with a 1/4-inch o.d. high-pressure

---

[24] A unit of activity is defined as 1 μmol of methyl viologen reduced per min.
[25] E. Morild, *Adv. Protein Chem.* **34**, 93 (1981).
[26] J. F. Miller, N. N. Shah, C. M. Nelson, J. M. Ludlow, and D. S. Clark, *Appl. Environ. Microbiol.* **54**, 3039 (1988).
[27] Recent evidence from our laboratory indicates that the proteolytic activity corresponds to a proteasome (unpublished result); however, this result awaits further confirmation. Thus, we shall continue to refer to this enzyme as a protease throughout this article.
[28] P. C. Michels and D. S. Clark, *Appl. Environ. Microbiol.* **63**, 3985 (1997).

connection-to-1/4-inch NPT (national pipe thread) coupling. This 1/4-inch o.d. high-pressure opening is sealed with a 1/4-in. high-pressure gland and plug. In addition, a small Teflon-coated magnet is included inside the reaction vessel (RV) and is stirred by a water-driven magnetic stirrer (GFS Chemicals, Columbus, OH) attached to the outside of the reactor. A schematic of this modified reactor system is shown in Hei and Clark.[29]

Activity assays are initiated by injecting (at ambient pressure) 100 μl of an enzyme stock solution followed by 1 ml of a preheated assay solution (50 mM HEPES, 250 mM CBz-Ala-Ala-Leu-p-nitroanilide, pH 6.5 at assay temperature and pressure) into 9 ml of assay solution preequilibrated at the desired temperature inside the reaction vessel. All liquid injections into the reaction vessel are accomplished through the low-pressure injection port. For high-pressure experiments, the reactor chamber is immediately pressurized with helium from the gas reservoir. Hyperbaric pressurization is usually accomplished using either helium or nitrogen. At 100° and 500 atm, the solubility in water is similar for both gases; the liquid mole fractions of helium and nitrogen are 0.003094 and 0.002707, respectively.[30,31] Samples are withdrawn at time intervals (e.g., at 1, 2, 4, 8, and 16 min) through the high-pressure sampling valve into microcentrifuge tubes and flash frozen in liquid nitrogen to quench the reaction. Frozen samples are thawed and spun down at 5000g, and the absorbance of the supernatant is measured at 405 nm. Amidase activities are determined from the initial slope of the reaction trajectory and are corrected for background hydrolysis of the substrate at each temperature and pressure.

Figure 4 shows the temperature profiles for amidase activity of the protease from $M. jannaschii$ at 10, 250, and 500 atm. The optimum temperature for activity of 116° at low pressure (10 atm) is one of the highest optimum temperatures for a proteolytic enzyme reported in the literature.[28] Figure 4 also shows that pressure substantially enhances the activity of the protease at each temperature; for example, application of 500 atm increases the maximum reaction rate about twofold and the rate at 130°C fivefold. The 400% enhancement of amidase activity at 130° translates to an overall activation volume, ΔV^*, of -106 ml mol^{-1}, as determined by the Johnson–Eyring equation [Eq. (2)]. The most negative activation volumes previously reported for serine proteases were -36 ml mol^{-1} for the digestion of casein by trypsin[32] and -33 ml mol^{-1} for the hydrolysis of Suc-Ala-Ala-pNA by α-chymotrypsin.[33] Michels and Clark also found that the protease is stabilized

---

[29] D. Hei and D. S. Clark, *Appl. Environ. Microbiol.* **60**, 932 (1994).
[30] H. L. Clever, *"Solubility Data Series,"* Vol. 1. Pergamon Press, Oxford, 1979.
[31] R. Battino, *"Solubility Data Series,"* Vol. 10. Pergamon Press, Oxford, 1982.
[32] This ΔV^* value was calculated by Morild[25] from data reported by D. Fraser and F. H. Johnson, *J. Biol. Chem.* **190**, 417 (1951).
[33] S. Makimoto and Y. Taniguchi, *Biochim. Biophys. Acta* **914**, 304 (1987).

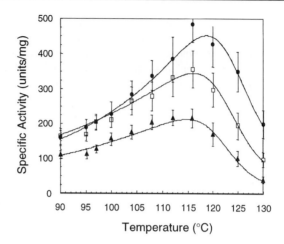

FIG. 4. Temperature profiles for amidase activity of *M. jannaschii* protease at 10 atm (▲), 250 atm (□), and 500 atm (●). Values represent averages of six separate experiments. A unit of activity is defined as 1 μmol of *p*-nitroaniline produced per min. Reprinted from Michels and Clark,[28] with permission; copyright 1997, American Society for Microbiology.

against thermoinactivation by pressure; for example, the thermal half-life at 125° increases 2.7-fold on application of 500 atm.[28]

The protease is not the only enzyme from *M. jannaschii* whose activity and stability are enhanced by pressure. As described above, pressure also increases the hydrogenase activity in crude extracts of *M. jannaschii*.[22] Likewise, the thermostability of hydrogenase is found to improve under pressure, increasing 4.8-fold at 90° on application of 500 atm.[29] These results suggest that the accelerated growth of *M. jannaschii* under pressure[26] may be a direct consequence of pressure-induced activation and/or stabilization of key enzymes.

High-Pressure Thermostability Measurements

Thermostability experiments are performed using the same high-pressure, high-temperature reactor system employed in the proteolytic activity assays described above. Thermoinactivation experiments are initiated by injecting at ambient pressure a few milliliters of the enzyme solution into the reaction vessel through the low-pressure injection port. It is important, especially for thermoinactivation temperatures at which the enzymatic half-lives are short, to measure the time required for the thermoinactivation solution to reach the desired temperature following injection. A shorter transient heating period can be achieved by preheating the enzyme solution at a subdenaturing temperature prior to injection. Alternatively, incubation buffer can be preheated in the reaction chamber and the experiment started by injecting a small volume (e.g., 0.5 ml) of concentrated enzyme solution followed by a 1-ml aliquot of incubation buffer to ensure that all of the enzyme solution enters the reactor chamber. For high-pressure thermoinactivation experiments, the reactor is immediately pressurized following solution injection using preheated helium from the gas reservoir. Preheating the helium in a large gas reservoir ensures that pressurization is rapid and that the pressure does not change after pressurization of the reactor chamber.

For thermal half-life determinations, samples are withdrawn from the reactor at appropriate time intervals through the high-pressure sampling valve. Prior to sampling, a small amount of enzyme solution ($\sim$0.4 ml) is removed to purge the sample line between the reactor and the outlet. Sampling is done slowly to prevent excessive foaming and agitation. Following sampling, the reactor pressure drops by about 5%; however, the reactor can be repressurized within a few seconds after sampling.

Thermoinactivation of Glutamate Dehydrogenase from Pyrococcus Furiosus

Glutamate dehydrogenase (GDH) from the hyperthermophile *P. furiosus* is a hexamer composed of six identical subunits, each containing 419 amino acids.[34] The enzyme has a melting temperature (T_m) for denaturation of 113°,[35] ranking it among the most thermostable GDHs studied to date. Thermoinactivation experiments using the high-pressure, high-temperature reactor system and the aforementioned methodology were performed on two GDHs: the native GDH (purified from *P. furiosus*) and a recombinant GDH mutant containing an extra tetrapeptide at the C terminus (rGDH^t).[36] Both GDHs are greatly stabilized by the application of pressure up to 750 atm. For example, as shown in Fig. 5, pressure stabilizes rGDH^t against thermoinactivation at 105°C, with thermal half-lives of 13 and 170 min at 5 and 275 atm, respectively. As described by Sun *et al.*,[36] the half-lives are determined by fitting the inactivation data to a model of thermal denaturation originally proposed by Singh *et al.*[37] for GDH from bovine liver,

$$M_6 \underset{k_{-1}}{\overset{k_1}{\rightleftarrows}} M_6' \overset{k_2}{\longrightarrow} 6U \; (\longrightarrow A) \tag{3}$$

where M_6 is the active, native form of the hexamer, M_6' is an inactive hexameric intermediate (referred to as a molten globule-like intermediate by Singh and co-workers), U is the unfolded monomer, and A is a final aggregated state. The best-fit

---

[34] K. S. P. Yip, T. J. Stillman, K. L. Britton, P. J. Artymiuk, P. J. Baker, S. E. Sedelnikova, P. C. Engel, A. Pasquo, R. Chiaraluce, and V. Consalvi, *Structure* **3**, 1147 (1995).
[35] H. Klump, J. DiRuggiero, M. Kessel, J. B. Park, M. W. W. Adams, and F. T. Robb, *J. Biol. Chem.* **267**, 22681 (1992).
[36] M. M. C. Sun, N. Tolliday, C. Vetriani, F. T. Robb, and D. S. Clark, *Protein Sci.* **8**, 1056 (1999).
[37] N. Singh, Z. Liu, and H. F. Fisher, *Biophys. Chem.* **63**, 27 (1996).

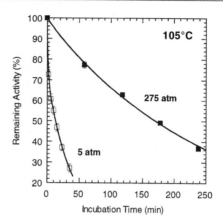

FIG. 5. Thermoinactivation trajectories of a recombinant GDH mutant (rGDH') from the hyperthermophile *Pyrococcus furiosus*. The experiments were performed at 105° with a GDH concentration of 50 μg/ml in 100 mM EPPS (pH 7.1 at 105°), and at 5 (□) and 500 (■) atm. Error bars represent mean deviations from two separate experiments.

curves (Fig. 5) correspond to the following values for the rate constants k_1, k_{-1}, and k_2 at 5 and 275 atm, respectively: 0.28, 0.58, 0.09 min^{-1} and 0.006, 0.03, 0.07 min^{-1}. According to the inactivation model [Eq. (3)] and the best-fit rate constants, pressure stabilizes GDH against thermoinactivation by decreasing the rate of conversion of the native form of the hexamer to the inactive hexameric intermediate (i.e., by decreasing k_1); pressure also reduces the rate constant k_{-1}, although to a lesser degree.

Stabilization of both native GDH and rGDH' is also achieved by adding glycerol in the thermoinactivation solution.[36] Glycerol has been shown to decrease both the volume and compressibility of protein interiors,[38] as well as reduce the flexibility and thermal backbone fluctuations of proteins.[39–41] Based on the known effects of glycerol on protein structure, the kinetic analysis of thermal inactivation, and the combined effect of glycerol and pressure, a mechanism of pressure-induced thermostabilization involving compression and/or rigidification of the native protein structure was proposed.[36]

GDH from *P. furiosus* is not the only enzyme stabilized by high pressure. Other enzymes found to be stabilized by high pressure in the authors' laboratory include GDHs from the hyperthermophiles *Thermococcus litoralis* and

[38] A. Priev, A. Almagor, S. Yedgar, and B. Gavish, *Biochemistry* **35**, 2061 (1996).
[39] P. Cioni and G. B. Strambini, *J. Mol. Biol.* **242**, 291 (1994).
[40] P. Cioni and G. B. Strambini, *J. Mol. Biol.* **263**, 789 (1996).
[41] S. L. Butler and J. J. Falke, *Biochemistry* **35**, 10595 (1996).

Pyrococcus endeavori (unpublished results[42]), hydrogenases from *Methanococcus jannaschii* and the extreme thermophile *Methanococcus igneus*,[29] α-glucosidase from *P. furiosus*,[29] glyceraldehyde-3-phosphate dehydrogenase from *Thermotoga maritima*,[29] and rubredoxin from *P. furiosus*.[29]

Concluding Remarks

The past decade has witnessed an explosion of research on hyperthermophily, with much of the attention focused on proteins and enzymes from hyperthermophiles. Many species of hyperthermophilic Archaea are found in the deep-sea environment near hydrothermal vents, where their proteins and enzymes must adapt not only to high temperatures but also to high pressures. Investigating the activity and stability of proteins from these microorganisms as a function of temperature and pressure has provided insights into possible mechanisms of protein stability beyond those that could be obtained from temperature studies alone. Moreover, pressure activation of enzymes at high temperatures points to a possible mechanism for barophilism in thermophilic and hyperthermophilic organisms.

The extreme conditions of high-pressure/temperature biochemical studies dictate that special equipment and methods be devised. The present contribution details some of the techniques and apparatus employed in our laboratory. High-temperature apparatus and methodologies can be found in the other contributions to this volume of *Methods in Enzymology*, and other examples of high-pressure devices and techniques are discussed elsewhere.[43,44] The best choice of equipment will depend on the specific requirements of the experiment. There are many options to choose from, some of which involve little more than a simple pressure vessel, a compressor, and a handful of valves and gauges. Thus, as the known biosphere expands to include a wider range of temperatures and pressures, the ability to employ high temperatures and pressures in laboratory studies of microbial behavior and biochemical phenomena should not be limiting.

Acknowledgments

Research in the authors' laboratory is supported by grants BES-9816490 and BES-9604561 from the National Science Foundation, the Schlumberger Fellowship of DSC, and by Kyowa Hakko Kogyo Co. Ltd. We thank Peter C. Michels for assistance in preparing the manuscript.

---

[42] M. M. C. Sun and D. S. Clark, 1998.
[43] C. Balny, P. Masson, and F. Travers, *High Press. Res.* **2**, 1 (1989).
[44] C. A. Royer, *Methods Enzymol.* **259**, 357 (1995).

[28] Thermodynamic Analysis of Hyperthermostable Oligomeric Proteins

By JAN BACKMANN and GÜNTER SCHÄFER

Introduction

(Hyper)thermostable proteins differ from their analogs in mesophilic organisms in different *intensive* parameters, such as packing density, fraction of buried apolar surface, the specific number of hydrogen bonds and of salt bridges, the extension of ion pair networks, and probably the specific amount of β structures.[1-3] The most important *extensive* parameter, however, seems to be the number of polypeptide chains per active protein molecule (The higher degree of oligomerization will, of course, influence the above-named intensive parameters.). Several proteins in hypothermophilic organisms were found to have a higher degree of oligomerization in comparison to their mesophilic analogs.[1,4-8] Like the other above-mentioned stabilizing structural differences between mesophilic and thermophilic proteins, oligomerization is an optional strategy used by evolution, but it is, we believe, one of the more important.

For studying the thermodynamic stability of oligomeric proteins, several aspects have to be taken into account: (i) the reversibility of unfolding, (ii) the type of the equilibrium between the native and the completely unfolded protein, (iii) the choice of the detection and denaturation method, and (iv) the specific contribution of the association between the monomers to the overall stability.

(i) For any thermodynamic analysis one has to make sure that the system in question is equilibrated before measurement, i.e., that all processes are reversible and that the system is given enough time to reach the equilibrium. In the case of protein folding the reversibility is checked by comparing the enzymatic activity or the circular dichroism (CD), fluorescence, or infrared spectrum after renaturation with that measured before denaturation. In the case of differential scanning calorimetry

---

[1] D. Maes, J. Ph. Zeelen, N. Thanki, N. Beaucamp, M. Alvarez, M.-H. Dao Thi, J. Backmann, J. A. Martial, L. Wyns, R. Jaenicke, and R. K. Wierenga, *PROTEINS: Struc. func. gen.* **37,** 44 (1999).
[2] G. Vogt and P. Argos, *Folding & Design* **2**(Suppl.), 40 (1997).
[3] G. Vogt, S. Woell, and P. Argos, *J. Mol. Biol.* **269,** 631 (1997).
[4] M. Kohlhoff, A. Dahm, and R. Hensel, *FEBS Lett.* **383,** 245 (1996).
[5] T. Dams, G. Böhm, G. Auerbach, G. Bader, H. Schurig, and R. Jaenicke, *Biol. Chem.* **379,** 367 (1998).
[6] J. Backmann, G. Schäfer, L. Wyns, and H. Bönisch, *J. Mol. Biol.* **284,** 817 (1998).
[7] V. Villeret, B. Clantin, C. Tricot, C. Legrain, M. Roovers, V. Stalon, N. Glansdorff, and J. Van Beeumen, *Proc. Natl. Acad. Sci. U.S.A.* **95,** 2801 (1998).
[8] R. Jaenicke and G. Böhm, *Curr. Opin. Struct. Biol.* **8,** 738 (1998).

(DSC) the thermograms of two successive scans are compared. Moreover, it must be confirmed that kinetic processes do not affect the measured signal, i.e., that the equilibrium is reached. Observing the signal over a sufficiently long interval of time can easily prove this. In the case of temperature gradients (DSC or optically monitored heat induced unfolding) one has to check that the shape and amplitude of the obtained thermogram or temperature profile does not depend on the scan rate.[9] As a rule of thumb, it can be considered that in the case of thermally induced unfolding the equilibrium is reached within minutes and for denaturant-induced unfolding it is reached overnight (at room temperature or below).

The concentration dependence of the temperature profile is another criterion for the irreversibility of a thermal unfolding, although this is not always unequivocal.[9] Whereas the shape of a temperature profile for a monomeric protein that unfolds reversibly should not depend on concentration at all, increased protein concentration of an oligomeric protein leads, in the case of reversibility, to a higher stability and thus to a higher melting temperature T_{trs} (see below). On the other hand, in the case of irreversible unfolding, the apparent melting temperature T_m decreases with higher protein concentration, which furthers intermolecular aggregation. One must also take into account that a higher degree of irreversibility of unfolding, even for proteins that unfold nearly (i.e., practically) reversibly at lower concentrations, can be induced by high protein concentration. This is especially important for those techniques that generally require higher concentrations in order to obtain a good signal-to-noise ratio, e.g., Fourier transform infrared (FTIR), near-UV CD, and DSC (especially in the case of older calorimeters). Generally, denaturant-induced unfolding tends to have a higher degree of reversibility in comparison to thermally induced.

(ii) In order to determine the free energy of unfolding, i.e., the thermodynamic stability of the native state, it is absolutely necessary to know which equilibrium conformations a protein can assume. In addition to the deviation from the two-state equilibrium that can occur in monomeric proteins, the unfolding of oligomeric proteins is complicated by the combination of monomer unfolding and oligomer dissociation. This problem and its practical implications are discussed below.

(iii) Once the type of the equilibrium is known a convenient detection method can be chosen. General aspects of the detection of protein conformational changes were discussed in detail by Schmid.[10] Special aspects for the case of oligomeric proteins are described below. Generally, highly thermostable, oligomeric proteins seem to unfold irreversibly upon increasing temperature.[11,12] One possible reason

---

[9] J. M. Sanchez-Ruiz, *Biophys. J.* **61**, 921 (1992).
[10] F. X. Schmid, in "A Practical Approach" (T. E. Creighton, ed.), p. 251. Oxford University Press, Oxford. (1989).
[11] H. Bönisch, J. Backmann, T. Kath, D. Naumann, and G. Schäfer, *Arch. Biochem. Biophys.* **333**, 75 (1996).
[12] B. S. McCrary, S. P. Edmondson, and J. W. Shriver, *J. Mol. Biol.* **264**, 784 (1998).

could be that the higher hydrophobicity of the protein interior makes it difficult to solvate the molecule when it unfolds. Another reason is probably that these proteins have to be brought to high temperatures for thermally induced unfolding to be observed. At such high temperatures the probability of destructive processes is generally higher. The latter can destroy the covalent structure even of proteins that are known to unfold reversibly.[13] This and the fact that melting occurs close to or above the boiling point of water limit the application of temperature gradients for the determination of thermodynamic parameters of hyperthermostable oligomeric proteins from DSC thermograms or spectroscopic thermal profiles.

(iv) The specific contribution of association between monomers to the overall stability cannot directly be measured in the case of a two-state equilibrium. An experimental determination of their contributions is possible only if dissociation and monomer unfolding are distinguishable.[14] Otherwise, only numerical methods based on the three-dimensional structure, if known, allow one to estimate the share of subunit interaction in overall thermodynamic parameters.[1,6,15] An interesting approach for analyzing the influence of oligomerization on stability has been described by Rietveld and Ferreira.[16]

Analysis of Equilibrium Type

The unfolding equilibrium of homooligomeric proteins comprises two processes, unfolding and dissociation, and can generally be presented in the form:

$$N_n \overset{K_1}{\rightleftharpoons} nN \overset{K_2}{\rightleftharpoons} nD \qquad (1)$$

N_n is the native oligomer, N the dissociated monomer with the same fold as the oligomerized state, D the fully denatured and dissociated state, and n the numbers of monomers in the native state. More complicated schemes are possible. For instance, if the monomer has more than one domain, it could be that one of them unfolds partly or fully before the oligomer disintegrates. Moreover, it is possible that in the dissociated state the monomer will have more than one equilibrium conformation. In this case one has to substitute nN by nI (intermediate) in the scheme above. A homogeneous oligomeric state D_n can be ruled out as an unfolded monomer will not have specific oligomerization sites. The equilibrium in Eq. (1) will be characterized by an equilibrium constant $K_n = K_1 K_2^n$ and the stability of the native state by $\Delta G_u = \Delta G_1 + n \Delta G_2$. In the case $K_1 \gg K_2$, the intermediate nN will be populated in a significant amount and the system will have a real

---

[13] S. E. Zale and A. M. Klibanov, *Biochemistry* **25**, 5432 (1986).
[14] K. E. Neet and D. E. Timm, *Protein Sci.* **3**, 2167 (1994).
[15] B. M. Baker and K. P. Murphy, *Methods in Enzymology* **295**, 294 (1998).
[16] A. W. Rietveld and S. T. Ferreira, *Biochemistry* **37**, 933 (1998).

three-state equilibrium. If $K_1 \ll K_2$, there will be no significant amounts of folded monomers and the system can be described sufficiently accurately with a two-state model:

$$N_n \rightleftharpoons nD \qquad (2)$$

The two-state model implies that quaternary interactions are necessary for the stabilization of the folded state and ΔG_u is equal to the dissociation free energy ΔG_{dis}.[14]

To scrutinize which type of model should be applied, an appropriate combination of at least two detection methods should be used, one sensitive for tertiary or quaternary structure and another for secondary structure. As the effective stability of an oligomeric protein depends on its concentration, both methods should ideally be applicable in the same protein concentration range. The combination of analytical ultracentrifugation (AUC) and spectropolarimetry (far-UV CD) fulfills these requirements. In both methods a protein concentration of 0.2–0.3 mg/ml can be applied. Before unfolding isotherms are measured spectroscopically, the spectra of the folded and the unfolded protein must be recorded and subtracted from each other in order to see at which wavelength a maximal signal related to protein unfolding can be obtained. In far-UV CD usually two extrema will be observed. In order to measure an unfolding isotherm using far-UV CD spectroscopy, the difference maximum at higher wavelength (205–230 nm) must be used. The difference maximum at lower wavelength (180–205 nm) cannot be detected because of the high absorbance of guanidine hydrochloride in this spectral range. (In the case of thermally induced unfolding and the absence of other absorbing substances, however, one tends to use the lower wavelength maximum as it often has a higher absolute value.) For sample preparation see refs. 10 and 17.

For the analysis of the guanidine hydrochloride–induced unfolding/dissociation equilibrium with AUC, it is recommended that the sedimentation–diffusion equilibrium mode be used, as this gives reliable values at least for the native state (approx. ±3%). The appropriate equipment (short column/3 channel cells) allows one to analyze 21 samples at the same time, so that an entire unfolding curve can be recorded in a single run. The determination of the molecular weight of the denatured state is a sophisticated task.[18] However, the analysis is much simpler if the primary structure of the oligomeric protein and thus its molecular mass is known. Using standard software, the ratio of the dissociated and oligomerized forms can be determined at the different guanidine hydrochloride concentrations. As this procedure is prone to error, one tends to confine oneself to the determination of the apparent molecular mass if this is sufficient to make a satisfying

---

[17] C. N. Pace, *Methods in Enzymology* **131**, 266 (1986).
[18] E. Reisler and H. Eisenberg, *Biochemistry* **8**, 4572 (1969).

conclusion on the denaturant-induced unfolding.[19,20] However, for a direct comparison of spectroscopic data and analytical ultracentrifugation results as well as for the thermodynamic analysis of the latter, fractions of (un)folded protein must be determined.

If two normalized curves (fraction versus denaturant concentration) from the different detection methods (e.g., far-UV CD and AUC) are superimposable within the experimental error, it can be concluded that the protein unfolds under the given conditions in a two-state equilibrium.

Extrapolation of Free Energy toward Zero Denaturant Concentration

A crucial problem for the extraction of thermodynamic parameters is the extrapolation of the free energy of unfolding toward zero concentration of the chaotropic agent (e.g., guanidine hydrochloride, urea). The simplest way is a linear extrapolation that is based on the assumption that the free energy of unfolding is directly proportional to the denaturant concentration, m being the factor of proportionality (which is a negative value).

$$\frac{\delta \Delta G_u}{\delta [D]} = m = \text{constant} \tag{3}$$

The validity of this linear extrapolation model (LEM) was proven for monomeric proteins in many cases.[21] Often the LEM is applied even without considering its applicability with the result that the m value has become generally recognized as a standard empirical parameter. The strength of the LEM lies in its simplicity. It is not necessary to calculate denaturant activities, which is an additional source of errors as in the case of the Tanford binding model.[22] Also, the LEM has only one parameter (m), whereas the binding model has two (Δn and k)* if one extrapolates the free energy toward zero denaturant concentration with the function

$$\Delta G = \Delta G(H_2O) - RT \Delta n \ln(1 + ka) \tag{4}$$

where a is the activity of the denaturant. Moreover, the m value has well-studied, straightforward empirical correlations to other experimentally accessible parameters like ΔC_p and ΔASA.[21] The binding model, however, may be more appropriate in some cases in which it fits better to the experimental data. On the other hand,

---

[19] T. Hansen, C. Urbanke, V. M. Leppanen, A. Goldman, K. Brandenburg, and G. Schäfer, *Arch. Biochem. Biophys.* **363**, 135 (1999).
[20] L. Carlini, U. Curth, B. Kindler, C. Urbanke, and R. D. Porter, *FEBS Lett.* **430**, 197 (1998).
[21] J. K. Myers, C. N. Pace, and J. M. Scholtz, *Protein Sci.* **4**, 2138 (1995).
[22] C. Tanford, *Adv. Protein Chem.* **24**, 1 (1970).
* These parameters have an disputed physical meaning, see ref. 37.

this model forces one to determine the extrapolation parameters (Δn, k) typically from four or five data points in a nonlinear regression analysis, whereupon, of course, rather weakly defined values are obtained.

Although it is not obvious that the LEM can be applied for multimeric proteins, it was proven to be valid in a number of cases.[14] The validity of LEM can be ascertained by combining urea- and guanidine hydrochloride–induced unfolding and comparing the resultant $\Delta G(H_2O)$ values. In the case of hyperthermostable oligomeric proteins, however, urea is often not a strong enough denaturant to measure a full unfolding isotherm.

Analysis of Unfolding Isotherm

Once we have measured a spectroscopic (or another) signal using different samples with the same protein concentration but different denaturant concentrations and plotted these signals versus denaturant concentration as shown in Fig. 1 (we will further call such a plot an unfolding isotherm), the results must be processed in an appropriate way in order to obtain thermodynamic parameters. In an unfolding isotherm pretransitional, transitional, and posttransitional regions can easily be identified. The processing of such data sets is based on the extrapolation of the post- and pretransitional regions over the entire concentration range. The nontransitional regions reflect an at least approximately linear dependence of the physical parameter y (CD, absorbance, fluorescence, or other) of the folded or

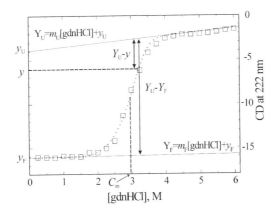

FIG. 1. Unfolding isotherm of the trimeric adenylate kinase from *Sulfolobus acidocaldarius* using CD spectroscopy. The squares represent the measured CD values at 222 nm versus the corresponding guanidinium chloride concentration at 40°. A linear dependence outside the transition region is assumed. The linear extrapolations of the pre- and posttransitional parts of the isotherms (solid lines) and the corresponding expressions are shown. The dotted line represents a fit using a function from Backmann *et al.*[6] for a trimeric protein.

unfolded protein on the denaturant concentration in the absence of conformational changes.† The distance of the measured value y from those extrapolated values Y_D and Y_U is correlated to the fraction of unfolded protein F_D by

$$F_D = \frac{Y_D - y}{Y_U - Y_F} \qquad (5)$$

(for explanation of symbols, see Fig. 1), which allows, in the case of a two-state transition, one to calculate the equilibrium constant

$$K_D = \frac{n^n C_t^{n-1} F_D^n}{1 - F_D} \qquad (6)$$

where C_t is the molar protein concentration expressed in oligomer equivalents and, consequently, the free energy $\Delta G = RT \ln K_D$ at the corresponding denaturant concentration. Once ΔG has been determined for different denaturant concentrations in the transition range, the free energy in absence of denaturant is obtained by extrapolation. Most researchers use a linear extrapolation (see above for the advantages and drawbacks of this model). For thermodynamic studies one will usually wish to measure a significant number of unfolding isotherms at different conditions such as temperature, protein concentration, ionic strength, and pH. In such a case data processing will be facilitated using appropriate fitting function to analyze the unfolding isotherms. Analytical expressions for monomeric, dimeric, and trimeric proteins are available.[6,23,24] For higher oligomers $\Delta G(H_2O)$ or C_m (the denaturant concentration at which the fraction of denatured protein F_D equals one-half, cf. Fig. 1) can be obtained only numerically or graphically. Fairman and co-workers[25] suggested a fitting procedure based on a numerical approach for higher oligomers. It is important to realize that C_m is always obtained with a significantly higher accuracy than ΔG. Thus, if the linear extrapolation model is applicable, procedures should be used that allow determination of C_m rather than $\Delta G(H_2O)$ because the latter can be calculated from the first with the expression

$$\Delta G(H_2O) = -m \left[C_m + \frac{RT}{m} \ln \left(\frac{n^n}{2^{n-1}} C_t^{n-1} \right) \right] \qquad (7)$$

where R is the gas constant, and T the temperature in degrees Kelvin.

---

† A linear dependence is preferable in the absence of strong counterevidence. More complex functions can lead to overparameterization. Anyway, the error of a relatively short linear extrapolation into the transition range is limited.

[23] M. M. Santoro and D. W. Bolen, *Biochemistry* **27**, 8063 (1988).

[24] H. Backes, C. Berens, V. Helbl, S. Walter, F. X. Schmid, and W. Hillen, *Biochemistry* **36**, 5311 (1997).

[25] R. Fairman, H.-G. Chao, L. Mueller, T. B. Lavoie, L. Shen, J. Novotny, and G. R. Matsueda, *Protein Science* **4**, 1457 (1995).

Determination of m Value and Its Possible Dependence on Temperature

In the case of hyperthermostable proteins, denaturant unfolding curves can be measured over a broad temperature range. Such an approach often provides the only reliable determination of the thermodynamic parameters, as in many cases thermally induced unfolding cannot be used for thermodynamic analysis because of the irreversibility of the thermal unfolding[11,12,19] and the difficulties in observing unfolding above the boiling point of water.* ΔH_{trs} and ΔC_p can only be obtained if reliable m-values are available (see below), but the m-value yielded from the fit of a single unfolding isotherm is usually subject to a large error. The best way to increase the reliability of the m-value is to average a significant number of measured values. This also includes m-values measured at different temperatures if, of course, there is evidence that m is constant over the entire temperature range. It is advisable to plot all m-values obtained with the corresponding error bars vs temperature and to determine whether there is a trend and whether all points lie within a reasonable distance from the mean represented by a horizontal line. When making conclusions as to the dependence or independence of the m-value on temperature, one must make sure that the changing lengths of the pre- and posttransitional regions (and consequently the number of data points) do not introduce artifactual temperature dependencies. The independence of m from temperature is important for the global fit of the C_m versus temperature with a modified Gibbs–Helmholtz equation as described below.

Residual Structure

It is possible that the measured m-value and ΔC_p fall below the theoretical expectation for a two-state equilibrium. Estimates can be obtained using parameterizations as suggested by Myers et al.[21] or Gomez et al.[26] that are based on the correlation of these parameters with the accessible surface area change upon unfolding. Even if the structure and thus the change in accessible surface area upon folding are not known, data sets such as those compiled by Miller et al.[27] Myers et al.,[21] and Pfeil[28] can be used to estimate approximate values of ΔC_p and m for the two-state equilibrium based on a rough correlation of the total molecular mass of the n-mer with these parameters.

---

* One can, of course, try to overcome these problems by carrying out the thermal denaturation experiments in presence of denaturants. In this case, however, uncertainties are introduced by the necessity of extrapolation of the measured values toward zero denaturant concentration.

[26] J. Gomez, V. J. Hilser, D. Xie, and E. Freire, *Proteins* **22,** 404 (1995).
[27] S. Miller, A. M. Lesk, J. Janin, and C. Chothia, *Nature* **328,** 834 (1987).
[28] W. Pfeil, "Protein Stability and Folding: A Collection of Thermodynamic Data." Springer-Verlag, Berlin (1998).

Strong deviations from the predicted values are an indication for more complex unfolding equilibria or the presence of residual structure in the "denatured" form. The latter possibility should certainly be taken into account in the case of large oligomeric proteins. A significant change of m-value with the pH can be an indication of a change of equilibrium type. However, one should avoid basing such arguments on too few determinations and biased data fittings, as a single m-value is determined with a relatively large error.

Determination of ΔH_{trs}, ΔC_p, and T_{trs} from Temperature Dependence of C_m

If an oligomeric protein has been shown to unfold reversibly in a two-state equilibrium, if the m-value does not depend on temperature and is reliably determined, and if all unfolding isotherms were measured at the same total protein concentration, then a plot of the C_m value yielded from all binding isotherms vs T can be fitted to a modified Gibbs–Helmholtz equation (cf. ref. 6)

$$C_m = -\frac{1}{m}\left[\Delta H(T_{trs})\frac{T_{trs} - T}{T_{trs}} - \Delta C_p(T_{trs} - T) + \Delta C_p T \ln\left(\frac{T_{trs}}{T}\right)\right] \quad (8)$$

where T_{trs} is the melting (transition) temperature (at the given concentration of the oligomeric protein), $\Delta H(T_{trs})$ is the enthalpy of unfolding at melting temperature, and ΔC_p is the heat capacity change upon unfolding. Here we consider m to be an independently determined constant and ΔC_p a fittable parameter. If both parameters are allowed to vary, then the fit cannot reasonably converge. This is easy to understand as both parameters are linearly (at least on an empirical basis) correlated to each other and to the accessible surface area change upon unfolding, ΔASA.[21] Consequently, one of the two parameters m or ΔC_p could be substituted using the empirical expression from Myers et al.[21] However, one must be aware that thereupon an additional condition for the validity (and source of errors) for the results is introduced, i.e., the applicability of the empirical expression by Myers and co-workers. Some interesting properties of function (8) should be mentioned here: The intersection with the temperature axis is the melting temperature of the oligomeric protein at the total protein concentration C_t at which the unfolding isotherm was measured. [Thus the transition temperature measured in absence of the denaturant can be used to improve the quality of the fit to Eq. (8).] Note that the function will shift with the protein concentration used in the measurement and thus is similar to the effective free energy (see below), differing from the latter only by a scaling factor, which is m.

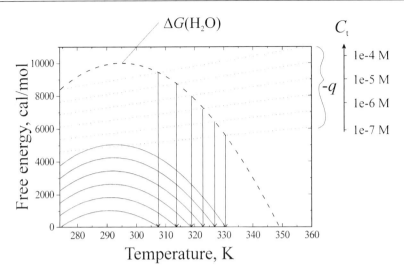

FIG. 2. Thermodynamic stability of a hypothetical dimeric protein with $T^° = 349$ K, $\Delta H^° = 126$ kcal/mol, and $\Delta C_p = 2138$ kcal/(mol K). The dashed line represents the intrinsic free energy $\Delta G(H_2O)$ as it would have been obtained experimentally. The dotted line shows the function $-q$ [cf. Eq. (10)] corresponding to different protein concentrations as approximately indicated by the concentration scale on the extreme right. If q is added to $\Delta G(H_2O)$, then ΔG_{eff} (solid curves) is obtained as shown by the arrows. The arrows also indicate the thermodynamic melting temperature T_{trs} corresponding to the total protein concentration C_t expressed in dimer equivalents.

Concentration Dependence of Melting Temperature and Denaturant Half-Concentration C_m

The dependence of melting temperature T_{trs} and denaturant half-concentration C_m on protein concentration must be taken into account if the results of different methods for the same protein are compared. In this paragraph we describe a simple approach to calculate the concentration dependence of T_{trs} and C_m.[†] Figure 2 shows the free energy function of a hypothetical dimeric protein. To obtain such a curve we recommend measuring denaturant-induced unfolding over the broadest possible temperature range, to determine C_m and the m-values for each unfolding curve, to average (if possible—see above) the m-values over all measurements, and to calculate $\Delta G(H_2O)$ for the different temperatures using Eq. (7). If a thermal unfolding profile was measured using the same protein concentration, the free

---

[†] The concentration dependence of T_{trs} of a reversible two-state oligomer unfolding should not be confused with the dependence of apparent melting on protein concentration in the case of irreversible unfolding.[9]

energy at the observed melting temperature can be calculated using

$$\Delta G(T_{trs}) = -RT_{trs} \ln(K_D(T_{trs})) = -RT_{trs} \ln\left(\frac{n^n}{2^{n-1}}C_t^{n-1}\right) \quad (9)$$

and can be added to the data set consisting of data pairs $\Delta G(H_2O, T)$ vs T, thus extending maximally the temperature range available for the fit. The values obtained are then plotted vs the thermodynamic temperature and fitted with the equation

$$\Delta G(H_2O) = \Delta H°\left(1 - \frac{T}{T°}\right) + \Delta C_p\left(T - T° - T \ln \frac{T}{T°}\right) \quad (10)$$

in order to obtain $\Delta H°$, $T°$, and ΔC_p [$T°$ does not equal the melting temperature at the given protein concentration, but is the temperature at which $\Delta G(H_2O)$ is zero]. With these parameters (i.e., $\Delta H°$, $T°$, ΔC_p) the $\Delta G(H_2O)$ values can be calculated for any temperature using Eq. (10). Thus a table displaying thermodynamic temperature with a sufficiently small interval (say 0.5 degree) with the corresponding $\Delta G(H_2O)$ can be compiled using spreadsheet software. In another column the function q of T and C_t

$$q = RT \ln\left(\frac{n^n}{2^{n-1}}C_t^{n-1}\right) \quad (11)$$

is calculated. Eventually in a fourth column one computes

$$\Delta G_{eff} = \Delta G(H_2O) + q. \quad (12)$$

These values are then used to obtain the effective (real) values of T_{trs} and $\Delta H(T_{trs})$ for any protein concentration by fitting them (ΔG_{eff} vs temperature T) with Eq. (10), where $T°$ becomes T_{trs} (the melting temperature) and $\Delta H°$ becomes $\Delta H(T_{trs})$ (calorimetrically measurable heat effect for the given protein concentration). The melting temperature T_{trs} can also be directly read out from the spreadsheet: it is the temperature at which ΔG_{eff} equals zero. Figure 2 illustrates this procedure. Provided that m is constant at least within the error of measurement over the relevant temperature range, division of ΔG_{eff} by the average m-value (in another column of the spreadsheet) gives the corresponding denaturant half-concentration C_m for any temperature at a particular protein concentration [cf. Eq. (8); see Fig. 3].

Concentration Normalization of the Free Energy Function of a Homooligomeric Protein

In the case of monomeric proteins the equilibrium constant for the two-state unfolding is dimension-less. For oligomeric proteins, however, the equilibrium

TABLE I
TOTAL PROTEIN CONCENTRATION EXPRESSED
IN n-MER EQUIVALENTS[a]

n	$C_t(q=0)$ (M)
2	0.500
3	0.384
4	0.314
5	0.267
6	0.232
7	0.206
8	0.185

[a] At which $\Delta G(H_2O)$ equals ΔG_{eff}.

constant $K_{\text{unf}} = [U]^n/[F_n]$ has the dimension M^{n-1}. Consequently, the free energy will be concentration-dependent and can only be normalized by defining standard conditions. The standard free energy of protein unfolding/dissociation equilibria for which the activity of all involved components equals unity corresponds to conditions that cannot be realized in practice. In Table I we show at which total protein concentrations $\Delta G(H_2O)$ will be equal to ΔG_{eff} [q should be zero; cf. Eq. (12)]. Under this condition $\Delta G(H_2O)$ of an oligomeric protein will be zero at melting temperature.

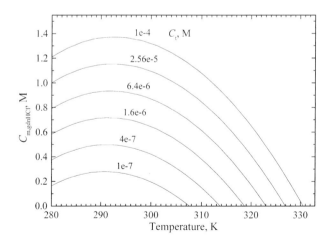

FIG. 3. The half-concentration of guanidinium chloride for the same hypothetical dimeric protein as shown in Fig. 2 with an m-value (for guanidinium chloride) of 3681 cal/(mol K).

Further Aspects in Stability Studies of Hyperthermostable Oligomeric Proteins

So far we have focused on the problem of how the thermodynamic parameters of a thermostable oligomeric protein can be determined. Some methods used to study kinetically controlled processes related to thermal stability will be briefly considered.

Characterization of Thermostable Proteins using H/D Exchange Monitored by FTIR Spectroscopy

The exchange of protein backbone amide protons, or H/D exchange, is an important indicator of stability and flexibility. It can provide very helpful information on specific properties of thermostable proteins, even if only the most slowly exchanging (i.e., structurally bound) amide protons are taken into account. Only NMR (with the exception of perhaps neutron diffraction[29]) allows site-specific measurements of H/D exchange. This method, however, requires relatively large amounts of protein, expensive equipment, and special expertise. FTIR (Fourier transform infrared spectroscopy) appears to be a satisfactory alternative if less specific information about the H/D exchange of slowly exchanging NH groups is required.[11,30,31] Infrared spectroscopy allows one to monitor H/D exchange by examining a temperature profile of the amide II band (1500–1600 cm^{-1}) or by measuring the rate of intensity change at constant temperature. One can normalize the signal using the area of the amide I band (1600–1700 cm^{-1}), which does not depend on H/D exchange. Such profiles are obtained without much effort as a by-product of the measurement of infrared temperature profiles. Proteins are often stored in the form of lyophilized powders, and samples can be prepared by dissolving the protein in a D_2O buffer. This is done at low temperature (e.g., 20°C or below) at which all protons not structurally bound will be exchanged. Before recording a temperature profile of unfolding in D_2O, one would usually preincubate the sample at elevated temperature below the unfolding region in order to ensure that the spectral changes, which will be observed later, can be unequivocally assigned to conformational changes and exchange-related effects can be precluded. If this necessary preincubation procedure is carried out directly in the cuvette at constant temperature, the exchange kinetic can be monitored. One can also heat the protein sample with a temperature gradient, thus allowing the exchange profile to be compared with those of other relevant proteins or mutants

---

[29] A. A. Kossiakoff, *Nature* **296**, 713 (1982).
[30] J. Backmann, C. Schultz, H. Fabian, U. Hahn, W. Saenger, and D. Naumann, *Proteins* **24**, 379 (1996).
[31] P. Zavodszky, J. Kardos, A. Svingor, and G. A. Petsko, *Proc. Natl. Acad. Sci. U.S.A.* **95**, 7406 (1998).

under the same conditions. The scan is stopped when full exchange is reached (which always occurs below the thermodynamic transition temperature, given a sufficiently low scan rate) and the protein is cooled down, whereupon another scan is carried out to obtain a temperature profile of unfolding/refolding. Temperature jump experiments can also be applied to study H/D exchange of proteins.[30] In the case when the exchange of structurally bound protons follows a global unfolding mechanism, it is theoretically possible to obtain thermodynamic stability parameters of the protein without heating the protein above the transition temperature,[32] which is especially relevant for highly thermostable oligomeric proteins.

Kinetics of Thermally Induced Isothermic Loss of Native Structure

A frequently used approach to study the thermal stability of proteins is to incubate a protein solution at an elevated constant temperature and to observe the change of certain physical parameters (e.g., CD, IR absorbance, enzyme activity) over periods of minutes or hours. Such measurements deliver precious information for the practical application of the protein in question. On the other hand, it is impossible to extract thermodynamic or structural parameters from such measurements, as they reflect the loss of native protein caused by a variety of processes. Irreversible thermal denaturation involves complex mechanisms and can lead to precipitation. The rates of such reactions depend on the concentration; the rate constants depend on temperature and solution conditions. The order of such reactions can vary from 1 to 2.[6,33] FTIR has the advantage that it at least allows clear identification of β-aggregation in the changes in the amide I band (1600–1700 cm^{-1}) of the infrared spectrum. The band component at around 1618 cm^{-1} reliably reflects the progress of β-aggregation.[11,34]

One should also keep in mind that most proteins have ligands or substrates or bind to other proteins. Such interactions will usually increase the thermodynamic and thermal stability of those proteins.[35,36] For instance, Hansen et al.[19] observed that Mg^{2+} ions caused an increase in operational melting temperature by 10° S. acidocaldarius pyrophosphatase and an increase of thermodynamic stability by 1.7 kcal/mol at 25°.

---

[32] Y. Bai, J. J. Englander, L. Mayne, J. S. Milne, and S. W. Englander, *Methods in Enzymology* **259**, 344 (1995).
[33] S. P. Roefs and K. G. De Kruif, *Eur. J. Biochem.* **226**, 883 (1994).
[34] H. H. Bauer, M. Müller, J. Goette, H. P. Merkle, and U. P. Fringeli, *Biochemistry* **33**, 12276 (1994).
[35] V. Rishi, F. Anjum, F. Ahmad, and W. Pfeil, *Biochem. J.* **329**, 137 (1998).
[36] E. Gazit and R. T. Sauer, *J. Biol. Chem.* **274**, 2652 (1999).
[37] Q. Zou, S. M. Habermann-Rottinghaus, and K. P. Murphy, *PROTEINS: Struc. Func. Gen.* **31**, 107 (1998).

Acknowledgments

We thank Dr. Claus Urbanke (Medizinische Hochschule Hannover, Germany) for giving us detailed consultations on the subject of analytical ultracentrifugation. This work was supported by F.W.O. (Foundation for Scientific Research), V.I.B. (Flemish Interuniversity Institute for Biotechnology) and the Deutsche Forschungsgemeinschaft (Grant Scha 125/17,1-3).

[29] Dynamics and Thermodynamics of Hyperthermophilic Proteins by Hydrogen Exchange

By S. WALTER ENGLANDER and REUBEN HILLER

Introduction

What strategies do hyperthermophilic proteins use to maintain their stability and activity at high temperature? The measurement of stability at very high temperatures and possibly correlated characteristics such as dynamic flexibility present special problems. Here we consider the application of hydrogen exchange approaches to these problems. Thermodynamic parameters and aspects of structural dynamics can, in many cases, be measured by hydrogen exchange methods at temperatures far below the melting temperature.

Thermal Stability

It is often assumed that the unusually high melting temperatures of thermophilic proteins reflect an unusually large stability, measured by the free energy for global unfolding. However, high melting temperature can be achieved in other ways as well. The stabilization free energy of a protein as a function of temperature, $\Delta G(T)$, is described by the Gibbs–Helmholtz equation [Eq. (1)].

$$\Delta G(T) = \Delta H_m + \Delta C_p(T - T_m) - T(\Delta S_m + \Delta C_p \ln T/T_m) \quad (1)$$

Alternative curves that fit Eq. (1) are shown in Fig. 1. The melting temperature, T_m, occurs where the difference in free energy between the native and unfolded states passes through zero. Figure 1 and Eq. (1) show that high melting temperature can be achieved in several different ways. The addition of stabilizing interactions will increase $\Delta G(T)$, shifting the melting curve vertically upward and increasing T_m. High T_m can also be achieved by a lateral shift in the free energy curve to higher temperature. This requires a change in either the melting enthalpy (ΔH_m) or entropy (ΔS_m), or in both. High T_m can also be achieved by flattening the curve so that it crosses the ΔG equals zero position at higher (and lower) temperature.

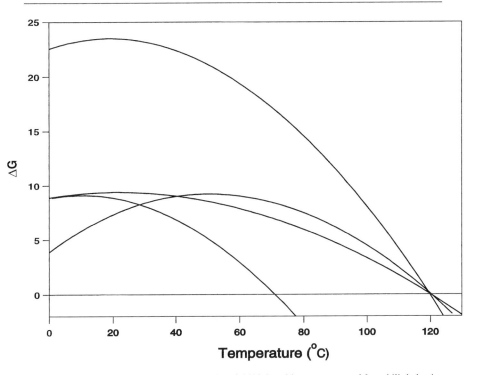

FIG. 1. Alternative thermal stability curves that yield high melting temperature. Mesophilic behavior (T_m of 76°) can be readjusted to higher T_m (120° shown) by altering the different parameters in Eq. (1) to raise, shift, or stretch the stability curve.

This requires changing the balance of hydrophobic interactions to decrease the specific heat parameter (ΔC_p).

For mesophilic proteins, one classically determines stability parameters and mutationally induced changes in stability by driving the protein through its thermal transition where the melting behavior can be measured.[1-3] The parameters determined can then be used together with Eq. (1) to infer the shape and positioning of the stability curve. The high T_m values of thermophilic proteins make these measurements problematic and impose additional uncertainty on the long extrapolations that this analysis requires. When melting is not fully reversible, as often occurs, these approaches fail.

Hydrogen exchange (HX) methods now make it possible to measure the stability parameters of native proteins, the distribution of stability and structural

[1] P. L. Privalov, *Adv. Protein Chem.* **33,** 167 (1979).
[2] P. L. Privalov, *Adv. Protein Chem.* **35,** 1 (1982).
[3] C. N. Pace, *CRC Crit. Rev. Biochem.* **3,** 1 (1975).

dynamics through the protein, and changes in these parameters due to experimentally imposed mutations, both locally and remotely.

Hydrogen Exchange and Protein Unfolding

Hydrogens on the polar groups of proteins engage in continual exchange with hydrogens in the aqueous solvent. These groups include the ionizable and neutral polar side chains and also the main chain peptide group NH hydrogen. Slowly exchanging protein hydrogens largely represent the main chain amide groups that participate extensively in secondary and tertiary structural H-bonding. Amide HX is rate-limited by a proton transfer reaction that requires H-bond formation to a solvent hydroxide ion (above pH 3). Because H-bonded NHs are sterically inaccessible to this process, their exchange requires transient H-bond breakage.

Hydrogen bonds can be broken in various kinds of opening reactions. These include small local fluctuations that break one H bond at a time, subglobal unfolding reactions (local unfolding) that break multiple H bonds in a cooperative partial unfolding reaction, and large whole-molecule global unfolding reactions. Thermodynamic principles require that protein molecules continually cycle through all of these high energy conformational states and that they populate each state according to the Boltzmann relationship in Eq. (2).

$$\Delta G_{op} = -RT \ln K_{op} \tag{2}$$

ΔG_{op} is the free energy of any given state (I or U) above the native state and K_{op} is the equilibrium constant for unfolding, given by I/N or U/N. In many cases, the most slowly exchanging hydrogens in a protein are so well protected that they can only exchange during the small fraction of time when the protein visits the globally unfolded state. The measurement of their exchange rates can directly evaluate the free energy for global unfolding under fully native conditions.[4,5] The faster exchange of hydrogens that are exposed by smaller opening reactions can provide structurally resolved information on local dynamics and energetics through the protein. HX experiments can be manipulated to distinguish and measure the different opening reactions.

Hydrogen Exchange Measurement and Analysis

HX has been measured by all combinations of H–H, H–D, and H–T approaches. At lowest resolution, HX can be measured at a whole molecule level using radioisotopic tritium counting[6] and by various spectroscopic methods that are sensitive

---

[4] S. N. Loh, K. E. Prehoda, J. Wang, and J. L. Markley, *Biochemistry* **32**, 11022 (1993).
[5] Y. Bai, J. S. Milne, L. Mayne, and S. W. Englander, *Proteins: Struct. Funct. Genet.* **20**, 4 (1994).
[6] S. W. Englander and J. J. Englander, *Methods in Enzymology* **49**, 24 (1978).

to H–D differences, including infrared[7] and ultraviolet[8] absorbance and Raman scattering.[9] At intermediate resolution, HX can be measured by tritium exchange labeling together with fragment separation methods,[10] and this approach has been extended to mass spectroscopic approaches.[11]

Most effectively, HX can be measured at the level of amino acid resolution by NMR spectroscopy using standard two-dimensional (2D) methods for relatively small protein molecules (<25 kDa). The NMR machine-time requirement can be reduced and faster HX rates can be measured by using ^{15}N-labeled protein together with heteronuclear nuclear magnetic resonance (NMR) methods, typically the heteronuclear single quantum COSY experiment. Alternatively, straightforward 1D proton NMR can be used, especially when only the slowest hydrogens are to be studied, since the prior exchange of the many faster hydrogens with deuterium simplifies the spectrum and can allow resolution of individual slowly exchanging NHs. This approach can even allow quite large proteins to be studied, especially for thermophilic proteins where the NMR measurements can be done at high temperatures that enhance molecular tumbling and sharpen NMR lines. Analogously, faster sites can be better resolved by using partially exchanged proteins prepared with the slower sites already deuterated.

HX can be initiated by placing a protein into D$_2$O under the conditions to be studied (pD, temperature, denaturant, etc.). Each hydrogen then begins to exchange with solvent deuterium at its characteristic rate. The exchange process can be followed by recording sequential NMR spectra in time. The measured first-order exchange rate (k_{ex}) can then be compared with the chemical rate expected for that site in the fully exposed condition (k_{ch}[cat]) in order to obtain the structural protection factor ($P = 1/K_{op}$) (where [cat] is the HX catalyst concentration, namely OH$^-$ for amide NHs above pH 3). From this the free energy for the dominant opening reaction that exposes the site to exchange can be computed [Eq. (3)].

$$\Delta G_{HX} = -RT \ln k_{ex}/k_{ch}[\text{cat}] = -RT \ln 1/P = -RT \ln K_{op} \quad (3)$$

Equation (3) assumes that exchange occurs only in the open state with the externally calibrated chemical rate constant, k_{ch}. The k_{ch} value depends on pH, temperature, local amino acid sequence, and other more minor factors[12,13] and can most easily be computed using spreadsheets that have been placed on the World Wide Web. When the measurement is not done at amino acid resolution, some averaged k_{ch} value might be used (e.g., characteristic for the Ala-NH-Ala proton).

---

[7] H. H De Jongh, E. Goormaghtigh, and J. M. Ruysschaert, *Biochemistry* **34**, 172 (1995).
[8] J. J. Englander, D. B. Calhoun, and S. W. Englander, *Anal. Biochem.* **92**, 517 (1979).
[9] P. Hildebrandt, F. Vanhacke, G. Heibel, and A. G. Mauk, *Biochemistry* **32**, 14158 (1993).
[10] J. J. Englander, J. R. Rogero, and S. W. Englander, *Anal. Biochem.* **147**, 234 (1985).
[11] Z. Zhang, C. B. Post, and D. L. Smith, *Biochemistry* **35**, 779 (1996).
[12] G. P. Connelly, Y. Bai, M.-F. Jeng, L. Mayne, and S. W. Englander, *Proteins: Struct. Funct. Genet.* **17**, 87 (1993).
[13] Y. Bai, J. S. Milne, L. Mayne, and S. W. Englander, *Proteins: Struct. Funct. Genet.* **17**, 75 (1993).

An important structural issue concerns the relationship between k_{ch} and opening–closing rates. The use of Eq. (3) assumes the relationship in Eq. (4).

$$k_{ex} = K_{op} k_{ch}[cat] \tag{4}$$

A more complete formulation uses Eq. (5).

$$k_{ex} = k_{ch}[cat] k_{op}/(k_{op} + k_{cl} + k_{ch}[cat]) \tag{5}$$

For stable structure, $k_{op} \ll k_{cl}$ so the k_{op} term in the denominator can be safely ignored. When $k_{cl} \gg k_{ch}[cat]$, the denominator in Eq. (5) then reduces to Eq. (4). Exchange then depends only on the preequilibrium opening and is said to be in the EX2 limit,[14] meaning that exchange follows second-order kinetics dependent on solvent catalyst concentration. When reclosing is slower than $k_{ch}[cat]$, HX is in the EX1 limit (monomolecular exchange) and yields the first-order opening rate constant ($k_{ex} = k_{op}$) rather than the opening equilibrium constant. A thorough discussion of these and related issues is given in ref. 15.

HX due to local fluctuations is unavoidably in the EX2 limit. Exchange due to unfolding reactions may press these limits, especially when measured under extreme conditions, e.g., with thermophilic proteins. Tests for EX2 behavior include the classical experiment for simple pH dependence,[14] the difficult and relatively noisy nuclear overhauser effect (NOE) correlation experiment,[16] and most easily the comparison of the rates measured for several NHs in a cooperative unfolding with their expected chemical rates.[5]

The effort to measure the slowest hydrogens of a protein in order to obtain the global unfolding free energy can be defeated by the long times necessary for exchange. For example, consider hydrogens that exchange by way of global unfolding in a protein that is stabilized by a free energy of 10 kcal/mol. The exchange time constant at pH 7 and 20° would be about 1 month, and it increases by 30-fold for each 2 kcal of further stabilization. The time required can be reduced and additional stability information can be obtained by performing HX experiments with added denaturant and/or at elevated temperature (below).

An Example

Stabilization free energy decreases with denaturant according to Eq. (6).

$$\Delta G_u(den) = \Delta G(0) - m[den] \tag{6}$$

The m value is the slope of a plot of the unfolding free energy (ΔG_u) against denaturant concentration. The m value proportions to the amount of buried protein

---

[14] A. Hvidt and S. O. Nielsen, *Adv. Protein Chem.* **21**, 287 (1966).
[15] S. W. Englander and N. R. Kallenbach, *Q. Rev. Biophys.* **16**, 521 (1984).
[16] H. Roder, G. Wagner, and K. Wuthrich, *Biochemistry* **24**, 7396 (1985).

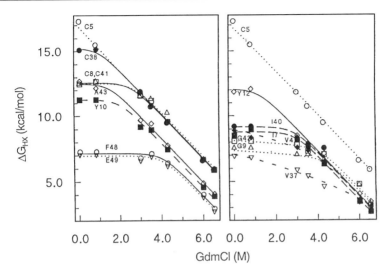

FIG. 2. HX results as a function of guanidinium chloride denaturant (GdmCl) for rubredoxin from *Pyrococcus furiosus*. Data were obtained by 1D NMR at 60° and computed using Eq. (3). The results distinguish HX that occurs by local fluctuations, a subglobal unfolding, and the highest free energy global unfolding. Reprinted with permission from R. Hiller, Z. H. Zhou, M. W. W. Adams, and S. W. Englander, *Proc. Natl. Acad. Sci. U.S.A.* **94**, 11329 (1997).

surface that is exposed to solvent in the unfolding reaction. An example is given in Fig. 2, which shows HX results as a function of guanidinium chloride denaturant (GdmCl), measured by 1D NMR at 60° for rubredoxin from *Pyrococcus furiosus* and computed as in Eq. (3).

The data in Fig. 2 show that various amide hydrogens exchange at low denaturant with essentially zero m value, therefore by way of small H-bond breaking fluctuations that expose little new surface. Increasing denaturant selectively promotes larger unfolding reactions (decreases ΔG_u). A large unfolding can then come to dominate the exchange of the hydrogens that it exposes [Eq. (6)]. This is seen as the merging of local fluctuation curves into an unfolding curve. The highest free energy global unfolding of hyperthermophilic rubredoxin, with ΔG_u of 17 kcal/mol at 60° and zero GdmCl, involves the disruption of four Cys NH hydrogen bonds that group at the crown of the molecule where a functional metal atom is held. The bulk of the protein unfolds with the somewhat lower free energy of 14 kcal/mol (zero GdmCl, 60°).

HX rate is also accelerated by increased temperature. The intrinsic k_{ch} increases about 10-fold for each 20° to 25° rise in addition to the effect of temperature on K_{op}. Figure 3 shows temperature-dependent results for rubredoxin. NMR experiments up to 120° maintained exchange in the EX2 regime so that the stabilization free energy could be computed.

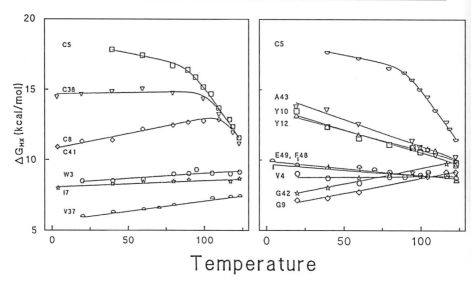

FIG. 3. HX results as a function of temperature for rubredoxin from *Pyrococcus furiosus*. Reprinted with permission from R. Hiller, Z. H. Zhou, M. W. W. Adams, and S. W. Englander, *Proc. Natl. Acad. Sci. U.S.A.* **94**, 11329 (1997).

The slope of the temperature-dependent curve gives the entropy for the determining unfolding reaction ($d\Delta G/dT = -\Delta S$). From ΔG and ΔS, the enthalpy of the unfolding reaction can be directly computed ($\Delta G = \Delta H - T\Delta S$). The melting temperature can be estimated by extrapolating the plot of ΔG vs T to zero ΔG (see Fig. 4). Thus HX measured as a function of temperature well below the T_m can provide all of the usual thermodynamic parameters. In addition, the denaturant dependence gives the structural parameter, m.

The behavior of the global unfolding reaction, taken from the Cys-5 behavior in Fig. 3, is shown in Fig. 4 and compared to two unusually stable mesophilic proteins of almost identical size. The hyperthermophilic rubredoxin attains an extraordinarily high melting temperature, close to 200°. The comparison in Fig. 4 shows that this is accomplished by an increase in unfolding free energy and perhaps also by a shift of the stability curve to higher temperature. The latter point, however, is somewhat ambiguous because of the difficulty of determining the true maximum of the stability curve.

Conformational Flexibility

The relationship of protein flexibility to stability and function has a long history.[17-19] The operational meaning of this term is ambiguous and its measurement is uncertain. How does measured "dynamics" relate to "flexibility"? Various aspects of the dynamic character of protein structure have been approached by

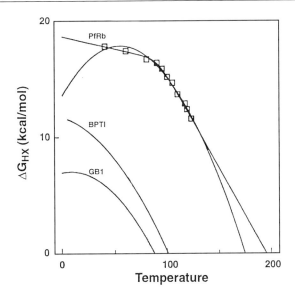

FIG. 4. Free energy of the global unfolding reaction, taken from the Cys-5 behavior in Fig. 3, compared to the analogous stability curves for two other unusually stable mesophilic proteins of almost identical size. The different curves through the rubredoxin (Rb) data indicate the uncertainty of extrapolation. The comparison can be viewed in the light of the type curves in Fig. 1. Reprinted with permission from R. Hiller, Z. H. Zhou, M. W. W. Adams, and S. W. Englander, *Proc. Natl. Acad. Sci. U.S.A.* **94,** 11329 (1997).

X-ray diffraction, nuclear magnetic resonance, and theoretical methods, as well as by hydrogen exchange. It seems likely that these different approaches access different motions that occur on very different time scales.

Thermophilic proteins may provide a good testing ground for the functional significance of protein flexibility and dynamics. The Petsko laboratory has studied a thermophilic enzyme that essentially loses its activity at lower temperature where its closely related mesophilic counterpart is maximally active.[20,21] They present data that suggests some general relationship between the flexibility or rigidity of a protein, measured by HX, and its functional temperature range.[22]

HX measurements access two kinds of dynamically occurring distortions that are much smaller than the global unfolding discussed above. Very local fluctuations

---

[17] K. U. Linderstrom-Lang and J. A. Schellman, *in* "The Enzymes" (P. D. Boyer, H. Lardy, and K. Myrback, eds.), pp. 443–510. Academic Press, New York, 1959.
[18] D. E. Koshland, G. Nemethy, and D. Filmer, *Biochemistry* **5,** 365 (1966).
[19] J. Monod, J. Wyman, and J. P. Changeaux, *J. Mol. Biol.* **12,** 88 (1965).
[20] A. Wrba, A. Schweiger, V. Schultes, R. Jaenicke, and P. Zavodsky, *Biochemistry* **29,** 7584 (1990).
[21] K. Hecht, A. Wrba, and R. Jaenicke, *Eur. J. Biochem.* **183,** 69 (1989).
[22] P. Zavodsky, J. Kardos, A. Svingor, and G. A. Petsko, *Proc. Natl. Acad. Sci. USA.* **95,** 7406 (1998).

can sever as few as one hydrogen bond at a time. The detailed motions involved are unknown and may represent, for example, some main chain twisting or bending that allows H-bond breakage and solvent access. It appears that the transient fluctuation necessary to bring an exchangeable hydrogen into fruitful H-bonding contact with solvent catalyst may require a fairly large separation of the protecting H-bond, perhaps by 4 Å or more.[23] Some hydrogens exchange by more extensive unfolding reactions involving entire secondary structural units, separately or together, mounting up to ultimate global unfolding.[24-28]

In applying HX methods to the flexibility problem, it will be necessary to distinguish the kind of underlying motions at work. Figures 2 and 3 show that the different kinds of fluctuations can be distinguished by their differing sensitivities to increasing denaturant and temperature. The slope of the free energy vs denaturant curve proportions to the amount of surface that is exposed in the opening reaction, indexed by the m parameter in Eq. (6). The slope of the free energy vs temperature curve gives the entropy of the responsible opening reaction. These parameters give information on the size of the structural distortion being measured in any particular case. In addition, it has been observed that neighboring hydrogens that exchange by way of small fluctuations by these criteria tend to exchange at very different rates. Neighboring hydrogens that appear to exchange by larger unfoldings should exchange with the same computed opening free energy, i.e., by way of the same larger scale opening reaction.

Summary

The naturally occurring hydrogen exchange of protein molecules can provide nonperturbing site-resolved measurements of protein stability and flexibility and changes therein. The measurement and understanding of these issues is especially pertinent to studies of thermophilic proteins. This chapter briefly reviews the considerations necessary for measuring hydrogen exchange and translating HX measurements into these detailed protein parameters.

Acknowledgment

This work was supported by a research grant from the National Institutes of Health USA.

---

[23] J. S. Milne, L. Mayne, H. Roder, A. J. Wand, and S. W. Englander, *Protein Sci.* **7,** 739 (1998).
[24] Y. Bai, T. R. Sosnick, L. Mayne, and S. W. Englander, *Science* **269,** 192 (1995).
[25] A. K. Chamberlain, T. M. Handel, and S. Marqusee, *Nature Struct. Biol.* **3,** 782 (1996).
[26] E. J. Fuentes and A. J. Wand, *Biochemistry* **37,** 3687 (1998).
[27] E. J. Fuentes and A. J. Wand, *Biochemistry* **37,** 3687 (1998).
[28] R. Hiller, Z. H. Zhou, M. W. W. Adams, and S. W. Englander, *Proc. Natl. Acad. Sci. U.S.A.* **94,** 11329 (1997).

[30] Nuclear Magnetic Resonance Analysis of Hyperthermophile Ferredoxins

By GERD N. LA MAR

I. Introduction

The single cubane cluster ferredoxins,[1] Fds, contain a [4Fe:4S] cluster typically coordinated by four cysteinyl residues. The Fds from three hyperthermophiles, the bacterium *Thermotoga maritima*,[2,3] *Tm*, which thrives optimally at 80°, and the archaea *Thermococcus litoralis*,[4] *Tl*, and *Pyrococcus furiosus*,[5] *Pf*, which thrive optimally at 90° and 100°, respectively, have been investigated by nuclear magnetic resonance (NMR). In each case, the purified Fds have been found to be hyperthermostable, exhibiting undetectable denaturation after 24 hr at 95°. As the sequence comparison shows in Fig. 1, they (in particular *Tm* and *Tl* Fd) exhibit significant homology to those of crystallographically characterized single cubane cluster Fds from the mesophiles *Desulfovibrio gigas*,[6] *Dg*, and *Desulfovibrio africanus*,[7] *Da*, and the thermophile *Bacillus thermoproteolyticus*,[8] *Bt* (sequence not shown for *Da* and *Bt* Fd). The *Pf* Fd exhibits somewhat less sequence homology to Fds from either mesophiles or the other hyperthermophiles. Alignments in Fig. 1 are based on the four ligands in the cluster-ligating consensus sequence, labeled **I–IV,** and occupied by all-Cys in both *Tm* and *Tl* Fds, as well as in *Bt* and *Da* Fds,[7,8] whereas *Pf* Fd has an Asp[9] at position **II**. These Fds often possess two additional Cys, designated Cys(**V**) and Cys(**VI**) in Fig. 1 (present in *Dg, Tm, Tl,* and *Pf* Fds, but absent in *Da* and *Bt* Fds), which can participate in a disulfide bond,[10–12] as shown schematically in Fig. 2 for *Pf* 4Fe Fd.

---

[1] R. Cammack, *Adv. Inorg. Chem.* **38,** 281 (1992).
[2] J. M. Blamey, S. Mukund, and M. W. W. Adams, *FEMS Microbiol. Lett.* **121,** 165 (1994).
[3] W. Pfeil, U. Gesierich, G. R. Kleemann, and R. Sterner, *J. Mol. Biol.* **272,** 591 (1997).
[4] S. Mukund and M. W. W. Adams, *J. Biol. Chem.* **268,** 13592 (1993).
[5] R. C. Conover, A. T. Kowal, W. Fu, J.-B. Park, S. Aono, M. W. W. Adams, and M. K. Johnson, *J. Biol. Chem.* **265,** 8533 (1990).
[6] C. R. Kissinger, L. C. Sieker, E. T. Adman, and L. H. Jensen, *J. Mol. Biol.* **219,** 693 (1991).
[7] A. Séry, D. Housset, L. Serre, J. Bonicel, C. Hatchikian, M. Frey, and M. Roth, *Biochemistry* **33,** 15408 (1994).
[8] K. Fukuyama, H. Matsubara, T. Tsukihara, and Y. Katsube, *J. Mol. Biol.* **210,** 383 (1989).
[9] L. Calzolai, C. M. Gorst, Z.-H. Zhao, Q. Teng, M. W. W. Adams, and G. N. La Mar, *Biochemistry* **34,** 11373 (1995).
[10] C. M. Gorst, Z.-H. Zhou, K. Ma, Q. Teng, J. B. Howard, M. W. W. Adams, and G. N. La Mar, *Biochemistry* **34,** 8788 (1995).
[11] H. Sticht, G. Wildegger, D. Bentrop, B. Darimont, R. Sterner, and P. Rösch, *Eur. J. Biochem.* **237,** 726 (1996).
[12] S. Macedo-Ribeiro, B. Darimont, R. Sterner, and R. Huber, *Structure* **4,** 1291 (1996).

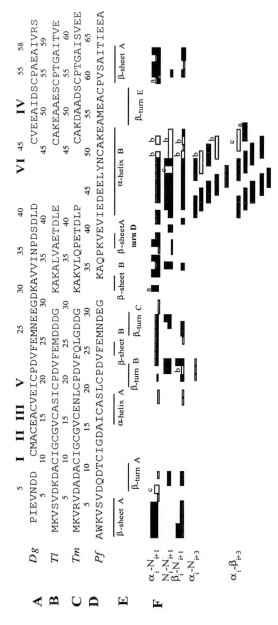

FIG. 1. Sequence alignment for structurally characterized cubane Fds with disulfide bonds from: (A) mesophilic *Desulfovibrio gigas*, *Dg*; (B) hyperthermophilic *Thermococcus litoralis*, *Tl*; (C) hyperthermophilic *Thermatoga maritima*, *Tm*; (D) hyperthermophilic *Pyrococcus furiosus*, *Pf*. The sequences are aligned on the basis of conserved ligands in the cluster consensus ligating sequence, **I–IV**, the disulfide bond Cys(**V**), Cys(**VI**), and tertiary contacts established herein. (E) Secondary structural elements identified for *Pf* Fd and also observed in the other hyperthermostable Fds. The *Dg* and *Tm* Fds have a β turn in place of the third strand of β sheet A. (F) Sequential and medium-range NOESY cross peaks[24] for: 4Fe Fd$_A^{ox}$ (with intact disulfide); (a) observed only in 50 msec mixing time NOESY spectrum, (b) too broadened by dynamic disulfide orientational heterogeneity to detect; (c) cross peak not detected due to near-degeneracy, position under solvent line, or paramagnetic influences. [Reprinted from P. L. Wang, L. Calzolai, K. L. Bren, Q. Teng, F. E. Jenney, Jr., P. S. Brereton, J. B. Howard, M. W. W. Adams and G. N. La Mar, *Biochemistry* **38**, 8167 (1999), with permission.]

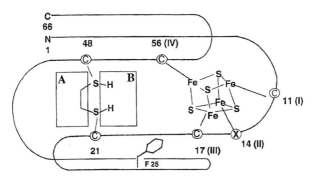

FIG. 2. Schematic representation of the molecular structure of a cubane single-cluster Fd. The cluster ligands are labeled by both the conserved ligation sequence **I–IV** and the sequence position in *Pf* Fd, 11, 14, 17 and 56, with position 14(**II**) occupied by Asp in WT *Pf* Fd. The alternate oxidation states of the two Cys-21(**V**), Cys-48(**VI**) remote from the cluster, two free sulfhydryl and a disulfide bond, are designated by the B and A form, respectively. [Reprinted from L. Calzolai, C. M. Gorst, K. L. Bren, Z.-H. Zhou, M. W. W. Adam, and G. N. La Mar, *J. Am. Chem. Soc.* **119,** 9341 (1997), with permission.]

II. General Properties Relevant to NMR

NMR can provide a detailed molecular model (structure) for the complete protein comparable to that obtained[13–15] by X-ray crystallography and can serve as a unique probe of important aspects of the electronic structure and magnetic properties of the invariably paramagnetic cluster,[16–18] as well as allow the characterization of a variety of relevant dynamic properties of either the cluster or the protein as a whole. Hyperthermostable Fds do not pose any experimental problems to the profitable application of NMR to study molecular and electronic structure of dynamic properties. Rather, they provide significant advantages in that molecular and electronic structures can be studied over an extremely wide temperature range and in the presence of organic solvents that generally denature Fds from mesophiles. This wide temperature range, moreover, broadens the range of rate processes that can be studied and facilitates more quantitative description of molecular conformational equilibria. The wide temperature range provides exceptional opportunities for determining if, and how, molecular structure differs at the temperature extremes.

---

[13] K. Wüthrich, "NMR of Proteins and Nucleic Acids." Wiley & Sons, New York, 1986.
[14] J. Cavanagh, W. J. Fairbrother, and A. G. Palmer III, and N. J. Skelton, "Protein NMR Spectroscopy." Academic Press, New York, 1996.
[15] I. Bertini, C. Luchinat, and A. Rosato, *Prog. Biophys. Mol. Biol.* **66,** 43 (1996).
[16] H. Cheng and J. L. Markley, *Annu. Rev. Biophys. Biomol. Struct.* **24,** 209 (1995).
[17] I. Bertini, S. Ciurli, and C. Luchinat, *Struct. Bond.* **83,** 2 (1995).
[18] B. J. Goodfellow and A. L. Macedo, *Annu. Repts. NMR Spectros.* **37,** 119 (1999).

III. Unusual Properties of *Pyrococcus furiosus* Fd

The *Tm* and *Tl* 4Fe Fds possess all-Cys ligation of the cubane cluster (Fig. 1) and a stable disulfide bond that is not cleaved by dithionite, such that their novel properties appear to be restricted to their hyperthermostability.[11,12,19] The *Pf* Fd deserves special attention in that it exhibits numerous unusual properties even among the other hyperthermostable Fds. In addition to the unusual Asp-14 at position **II**,[20] *Pf* Fd exhibits a predominant $S = 3/2$ ground state in its reduced form[5,21] at *cryogenic temperatures* rather than the conventional $S = 1/2$ ground state for all-Cys ligated clusters,[1] ligation of one reduced cluster position by exogenous ligand,[5,22,23] a facile 3Fe $\rightleftharpoons$ 4Fe cluster interconversion,[5] a redox-active disulfide bridge,[10] extraordinarily slow electron self-exchange,[20] and an equilibrium disulfide orientational heterogeneity.[24] The latter seriously complicates the detailed description of the fascinating properties of this Fd, but at the same time provide numerous challenges that NMR has been able to master more effectively than other spectroscopic approaches. On the other hand, the slow formation and cleavage of the disulfide bridge enables two forms of the protein to be isolated independently of the cluster redox state. These are designated as the A and B forms (as in Fd$_A$ and Fd$_B$), which have the disulfide bridge and two free thiols, respectively.[10] These are discussed further in Section VII.

IV. Cluster Magnetic Properties

The electronic structure of the cubane clusters can be understood only on the basis of pairwise iron valence delocalization of solely high-spin ferrous ($S = 2$) and ferric ($S = 5/2$) ions. The four (or three) iron of the cluster, numbered 1–4, exhibit pairwise valence delocalization that results in a near "ferromagnetic" coupling for the pairs, $\vec{S}_A = \vec{S}_1 + \vec{S}_2$ and $\vec{S}_B = \vec{S}_3 + \vec{S}_4$, with the net cluster spin, resulting from antiferromagnetic coupling for the net spins for the two *pairs* of

---

[19] P. L. Wang, A. Donaire, Z. H. Zhou, M. W. W. Adams, and G. N. La Mar, *Biochemistry* **35**, 11319 (1996).

[20] L. Calzolai, Z. H. Zhou, M. W. W. Adams, and G. N. La Mar, *J. Am. Chem. Soc.* **118**, 2513 (1996).

[21] J. Telser, H. Huang, H.-I. Lee, M. W. W. Adams, and B. M. Hoffman, *J. Am. Chem. Soc.* **120**, 861 (1998).

[22] L. Calzolai, C. M. Gorst, K. L. Bren, Z. H. Zhou, M. W. W. Adams, and G. N. La Mar, *J. Am. Chem. Soc.* **119**, 9341 (1997).

[23] J. Telser, E. T. Smith, M. W. W. Adams, R. C. Conover, M. K. Johnson, and B. M. Hoffman, *J. Am. Chem. Soc.* **117**, 5133 (1995).

[24] P.-L. Wang, L. Calzolai, K. L. Bren, Q. Teng, F. E. Jenney, Jr., P. S. Brereton, J. B. Howard, M. W. W. Adams, and G. N. La Mar, *Biochemistry* **38**, 8167 (1999).

iron $S_T = \vec{S}_A + \vec{S}_B$. We provide here only a brief qualitative description of the spin coupling. Detailed descriptions have been summarized.[17,25,26]

A. Four-Iron Clusters

The four-iron oxidized cluster, 4Fe Fd^{ox}, is represented by [4Fe:4S]$^{2+}$, indicating the presence of 2Fe^{2+} and 2Fe^{3+}. However, Mössbauer spectroscopy shows that the four iron are oxidation-state equivalent,[27] such that a more appropriate representation is 2Fe$^{2.5+}$:2Fe$^{2.5+}$, where the cluster consists of two valence-delocalized iron pairs with individual iron properties intermediate between ferrous and ferric ions. The two spins (one $S = 5/2$ than other $S = 2$) in each 2Fe$^{2.5+}$ pair[17,25,26] couple to lead to $S_A = S_B = 9/2$. The antiferromagnetic coupling of $\vec{S}_A$ and $\vec{S}_B$ leads to $\vec{S}_T = 0$ ground state, but the excited $S = 1, 2, 3, \ldots$, states are appreciably populated at ambient temperature. The reduced four-iron cluster, 4Fe Fd^{red}, is represented by [4Fe:4S]$^{1+}$, and reflects the presence of 1Fe^{3+} and 3Fe^{2+}. However, Mössbauer spectroscopy[27] shows that a more appropriate description is 2Fe$^{2.5+}$:2Fe$^{2.0+}$, where one of the valence-delocalized iron pair is reduced by one electron relative to that 4Fe Fd^{ox}. The diferrous pair has $\vec{S}_B = \vec{S}_3 + \vec{S}_4 = 4$, such that the antiferromagnetic coupling[17,25,26] between $\vec{S}_A = 9/2$ and $\vec{S}_B = 4$ generally yields a $S_T = 1/2$ ground state (see, however, evidence[5,21] for $S = 3/2$ ground state for *Pf* 4Fe Fd^{red} at 4 K).

B. Three-Iron Clusters

The 4Fe cluster in the 4Fe Fds can be converted *in vitro* by ferricyanide oxidation to a cubane [3Fe:4S] cluster by the loss of one iron atom. The oxidized cluster is represented by [3Fe:4S]$^{1+}$, which indicates that all three iron are ferric. The observed total $S_T = 1/2$ has been rationalized[28] on the basis of two of the ferric ion coupling to yield $\vec{S}_B = \vec{S}_2 + \vec{S}_2 = 2$, which couples antiferromagnetically with the remaining ferric ion $\vec{S}_A = 5/2$ to yield $\vec{S}_T = 1/2$. The reduced three iron cluster, [3Fe:4S]^0, formally consists of 2Fe^{3+} and 1Fe^{2+}, but Mössbauer has shown[29] that, in fact, the appropriate description is 2Fe$^{2.5+}$:1Fe$^{3.0+}$, i.e., a valence-delocalized pair and a ferric ion. The 2Fe$^{2.5+}$, as in the case of the 4Fe cluster, has

---

[25] L. Noodleman, T. L. Chen, D. A. Case, C. Giori, G. Rius, J.-M. Mouesca, and B. Lamotte, *in* "Nuclear Magnetic Resonance of Paramagnetic Macromolecules" (G. N. La Mar, ed.), pp. 339–367. Kluwer Press, Dordrecht, 1995.
[26] L. Noodleman, C. Y. Peng, J.-M. Mouesca, and D. A. Case, *Coord. Chem. Rev.* **144**, 199 (1995).
[27] P. Middleton, D. P. E. Dickson, C. E. Johnson, and J. D. Rush, *Eur. J. Biochem.* **88**, 135 (1978).
[28] T. A. Kent, B. H. Huynh, and E. Münck, *Proc. Natl. Acad. Sci. U.S.A.* **77**, 6574 (1980).
[29] E. Münck, V. Papaefthymiou, K. K. Surerus, and J. J. Girerd, *in* "Metal Clusters in Proteins" (L. Que, ed.), pp. 302–325. ACS, Washington, D.C., 1988.

$\vec{S}_A = \vec{S}_1 + \vec{S}_2 = 9/2$, which couples antiferromagnetically to the ferric ion with $\vec{S}_B = 5/2$ to yield $\vec{S}_T = 2$.

In the above descriptions, it is noted that the pair of iron (or single iron) that is described by $\vec{S}_A$ has a *magnetic moment larger* than that of the other pair of iron (or single iron) described by $\vec{S}_B$. When in the ground state, the total spin $\vec{S}_T$ is aligned parallel to the applied magnetic field B_0, $\vec{S}_A$ is *necessarily aligned parallel* with B_0, while $\vec{S}_B$ is *necessarily aligned antiparallel* to the field.[16–18] The location of the unique valence-delocalized pair in 4Fe Fd^{red} or 3Fe Fd^{red} reflects the different micropotential of the iron pairs. The investigation of the structural bases of both the micropotentials that select among the pair of irons that become reduced, as well as the variation in overall cluster reduction potential, is currently a very active research area.[30]

V. NMR Spectral Parameters

Because all cubane clusters exist in either paramagnetic ground states (3Fe Fd^{ox}, 3Fe Fd^{red}, 4Fe Fd^{red}), or significantly populate paramagnetic excited states (4Fe Fd^{ox}), the effect of paramagnetism on the NMR spectral parameters must be considered.[16,18,31–35]

A. Relaxation

The cluster paramagnetism leads to increased relaxation which, in the limit of the point-dipole approximation,[36] can be described by the T_1 for nucleus i at a distance R_{q-i} from iron q in a cluster of n iron, as[19,31,37]

$$T_{1i}^{-1} = \sum_{q=1}^{n} D_q R_{q-i}^{-6} \qquad (1)$$

where D is a constant that is proportional to the spin-magnetization on the Fe_q, $\langle S_z \rangle_q$, and the electron spin relaxation time, τ_e. For 4Fe Fd^{ox}, D_q is identical for the four iron; however, for the other three cluster states, the D_q depends whether the iron is on a valence-delocalized pair or not. The nature of the constant D is

---

[30] P. J. Stephens, D. R. Jollie, and A. Warshel, *Chem. Rev.* **96**, 2491 (1996).
[31] L. Banci, I. Bertini, and C. Luchinat, "Nuclear and Electron Relaxation." VCH, Weinheim, 1991.
[32] A. Xavier, D. Turner, and H. Santos, *Methods in Enzymology* **227**, 1 (1993).
[33] I. Bertini, P. Turano, and A. J. Vila, *Chem. Rev.* **93**, 2833 (1993).
[34] L. Banci, I. Bertini, and C. Luchinat, *Methods in Enzymology* **239**, 485 (1994).
[35] I. Bertini and C. Luchinat, *Coord. Chem. Rev.* **150**, 1 (1996).
[36] S. J. Wilkens, B. Xia, B. F. Volkman, F. Weinhold, J. L. Markley, and W. M. Westler, *J. Phys. Chem. B* **102**, 8300 (1998).
[37] I. Bertini, M. M. J. Couture, A. Donaire, L. D. Eltis, I. C. Felli, C. Luchinat, M. Piccioli, and A. Rosato, *Eur. J. Biochem.* **241**, 440 (1996).

such that T_1 values are typically 3 ms for protons ~ 4 Å from the closest iron, while the paramagnetic effect is largely suppressed to $T_{1e} \sim 100$ ms when the closest $R_{q-i} \sim 6$ Å and is completely negligible for proton $R_{q-i} > 8$ Å.

B. Hyperfine Shifts

The observed shift, referenced to 2,2-dimethyl-2-silapentane 5-sulfonate (DSS), δ_{DSS^i}(obs), for nucleus i is given by:[35]

$$\delta^i_{DSS}(obs) = \delta^i_{DSS}(dia) + \delta^i_{hf} \tag{2}$$

where δ_{DSS^i}(dia) is the shift for the same nucleus in an *isostructural diamagnetic complex*, and the hyperfine shift, δ_{hf}, is given by the sum of the contact and dipolar shifts, i.e.,

$$\delta^i_{hf} = \delta^i_{con} + \delta^i_{dip} \tag{3}$$

The dipolar shift results from a through-space interaction with a *magnetically anisotropic* cluster magnetic moment and is given by:

$$\delta_{dip^i} = (24\pi)^{-1}\left[2\Delta\chi_{ax}(\cos^2\theta_i - 1)R_i^{-3} + 3\Delta\chi_{rh}(\sin^2\theta_i \cos 2\Omega_i)R_2^{-3}\right] \tag{4}$$

where R, θ, Ω are the polar coordinates of nucleus i in the *magnetic coordinate system* where the *anisotropic* paramagnetic susceptibility tensor, χ, is diagonal and $\Delta\chi_{ax}$ and $\Delta\chi_{rh}$ are the axial and rhombic anisotropies of the tensor. In the case of cubane Fd, there is no evidence[16–18] for significant magnetic anisotropy in any of the three NMR characterized clusters forms 3Fe Fd^{ox}, 4Fe Fd^{ox}, and 4Fe Fd^{red}, and therefore δ_{dip} may be ignored for the present considerations.

Hence only the contact shift is important in interpreting the ^{1}H NMR spectra of cubane cluster Fd. The contact shift for nucleus i, ligated to iron q, can be written in general as:[17,35]

$$\delta_{con^i} = (A^i/\hbar)B_o\langle S\rangle_q \tag{5}$$

where $A^i/\hbar$ is the contact coupling constant that is related to the amount of delocalized spin density, ρ_i, on nucleus i and $\langle S_z\rangle_q$ is the spin-magnetization of iron q to which the ligand with nucleus i is ligated; $\langle S_z\rangle_q$ exhibits a sign that indicates whether the spin is aligned parallel or antiparallel to the applied field and is strongly temperature-dependent. The result of the paramagnetism on the ^{1}H NMR spectra of Fds in general is to yield a number of strongly relaxed single proton resonances outside the 0–10 ppm where "diamagnetic" proton signals resonate, and whose chemical shifts are strongly temperature dependent.[16–18,33] Since spin delocalization can occur only for ligands coordinated to the iron, the resonances that are resolved, strongly relaxed, and whose shifts are strongly temperature-dependent necessarily originate from the nuclei of the cluster ligands. The ^{1}H NMR spectra

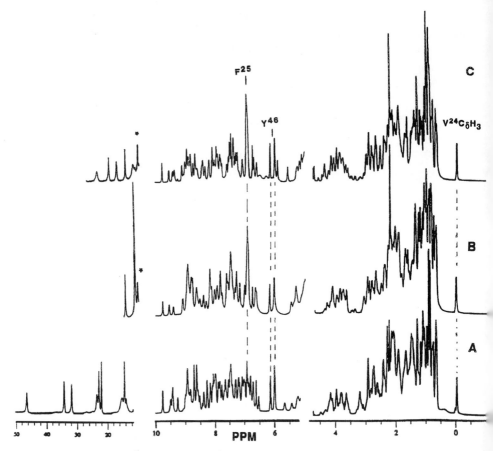

FIG. 3. 500-MHz ^{1}H NMR spectra in ^{1}H$_2$O, pH 8.0, at 30°C of (A) as-isolated, reduced Pf 4Fe Fd with cleaved disulfide, designated Pf 4Fe Fd$_B^{red}$, (B) the oxidized form with intact disulfide, designated Pf 4Fe Fd$_A^{ox}$, and (C) the oxidized form of Pf 3Fe Fd with intact disulfide, designated Pf 3Fe Fd$_A^{ox}$, a$_i^{ox}$. Labile proton peaks are labeled by asterisks. The regions 50–10 ppm and 10–5 ppm are expanded vertically by factors of 3 and 2, respectively, when compared with the 5–0 ppm window. [Reprinted from C. M. Gorst, Z. H. Zhou, K. Ma, Q. Teng, J. B. Howard, M. W. W. Adams, and G. N. La Mar, *Biochemistry* **34**, 8788 (1995), with permission.]

of Pf 3Fe Fd^{ox}, 4Fe Fd$_A^{ox}$, and 4Fe Fd$_B^{red}$, the only three cubane cluster oxidation states characterized in detail to date, are illustrated[10] in Figs. 3A–C, respectively. It is noted that the 0–10 ppm "diamagnetic envelope" is essentially identical for the three forms, whereas the 0–50 ppm "hyperfine shifted window," which contains the majority of the cluster ligand signals, differs dramatically and highly characteristically for the three cluster states.[16–18,33]

The hyperfine coupling constant, A^i/h, possesses valuable structural information in the dependence of A^i/h for individual Cys $C_\beta Hs$ on the Fe–S–C_β–H dihedral angle,[17,25,38–40] ϕ, i.e.,

$$(A^i/h) = Q f(\phi) \tag{6}$$

where Q is a constant and $f(\phi)$ was initially approximated[38] by $\cos^2 \phi$, although more complex relationships have been deduced[40] by comparing the observed δ_{con} to the experimental ϕ in crystallographyically characterized Fds. For *a single paramagnetic iron* such as a rebredoxin, the spin magnetization can be approximated by:

$$\langle S_z \rangle = -\frac{S(S+1)}{3kT} B_o \tag{7}$$

such that low-field δ_{con} result that exhibit the Curie law (i.e., δ_{con} decreases with increasing temperature, $\delta_{con} \propto 1/T$).

C. Temperature Behavior of Cluster Ligand Contact Shifts

Compared to a single iron atom, a more complex pattern of temperature dependence is expected in spin-coupled systems[17,25,26,41] that reflects the population of different excited spin states and the position of the individual iron in the spin-coupling hierarchy qualitatively described in Section IV. The simplest state to interpret is the 4Fe Fd^{ox}, where the four iron are oxidation-state equivalent and the ground state has $\vec{S} = 0$ (i.e., $\delta_{con} = 0$ in the ground state). Hence, δ_{con} becomes nonzero as excited spin states are populated, with the result that all nuclei will exhibit anti-Curie behavior[16–18,25,33] (i.e., δ_{con} increases with increasing temperature), as shown[42] for *Tl* 4Fe Fd^{ox} in Fig. 4A.

The other three cluster states, 4Fe Fd^{ox}, 3Fe Fd^{ox}, and 3Fe Fd^{red}, exhibit much more interesting and informative temperature behavior for the δ^i_{con}, and the relationship of this temperature dependence and the electronic structure of the individual iron can be represented qualitatively[43] in Fig. 5. Equation (5) shows that the *direction* of the contact shifts (downfield or upfield) for a cluster ligand depends on the *sign* of $\langle S_z \rangle_q$. Thus, in the *low-temperature limit* only the ground state is populated (i.e., region 3 in Fig. 5), S_T, and hence the larger $\vec{S}_A$ (in a cluster coupled to a

---

[38] M. Poe, W. D. Phillips, C. C. McDonald, and W. Lovenberg, *Proc. Natl. Acad. Sci. U.S.A.* **65**, 797 (1970).

[39] S. C. Busse, G. N. La Mar, and J. B. Howard, *J. Biol. Chem.* **266**, 23714 (1991).

[40] I. Bertini, F. Capozzi, C. Luchinat, M. Piccioli, and A. J. Vila, *J. Am. Chem. Soc.* **116**, 651 (1994).

[41] W. R. Dunham, G. Palmer, R. H. Sands, and A. J. Bearden, *Biochim. Biophys. Acta* **253**, 373 (1971).

[42] A. Donaire, C. M. Gorst, Z. H. Zhou, M. W. W. Adams, and G. N. La Mar, *J. Am. Chem. Soc.* **116**, 6841 (1994).

[43] S. C. Busse, G. N. La Mar, L. P. Yu, J. B. Howard, E. T. Smith, Z. H. Zhou, and M. W. W. Adams, *Biochemistry* **31**, 11952 (1992).

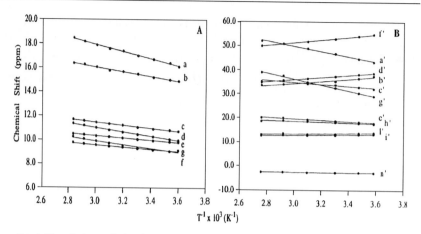

FIG. 4. Plot of observed chemical shifts relative to DSS versus reciprocal absolute temperature (Curie plot) for the assigned resolved contact shifted Cys resonances from (A) Tl 4Fe Fd^{ox} and (B) Tl 4Fe Fd^{red}. Resonances are assigned[42] as d, f, n and d', f', n' for Cys10(**I**); a, g, e and a', g', e' for Cys13(**II**); b, l, i and b', l', i' for Cys16(**III**); and c, h, m and c', h', m' for Cys51(**IV**). [Reprinted from A. Donaire, C. M. Gorst, Z. H. Zhou, M. W. W. Adams, and G. N. La Mar, *J. Am. Chem. Soc.* **116,** 6841 (1994), with permission.]

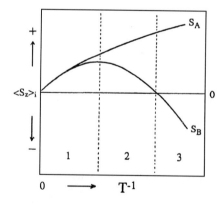

FIG. 5. Qualitative schematic representation of the temperature dependence with spin magnetization (S_z) for the two local or intermediate spin states S_A and S_B, which couple to yield a total cluster spin state S_T. The qualitative behavior shown is that for $|S_A| > |S_B|$. The three regions are (1) high temperature limit, where both $\langle S_z \rangle_A$ and $\langle S_z \rangle_B$ exhibit Curie-like behavior with the same sign, (2) intermediate temperature where $\langle S_z \rangle_A$ exhibits Curie-like and $\langle S_z \rangle_B$ anti-Curie behavior, both with the same sign, and (3) low temperature limit where $\langle S_z \rangle_A$ exhibits Curie-like behavior and $\langle S_z \rangle_B$ exhibits Curie-like behavior but with opposite signs from that of $\langle S_z \rangle_A$. The observed behavior for both 3Fe Fd^{ox} and 4Fe Fd^{red} is that shown for region (2). [Reprinted from S. C. Busse, G. N. La Mar, L. P. Yu, J. B. Howard, E. T. Smith, Z. H. Zhou, and M. W. W. Adams, *Biochemistry* **31,** 11952 (1992), with permission.]

smaller $\vec{S}_B$) is necessarily aligned parallel to the applied magnetic field (positive $\langle S_z \rangle$) and hence must experience low-field δ_{con}. The smaller $\vec{S}_B$, since it is antiferromagnetically coupled to the larger $\vec{S}_A$, must therefore necessarily be aligned antiparallel to the field (negative S_z) and hence exhibit *upfield* δ_{con}. In the *high temperature limit*, the $\vec{S}_A$ and $\vec{S}_B$ are largely "uncoupled," and both show some form of "Curie" behavior with positive $\langle S_z \rangle$ and low-field shifts, as shown in region 1 in Fig. 5.

The logical connection of the extreme low and high temperature limits in Fig. 5, region 2, which is where the NMR experiments are generally carried out, thus predicts Curie-type behavior for the nuclei on cluster ligands to the iron pair (or single iron) with the *larger* $\vec{S}_A$ (i.e., $2Fe^{2.5+}$ in 4Fe Fd^{red} and 3Fe Fd^{red}, or the single ferric iron in 3Fe Fd^{ox}). Conversely, the nuclei ligated to the iron pair with the smaller $\vec{S}_B$ will exhibit either low-field, anti-Curie or high-field, pseudo-Curie behavior.[16–18,33] The high-field pseudo-Curie behavior for the valence-delocalized iron pair is observed in the reduced high potential iron sulfur proteins, HIPIP,[16–18,33] whereas only the anti-Curie behavior is observed in Fd^{red}. Indeed, the predicted Curie and anti-Curie temperature behavior δ_{con} for cluster ligand to the valence-delocalized $2Fe^{2.5+}$ and diferrous pair, $2Fe^{2.0+}$, respectively, in reduced 4Fe cubane cluster Fd has been observed for each NMR characterized Fd with all-Cys ligation, as illustrated[42] by the data for *Tl* 4Fe Fd^{red} in Fig. 4B. One of the major contributions of NMR spectroscopy to the understanding of the electronic structure of iron–sulfur clusters is that this technique is unique in its ability to identify the oxidation state of an iron ligated to a particular cluster ligand in the sequence. However, although it has been possible to identify the cluster ligands to the valence-delocalized pair in 4Fe Fd^{red} in numerous cases, the structural basis for the preferences for the reducing electron to be shared by a particular iron pair is not yet understood.

VI. NMR Experiments

A. *Sample Conditions*

The sample size requirement for NMR depends on the instrumentation available. High magnetic fields such as those on 600–800 MHz spectrometers provides optimal resolution and sensitivity and allow effective 2D NMR molecular structural characterization of samples as small as ~ 1 mM Fd in 0.5 ml in ^1H_2O. The relatively small size of Fds, however, do not present serious resolution problems so that similarly effective ^1H 2D NMR studies can be carried out on lower-field spectrometers, such as at 300–400 MHz, provided a more concentrated sample, i.e., $\sim 4–6$ mM in 0.5 ml H_2O, is available. The hyperthermostability of Fds is reflected in the very slow exchange of peptide protons with bulk water, such that the crucial sequence-specific assignments can be carried out at much more alkaline pH (7–8) than for other such small proteins. As is the case for ^{1}H NMR studies of any protein, buffers must be deuterated to avoid damaging artifacts in the 2D NMR spectra; the ideal buffer is phosphate. Whereas the sequence-specific assignment of nonligated residues must be carried out in ^{1}H$_2$O, the assignment of the strongly relaxed and hyperfine shifted cluster ligand signals are optimally assigned in ^{2}H$_2$O solution.

A typical ∼3–4 mM 0.50 ml sample on a modern 500 MHz spectrometer may require 6–12 hr for an informative TOCSY or NOESY map. Measurement of steady-state NOEs generally require less than 1 hr of spectrometer time. Considering the range of relaxation times that are encountered and the need for optimizing parameters for each range of T_1 values, between 6 and 10 2D maps are required to achieve a reasonably robust structure by 2D NMR, while the sequence-specific assignment of the cluster ligand could be achieved on the basis of as few as 3 or 4 2D maps.

B. Reference Spectra

The experiments needed to assign the resonances, determine the molecular structure, and characterize the cluster electronic structure are the same as those employed for paramagnetic metalloproteins in general,[35,44] which, in turn, are largely the same as those employed for comparable diamagnetic proteins,[14] with the exception that faster repetition rates and much shorter mixing times are necessary to characterize the cluster environment. Since nuclear relaxation times can vary from 1 ms to several seconds in a cubane Fd,[16–18,33] a range of experimental conditions are needed to optimally characterize the complete protein. Moreover, since strongly relaxed peaks have much less intensity than weakly or inconsequentially relaxed peaks, the former are frequently not detectable in a "normal" spectrum. The relaxed peaks, however, can be "emphasized" by recoding spectra with fast repetition rates and/or with a WEFT pulse sequence.[35,44,45] The dramatic "resolution" of strongly relaxed peaks within the diamagnetic envelope for Tl 4Fe Fd^{ox} is illustrated[42] in Fig. 6.

C. Relaxation Times

The T_1 values for resolved resonances are readily obtained by the standard inversion–recovery experiment. For nonresolved but paramagnetically influenced nuclei, a 2D TOCSY or COSY experiment within an inversion–recovery preparation pulse sequence will similarly allow reasonable T_1 estimates[46,47] for nonresolved, spin-coupled protons.

---

[44] G. N. La Mar and J. S. de Ropp, *Biol. Magn. Reson.* **18**, 1 (1993).
[45] R. K. Gupta, *J. Magn. Reson.* **24**, 461 (1976).
[46] A. S. Arseniev, A. G. Sobol, and V. F. Bystrov, *J. Magn. Reson.* **70**, 427 (1986).
[47] J. G. Huber, J.-M. Moulis, and J. Gaillard, *Biochemistry* **35**, 12705 (1996).

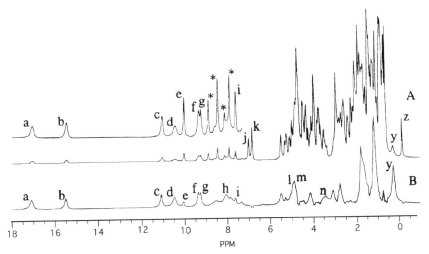

FIG. 6. (A) 500-MHz ^{1}H NMR spectrum of 4Fe Tl Fd^{ox} in ^{2}H$_2$O, pH 7.5, at 30°, collected under nonsaturating conditions.[42] Contact shifted Cys resonances are labeled $a–g$, the Phe 24 aromatic ring protons are labeled j, k, and two upfield methyls, one strongly relaxed and the other assigned to Val 23, are labeled y, z, respectively. Slowly exchanging peptide protons in the low-field window are marked by asterisks. (B) Super-WEFT trace of the same sample collected with a repetition rate of 20 s^{-1} and a delay time of 20 ms, which emphasizes very rapidly relaxed (T_1 < 10 ms) resonances in both the resolved ($a-i$, z) and unresolved (peaks l, m, n) spectral window. [Reprinted from A. Donaire, C. M. Gorst, Z. H. Zhou, M. W. W. Adams, and G. N. La Mar, *J. Am. Chem. Soc.* **116**, 6841 (1994), with permission.]

D. Spin Connectivity

^{1}H–^{1}H scalar correlation is optimally mapped by TOCSY, although magnitude COSY, MCOSY, has also been used successfully.[35,44] For the major "diamagnetic" portion of the protein (i.e., T_1 > 200 ms), 0.5 to 1.0 s^{-1} repetition rates and 150–250 ms mixing times are appropriate, as in comparably sized diamagnetic proteins.[14] At the other extreme are the cluster ligands with T_1 values of 2–25 ms, where minimally the geminal, and in some cases the vicinal, coupling can be detected in a TOCSY spectrum with 3–15 ms mixing time collected at 10 s^{-1}. The TOCSY connections[42] for the cluster ligands in Tl 4Fe Fd^{ox} are illustrated in Fig. 7. Scalar connection among protons with intermediate T_1 values of 30–100 ms require intermediate mixing times and repetition rates.[48]

Because of the small size of Fds, heteronuclear ^{15}N and/or ^{13}C labeling has not been necessary to achieve assignments. Hence, only limited heteronuclear labeling of hyperthermostable Fds has been reported[24] and consisted of

---

[48] C. M. Gorst, T. Yeh, Q. Teng, L. Calzolai, J. H. Zhou, M. W. W. Adams, and G. N. La Mar, *Biochemistry* **34**, 600 (1995).

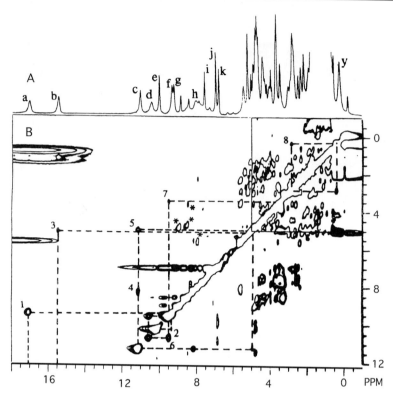

FIG. 7. (A) Reference trace for 4Fe *Tl* Fd^{ox} in ^{2}H$_2$O, pH 7.5, at 30°, collected at a repetition rate of 4.4 s^{-1} to emphasize relaxed resonances.[42] (B) 12.4 ms mixing time TOCSY spectrum illustrating the two cross peaks that correlate the scalar connectivities for the four-three-spin systems for the ligated cysteines. Cross peaks 1, 2 correlate *a*, *g*, and *e* [Cys-13(**II**)]; cross peak 3 correlates *b* and *l* [Cys-16(**III**)]; cross peaks 4, 5 correlate *c*, *h*, and *m* [Cys-51(**IV**)]; and cross peaks 6, 7 correlate *d*, *f*, and *n* [Cys-10(**I**)]. [Reprinted from A. Donaire, C. M. Gorst, Z. H. Zhou, M. W. W. Adams, and G. N. La Mar, *J. Am. Chem. Soc.* **116,** 6841 (1994), with permission.]

^{1}H-detected [^{1}H-^{15}N]-HSQC[14] in *Pf* 4Fd Fd$_A^{ox}$ where the peptide ^{15}N chemical shifts were reported for 44 of the 66 residues.

E. Spatial Proximity

Spatial or dipolar contacts for the resolved and strongly relaxed (such as cluster ligand) resonances are effectively pursued by steady-state NOEs,[35,44] η_{j-i}, to nuclei *i* upon saturating nucleus *j* according to:

$$\eta_{i \to j} = \sigma_{ij} T_{1i} \qquad (8)$$

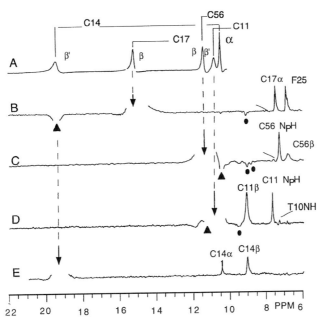

FIG. 8. (A) Low-field portion of the 500 MHz ^{1}H NMR spectrum of PfD14C-Fd$_A^{ox}$. (B)–(E): Steady-state NOE difference traces obtained by saturating the resonance indicated by the vertical arrow[22], the position of the irradiation in the reference trace is indicated by a solid triangle and NOEs due to the reference trace are marked with solid ovals; (B) saturate Cys-17 C$_\beta$H and observe NOEs to Cys-17 C$_\alpha$H and Phe-25 ring; (C) saturate Cys-56 C$_\beta$H and observe NOEs to Cys-56 C$_\beta$H and N$_P$H; (D) saturate Cys-11 C$_\beta$H′ and observe NOEs to Cys-11 C$_\beta$H and N$_P$H and Thr-10 N$_P$H; (E) saturate Cys-14 C$_\beta$H′ and observe NOEs to Cys-14 C$_\alpha$H and C$_\beta$H. [Reprinted from L. Calzolai, C. M. Gorst, K. L. Bren, Z.-H. Zhou, M. W. W. Adams, and G. N. La Mar, *J. Am. Chem. Soc.* **119**, 9341 (1997), with permission.]

where the cross-relaxation rate, σ_{ij}, is given by

$$\sigma_{ij} = 0.1\gamma^2\hbar^4 r_{ij}^{-6}\tau_r \qquad (9)$$

with r_{ij} as the i–j interproton separation and τ_r as the molecular tumbling time, which for the ∼7 kDa Fd is ∼3 ns. An example of the assignment of Cys residues in mutant Pf 4Fe D14C-Fd$_A^{ox}$ is illustrated[22] in Fig. 8. For the less strongly or inconsequentially relaxed proton, NOESY spectrum at either ∼50 ms mixing time and 3–5 s^{-1} repetition rates, or ∼150–250 ms mixing time and 0.5 s^{-1} repetition rate, respectively, have been found most effective.

F. Magnetization Transfer between Fd^{ox} and Fd^{red}

When the rate of interconversion between two alternate cluster oxidation states is comparable to T^{-1}, saturation of a peak for one oxidation state (designated Y)

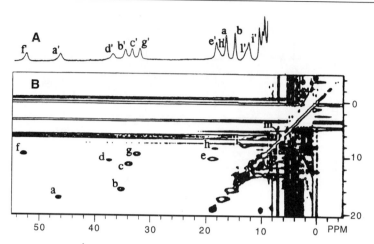

FIG. 9. (A) 500-MHz ^{1}H NMR reference trace of a $\sim$1 : 1 mixture of Tl 4Fe Fd^{ox} and Fd^{red} in ^{2}H$_2$O, pH 7.5, at 30°. The resonances for Fd^{ox} are labeled $a-m$, as in Fig. 7, whereas those for the Fd^{red} are labeled $a'-m'$, with the same label reserved for cross-correlated resonances. (B) Portion of the 3 ms mixing time NOESY (EXSY) spectrum illustrating the cross correlation of resonances between 4Fe Tl Fd^{ox} and Fd^{red}; the intense cross peaks are labeled with the letter for the correlated peak.[42] [Reprinted from A. Donaire, C. M. Gorst, Z. H. Zhou, M. W. W. Adams, and G. N. La Mar, *J. Am. Chem. Soc.* **116,** 6841 (1994), with permission.]

leads to saturation (magnetization)-transfer[49] for this same proton in the alternate oxidation state (designated X). Hence the assignments for the cluster in one oxidation state can be "transferred" to the other oxidation state, either by steady-state 1D methods[9,22] or with 2D EXSY, as illustrated[42] for Tl 4Fe Fd in Fig. 9.

Qualitative analysis of the magnetization transfer, moreover, allows the determination of the electron self-exchange rate for the reaction:

$$[4Fe^*; 4S^*]^{1+} + [4Fe; 4S]^{2+} \rightarrow [4Fe^*; 4S^*]^{2+} + [4Fe; 4S]^{1+} \tag{10}$$

The fractional intensity change, F_x, for X on saturating $Y(F_x = (X_o - X_\infty)X_o^{-1}$, where X_∞ and X_o are the peak intensities with, and without, saturating Y, respectively), and the intrinsic relaxation rate for X is T_{1X}^{-1}, yield[49] an observed rate k_{xy}

$$k_{xy} = F_x(1 - F_x)^{-1}(T_{1X})^{-1} \tag{11}$$

and an electron self-exchange rate, k_{se}:

$$k_{se} = k_{xy}/[Y] \tag{12}$$

---

[49] J. Sandström, "Dynamic NMR Spectroscopy." Academic Press, New York, 1982.

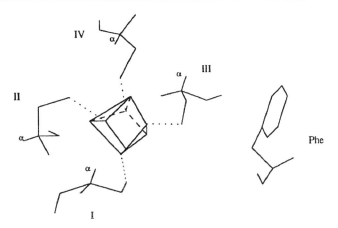

FIG. 10. Schematic representation of the coordination of a Fd cubane clusters. Cys(**II**) is either replaced or not coordinated in 3Fe ferredoxin. The position of the highly conserved Phe near Cys(**III**) is shown. [Reprinted from A. Donaire, C. M. Gorst, Z. H. Zhou, M. W. W. Adams, and G. N. La Mar, *J. Am. Chem. Soc.* **116,** 6841 (1994), with permission.]

G. Sequence-Specific Assignments

The majority of the residues that are minimally influenced by cluster paramagnetism are assigned sequentially in a manner that is identical to that used for diamagnetic proteins,[13] and hence is not elaborated here. The necessary ^{1}H NMR data to effect such assignments for *Pf* 4Fe Fd$_A^{ox}$ are illustrated schematically in Fig. 1E. For the cluster ligands, it has been shown[48] that the four Cys in cubane Fds ligate in characteristic orientations relative to the cluster[6–8,12,50–52] (as illustrated[42] in Fig. 10), such that two Cys [Cys(**II**), and Cys(**III**)] always have the C$_\alpha$H relatively remote from the closest Fe ($R_{Fe} \sim 5$ Å, $T_1 \sim 30$ ms), with the NH strongly relaxed ($R_{Fe} \sim 3$ Å, $T_1 \sim 1$ ms), while the other two Cys [Cys(**I**), and Cys(**IV**)] always have the C$_\alpha$H strongly relaxed ($R_{Fe} \sim 4$ Å, $T_1 \sim 2$ ms) but the NH is relatively remote from the iron ($R_{Fe} \sim 6$ Å, $T_1 \sim 50$ ms).

The orientation of all Cys in cubane clusters, moreover, is such that one C$_\beta$H is close to the iron ($R_{Fe} \sim 3$ Å, $T_1 \sim 2$–5 ms, hereafter labeled C$_\beta$H$'$) and the other is more remote ($R_{Fe} \sim 4.2$ Å, $T_1 \sim 10$–15 ms, hereafter labeled C$_\beta$H), such that stereospecific assignments[39,53] are generated. Hence TOCSY locates two C$_\beta$H$_2$-C$_\alpha$H and two C$_\beta$H$_2$ fragments (as shown[42] for *Tl* 4Fe Fd^{ox} in Fig. 7)

---

[50] E. T. Adman, L. C. Sieker, and L. H. Jensen, *J. Biol. Chem.* **251,** 3801 (1976).
[51] C. D. Stout, *J. Biol. Chem.* **268,** 25920 (1993).
[52] Z. Dauter, K. S. Wilson, L. C. Sieker, J. Meyer, and J. M. Moulis, *Biochemistry* **36,** 16065 (1997).
[53] S. L. Davy, M. J. Osborne, J. Breton, G. R. Moore, A. J. Thomson, I. Bertini, and C. Luchinat, *FEBS Lett.* **363,** 199 (1995).

that can be assigned to the Cys(II), Cys(III) and Cys(I), Cys(IV) pairs, respectively. The NHs for the latter pair of Cys are readily observed in steady-state NOEs from the C$_\beta$Hs, as shown[9] in Fig. 8 for Pf 4Fe D14C-Fd$_A^{ox}$. Cys(I) and Cys(IV) are readily distinguished on the basis that only the former exhibits a NOESY cross peak[42,48] (or steady-state NOE) to the N$_P$H of the residue adjacent on the N-terminal side of Cys(I) (shown in Pf 4Fe D14C-Fd$_A^{ox}$ in Fig. 8D), while Cys(II) and Cys(III) are distinguished by only the latter residues C$_\beta$H exhibiting strong NOEs[22,42,48] to the unique and highly conserved Phe in the hydrophobic core, as illustrated in Fig. 8B for Pf 4Fe Fd$_A^{ox}$.

H. Molecular Structure Determination

The complete solution molecular structure of a ferredoxin (or other small to moderate-sized paramagnetic metalloproteins) can be determined by NMR using distance geometry or simulated annealing in spite of the paramagnetism.[15,18] The procedures are, for all practical purposes, the same as for a comparably sized diamagnetic protein, except that fewer conventional dipolar constraints will be available for residues near the cluster. The increased relaxation will diminish NOEs and NOESY cross peak intensities near the cluster, some to the point of nonobservability. However, the relaxation data [via Eq. (1)], as well as the pattern of cluster ligand Cys δ_{con} via Eq. (7), provides alternate constraints[15,18,19,37,54] that ultimately lead to relatively robust molecular models. In some cases, it has been possible to use a homology model for the cluster environment[11] to generate a robust solution structure.

VII. Structural Isomerism in Pf Fd

Even the earliest ^{1}H NMR studies[43] of Pf Fd revealed that the protein can exist in multiple forms, in addition to its alternate cluster redox states, Fd^{ox} and Fd^{red}, with Pf Fd capable of existing in no less than six discrete molecular states.[9,10,22,24,43] Some of these states are only metastable,[9,10,22] while others are clearly equilibrium[24,43] structural heterogeneities. Similar molecular heterogeneity was not detected in the ^{1}H NMR spectra of either $Tm^{19,42}$ or $Tl^{11,55}$ Fds. A qualitative understanding of the molecular bases of these heterogeneities is necessary for interpreting either the magnetic coupling topology of the cluster or the molecular structure of any one Fd form.

---

[54] S. L. Davy, M. J. Osborne, and G. R. Moore, J. Mol. Biol. **277**, 683 (1998).
[55] G. Wildegger, D. Bentrop, A. Ejchart, M. Alber, A. Hage, R. Sterner, and P. Rösch, Eur. J. Biochem. **229**, 658 (1995).

A. Redox-Active Disulfide Bond

Standard cycling between the expected alternate cluster oxidation states, [4Fe:4S]$^{2+}$, (Fd^{ox}), and [4Fe:4S]$^{1+}$, (Fd^{red}), using O$_2$/dithionite as oxidant/reductant), revealed[10] that *two distinct states* could be prepared with ^{1}H NMR spectra characteristic of [4Fe:4S]$^{2+}$ and [4Fe:4S]$^{1+}$, as illustrated in Fig. 11. Thus, the "as isolated" reduced *Pf* 4Fe Fd exhibited the low-field ^{1}H NMR spectrum in Fig. 11A with obvious resolved cluster ligand proton signal with T_1 values and temperature-dependent shift characteristic of the reduced cluster [4Fe:4S]$^{1+}$, and is defined as Fd$_B^{red}$. Oxidation with O$_2$ afforded a new species (Fig. 11B) with chemical shifts, T_1 values, and temperature dependence diagnostic of oxidized cluster, [4Fe:4S]$^{2+}$, defined as Fd$_B^{ox}$. However, in the presence of excess O$_2$, the initially formed Fd$_B^{ox}$ converted to another form (Figs. 11A–11C) that also possesses an oxidized cluster that was labeled Fd$_A^{ox}$, with ^{1}H NMR spectra as shown in Fig. 11C. Lastly, reduction of Fd$_A^{ox}$ yielded a species with the ^{1}H NMR spectrum in Fig. 11E, which is diagnostic of a *reduced cluster,* which is, however, *not identical* to that of the "as isolated" Fd$_B^{red}$, and defined as Fd$_A^{red}$ (compare Figs. 11A and 11E). However, in the presence of excess reductant, Fd$_A^{red}$ converts to the original starting material, Fd$_B^{red}$, as shown in Fig. 11E, 11F. The conversion Fd$_B^{ox}$ → Fd$_A^{ox}$ (Figs. 11B–11D)

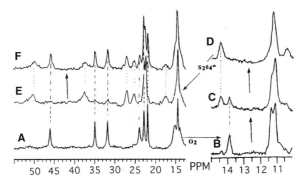

FIG. 11. Hyperfine shifted portions (to the low field of 10 ppm) of the 500-MHz ^{1}H NMR spectra of *Pf* 4Fe Fd, as a function of the presence of oxidant (O$_2$), reductant (S$_2$O$_4^{2-}$), and time.[10] (A) As-isolated *Pf* 4Fe Fd$_B^{red}$; (B) species formed immediately upon exposing 4Fe Fd$_B^{red}$ to air at 30°C, pH 8.0, designated *Pf* 4Fe Fd$_B^{ox}$; (C) the same sample as in panel B after 1 week open to the atmosphere at 30°, pH 8.0; (D) the same sample as in panel B 5 weeks after opening to the atmosphere at 30°, pH 8.0, when now only a single species is present, designated *Pf* 4Fe Fd$_A^{ox}$; (E) addition of excess S$_2$O$_4^{2-}$ to the sample in panel D leads to appearance of a new species we designate *Pf* 4Fe Fd$_A^{red}$; (F) after 16 hr at 30°, pH 8.0, and in the presence of excess S$_2$O$_4^{2-}$, the starting material *Pf* 4Fe Fd$_B^{red}$ is slowly regenerated (whose spectrum in pure form is shown in trace A). [Reprinted from C. M. Gorst, Z. H. Zhou, K. Ma, Q. Teng, J. B. Howard, M. W. W. Adams, and G. N. La Mar, *Biochemistry* **34**, 8788 (1995), with permission.]

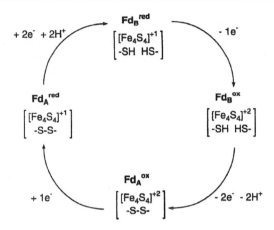

FIG. 12. Proposed cycle for the four redox states of *Pf* Fd. [Reprinted from C. M. Gorst, Z. H. Zhou, K. Ma, Q. Teng, J. B. Howard, M. W. W. Adams, and G. N. La Mar, *Biochemistry* **34**, 8788 (1995), with permission.]

required oxidant, and the conversion $Fd_A^{red} \rightarrow Fd_B^{red}$ (Figs. 11E, 11F) required reductant, demonstrating the presence of *four* distinct oxidation states.[10]

The additional redox chromophore in *Pf* Fd was traced to the Cys(**V**), Cys(**VI**) pair by titrations with 5,5′-dithiobis(2-nitrobenzonic acid) (DTNB),[10] which revealed a complete redox cycle of *Pf* 4Fd is shown schematically in Fig. 12. Hence the subscript A and B to the Fd with oxidized or reduced cluster refer to the oxidation state of the Cys(**V**), Cys(**VI**) pairs, with subscript A denoting an intact disulfide and subscript B denoting free thiols. The ^{1}H NMR spectra of the 3Fe Fd_A^{ox} and 3Fe Fd_B^{ox} have been similarly characterized.[10]

Whereas the change in solely the cluster oxidation state can be effected within milliseconds,[20,42] redox reactivity of the disulfide was found extraordinarily sluggish[10] at 30°, some ~70° below the physiological temperature for *Pf*. A disulfide bond that is redox active at approximately the same potential as the cluster has been reported for the 3Fe Fd from the mesophile *Dg*, but the disulfide reactivity is rapid such that only very small amounts of the intermediate states are populated at equilibrium,[56] and hence no structural information could be obtained on the two intermediate oxidation states. For *Pf* Fd, the sluggish disulfide reactivity, likely because of its remarkable hyperthermostability, allows the preparation of the two metastable intermediate oxidation states, Fd_B^{ox} and Fd_A^{red}, in relatively pure conditions and over times sufficient to carry out structural characterization of the four distinct redox states.[9,10,22,24]

[56] A. L. Macedo, I. Moura, K. K. Surerus, V. Papaefthymiou, M. Y. Liu, J. Legall, E. Münck, and J. J. G. Moura, *J. Biol. Chem.* **269**, 8052 (1994).

B. Disulfide Bond Orientational Isomerism in Fd_A

All ^{1}H NMR characterized forms of Pf Fd with an intact disulfide bond (i.e., Fd_A^{ox} or Fd_A^{red}) exhibit a large number of resonances that are severely broadened [9,10,22,24,43] (as shown for Pf 3Fe Fd_A^{ox} and Fd_B^{ox} in Fig. 14), for which the temperature and magnetic field dependence of the linewidth established that they arise from an *equilibrium dynamic*, conformational equilibrium. Similar line broadening is completely absent from all Pf Fd with cleaved disulfide [9,10,22,24] (i.e., Fd_B^{ox}, Fd_B^{red}). Generally, the most severely affected resonances are for the cluster ligands, seen most dramatically [10,24,43] in the ^{1}H NMR spectra of 3Fe Fd_A^{ox}, but also observed in the 4Fe Fd_A^{red} states.[9,22] Thus, the five resolved assigned Cys $C_\beta H$ signals in Pf 3Fe Fd_B^{ox} are broad at $\sim 30°$, and split into two distinct molecular species with separate sets of signals labeled with subscript S and R, respectively,[24] as shown in Fig. 13. Because this equilibrium heterogeneity manifests itself only in the presence of the disulfide bond between Cys-21(**V**) and Cys-48(**VI**), it is reasonable to assume that the heterogeneity is centered at the disulfide bond.

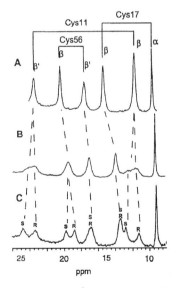

FIG. 13. Low field portions of the 500 MHz ^{1}H NMR spectra with the assigned Cys ligand $C_\beta H$ resonances of 3Fe Fd_A^{ox} at (A) 50°; (B) 20°; and (C) 5°, showing that two nonequivalent cluster environments can be resolved because of an equilibrium dynamic heterogeneity in the configuration about the Cys-21(**V**) to Cys-48(**VI**) disulfide bond.[24,43] The split resonances are labeled S or R for the resolved peaks with "S," "R" corresponding to the deduced chirality of the disulfide bond for the two "frozen out" forms, respectively. [Reprinted from P. L. Wang, L. Calzola, K. L. Bren, Q. Teng, F. E. Jenney, Jr., P. S. Brereton, J. B. Howard, M. W. W. Adams, and G. N. La Mar, *Biochemistry* **38**, 8167 (1999), with permission.]

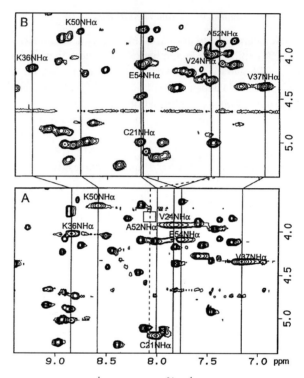

FIG. 14. The 60 ms mixing time ^{1}H TOCSY map^{24} in ^{1}H$_2$O, pH 7.6 at 30° for (A) Pf 4Fe Fd$_A^{ox}$ (with intact Cys-21(**V**)–Cys-48(**VI**) disulfide bond); and (B) ~75% 4Fe Fd$_B^{ox}$ (with cleaved disulfide) and ~25% 4Fe Fd$_A^{ox}$. The resonance position of the NHs that exhibit anomalous line broadening in 4Fe Fd$_A^{ox}$ due to disulfide orientational isomerism are connected to those in 4Fe Fd$_B^{ox}$ where the linewidths are "normal." The dashed line and box in A show where the Ala-52 NH shows up at 50°. [Reprinted from P. L. Wang, L. Calzolai, K. L. Bren, Q. Teng, F. E. Jenney, Jr., P. S. Brereton, J. B. Howard, M. W. W. Adams, and G. N. La Mar, *Biochemistry* **38**, 8167 (1999), with permission.]

Disulfide bonds generally adopt one of two orientations[57] ("chiralities"), left-handed R, or right-handed S helical orientation, of which only the former is observed in the crystal structure of Fds from both hyperthermophiles[12,19] and mesophiles.[6] A number of nonligated residues, including both some close to and very remote from the cluster, also exhibit line broadening due to the equilibrium conformation,[24] as illustrated in the peptide proton window of the ^{1}H NMR TOCSY spectra of Pf 4Fe Fd$_A^{ox}$ in Fig. 14A (this line broadening is again absent in Fd$_B^{ox}$, as shown in Fig. 14B). The pattern of the effected resonances radiates from the Cys-21(**V**)–Cys-48(**VI**) disulfide bond in a manner consistent with the

---

[57] J. S. Richardson, *Adv. Protein Chem.* **34**, 167 (1981).

population of alternate S or right-handed, and R or left-handed helical orientation of the disulfide bond.[24] The population of the R structure in Pf, but not Tl or Tm, Fds can be rationalized by their differences in the length of a helix (see Section IX). The significant effect of the disulfide orientation on the cluster ligand δ_{con} in 3Fe Fd^{ox} (see Fig. 13) and on residues to as far as 20 Å from the cluster[24] indicate that the consequences of the disulfide conformational heterogeneity is global rather than local (see Section IX), as observed in other proteins of similar size.[58] The presence in solution of comparable amounts of two interconverting structures in solution may be the basis for finding the Fd disordered in the 1:1 oxidoreductase:Fd crystal structure.[59]

VIII. Cluster Electronic Structure

A. Oxidized Three-Iron Cluster

The resolved, hyperfine-shifted portions of the ^{1}H NMR spectra of Tl and Pf 3Fe Fd^{ox} (Fig. 13A)[43] are similar to that of the 3Fe Fd from mesophilic Dg.[60,61] ^{1}H NMR spectra for hyperthermostable 3Fe Fd^{red} have not been reported. The three cluster ligated Cys have been identified[43,48] in both Tl and Pf 3Fe Fd, with one Cys exhibiting Curie, and two Cys displaying anti-Curie behavior, as shown in Fig. 15 for the latter protein, consistent with the magnetic complex[62] of a $S = 2$ diferric pair with a single ferric $S = 5/2$. The "unique" Cys in the three Fd, however, is *not* at the same position in the consensus ligating sequence.[42,43,48,60,61] The structural basis for the "unique" iron is not understood.

For Pf 3Fe Fd$_A^{ox}$, the Cys contact shift pattern, and hence magnetic coupling, depends on pH, with the apparent pK corresponding to the titrating of the unligated Asp-14(**II**).[48] The protonation of Asp-14(**II**) also abolishes the 2:1 magnetic inequivalent such that all three Cys exhibit very similar variable temperature behavior, as shown in Fig. 15. Moreover, the degree of magnetic asymmetry in 3Fe Fd$_A^{ox}$ depends on the orientation of the Cys-21(**V**)–Cys-48(**VI**) disulfide bond, with Cys-11(**I**) exhibiting stronger anti-Curie behavior for the R than S disulfide chirality (M. Webba da Silva, S. Sham, M. W. W. Adams, and G. N. La Mar, submitted for publication). The detailed interpretation of the coupling of the cluster and disulfide redox chromophores must await the determination of a robust molecular structure for Pf Fd.

---

[58] G. Otting, E. Liepinsh, and K. Wüthrich, *Biochemistry* **32**, 3571 (1993).
[59] L. Yu, S. Faham, R. Roy, M. W. W. Adams, and D. C. Rees, *J. Mol. Biol.* **286**, 899 (1999).
[60] A. L. Macedo, I. Moura, J. J. G. Moura, J. LeGall, and B. H. Huynh, *Inorg. Chem.* **32**, 1101 (1993).
[61] A. L. Macedo, P. N. Palma, I. Moura, J. LeGall, V. Wray, and J. J. G. Moura, *Magn. Reson. Chem.* **31**, S59 (1993).
[62] E. P. Day, J. Peterson, J. J. Bonovoisin, I. Moura, and J. J. G. Moura, *J. Biol. Chem.* **263**, 3684 (1988).

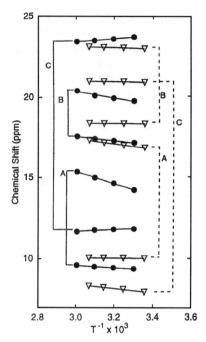

FIG. 15. Plot of chemical shift vs reciprocal absolute temperature[48] (Curie plot) of the resolved $C_\beta H$, $C_\beta 'H$, and $C_\alpha H$ peaks for Cys A [Cys-17(**III**)], B [Cys-56(**IV**)], and C [Cys-11(**I**)] for Pf 3Fe Fd^{ox} at pH 8.0 (solid markers, solid lines) and at pH 3.4 (open markers, dashed lines) in ^2H_2O at 30°. Note that while Cys A and B exhibit anti-Curie (negative slope) and Cys C display Curie-like (positive slope) behavior at pH 8.0, all three Cys exhibit similarly weak anti-Curie behavior at low pH. [Reprinted from C. M. Gorst, Y.-H. Yeh, Q. Teng, L. Calzolai, Z-H. Zhou, M. W. W. Adams, and G. N. La Mar, *Biochemistry* **34**, 600 (1995), with permission.]

B. Oxidized Four-Iron Cluster

1. All-Cys Ligation. Complete $C_\beta H$ assignment in the all Cys-ligated wild-type Tm Fd^{ox}[55], WT Tl Fd^{ox}[42], and Pf[22] mutant D14C-Fd_A^{ox} and D14C-Fd_B^{ox} reveal very similar δ_{con} for the four Cys and uniform anti-Curie temperature behavior (shown for Tl Fd in Fig. 4A), as expected[16–18] for the four oxidation-state equivalent $Fe^{2.5+}$. The pattern of δ_{con} and T_1 values for the $C_\beta H$, $C_\beta H'$ in each case is consistent with a highly conserved cluster geometry and ligand orientations among both the three hyperthermostable Fd, as well as other, less thermostable cubane Fds.[16–18,33]

2. Effect of Non-Cys Ligands. Pf Fd is unique in providing a very stable protein for which the ligation of Asp-14(**II**) in WT[9] and a Ser in the D14S-Fd^{22} could be unequivocally demonstrated in both presence (Fd_A^{ox}) and absence (Fd_B^{ox}) of the disulfide bond. The conserved oxidation-state equivalent environments of the four iron is supported by the observation of Cys $C_\beta H$ δ_{con} and variable temperature behavior (uniformly anti-Curie) that are conserved relative to all Cys-ligated

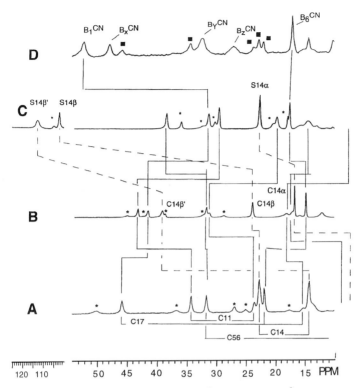

FIG. 16. Resolved low-field portions of the 500 MHz ^{1}H NMR spectra in ^{2}H$_2$O, 50 mM phosphate at 30° for (A) WT Pf Fd$_B^{red}$; (B) Pf D14C-Fd$_B^{red}$; (C) Pf D14S-Fd$_B^{red}$; and (D) WT Pf Fd$_B^{red}$-CN[22]; positions of unreacted WT Fd$_B^{red}$ peaks are indicated by filled squares. Positions of the minor Fd$_A^{red}$ form in each case are indicated by asterisks. Peak assignments are given for WT, and similarly labeled and assigned peaks for Cys11, 17, and 56 in (B)–(D) are connected by solid lines. The variable ligand 14 resonances are connected by dashed lines. Only Cys-17(III) could be unambiguously assigned in (D). [Reprinted from L. Calzolai, C. M. Gorst, K. L. Bren, Z.-H. Zhou, M. W. W. Adam, and G. N. La Mar, *J. Am. Chem. Soc.* **119,** 9341 (1997), with permission.]

clusters. The magnitude of the unique cluster-ligand II C$_\beta$H δ_{con} varies,[22] with Ser exhibiting the largest, and Asp exhibiting the smallest, with the relative δ_{con} for the three ligands in the approximate ratios found in model compounds.[63] For the oxidized, four-iron cluster, the cluster ligand C$_\beta$H δ_{con} exhibited[9] only weak dependence on either the presence of the disulfide (Fd$_A^{ox}$ vs Fd$_B^{ox}$), or the orientation (S vs R) of the disulfide bond in Fd$_A^{ox}$.

C. Reduced Four-Iron Cluster

The hyperfine-shifted region of the ^{1}H NMR spectra of Pf WT, D14C-, D14S-, and cyanide-ligated WT 4Fe Fd$_B^{red}$ are reproduced[22] in Fig. 16A–16D, respectively.

[63] J. A. Weigel and R. H. Holm, *J. Am. Chem. Soc.* **113,** 4184 (1991).

1. *All-Cys Ligation.* Cluster ligated assignments have been carried out on Tl^{42} and the *Pf* D14C-Fd^{22} with intact (D14C Fd$_A^{red}$) and cleaved disulfide (D14C-Fd$_B^{red}$). The pattern of δ_{con}, T_1 values, and temperature behavior for Cys C$_\beta$Hs in each of the complexes are typical of that observed for Fd^{red} from mesophiles,[16–18,33] and that expected for the cluster state, namely two Cys exhibiting Curie-like (2Fe$^{2.5+}$) and two displaying anti-Curie behavior (2Fe$^{2.0+}$), such as those for *Tl*; for 4Fe Fd^{red} shown in Fig. 4B; and for *Pf* 4Fe D14C-Fd^{red} in Fig. 17. In each case the same two Cys (**I** and **III**) exhibited the Curie-like behavior of the 2Fe$^{2.5+}$ pair,

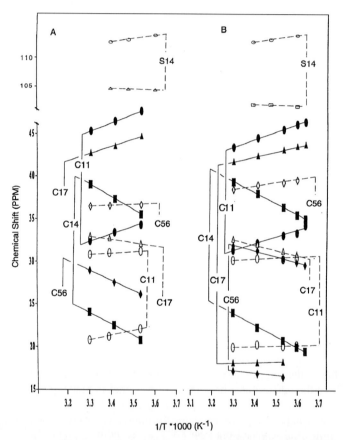

FIG. 17. Curie plots[22] for cluster ligand resonances for (A) D14C-Fd$_A^{red}$ (closed markers) and D14S-Fd$_A^{red}$ (open markers); each with intact disulfide bonds and (B) D14C-Fd$_B^{red}$ (closed markers) and D14S-Fd$_B^{red}$ (open markers), each with cleaved disulfide, all in ^{2}H$_2$O, 50 m*M* in phosphate, pH 7.2. The proton peaks for Cys-11, ligand 14, Cys-17, and Cys-56 are shown by circles, squares, triangles, and diamonds, respectively. [Reprinted from L. Calzolai, C. M. Gorst, K. L. Bren, Z.-H. Zhou, M. W. W. Adam, and G. N. La Mar, *J. Am. Chem. Soc.* **119**, 9341 (1997), with permission.]

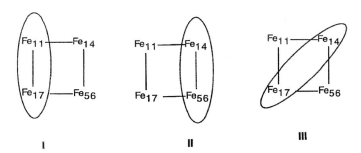

FIG. 18. Schematic representation of a [Fe$_4$S$_4$]$^{1+}$ cluster with the location of the mixed valence pair 2Fe$^{2.5+}$ (encircled Fe) ligated by: (**I**), Cys-11(**I**) and Cys-17(**III**); (**II**), Ligand 14(**I**) and Cys56(**IV**); and (**III**), Ligand 14(**II**) and Cys 17(**III**). [Reprinted from L. Calzolai, C. M. Gorst, K. L. Bren, Z.-H. Zhou, M. W. W. Adam, and G. N. La Mar, *J. Am. Chem. Soc.* **119,** 9341 (1997), with permission.]

as also found[40] in Fd^{red} from less thermostable cubane-cluster Fd^{red}. The structural basis of this discrimination of the micropotentials is not yet fully understood. The distinct difference in the oxidation state of the two pairs of irons is manifested in both the different temperature dependence and T_1 values of the ligated Cys. As may be expected because of the larger $\langle S_z \rangle$ for the valence-delocalized ($S = 9/2$) than diferrous ($S = 4$) iron pair,[17,22,25] the ligated Cys C$_\beta$Hs exhibit stronger paramagnetic relaxation than the Cys C$_\beta$Hs ligated to the diferrous pair.

Interestingly, for *Pf* D14C-Fd^{red}, Cys-11(**I**) and Cys-7(**III**) C$_\beta$Hs exhibit larger δ_{con} and steeper Curie-like behavior in the presence (D14C-Fd$_A^{red}$) than in the absence (D14C-Fd$_B^{red}$) of the disulfide bond,[22] as shown in Fig. 17. These results suggest that, in fact, the Fd^{red} cluster exists in an equilibrium of two electronic structures with the 2Fe$^{2.5+}$ ligated by Cys(**I**), and Cys(**III**) or Cys(**II**), and Cys(**IV**) as depicted in structures **I** and **II** in Fig. 18. In each case, structure **I** is more strongly populated than **II**, but the preference is *stronger* in the presence than in the absence of the disulfide bridge. Similar equilibrium among cluster electronic structures with variable location of the valence-delocalized iron pair have been characterized in oxidized HIPIP.[17,33]

2. *Effect of Non-Cys Ligands.* Asp ligation to a cubane cluster has been established only for *Pf* 4Fe Fd.[9] The ^{1}H NMR spectra of WT *Pf* Fd$_A^{red}$ and Fd$_B^{red}$ (Fig. 16A) reveal a pattern of ligated C$_\beta$Hs δ_{con}, T_1 values, and variable temperature behavior (mixed Curie, anti-Curie; see Fig. 19) that are very similar to those characteristic in all-Cys ligated 4Fe Fd^{red}. Since both δ_{con} and T_1^{-1} should increase dramatically with ground state S, the similar δ_{con}, T_1, and VT patterns[22] dictate for WT and D14C-Fd^{red} that in solution at ambient temperature, the cluster ground state is $S = 1/2$, rather than the $S = 3/2$ observed by EPR at cryogenic temperatures.[5,21] Although it is now clear that this difference in spin ground state is real, the environmental origin of the differences is obscure. The temperature

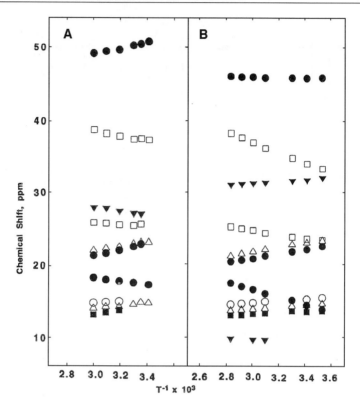

FIG. 19. Curie plots for low-field hyperfine shifted and relaxed resonances for (A) WT *Pf* Fd$_A^{red}$ ([Fe$_4$S$_4$]$^{1+}$ with free thiols for Cys-21, 48) and (B) *Pf* Fd$_B^{red}$ ([Fe$_4$S$_4$]$^{1+}$ with disulfide bridge between Cys-21, 48). The resonances belonging to the four ligands are given by the same symbols (□ for Cys-11, △ for Asp-14, ● for Cys-17, and ▽ for Cys-56) in both panels. Positive and negative slopes are designated Curie and anti-Curie, respectively. [Reprinted from L. Calzolai, C. M. Gorst, Z.-H. Zhao, Q. Teng, M. W. W. Adams, and G. N. La Mar, *Biochemistry* **34**, 11373 (1995), with permission.]

behavior of the ligand C$_\beta$Hs of the unique Asp-14(**II**)-ligated *Pf* 4Fe Fd^{red} has been interpreted[22] in terms of a cluster with an ∼1 : 1 equilibrium between species with a valence-delocalized iron pair ligated to Asp-14(**II**) and Cys-56(**IV**) (structure **II** in Fig. 18) and an iron pair ligated to Asp-14(**I**) and Cys-17(**II**) (structure **III** in Fig. 18). This finding that Asp-14 is bound to a Fe$^{2.5+}$ is consistent with EPR data.[23]

The introduction of a ligated Ser in D14S-Fd^{red} yields a Cys C$_\beta$H δ_{con} pattern shown in Fig. 16C that is similar to D14C-Fd^{red} (Fig. 16B) or WT Fd^{red} (Fig. 16A). The ligated Ser, in contrast, appear in the unprecedented low-field window near 100 ppm,[22] as shown in Fig. 16C. Relative to D14C-Fd^{red}, D14S-Fd^{red} exhibits similar δ_{con} and less steep Curie-like behavior for Cys-11(**I**) and Cys-17(**III**), and larger δ_{con} and weaker slopes for Cys-56(**IV**), as shown in Fig. 17. The changes in slopes are consistent with a cluster which is an ~1:1 equilibrium between structure **I** and **II** in Fig. 18. Thus, the higher oxidation state is favored by Ser compared to Cys ligation. Similar conclusions have been reached for Cys→Ser mutants of Hipip.[64]

In the presence of excess cyanide, WT 4Fe Fd^{red} ligates cyanide,[5,21,22] presumably by replacing Asp-14(**I**). The ^{1}H NMR spectrum of the CN-ligated WT 4Fe Fd$_B^{red}$ (Fig. 16D) is very similar to that of either WT (Fig. 16A) or D14C-Fd^{red} (Fig. 16B). However, the Cys C$_\beta$H resonances exhibit factors ~5–8 shorter T_1 values[22] than WT, D14C-, or D14S-Fd^{red}. The increased relaxation relative to WT and D14C-Fd^{red} allowed the assignment[22] of only the Cys-17(**III**) C$_\beta$H and C$_\beta$H. However, based on the variable temperature behavior of the remaining C$_\beta$H resonances, an assignment was proposed that located Cys-11(**I**) and Cys-17(**III**) as being ligated to 2Fe$^{2.5+}$, with Cys-56(**IV**) and CN$^-$ ligated to 2Fe$^{2.0+}$ (i.e., structure **I** in Fig. 18). These conclusions are consistent with ENDOR results.[5,21]

3. *Effect of Cys-21(V), Cys-48(IV) Redox States.* The pattern of Cys C$_\beta$H δ_{con} for a given Fd^{red} is similar for Fd$_A^{red}$ (intact disulfide) and Fd$_B^{red}$ (cleaved disulfide), but differs characteristically in the magnitude of δ_{con} and the Curie plot slope.[22] The difference in the δ_{con} and Curie slopes show that the free sulfhydryls in Fd$_B^{red}$ slightly favor structure **II** over **I** (Fig. 18) compared to that in the presence of the disulfide (Fig. 17) for both D14C-Fd^{red} and D14S-Fd^{red}. The apparent significant interaction between the disulfide and cluster redox chromophores is to be expected, since the available structures of cubane Fds[6,12,19] show that the peptide NH of Cys(**V**) invariably serves as a H-bond donor to the carbonyl of cluster ligand Cys(**III**). The ability to characterize this interaction between the two chromophores is *the direct consequence of the significant thermostability of the Pf Fd structures* that allow the characterization of the metastable redox states, Fd$_B^{ox}$ and Fd$_A^{red}$, for each Fd.

IX. Cluster Electron Exchange Rates

The rate of electron transfer into and out of the cluster can be measured by self-exchange, [i.e., Eq. (10)]. This rate is independent of potential and measures the reorganization energy[65] for electron transfer i.e., the degree to which the molecular structural environment of the cluster changes on adding or losing an electron). The electron self-exchange rate for *Tl* Fd has been determined[42] and found to be very similar to that for other less thermostable cubane cluster Fds.[66] Hence,

---

[64] E. Babini, I. Bertini, M. Borsari, F. Capozzi, A. Dikiy, L. D. Eltis, and C. Luchinat, *J. Am. Chem. Soc.* **118**, 75 (1996).
[65] G. McLendon, *Acc. Chem. Res.* **21**, 160 (1988).
[66] I. Bertini, F. Briganti, C. Luchinat, L. Messori, R. Monnanni, A. Scozzafava, and G. Vallini, *Eur. J. Biochem.* **204**, 831 (1992).

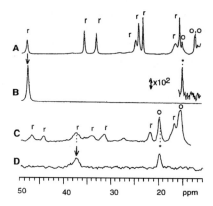

FIG. 20. Hyperfine-shifted portion of the 500 MHz ^{1}H NMR spectrum[20] of (A) ~7 mM solution of ~90% Pf WT 4Fe Fd$_B^{red}$ (peaks labeled r) and ~10% WT 4Fe Fd$_B^{ox}$ (peaks labeled O) at 30°; and (B) difference spectrum at 30° on saturating the low-field peak marked with a vertical arrow for Pf WT 4Fe Fd$_B^{red}$. (C) Hyperfine-shifted portion of the 500 MHz ^{1}H NMR spectrum of ~30% Pf 4Fe D14C-Fd$_B^{red}$ (peaks labeled r) and ~70% Pf 4Fe D14C-Fd$_B^{ox}$ (peaks labeled O) at 10°; and (D) difference spectrum at 10°C on saturating the low-field peak marked with the vertical arrow for Pf 4Fe Fd$_B^{red}$. Note $>10^2$ difference between WT and D14C Fd in the degree of magnetization transfer to the analogous Fd$_B^{ox}$ complex. [Reprinted from L. Calzolai, Z. H. Zhou, M. W. W. Adams, and G. N. La Mar, *J. Am. Chem. Soc.* **118,** 2513 (1996), with permission.]

hyperthermostability does not *necessarily* alter the rate of electron transfer, and hence, does not *necessarily* increase the reorganization energy.

However, for WT Pf Fe with the ligated Asp-14(**I**), this electron transfer rate is highly anomalous. Magnetization transfer experiments[20] on WT 4Fe Fd$_B$ at 30° (Figs. 20A and 20B) and on D14C-Fd$_B$ (Figs. 20C and 20D) show that the saturation factor for WT is much smaller (by $>10^2$) in WT Fd$_B$ at 30° than D14C-Fd$_B$ at 10°. These data reflects an electron self-exchange rate [determined via Eqs. (11) and (12)] that is $\sim 10^3$ *slower for ligated Asp-14(**II**) than for Cys-14(**II**)*. The k_{se} for Pf D14C-Fd is the same as for Tl^{42} and other cubane cluster Fds,[66] indicating that ligated Asp must provide some "gating" mechanism.[20] The increased reorganization energy that must account for the $\sim 10^3$ slower rate in WT must come from some coupling between the thermostable folding topology and the cluster geometry. One proposal that could account for such an effect is that the unique Asp-14(**II**) ligates as monodentate in one, and as bidentate ligand in the other cluster oxidation state.[20] Such bidentate ligation is not possible for either Cys or Ser.

X. Molecular Structure

The simple cubane cluster Fds share high sequence homology, and crystal structures of such Fds from three mesophiles, Dg,[6] Da,[7] and Bt,[8] have revealed a highly conserved fold that consists of two antiparallel, double-stranded β sheets [one that includes the two termini (β sheet A) and a three-residue internal β sheet B], two α helices [a short helix A initiated by cluster ligand Cys(**III**), and a long (10 member) helix B that terminates just before Cys(**IV**)] and five turns A [just before Cys(**I**), B initiated by the conserved Pro residue to the C terminus of Cys(**III**), C (connecting the two strands of β sheet B), D (connecting β sheet B and helix B), and E (initiated by the Pro after Cys(**IV**)].

NMR has addressed the solution structures of Tm, Tl, and Pf Fds. Robust solution molecular models have been reported for $Tm^{11,67}$ and Tl^{19} 4Fe Fds, the former using a homology model[11] to generate the cluster structure, and the latter[19] relying solely on NMR constraints. A high-resolution crystal structure[12] has been solved only for Tm 4Fe Fd^{ox}. Pf 4Fe Fd failed to provide usable crystals, and in a high-resolution crystal structure of the 1:1 complex between the 4Fe Fd^{ox} and its oxidoreductase, the Fd was found disordered.[59] Thus, for Pf Fd, only the secondary structures and tertiary contacts have been identified by NMR.[24,68] Each of the hyperthermostable Fds was found to exhibit essentially the same fold as found common to less thermostable Fds.[6–8] This can be viewed as surprising in view of the enormous differences in thermostability. However, similar structures are consistent with the high sequence homology among Fds from both mesophiles and hyperthermophiles. Thus, Tm and Dg exhibit sequence similarity and sequence identity of 61% and 39%, respectively, to the mesophiles Dg Fd, 83% and 73%, respectively, to Tl Fd, and 83% and 45%, respectively, to Pf Fd. The secondary structural elements for the three hyperthermostable Tm,[11,12] Tl,[19] and $Pf^{19,68}$ Fds are compared schematically in Fig. 1 to those reported for Dg Fd.[6]

A. *Tm 4Fe Fd^{ox}*

A molecular model was generated[11] based on the assignment of 51 of the 60 residues (including the four cluster ligands[55]), using the cluster geometry from crystallographically characterized Fds. The 20 generated structures displayed excellent backbone rmsd, as illustrated in Fig. 21. Calculations with and without the disulfide bond between Cys-20(**V**) and Cys-43(**II**) showed that the structure was not influenced by the disulfide bond.[11] This result is consistent with previous observations that essentially the same contacts in that part of the structure are observed in Fds from mesophiles with,[6] and without,[7] such a disulfide bond.

The secondary structural elements clearly identified were helix B, β sheet A, and four (A, B, D, E) of the five turns.[11] Limited NMR constraints suggested, but

---

[67] H. Sticht and P. Rösch, *Prog. Biophys. Molec. Biol.* **70**, 95 (1998).
[68] Q. Teng, Z. H. Zhou, E. T. Smith, S. C. Busse, J. B. Howard, M. W. W. Adams, and G. N. La Mar, *Biochemistry* **33**, 6316 (1994).

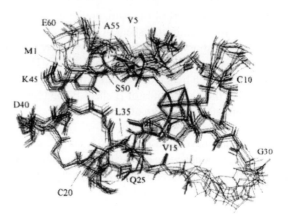

FIG. 21. Superposition of 10 backbone structures of *Tm* 4Fe Fd^{ox}. The 10 structures were obtained[11] from restrained MD calculations and showed the lowest internal energy and the smallest number of violations of the experimental data. For clarity, only a subset of the amino acid positions is labeled. [Reprinted from H. Sticht, G. Wildegger, D. Bentrop, B. Darimont, R. Sterner, and P. Rösch, Eur. *J. Biochem.* **237**, 726 (1996), with permission.]

could not confirm, the existence of the short helix A. The presence of neither β sheet B nor turn C could be confirmed.[11,55] Although two residues in this portion of the structure were not assigned,[55] the failure to uniquely form these two secondary structural elements is supported by rapid backbone labile proton exchange. Moreover, two sets of resonances were resolved for Val-34 and Leu-35. Both the β sheet B and turn C appear well formed in the crystal.[12] It is likely that this portion of *Tm* Fd exhibits a dynamic conformational heterogeneity in solution. Only 11 residues exhibited very slow peptide NH exchange times characteristic of strong backbone H-bonds.[11,55] Turn D, consisting of Pro-37–Thr-39, is homologous to that in *Dg* Fd, but the crystal structure showed[12] that it makes more backbone H-bonds to the N-terminal strand of β sheet A than in *Dg* Fd.[6]

B. *Tl* 4Fe Fd^{ox}

The molecular model was generated[19] using only NMR data based on the assignment of at least a major portion of all 59 residues.[42] The cluster structure, moreover, was generated as part of the NMR structure. In addition to the conventional NOESY, NHC$_\alpha$H spin coupling and H-bond (based on NH lability) constraints, the structure of the cluster and its environment benefited from the use of relaxation times [Eq. (1)] and the Cys Fe–S–C$_\beta$–H dihedral angle, ϕ, deduced from the contact shift pattern via Eq. (6), as well as short mixing time NOESY and steady-state NOEs, as structural constraints.[19] The nature of the constraints

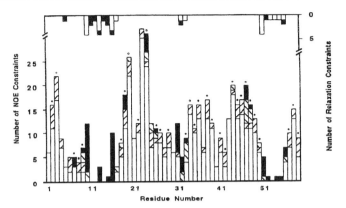

FIG. 22. Schematic representation[19] of the structural constraints used to generate the molecular model of 4Fe Tl Fd^{ox}. On the lower portion are shown the number of NOE constraints (left margin) from inter-(O) and intraresidue (▨) long (300 ms) mixing time NOESY, short mixing time (50 ms) NOESY (■), and steady state NOEs (n). The residues with estimates for $^3J(NHC\alpha H)$ are shown by asterisks. In the upper portion are shown the number of relaxation constraints (right margin) with both upper and lower R_{Fe} bounds (O), and with only upper bounds to R_{Fe} (n). [Reprinted from P.-L. Wang, A. Donaire, Z. H. Zhou, M. W. W. Adams, and G. N. La Mar, *Biochemistry* **35**, 11319 (1996), with permission.]

used are depicted schematically in Fig. 22. The superposition of the 10 accepted structures is shown in Fig. 23A and reflects a robust structure with excellent main chains on rmsd (root means square deviation). Even the cluster geometry is remarkably well-defined, as shown in Fig. 23B, in spite of the relative paucity of NMR constraints for this part of the structure.

The structure confirmed the presence of β sheets A and B. However, β sheet A involving the termini was expanded from a double to a *triple stranded* one by incorporating the third strand consisting of Val-40–Glu-42 (Fig. 1), as shown schematically in Fig. 24. The presence of both helices A and B were directly confirmed, with short mixing time NOESY and steady-state NOEs crucial to the characterization of the short helix A.[42] Four of the turns, A, B, C, E, were observed, but the turn D present in Fds from mesophiles[6-8] and in Tm Fd^{12} was replaced by the short extended segment Val-40–Glu-42 that comprises the third strand of β sheet A, as shown in Figs. 1 and 24. Nineteen very slowly exchanging peptide NHs were identified[42] and could be correlated with strong H-bonds within secondary structural elements.[19] The exceptional stability of the β sheet A near its termini is reflected in the very slow peptide NH exchange for both Lys-2 and Glu-59, and the importance of the stabilizing effect of the third strand to β sheet A is reflected in the slow exchange of the Val-3 and Thr-39 NHs due to interstrand H-bonds.

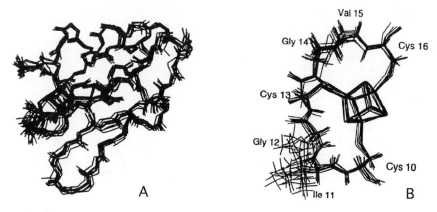

FIG. 23. (A) Superposition of backbones for 10 accepted structures of Tl 4Fe Fd^{ox} from the structural calculations. (B) Superimposition of the cluster binding loop, Cys-10–Cys-16, for the 10 accepted structures for portions of the structure with minimal constraints.[19] [Reprinted from P.-L. Wang, A. Donaire, Z. H. Zhou, M. W. W. Adams, and G. N. La Mar, *Biochemistry* **35,** 11319 (1996), with permission.]

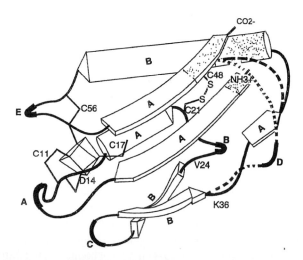

FIG. 24. Schematic structure of Pf 4Fe Fd generated by simply extending the α helix B by four residues and β sheet A by 2 residues on the N terminus, and by 1 residue at the C terminus, relative to the Tl Fd structure.[24] Solid lines illustrate the relative positions of the secondary elements, the cluster, helices A and B, β sheets A and B, and four β turns A–C, D. The position of the disulfide link involving Cys-21(**V**) and Cys-48(**VI**) is also shown. The portion between β sheet B and helix B retained in the other hyperthermostable T_m and Tl Fds is shown as a dashed line. The portion of the structure between β sheet B and helix B retained in the Fd from the Dg mesophile is shown as a dotted line. The shaded portion of the structure represents the parallel "modular" extension of helix B and β sheet A that results from the 6,7 extra residue in Pf relative to Tm or Tl Fd. [Reprinted from P.-L. Wang, A. Donaire, Z. H. Zhou, M. W. W. Adams, and G. N. La Mar, *Biochemistry* **35,** 11319 (1996), with permission.]

C. Pf Fd Solution

A molecular model of *Pf* Fd has not yet been determined, in large part because of the complications of its multiple molecular heterogeneities discussed in Section VII, which likely also complicate crystallization of the Fd. However, nearly complete ^{1}H NMR assignments for a variety of derivatives,[24,68] including 3Fe Fd$_A^{ox}$, 4Fe Fd$_A^{ox}$, 4Fe Fd$_B^{ox}$, 4Fe D14C-Fd$_A^{ox}$, as well as ^{15}N assignments in 4Fe Fd$_A^{ox}$ and 4Fe Fd$_B^{ox}$, together with the detection of the secondary structural elements and their tertiary contacts, allow at least a preliminary discussion of the unique structural features of this Fd relative to Fds from both mesophiles and other hyperthermophiles. The presence of the same secondary structural elements and tertiary contacts in *Pf* Fd could be directly established[24,68] as reported for *Tl* 4Fe Fd,[19] including the "conversion" of the turn D in the Fd from mesophiles and *Tm* Fd into a short, extended strand that expands β sheet A from antiparallel, double into a triple-stranded β sheet, as shown in Figs. 1 and 24. The lengths of the C- and N-terminal strands in β sheet A are also increased over that for either *Tl* and *Tm* Fds. Lastly, the apparent insertion of four residues in the sequence in *Pf* relative to *Tm* and *Tl* Fds (Fig. 1) manifests itself clearly as a four-residue (∼one turn) extension of the long helix B[24,68] (Figs. 1 and 24). The extension of helix B and β sheet A form close contacts so as to represent a "modular" extension of the molecular structure, as depicted schematically by the shaded portion[24] in the homology model shown in Fig. 24.

1. Effect of Cleaving Disulfide Bond. Although the NOESY cross peak pattern for 4Fe Fd$_B^{ox}$ [with free thiols for Cys-21(**V**) and Cys-48 (**VI**)] is very similar to that of 4Fe Fd$_A^{ox}$ (intact disulfide between Cys-21 (**V**) and Cys-48(**VI**), there are large chemical shift differences for the peptide NHs, as shown in Fig. 25A, and these differences are very widely distributed over the whole protein.[24] The large NH shift differences occur for Cys-21, Val-24, Lys-36, Val-37, Lys-50, Ala-52 and Glu-54, among others, indicating structural accommodation that involves both helix B and the loop linking β-sheet B to β-sheet A. The linkage among resonances with proton peaks experiencing significant shift difference upon cleaving the disulfide are shown in bold in the schematic structure of β-sheets A and B in Figure 26. The exact nature of the structural differences is yet to be determined.

2. Effect of Disulfide Orientation in Fd$_A^{ox}$. The structure of *Pf* Fd$_A^{ox}$ was shown to be heterogeneous in the presence of a disulfide bridge[9,22,24,43] (as described in Section VI) and the extensive line broadening of key residue NHs (Cys-21, Val-24, Lys-36, Val-37, Lys-50, Ala-52, Glu-54) in Fd$_A^{ox}$ complicated both the detection and interpretation of NOEs involving these residues.[24] Interestingly, precisely the same residues have their NHs broadened by the disulfide orientational heterogeneity as those experiencing the largest shift differences upon cleaving the disulfide (see above section). Hence, it is concluded that *one* of the two interconverting forms of Fd$_A^{ox}$ strongly resembles the structure of 4Fe Fd$_B^{ox}$ without the disulfide bond in which helix B is intact. The disulfide orientation heterogeneity was concluded to

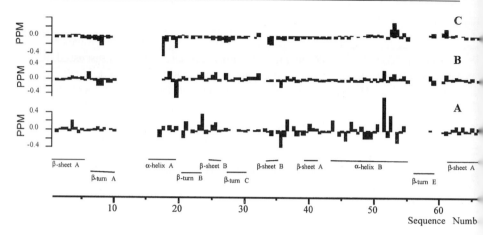

FIG. 25. Backbone proton chemical shift difference (N_pH closed bars; $C_\alpha H$ open bars) in ppm at 30°, among the four complexes of Pf Fd using 4Fe Fd_A^{ox} as reference.[24] (A) ^1H δ_{DSS}(4 Fe Fd_A^{ox})–δ_{DSS}(4Fe Fd_B^{ox}); (B) ^1H δ_{DSS} (4Fe Fd_A^{ox})–δ_{DSS}(3Fe Fd_A^{ox}); and (C) ^1H δ(4Fe Fd_A^{ox}–δ_{DSS} (4Fe D14C Fd_A^{ox}). Blanks indicate that the assignment is not available in one or the other derivative. The secondary structural element on which a residue resides is shown at the bottom. [Reprinted from P.-L. Wang, A. Donaire, Z. H. Zhou, M. W. W. Adams, and G. N. La Mar, *Biochemistry* **35**, 11319 (1996), with permission.]

be a natural consequence of the extension of helix B by one turn, which destabilizes the *S* or right-hand orientation of the disulfide observed in *Dg*, *Tm*, and *Tl* Fds.[6,11,12,19] The wide distribution of residues whose NHs are broadened by the dynamic disulfide orientational heterogeneity dictates that the structural consequences for this heterogeneity are global (see Figs. 25 and 26) rather than local as found in other small proteins with disulfide orientational heterogeneity.[58] Such a global response to the heterogeneity could account for the disordered Fd observed in the 1:1 Co-crystal with its oxidoreductase.[59]

It was concluded that the helix B is strongly bent near Ala-52 in one of the forms of Fd_A^{ox} with the *S* or right-handed disulfide conformation. Interestingly, while the alternate disulfide orientation are comparably populated at ambient temperature,[24,43] the species with the *S* orientation disappears at highly elevated temperatures where the *Pf* organisms thrive. Thus, this disulfide orientational heterogeneity reflects a "low temperature effect" on a protein that is largely homogenous at the physiological temperature of the organism. The prospects for the eventual solving of the solution NMR structure of *Pf* Fd_A^{ox} for both the *S* and *R* disulfide chiralities are improved by the observation that all of the backbone NH chemical shifts, and therefore likely the structure, of the 4Fe Fd_A^{ox} species with the *R* disulfide orientation are very similar to those of *Pf* 4Fe Fd_B^{ox} where the disulfide bond is cleaved.[24] Thus, a molecular model for *Pf* 4Fe Fd_B^{ox} is a high priority.

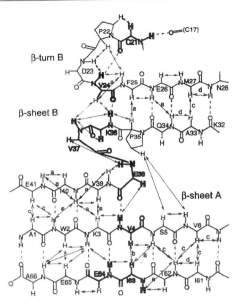

FIG. 26. Schematic representation[24] of the two β sheet of Pf 4Fe Fd showing observed NOESY cross peaks (double-sided arrows) and inferred backbone hydrogen bonds (dashed lines). (a) Cross peak is stronger in 4Fe Fd$_A^{ox}$ than in 4Fe Fd$_B^{ox}$; (b) cross peak is stronger in 4Fe Fd$_B^{ox}$ than 4Fe Fd$_A^{ox}$; (c) expected cross peaks not detected due to near degeneracy, signal under the solvent line, or paramagnetic influences; (d) observed only in a 50 ms mixing time NOESY spectrum. Residues whose chemical shifts either are affected most strongly by cleaving the disulfide bridge or are broadened most seriously by the dynamic orientational isomerism of the Cys-21(**V**)–Cys-48(**VI**) disulfide bond are shown in bold. [Reprinted from P.-L. Wang, L. Calzolai, K. L. Bren, Q. Teng, F. E. Jenney, Jr., P. S. Brereton, J. B. Howard, M. W. W. Adams, and G. N. La Mar, *Biochemistry* **38**, 8167 (1999) with permission.]

3. Effect of Cluster Architecture and Ligand Mutation. In contrast to the widely distributed NH shift changes upon cleaving the disulfide bond (Fig. 25A), the conversion of 4Fe to 3Fe (Fig. 25B) and 4Fe-D14C mutation (Fig. 25D) revealed significant shift changes primarily near the cluster, reflecting only local structural accommodation. However, at least part of helix B is coupled to the cluster environments and thus may account for the "gating" of electron transfer rate with Asp-14(**II**) relative to that with Cys-14(**II**)[20] (see Section **VIII**).

D. *Structural Implications for Hyperthermostability*

Each of the proposed mechanisms for increasing thermostability in proteins, increased hydrophobic interactions, greater packing effectiveness, more salt bridges, more H-bonds, stabilization of α helices by Ala, loop stabilization, and decreased thermolability of residues (i.e., discrimination against Asn, Gln, Cys), have been found operative in one or another hyperthermostable protein, with different mechanisms dominating in different large, and largely oligomeric, proteins.[69,70] The prospects for elucidating the "secrets" of hyperthermostability in small, monomeric proteins were dashed by the early reports of the solution and crystal structures of the ~50 residue, monomeric rubredoxin,[71–73] which, except for a one- to two-residue extension of the antiparallel, triple-stranded β sheet involving the termini, were found essentially superimposable on the structures for the rubredoxin from the mesophile *Clostridium pasteurianum*.[74]

The available solution structures[11,19,24] of hyperthermostable Fds have not provided any significant insight into the structural origin of hyperthermostability. The *Tm* Fd crystal structure has, however, been more useful.[12,75] Clearly there is no increased number of salt bridges and the packing density appears ordinary. There is an increased number of backbone H-bonds, including those between turn D and β sheet A in *Tm* Fd,[12,75] or those due to the incorporation of a third strand into β sheet A in *Tl*[19] and *Pf* Fd.[24,68] In all three Fds, there seems to be stronger interstrand H-bonds that deter "fraying" of the termini, as also observed in rubredoxin[71–73] and the long helical B does generally have more Ala (3) than Fds from mesophiles. Although the number of thermolabile residues such as Asn and Gln (2-3) in *Tm* and *Tl* Fds is comparable to that in Fds from mesophiles, this number is *increased* (5) in *Pf* Fd. Replacement of bulky side chains in the cluster loops reduces strain,[12,75] and both *Tm* and *Tl* Fd introduce two Gly in the loop. *Pf* Fd, on the other hand, has only a single Gly in the cluster loop (Fig. 1).

The optimal growth temperatures increase in the order *Tm* (80°) < *Tl* (88°) < *Pf* (100°). However, it is clear that the thermostabilities are more comparable.[3] More detailed comparisons and generalization must await a higher resolution solution structure for *Tl* Fd and the generation of a robust molecular model for *Pf* Fd. *Pf* Fd appears to reflect the most extensive expansion of secondary structure relative to Fds from mesophiles, with β sheet A extended to incorporate a third strand and lengthened by one to two residues, and helix B extends by four residues (~one full turn) and the two extensions strongly interact via the Trp-2 (β sheet A extension) and Tyr-46 (helix B extension).[24] However, these extensions appear to *function more to destabilize the disulfide bond*[24] *and make it reducible*[10,76] *rather*

[69] C. Vieille and J. G. Zeikus, *Trends Biotech.* **14**, 183 (1996).
[70] R. Jaenicke and G. Böhm, *Curr. Opin. Struct. Biol.* **8**, 738 (1998).
[71] P. R. Blake, M. W. Day, B. T. Hsu, L. Joshua-Tor, J.-B. Park, Z. H. Zhou, D. R. Hare, M. W. W. Adams, D. C. Rees, and M. F. Summers, *Protein Sci.* **1**, 1522 (1992).
[72] M. W. Day, B. T. Hsu, L. Joshua-Tor, J.-B. Park, Z. H. Zhou, M. W. W. Adams, and D. C. Rees, *Protein Sci.* **1**, 1494 (1992).
[73] P. R. Blake, J.-B. Park, Z. H. Zhou, D. R. Hare, M. W. W. Adams, and M. F. Summers, *Protein Sci.* **1**, 1508 (1992).
[74] K. D. Watenpaugh, L. C. Sieker, and L. H. Jensen, *J. Mol. Biol.* **131**, 509 (1979).
[75] S. Macedo-Ribeiro, B. Darimont, and R. Sterner, *Biol. Chem.* **378**, 331 (1999).
[76] L. F. Kress and M. Laskowski Sr., *J. Biol. Chem.* **242**, 4925 (1967).

than to stabilize the protein. Clearly more studies would be needed to define the common stabilizing interactions in hyperthermostable Fds.

Acknowledgments

The author's research on hyperthermostable Fds has been carried out in collaboration with Prof. M. W. W. Adams at the University of Georgia and was supported by grants from the National Science Foundation DMB 91-04018 and MCB 96-00759.

[31] Calorimetric Analyses of Hyperthermophile Proteins

By JOHN W. SHRIVER, WILLIAM B. PETERS, NICHOLAS SZARY, ANDREW T. CLARK, and STEPHEN P. EDMONDSON

Introduction

The enhanced stability of hyperthermophile proteins relative to those of mesophiles is the result of an interplay of many different forces. It would appear that nature uses all means at her disposal. In some thermophile proteins there is an intrinsic increase in stability through increased numbers of electrostatic interactions (including salt bridges, ion pair networks, and hydrogen bonds) and more efficient hydrophobic core packing.[1-3] However, in some the intrinsic stability is only marginal at the physiological growth temperature,[4,5] and these are stabilized by ligand binding and modification of solvent structure through increased salt and osmolyte concentrations.[6-9] A determination of the relative importance of any one of these factors requires a quantitative measure of its contribution.

A quantitative description of the individual forces contributing to protein folding and ligand binding requires determination of the thermodynamic state functions

---

[1] M. W. W. Adams, *Ann. Rev. Microbiol.* **47**, 627 (1993).
[2] C. Vieille and J. G. Zeikus, *TIBTECH* **14**, 183 (1996).
[3] R. Jaenicke and G. Böhm, *Curr. Opin. Struct. Biol.* **8**, 738 (1998).
[4] B. S. McCrary, S. P. Edmondson, and J. W. Shriver, *J. Mol. Biol.* **264**, 784 (1996).
[5] W.-T. Li, R. Grayling, K. Sandman, S. Edmondson, J. W. Shriver, and J. N. Reeve, *Biochemistry* **37**, 10563 (1998).
[6] R. Hensel and H. König, *FEMS Microbiol Lett.* **49**, 75 (1988).
[7] T. Oshima, *in* "The Physiology of Polyamines" (U. Bachrach and, Y. Heimer, eds.), pp. 35–46 (CRC Press, Boca Raton, FL, 1989).
[8] L. Martins, R. Huber, H. Huber, K. O. Stetter, M. S. Da Costa, and H. Santos, *Appl. Env. Microbiol.* **63**, 896 (1997).
[9] K. Hamana, H. Hamana, M. Niitsu, K. Samejima, and T. Itoh, *Microbios* **87**, 69 (1996).

associated with these processes for native and site-directed mutants.[10] Calorimetry is often the easiest and most reliable method for determining free energy, enthalpy, and heat capacity changes accompanying biochemical reactions. The Gibbs free energy change quantitatively describes protein stability in the case of protein folding and ligand affinity in the case of association. Most importantly for thermophile studies, the state functions also provide the temperature dependence of these processes. The temperature dependence of the free energy is determined by the enthalpy change associated with the reaction through the van't Hoff relation. The enthalpy change is also usually temperature dependent as determined by the change in heat capacity.

In more quantitative terms, we know that the equilibrium constant for a reaction at any temperature T is related to the Gibbs free energy change $\Delta G°$ by[11]

$$K(T) = e^{-\Delta G°(T)/RT} \tag{1}$$

where R is the gas constant and T is the temperature in degrees Kelvin. The free energy change at temperature T is given by

$$\Delta G°(T) = \Delta H°(T) - T\Delta S°(T) \tag{2}$$

or

$$\Delta G°(T) = \Delta H°(T_m)\left(\frac{T_m - T}{T_m}\right) - (T_m - T)\Delta C_p + T\Delta C_p \ln\left(\frac{T_m}{T}\right) \tag{3}$$

where T_m is the temperature where $K = 1$ (i.e., $\Delta G° = 0$), $\Delta H°(T)$ is the enthalpy change at temperature T, $\Delta S°(T)$ is the entropy change, and ΔC_p is the change in heat capacity.[12] Equation (3) is sometimes referred to as a modified Gibbs–Helmholtz equation. The temperature dependence of the protein stability (at a given pH, salt, and ligand concentration, if applicable) is referred to as the *protein stability curve*.[13]

One of the principal goals of calorimetry of proteins is to define the protein stability curve (Fig. 1), i.e., to define T_m, $\Delta H°(T_m)$, and ΔC_p for protein unfolding. Because ΔC_p is positive for protein unfolding,[14] the curve is concave downwards, and $\Delta G°$ is zero at two temperatures (assuming more than marginal stability). These are the heat denaturation (T_m) and cold denaturation midpoint temperatures.[12] $\Delta H°$ controls the steepness of the curve at any temperature, and

---

[10] E. di Cera, "Thermodynamic Theory of Site-Specific Binding Processes in Biological Macromolecules." Cambridge University Press, Cambridge, UK, 1995.
[11] G. N. Lewis, M. Randall, K. S. Pitzer, and L. Brewer, "Thermodynamics." McGraw-Hill, New York, 1961.
[12] P. Privalov, Y. Griko, S. Venyaminov, and V. Kutyshenko, *J. Mol. Biol.* **190**, 487 (1986).
[13] W. Becktel and J. Schellman, *Biopolymers* **26**, 1862 (1987).
[14] P. Privalov, *Adv. Protein. Chem.* **33**, 167 (1979).

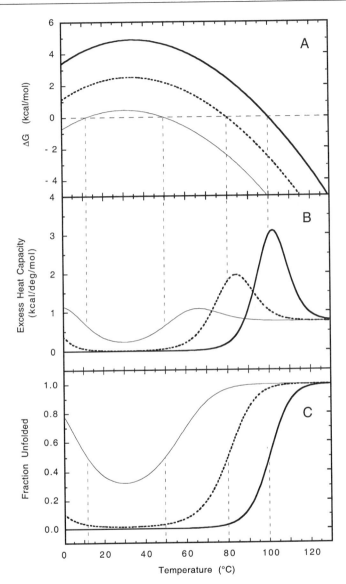

FIG. 1. Simulated protein stability curves (A) compared with thermal unfolding followed by DSC (B) and unfolding progress as might be observed by a spectroscopic probe (C). The curves in the lower two panels demonstrate the effects of varying overall stability on the data that might be obtained by varying pH. Decreasing stability leads to a decrease in $\Delta H°$, and therefore a broadening of the transition. Lower stability makes it possible to observe cold denaturation and the temperature of maximal stability. The data were simulated using Eq. (3) (A), Eq. (11) (B), and Eq. (6) (C) with $\Delta C_p = 750$ cal/deg/mol. The bold curves corresponds to a $\Delta H°$ of 54 kcal/mol, $T_m = 100°$; the dotted curves to $\Delta H = 37.5$ kcal/mol, $T_m = 80°$; and thin solid curves to $\Delta H° = 15$ kcal/mol, $T_m = 50°$.

ΔC_p determines the degree of curvature. Defining the protein stability curve facilitates a comparison of the stabilities of different proteins. For example, increasing the T_m of a hyperthermophile protein can be accomplished by shifting the protein stability curve vertically or shifting it laterally to higher temperature, or by decreasing the curvature.[4,15,16] Understanding the structural factors that control the magnitude of these parameters, and therefore the shape and position of the curve, is central to a structural understanding of hyperthermophile protein stability.

Similarly, the goal of studying binding by calorimetry is to define not only the binding affinity, but also its temperature dependence. The change in heat capacity is of particular importance in structural interpretations. Indeed, many of the same factors responsible for a ΔC_p in protein folding also play an important role in ligand binding, e.g., changes in solvent accessible surface areas.[17,18]

The thermodynamic parameters associated with heat-induced transitions (e.g., unfolding) are readily determined by differential scanning calorimetry (DSC), whereas those associated with ligand-induced changes (binding) are most reliably obtained from isothermal titration calorimetry (ITC). Van't Hoff methods that derive an enthalpy from the temperature dependence of an apparent equilibrium constant are often unreliable. Fortunately, current calorimeters are very sensitive instruments, making it straightforward to directly obtain accurate measurements of enthalpies and heat capacities. Calorimetry is especially valuable for studies of hyperthermophile proteins, since measurements at the physiological growth temperature are often readily accomplished.

The purpose of this chapter is to describe the practical use of DSC and ITC to measure protein stability and binding constants at high temperature. The goal is to go beyond simple unfolding of monomeric proteins and describe the tools needed to characterize the effects of pH, ligand binding, and oligomerization on protein stability. The advantages and disadvantages of the techniques are presented and, where possible, pitfalls are described, especially in relation to linkage of stability and binding. The emphasis is on practical applications and not on interpretation of data. The reader is referred to a number of excellent reviews for discussions of the current interpretation of thermodynamic parameters.[18-23]

It must be recognized that nature often places limitations on our ability to obtain thermodynamic information. This is especially true of hyperthermophile proteins.

[15] H. Nojima, A. Ikai, T. Oshima, and H. Noda, *J. Mol. Biol.* **116**, 429 (1977).
[16] D. C. Rees and M. W. W. Adams, *Structure* **3**, 251 (1995).
[17] P. R. Connelly, *Curr. Opin. Biotech.* **5**, 381 (1994).
[18] I. Luque and E. Freire, *Methods in Enzymology* **295**, 100 (1998).
[19] E. Freire, O. L. Mayorga, and J. M. Sanchez-Ruiz, *Ann. Rev. Biophys. Biophys. Chem.* **19**, 159 (1990).
[20] K. J. Breslauer, E. Freire, and M. Straume, *Methods in Enzymology* **211**, 533 (1992).
[21] E. Freire, *Methods in Enzymology* **259**, 144 (1995).
[22] G. Makhatadze and P. L. Privalov, *Adv. Prot. Chem.* **47**, 308 (1995).
[23] T. Lazaridis, G. Archontis, and M. Karplus, *Adv. Prot. Chem.* **47**, 231 (1995).

Accurate determination of thermodynamic changes for a chemical reaction (i.e., folding or binding) can only be obtained for reversible reactions under equilibrium conditions. Even application of hydrogen exchange data to obtain free energies of unfolding requires reversible unfolding.[24] Experimentally, a free energy change is derived from an equilibrium constant [Eq. (1)]. Irreversible reactions invariably contain a nonequilibrium, kinetic component that is difficult, if not impossible, to isolate from the reaction of interest.

It should also be recognized that reliance on a single experimental technique is often unwise. Interpretation of a result obtained under one set of conditions is even more risky. Interpretation of data collected at a single pH, ligand concentration, and temperature can be difficult, especially if the process of interest is linked to other processes that are unknown. In general, calorimetric data should be collected over a wide range of conditions and the results compared to other experimental data, e.g., folding and binding followed spectroscopically.

Differential Scanning Calorimetry

DSC has been used extensively to study mesophile protein stability.[21,22] There have been relatively few DSC studies of hyperthermophile proteins.[4,5,25-31] DSC is ideally suited for hyperthermophile protein stability studies because of the ability to collect data to 130° (and possibly to 150°).

A DSC instrument contains two metal cells, sample and reference, enclosed within a thermal shield. The temperature of both is raised simultaneously with an external heater at a rate slow enough to ensure thermal and chemical equilibrium within the cells at all temperatures. Peltier thermoelectric units associated with the sample cell maintain its temperature the same as the reference cell. The raw data is the heat flow from the Peltier device. The heat capacity for a protein as a function of temperature is typically linear below the unfolding transition (Fig. 1B). As the temperature of the sample enters a range where an unfolding transition occurs, the sample takes on an anomalous (or excess) heat capacity since a portion of the heat transferred to the sample cell drives the transition. On completion of the transition

---

[24] R. Hiller, Z. Zhou, M. W. W. Adams, and S. W. Englander, *Proc. Natl. Acad. Sci. U.S.A.* **94**, 11329 (1997).
[25] A. Wrba, A. Schweiger, V. Schultes, R. Jaenicke, and P. Zavodszky, *Biochemistry* **29**, 7584 (1990).
[26] H. Klump, J. Di Ruggiero, M. Kessel, J.-B. Park, M. W. W. Adams, and F. T. Robb, *J. Biol. Chem.* **267**, 22681 (1992).
[27] K. Laderman, B. Davis, H. Krutzsch, M. Lewis, Y. Griko, P. Privalov, and C. Anfinsen, *J. Biol. Chem.* **268**, 24394 (1993).
[28] H. H. Klump, M. W. W. Adams, and F. T. Robb, *Pure Appl. Chem.* **66**, 485 (1994).
[29] J. McAfee, S. Edmondson, P. Datta, J. Shriver, and R. Gupta, *Biochemistry* **34**, 10063 (1995).
[30] S. Knapp, A. Karshikoff, K. D. Berndt, P. Christova, B. Atanasov, and R. Ladenstein, *J. Mol. Biol.* **264**, 1132 (1996).
[31] B. S. McCrary, J. Bedell, S. P. Edmondson, and J. W. Shriver, *J. Mol. Biol.* **276**, 203 (1998).

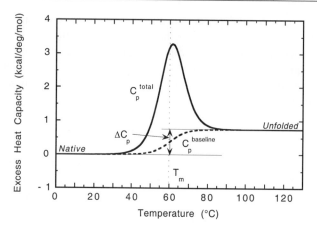

FIG. 2. Simulated DSC endotherm, C_p^{total} (bold line), for a monomeric protein unfolding reversibly via a two-state transition ($T_m = 60°$, $\Delta H° = 50$ kcal/mol, $\Delta H_{cal}/\Delta H_{vh} = 1.0$). The sigmoidal heat capacity baseline, $C_p^{baseline}$ (dashed line), shifts from the native to unfolded due to a ΔC_p of 750 cal/deg/mol. The T_m (indicated with the dotted line) differs slightly from T_{max} because of the shifting baseline. The ΔC_p of unfolding is the difference between the extrapolated native and unfolded baselines at the T_m. The area between the two curves, C_p^{total} and $C_p^{baseline}$, is the calorimetrically determined enthalpy change, ΔH_{cal}.

at higher temperature, the baseline does not necessarily return to the pretransition level since the heat capacity of the unfolded protein is usually higher than that of the native protein. A typical DSC scan (Fig. 2) is therefore composed of a linear pretransition baseline, a peak or endotherm centered at a midpoint temperature T_m with an intensity dependent on the heat (i.e., enthalpy) of the reaction, and a displaced posttransition baseline. The difference in the extrapolated baselines at the T_m represents the change in heat capacity, ΔC_p, on unfolding.

Instrument Design and Performance

The designs of modern differential scanning calorimeters have been described in detail elsewhere.[32,33] There are currently two manufacturers of high-sensitivity DSC instruments appropriate for biological calorimetry, MicroCal (Northampton, MA) and Calorimetry Sciences (Provo, UT). Their performance is comparable, although there can be significant variation among instruments even from the same manufacturer. Modern instruments can collect data during both heating and cooling. The short-term and long-term noise levels, as well as baseline reproducibility on repetitive scans and reloading, are of paramount importance. Optimum performance is obtained if the instrument is powered by an isolated electrical circuit with

---

[32] G. Privalov, V. Kavina, E. Freire, and P. L. Privalov, *Anal. Biochem.* **232,** 79 (1995).
[33] V. Plotnikov, J. M. Brandts, L.-N. Lin, and J. F. Brandts, *Anal. Biochem.* **250,** 237 (1997).

a true isolated ground. The DSC should never be left idle but should always be scanning between the lower and upper temperature limits even when not collecting data.

The cells of a DSC instrument are typically gold or tantalum and can be cylindrical, lollipop-shaped, or capillaries. Capillary cells can eliminate baseline distortions associated with precipitate formation after irreversible unfolding, although the advantage is largely cosmetic since it is difficult to analyze irreversible data.

Cell Cleaning. It is important to be aware of the chemical reactivities of the materials used in construction of the calorimeter cells and attachments (capillaries, seals, pressure-sensitive devices). Exhaustive cleaning is necessary to obtain the most reproducible data. We routinely use the following method suggested by Calorimetry Sciences. The cells are incubated for 3 hr at 30° with a solution of pepsin (1 mg/ml) in 0.5 M NaCl and 0.1 M acetic acid, washed with 100 ml of deionized water (vacuum-assisted washing devices supplied by the manufacturer permit rapid flushing of the cells with large volumes), incubated for 20 min with 50% (v/v) formic acid at 65°, washed with 100 ml of deionized water, incubated overnight at 90° with 4 M NaOH, washed with 100 ml of deionized water, incubated for 20 min at 50° with high-performance liquid chromatography (HPLC) grade tetrahydrofuran, then washed with 100 ml of 1% sodium dodecyl sulfate (SDS) solution and finally with 500 ml of deionized water. Deionized, degassed water is scanned to 130° under 3 atm pressure, cooled, and the water in both cells is discarded. Following this procedure, repetitive scans of water should yield identical baselines to within a microwatt (μW) (Fig. 3).

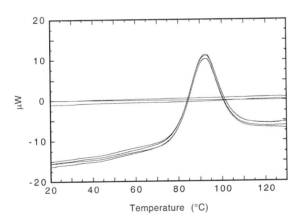

FIG. 3. Raw DSC data showing the reproducibility obtained with successive scans. The only data manipulation performed is subtraction of the first buffer vs buffer scan from all succeeding scans. The buffer vs buffer baselines centered on 0 μW are three repetitive scans (without reloading) on 10 mM potassium acetate (pH 5.7) and 0.3 M KCl. The four protein vs buffer scans were collected with four separate loadings of the same sample (Sac7d, 1.6 mg/ml in the above buffer). The deviation observed is on the order of 1 μW and is comparable to that observed for repetitive scans of buffer. Data were collected on a Calorimetry Sciences Nano II instrument with gold cylindrical cells.

Instrument Calibration. Heat flow calibration is accomplished using an internal heater under software control. Calibration pulses of various magnitudes can be applied during buffer vs buffer scans. Temperature accuracy can be checked using standards supplied in sealed ampoules by the manufacturer of the instrument. Performance can also be checked with RNase A (Sigma, St. Louis, MO). At pH 2.2 (0.1 M KCl, 0.02 M glycine) RNase A (ε_{280} of 0.69 ml/mg • cm, molecular weight 13,700) unfolds with a T_m of 36.0° and a ΔH of 74.5 kcal/mol ($\Delta H_{cal}/\Delta H_{vh}$ ratio of 1.00 ± 0.01).[34]

Data Collection

Buffers. Sample buffers should not interact with the protein and should have negligible change in pK with temperature, i.e., the heats of protonation should be small (tabulations can be found in ref. 35). Glycine and acetate are often used since they compensate protonation changes associated with protein unfolding.[14] Amines (including Tris) should be avoided because of high heats of protonation. Buffers such as glycine and acetate (even ultrapure grades) often contain impurities that create baseline artifacts, especially at high temperature. In our experience such effects are not significant after the first scan and usually are not noticeable in the presence of protein.

Sample Preparation. The behavior of protein samples should be investigated outside the calorimeter at the highest temperature expected in a DSC scan prior to running an experiment. Samples that precipitate should not be used since it is doubtful that useful information can be obtained because of irreversibility, and removal of precipitate can be difficult (especially if the cells are capillaries). Protein samples are dialyzed against the desired buffer at the required concentration, with three changes to ensure equilibration. Protein concentrations are typically on the order of 1 mg/ml, although current instruments can be used with 0.1 mg/ml for larger proteins. With a sample cell volume of 0.3 ml, this only requires about 30 ng of protein per scan. In practice about 2 ml is needed to facilitate cell loading and measuring concentration. Following dialysis, the sample and a portion of the last dialysis buffer are degassed by gentle stirring under vacuum for 10 min. A portion of the degassed sample is loaded in the DSC cell (see below), and the remainder is used to measure the sample concentration spectrophotometrically. The accuracy of the spectrophotometer should be checked using K_2CrO_4 standard solutions.[36]

Sample Loading. DSC cells must be preconditioned with the desired buffer by scanning to the upper temperature limit. The buffer in both cells is discarded on reaching 30° on the downscan, and the cells are reloaded with freshly degassed

---

[34] E. Tiktopulo and P. Privalov, *Biophys. Chem.* **1**, 349 (1974).

[35] J. J. Christensen, L. D. Hansen, and R. M. Izatt, "Handbook of Proton Ionization Heats." John Wiley & Sons, New York, 1976.

[36] A. Gordon and R. Ford, "The Chemist's Companion: A Handbook of Practical Data, Techniques, and References." John Wiley, New York, 1972.

buffer. After establishment of reproducible buffer vs buffer scans, degassed protein solution is loaded on the next downscan at 30°. The best results are obtained if degassed buffer is loaded at the same time in the reference cell.

Both buffer and sample must be loaded with care to prevent introducing air bubbles into the cells, which can lead to erratic baselines. Samples are loaded into the syringes slowly to prevent cavitation and then slowly injected into the cells. Both sample and reference cells are overloaded so that sufficient solution accumulates at the top to permit withdrawal of approximately 100 μl of solution without introducing air into the capillaries or cells. Rapid pulses of solution in and out of the cells created by oscillation of the syringe barrel back and forth quickly between the forefinger and thumb should dislodge and expel air bubbles trapped in the cell and loading capillaries.

Data Collection. The cells are capped, and 3 atm pressure is applied to prevent degassing and boiling up to 130°. Scans are performed over a temperature range sufficient to define not only the transition, but also the baselines below and above the transition. The sample should never be allowed to freeze, as this may damage the cells. The scan rate should be sufficient to achieve an adequate signal-to-noise ratio (signal-to-noise increases with scan rate), but not so fast as to exceed the unfolding rate. A rate of 1 deg/min is normal, but various rates should be checked to ensure that there is no scan rate dependence in the data (Fig. 4). The data collection rate is typically on the order of 10 points/min, providing 1000 data points in a 0° to 100° scan at 1 deg/min.

Reversibility. Unfolding must be reversible and is monitored by determining the reproducibility of repetitive scans on the same sample. For proteins that unfold at moderate temperatures (e.g., 60° to 80°), this can be accomplished by comparing

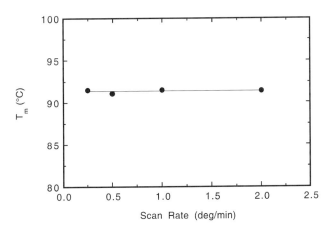

FIG. 4. The scan rate dependence of the T_m observed for DSC of Sac7d. The negligible scan rate dependence (0.19° standard deviation) indicates that the unfolding of Sac7d is at equilibrium during the DSC scan at rates at least up to 2 deg/min (the maximal scan rate available).

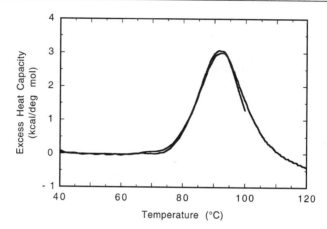

FIG. 5. The reversibility of the unfolding of Sac7d demonstrated by DSC with termination of the first DSC scan at 100°, rapid cooling (2 deg/min), and rescanning to 120°. The two scans coincide to within the error expected for repetitive scans (see Fig. 2). The pretransition slopes have been adjusted to zero since rapid cooling leads to baseline perturbation.

the ΔH_{cal} values for repetitive scans. For hyperthermophile proteins with T_m values greater than 90°, this can be difficult because of lengthy exposure of the unfolded protein to high temperature. For example, the protein will be above 100° for about 1 hr during scanning to 130° at a rate of 1 deg/min. It is not normally necessary to go beyond the T_m to demonstrate reversibility. If the relaxation time for the unfolding reaction at the T_m is less than 1 sec, the protein unfolds and refolds on the order of a hundred times per minute. A protein at the T_m will have unfolded and refolded more than a thousand times at a scan rate of 1 deg/min. Thus, duplication of successive scans to the midpoint, or slightly above, should be sufficient (Fig. 5).

Data Analysis

The raw data in DSC is the heat flow into the sample cell per unit time (e.g., µJ/sec or µW). Division by the scan rate (e.g., 1 deg/min) provides the heat flow per degree change and represents the temperature dependence of the heat capacity of the solution in the sample cell.

Absolute Heat Capacity. The absolute partial molar heat capacity of the protein can be calculated[37] using the temperature dependence of the heat capacity of the protein solution (i.e., a protein DSC scan), the heat capacity of the buffer (i.e., a buffer scan), the sample cell volume, the protein concentration, and the partial specific volume of the protein (calculated from the amino acid composition).[38]

---

[37] P. Privalov and S. Potekin, *Methods in Enzymology* **131**, 4 (1986).
[38] G. Makhatadze, V. N. Medvedkin, and G. Privalov, *Biopolymers* **30**, 1001 (1990).

Routines for performing this calculation are incorporated into commercial DSC software. Division by the molecular weight provides the absolute partial specific heat capacity.

Note that calculation of the absolute heat capacity requires two scans: one of buffer vs buffer and one of protein vs buffer. The buffer scan is usually the scan of buffer collected immediately prior to loading the protein. The two must be highly reproducible for the result to be meaningful. Attention must be paid to cell cleaning, preconditioning, and sample loading.

The partial specific heat capacity of the folded state is typically[39] on the order of 0.3 cal K^{-1} g^{-1} at 25° with a linear temperature dependence of 0.001 to 0.002 cal K^{-1} g^{-1}. The partial specific heat capacity of the unfolded state is approximately 0.5 cal K^{-1} g^{-1} and varies in a nonlinear fashion with temperature, reaching a plateau above about 70°. It can be calculated from the amino acid composition of the protein if there is no residual structure.[39,40] In practice, the baseline above the transition contains contributions from effects other than the heat capacity of the unfolded protein (e.g., chemical modification of the unfolded chain and aggregation[41]) so that the experimental measurement of the absolute heat capacity of the unfolded chain can be unreliable. Thus, although it might be expected that DSC should be an ideal method to determine differences in heat capacities, in practice the measurement obtained from a difference between the extrapolated baselines is not used.

Excess Heat Capacity. The total absolute heat capacity of the sample is the sum of the excess heat capacity and the heat capacity baseline, $C_p^{baseline}$, which represents a progress curve from folded to unfolded species[42] (Fig. 2):

$$C_p^{total} = C_p^{excess} + C_p^{baseline} \tag{4}$$

For a two-state unfolding reaction $N \rightleftharpoons U$, the baseline is a weighted sum of two curves representing the heat capacities of the native and unfolded species, with the weights determined by the relative populations of the native and unfolded states, P_N and P_u, respectively:

$$C_p^{baseline} = P_N(T)(A + BT) + P_U(T)(C + DT) \tag{5}$$

$$C_p^{baseline} = [1 - \alpha(T)](A + BT) + \alpha(T)(C + DT) \tag{6}$$

$\alpha(T)$ is the progress of the reaction as a function of temperature with values between 0 and 1, given by

$$\alpha(T) = \frac{[U]}{[N] + [U]} = \frac{K_{un}(T)}{1 + K_{un}(T)} \tag{7}$$

---

[39] G. Makhatadze, *Biophys. Chem.* **71**, 1 (1998).
[40] M. Häckel, H.-J. Hinz, and G. R. Hedwig, *Thermochim. Acta* **308**, 23 (1998).
[41] T. Ahern and A. Klibanov, *Science* **228**, 1280 (1985).
[42] J. Sturtevant, *Ann. Rev. Phys. Chem.* **38**, 463 (1987).

$C_p^{baseline}$ is a sigmoidal curve not unlike that observed with spectroscopic probes. If T is expressed relative to the midpoint temperature, T_m, where $K_{un} = 1 (\Delta G_{un}^\circ = 0$ and $\alpha = 0.5$), then the term C in Eqs. (5) and (6) is A plus the change in heat capacity (ΔC_p) at T_m. Note that for a two-state reaction, T_m represents the maximum in C_p^{excess} but not C_p^{total}. Subtraction of $C_p^{baseline}$ from C_p^{total} provides C_p^{excess}.

DSC provides two independent measures of the enthalpy change. The calorimetric enthalpy is the total excess heat absorbed, which is the area under the endotherm after removal of $C_p^{baseline}$:

$$\Delta H_{cal} = \int C_p^{excess} dT \tag{8}$$

Since ΔH is temperature dependent, this is actually ΔH° at T_m. DSC also provides the progress of the reaction as a function of temperature by comparing the percentage of the endotherm completed at any temperature to the total endotherm. Thus, DSC provides the temperature dependence of K_{un}, from which an enthalpy can also be obtained from the van't Hoff relation[11] and is called the van't Hoff enthalpy:

$$\Delta H_{vh} = \frac{-R d(\ln K_{un})}{d(1/T)} \tag{9}$$

This is best obtained from a direct fitting of the data, although an effective van't Hoff enthalpy, ΔH_{vh}^{eff}, has been defined by Privalov and Khechinashvili[43]:

$$\Delta H_{vh}^{eff} = \frac{4RT_m^2 C_{p,max}}{\Delta H_{cal}} \tag{10}$$

This latter function is overly dependent on the accuracy of a single data point, i.e., $C_{p,max}$, and therefore is less reliable than a direct fit of the entire data set.

The calorimetric and van't Hoff enthalpies should be the same for a two-state reaction ($N \rightleftharpoons U$) that proceeds from completely folded to unfolded protein. It is often argued that the $\Delta H_{cal}/\Delta H_{vh}$ ratio indicates whether or not a reaction is two-state. However, care must be taken in application of this criterion.[31,44] Marginally stable proteins may not be completely folded at the beginning of a thermal melt (Fig. 1, thin solid line), and then the measured ΔH_{cal} will be less than ΔH_{vh} even if the unfolding is two-state. Similarly, such data can yield an apparent T_m that is not an accurate reflection of where $\Delta G_{obs}^\circ = 0$ (see also ref. 45).

Thermodynamic Parameter Estimation by Nonlinear Regression. Although commercial DSC software contains routines for extracting thermodynamic parameters from data, we have found these too restrictive. In general, users must

---

[43] P. Privalov and N. Khechinashvili, *J. Mol. Biol.* **86**, 665 (1974).
[44] E. Freire, in "Protein Stability and Folding" (B. Shirley, ed.). Humana Press, Totowa, NJ, 1995.
[45] D. T. Haynie and E. Freire, *Anal. Biochem.* **216**, 33 (1994).

write their own data analysis software to achieve the necessary flexibility required to fit data with the most appropriate model. Two approaches are commonly used to derive thermodynamic parameters from excess heat capacity data: (1) statistical mechanical deconvolution[46,47] and (2) nonlinear regression.[42] Although the first method is quite elegant, it cannot provide error limits for the derived parameters, and in actual practice the final parameters are further refined by nonlinear regression.[44]

Monomeric Protein Stability. DSC data for a protein that is suspected or known to be monomeric is initially fit by assuming a two-state transition, i.e., $\Delta H_{cal}/\Delta H_{vh} = 1$, with a nonzero ΔC_p. If the fit is not satisfactory, ΔC_p (i.e., $C_p^{baseline}$) is removed from the data as described below, and the data is fit by allowing ΔH_{cal} and ΔH_{vh} to float independently.

Two-state unfolding transition ($\Delta H_{cal} = \Delta H_{vh}$) *with a nonzero* ΔC_p. Given initial values of $\Delta H°(T_m)$, T_m, and ΔC_p, one can calculate $\Delta G°$ and therefore K and $\alpha(T)$ at any temperature [Eqs. (1), (3), and (7)]. The excess heat capacity in a DSC scan is given by the incremental change in α at temperature T multiplied by the molar enthalpy change.[42]

$$C_p^{excess} = \Delta H°(T)\frac{d\alpha(T)}{dT} \quad (11)$$

where $\Delta H°$ is temperature dependent:

$$\Delta H°(T) = \Delta H°(T_m) + \int_{T_m}^{T} \Delta C_p \, dT \quad (12)$$

$$\Delta H°(T) = \Delta H°(T_m) + (T - T_m)\Delta C_p \quad (13)$$

DSC data can be fit to Eq. (11) with three adjustable parameters [$\Delta H°(T_m)$, T_m, ΔC_p], plus three additional parameters from Eq. (6), A, B, and D. Fitting of experimental data collected on the hyperthermophile protein Sac7d using this approach is demonstrated in Fig. 6. An overlay of $C_p^{baseline}$ using the derived parameters is provided. For more thermostable proteins the posttransition baseline may not be defined by the data. In such cases only A, B, $\Delta H°(T_m)$, and T_m can be fitted.

Non-two-state unfolding transition with $\Delta H_{cal} \neq \Delta H_{vh}$. The simultaneous unfolding of multiple independent domains with similar midpoint temperatures, such that only one endotherm is observed, results in the observed ΔH_{cal} being greater than ΔH_{vh}. A simple example might be the overlapping of two two-state transitions with identical ΔH_{cal} values; as a result, the area of the endotherm (i.e., the total observed ΔH_{cal}) would be twice that expected from the width of the

---

[46] E. Freire and R. Biltonen, *Biopolymers* **17**, 463 (1978).
[47] S.-I. Kidokoro and A. Wada, *Biopolymers* **26**, 213 (1987).

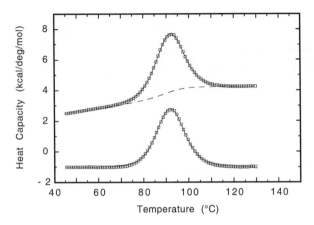

FIG. 6. Fitting of DSC of Sac7d by nonlinear regression. Only one-tenth of the data points fitted are indicated by the open squares. The upper curve shows a fit (solid line) of the absolute molar heat capacity and the derived baseline (dashed line) assuming a two-state transition ($\Delta H_{cal}/\Delta H_{vh} = 1.0$) with six fitting parameters: $\Delta H°(T_m)$, T_m, ΔC_p, the slope and y intercept of the pretransition baseline, and the slope of the posttransition baseline (the y intercept of the posttransition baseline is the sum of the pretransition baseline y intercept and ΔC_p). The parameters from the fit are $\Delta H°(T_m) = 62.5$ kcal/mol, $T_m = 90.5°$, and $\Delta C_p = 665$ cal/deg/mol. The lower curve shows a fit (soild line) of the same data after removal of the $C_p^{baseline}$ (the curve is displaced by -1000 cal/deg/mol for clarity). The three parameter nonlinear regression gives $\Delta H(T_m) = 64.2$ kcal/mol, $T_m = 92.0°$, and $\Delta H_{cal}/\Delta H_{vh} = 0.97$.

transition. If

$$\beta = \frac{\Delta H_{cal}}{\Delta H_{vh}} \tag{14}$$

then

$$C_p^{excess} = \beta \, \Delta H°(T) \frac{d\alpha(T)}{dT} \tag{15}$$

The magnitudes of the ΔC_p values for unresolved transitions cannot be defined by the data. It is not possible to use the complete expression for $\Delta G°$ [Eq. (3)] in fitting the data. Therefore, the ΔC_p, or more specifically $C_p^{baseline}$, must be removed prior to nonlinear regression as follows.

The progress of the reaction at any temperature T is obtained by calculating a progress function, $I(T)$, by taking the ratio of the integral of C_p^{total} from any arbitrary temperature, T_i, preceding the transition to any temperature, T, divided by the total integral over the transition. Numerically, $I(T)$ is defined as follows:

$$I(T) = \frac{\sum_{t=T_i}^{T} C_p^{total}(T)}{\sum_{t=T_i}^{T_i} C_p^{total}(T)} \tag{16}$$

$I(T)$ is linear before and after the transition, with a sigmoidal transition in between. An equilibrium constant is obtained from the progress curve at any temperature T as commonly done with spectroscopic data[48] [see Eq. (18) below; a two-state reaction can be assumed for this purpose without introducing significant error], and $\alpha(T)$ is calculated using Eq. (7). $C_p^{baseline}$ is calculated according to Eq. (5) using $\alpha(T)$ and values for A, B, C, and D obtained from fitting segments of the baseline before and after the transition in the DSC scan. $C_p^{baseline}$ can be smoothed by nonlinear regression. The fitted $C_p^{baseline}$ curve is subtracted from the original data, i.e., C_p^{total}, to obtain C_p^{excess}, which is then fitted to obtain ΔH_{cal}, ΔH_{vh}, and T_m. An example of fitting data in this manner is also shown in Fig. 6.

Determination of ΔC_p. As indicated above, determination of ΔC_p from the difference in the extrapolated DSC baselines at the T_m is unreliable and is never done. The most common method takes advantage of the linkage of pH to stability (see Linkage section). Decreasing the T_m by decreasing pH (typically between pH 2 and 4) leads to a decrease in ΔH (see Fig. 1) according to the Kirchhoff relation[11]:

$$\Delta C_p = \frac{\partial (\Delta H^\circ)}{\partial T} \qquad (17)$$

A plot of the resulting ΔH° vs T_m (i.e., a Kirchhoff plot) is commonly linear and the slope provides ΔC_p.[14] Although this approach is quite common, it must be realized that it can give erroneous results in the presence of ion binding at low pH.[31] In this case, the ΔC_p must be obtained from a global analysis of data collected as a function of pH and salt concentration (see Linkage section).

An alternative procedure is to fit a family of DSC curves measured at different pH values globally to a common ΔC_p and $\Delta H^\circ(T_m)$. The number of variables fitted for n DSC curves decreases from $2n + 2$ to $n + 2$, excluding any baseline parameters. This significantly increases the precision of determining ΔC_p and the protein stability curve.

Another reliable method of determining ΔC_p is available if cold denaturation occurs.[12,49] When both cold and heat denaturation are observable in the same scan, the curvature of the protein stability curve is defined by a single DSC scan (Fig. 1B, thin solid line). Since ΔC_p determines the curvature, fitting the DSC data over both cold and heat denaturation regions using Eqs. (3), (6), (7), and (11) accurately specifies ΔC_p.

ΔC_p can also be obtained from thermal denaturation followed by spectroscopic probes if cold denaturation can be observed[5] (Fig. 1C, thin solid line). Thermal denaturation curves followed by CD or fluorescence can be fit by nonlinear regression

---

[48] M. Eftink, *Methods in Enzymology* **259**, 487 (1995).
[49] J. M. Scholtz, *Protein Sci.* **4**, 35 (1995).

using Eqs. (1), (3), and (7) above where K_{un}

$$K_{un} = \frac{y_{obs} - (y_n + m_n T)}{(y_u + m_u T) - y_{obs}} \qquad (18)$$

y_{obs} is the observed signal intensity, y_n and m_n are the y intercept and slope of the linear baseline prior to the transition, and y_u and m_u are the y intercept and slope of the baseline following the transition.

Irreversible Unfolding Compared to Reversible Unfolding of Monomer. The unfolding of many hyperthermophile proteins has proven to be irreversible. Although it is not possible to derive thermodynamic information from data on irreversible systems, simulations can provide insight into the limits that irreversibility imposes. We consider here the irreversible unfolding of a monomer. The results can be extrapolated to more complicated systems.[50] The irreversible thermal unfolding of a protein might be described as follows[51]:

$$N \rightleftharpoons U \to I \qquad (19)$$

where N unfolds reversibly to U, which then proceeds irreversibly to I. Irreversibility may be due to chemical modification of the unfolded protein, aggregation, etc.[41] The thermostability of such a system reflects not only the equilibrium constant between native and folded forms, but also the kinetics of the succeeding step (or steps). Thus, irreversibility can be more a reflection of the unfolded protein than of the native, so that attempting to understand the thermostability of the protein in terms of its native structure can be risky.

The presence of the irreversible step can significantly distort a DSC endotherm. The equilibrium between N and U is dynamic, and at any temperature there is a finite probability that N will unfold and quickly refold well below the T_m of the transition. Heat is absorbed and then given back to the surrounding lattice with no net change. As the temperature increases, the rates of these reactions increase, including the rate of the second step in irreversible systems. At sufficiently high temperature the irreversible step will effectively compete with refolding, and the heat absorbed on unfolding is not regained. The result is a shifting of the endotherm to lower temperature and a distortion of its shape (Fig. 7). Because of the kinetic aspect of the distortion, the degree to which the peak is shifted is scan rate dependent. Thus, one criterion for reversibility is the demonstration of scan rate independence of the T_{max}.

It should be recognized that the second reaction is not necessarily "silent" thermodynamically, i.e., significant heat changes may be associated with the chemical events leading to irreversibility. An endotherm for the unfolding process can be

---

[50] B. I. Kurganov, A. E. Lyubarev, J. M. Sanchez-Ruiz, and V. L. Shnyrov, *Biophys. Chem.* **69**, 125 (1997).

[51] R. Lumry and H. Eyring, *J. Phys. Chem.* **58**, 110 (1954).

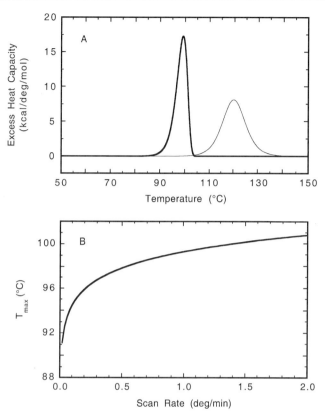

FIG. 7. The effect of irreversibility on DSC data. (A) The excess heat capacity was simulated (bold line) using the expressions given in ref. 19 assuming that unfolding can be described by $N \rightleftharpoons U \rightarrow I$ (see text). The unfolding reaction is described by a $T_m = 120°$ and $\Delta H = 100$ kcal/mol. The kinetics of the second step is described by a frequency factor $\ln A = 40$, an activation energy $E_a = 28$ kcal/mol, and a scan rate of 1 deg/min. For comparison, the DSC curve for reversible unfolding $N \rightleftharpoons U$ is shown (thinner line) with identical T_m and ΔH values. Only C_p^{excess} is shown to clearly indicate the distortion in the shape of the endotherm caused by irreversibility. (B) The temperature of the maximum of the DSC endotherm, T_{max}, observed for irreversible unfolding is scan rate dependent.

increased or diminished, and its midpoint temperature shifted, because of overlap with an endothermic or exothermic irreversible process (or processes) that occurs essentially simultaneously. There are no reliable methods that permit the unambiguous deconvolution of such transitions.

Oligomer Stability. Many thermophile proteins are oligomers that require some special considerations—specifically the need to specify concentration when discussing stability. Total protein concentration does not enter into the expressions for the stability of monomeric proteins. This is not the case for oligomers because of the increase in the number of particles upon unfolding. Le Chatelier's principle requires that unfolding be promoted by decreasing concentration.

The expressions for fitting monomeric protein data cannot be applied to oligomeric systems. We describe here the unfolding of a dimer to demonstrate the differences, and the results can be extended to higher order oligomers.[52] In the simplest case, a dimer unfolds to give two identical random coil monomers:

$$N \rightleftharpoons 2U \qquad (20)$$

The total concentration of protein, $[P]_t$, can be expressed in terms of monomer chains such that

$$[P]_t = [U] + 2[N] \qquad (21)$$

The equilibrium constant for unfolding is given by

$$K_{un} = \frac{[U]^2}{[N]} = \frac{[U]^2}{0.5([P_t] - [U])} = \frac{2[U]^2}{[P_t] - [U]} \qquad (22)$$

Solving for $[U]$ provides an expression for the concentration of unfolded monomer in terms of the equilibrium constant K_{un} and the total protein concentration[53]:

$$[U] = \frac{-K_{un}}{4} + \frac{K_{un}}{4}\left(1 + \frac{8[P_t]}{K_{un}}\right)^{1/2} \qquad (23)$$

The equilibrium constant is determined by the free energy of unfolding the dimer at any temperature T by Eq. (1).

The free energy of unfolding the dimer is given by:

$$\Delta G°(T) = \Delta H°(T°)\left(\frac{T° - T}{T°}\right) - (T° - T)\Delta C_p + T\Delta C_p \ln\left(\frac{T°}{T}\right) \qquad (24)$$

where $\Delta H°(T°)$ is the standard state enthalpy of unfolding the dimer at $T°$, ΔC_p is the change in heat capacity for unfolding the dimer, and $T°$ is the temperature at which the free energy of unfolding of a 1.0 M protein solution is zero (the standard state is defined as 1 M total protein). T_m is $T°$ only at 1 M protein. Note that the standard state is not normally accessible experimentally; it is a reference state obtained by fitting the data.

The progress of the unfolding reaction as a function of temperature is given by the temperature dependence of the fraction of protein that exists as unfolded monomer:

$$\alpha(T) = \frac{[U]}{2[N] + [U]} = \frac{[U]}{[P_t]} = \frac{(K_{un}[N])^{1/2}}{[P_t]} \qquad (25)$$

[52] E. Freire, *Comments Mol. Cell. Biophys.* **6**, 123 (1989).
[53] C. Steif, P. Weber, H.-J. Hinz, J. Flossdorf, G. Cesareni, and M. Kokkinidis, *Biochemistry* **32**, 3867 (1993).

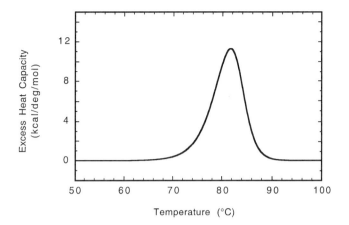

FIG. 8. The effect of oligomerization on DSC data. A simulation of C_p^{excess} for the reversible unfolding of a 1 μM dimer is shown with $T° = 100°$, $\Delta H = 100$ kcal/mol, and $\Delta C_p = 1000$ cal/deg/mol. Only C_p^{excess} is shown to clearly indicate the distortion in the shape of the endotherm caused by oligomerization.

A DSC endotherm is proportional to the partial derivative of α with respect to T [Eq. (11)], and the thermodynamic parameters are obtained by nonlinear regression.

Simulations of DSC data using these expressions demonstrate that unfolding of an oligomer leads to an asymmetric endotherm (Fig. 8). Because of the asymmetry of the transition, the temperature of the midpoint of the transition, T_m, is not the same as the maximum of the DSC endotherm or the inflection point of the progress curve.[52] By definition, $\alpha(T)$ is equal to 0.5 at the T_m, and since

$$K_{un} = \frac{2\alpha^2 [P_t]}{1 - \alpha} \quad (26)$$

the standard state free energy of unfolding is not zero at $\alpha(T_m) = 0.5$, but [53]

$$\Delta G° = -RT_m \ln[P]_t \quad (27)$$

which is only zero when $[P]_t = 1 M$. The free energy of unfolding is concentration dependent, and therefore so is T_m and the endotherm maximum. Both increase logarithmically without limit with increasing protein concentration (Fig. 9). Because of the concentration dependence, thermodynamic parameters for oligomers cannot be compared at different concentrations. Comparisons are typically done at $1 M$ protein, the extrapolated reference state.

Error Analysis. Error estimates for all parameters derived from nonlinear regression of DSC data are obtained using standard techniques that characterize the

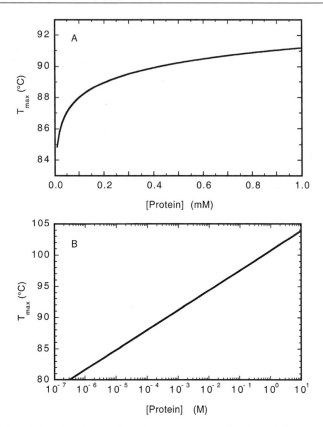

FIG. 9. The stability of an oligomer depends on concentration. (A) The variation of T_{max} with concentration is shown for a dimer with $T° = 100°C$, $\Delta H = 100$ kcal/mol, and $\Delta C_p = 1000$ cal/deg/mol. The concentration dependence is logarithmic (B) and does not plateau at high concentrations. The $T°$ is defined as the T_m at 1.0 M protein, 100°, which differs slightly from the T_{max} that would be observed at 1.0 M protein because of the asymmetry of the endotherm.

shape of the chi-squared surface in the vicinity of the minimum.[54,55] The Monte Carlo method of confidence interval estimation[56] is especially attractive for this analysis. Errors in thermodynamic functions such as $\Delta G°$ are derived from fitted parameters using standard error propagation formulas.[4,54,57]

[54] P. R. Bevington and D. K. Robinson, "Data Reduction and Error Analysis for the Physical Sciences." McGraw-Hill, New York, 1992.
[55] M. Johnson and L. Faunt, *Methods in Enzymology* **210,** 1 (1992).
[56] W. Press, B. Flannery, S. Teukolsky, and W. Vetterling, "Numerical Recipes: The Art of Scientific Computing (Fortran Version)." Cambridge University Press, Cambridge, UK, 1989.
[57] L. Swint-Kruse and A. D. Robertson, *Biochemistry* **34,** 4724 (1995).

Isothermal Titration Calorimetry

There has been little application of ITC to studies of hyperthermophile proteins,[58,59] and none of these used ITC to characterize binding at physiologically relevant temperatures. This is surprising because the technique is especially useful for measuring binding parameters at temperatures up to 110°.

Similar to DSC, an ITC instrument contains two metal (Hastelloy) cells, sample and reference, contained within a thermal shield. The temperatures of the cells are kept constant during titration of concentrated ligand into the sample cell from a computer controlled motor driven syringe. Nearly all reactions are associated with heat changes,[60] and binding of the ligand leads to either absorption or release of heat depending on whether the binding is endothermic or exothermic. The temperature of the surrounding thermal bath is maintained a few degrees lower than the sample and reference cells, which are regulated at the desired temperature with a heater. The temperature differential permits the observation of both exothermic and endothermic reactions, since the heat change associated with binding modifies the amount of heat required to maintain the differential (exothermic reactions lead to a decrease in heat required and endothermic reactions require more). The deviation from the baseline level of heat input into the cells provides the heat associated with binding of the injected ligand.

Instrument Design and Performance

The design of an ITC instrument is described in detail elsewhere.[61] In the conventional configuration supplied by one manufacturer (Calorimetry Sciences, Provo, UT), there are two removable cylindrical cells with volumes of about 1.3 ml. Each cell has three capillary access ports. A motor-driven syringe is inserted into one port of the sample cell, stirring is accomplished with a continuously rotating propeller inserted through another port, and sample loading is accomplished through the third port. Best results are obtained if data are collected with slightly overfilled cells. The reference cell is typically filled with water. Optimum performance is obtained if the instrument is powered by an isolated electrical circuit with a true isolated ground. The upper temperature limit of currently available instruments appropriate for biochemical studies ranges from 80° to 100°.

Cell Cleaning. Cells are cleaned by disassembly and sonicating in a detergent mixture, followed by thorough rinsing with distilled water. Cell parts are dried in an oven at 110° for 30 min prior to reassembly. It is essential that the gaskets be dried thoroughly, given the high heat of vaporization of water. Evaporation of

---

[58] T. Lundbäck and T. Härd, *J. Phys. Chem.* **100**, 17690 (1996).
[59] T. Lundbäck, H. Hansson, S. Knapp, R. Ladenstein, and T. Härd, *J. Mol. Biol.* **276**, 775 (1998).
[60] H.-J. Hinz, "Thermodynamic Data for Biochemistry and Biotechnology." Springer-Verlag, Berlin, 1986.
[61] T. Wiseman, S. Williston, J. F. Brandts, and L.-N. Lin, *Anal. Biochem.* **179**, 131 (1989).

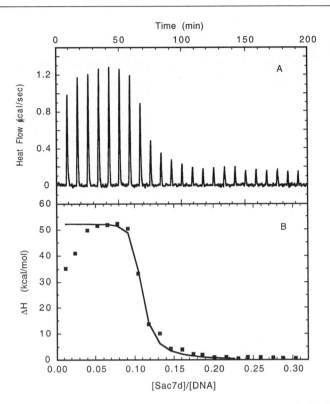

FIG. 10. Isothermal titration calorimetry of the binding of Sac7d to DNA at 80° showing the heat flow (A) following a series of 10 μl injections of Sac7d (0.55 mM) into poly(dG-dC) • poly(dG-dC) (0.34 mM) in 10 mM KH$_2$PO$_4$ (pH 6.8) and 25 mM KCl. The baseline has been flattened by subtraction of a polynomial fit of the raw data. A fit of the integrated heats per mole of injected protein (B) with the McGhee–von Hippel model is shown with the solid curve through the data ($K = 1.04 \times 10^7 M^{-1}$, site size = 4.2 base pairs, $\Delta H° = -52.5$ kcal/mol). The first two points are inaccurate because of diffusion from the injection needle during equilibration and were not included in the fit.

1 μl of water is associated with the absorption of 1000 μcal of heat, which is significantly greater that the heat observed in a typical ITC injection (Fig. 10).

Calibration. Calibration is accomplished by using internal heater pulses similar to that described above for the DSC. Note that it is important to calibrate the instrument at each temperature in variable temperature work. Most ITC experiments are run with overfilled cells, and the effective volume of the calorimeter is defined using a reaction of known heat (e.g., titration of 18-*crown*-6 ether with barium chloride[62]). The binding of 2'-CMP to RNase A is commonly used as a benchmark test for evaluating the performance of calorimeters.[61]

[62] P. Wang, R. M. Izatt, S. E. Gillespie, J. L. Oscarson, X. X. Zhang, C. Wang, and J. D. Lamb, *J. Chem. Soc. Faraday Trans.* **91,** 4207 (1995).

Data Collection

Experimental Design. The product $[P]_t K$ must be in the range of 50 to 1000 to obtain useful data to define both the binding affinity and heat of binding.[61] Higher values lead to data that cannot define the binding affinity, but reliable heats can still be measured. Note that for very tight binding it is necessary to lower the protein concentration to where the signal-to-noise ratio becomes too low. The concentration of the protein and ligand must be adjusted so that the midpoint of the titration appears about midway through the titration to define the titration curve before and after equivalence. The stock titrant concentration in the syringe is typically on the order of 100 times more concentrated than the solution being titrated; thus, solubility limits may dictate the macromolecule concentration. An estimate of the binding affinity and $\Delta H°$ can be used with simulation programs to determine the optimal concentrations of protein and ligand.

Sample Preparation. The protein and ligand solutions must be closely matched to prevent large heats of dilution. Although these can be removed in the data analysis, they are best kept to a minimum. If possible, the protein and ligand should be dialyzed against the same buffer solution. The ligand can be dissolved in the buffer that the protein was last dialyzed against if it is too small to dialyze.

Buffer considerations are as described above for DSC. Heats of protonation must be considered if binding is associated with ionization changes, and methods for removing the effects of protonation using linkage analysis have been described.[63]

Cell Loading. Sample and reference buffer are loaded into the cells using the same procedure described above for DSC to eliminate air bubbles. Erratic performance usually indicates either bubbles or leaking cells. The cells are overfilled outside the calorimeter to about 2 mm above the tip of the loading capillary and then inserted stepwise into the calorimeter to gradually equilibrate the cells and minimize perturbation of the calorimeter. The syringe is put into place at the end of the equilibration to minimize the effects of leakage from the tip of the injection needle. Total equilibration requires approximately 2 hr.

Data Collection. Data collection is under computer control with typically 20 to 25 injections of 5 to 10 μl each (Fig. 10). Each injection is accompanied by a heat change until the binding sites are saturated. A small heat change is also typically associated with dilution of the ligand and can be seen at the end of the injection series after saturation. There must be sufficient time between injections to permit the heat flow to return to baseline. A typical titration requires about 2 hr.

Data Analysis

Software supplied by the ITC manufacturer can be used to perform straightforward binding analysis. More complicated models can also be added as

---

[63] B. M. Baker and K. P. Murphy, *Biophys. J.* **71**, 2049 (1996).

user-defined external modules. However, as with DSC, the flexibility achieved from writing one's own software is often essential. This is especially true if closed form expressions cannot be written for the concentration of complex in terms of ligand concentration, binding affinity, and number of sites.

The binding of a ligand L to a macromolecule N is studied by injecting L into N. The cell is overfilled, and therefore the reaction volume observed by the calorimeter remains constant. With the ith injection of L of volume Δv into a constant cell volume c, the concentration of N decreases because of displacement of a volume equal to the injection volume, so that the concentration of macromolecule after the ith injection is given in discrete numerical form by;

$$N_i = N_{i-1} - \frac{\Delta v}{c} N_{i-1} \tag{28}$$

In differential form this becomes:

$$\frac{\partial N}{N} = -\frac{\partial v}{c} \tag{29}$$

which can be integrated to give[64]:

$$N_i = N_0 e^{-i\Delta v/c} \tag{30}$$

where N_0 is the starting concentration of macromolecule in the cell. During the titration, the concentration of the ligand increases with each injection i, but a slight amount is also lost due to displacement of the cell contents with each injection. The concentration of ligand after the ith injection is given by:

$$L_i = L_{i-1} + L_0 \frac{\Delta v}{c} - L_{i-1} \frac{\Delta v}{c} \tag{31}$$

where L_0 is the stock ligand concentration in the syringe. In differential form this is:

$$\frac{\partial (L - L_0)}{(L - L_0)} = -\frac{\partial v}{c} \tag{32}$$

so that:

$$L_i = L_0 (1 - e^{-i\Delta v/c}) \tag{33}$$

The amount of heat generated with each injection is proportional to the amount of new $N \cdot L$ generated, i.e., $N \cdot L_i - N \cdot L_{i-1}$, plus the amount of complex lost because of extrusion that accompanies the injection. Since the displacement does not take place instantaneously, and efficient mixing is continuously occurring, the amount lost is taken as the average of the concentrations at the beginning and end of the injection:

$$\Delta N \cdot L_i = N \cdot L_i - N \cdot L_{i-1} + 0.5 \frac{\Delta v}{c} (N \cdot L_{i-1} + N \cdot L_i) \tag{34}$$

[64] D. R. Bundle and B. W. Sigurskjold, *Methods in Enzymology* **247**, 288 (1994).

The heat generated in the cell is given by the cell volume times the molar enthalpy times the incremental change in complex plus a heat of dilution[64]:

$$Q_i = c\Delta H° \Delta N{\cdot}L_i + q_i \tag{35}$$

The heat generated can be normalized by the concentration increment of the ligand titrated. The change in complex per incremental increase in total ligand concentration is given by:

$$\frac{\Delta N{\cdot}L_i}{\Delta L_i} = \left[N{\cdot}L_i - N{\cdot}L_{i-1} + 0.5\frac{\Delta v}{c}(N{\cdot}L_{i-1} + N{\cdot}L_i)\right] \Big/ \left[L_i - L_{i-1} + 0.5\frac{\Delta v}{c}(L_{i-1} + L_i)\right] \tag{36}$$

so that the molar heat change is given by:

$$\bar{Q}_i = \Delta H° \frac{\Delta N{\cdot}L_i}{\Delta L_i} + q_i \tag{37}$$

The fitting of ITC data requires a model to calculate the change in complex concentration for each incremental increase in ligand concentration. For example, for single-site binding with a binding constant K, the concentration of complex after the ith injection is given by:

$$N{\cdot}L_i = \frac{1 + KN_i + KL_i - \sqrt{(1 + KN_i + KL_i)^2 - 4K^2 N_i L_i}}{2K} \tag{38}$$

Thus, ITC data for single-site binding can be fit by nonlinear regression using Eqs. (35) and (38) with adjustable parameters $\Delta H°$, K, and q_i.

Binding Studies at Physiologically Relevant Temperatures. As indicated above, ITC is especially well suited for binding studies of hyperthermophile proteins at high temperature, although to our knowledge this has never been done. The binding of Sac7d to duplex DNA at 80° is shown in Fig. 10, demonstrating the high quality of data that can be obtained at this temperature. The heats associated with the first two injections (and possibly the third) are low because of diffusion of the titrant out of the syringe needle during thermal equilibration. At lower temperature only the first point is typically in error. Fitting of the data to the McGhee–von Hippel model[65] for nonspecific binding to an infinite lattice gives a binding constant of $1.04 \times 10^7 M^{-1}$ and a site size of 4.2 base pairs. The binding is exothermic with an enthalpy change of -12.5 kcal/mol. This is a clear case of why a van't Hoff analysis can be unreliable, since there is little change in binding affinity with temperature [which implies an apparent enthalpy change near zero according to Eq. (9)]. The small dependence of K on temperature arises from a significant negative ΔC_p associated with binding, so that at 25° the binding is endothermic

---

[65] J. McGhee and P. von Hippel, *J. Mol. Biol.* **86**, 469 (1974).

with an enthalpy change of approximately 10 kcal/mol.[59] The large ΔC_p may be due to the unwinding and bending induced in DNA by Sac7d.[59,66]

Linkage of Folding and Binding

We now proceed to the combined effects of coupled folding and binding reactions. The effect of linked reactions cannot be overstated in the analysis of thermodynamic data.[63,67,68] Simple examples of the effect of linkage include the stabilization of a protein by metal ion binding, or the effect of pH on thermal stability. Failure to account for these reactions can lead to unknown (or unintended) contributions to the thermodynamic parameters. The purpose of linkage analysis is to separate the individual contributions to obtain the thermodynamic parameters associated with the intrinsic reaction of interest. Such an analysis is particularly valuable for interpretation of thermodynamic parameters in terms of structure.

The effects of linkage can be made more concrete by consideration of the following general model that includes the effects of both pH and ligand concentration.[31] The intrinsic unfolding reaction $N \rightleftharpoons U$ is linked to a single protonation reaction and a single ligand binding reaction (Scheme I)

$$
\begin{array}{ccccc}
& K_l & & K_{un} & \\
N \cdot L & \rightleftharpoons & N & \rightleftharpoons & U \\
K_c \updownarrow & & K_n \updownarrow & & K_u \updownarrow \\
N \cdot L^+ & \rightleftharpoons & N^+ & \rightleftharpoons & U^+
\end{array}
$$

SCHEME 1.

($N \cdot L$ represents the ligand bound form, and the + superscript indicates a relative change in charge due to protonation). The intrinsic unfolding reaction of the unprotonated form is characterized by an equilibrium constant $K_{un} = [U]/[N]$. Ligand binding to the unprotonated protein is characterized by a binding constant $K_l = [N \cdot L]/[N][L]$. The protonation reactions are described by proton binding constants $K_n = [N^+]/([N][H^+])$, $K_u = [U^+]/([U][H^+])$, and $K_c = [N \cdot L^+]/([N \cdot L][H^+])$. Changes in protonation of the ligand are not included here but may be necessary in some situations. Direct interconversion between all species is possible but are not explicitly drawn above for clarity (e.g., unfolding of $N \cdot L$ to U). Ligand binding to the unfolded chain is considered to be negligible. The temperature

---

[66] H. Robinson, Y.-G. Gao, B. S. McCrary, S. P. Edmondson, J. W. Shriver, and A. H.-J. Wang, *Nature* **392**, 202 (1998).

[67] J. Wyman and S. J. Gill, "Binding and Linkage: Functional Chemistry of Biological Macromolecules." University Science Books, Mill Valley, CA, 1990.

[68] M. Straume and E. Freire, *Anal. Biochem.* **203**, 259 (1992).

dependence of each of the reactions in the model is specified by the appropriate changes in enthalpies and heat capacities.

The overall equilibrium constant for the unfolding reaction, K_{obs}, is given by the sum of the concentrations of all of the unfolded species divided by the sum of the concentrations of all of the native species:

$$K_{obs} = \sum U_i / \sum N_i = ([U] + [U^+])/([N] + [N^+] + [N \cdot L] + [N \cdot L^+]) \quad (39)$$

and is a function of temperature, ligand concentration, and pH:

$$K_{obs}(T, [L], [H]) = K_{un}(1 + K_u[H])/(1 + K_n[H] + K_l[L] + K_c[H]K_l[L]) \quad (40)$$

where [H] is the hydrogen ion activity and [L] is the activity of the free ligand. The fraction of each state in the ensemble of four native states is:

$$\begin{aligned} f_{nn}(T, [L], [H]) &= 1/Q_n \\ f_{nn+}(T, [L], [H]) &= K_n[H]/Q_n \\ f_{nnl}(T, [L], [H]) &= K_l[L]/Q_n \\ f_{nnl+}(T, [L], [H]) &= K_l K_n[L][H]/Q_n \end{aligned} \quad (41)$$

where

$$Q_n(T, [L], [H]) = 1 + K_n[H] + K_l[L] + K_l K_c[L][H] \quad (42)$$

(Note: these are fractions of each within the ensemble of native states, not all states.) The fraction of each state within the ensemble of unfolded states is

$$\begin{aligned} f_{uu}(T, [L], [H]) &= 1/Q_u \\ f_{uu+}(T, [L], [H]) &= K_U[H]/Q_u \end{aligned}$$

where

$$Q_u(T, [L], [H]) = 1 + K_u[H] \quad (43)$$

The observed heat of unfolding includes the intrinsic heat of unfolding, the heats associated with adjustment of the relative populations within the folded and unfolded ensemble of states, and the heat of protonation of the buffer:

$$\begin{aligned} \Delta H^\circ_{obs}(T, [L], [H]) = {} & \Delta H^\circ_{un} + f_{uu+}\Delta H^\circ_u - f_{nn+}\Delta H^\circ_n - f_{nnl}\Delta H^\circ_l \\ & - f_{nnl+}(\Delta H^\circ_c + \Delta H^\circ_l) - \Delta n\,\Delta H^\circ_b \end{aligned} \quad (44)$$

where $\Delta n = f_{uu+} - f_{nn+} - f_{nnl+}$

If the heats of protonation of the ionizing group in the U, N and $N \cdot L$ states are identical (i.e., $\Delta H^\circ_n = \Delta H^\circ_u = \Delta H^\circ_c$), then

$$\Delta H^\circ_{obs}(T, [L], [H]) = \Delta H^\circ_{un} + \Delta n\,\Delta H^\circ_n - \Delta n\,\Delta H^\circ_b - (f_{nnl} + f_{nnl+})\Delta H^\circ_l \quad (45)$$

If the buffer is chosen such that its heat of protonation matches that of the heats of protonation of the titrating groups of N, then the middle two terms on the right-hand side of Eq. (45) cancel, and the observed heat of unfolding is given by the intrinsic heat of unfolding *minus a contribution due to ligand binding*.

$$\Delta H^\circ_{obs}(T, [L], [H]) = \Delta H^\circ_{un} - (f_{nnl} + f_{nnl+})\Delta H^\circ_l \qquad (46)$$

The observed heat of unfolding may differ significantly from the intrinsic heat of unfolding, depending on the magnitude of the heat of ligand binding and the change in binding stoichiometry upon unfolding. The contributions from ligand binding must be determined to obtain the parameters for the intrinsic unfolding process.

The DSC endotherm is given by

$$C_p(T, [L], [H]) = \Delta H^\circ_{obs}(T, [L], [H]) \; d\alpha_u(T, [L], [H])/dT \qquad (47)$$

and the temperature dependence of the unfolding progress is

$$d\alpha_u(T, [L], [H])/dT = dK_{obs}/dT/(1 + K_{obs})^2 \qquad (48)$$

where

$$\begin{aligned}dK_{obs}(T, [L], [H])/dT = &\{Q_n[K_{un}\Delta H^\circ_{un}/(RT^2) + K_{un} K_u \Delta H^\circ_u[H]/(RT^2) \\&+ K_u[H]K_{un} \Delta H^\circ_{un}/(RT^2)] \\&- (K_{un} + K_{un} K_u[H])dQ_n/dT\}/Q_n^2 \end{aligned} \qquad (49)$$

and

$$\begin{aligned}dQ_n(T, [L], [H])/dT = &K_n \Delta H^\circ_n[H]/(RT^2) + K_l \Delta H^\circ_l[L]/(RT^2) \\&+ K_l[L]K_c[H]\Delta H^\circ_c/(RT^2) \\&+ K_l[L]K_c[H]\Delta H^\circ_l/(RT^2) \end{aligned} \qquad (50)$$

The apparent T_m occurs at the maximum of $d\alpha_u/dT$ and can be obtained numerically. The above relations have been implemented in a *Mathematica* Notebook (available at http://w3.biochem.siu.edu/thermo/thermoware.html). We use this model below to first interpret separately the effects of pH and ligand binding on protein stability, and then combine them to interpret the effects of pH and anion concentration on Sac7d folding.

Effect of pH on Protein Stability. Electrostatic interactions are thought to be important in conferring increased stability in thermophile proteins.[3] The effect of pH on stability can be used to quantitatively probe the location and importance of these interactions, since ionic interactions must perturb the pK values of the groups involved.[69,70]

---

[69] A.-S. Yang and B. Honig, *J. Mol. Biol.* **231**, 459 (1993).
[70] B. Garcia-Moreno, *Methods in Enzymology* **259**, 512 (1995).

A simple linkage model (Scheme II) can be used to demonstrate the effect of pH on thermostability of proteins.

$$N \rightleftharpoons U$$
$$\Updownarrow \qquad \Updownarrow$$
$$N^+ \rightleftharpoons U^+$$

SCHEME 2.

(Protonation is actually binding of proton, but we separate pH and ionization effects from ligand binding for clarity.) For simplicity, we assume initially that a single protonation site is linked to folding. This does not imply that there is only a single ionizing group on the protein, but that only one significantly affects protein stability. Such might be the case below pH 7 for a carboxyl group involved in a salt bridge.

As with any thermodynamic box, knowledge of the free energies for any three reactions specifies the free energy change for the fourth. If we specify the unfolding free energy for $N \rightleftharpoons U$ along with the pK values for protonation of N and U, the free energy for unfolding N^+ to U^+ is also specified. If the pK values are identical, pH is not linked to folding, and a change in pH will have no effect on stability (Fig. 11). pH can only affect stability if the pK of an ionizing group is linked to folding, which requires that pK values must differ in the folded and unfolded states. A simulation of the effect of pH on the T_m for a protein with a single ionizing group linked to folding is shown in Fig. 11.

Note that the pH dependence of T_m is not a titration curve.[69,71] Rather, the upper and lower limits of the sigmoidal transition are defined by the two pK values. The altered pK *of* the titrating group in the folded protein indicates that an electrostatic interaction affects the stability of the protein. The steepness of the sigmoidal transition reflects the number of protonation reactions linked to folding since

$$\frac{\partial T_m}{\partial pH} = \frac{2.303 R T_m^2 \Delta n}{\Delta H^\circ} \tag{51}$$

where Δn is the change in protonation upon unfolding.[14] For comparison, the pH dependence of the T_m for a protein with two protonation sites is shown in Fig. 11. We have assumed here that the two groups are identical. If they have different pK values, then the upper and lower limits for the sigmoidal transition will be spread out, and the steepness of the curve will be decreased. The steepness at any point can only indicate the change in protonation at that temperature on unfolding, which will be less than or equal to the number of protonation sites linked to folding.

---

[71] M. Oliveberg, S. Vuilleumier, and A. Fersht, *Biochemistry* **33**, 8826 (1994).

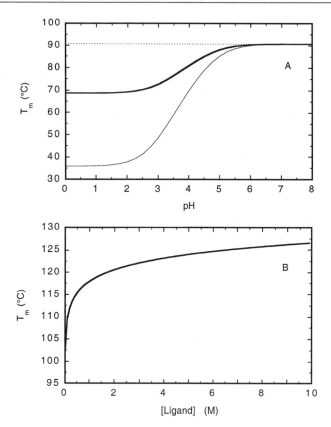

FIG. 11. The effect of pH and ligand binding on the thermal stability. (A) The T_m depends on pH due to linkage of folding to protonation. If the pK values for all of the ionizing groups are unchanged upon unfolding, there is no pH dependence (dotted line). The bold line is that expected for the linkage of one ionizing group, and the thinner solid line for two identical groups to the unfolding of a monomeric protein ($T_{m,un} = 91°$, $\Delta H_{un} = 60$ kcal/mol, $\Delta C_{p,un} = 725$ cal/deg/mol, $pK_n = 3$, $pK_u = 5$, $\Delta H_n = \Delta H_u = -1.5$ kcal/mol, $\Delta C_{p,n} = \Delta C_{p,u} = 25$ cal/deg/mol). (B) The binding of a ligand increases the stability with a logarithmic dependence of T_m on the concentration of ligand. The dependece of T_m for a monomeric protein on ligand concentration is shown with $T_{m,un} = 91°$, $\Delta H_{un} = 60$ kcal/mol, $\Delta C_{p,un} = 725$ cal/deg/mol, and a single binding site with $pK_l = 3$ ($K_d = 0.0001M$), $\Delta H_l = -1.5$ kcal/mol, $\Delta C_{p,l} = 25$ cal/deg/mol. The binding site is saturated at 2 mM ligand although the T_m (and ΔG_{obs}) continues to increase.

Effect of Ligand Binding on Protein Stability. As indicated above, many thermophile proteins are stabilized by ligand binding. Indeed, in some the intrinsic stability is only marginal, and they must be stabilized *in vivo* by extrinsic factors.[4,5] The high salt concentrations found intracellularly in many hyperthermophiles may promote folding not only by inducing changes in the water structure (i.e., enhancing the hydrophobic effect), but also by specific ion binding.

As with oligomer unfolding, discussions of protein stability in the presence of ligand requires specification of concentrations. The effect of ligand binding needs to be explicitly treated in the analysis of protein stability data if binding occurs. Working at saturating concentrations of ligand does not suffice; it is impossible to saturate the effect of ligand binding on protein stability.

The linkage model derived above can also be used to demonstrate the effects of binding. We ignore for the moment the effect of protonation and consider only ligand binding (Scheme III):

$$N \rightleftharpoons U$$
$$\updownarrow \quad \updownarrow$$
$$N \cdot L \rightleftharpoons U \cdot L$$

SCHEME 3.

If L binds preferentially to the N state, increasing ligand concentration must increase stability by the law of mass action. As shown in Fig. 11B, the increase in T_m with ligand concentration does not reach a limiting value, even at concentrations well beyond the saturation of available binding sites. The increase in free energy is logarithmic, similar to that shown above for the concentration dependence of dimer unfolding free energy, and does not plateau.

Linkage of Ligand Binding and pH to Protein Stability. It is often difficult to design an experiment that allows one to isolate the effects of ligand binding and protonation from protein stability. For example, if ion binding is associated with changes in ionization, heats of protonation will be included in the thermodynamics of binding.[63] Variation of pH is often used to destabilize a protein to perform a Kirchhoff analysis to obtain a ΔC_p for unfolding. This will require a global analysis that includes unfolding, ionization, and binding energetics if ligand binding is pH dependent.[31] The latter situation is described here with the linkage of anion binding and protonation to protein folding.

Sac7d is unusually well behaved for thermodynamic studies over a wide range of pH (pH 0 to 10), salt concentration (0 to 0.3 M), and temperature (0° to 130°).[4,31] Such a wide window of experimental conditions has permitted the observation of linkage effects that would not normally be apparent. For example, upon decreasing the pH below 4 in the presence of 0.3 M NaCl, the T_m of Sac7d decreases similarly to that shown in Fig. 11. However, the T_m increases below pH 2, an effect not expected by simple linkage of protonation and folding. Further, in the absence of salt, the DSC endotherm shifts to low temperature with decreasing pH and disappears entirely at pH 2, but then it reappears with a further decrease in pH and shifts back to higher temperature. Although there is little or no endotherm at pH 2, circular dichroism shows that about 25% of the protein is folded at pH 2 and that heating leads to a cooperative transition.

The pH dependence of folding can also be followed by CD at constant temperature. At 25°C, decreasing pH below 4 leads to unfolding of the protein with about 75% unfolded at pH 2. Decreasing the pH further leads to refolding to the native state at pH 0. The protein can also be refolded at pH 2 by adding salt. The effectiveness of various salts shows no variation with cation but rather depends on the type of anion, the order of effectiveness being sulfate > perchlorate > chloride.

These data indicate that the folding of Sac7d is linked to protonation and anion binding. The minimal model that can explain the data is the linkage of two protonation and two anion binding reactions to protein folding.[31] The dependence of free energy, heat capacity, and extent of folding on temperature is more complicated that that shown in Fig. 1, and these are now described by four-dimensional surfaces (e.g., $\Delta G°$ vs T, pH, and salt). The surfaces can be visualized by simulations of three-dimensional surfaces at two salt concentrations (Fig. 12). A decrease in stability is seen in all of the surfaces in the vicinity of pH 2. The effect is most pronounced at low salt, where the free energy surface dips below the zero plane in the vicinity of pH 2, and only about 25% of the protein remains folded. The protein stability increases below pH 2 due to the binding of the anion contributed by the acid.

It is interesting to note that the simulations in Fig. 12 demonstrate that the absence of a DSC endotherm does not necessarily imply the lack of a cooperative transition (see also ref. 72). $\Delta H°_{obs}$ of unfolding decreases with temperature due to a positive $\Delta C°_p$ and tends toward zero near the temperature of maximal stability, which for many proteins occurs near 10 to 20°.[14] Thus, any perturbation that shifts the T_m to this temperature will display a negligible endotherm in the DSC, even though a thermal melting transition can be observed with spectroscopic methods. This is an excellent example of the advantage of using more than one experimental technique.

Effect of Linkage on Determining ΔC_p of Unfolding. The ΔC_p for unfolding Sac7d obtained from a Kirchhoff plot of DSC data (namely 498 cal/deg/mol) is not the intrinsic ΔC_p for unfolding, since the enthalpies contain contributions from ion binding [Eq. (46)]. This is indicated by the observation of a maximum in the unfolding progress curves followed by CD between 10° and 20° observed under low salt conditions and low pH.[31] The position of the maximum defines an intrinsic ΔC_p of 725 cal/deg/mol. Anion binding is significant when pH is used as a perturbant, and the low ΔC_p value obtained from a Kirchhoff plot of the DSC data can be explained by an increased contribution from the heat of anion binding at low pH. In this case the intrinsic ΔC_p, and therefore the intrinsic free energy of unfolding, can only be obtained from a global analysis.

Linkage Analysis and Global Nonlinear Regression. Reliable extraction of thermodynamic parameters for systems that involve linkages requires multidimensional data sets, e.g., DSC as a function of pH and salt concentration, and the extent

---

[72] D. Haynie and E. Freire, *Proteins* **16**, 115 (1993).

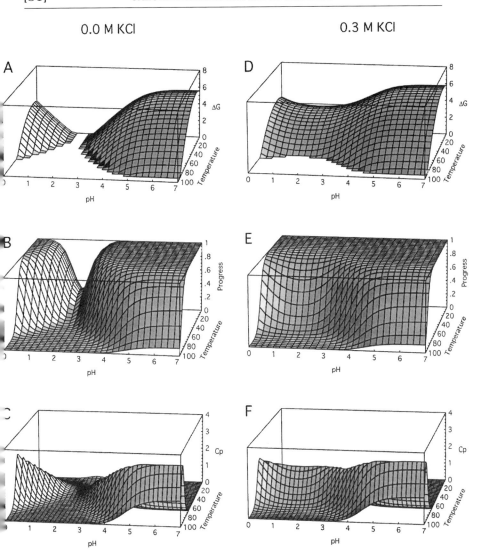

FIG. 12. The dependence of Sac7d protein stability, folding progress, and excess heat capacity on temperature, pH, and salt concentration. The surfaces are obtained from a fit of multidimensional DSC and CD data with a linkage model similar to that of Scheme I, where unfolding is linked to the protonation of two carboxyl groups and two anion binding sites.[31] The intrinsic unfolding reaction is characterized by $T_m = 91°$, $\Delta H_{un} = 60$ kcal/mol, and $\Delta C_p = 725$ cal/deg/mol. The pK for the carboxyls in the unfolded species is 4.6 at low salt and 4.4 at high salt, and in the N state they are 0.0. Anion binding is characterized by a pK_l of -0.3, but increases to 2.4 in low salt for binding of the second ligand. The pK of the carboxyls in the ligand bound states is 2.7, except in high salt one increases to 4.5.

of folding followed spectroscopically as a function of temperature, pH, and salt concentration. Each independent reaction in the linkage model (including buffer ionization) is characterized by ΔH, ΔC_p, and either an equilibrium constant at a reference temperature or the T_m. The number of parameters even in a simple model can be quite large, but many parameters can be fixed to minimize the complexity of the problem. Thermodynamic parameters for the buffer can be set to literature values; ΔH and ΔC_p for the ionizing groups and the pK values in the unfolded state can be set to values for model compounds; the pK values in the native state can be determined by multidimensional NMR; and the enthalpies and heat capacities of binding can be characterized by ITC. An advantage of an explicit linkage model is that it clearly indicates the assumptions made in the analysis. In the presence of linkages, it is certainly better than modeling the unfolding as $N \rightleftharpoons U$.

Multidimensional data can be fitted to a linkage model using global nonlinear regression.[68,73] An adaptive grid refinement global optimization algorithm (Loehle Enterprises, Naperville, IL) is especially useful for searching parameter space for the best parameters. This method has the advantage of locating all minima in the chi squared (χ^2) surface, and thus the common problem of becoming trapped in local minima is eliminated. It also defines the precision of the derived parameters and explicitly includes covariances. Unfortunately, the method is computationally intensive and so is best used with groups of three to five parameters at a time. For the final fits, all of the parameters can be refined simultaneously using a simplex optimization procedure.

Concluding Remarks

Calorimetry is one of the few biophysical techniques that permits a direct characterization of the energetics of hyperthermophile proteins at physiologically relevant temperatures. No other method directly measures enthalpy changes, and as such it should be viewed as an essential tool for thermodynamic studies. However, calorimetry is not selective. It "sees" everything, and therefore interpretation in the presence of linked reactions can be difficult or deceptive. Thermodynamic descriptions often resemble a Russian Matryoshka doll in that we often think we understand a system, only to discover that there is another unforeseen little doll hidden within. A global analysis of DSC, ITC, and spectroscopic data can often resolve these difficulties and make it more likely to achieve an accurate representation of the stability and function of hyperthermophile proteins.

Acknowledgment

This work was supported by the National Institutes of Health (GM 49686).

[73] J. M. Beechem, *Methods in Enzymology* **210**, 37 (1992).

[32] Crystallographic Analyses of Hyperthermophilic Proteins

By Douglas C. Rees

Following the first structure determination in 1992 of a protein from a hyperthermophilic organism, the rubredoxin from *Pyrococcus furiosus*,[1,2] structures of hyperthermophilic proteins have appeared at an ever increasing rate, with more than 50 distinct structures having been submitted to the Protein Data Bank[3,4] as of October 1999 (Table I). These structure determinations have been undertaken for many of the reasons discussed elsewhere in this volume—to understand the structural basis of thermostability; to more fully characterize proteins with unusual cofactors or active centers; for biotechnological purposes; as homologs of eukaryotic proteins, especially those involved in some aspect of macromolecular biosynthesis; and as part of structural genomics efforts, because they either do, or do not, look like some other proteins. As there is an expectation that hyperthermostable proteins should be more robust and potentially better suited for withstanding the solution conditions and time scales of crystallization experiments, these proteins have provided attractive targets for structural studies. The focus of this chapter is to review the status and some general implications of crystallographic studies on hyperthermophilic proteins, with emphasis on relevant technical aspects of the structure determination, quality of the crystals, and conclusions about thermal stability and other properties of these fascinating proteins.

Crystal Structure Determination of Hyperthermophilic Proteins

The basic sequence of steps in the crystallographic analysis of a hyperthermophilic protein is unaltered from that used for more "conventional" proteins; one must (1) obtain sufficient quantities of pure material; (2) produce "diffraction quality" crystals; (3) determine the phases; and finally (4) complete the structure determination by building and refining a molecular model. Increasingly, the rate-determining step in any structure analysis, including hyperthermophilic proteins, involves obtaining suitable samples for crystallization trials.

---

[1] M. W. Day, B. T. Hsu, L. Joshua-Tor, J.-B. Park, Z. H. Zhou, M. W. W. Adams, and D. C. Rees, *Prot. Sci.* **1**, 1494 (1992).
[2] P. R. Blake, J.-B. Park, Z. H. Zhou, D. R. Hare, M. W. W. Adams, and M. F. Summers, *Prot. Sci.* **1**, 1508 (1992).
[3] J. Sussman, D. Lin, J. Jiang, N. Manning, J. Prilusky, O. Ritter, and E. Abola, *Acta Crystallogr.* **D54**, 1078 (1998).
[4] Research Collaboratory for Structural Bioinformatics PDB, http://www.rcsb.org/pdb/.

TABLE I
HYPERTHERMOPHILIC PROTEINS AVAILABLE IN PROTEIN DATA BANK[a]

Year	PDB ID	Res (Å)	R-factor	Protein	Source
1992	1CAA	1.8	0.178	P. furiosus rubredoxin[1,2]	wt
	1BRF	0.95	0.132	P. furiosus rubredoxin (high resolution-1998)[37]	wt
	1BQ8	1.1	0.119	P. furiosus rubredoxin (recomb.met-0 mutant-1998)[37]	rec
1995	1HDG	2.5	0.166	T. maritima glyceraldehyde-3-phosphate dehydrogenase[38]	rec
	1AOR	2.3	0.155	P. furiosus aldehyde ferredoxin oxidoreductase[39]	wt
	1IGS	2.0	0.177	S. solfataricus indole-3-glycerolphosphate synthase[40]	rec
1996	1GTM	2.2	0.174	P. furiosus glutamate dehydrogenase[41]	wt
	1NSJ	2.0	0.192	T. maritima phosphoribosylanthranilate isomerase[40]	rec
	1XER	2.0	0.173	Sulfolobus sp. ferredoxin[42]	wt
	1GOW	2.6	0.219	S. solfataricus β-glycosidase[43]	wt
	1PCZ	2.2	0.218	P. woesei TATA binding protein[44]	rec
1997	1VJW	1.75	0.159	T. maritima ferredoxin[45]	rec
	1B8A	1.90	0.168	P. kodakaraensis aspartyl tRNA synthetase[46]	rec
	1AIS	2.1	0.212	P. woesei TBP + TFIIB + DNA[47]	rec
	1XGS	1.75	0.187	P. furiosus methionine aminopeptidase[48]	wt
	1VPE	2.0	0.198	T. maritima phosphoglycerate kinase[49]	rec
	1AJ8	1.9	0.191	P. furiosus citrate synthase[50]	rec
	1TMY	1.9	0.186	T. maritima CheY[51]	rec
	1FSZ	2.8	0.199	M. jannaschii cell division protein Ftsz[52]	rec
	1FTR	1.7	0.198	M. kandleri formylmethanofuran: tetrahydromethanopterin formyltransferase[53]	rec
	1AZP	1.6	0.211	S. acidocaldarius Sac7d + DNA[54]	rec
	1A0E	2.7	0.180	T. neapolitana xylose isomerase	rec
	1A1S	2.7	0.226	P. furiosus ornithine carbamoyltransferase[5]	rec
1998	1A2Z	1.73	0.18	T. litoralis pyrrolidone carboxyl peptidase[55]	rec
	1A5Z	2.1	0.200	T. maritima lactate dehydrogenase[56]	rec
	1A7W	1.55	0.198	M. fervidus histone HmfB	rec
	1HTA	1.55	0.198	M. fervidus histone HmfA	rec
	1A76	2.0	0.214	M. jannaschii FLAP endonuclease-1[57]	rec
	1A79	2.28	0.200	M. jannaschii tRNA splicing endonuclease[58]	rec
	1A8L	1.9	0.192	P. furiosus protein disulfide oxidoreductase[59]	rec
	1NKS	2.57	0.160	S. acidocaldarius adenylate kinase[60]	rec
	1EIF	1.9	0.205	M. jannaschii translation initiation factor 5a[61]	rec
	1SHS	2.9	0.216	M. jannaschii small heat shock protein[62]	rec
	1BNZ	2.0	0.168	S. acidocaldarius Sso 7d protein + DNA[63]	rec
	1MJH	1.7	0.210	M. jannaschii hypothetical protein Mj0577[64]	rec
	1BVU	2.5	0.192	T. litoralis glutamate dehydrogenase[65]	wt
	1SSS	2.3	0.166	S. solfataricus superoxide dismutase[66]	rec
	1B25	1.85	0.174	P. furiosus formaldehyde ferredoxin oxidoreductase[67]	wt

TABLE I (continued)

Year	PDB ID	Res (Å)	R-factor	Protein	Source
Hold	1THF			T. maritima imidazole glycerolphosphate synthase-cyclase[68]	
	1B06			S. acidocaldarius superoxide dismutase[69]	
	1B26			T. maritima glutamate dehydrogenase[70]	
	1B3Q			T. maritima CheA histidine kinase domain[71]	
1999	1B7G	2.05	0.226	S. solfataricus glyceraldehyde-3-phosphate dehydrogenase[72]	rec
	1QEZ	2.7	0.197	S. acidocaldarius inorganic pyrophosphatase[73]	rec
	1QC7	2.2	0.235	T. maritima FliG[74]	rec
	1QDL	2.5	0.226	S. solfataricus anthranilate synthase[75]	rec
	1COJ	1.9	0.170	A. pyrophilus superoxide dismutase[76]	rec
	1QVB	2.4	0.210	T. aggregans β-glycosidase[77]	rec
	1C3P	1.8	0.198	A. aeolicus histone deacetylase-like protein[78]	rec
	1QLM	2.0	0.198	M. kandleri methenyltetrahydromethanopterin cyclohydrolase[79]	rec
hold	1B43			P. furiosus FEN-1[80]	
	1MGT			P. kodakaraensis O^6-methylguanine-DNA transferase[81]	
	1FCI			M. fervidus glyceraldehyde-3-phosphate dehydrogenase[82]	
	1B73			A. pyrophilus glutamate racemase[83]	
	1B8Z			T. maritima HU	
	1B9B			T. maritima triose-phosphate isomerase	
	1CJS			M. jannaschii ribosomal protein L1	
	1FBN			M. jannaschii fibrillarin homolog	
	1C3C			T. maritima adenylosuccinate lyase	
	1CZ3			T. maritima dihydrofolate reductase	
	1DO6			P. furiosus superoxide reductase	

[a] [PDB from J. Sussman, D. Lin, J. Jiang, N. Manning, J. Prilusky, O. Ritter, and E. Abola, *Acta Crystallogr.* **D54**, 1078 (1998); Research Collaboratory for Structural Bioinformatics PDB, http://www.rcsb.org/pdb/] as of October 1999. The year of submission of the coordinates (year), the PDB accession number (PDB ID), resolution, and *R*-factor of the crystal structure refinement (Res and *R*-factor, respectively); the name and host organism of the protein and the primary structure reference (if published); and the source of the protein [i.e., either wild-type (wt) or recombinant (rec)], are indicated for each structure. Coordinate sets on hold are listed at the end of the year they were submitted. Structures were found via keyword searches that should identify most, but perhaps not all, of the relevant sets of hyperthermophilic proteins in the PDB.

Protein Expression and Purification

The two basic approaches for obtaining hyperthermophilic proteins are either to grow the organism and isolate the desired protein, or to clone the gene for a desired protein and express it in a suitable mesophilic host. As emphasized in Table I, the latter route has been more widely utilized to produce protein for structural studies. In general, genes from hyperthermophiles have been expressed in *Escherichia coli;* the one exception to date used in a published structural study is the *P. furiosus* ornithine transcarbamoylase that was expressed in yeast.[5] A potential advantage with recombinantly expressed hyperthermophilic proteins is that they are often substantially more thermostable than all host proteins, so that a heat step provides a convenient and effective purification method. Because it has been difficult to develop a mutagenesis system in a hyperthermophile, the ability to produce recombinant versions of hyperthermostable proteins currently represents the most direct way to apply mutagenesis methods to the study of fundamental issues of stability, as well as to confer proteins with more technologically desirable properties. With the increasing availability of genome sequences and the development of molecular biology methods, recombinant approaches to protein expression and purification will undoubtedly become more and more powerful.

Recombinant methods have not completely replaced "traditional" methods of isolating enzymes from the host organism, however, because not all proteins have been successfully produced recombinantly. For example, some proteins appear to be unable to fold into the active form at mesophilic temperatures when expressed in a mesophilic host, whereas others contain complex cofactors that may be difficult to incorporate in recombinant systems. For these reasons, recombinant techniques have not entirely superseded more traditional methods of protein purification. Indeed, it is striking, at least at present, that even recombinantly expressed proteins are predominantly from biochemically better characterized host organisms such as *P. furiosus* or *Thermotoga maritima,* rather than from organisms such as *Methanococcus jannaschii* for which the complete genome sequence was first available,[6] but which has been less extensively characterized biochemically.

Because many hyperthermophiles require anaerobic growth conditions, it is not surprising that some of the proteins isolated from these organisms are oxygen sensitive and must be purified under strictly anaerobic conditions. Systems for manipulating biochemical samples under anaerobic conditions have been devised and typically include both Schlenk manifolds[7] for benchtop operations (important for protein purification) and anaerobic chambers for more detailed manipulations (such as crystallization) and storage. The Schlenk manifold contains two lines: vacuum and inert gas. The latter typically consists of either Ar or N_2 that have been scrubbed free of O_2 through a supported copper catalyst.[7] Samples are stored in containers sealed with rubber septa that are attached to the manifold via needles and butyl rubber tubing. When a solution is connected to the manifold, oxygen can be removed by multiple cycles of degassing under vacuum and purging with inert gas, followed by addition of several millimolar sodium dithionite (assuming

---

[5] V. Villeret, B. Clantin, C. Tricot, C. Legrain, M. Roovers, V. Stalon, N. Glansdorff, and J. Van Beeumen, *Proc. Natl. Acad. Sci. U.S.A.* **95,** 2801 (1998).

[6] C. J. Bult, O. White, G. J. Olsen, L. Zhou, R. D. Fleischmann, G. G. Sutton, J. A. Blake, L. M. FitzGerald, *et al., Science* **273,** 1058 (1996).

[7] D. F. Shriver and M. A. Drezdzon, "The Manipulation of Air-Sensitive Compounds," 2nd ed. John Wiley & Sons, New York, 1986.

the protein sample and/or buffer components can tolerate this reductant). Once the solutions are prepared, the columns, pumps, sample injectors, etc., can be made anaerobic by passing oxygen-free solutions through the chromatography system. With practice and proper manifold design, it is possible to do protein purifications under anaerobic conditions using the same general procedures developed for oxygen stable samples.

A variety of anaerobic chambers may be utilized for handling oxygen sensitive biochemical materials. A Coy Industries (Ann Arbor, MI) anaerobic chamber has been employed in the author's laboratory, but glass and metal designs favored by synthetic chemists can also be used. An important feature for crystal screening is the installation of a viewing port for a stereomicroscope.

Crystallizations

The principles of crystallization for hyperthermostable proteins are identical to those for "ordinary" macromolecules: namely, macromolecular solutions are brought to the point of supersaturation in such a manner that the macromolecule comes out of solution as crystals, rather than as precipitate. The most important factor for a successful crystallization effort is to have sufficient quantities of pure and homogenous material to screen for suitable conditions. Since crystallization methods have been extensively reviewed in these volumes and elsewhere (see refs. 8, 9), the technical aspects of this process will not be detailed here. There is, however, one unique property of hyperthermophilic proteins that could potentially be exploited in screening crystallization conditions: their thermostability. The crystallization temperature is an important parameter to optimize, and the use of hyperthermophilic proteins extends the temperature range available for crystallization trials. The successful use of elevated temperatures in crystallization trials does not yet appear to have been achieved, but should be considered when working with hyperthermophilic proteins.

Oxygen-sensitive proteins are crystallized in the same fashion as oxygen-stable proteins, except that the manipulations must obviously be performed under anaerobic conditions. We have found that several variants of the melting-point capillary method[10,11] can work well for protein crystallization in a glove box. Typically, 10–20 μl of precipitant solution is introduced into a melting-point capillary (diameter ∼1–1.5 mm, generally treated with AquaSil™ (Pierce, Rockford, IL) siliconizing solution) and separated by several millimeters from an approximately

---

[8] A. MacPherson, "Preparation and Analysis of Protein Crystals." John Wiley & Sons, New York, 1982.
[9] C. W. J. Carter and R. M. Sweet, eds., "Macromolecular Crystallography, Part A." *Methods Enzymol.*, Vol. 276, Academic Press, San Diego, 1997
[10] J. Drenth, W. G. J. Hol, and R. K. Wierenga, *J. Biol. Chem.* **250**, 5268 (1975).
[11] M. M. Georgiadis, H. Komiya, P. Chakrabarti, D. Woo, J. J. Kornuc, and D. C. Rees, *Science* **257**, 1653 (1992).

equal volume of the protein solution. The capillary is closed with sealing wax and the solutions are brought together by either manual shaking or centrifugation. The capillaries are then left undisturbed in the glove box for a suitable period of time before checking for crystals. Many variants of this general approach are possible: the protein solution may be replaced with a 50:50 mixture of protein and precipitant solution; a gap may be left between protein and precipitant solutions (so that this more resembles a vapor diffusion method); or the protein solution may be gently layered on the precipitant solution without resorting to shaking or centrifugation. As with any crystallization study, the exact protocols, including volumes and concentrations of solutions, must be empirically established.

In our experience, the manipulations required for these capillary setups are more manageable in a glove box than for sitting drops or hanging drops. In addition, capillaries can be stored more efficiently than trays in the glove box, and they can be removed from the chamber, if necessary, for detailed analysis under a microscope, photography, or storage at different temperatures (unless strict anaerobic conditions are required).

Although the conventional wisdom suggests that hyperthermophilic proteins should be more amenable for crystallizations, controlled tests of this hypothesis have not been rigorously conducted. In the author's laboratory, crystallization trials for nine hyperthermophilic proteins resulted in useful crystals for five proteins; one protein gave thin, plate-like crystals that diffracted well but were too mosaic to provide useful data; two proteins gave big, beautiful and poorly diffracting crystals that heartbreakingly defied all efforts to improve the diffraction quality; and one protein resisted crystallization. This outcome seems comparable to what would be expected for crystallization trials with mesophilic proteins that are available in sufficient quantities of pure, homogenous material. As with any unsuccessful crystallization study, there are many possibilities for the failure to obtain suitable crystals, including lack of success at finding the right conditions, sample heterogeneity, and inherent difficulty in crystallization (perhaps due to flexibility, or lack of appropriately positioned residues to form lattice contacts). Although hyperthermophilic proteins may not necessarily be more suitable for crystallizations, the unshakeable conclusion from the very first protein structure determination (myoglobin[12]) is that the more proteins from different organisms that are screened for crystallizations, the more likely it is that one of them will work well. Consequently, it is a prudent strategy to include hyperthermophilic homologs when attempting to crystallize a particular protein.

Crystallographic Analysis

Methods for the crystal structure determination for hyperthermophilic proteins are no different than for other types of macromolecules, which have been

---

[12] J. C. Kendrew and R. G. Parrish, *Proc. Roy. Soc.* **A238**, 305 (1957).

reviewed in detail in these volumes[9,13] and elsewhere.[14,15] For tungsten containing enzymes that are predominantly found in hyperthermophilic organisms, the W edge would be ideal for the application of multiwavelength anomalous diffraction (MAD) phasing methods, although this approach has not been utilized to date.

One technical problem of potential relevance to hyperthermophilic proteins concerns the application of cryocrystallographic methods[16,17] to oxygen-sensitive crystals. Normally (i.e., with non-oxygen-sensitive proteins), the crystallization chambers are opened in air and the crystals are picked up on fiber loops for transfer to the goniometer under a cold (~100 K) nitrogen stream. This step must be done quickly, since water evaporates rapidly from the loop, which alters the composition of the solution bathing the crystal and leads to loss of crystallinity. Not uncommonly, it is necessary to modify the crystallization solution by the addition of cryoprotectants that allow the solution to form a glass under cryogenic conditions, rather than to form ice crystals that can destroy a crystal. For oxygen-sensitive crystals, the addition of oxygen-free cryoprotectants can be performed in the glove box. Because of the time involved, it is not possible to mount a crystal on a fiber loop in the glove box, remove it from the glove box, transfer it to the cold stream, and still have the crystal remained ordered. We have found it possible to remove crystallization capillaries from the glove box, quickly open them in air, and immediately scoop crystals in the loop for transfer to the cold nitrogen stream; since gas diffusion through crystals takes on the order of minutes,[18] with suitable agility the crystals can be cooled in a fraction of that time and henced maintained in an anaerobic state. Ideally, the crystals should be mounted on the loop and cooled to cryogenic temperatures in the glove box, and then removed while being maintained at low temperature; although it is more difficult to work with liquid nitrogen in the glove box (since it experiences a substantial volume increase upon warming), cryocrystallization capabilities in anaerobic chambers have been described.[19,20]

---

[13] C. W. J. Carter and R. M. Sweet, eds., "Macromolecular Crystallography, Part B." *Methods Enzymol.*, Vol. 277, Academic Press, San Diego, 1997.

[14] J. P. Glusker, M. Lewis, and M. Rossi, "Crystal Structure Analysis for Chemists and Biologists." VCH Publishers, New York, 1994.

[15] C. Giacovazzo, ed., "Fundamentals of Crystallography." Oxford University Press, Oxford, 1992.

[16] E. F. Garman and T. R. Schneider, *J. Appl. Crystallogr.* **30**, 211 (1997).

[17] D. W. Rodgers, *Methods in Enzymology* **276**, 183 (1997).

[18] S. M. Soltis, M. H. B. Stowell, M. C. Wiener, G. N. Phillips, and D. C. Rees, *J. Appl. Crystallogr.* **30**, 190 (1997).

[19] K. Valegård, A. C. Terwisscha van Scheltinga, M. D. Lloyd, T. Hara, S. Ramaswamy, A. Perrakis, A. Thompson, H.-J. Lee, J. E. Baldwin, C. J. Schofield, J. Hajdu, and I. Andersson, *Nature* **394**, 805 (1998).

[20] X. Vernède and J. C. Fontecilla-Camps, *J. Appl. Crystallogr.*, **32**, 505 (1999).

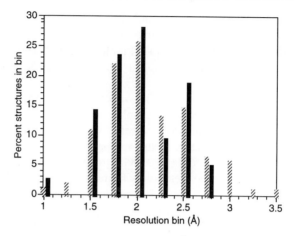

FIG. 1. Resolution histogram in resolution bins of 0.25 Å for the hyperthermophilic proteins (solid bars) listed in Table I and for the entire Protein Data Base (PDB) [cross-hatched bars; J. Sussman, D. Lin, J. Jiang, N. Manning, J. Prilusky, O. Ritter, and E. Abola, *Acta Crystallogr.* **D54**, 1078 (1998); Research Collaboratory for Structural Bioinformatics PDB, http://www.rcsb.org/pdb/], as estimated in October, 1999 from http://db2.sdsc.edu/moose/images/summary.gif.

Structural Properties of Hyperthermophilic Proteins

Crystal Quality: Resolution and B Factors

As a consequence of their extreme thermostability, the common perception is that hyperthermophilic proteins are "rocks," which further suggests that their crystals should be well-ordered, diffract to high resolution, and have lower temperature (B) factors. Although the structural database of hyperthermophilic proteins is still relatively small, it is of sufficient size to begin to test these ideas. Somewhat surprisingly, the average resolutions of structures in the complete Protein Data Base ($\sim$2.1 Å resolution[21]) and the subset of hyperthermophilic proteins ($\sim$2.0 Å resolution) are very similar, as are the resolution distributions for the two sets of structures (Fig. 1). When the variation of average main chain B with resolution for hyperthermophilic proteins is compared to a representative set of nonhyperthermophilic proteins, there is qualitatively little distinction (Fig. 2). Comparison of B factors between different structures is problematic, however, because of the large variation normally observed in B factors due to differences in resolution, refinement programs, etc.[22] Finally, the R-factors after refinement of hyperthermophilic proteins (Table I) do not appear significantly different than would be expected for "ordinary" proteins. At this stage, the general conclusion must be that there is

---

[21] http://db2.sdsc.edu/moose/images/summary.gif.
[22] S. Parthasarathy and M. R. N. Murthy, *Prot. Sci.* **6**, 2561 (1998).

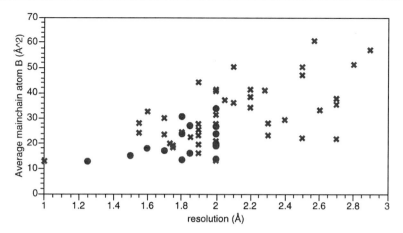

FIG. 2. Variation in average main chain temperature B factor as a function of resolution for the hyperthermophilic proteins listed in Table I (designated with an "✖") and a representative group of mesophilic proteins from the PDB (designated with "●"). Coordinates for the latter set of proteins were selected as having been submitted to the Brookhaven PDB [J. Sussman, D. Lin, J. Jiang, N. Manning, J. Prilusky, O. Ritter, and E. Abola, *Acta Crystallogr.* **D54,** 1078 (1998)] since January 1998, containing more than 200 residues and diffracting to at least 2.0 Å resolution. These structures consisted of PDB accession numbers 1A28, 1A6Q, 1A6Q, 1A7T, 1A9T, 1BA1, 1BF2, 1BFD, 1BN8, 1BOL, 1BRT, 1PRX, 2BLS, 2DOR, 2GAR, 2WEA.

nothing distinctive about the diffraction properties of crystals of hyperthermophilic proteins.

Analysis of Thermostability

An important motivation for the structure determination of hyperthermophilic proteins is to identify and characterize structural determinants of protein stability. Typically, the types of interactions implicated as contributing to stability include salt bridges, favorable packing interactions (which implies fewer cavities and improved hydrophobic interactions), increased secondary structure stability, improved hydrogen bonding, replacement of strained residues with glycines, and cross-links formed by disulfides and other groups. The literature on protein stability is vast, and reviews addressing issues of hyperthermostability provide excellent entry points to more detailed discussions.[23–28] Ideally, the most sensitive types

---

[23] D. C. Rees and M. W. W. Adams, *Structure* **3,** 251 (1995).
[24] A. Goldman, *Structure* **3,** 1277 (1995).
[25] R. Jaenicke and G. Bohm, *Curr. Opin. Struct. Biol.* **8,** 738 (1998).
[26] R. J. Russell and G. L. Taylor, *Curr. Opin. Biotechnol.* **6,** 370 (1995).
[27] R. Jaenicke, *Adv. Prot. Chem.* **48,** 181 (1996).
[28] F. T. Robb and D. L. Maeder, *Curr. Opin. Biotechnol.* **9,** 288 (1998).

of analyses involve comparison of homologous proteins with different thermodynamic stabilities, with the objective of correlating stability with quantitative changes in specific types of interactions. As indicated in Table I, this is now possible for several families of enzymes, including electron transfer proteins, glycolytic enzymes, glutamate dehydrogenase, and citrate synthase, that have structurally characterized representatives from organisms growing under a range of thermal conditions.

A summary of the possible involvement of these different energetic contributions to hyperthermostability in structurally characterized proteins is provided in Table II. The most general conclusion from these studies is that hyperthermostability can be achieved without requiring any new types of interactions to stabilize the folded conformation. Rather than being the consequence of any one dominant type of interaction, the increased stability of these proteins reflects a number of subtle interactions: i.e., hyperthermophilic proteins are stabilized in the same ways as mesophilic proteins, except that they just do a little better at remaining in the native conformation at elevated temperatures. Increased numbers of salt bridges, particularly involving multiple groups, are commonly reported in hyperthermophilic proteins, as are evidences of improved packing or hydrophobic interactions, shortened and/or immobilized loops and termini, and enhanced secondary structure stability. Not uncommonly, the increased numbers of salt bridges and/or improved hydrophobic interactions occur at subunit–subunit interfaces, so that oligomerization effects have been associated with increased thermal stability. Somewhat surprisingly, in view of prevailing ideas prior to structure determinations, there is little evidence for an important role for cross-links in contributing to hyperthermostability. Table II also demonstrates that after a major emphasis on protein stability in the first papers to describe the structures of hyperthermophilic proteins, more recent papers have tended to ignore this issue, undoubtedly reflecting the subtleties and difficulties of this problem.

Thermal Expansion of Hyperthermophilic Proteins

With the exception of a few odd compounds, such as water under certain conditions, matter typically expands with increasing temperature, and proteins provide no exception to this behavior. Values of the coefficient of thermal expansion

$$\alpha = \frac{1}{V}\frac{dV}{dT}$$

have been reported for proteins of $\sim 10^{-4}$ K^{-1} (ref. 29). As the temperature of the protein environment increases, the size and number of interior cavities should increase as a consequence of this thermal expansion. Cavities will facilitate internal

---

[29] K. Heremans and L. Smeller, *Biochim. Biophys. Acta* **1386**, 353 (1998).

TABLE II
PROPOSED CONTRIBUTIONS OF VARIOUS ENERGETIC INTERACTIONS TO HYPERTHERMOSTABILITY[a]

PDB ID	Salt bridges/ hydrogen bonds	Packing/ hydrophobicity	Loops/ termini	2° Structure stabilization	Other interactions/ comments
1CAA	x		x		
1HDG	x				
1AOR	x	x			
1IGS	x	x	x	x	
1GTM	x				
1XER					metal binding
1NSJ	x	x	x		
1GOW	x				buried solvent networks
1PCZ	x	x			disulfide bridges
1VJW	x		x	x	strained residues replaced by glycine
1XGS	x		x	x	buried solvent network; strained residues replaced by glycine
1VPE	x		x	x	
1AJ8	x	x	x		
1TMY					nothing obvious
1FTR	x	x			highly charged surface for salt-dependent thermostability; oligomerization
1A1S		x			oligomerization
1A2Z		x			disulfide bridge
1A5Z	x	x		x	
1NKS					oligomerization
1BVU	x				
1SSS	x				
1B26	x	x			
1B7G	x				disulfide bridge
1QEZ	x				oligomerization
1QDL			x		
1COJ	x	x			
1QVB	x	x			oligomerization buried solvent network
1QLM		x	x		oligomerization

[a] For the proteins listed in Table I, as identified by the authors in the indicated references. Salt bridges/hydrogen bonds indicate that increased numbers of these interactions are observed in a hyperthermophilic protein; packing/hydrophobic interaction indicate that more effective packing or better burial of hydrophobic surface was noted in a hyperthermophilic protein, and is often associated with subunit–subunit interfaces in oligomers; loops/termini denotes observations of shortened loops or fixed polypeptide chain termini in a hyperthermophilic protein; and 2° structure stabilization indicates evidence for enhanced stability of α helices or β sheets. Structures for which hyperthermostability was not significantly addressed in the original structure determination paper are omitted from this table.

rearrangements that may precede unfolding, and these events are reflected in a positive volume of activation, $\Delta V^\ddagger$, for protein unfolding. For staph nuclease, $\Delta V^\ddagger$ has been measured to be ~30 Å^3/molecule,[30] which can be generated by a 20° temperature increase for $\alpha = 10^{-4}$ K^{-1} and $V = 1.5 \times 10^4$ Å^3. This behavior parallels the substantial positive volume increase observed in solids near their melting temperature[31] that allows the rearrangements necessary for melting.

Thermal expansion should be increasingly important for proteins that exist at high temperatures, since they would be expected to have increased numbers and sizes of cavities under physiological conditions, which would be destabilizing. To minimize this effect, it is possible that hyperthermophilic proteins are constructed to have a smaller value of α to compensate for their high environmental temperature. A similar proposal has been explored by Palma and Curmi,[32] who used molecular dynamics simulations to examine the correlation between surface thermal expansion and protein stability and found that more stable proteins are calculated to exhibit smaller surface thermal expansion.

Thermal expansion properties of proteins can be studied by examination of structures determined at different temperatures.[33] Most commonly, the temperature range is between room temperature (~298 K) and the temperature of liquid nitrogen (~100 K), but hyperthermophilic proteins provide the opportunity to extend this range up to ~373 K. Figure 3 illustrates the dependence of volume/buried atom on temperature for several proteins. The volume is calculated as the average volume of a Voronoi polyhedron[34] surrounding every buried atom (defined as having zero accessible surface area) in a structure. For mesophilic structures, a series of structures at different temperatures have been reported for lysozyme and ribonuclease.[35,36] For hyperthermophilic enzymes, the temperature-dependent data are more scarce, although a preliminary refinement of *P. furiosus* rubredoxin at 95° has been performed (M. K. Chan, M. W. Day, M. W. W. Adams, and

---

[30] G. J. A. Vidugiris, J. L. Markley, and C. A. Royer, *Biochemistry* **34,** 4909 (1995).

[31] A. Bondi, "Physical Properties of Molecular Crystals, Liquids, and Glasses." John Wiley & Sons, New York, 1968.

[32] R. Palma and P. M. G. Curmi, *Prot. Sci.* **8,** 913 (1999).

[33] H. Frauenfelder, H. Hartmann, M. Karplus, I. D. J. Kuntz, D. Ringe, R. F. J. Tilton, M. L. Connolly, and L. Max, *Biochemistry* **26,** 254 (1987).

[34] F. M. Richards, *Ann. Rev. Biophys. Bioeng.* **6,** 151 (1977).

[35] I. V. Kurinov and R. W. Harrison, *Acta Crystallogr.* **D51,** 98 (1995).

[36] R. F. Tilton, J. C. Dewan, and G. A. Petsko, *Biochemistry* **31,** 2469 (1992).

[37] R. Bau, D. C. Rees, D. M. J. Kurtz, R. A. Scott, H. Huang, M. W. W. Adams, and M. K. Eidsness, *J. Biol. Inorg. Chem.* **3,** 484 (1998).

[38] I. Korndorfer, B. Steipe, R. Huber, A. Tomschy, and R. Jaenicke, *J. Mol. Biol.* **246,** 511 (1995).

[39] M. K. Chan, S. Mukund, A. Kletzin, M. W. W. Adams, and D. C. Rees, *Science* **267,** 1463 (1995).

[40] M. Hennig, R. Sterner, K. Kirschner, and J. N. Jansonius, *Biochemistry* **36,** 6009 (1997).

[41] K. S. P. Yip, T. J. Stillman, K. L. Britton, P. J. Artymiuk, P. J. Baker, S. E. Sedelnikova, P. C. Engel, A. Pasquo, R. Chiaraluce, V. Consalvi, R. Scandurra, and D. W. Rice, *Structure* **3,** 1147 (1995).

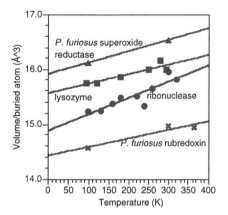

FIG. 3. Dependence of the average volume per buried atom on the temperature of the structure determination. Structure analyses are for RNase [●; PDB data sets 1RAT, 2RAT, 3RAT, 4RAT, 5RAT, 6RAT, 7RAT, 8RAT, 9RAT; R. F. Tilton, J. C. Dewan, and G. A. Petsko, *Biochemistry* **31,** 2469 (1992)], lysozyme[35] [■; PDB data sets 1LSA, 1LSB, 1LSC, 1LSD, 1LSE; I. V. Kurinov and R. W. Harrison, *Acta Crystallogr.* **D51,** 98 (1995)]; *P. furiosus* rubredoxin (✖; PDB data set 1CAA [100 K; M. W. Day, B. T. Hsu, L. Joshua-Tor, J.-B. Park, Z. H. Zhou, M. W. W. Adams, and D. C. Rees, *Prot. Sci.* **1,** 1494 (1992)], and at 298 K and 368 K [M. K. Chan, M. W. Day, M. W. W. Adams, and D. C. Rees, unpublished results (1995)]; and *P. furiosus* superoxide reductase (▲; A. P. Yeh, Y. Hu, F. E. Jenney Jr., M. W. W. Adams, and D. C. Rees, *Biochemistry* **39,** 2499 (2000)).

D. C. Rees, unpublished results, 1995) that complements structures at low and room temperatures.[1] The slopes of these lines divided by the atomic volume yield the coefficient of thermal expansion, which is $\sim 10^{-4}\,\mathrm{K}^{-1}$ for these proteins, similar to the results of solution studies. To first order, there appears little difference in the thermal expansion behavior of mesophilic and hyperthermophilic proteins, at least between liquid nitrogen and room temperatures. There is a hint that the volume of buried atoms in the *P. furiosus* rubredoxin may not increase as rapidly at higher temperatures, but this remains to be established confidently. Undoubtedly, further work is needed to address the thermal properties of hyperthermophilic proteins and whether this has any relationship to stability.

[42] T. Fujii, Y. Hata, H. Moriyama, T. Wakagi, N. Tanaka, and T. Oshima, *Nature Struct. Biol.* **3,** 834 (1996).
[43] C. F. Aguilar, I. Sanderson, M. Moracci, M. Ciaramella, R. Nucci, M. Rossi, and L. H. Pearl, *J. Mol. Biol.* **271,** 789 (1997).
[44] B. S. DeDecker, R. Obrien, P. J. Fleming, J. H. Geiger, S. P. Jackson, and P. B. Sigler, *J. Mol. Biol.* **264,** 1072 (1996).
[45] S. Macedo-Ribeiro, B. Darimont, R. Sterner, and R. Huber, *Structure* **4,** 1291 (1996).
[46] E. Schmitt, L. Moulinier, S. Fujiwara, T. Imanaka, J. C. Thierry, and D. Moras, *EMBO J.* **17,** 5227 (1998).
[47] P. F. Kosa, G. Ghosh, B. S. DeDecker, and P. B. Sigler, *Proc. Natl. Acad. Sci. U.S.A.* **94,** 6042 (1997).

Future Developments

We are still in the early days for the crystallographic analysis of hyperthermophilic proteins. Even so, there has been widespread interest in this area, as proteins studied so far range from homologs of well-characterized proteins to novel enzymes found primarily in these organisms. One surprising omission, given the biotechnological relevance, is the absence of polymerase structures (at least, that have been publicly described or deposited in the PDB as of October, 1999), but presumably this situation will change shortly. With the increasing availability of genomic sequences for hyperthermophiles, the number of structurally characterized proteins will explode; undoubtedly, even by the time this review appears, the list of structures in Table I will be woefully incomplete. Beyond continuing structural studies, hyperthermophilic proteins provide exciting opportunities, particularly in characterizing the response of protein structure to physiologically extreme conditions of temperature and pressure, and in the currently underexploited analysis of hyperthermophilic membrane proteins. Together, these diverse structural efforts

[48] T. H. Tahirov, H. Oki, T. Tsukihara, K. Ogasahara, K. Yutani, K. Ogata, Y. Izu, S. Tsunasawa, and I. Kato, *J. Mol. Biol.* **284**, 101 (1998).

[49] G. Auerbach, R. Huber, M. Grattinger, K. Zaiss, H. Schurig, R. Jaenicke, and U. Jacob, *Structure* **5**, 1475 (1997).

[50] R. J. M. Russell, J. M. C. Ferguson, D. W. Hough, M. J. Danson, and G. L. Taylor, *Biochemistry* **36**, 9983 (1997).

[51] K. C. Usher, A. F. A. De la Cruz, F. W. Dahlquist, R. V. Swanson, M. I. Simon, and S. J. Remington, *Protein Sci.* **7**, 403 (1998).

[52] J. Lowe and L. A. Amos, *Nature* **391**, 203 (1998).

[53] U. Ermler, M. C. Merckel, R. K. Thauer, and S. Shima, *Structure* **5**, 635 (1997).

[54] H. Robinson, Y. G. Gao, B. S. McCrary, S. P. Edmondson, J. W. Shriver, and A. H. J. Wang, *Nature* **392**, 202 (1998).

[55] M. R. Singleton, M. N. Isupov, and J. A. Littlechild, *Struct. Fold. Des.* **7**, 237 (1999).

[56] G. Auerbach, R. Ostendorp, L. Prade, I. Korndorfer, T. Dams, R. Huber, and R. Jaenicke, *Structure* **6**, 769 (1998).

[57] K. Y. Hwang, K. Baek, H. Y. Kim, and Y. Cho, *Nature Struct. Biol.* **5**, 707 (1998).

[58] H. Li, C. R. Trotta, and J. Abelson, *Science* **280**, 279 (1998).

[59] B. Ren, G. Tibbelin, D. de Pascale, M. Rossi, S. Bartolucci, and R. Ladenstein, *Nature Struct. Biol.* **5**, 602 (1998).

[60] C. Vonrhein, H. Boenisch, G. Schaefer, and G. E. Schulz, *J. Mol. Biol.* **282**, 167 (1998).

[61] K. K. Kim, L. W. Hung, H. Yokota, R. Kim, and S. H. Kim, *Proc. Natl. Acad. Sci. U.S.A.* **95**, 10419 (1998).

[62] K. K. Kim, R. Kim, and S. H. Kim, *Nature* **394**, 595 (1998).

[63] Y. G. Gao, S. Y. Su, H. Robinson, S. Padmanabhan, L. Lim, B. S. McCrary, S. P. Edmondson, J. W. Shriver, and A. H. J. Wang, *Nature Struct. Biol.* **5**, 782 (1998).

[64] T. I. Zarembinski, L. W. Hung, H. J. Mueller-Dieckmann, K. K. Kim, H. Yokota, R. Kim, and S. H. Kim, *Proc. Natl. Acad. Sci. U.S.A.* **95**, 15189 (1998).

[65] C. Vetriani, D. L. Maeder, N. Tolliday, K. S. P. Yip, T. J. Stillman, K. L. Britton, D. W. Rice, H. H. Klump, and F. T. Robb, *Proc. Natl. Acad. Sci. U.S.A.* **95**, 12300 (1998).

on hyperthermophilic proteins will significantly contribute to advances ranging from fundamental aspects of protein structure and stability to new technological applications.

Acknowledgments

Heated discussions with co-workers and collaborators, particularly M. W. W. Adams and J. B. Howard, are gratefully acknowledged. This work was supported by the Howard Hughes Medical Institute and the National Institutes of Health (GM45162).

[66] T. Ursby, B. S. Adinolfi, S. Al-Karadaghi, E. De Vendittis, and V. Bocchini, *J. Mol. Biol.* **286,** 189 (1999).
[67] Y. Hu, S. Faham, R. Roy, M. W. W. Adams, and D. C. Rees, *J. Mol. Biol.* **286,** 899 (1999).
[68] R. Thoma, G. Obmolova, D. A. Lang, M. Schwander, P. Jeno, R. Sterner, and M. Wilmanns, *FEBS Lett.* **454,** 1 (1999).
[69] S. Knapp, S. Kardinahl, N. Hellgren, G. Tibbelin, G. Schafer, and R. Ladenstein, *J. Mol. Biol.* **285,** 689 (1999).
[70] S. Knapp, W. M. deVos, D. Rice, and R. Ladenstein, *J. Mol. Biol.* **267,** 916 (1997).
[71] A. M. Bilwes, L. A. Alex, B. R. Crane, and M. I. Simon, *Cell* **96,** 131 (1999).
[72] M. N. Isupov, T. M. Fleming, A. R. Dalby, G. S. Growhurst, P. C. Bourne, and J. A. Littlechild, *J. Mol. Biol.* **291,** 651 (1999).
[73] V.-M. Leppanen, H. Nummelin, T. Hansen, R. Lahti, G. Schafer, and A. Goldman, *Protein Sci.* **8,** 1218 (1999).
[74] S. A. Lloyd, F. G. Whitby, D. Blair, and C. P. Hill, *Nature* **400,** 472 (1999).
[75] T. Knoechel, A. Ivens, G. Hester, A. Gonzalez, R. Bauerle, M. Wilmanns, K. Kirschner, and J. N. Jansonius, *Proc. Natl. Acad. Sci. U.S.A.* **96,** 9479 (1999).
[76] J. H. Lim, Y. G. Yu, Y. S. Han, S. J. Cho, B. Y. Ahn, S. H. Kim, and Y. J. Cho, *J. Mol. Biol.* **270,** 259 (1997).
[77] Y. I. Chi, L. A. Martinez-Cruz, J. Jancarik, R. V. Swanson, D. E. Robertson, and S.H. Kim, *FEBS Lett.* **445,** 375 (1999).
[78] M. S. Finnin, J. R. Donigian, A. Cohen, V. M. Richon, R. A. Rifkind, P. A. Marks, R. Breslow, and N. P. Pavletich, *Nature* **401,** 188 (1999).
[79] W. Grabarse, M. Vaupel, J. A. Vorholt, S. Shima, R. K. Thauer, A. Wittershagen, G. Bourenkov, H. D. Bartunik, and U. Ermler, *Structure* **7,** 1257 (1999).
[80] D. J. Hosfield, C. D. Mol, B. H. Shen, and J. A. Tainer, *Cell* **95,** 135 (1998).
[81] H. Hashimoto, T. Inoue, M. Nishioka, S. Fujiwara, M. Takagi, T. Imanaka, and Y. Kai, *J. Mol. Biol.* **292,** 707 (1999).
[82] C. Charron, F. Talfournier, M. N. Isupov, G. Branlant, J. A. Littlechild, B. Vitoux, and A. Aubry, *Acta Crystallogr.* **D55,** 1353 (1999).
[83] K. Y. Hwang, C. S. Cho, S. S. Kim, H. C. Sung, Y. G. Yu, and Y. J. Cho, *Nature Struct. Biol.* **6,** 422 (1999).

[33] Thermostability of Proteins from *Thermotoga maritima*

By RAINER JAENICKE and GERALD BÖHM

Introduction

Proteins, independent of their mesophilic or thermophilic origin, consist exclusively of the 20 canonical natural amino acids. In the multicomponent system of the cytosol, these are known to undergo covalent modifications at the upper limit of temperature observed in the biosphere (deamidation, β-elimination, disulfide exchange, oxidation. Maillard reactions, hydrolysis, etc.[1]). Extremophiles must compensate for these degradation processes either by using compatible protectants or by enhanced synthesis and repair. Little is known about the chemistry involved, e.g., in the hydrothermal decomposition of proteins, and even less about protection and repair, with one exception: In the hyperthermophilic bacterium *Thermotoga maritima* (*Tm*). a superactive repair enzyme, L-isoaspartyl methyltransferase, has been isolated that catalyzes the transfer of the methyl group from S-adenosylmethionine (SAM) to the α-carboxyl group of L-isoaspartyl residues resulting from the deamidation of Asn and the isomerization of Asp. Interestingly, its specific activity under physiological conditions is ≈20-fold higher than the maximal activities of mesophilic homologs at 37°.[2] Because deamidation is one of the most important degradation reactions in the temperature regime of hydrothermal vents,[3] the enzyme may be essential for initiating repair of proteins at the borderline of viability (T_{max} ≈ 90°). At 80°, the optimum growth temperature of *Thermotoga maritima*,[4] no significant damage is expected to occur. The same holds for the essential metabolites and coenzymes, because the temperature at which ATP hydrolysis becomes limiting for viability lies between 110° and 140°[5]; at T_{opt} and pH 5–7, the half-time of hydrolysis is of the order of 6–7 hr. The hydrothermal oxidation of NADH can be ignored for cytosolic proteins in a strictly anaerobic environment. Similarly, the aqueous solvent still sustains its characteristics so that the two essential requirements for the self-organization of proteins, the integrity of the natural amino acids and the formation of the hydrophobic core

---

[1] D. B. Volkin, H. Mach, and C. R. Middaugh, *in* "Protein Stability and Folding: Theory and Practice" (B. A. Shirley, Ed.), *Meth. Mol. Biol.* **40,** 35 (1995).
[2] J. K. Ishikawa and S. Clarke, *Arch. Biophys. Biochem.* **358,** 222 (1998).
[3] R. Jaenicke, *Biochemistry (Moscow)* **63,** 312 (1998).
[4] R. Huber, T. A. Langworthy, H. König, M. Thomm, C. R. Woese, U. B. Sleytr, and K. O. Stetter, *Arch. Microbiol.* **144,** 324 (1986).
[5] E. Leibrock, P. Bayer, and H.-D. Lüdemann, *Biophys. Chem.* **54,** 175 (1995).

in the process of protein folding, are still operative.[6] Thus, in the following it is possible to discuss the problems of protein stability and protein folding over the whole temperature range from the freezing to the boiling point of water.

Many studies on the thermal stability of proteins have focused on enzymes from hyperthermophiles. In this context, *Thermotoga maritima,* with its temperature range of growth between 55° and 90°, has become one of the favorite organisms, because it is widespread in marine geothermal vents as well as in low-salinity solfataric springs, it can be cultivated in large-scale fermentations, and it has been characterized in detail with respect to its metabolic requirements.[7] The bacterium is a strictly anaerobic fermentative organotroph that grows on various sugars, cellulose, starch, and glycogen as the carbon source: peptides are required for growth on carbohydrates because the organism does not utilize ammonia or free amino acids as the N source.[6,7]

Fundamentals of Protein Stability

Globular proteins exhibit marginal stabilities that are equivalent to only a small number of weak intermolecular interactions.[8–13] In this respect, proteins from thermophiles and hyperthermophiles do not differ significantly from their mesophilic counterparts. In most cases, thermophilic adaptation is accompanied by a small increase in the free energy of stabilization that is generated either by a flattening or lowering of the $\Delta G_{N \to U}$ vs temperature profiles, or by a shift to a higher stability maximum. Because of thermal effects on the heat capacity ΔC_P of proteins, the temperature profiles show the characteristics of skewed parabolas, defining denaturation transitions at both high and low temperature (Fig. 1).[14–17]

No general strategy of thermal adaptation has yet been established. This holds for two reasons: First, there is a hierarchical order of incremental contributions to

---

[6] R. Jaenicke, H. Schurig, N. Beaucamp, and R. Ostendorp, *Adv. Protein Chem.* **48,** 181 (1996).
[7] M. W. W. Adams, J.-B. Park, S. Mukund, J. Blamey, and R. M. Kelly, *In* "Biocatalysis at Extreme Temperature" (M. W. W. Adams and R. M. Kelly, Eds.), ACS Symp. Ser. 498, 2. ACS, Washington, DC, 1992.
[8] K. A. Dill, *Biochemistry* **29,** 7133 (1990).
[9] R. Jaenicke, *Eur. J. Biochem.* **202,** 715 (1991).
[10] S. K. Burley and G. A. Petsko, *Adv. Protein Chem.* **39,** 125 (1988).
[11] P. L. Privalov, *in* "Protein Folding" (T. E. Creighton, Ed.), P. 83. W. H. Freeman, New York, 1992.
[12] B. W. Matthews, *FASEB J.* **10,** 35 (1996).
[13] W. Pfeil, "Protein Stability and Folding: A Collection of Thermodynamic Data," Springer, Berlin, 1998.
[14] J. F. Brandts, J. Fu, and J. H. Nordin, *in* "The Frozen Cell" (G. E. W. Westenholme and M. O'Connor, Eds.), P. 189. J. & A. Churchill, London, 1970.
[15] F. Franks, *Adv. Protein Chem.* **46,** 105 (1995).
[16] P. L. Privalov and S. J. Gill, *Adv. Protein Chem.* **39,** 191 (1988).
[17] R. Jaenicke, *Phil. Trans. Roy. Soc. Lond. B* **326,** 535 (1990).

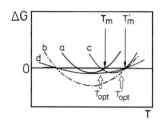

FIG. 1. Hypothetical temperature profiles of the free energy of (a) mesophilic and (b–d) thermophilic proteins. ΔG is defined as the difference in the free energies between the native and denatured states. T_m and T'_m are the melting temperatures of the mesophilic and thermophilic variants, respectively. The minimum of the ΔG parabola for a given protein (i.e., maximum stability) is observed at a temperature that is much below the optimal temperature (T_{opt} and T'_{opt}) of the respective mesophilic or thermophilic organism.

the *intrinsic* stability of polypeptide chains or their assemblies; second, *extrinsic factors*, not encoded in the amino acid sequence, may be of major importance. Contributions to the latter may come from ligands or conjugating compounds such as metabolites, coenzymes, and compatible solvent components, but also from other biopolymers. In the case of *Thermotoga maritima*, extrinsic effects have been reported for a number of proteins, e.g., Ca^{2+}/Mg^{2+} for Tm enolase, oligonucleotides for Tm cold shock protein, NAD for glyceraldehyde-3-phosphate dehydrogenase, and NADP/methotrexate for dihydrofolate reductase. Intrinsic stability seems to be a common feature of the whole protein inventory of the cell, including its surface layer, the *toga*.[4,18] Interestingly, this holds not only for the proteins in their functional state at the optimum growth temperature of the bacterium, but also for the process of self-organization: Denaturation/renaturation experiments at temperatures between 20° and 85° yield the native starting material[19] (see below).

The above-mentioned incremental nature of protein stability has been established by detailed studies on model proteins such as phage T4 lysozyme,[9,20] and by rational protein design.[21] As mentioned, stabilization may involve all levels of the hierarchy of protein structure, local packing of the polypeptide chain, secondary and supersecondary structural elements, domains, and subunits.[9] Approaches to assign specific structural alterations to changes in stability are summarized in Table I.

As proteins are fundamentally multifunctional, combining the capacity of folding, working, and feeding at the same time, evolution had (and still has) to compromise between rigidity or stability, on the one hand, and flexibility or function, regulation, and turnover, on the other. The way nature reaches this goal is

---

[18] R. Rachel, A. M. Engel, R. Huber, K. O. Stetter, and W. Baumeister, *FEBS Lett.* **262**, 64 (1990).
[19] V. Rehaber and R. Jaenicke, *J. Biol. Chem.* **267**, 10999 (1992).
[20] B. W. Matthews, *Adv. Protein Chem.* **46**, 249 (1995).
[21] B. van den Burg, G. Vriend, O. R. Veltman, G. Venema, and V. G. H. Eijsink, *Proc. Natl. Acad. Sci. U.S.A.* **95**, 2056 (1998).

TABLE I
MOLECULAR BASIS OF PROTEIN STABILITY

Experimental approaches	Example	Reference
Selection of ts mutants[a]	Trp repressor (trpR)	b
	Bacteriophage f1 gene V protein	c
	Phage P22 tailspike/capsid protein	d,e
Variations of amino acid residues in model proteins	Ubiquitin	f
	Lambda repressor	g
	Streptococcus protein Gβ1 domain	h
	Repressor of primer (Rop)	i
	Barnase	j
	T4 lysozyme	12, 20, 22
Fragmentation of domain Ps, modifications of connecting peptides between domains	βγ Crystallins	28
	Phosphoglycerate kinase	k, 28
	Circular permutants (DHFR and others)	l, 28
	Single-chain antibodies (scFv)	m
Modification of subunit interfaces, hybridization, or solvent perturbation	Insulin	n
	Hemoglobin	o
	Glutathione transferase	p

[a] ts, Temperature sensitive, J. A. Schellman, *Biophys. J.* **73**, 2960 (1997).
[b] L. Jin, J. W. Fukayama, I. Pelczer, and J. Carey, *J. Mol. Biol.* **285**, 361 (1999).
[c] W. S. Sandberg, P. M. Schlunk, H. B. Zabin, and T. C. Terwilliger, *Biochemistry* **34**, 11970 (1995).
[d] M. L. Gallisteo, C. L. Gordon, and J. King, *J. Biol. Chem.* **270**, 16595 (1995).
[e] S. D. Betts and J. King, *Protein Sci.* **7**, 1516 (1998).
[f] G. A. Lazar, J. R. Desjarlais, and T. M. Handel, *Protein Sci.* **6**, 1167 (1997).
[g] W. A. Lim, A. Hodel, R. T. Sauer, and F. M. Richards, *Proc. Natl. Acad. Sci. U.S.A.* **91**, 423 (1994).
[h] S. M. Malakauskas and S. L. Mayo, *Nature Struct. Biol.* **5**, 470 (1998).
[i] M. Munson, S. Balasubramanian, K. G. Fleming, A. D. Nagi, R. O'Brien, J. M. Sturtevant, and L. Regan, *Protein Sci.* **5**, 1584 (1996).
[j] C. K. Vaughan, A. M. Buckle, and A. R. Fersht, *J. Mol. Biol.* **286**, 1487 (1999).
[k] K. Zaiss and R. Jaenicke, *Biochemistry* **38**, 4633 (1999).
[l] M. Iwakura and T. Nakamura, *Protein Eng.* **11**, 707 (1998).
[m] Y. Tang, N. Jiang, C. Parakh, and D. Hilvert, *J. Biol. Chem.* **271**, 15682 (1996).
[n] J. Brange, U. Ribel, J. F. Hansen, G. Dodson, C. M. T. Hansen, S. Havelund, S. G. Melberg, F. Norris, K. Norris, L. Snel, A. R. Sørensen, and H. O. Voigt, *Nature* **333**, 679 (1988).
[o] G. J. Turner, F. Galacteros, M. L. Doyle, B. Hedlund, D. W. Pettigrew, W. B. Turner, F. R. Smith, W. Moo-Penn, D. L. Rucknagel, and G. K. Ackers, *Proteins* **14**, 333 (1992).
[p] J. M. Stevens, J. A. Hornby, R. N. Armstrong, and H. W. Dir, *Biochemistry* **37**, 15534 (1998).

that the attractive and repulsive interatomic interactions that determine the three-dimensional structure of proteins more or less compensate each other so that $\Delta G_{N \to U}$ represents a small difference between large numbers. The total energy that holds the polypeptide together in its densely packed native state and the Gibbs

[22] B. W. Matthews, *Annu. Rev. Biochem.* **62**, 139 (1993).

free energy of stabilization are of the order of 10^7 kJ/mol and 50 ± 20 kJ/mol, respectively.[9,23] Available data for $\Delta G_{N \to U}$ of ultrastable proteins from *Thermotoga maritima* do not exceed 120 kJ/mol,[24-27] confirming the previous discussion even for the protein repertoire of hyperthermophiles.

The stability of proteins refers to the maintenance of a defined three-dimensional structure with specific thermodynamic and functional properties. High-resolution structures in the crystalline state and in solution have reached a stage at which the atomic coordinates of proteins can be compared with an accuracy down to root mean square deviation (r.m.s.d.) values less than 1 Å. However, even this precision does not allow the free energy of stabilization to be calculated from the coordinates, nor does it allow predictions with respect to the dynamics of functionally relevant local interactions in active or regulatory sites of homologous proteins. The fluctuations between preferred conformations of native proteins involved in such functionally important motions may very well show amplitudes and angles of up to 50 Å and 20°, respectively.[28]

In most cases when thermophiles or hyperthermophiles are compared with their mesophilic counterparts, evolutionary adaptation of proteins is nothing but the conservation of the functionally significant flexibility. This means that, under altering environmental conditions, homologous proteins are in "corresponding states." For example, pyruvate dehydrogenase from *Escherichia coli* at 30° exhibits the same specific activity as the enzyme from the moderately thermophilic *Bacillus stearothermophilus* at 60°.[29] Correspondingly, the glycolytic enzymes from *Thermotoga maritima* are practically inactive at room temperature; however, at the optimum growth temperature of the bacterium, at 80°, the specific activity reaches or exceeds the level of the mesophilic homologs at their physiological temperature around 30°.[4,6] Apparent exceptions have been reported for enzymes with highly unstable substrates or important repair functions, e.g., *Tm* isoaspartyl methyltransferase[2] (see above) or *Tm*PRAI[30] (cf. ref. 30a); actually, in such cases the only difference is that the principle of corresponding states holds here also for the kinetic competition of the hydrothermal and the enzymatic reactions.

---

[23] R. L. Baldwin and D. Eisenberg, *in* "Protein Engineering" (D. E. Oxender and C. F. Fox, Eds.), p. 127. Liss, New York, 1987.
[24] W. Pfeil, U. Gesierich, G. R. Kleemann, and R. Sterner, *J. Mol. Biol.* **272**, 591 (1997).
[25] M. Grättinger, A. Dankesreiter, H. Schurig, and R. Jaenicke, *J. Mol. Biol.* **280**, 525 (1998).
[26] K. Zaiss and R. Jaenicke, *Biochemistry* **38**, 4633 (1999).
[27] T. Dams and R. Jaenicke, *Biochemistry* **38**, 9169 (1999).
[28] R. Jaenicke, *Progr. Biophys. Mol. Biol.* **71**, 155 (1999).
[29] R. Jaenicke and R. N. Perham, *Biochemistry* **21**, 3378 (1982).
[30] R. Sterner, G. R. Kleemann, H. Szadkowski, H. Lustig, M. Hennig, and K. Kirschner, *Protein Sci.* **5**, 2000 (1996).
[30a] R. Sterner, A. Marz, R. Thoma, and K. Kirschner, *Methods in Enzymology* **331**[23], (2001).

Forces and Mechanisms Involved in Protein Stability and Stabilization

The spatial structure of proteins is determined by electrostatic forces between polar and ionized groups and by hydrophobic interactions involving nonpolar residues.[8-10,31] The electrostatic forces include ion pairs, hydrogen bonds, weakly polar interactions, and van der Waals forces. Studies involving single surface-located ion pairs have shown that their contribution to the stability of a given protein is only marginal (<4 kJ/mol), because the gain in free energy is about equal to the entropic cost of dehydration and the reduction of the conformational freedom.[9,12,32,33] However, extensive networks of charged groups may play an important role in maintaining enzyme stability at extreme temperatures: (1) because each extra ion pair added to the network requires the desolvation and localization of only a single residue; (2) because networks are often located in cavities and at interfaces where their conformational freedom is already restricted. As a consequence, part of the entropic cost has already been provided during the folding of the protein; and (3) because hydration effects play a minor role at high temperature, and the dielectric constant decreases with temperature, resulting in an increased electrostatic energy on ion-pair formation. The classical observation[34] that enhanced ion pairing correlates with thermal stability may be rationalized on the basis of these arguments (for examples, see below).

The significance of hydrogen bonds as the dominant stabilizing force in protein folding has been controversial for more than 40 years.[35-41] Estimates of the energy increment inherent to an intrachain H-bond relative to the H-bond with the aqueous solvent are of the order of 3.2 ± 0.8 kcal/mol.[40,41] Thus, its contribution to the overall Gibbs free energy is again close to the thermal energy kT, in accordance with the common notion that the thermal denaturation of proteins can be ascribed to the breakage of hydrogen bonds. However, considering a 10 kDa protein, about 440 polar sites are exposed on denaturation; half of them are involved in internal H-bonds in the native state. Therefore, even a marginal difference in H-bond strength between protein–water and water–water will be magnified to a large energy

---

[31] R. Jaenicke, *Biochemistry* **30**, 3147 (1991).
[32] A. Horovitz, L. Serrano, B. Avron, M. Bycroft, and A. R. Fersht, *J. Mol. Biol.* **216**, 1031 (1990).
[33] S. Dao-pin, U. Sauer, H. Nicholson, and B. W. Matthews, *Biochemistry* **30**, 7142 (1991).
[34] M. F. Perutz and H. Raidt, *Nature* **255**, 256 (1975).
[35] A. E. Mirsky and L. Pauling, *Proc. Natl. Acad. Sci. U.S.A.* **22**, 439 (1936).
[36] J. A. Schellman, *Compt. Rend. Lab. Carlsberg, Sér. Chim.* **29**, 223 (1995).
[37] J. D. Bernal, *Disc. Faraday Soc.* **25**, 7 (1958).
[38] W. Kauzmann, *Adv. Protein Chem.* **14**, 1 (1959).
[39] C. Tanford, *Adv. Protein Chem.* **23**, 121 (1968).
[40] A. R. Fersht, *TIBS* **12**, 301 (1987).
[41] C. N. Pace, U. Heinemann, U. Hahn, and W. Saenger, *Angew. Chem., Int. Ed. Engl.* **30**, 343 (1991).

change. Site-directed mutagenesis has shown that additional H-bonds or H-bond networks may contribute to the stability of thermophilic proteins.[40–43]

The physical nature of hydrophobic effects was previously considered to be entropic.[38,44,45] Based on this hypothesis, it has often been claimed that the thermal stabilization of proteins in thermophiles may be correlated with an increase in the number of hydrophobic residues. A critical analysis proved the differences to be statistically insignificant[46]; the recent dramatic increase in sequence data from complete genomes of mesophilic and (hyper-) thermophilic bacteria and archaea clearly confirmed this finding[47] (see below). Considering the real meaning of the word *hydrophobic*, it is clear that the aversion of nonpolar solutes to water becomes more ordinary and less entropy-driven at extreme temperatures, whereas in the mesophilic temperature regime the hydrophobic effect is indeed entropic. Maximum aversion arises at the temperature at which the free energy of transfer of nonpolar solutes into water shows its maximum. Under this condition, the entropy (i.e., the temperature derivative of ΔG) equals zero, so that the hydrophobic effect must be driven by enthalpic contributions, attributable to van der Waals forces in the core of the protein.[8,16,48]

Strategies for Thermal Stabilization

Amino Acid Composition: Lessons from Complete Genomes

When available genome sequences and protein structures are inspected, there seems to be no way to unambiguously correlate extreme thermal stability with either the amino acid composition or the three-dimensional structure of a given set of homologous proteins. In the present context, the use of a variety of marker proteins (apart from 16S ribosomal RNA) for evolutionary studies allowed three conclusions to be drawn: First, the bacterial and archaeal (hyper-) thermophiles are positioned at short branches of the phylogenetic tree, supporting the idea that in the present diversity of organisms thermophiles seem to precede their mesophilic counterparts. It is important to note that such phylogenetic relationships say nothing about the temperature at which life started.[49,50] Second, available complete genome

---

[42] M. W. Day, B. T. Hsu, L. Joshua-Tor, J. B. Park, Z. H. Zhou, M. W. Adams, and D. C. Rees, *Protein Sci.* **1,** 1494 (1992).
[43] S. Macedo-Ribeiro, B. Darimont, and R. Sterner, *Biol. Chem.* **378,** 331 (1997).
[44] C. Tanford, "The Hydrophobic Effect," 2nd Ed. Wiley. New York, 1980.
[45] F. H. Stillinger, *Science* **209,** 451 (1980).
[46] G. Böhm and R. Jaenicke, *Int. J. Peptide Prot. Res.* **43,** 97 (1994).
[47] R. Jaenicke and G. Böhm, *Curr. Opin. Struct. Biol.* **8,** 738 (1998).
[48] R. L. Baldwin, *Proc. Natl. Acad. Sci. U.S.A.* **83,** 8069 (1986).
[49] E. V. Koonin, R. L. Tatusov, and M. L. Galperin, *Curr. Opin. Struct. Biol.* **8,** 355 (1998).
[50] S. L. Miller and A. Lazcano, *J. Mol. Evol.* **41,** 689 (1995).

sequences clearly show that the phylogenetic tree has more complex roots than assumed previously based on the rRNA tree. Considering different genomes, single genes may show different phylogenetic relationships or, even more perplexing, they turn out to contain a mix of DNAs, some close to Archaea, others close to Bacteria. It looks as if each gene has its own history, due to lateral gene transfer or "swapping of genes" among organisms.[51,52] Third, among the 15 bacterial and archaeal genome sequences unravelled until 1998, six belong to hyperthermophiles. A comparison of these genomes with respect to specific genes from mesophiles provides a data set sufficiently large to extract significant trends in amino acid usage: Compared to mesophiles, genomes of thermophiles encode higher levels of charged amino acids (29.8 vs 24.1%), primarily at the expense of uncharged polar residues (26.8 vs 31.2%); there seems to be no preference for hydrophobic residues in thermophiles (43.4 vs 44.7%). Glutamine seems to be significantly discriminated against in hyperthermophiles, possibly because of an increased rate of deamidation of this residue at high temperatures; surprisingly, asparagine does not appear to be subject to the same discrimination.[51]

Comparison of Homologous Proteins

The picture of the adaptive mechanisms of thermophiles emerging from the statistical analysis of their complete genomes is at best vague, mainly due to the fact that the small differences in $\Delta G_{N \rightarrow U}$ between homologous proteins form thermophiles and mesophiles allow an astronomical number of combinations of small structural variations. There is a repertoire of different strategies of thermal adaptation that is "used" by different proteins in an individual fashion. To illustrate this, in the following, we focus on a set of proteins from *Thermotoga maritima* that have been characterized in sufficient detail to allow a correlation of their thermodynamic, kinetic, and structural properties. All of them exhibit high intrinsic stability up to the maximum growth temperature of *Thermotoga maritima:* most have been cloned and overproduced to high expression levels in *E. coli* as authentic proteins. In addition, their folding topologies were found to be highly similar to those known for their mesophilic counterparts. The most significant difference is their low conformational flexibility at room temperature. This is correlated with an increase in compactness of the hydrophobic core (i.e., a decreased cavity volume) and with an increased number of peripheral and/or intersubunit ion pairs and hydrogen bonds. According to the principle of corresponding states, the anomalous rigidity disappears at elevated temperature. One further distinction between thermophiles and mesophiles, which again reflects the extreme environment, is

---

[51] G. Deckert, P. V. Warren, T. Gaasterland, W. G. Young, A. L. Lenox, D. E. Graham, R. Overbeek, M. A. Snead, M. Keller, M. Aujay, R. Huber, R. A. Feldman, J. M. Short, G. J. Olsen, and R. V. Swanson, *Nature* **392**, 353 (1998).

[52] R. F. Doolittle, *Nature* **392**, 339 (1998).

the integration of enzymes such as hydrolases and oxidoreductases into the cell membrane. Evidently, in a turbulent hydrothermal vent or in fumaroles, physical contact of the enzymes with the nutrient substrate makes sense. In the case of *Thermotoga maritima,* the activities of α- and β-glucoamylases as well as xylanase are pelleted together with the *toga:* the amylases are not secreted into the outside medium, nor are they detectable in the cytosolic fraction.[53-55] For AmyA, one of the amylose gene products, the reason may be that the enzyme has been shown to be a lipoprotein.[56] In the case of 1,4-D-xylanase, the C-terminal domain of the enzyme seems to be involved in cellulose binding.[57,58] In accordance with the localization in the cell membrane, the expression of the previously mentioned hydrolases in the bacterium is extremely low. In their recombinant form, some of them were found to be extremely stable, with transition temperatures between 90° and >105°.[53-55,58]

Obviously, there are also cytosolic hydrolases in *Thermotoga.* For example, a 670 kDa homomultimeric protease has been identified that shows both tryptic and chymotryptic activity with an optimum at 90°C and a half-life of 36 min at 95°C; its sequence homology suggests antifungal proteolytic activity.[59] In this context, it is worth mentioning that *Thermotoga maritima* also contains a chaperone complex showing certain structural and functional similarities to the Hsp60 (GroEL) complex from *Escherichia coli:* sevenfold symmetry of a homooligomeric two-layer ring system with 60–70 kDa subunits and ATPase activity with a temperature optimum at 70–90°, paralleled by reduced reactivation of Tm lactate dehydrogenase as a model substrate.[60,61]

Selected Examples

Triose-Phosphate Isomerase and Phosphoglycerate Kinase

Triose-phosphate isomerase (TIM) (EC 5.3.1.1) and phosphoglycerate kinase (PGK) (EC 2.7.2.3) are two ubiquitous enzymes in the major pathways of carbohydrate metabolism (glycolysis, gluconeogenesis, and pentose phosphate pathway), catalyzing the interconversion of dihydroxyacetone phosphate and D-glyceraldehyde 3-phosphate, and the phospho-group transfer between

---

[53] J. Schumann, A. Wrba, R. Jaenicke, and K. O. Stetter, *FEBS Lett.* **282,** 122 (1991).
[54] C. Winterhalter and W. Liebl, *Appl. Environ. Microbiol.* **61,** 1810 (1995).
[55] S. A. Käslin, S. E. Childers, and K. M. Noll, *Arch. Microbiol.* **170,** 297 (1998).
[56] W. Liebl, I. Stempfinger, and P. Ruile, *J. Bacteriol.* **179,** 941 (1997).
[57] C. Winterhalter, P. Heinrich, A. Candussio, G. Wich, and W. Liebl, *Mol. Microbiol.* **15,** 431 (1995).
[58] D. Wassenberg, H. Schurig, W. Liebl, and R. Jaenicke, *Protein Sci.* **6,** 1718 (1997).
[59] P. M. Hicks, K. D. Rinker, J. R. Baker, and R. M. Kelly, *FEBS Lett.* **440,** 393 (1998).
[60] R. Ostendorp, Ph.D. Thesis, University of Regensburg (1996).
[61] K. Lohmüller, Thesis, University of Regensburg (1996).

1,3-bisphosphoglycerate and ATP. Both enzymes show high structural similarity when homologs from various sources, including meso-, thermo-, and hyperthermophiles, are compared. In contrast to the common dimeric quaternary structure of mesophilic TIMs. TmTIM does not occur as a distinct monofunctional entity, although its backbone shows the prototype eight-stranded α/β-barrel topology. TmPGK is monomeric and folds into two distinct lobes of approximately equal size, with the active site positioned in a cleft between the domains. On screening for the activities of the two enzymes in crude extracts of *Thermotoga maritima*, one peak with TIM activity (at 350 kDa) and two well-separated peaks with PGK activity (at 45 and 350 kDa) were eluted by gel-permeation chromatography, the first belonging to a "normal" monomeric PGK, the second to a tetrameric bifunctional PGK–TIM fusion protein, with TIM as the C-terminal part of each of the four polypeptide chains.[62] As taken from the complete sequence of the *gap* operon, the fusion protein is not part of a glycolytic multienzyme complex: the *gap* gene (coding for glyceraldehyde-3-phosphate dehydrogenase, GAPDH) is regulated by a different promoter and *precedes* the PGK and TIM genes. Within the *gap* gene cluster, the TIM gene (*tpi*) lacks a start codon as well as promoter elements and the ribosomal binding site, so that *tpi* is expressed together with the preceding *pgk* gene. Moreover, both *pgk* and *tpi* are not fused in frame; obviously, a mechanism such as a programmed reading frameshift or transcription slippage must be in effect upstream of the PGK stop codon.[6,62,62a] Cloning and expression of the *Tm pgk-tpi* gene in *E. coli* clearly proves that the fusion mechanism is not specific for *Thermotoga:* both recombinant enzymes are found to be authentic, with the only difference that the ratio of the monomeric PGK and the fusion protein is different.[63]

In order to study how the covalent joining of the two enzymes and the anomalous state of association effects the structure, stability, and function of TmTIM, the *pgk-tpi* gene was dissected, and the isolated enzyme cloned and expressed in *E. coli*.[64] With a molecular mass of 104 kDa, the purified protein was found to be tetrameric, in contrast to the dimeric enzyme from mesophilic sources. The tetrameric state of association has also been reported for the enzyme from two other hyperthermophiles, *Thermoproteus tenax*[65] and *Pyrococcus woesei*.[66] As taken from ultracentrifugal analysis at low concentrations, all three seem to form stable dimers of dimers with an exceedingly high association constant (R. Jaenicke,

---

[62] H. Schurig, N. Beaucamp, R. Ostendorp, R. Jaenicke, E. Adler, and J. R. Knowles, *EMBO J.* **14**, 442 (1995).

[62a] D. Wassenberg, M. Wuhrer, N. Beaucamp, H. Schurig, M. Wozny, D. Reusch, S. Fabry, and R. Jaenicke, *Biol. Chem.* in press (2001).

[63] N. Beaucamp, R. Ostendorp, H. Schurig, and R. Jaenicke, *Prot. Peptide Lett.* **2**, 281 (1995).

[64] N. Beaucamp, A. Hofmann, B. Kellerer, and R. Jaenicke, *Protein Sci.* **6**, 2159 (1997).

[65] M. Kohlhoff, A. Dahm, and R. Hensel, *FEBS Lett.* **383**, 245 (1996).

[66] G. S. Bell, R. J. M. Russell, M. Kohlhoff, R. Hensel, M. J. Danson, D. W. Hough, and G. L. Taylor, *Acta Crystallogr.* **D54**, 1419 (1998).

unpublished, 1998). The crystal structure of the *Thermotoga* enzyme clearly supports this result, allowing correlation of the pairwise topology to a specific hydrophobic patch of residues in loop 5 and helix 5, apart from a network of ionpairs.[67]

The tetrameric quaternary structure of *Tm*TIM suggests that the wild-type PGK–TIM fusion protein consists of a core of two TIM dimers covalently linked to four PGK chains via the protruding N-terminal end of the polypeptide chain. A large hydrophobic patch on the surface of the *Tm*PGK moiety represents a good candidate for the attachment site (see below). Successful crystallization screens and crystallographic refinement procedures will allow the detailed topology of the complete PGK–TIM fusion protein to be elucidated in the not too distant future.[68]

When the physicochemical and catalytic properties of the isolated *Tm*TIM moiety are compared with the enzyme integrated into the bifunctional PGK–TIM fusion protein, the separate dimer shows significant changes, reflected not only by an increase in intrinsic stability, but also by a drastic enhancement of the catalytic efficiency of the enzyme. In accordance with the principle of corresponding states, the catalytic properties of the enzyme at $\sim 80°$ are similar to those of its mesophilic counterparts at their respective physiological temperatures. Under this condition, the *Tm*TIM within the fusion protein shows long-term stability, whereas the separate recombinant enzyme exhibits a denaturation half-time of 2 hr. At neutral (cellular) pH, the thermal denaturation transitions of the fusion protein and its constituent parts are irreversible so that a sound thermodynamic analysis is rendered impossible. Obviously, misfolding and/or wrong domain interactions interfere with correct folding and/or association. Therefore, guanidinium chloride-induced (GdmCl) denaturation/renaturation was applied in order to get insight into the stability of *Tm*TIM and *Tm*PGK and their mutual stabilization within the PGK–TIM fusion protein. As illustrated in Fig. 2, the TIM part of the fusion protein is significantly more stable than the PGK part: the onset of deactivation of PGK is found to coincide with the first step of the bimodal CD transition; the second step parallels the denaturation of TIM, proving the independent unfolding/folding of the two enzymes linked in the bienzyme complex. With increasing temperature, the half-concentration of GdmCl is shifted to lower concentrations; at the same time, the transitions change from bimodal to unimodal, accompanied by a marked decrease in the yield of reactivation.[69]

In contrast to TIM, which is not available as a single gene product, PGK is expressed both as bifunctional PGK–TIM fusion protein and as separate monomeric entity.[62] The comparison of the physicochemical data for both forms

---

[67] D. Maes, J. P. Zeelen, N. Thanki, N. Beaucamp, M. Alvarez, M. H. D. Thi, J. Backmann, J. A. Martial, L. Wyns, R. Jaenicke, and R. K. Wierenga, *Proteins: Struct. Funct. Genet.* **37**, 441 (1999).
[68] T. Dams, G. Auerbach, G. Bader, and R. Huber, work in progress.
[69] N. Beaucamp, H. Schurig, and R. Jaenicke, *Biol. Chem.* **378**, 679 (1997).

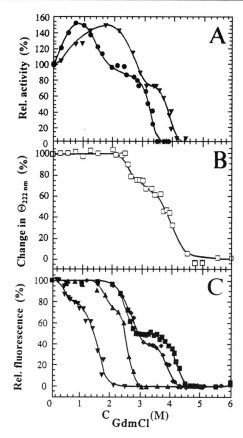

FIG. 2. GdmCl-induced equilibrium transitions of the PGK–TIM fusion protein in 50 mM sodium phosphate, pH 7.5, after 48 hr incubation at 40°. (A) Deactivation of the PGK portion (●) and the TIM portion (▼) of the fusion protein (20 μg/ml). (B) Denaturation transition of the PGK–TIM fusion protein (50 μg/ml) monitored by circular dichroism at 222 nm (□). (C) Temperature dependence of the PGK–TIM fusion protein (20 μg/ml) monitored by fluorescence emission at 320 nm (λ_{exc} = 280 nm): (■) 20°, (♦) 40°, (▲) 60°, (▼) 80°.[69]

of the enzyme is summarized in Table II. As shown by thermal and chemical denaturation/deactivation [e.g., at high GdmCl concentration], the fusion does not contribute significantly to the intrinsic stability. At neutral pH, the GdmCl- and temperature-induced denaturation transitions reveal two-state behavior with high cooperativity (Fig. 3A). As taken from the temperature dependence of the free energy of unfolding at zero GdmCl concentration and pH 7, optimum stability is observed at ca 30° (Fig. 3B). The difference in the free energies of stabilization for the enzymes from mesophiles and *Thermotoga* amounts to $\Delta\Delta G$ = 85 kJ/mol.

TABLE II
PHYSICOCHEMICAL AND CATALYTIC PROPERTIES OF TmPGK–TIM FUSION PROTEIN
AND ITS CONSTITUENT PARTS TmPGK AND TmTIM[a]

Property	PGK–TIM fusion protein	PGK	Separated TIM
Molecular mass			
Monomer (calculated)	71.6 kDa	43.5 kDa	28.5 kDa
Monomer (SDS–PAGE)	71 ± 3 kDa	43 ± 3 kDa	28 ± 4 kDa
Native (UC)[b]	286 ± 10 kDa	45 ± 4 kDa	104 ± 3 kDa
Native (CL)[c]	300 ± 20 kDa		
Absorbance[d]	0.56 ± 0.03	0.51 ± 0.03	0.61 ± 0.03
Catalysis			
pH optimum	5.2–6.0 (PGK)	5.2–6.0	
	7.5–9.0 (TIM)		8.0
$K_{m,app}$ (40°)[e]	0.7 mM (PGK)	0.3 mM	
	0.4 mM (TIM)		1.2 mM
Thermal transition	100° (PGK)	100°	
	90° (TIM)		82°

[a] For details, cf. refs.[62–64,67,69].
[b] Ultracentrifugation: no dissociation/association at 0.1–1.0 mg/ml protein concentration.
[c] Cross-linking with glutardialdehyde.
[d] Absorption coefficient $A_{280\,nm}$ (0.1%, 1 cm).
[e] $K_{m,app}$ for PGK and TIM refer to 3-phosphoglycerate and dihydroxyacetone phosphate, respectively.

The extrapolated temperatures of cold and heat denaturation are about −10° and 85°. This indicates that the stability profile of TmPGK is shifted to higher free energy values and, at the same time, broadened over a wider temperature range, compared to that observed for PGKs from mesophiles or moderate thermophiles. In order to achieve cold or heat denaturation, GdmCl concentrations of ∼1.8 or ∼0.9 M are required. Because of a kinetic intermediate on the pathway of cold denaturation, equilibration in the transition range takes exceedingly long.[25]

If we bear in mind that PGK is a two-domain protein and the reaction of the substrate with the nucleotide requires hinge motions over distances of the order of 10 Å, the two-state assumption is by no means trivial. Whereas GdmCl-induced unfolding transitions of yeast or *Thermus thermophilus* PGK are highly cooperative, for the enzyme from horse muscle the quantification of unfolding/refolding transitions require at least a three-state mechanism.[28] In the case of *Thermotoga* PGK, calorimetric studies are hampered by the combination of extreme thermal stability and limited reversibility of the unfolding transition (stability up to 100°, $t_{1/2,deactivation}$ at 100°C is 2 hr).[62] At neutral pH in the absence of denaturants, no thermal denaturation can be induced at 0–100°. Therefore, in order to destabilize the protein, moderate GdmCl concentrations or low pH had to be applied; at pH

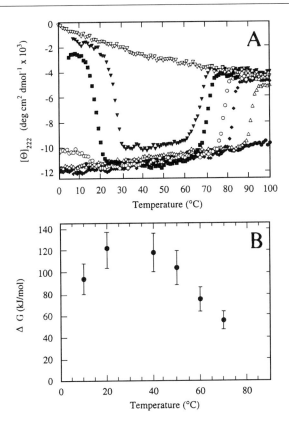

FIG. 3. Thermal unfolding of *Tm*PGK (0.2 mg/ml) in 10 m*M* sodium phosphate, pH 7.0 at varying GdmCl concentration, monitored by dichroic absorption at 222 nm. (A) (●) 0 *M*, (△) 1.1 *M*, (◆) 1.6 *M*, (○) 1.85 *M*, (■) 2.3 *M*, (▼) 2.6 *M*, (▽) 3.2 *M* GdmCl. (B) Temperature dependence of the free energy of unfolding of *Tm*PGK extrapolated to 0 *M* GdmCl [$\Delta G(H_2O)$] at pH 7.0. Error bars refer to repeated measurements (cf. Ref. 25).

3–4, full reversibility was accomplished up to 100°. Making use of these conditions, a detailed spectral and calorimetric analysis of *Tm*PGK and its isolated N- and C-terminal domains showed that the changes in heat capacity in the case of the intact two-domain protein reflect, first, the disruption of the domain contacts, and second, their sequential unfolding, *in toto* obeying at least a three-state mechanism. Unexpectedly, the domains also unfold via partially structured intermediates. To what extent they differ in intrinsic stability cannot be derived from the observed differences in their thermodynamic properties, because it still needs to be show that their structure as separate entities and within the two-domain protein is identical.[26]

The 2.0 Å X-ray structure of the full-length TmPGK[70,71] exhibits high topological similarity to PGKs from other sources, except for local differences attributable to the anomalously high thermal stability of the hyperthermophilic protein: (i) The interdomain angle in TmPGK is significantly smaller than in any other PGK structure solved so far; (ii) the flexibility of the polypeptide chain in TmPGK is drastically reduced; (iii) the helical content is increased, the length of the adjacent loops is reduced, and the caps of helices are stabilized; (iv) there are additional hydrogen bonds and a substantial increase in the number of ion pairs; (v) the value for the ratio of surface area to volume is low; and (vi) in the case of the fusion protein, as in other thermophilic assemblies, the formation of the anomalous tetramer contributes to thermal stability.[47] Surface area calculations of the TmPGK crystal structure show a smooth noncharged surface area that is presumably involved in the packing of the fusion protein, in accordance with this assumption.[71]

Glyceraldehyde-3-phosphate Dehydrogenase

Structure and Stability of the Wild-Type Enzyme. Among the cytosolic enzymes that have been screened in *Thermotoga maritima,* glyceraldehyde-3-phosphate dehydrogenase (GAPDH) (EC 1.2.1.12) shows the highest expression level. The enzyme catalyzes the oxidative phosphorylation of GAP to 1,3-BPG and has been the paradigm for the structure–function relationship of thermophilic proteins, as it was among the first to be investigated at the sequence and crystallographic level.[72,73] Eukaryotic and prokaryotic homologs (including examples from thermophiles) are topologically identical, sharing a strictly conserved catalytic mechanism. In the case of TmGAPDH, marked differences between the thermophilic and mesophilic enzyme are summarized in Table III.[74] The data reflect the high rigidity and packing density and the tight subunit interactions in the case of the hyperthermophilic enzyme compared to its mesophilic counterpart. The amino acid sequences of the Tm enzyme and its homologs from *Thermus aquaticus* and *B. stearothermophilus* show identities of >60%, with only 8% nonconservative exchanges.[75] In spite of that, still about 100 out of 330 residues are left as candidates to explain the stabilization of the enzyme; which of these are responsible for the change in ΔG_{stab} is obviously not a trivial question. The crystal structure of the holoenzyme (solved to a resolution of 2.5 Å, with an overall mean positional error of 0.26 Å) does not

---

[70] G. Auerbach, U. Jacob, M. Grättinger, H. Schurig, and R. Jaenicke, *Biol. Chem.* **378,** 327 (1997).
[71] G. Auerbach, R. Huber, M. Grättinger, K. Zaiss, H. Schurig, R. Jaenicke, and U. Jacob, *Structure* **5,** 1475 (1997).
[72] J. I. Harris and J. E. Walker, in "Pyridine Nucleotide-Dependent Dehydrogenases" (H. Sund. Ed.), p. 43 de Gruyter, Berlin, 1977.
[73] T. Skarzynski, P. C. Moody, and A. J. Wonacott, *J. Mol. Biol.* **193,** 171 (1987).
[74] A. Wrba, A. Schweiger, V. Schultes, R. Jaenicke, and P. Závodszky, *Biochemistry* **29,** 7584 (1990).
[75] V. Schultes, R. Deutzmann, and R. Jaenicke, *Eur. J. Biochem.* **192,** 25 (1990).

TABLE III
PROPERTIES OF HOMOLOGOUS MESOPHILIC AND THERMOPHILIC GAPDHs[a]

Property	GAPDH	
	T. maritima	Yeast
Molecular mass (tetramer/monomer, kDa)	145/37	144/36
Change in $s_{20,W}$ (holo vs apo)[b]	3.5%	5.3%
Thermal transition (T_m of holo-GAPDH, °)[c]	109	40
Denaturation transition in GdmCl ($c_{1/2}$ 20°)	2.1 M	0.5 M
Relative activation at 0.5 M GdmCl (70°)	300%	0%
Relative H–D exchange rate (B_{rel} 25°C)[d]	64	100
Specific activity (U/mg)	200 (85°)	70 (20°)
K_m (μM substrate)	400 (60°)	160 (25°)
K_m (μM NAD$^+$)	79 (60°)	44 (25°)

	T. maritima	B. stear thermophilus	Homarus americanus
Peripheral ion pairs			
Intrasubunit	78	61	55
Intersubunit	16	23	13
Surface area (Å^2)[e]			
Total	14,641	15,543	13,807
Hydrophobic residues	7,628	7,551	5,789
Charged residues	3,693	5,028	3,220

[a] For details, cf. refs. 3, 6, 74, 76.
[b] Change in $s_{20,w}$ in the presence of 1 mM NAD$^+$, relative to $s_{20,w}$ (apo-GAPDH).
[c] Irreversible denaturation monitored by differential scanning calorimetry.
[d] Exchange rate constant according to EX$_{II}$ mechanism.[74]
[e] Accessible surface area buried in tetramer contacts.

give the answer. As expected from the sequence homology, the structures of the *Bacillus* and the *Thermotoga* enzyme are closely similar. When the catalytic and the NAD$^+$-binding domains are fixed separately, the r.m.s.d. of superimposed C^α positions is 0.57 and 0.83 Å, respectively. Insertions and deletions in the primary structure are found in loops on the surface of the molecules. The relatively high overall r.m.s.d. of 2.56 Å for the entire monomer is attributable to a rigid body rotation of the two domains relative to each other. Whether this causes a significant improvement of van der Waals contacts in the dimer-of-dimers assembly of the enzyme needs further investigation.[76] When the enzyme from lobster (physiological temperature <20°) is included in the comparison, it becomes obvious that the increase in thermal stablity correlates with the number of ion pairs in the

[76] I. Korndörfer, B. Steipe, R. Huber, A. Tomschy, and R. Jaenicke, *J. Mol. Biol.* **246,** 511 (1995).

periphery of the tetramer (Fig. 4), in agreement with earlier findings.[34] In contrast to other oligomeric proteins, there is no significant alteration in the intersubunit charge pattern. The number of charges in TmGAPDH is even decreased to a value close to that of the lobster enzyme; evidently, not all charges in the interfaces form stabilizing intersubunit ion pairs. On the other hand, the increase in hydrophobic surface area suggests hydrophobic effects to be involved in the stabilization of the quaternary structure.[76] Further increments of stabilization of the overall structure may be attributed to van der Waals interactions optimized in the compact core of the hyperthermophilic protein. The increased packing density is clearly indicated by three observations: (i) the decrease in the H–D exchange rates, (ii) the relatively small decrease in the hydrodynamic volume on NAD^+-binding, and (iii) the activation of the enzyme at low GdmCl concentrations.[19,74] As has been mentioned, the central issue in the evolutionary adaptation of proteins is the conservation of their functional state. Considering the marginal extra free energy accompanying the transition between mesophiles and thermophiles, evidently the enhanced or decreased stability can hardly be correlated with a clear-cut "mechanism of thermal adaptation." The repertoire of increments to $\Delta\Delta G_{stab}$ is the same as summarized in connection with TmPGK.

Recombinant Enzyme, Mutants, and Constructs. Attempts to confirm hypothetical mechanisms of stabilization made use of mutants and constructs of TmGAPDH expressed in *E. coli*. The cloning, expression, and purification of the authentic recombinant protein have been reported.[77,78] In order to correlate the anomalous stability of TmGAPDH with specific amino acid substitutions, selection for temperature-sensitive mutants was performed using random mutagenesis of the *E. coli* strain W3CG/pAT1.[78] When a number of clones were sequenced point mutations were observed either in the N-terminal tripeptide Ala-Arg-Val [Ala1Val, Ala1Thr, Val3Ile], or in the ribosomal binding site. Since the N-terminal sequence of all known GAPDHs has been shown to be insignificant for the structure and stability of the enzyme,[72,76,79] this approach was not pursued further. Site-directed mutations were hampered by the fact that the choice of suitable candidates is highly ambiguous because of the high number of amino acid exchanges, even in going from Tm- to BsGAPDH. To reduce the number of candidates, experiments focused on ion pairs.[34] The following three types of ion pairs were selected: (i) Nonconserved charges in TmGAPDH that are not present in BsGAPDH, (ii) highly conserved residues involved in the Arg-10 charge cluster in the N-terminal (NAD^+-binding) domain, (iii) ion pairs in the "S-loop" of the substrate-binding domain that is involved in tertiary and quaternary contacts in the wild-type enzyme.

[77] A. Tomschy, R. Glockhuber, and R. Jaenicke, *Eur. J. Biochem.* **214**, 43 (1993).
[78] A. Tomschy, G. Böhm, and R. Jaenicke, *Protein Eng.* **7**, 1471 (1994).
[79] J. I. Harris and M. Waters, in "The Enzymes," 3rd ed. (P. D. Boyer, Ed.), Vol. 13, P. 1. Academic Press, New York, 1975.

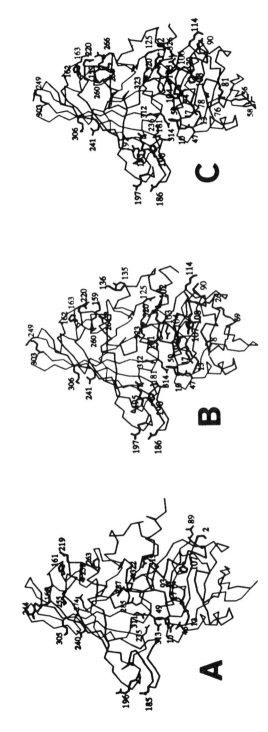

FIG. 4. Structure of homologous GAPDHs at 2.5 Å resolution. C$^\alpha$ backbones of the subunit of GAPDH from *Homarus americanus* (*Ha*) (A), *B. stearothermophilus* (*Bs*), B), and *Tm* (C).[76] Residues involved in ion pairs are numbered. Those specific for *Ha* are: 2, 244, 255, 259, 387 (positive); 89, 168, 235, 274, 285 (negative); those for *Bs*: 69, 101, 102, 107, 114, 136, 190, 195, 303 (positive); 26, 78, 90, 103, 104, 125, 135, 163, 181, 249 (negative); those for *Tm*: 20, 56, 81, 102, 104, 114, 159, 190, 195, 266, 303 (positive); 58, 76, 78, 90, 103, 106, 125, 163, 181, 192, 236, 249, 323, 326 (negative).

The first category was found to have no effect: Neither the thermal stability nor the 3D structure showed significant alterations. In the second case, substitutions referred to two arginine charge clusters around Arg10 and Arg20. Arg10 involves one intersubunit and two intrachain ion pairs; in addition, the backbone imino group froms one out of eight hydrogen bonds responsible for coenzyme binding. To keep the backbone in place and to explore how critical the charge distances are, the arginine residue was replaced by Met and Lys. Both mutations led to the uncoupling of GdmCl-induced deactivation (due to the release of the coenzyme) and denaturation. Deactivation occurred at exceedingly low denaturant concentration, where the change in fluorescence emission remained unaltered, proving that the mutations have no significant effect on the overall protein stability. In substituting Arg20 by Ala and Asn, no differences in thermodynamic stability compared to the wild-type protein were detected; instead, the unfolding kinetics revealed that the resistance against thermal denaturation is strongly diminished. This is reflected by a decrease in free energy of activation by ∼4 kJ/mol at 100° and a shift of the temperature of half-denaturation after 1 hr incubation from 96° to 89° for both mutant enzymes.[80] Due to a large decrease in activation enthalpy, the effects are temperature-dependent and become even more significant at the optimum growth temperature of *Thermotoga maritima*. The third case involved the "S-loop" (residues 178–201), which in mesophilic GAPDHs is highly conserved. In thermophilic GAPDHs, there is a unique ion pair, Asp181–Arg195, which in the case of *Bs*GAPDH has been suggested to play a role in the stabilization of the tertiary and quaternary structure.[73] A single point mutation, Arg195Asp, and a double mutant, Asp181Lys–Arg195Asp, were constructed in order to test this hypothesis. The latter was designed as a suppressor mutant where, for steric reasons, Arg was replaced by Lys. Both mutants showed reduced specific activity, due to the involvement of Arg195 in the binding of the "substrate" phosphate ion. There was also a significant decrease in stability as well as a change in the denaturation mechanism from two-state to three-state, but no changes in the tetramer→monomer transition, even at low temperature.[78]

In summary, there is no simple correlation between the occurrence of peripheral ion pairs and thermal stabilization. Even after careful selection of specific loci for site-directed mutagenesis on the basis of multiple sequence alignments and homology modeling, the results are not clearly predictable. If we recall that the overall stability of proteins is a minute difference between large contributions of attractive and repulsive forces and the corresponding entropy contributions, this result is not surprising.

As has been discussed in connection with modular proteins, single domains may be used to determine increments of stabilization.[28] In this context, the N-terminal domain of *Tm*GAPDH has been shown to bind NAD^+ and NADH

---

[80] G. Pappenberger, H. Schurig, and R. Jaenicke, *J. Mol. Biol.* **274,** 676 (1997).

with high affinity and to exhibit the same GdmCl unfolding transition as the native tetramer, thus indicating that the excised domain preserves its native structure and that the constituent parts of TmGAPDH share the high intrinsic stability of the parent molecule.[81] In the construct, the native domain was mimicked by combining the compact N-terminal 146 residues with the C-terminal α_3 helix (residues 313–333). It would be interesting to know how this structural element is involved in the stabilization of the core of the domain. Obviously, during protein synthesis interactions with the C-terminal domain connot occur until the entire polypeptide chain has been synthesized. Because monomer folding and oligomerization are coupled processes, it is conceivable that locking the domain by means of its C-terminal helix is an important step in the assembly of the enzyme.

Enolase

Enolase (EC 4.2.1.11) catalyzes the dehydration of 2-phospho-D-glycerate (2-PG) to form phosphoenol pyruvate. It is one of the most abundant proteins in *Thermotoga maritima* and can be easily purified to homogeneity.[82] Like enolases from other sources, *Tm* enolase is a metalloenzyme with tightly bound divalent structural and catalytic metal ions such as Mg^{2+} or Zn^{2+}. Regarding their state of association, commonly enolases have been shown to be 90 kDa homodimers. The fact that *Tm* enolase forms homooctamers (with a tetramer-of-dimers topology) is not necessarily correlated with its high intrinsic stability, as nonthermophilic bacterial enolases have also been reported to be homooctameric.[6] Apart from N-terminal sequences (which show a high degree of similiarity to other enolases), no sufficient sequence information is at hand to deduce principles of stabilization from sequence alignments. In the present context, three observations are important: (i) *Tm* enolase shares its high intrinsic stability with the other enzymes, with an additional extrinsic contribution from its divalent metal ligands: Mg^{2+} shifts the thermal transitions from 90 to 94° (Fig. 5A). (ii) Regarding its catalytic activity, at the optimum growth temperature, the specific activity of *Tm* enolase reaches $\sim$2000 U/mg (Fig. 5B); corresponding values for the enzymes from *Bacillus megaterium, E. coli,* and *Thermus aquaticus* are $\sim$70, $\sim$160, and $\sim$700 U/mg, respectively. Obviously, during evolution, the *Tm* enzyme succeeded in optimizing both catalytic efficiency and stability at the upper limit of the physiological temperature regime. (iii) K_m values of *Tm* enolase for 2-PG and Mg^{2+} show a pronounced temperature dependence: raising the temperature from 13° to 75° leads to a decrease of K_m by a factor of 2 and 30, respectively. Beyond 45°, K_m remains practically constant, and at physiological temperature both reach plateau values close to the levels observed for the enzyme at room temperature (Fig. 5C). Again, the previously mentioned "corresponding states" situation seems to hold.[82]

[81] M. Jecht, A. Tomschy, K. Kirschner, and R. Jaenicke, *Protein Sci.* **3,** 411 (1994).
[82] H. Schurig, K. Rutkat, R. Rachel, and R. Jaenicke, *Protein Sci.* **4,** 228 (1995).

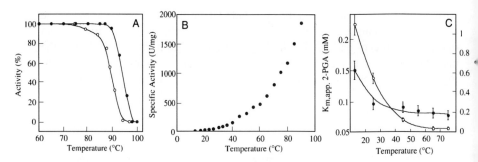

FIG. 5. Thermal stability and catalysis of Tm enolase. (A) Extrinsic stabilization by Mg^{2+} ions (HEPES buffer, pH 7.5, monitored after 2 hr incubation at given temperatures). Equilibrium transitions in the absence (○) and in the presence (●) of 5 mM $MgCl_2$. (B) Temperature dependence of the specific activity for the 2-PG dehydration reaction, measured in Tris buffer, pH 7.5 at optimum Mg^{2+} concentration. (C) Effect of temperature on $K_{m,app}$ for 2-PG (○) and Mg^{2+} (●).[82]

Lactate Dehydrogenase

Lactate dehydrogenase (LDH) (EC 1.1.1.27) catalyzes the NAD-dependent redox interconversion of pyruvate and lactate. In connection with the correlation of sequence homology and stability, LDH has been one of the most intensively studied objects, using mainly the enzymes from *B. stearothermophilus* and *B. megaterium*.[83] In summarizing the results, the weak correlation of the stabilizing or destabilizing effects illustrates the practically unlimited structural and functional adaptability of proteins without unveiling the molecular basis of their stability.

The expression of LDH in *Thermotoga maritima* is extremely low.[84] Cloning by complementation of an *E. coli* pyruvate formate–lyase⁻/LDH⁻ double mutant yielded the active, authentic enzyme.[85] Its physicochemical and catalytic properties have been studied in detail[83–85] and compared with those of moderately thermophilic and mesophilic homologs.[86] Sequence comparisons of *Tm*LDH with the enzymes from *Thermus aquaticus* and *B. stearothermophilus* yield high identities (similarities) between 48 and 40% (62 and 60%), close to the values observed for GAPDH.[6,75,85] Correspondingly, both the three-dimensional fold and the catalytic mechanism of all three homologs are strictly conserved. By making use of 2.0–3.0 Å resolution crystal structures,[86] a variety of structural differences can be pinned down that clearly confirm observations in other enzymes from *Thermotoga*

---

[83] H. Zuber, *Biophys. Chem.* **29,** 171 (1988).
[84] A. Wrba, R. Jaenicke, R. Huber, and K. O. Stetter, *Eur. J. Biochem.* **188,** 195 (1990).
[85] R. Ostendorp, W. Liebl, H. Schurig, and R. Jaenicke, *Eur. J. Biochem.* **216,** 709 (1993).
[86] G. Auerbach, R. Ostendorp, L. Prade, I. Korndörfer, T. Dams, R. Huber, and R. Jaenicke, *Structure* **6,** 769 (1998).

TABLE IV
STRATEGIES FOR INTRINSIC PROTEIN STABILIZATION GAINED FROM CRYSTAL STRUCTURE OF TmLDH[a]

Parameter	TmLDH	BsLDH	LcLDH	SsLDH	SaLDH
Ion pairs within (4Å/6Å)					
per monomer	17/32	16.5/28.9	8/20	10.5/20	13/25
per residue	0.05/0.10	0.05/0.09	0.03/0.06	0.03/0.06	0.04/0.07
Secondary structure (%/ residues involved)					
α helices	46/144	42/134		42/139	40/131
β strands	23/72	22/71		19/58	19/63
Accessible surface area (% ASA/ASA in Å^2)					
Monomer					
Total ASA	100/14,673	100/13,714	100/15,519	100/16,050	100/16,743
Hydrophobic	26/3873	27/3734	29/4604	33/5438	32/5424
Polar	27/4011	31/4299	29/4569	28/4551	29/5022
Charged	46/6852	41/5680	40/6345	37/6060	37/6296
Hydrophobic/charged	0.57	0.66	0.73	0.89	0.86
Tetramer					
Total ASA	100/46,891	100/39,683	100/45,665	100/42,189	100/44,730
Hydrophobic	22/10,566	22/8943	24/11,340	26/11,044	24/11,072
Polar	27/13,086	61/12,445	27/12,669	29/12,385	29/13,332
Charged	50/23,485	46/18,294	47/21,654	44/18,759	45/20,325
Hydrophobic/charged	0.44	0.48	0.51	0.59	0.53
Cavities					
Number of cavities	2	4			
$V_{\text{single cavities}}$ (Å^3)	19.8, 19.9	29.7–41.2			
$V_{\text{cavities total}}$ (Å^3)	39.8	109.6			
% of monomer volume	0.08	0.27			

[a] At 2.1 Å.[86] Bs, B. stearothermophilus; Bl, Bifidobacterium longum; Lc, Lactobacillus casei; Ss, Sus scrofa (pig); Sa, Squalus acanthias (dogfish).

and from other hyperthermophiles: an increased number of intrasubunit ion pairs; increased hydrophobicity in the core of the molecule, especially in the subunit interfaces; a decreased ratio of hydrophobic to charged surface (mainly caused by an increased number of arginine and glutamate side chains on the protein surface); an increased secondary structure content, including an additional unique *thermohelix;* more tightly bound intersubunit contacts mainly based on hydrophobic interactions; and a decrease in both the number and the total volume of internal cavities (Table IV). Contributions to the intrinsic stability from additional proline residues in loop regions (due to the entropy-driven destabilization of the denatured state)[87] cannot be of importance because there are no additional proline residues in TmLDH.[6]

[87] Y. Suzuki, *Proc. Japan Acad., Ser. B. Phys. Biol. Sci.* **65,** 146 (1989).

Glutamate Dehydrogenase

Glutamate dehydrogenase (GluDH) (L-glutamate: NAD$^+$ oxidoreductase, EC 1.4.1.2) catalyzes the oxidative deamination of glutamate to α-ketoglutarate. The *Thermotoga* enzyme (*Tm*GluDH) was cloned by complementation, expressed in *E. coli* as an active enzyme, and characterized regarding its catalytic and phylogenetic properties. It forms a homohexamer of 46 kDa subunits retaining full catalytic activity up to 80°.[88] The three-dimensional structure shows the typical features of hexameric GluDHs with six subunits arranged in 32 symmetry.[89] Each subunit consists of two domains connected by a flexible hinge region. A structural comparison of the enzymes from the hyperthermophiles *Tm* and *Pyrococcus furiosus* (*Pf*) (sequence identity/similarity: 55/74%) with the enzyme from the mesophilic bacterium *Clostridium symbiosum* (Cs) (35/54%) revealed that the overall topology of all three homologs (with 50% α helices and 16% β sheet) is closely similar: The C$^\alpha$ positions of the nucleotide-binding domains (residues 188–339) of *Tm*- and *Pf*GluDH differ with an r.m.s.d. of 3.5 Å, and the structure from *Cs*GluDH with an r.m.s.d. of 4.3 Å. The corresponding values of the N-terminal domain (residues 4–188 and 339–412) are 2.2 Å (*Pf*) and 3.2 Å (*Cs*). The catalytic residues of the active sites of all three enzymes are conserved, differing in the r.m.s.d. values by no more than 0.3–0.7 Å.[89]

Assuming ion pairs and ion pair networks to play a key role in protein stabilization at elevated temperature, *Cs*-, *Tm*-, and *Pf*GluDH were used as models for ion pair statistics and site-directed mutagenesis experiments.[89,90] In agreement with earlier findings, the total number and the sizes of intrasubunit ion pair networks are found to increase with the optimum growth temperature; the same holds for the "melting points" of the enzymes, whereas the volume of intrasubunit cavities is decreased. For example, compared to the *Cs*GluDH, the *Tm* enzyme contains ion pair networks with as many as seven charged residues; in *Pf*GluDH, the largest network consists of 18 residues (Table V). Similarly to *Tm*GAPDH, there is no increase in intersubunit ion pairs.[89]

In discussing these findings in more detail, one problem is the distinction between species-specific variations and real contributions to thermal stability. There are two ways to circumvent this problem, either by extending the comparison to a larger sample size, or by site-directed mutagenesis. In connection with the first approach, structure-based sequence alignments have shown that the increase in thermal stability of various GluDHs is correlated with (i) an increase in rigidity by a reduction of glycine residues, (ii) an increase of hydrophobic contacts, (iii) a

---

[88] R. Kort, W. Liebl, B. Labedan, P. Forterre, R. I. Eggen, and W. M. de Vos, *Extremophiles* **1**, 52 (1997).

[89] S. Knapp, W. M. de Vos, D. Rice, and R. Ladenstein, *J. Mol. Biol.* **267**, 916 (1997).

[90] J. H. Lebbink, S. Knapp, J. van der Oost, D. Rice, R. Ladenstein, and W. M. de Vos, *J. Mol. Biol.* **280**, 287 (1998).

TABLE V
STRUCTURAL AND PHYSICAL PROPERTIES OF GluDHs[a]

Parameter	Cs	Tm	Pf
Structural and functional characteristics			
Molecular mass of hexamer (kDa)	295.0	271.8	282.0
Sequence identity to TmGluDH (%)	35	100	55
Sequence similarity to TmGluDH (%)[b]	54	100	74
$T_{\text{opt growth}}$ of organism (°C)	37	80	100
Melting temperature (°C)	55[c]	<98[d]	113
% residues in helical conformation	50	50	50
% residues in β sheet conformation	15	16	16
Ion pair statistics[e]			
No. of ion pairs per hexamer	188(308)	223(362)	288(462)
No. of ion pairs per residue	0.07(0.11)	0.09(0.15)	0.11(0.18)
% residues forming ion pairs	48(70)	49(73)	58(82)
% ion pairs formed by Arg/Lys/His	47/33/20	58/35/7	61/35/4
% ion pairs formed by Glu/Asp	56/44	58/42	53/47
% of all Arg forming ion pairs	61(78)	68(88)	90(100)
No. residues forming two ion pairs	40(83)	78(98)	118(210)
No. residues forming three ion pairs	10(34)	18(49)	24(74)
No. 2/3/4 member ion pair networks	72/24/12	66/33/18	54/24/12
No. 5/6/7/18 member ion pair networks	0/0/0/0	0/0/6/0	12/6/0/3
No. intersubunit ion pairs	36(54)	34(73)	54(90)

[a] From Ref. 89.
[b] Calculated based on comparison tables using the GCG program package.
[c] Measured by thermal unfolding at pH 7 using CD spectroscopy.
[d] Measured by thermal unfolding at pH 5.5–7.5 using DS calorimetry.
[e] Number of ion pairs were evaluated with a charge–charge distance <4 Å (or 6 Å for the numbers in parentheses).

decrease in the sulfur content, and (iv) a modulation of the subunit flexibility via domain movements. Evidently, these conclusions cannot be generalized, because for other Tm enzymes (such as GAPDH and LDH) only part of the strategies have been confirmed.[75,85] Much the same holds for variations in the secondary and tertiary structure; in contrast to TmLDH (where an additional helix was found), no significant differences in secondary structure were discovered that occurred exclusively in (hyper-) thermophilic GluDHs.[89,91] Regarding cavities and accessible surface areas, again, apparent differences vanish if the areas are normalized to the size of the enzymes; thus, normalized accessible surface areas do not correlate with thermal stability.[89] When the surfaces buried on hexamer formation are analyzed on a residue per residue basis rather than an atom per atom basis,[91] major structural

[91] K. S. P. Yip, T. J. Stilman, K. L. Britton, P. J. Artymiuk, P. J. Baker, S. E. Sedenikova, P. C. Engel, A. Pasquo, R. Chiaraluce, R. Consalvi, R. Scandurra, and D. W. Rice, *Structure* **3**, 1147 (1995).

differences emerge: Whereas in *Pf*GluDH most of the buried surface arises from charged residues, in *Cs*- and *Tm*GluDH the major contributions are due to hydrophobic interactions.[89] Thus, the two thermophilic enzymes differ clearly in the way they stabilize the contacts between the two GluDH trimers. The accumulation of hydrophobic residues at the dimer of trimers interface in *Tm*GluDH indicates a stabilization by van der Waals interactions and by the removal of nonpolar groups from the solvent. In *Pf*GluDH, an 18-residue network in this area suggests a stabilizing effect by electrostatic interactions.[91] The attempt to transplant a unique ion pair network from *Pf*GluDH into the less thermostable *Tm* enzyme was unsuccessful. In spite of the fact that the presence of the additional charge cluster was confirmed by X-ray analysis, no stabilizing effect was detectable; rather, the specific activity of the enzyme and the temperature at which it exhibits optimum activity were affected.[90]

Dihydrofolate Reductase

Dihydrofolate reductase (DHFR) (EC 1.5.1.3) catalyzes the NADPH-dependent reduction of dihydrofolic acid, thereby restoring THFA as a central cofactor in C_1 transfer reactions. Commonly the enzyme has been found to be monomeric, with a molecular mass around 20 kDa. It possesses an α/β fold and does not contain a cofactor. Phylogenetically, the relationship of the enzyme to eukaryotic homologs seems to be closer than to other bacterial DHFRs. Since the expression of the natural enzyme in *Tm* is extremely low, studies focused on the structure, stability, and folding of the recombinant protein. For its preparation and general characterization, cf. ref. 91a.

*Tm*DHFR forms a stable homodimer exhibiting extreme intrinsic stability, with denaturation temperatures between 80° and 100°, and half-concentrations of chaotrophic denaturation around 3 *M* GdmCl and 6 *M* urea, respectively.[92,93] Dissociation is accompanied by denaturation and deactivation so that under no conditions, including the folding/unfolding pathway, are structured monomers detectable. Thus, the thermodynamic characterization is hampered by the superposition of folding and association, obeying either the sequential mechanism

$$2U \rightarrow 2I \rightarrow \rightarrow \rightarrow I_2 \rightarrow N_2 \tag{1}$$

(with U, I, and N_2 as unfolded, intermediate, and native states, respectively), or a parallel "multiple-pathway" scheme. Spectral data gave evidence that the unfolding reaction can be quantified by the two-state model ($N_2 \rightleftharpoons 2U$), with an extremely high free energy of stabilization $\Delta G_{N \rightarrow U} = 142 \pm 10$ kJ/mol at 15°. Maximum stability is observed at ca. 35°. There is no flattening of the ΔG vs T

[91a] T. Dams and R. Jaenicke, *Methods in Enzymology* **331** [27], (2001).
[92] T. Dams, Ph.D. Thesis, University of Regensburg (1998).
[93] T. Dams, G. Böhm, G. Auerbach, G. Bader, H. Schurig, and R. Jaenicke, *Biol. Chem.* **379**, 367 (1998).

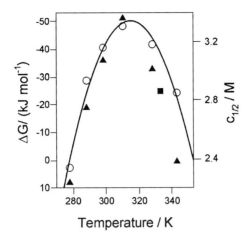

FIG. 6. Thermodynamic stability of TmDHFR in 2.9 M GdmCl.[92] (○) $\Delta G_{2.9\ M\ \text{GdmCl}}$ calculated according to Backmann et al.[94] (with $\Delta H_m = -466$ kJ/mol, $T_m = 333 K$, $\Delta C_p = 22.4$ kJ/(mol K) and $P_t = 5 \times 10^{-7} M$ for the enthalpy of melting, the melting temperature, the change in heat capacity between the native and unfolded states, and the monomer concentration, respectively. (▲) Denaturant concentration of half-denaturation; (■) melting point in 2.9 M GdmCl.

profile (Fig. 6). Rather, the enhanced stability is characterized by shifts toward higher overall stability and a higher temperature of maximum stability. Because of the extremely high association constant of the dimer, unfolding occurs exceedingly slowly without intermediates; accordingly, reassociation is not rate-limiting, in agreement with the absence of the structured monomer on the folding pathway.[95] The presently available model structure of the monomeric enzyme (obtained from sequence alignments and homology modeling[93]) is not sufficient to propose mechanisms of stabilization for TmDHFR and its homologs. High-resolution crystal structures of the apoenzyme as well as its complexes with substrate (DHF), coenzyme (NADP/NADPH), and inhibitor (methotrexate) will provide the information that is required to explain the anomalously high stability of the dimer in comparison with its mesophilic counterparts.[96]

Phosphoribosylanthranilate Isomerase

Phosphoribosylanthranilate isomerase (PRAI) (EC 2.4.2.18) catalyzes the penultimate step in tryptophan biosynthesis. In contrast to the enzyme from mesophiles, TmPRAI is again a homodimer showing thermal stability up to 95°.[97] Under

[94] J. Backmann, G. Schäfer, L. Wyns, and H. Bönisch, *J. Mol. Biol.* **284**, 817 (1998).
[95] T. Dams and R. Jaenicke, *Biochemistry* **38**, 9169 (1999).
[96] T. Dams, G. Auerbach, G. Bader, T. Ploom, R. Huber, and R. Jaenicke, *J. Mol. Biol.* **297**, 659 (2000).
[97] R. Sterner, G. R. Kleemann, H. Szadkowski, A. Lustig, M. Hennig, and K. Kirschner, *Protein Sci.* **5**, 2000 (1996).

physiological conditions, the enzyme exhibits about 35-fold higher specific activity than its mesophilic counterpart from *E. coli* at the corresponding temperature of ca 37°, mainly due to its lower K_m value for the substrate. The high catalytic efficiency allows the enzyme to compete with the rapid spontaneous hydrolysis of the thermolabile substrate PRA at 80°.[98] The crystal structure of *Tm*PRAI at 2.0 Å resolution showed that the two subunits of the TIM-barrel protein associate via the N-terminal faces of their central β barrels. The main features contributing to the high thermal stability of the *Tm* enzyme are additional hydrophobic interactions involving two long loops that protrude into cavities of the neighboring subunit; apart from that, the N-terminal Met and the C-terminal Leu residues of both subunits are immobilized in a hydrophobic cluster, and the number of ion pairs is increased.[98] Protein engineering experiments were devised to prove the given mechanisms of stabilization; eliminating residues contributing to the various hydrophobic and charge clusters led to the fully active monomer, at the same time reducing the stability drastically.[99]

Cold Shock Protein

Thermotoga maritima contains a monomeric 66-residue protein that shows high sequence identity to the cold shock protein CspB from *B. caldolyticus* and high similarity to other known Csp's.[100,101] Computer-based homology modeling allowed the prediction that *Tm*Csp represents a β barrel similar to CspA from *E. coli* and CspB from *B. subtilis;* preliminary NMR data confirm this prediction.[102] Csp's and their homologous Y-box domains in eukaryotes belong to the most conserved proteins presently known. Their strong affinity to single-stranded DNA suggests that they may be involved in cell regulation at the transcription level.[101] In *Tm*, the protein exhibits an extremely low expression level; its function is still unresolved.

The thermal equilibrium transition of *Tm*Csp at 87° exceeds the maximum growth temperature of *Tm* and represents the maximal T_m value reported for Csp's so far. As taken from GdmCl-induced equilibrium transitions, *Tm*Csp is very stable compared to its homologs from thermophiles and mesophiles (Fig. 7a); with $\Delta G_{N \to U} = 26$ kJ/mol, its free energy of stabilization is increased more than twofold. Interestingly, all sequence variations that stabilize *Tm*Csp relative to its mesophilic counterpart are located in (or near) the surface of the β barrel.[103] In attempting to correlate hydrogen bonds or charges to the enhanced stability, no

---

[98] M. Hennig, R. Sterner, K. Kirschner, and H. N. Jansonius, *Biochemistry* **36**, 6009 (1997).
[99] R. Thoma and K. Kirschner, personal communication (1999).
[100] C. Welker, G. Böhm, H. Schurig, and R. Jaenicke, *Protein Sci.* **8**, 394 (1999).
[101] P. Graumann and M. A. Marahiel, *Bioessays* **18**, 309 (1996).
[102] S. Harrieder, Thesis, University of Regensburg (1998).
[103] D. Perl, C. Welker, T. Schindler, K. Schröder, M. A. Marahiel, R. Jaenicke, and F. X. Schmid, *Nature Struct. Biol.* **5**, 229 (1998).

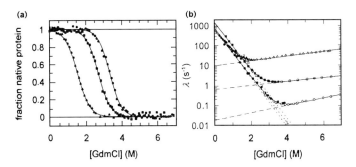

FIG. 7. Conservation of the unfolding/folding mechanism of cold-shock proteins (Csp) from *B. subtilis* (*Bs*), *B. caldolyticus* (*Bc*), and *Thermotoga maritima* (*Tm*). (a) Equilibrium unfolding transitions of Csp from *Bs* (▲), *Bc* (■), and *Tm* (●) induced by GdmCl at 25° and monitored by intrinsic fluorescence. Least-squares fit analyses based on the two-state model yield stabilization energies ΔG_{stab} = 11.3, 20.1, and 26.2 kJ/mol for Csp from *Bs*, *Bc*, and *Tm*, respectively. (b) Kinetics of unfolding (open symbols) and refolding (closed symbols) of *Bs* (△,▲), *Bc* (□,■) and T_m Csp (○,●), respectively. The apparent rate constants, λ, are plotted against the GdmCl concentration. The fits are on the basis of the linear two-state model.[103]

obvious statistical differences in the number of ion pairs or hydrogen bonds are found. In accordance with the calculated high isoelectric point (p*I* 7.7), the overall surface charge is found to be drastically increased compared to CspB. A putative nucleic acid-binding motif is found to cluster in an exposed region on one side of the protein; its size may be correlated with the high affinity of *Tm*Csp to a model oligonucleotide (CTT GAG GTT AAT CCA: K_d < 10 n*M*, at pH 7.0, 20°; unpublished result).[100,103] *In vivo*, binding to the target nucleic acid is assumed to cause significant stabilization of *Tm*Csp. In the case of *Bs*CspB, three peripheral phenylalanine residues (two of them conserved in *Tm*Csp) are known to be essential for nucleic acid binding. Because Phe side chains are usually buried in the interior of globular proteins, it is tempting to ascribe the low conformational stability of Csp's to the exposure of the aromatic residues. However, individually replacing Phe 15, Phe 17, and Phe 27 by Ala lead to destabilization, indicating that the aromatic side chains in the binding site of Csp are essential for both binding nucleic acids and conformational stability. Model calculations showed that all three Phe residues lose accessible surface on folding, thus favoring the native state.[104]

Thermal Stability and Folding

Stability and folding of proteins are closely connected: both involve the complex molecular recognition of structural elements along the polypeptide chain,

---

[104] T. Schindler, D. Perl, P. Graumann, V. Sieber, M. A. Marahiel, and F. X. Schmid, *Proteins: Struct. Funct. Genet.* **30**, 401 (1998).

which is determined by the cooperative action of many weak intermolecular interactions. As has been exemplified for a number of individual proteins, the repertoire of forces is small, but their combination in each single polypeptide, with its specific composition and sequence, allows an astronomical variety of solutions. For a detailed discussion of the present state of the art in the field of protein folding, reference is made to a selected number of recent reviews.[105–110] In the present context, mainly two aspects need consideration: (i) the extent to which mechanisms of protein folding are affected by temperature, and (ii) how homologous proteins from mesophiles and thermophiles differ in their folding characteristics. Regarding the first question, the most complex case, i.e., the folding and association of oligomeric proteins, has been investigated in some detail. It has the advantage that the regaining of biological activity may be used to quantify the associated native state, since usually reactivation and reassociation run parallel.[111] The overall reaction has been shown to obey a sequential uni-bimolecular mechanism with one rate-limiting first-order and one concentration-dependent second-order reaction, and competing side reactions due to the "kinetic partitioning" between folding/association, on the one hand, and misfolding/aggregation, on the other:

$$n\text{U} \xrightarrow{k_1} n{M} \xrightarrow{k_1'} n\text{M} \xrightarrow{k_2} n/2\ \text{M}_2 \xrightarrow{k_2'} n/4\ \text{M}_4 \cdots \quad (2)$$
$$\downarrow \qquad \downarrow \qquad \downarrow$$
$$\text{A} \qquad \text{A} \qquad \text{A}$$

Here, U, M, and M refer to unfolded, collapsed, and folded monomers, A to aggregates, and k_1, k_2 to the first- and second-order rate constants of folding and association, respectively. If the side reactions win, the protein will be degraded or precipitated, or chaperones will take care of it.[106,111] Each step along the sequential uni-bimolecular mechanism may contribute to the overall kinetics, depending on the temperature dependence of each single first- or second-order rate constant. Evidently, productive folding to the native state depends on the formation of the correct complementary recognition sites on the surface of domains and subunits as the prerequisite for proper docking and association; thermally unstable intermediates at any stage along the pathway will favor the side reactions and interfere with proper structure formation.

For GAPDHs from mesophiles and (hyper-)thermophiles, the above kinetic mechanism was confirmed up to a certain concentration limit, at which aggregation

---

[105] K. A. Dill and H. S. Chan, *Nature Struct. Biol.* **4,** 10 (1997).
[106] R. Jaenicke, *Curr. Top. Cell. Regulation* **34,** 209 (1996).
[107] D. V. Laurents and R. L. Baldwin, *Biophys. J.* **75,** 428 (1998).
[108] C. M. Dobson, A. Sali, and M. Karplus, *Angew. Chem., Int. Ed. Engl.* **37,** 868 (1998).
[109] A. R. Fersht, "Structure and Mechanism in Protein Science: A Guide to Enzyme Catalysis and Protein Folding." W. H. Freeman, New York, 1998.
[110] R. L. Baldwin and G. D. Rose, *TIBS* **24,** 26, 77 (1999).
[111] R. Jaenicke, *Progr. Biophys. Mol. Biol.* **49,** 127 (1987).

takes over. These results proved unambiguously that catalytic activity is strictly limited to the tetrameric quaternary structure.[19,112] Bearing in mind that GAPDH is a dimer-of-dimers in which pairs of subunits share active sites, this is not trivial. For the yeast enzyme, incubation at $0°$ in the presence of ATP was shown to yield native-like monomers capable of NAD^+ binding. Their thermal reconstitution again obeys the above equation.[113] TmGAPDH does not undergo cold inactivation. Both the rate and yield of reactivation show a complex temperature dependence: at $30-85°$, reactivation after preceding GdmCl denaturation reaches 90%. At $0°$, no regain of activity is detectable: beyond $85°$, thermal denaturation of folding intermediates competes with refolding, but, even at $100°$, reactivation still amounts to 30–40%. Obviously, the high thermal stability refers to all species included in the above equation as hypothetical intermediates. The fact that k_1 and k_2 are sufficient to quantify the kinetics clearly indicates that unimolecular shuffling at the tetramer level does not contribute significantly to the kinetics of reconstitution. However, there is a transition from the uni-bimolecular to a unimolecular mechanism at >3 μg/ml. Beyond this concentration, association becomes diffusion-controlled so that the overall reaction is exclusively determined by the folding of the monomer. Since reactivation kinetics monitor only those enzyme molecules that are accessible to reconstitution, irreversible side reactions do not interfere with the quantitative kinetic analysis.[111]

Considering the temperature dependence of reactivation together with the observation that there is no cold denaturation detectable, the question arises what is the physical nature of the "cold intermediate" ($I_{0°}$) trapped upon reactivation at $0°$. As it is stable, its characterization is feasible. Temperature shift experiments show that $I_{0°}$ reaches the native state within manual mixing time, proving that the zero level of reactivation is attributable to the formation of an inactive intermediate on the reconstitution path rather than irreversible side reactions. Ultracentrifugal analysis at $0°$ yields a dissociation/association equilibrium with average and minimum molecular masses of 110 and 36 kDa (the monomer), respectively. Shifting the temperature back to $25°$ leads to the homogeneous 144 kDa tetramer. When the kinetics of the $I_{0°} \rightarrow N$ reconstitution is monitored at $50°$, fluorescence emission shows a biphasic overshoot reaction reflecting the fast collapse of the hydrophobic core and subsequent reshuffling. The change in CD at 222 nm parallels the fast collapse, leading to native-like secondary structure. The near-UV CD clearly points to alterations of tertiary contacts in the environment of aromatic residues.[114] Thus, $I_{0°}$ represents an "assembled molten globule" with the secondary structure close to the native state, but the aromatic residues still exposed to the aqueous solvent. In contrast to the common molten globule, $I_{0°}$ shows a highly cooperative denaturation

---

[112] R. Jaenicke and R. Seckler, *Adv. Protein Chem.* **50**, 1 (1997).
[113] P. Bartholmes and R. Jaenicke, *Eur. J. Biochem.* **87**, 563 (1978).
[114] V. Schultes and R. Jaenicke, *FEBS Lett.* **290**, 235 (1991).

transition with the characteristics of the native enzyme. The differences in the limiting GdmCl concentrations for the fluorescence and far-UV CD denaturation profiles clearly indicate that full exposure of the fluorophores precedes the helix → coil transition.[115]

The observation that folding at low temperature yields native-like enzyme has been confirmed by the fact that recombinant *Tm* proteins can be expressed in *E. coli* about 60° below the physiological temperature in *Tm*. When minus strains of the host are used (lacking the gene of the respective gene), *E. coli* shows normal growth. The detailed characterization proved the recombinant proteins to be authentic regarding their physical and biochemical properties.[6]

To answer the question whether homologous proteins from mesophiles and thermophiles differ in their folding characteristics, homologous cold shock proteins were used as a paradigm.[103] In a comparison of Csp's from mesophilic *Bacillus subtilis* (*Bs*), thermophilic *Bacillus caldolyticus* (*Bc*) and hyperthermophilic *Thermotoga maritima*, all three all-β proteins were shown to fold extremely fast in a simple N $\rightleftharpoons$ U two-state reaction (Fig. 7b). In spite of the large difference in stability for the three Csp's, the time constants are close (about 1 ms at 25°), indicating that there is no correlation between conformational stability and fast folding. Thus, the absence of intermediates in the folding of *Bs*CspB is not a corollary of its low stability; rather, two-state folding and an unusually native-like activated state of folding seem to be inherent properties of these small all-β proteins. There is no link between stability and folding rate, and numerous sequence positions exist that can be varied to modulate the stability without affecting the rate and mechanism of folding.[103] Whether these conclusions also apply for other, more complex proteins needs further investigation.

Conclusions

Studies on *Thermotoga maritima* have been focusing mainly on evolution, metabolism, and the correlation of protein structure and stability. It has been established that viability of the hyperthermophile in its hostile natural habitats is based on the inherent thermostability of its whole cell inventory. In some cases this is assisted by extrinsic components such as the cell membrane or specific ligands and compatible solutes.[18,53,54,116] As a response to short-term temperature shocks, chaperones have been found to serve as additional safeguards. For a given protein, the enhancement of stability may originate from a variety of mutational changes, which altogether accumulate to no more than marginal $\Delta\Delta G_{stab}$ values. No unequivocal strategy of thermal stabilization has been extracted from a large

---

[115] V. Rehaber and R. Jaenicke, *FEBS Lett.* **317**, 163 (1993).
[116] L. O. Martins, L. S. Carreto, M. S. da Costa, and H. Santos, *J. Bacteriol.* **178**, 5644 (1996).

amount of data. Obviously, each protein shows individual characteristics. Tentative generalizations refer to (i) the principle of corresponding states reflected by the low structural flexibility around 20°, (ii) the types of weak interactions involved mainly in ion pairs or ion-pair clusters, hydrogen-bond networks, and enhanced hydrophobic packing through van der Waals forces, (iii) an increase in the hydrophobic surface area of folded monomers buried on assembly, and (iv) the distinction between functional and structural amino acids optimized for *flexibility* (catalytic function), on the one hand, and *stability*, on the other. A broader database (now accessible from the complete genome sequence,[117] as well as directed evolution experiments[118]) may provide more general conclusions and finally help in elucidating general strategies of protein stabilization.

Acknowledgments

Work in the authors' laboratory was supported by Grants of the Deutsche Forschungsgemeinschaft, the Fonds der Chemischen Industrie, and the European Community.

---

[117] K. E. Nelson, R. A. Clayton, S. R. Gill, M. L. Gwinn, R. J. Dodson, D. A. Haft, E. K. Hickey, J. D. Peterson, *et al., Nature* **399**, 323 (1999).
[118] W. P. Stemmer, *Nature* **370**, 389 (1994).

[34] Structural Basis of Thermostability in Hyperthermophilic Proteins, or "There's More Than One Way to Skin a Cat"

By GREGORY A. PETSKO

The great American songwriter Cole Porter was once asked if he wrote the words first or the music first. He said, "Yes." Consider the question of what factor is responsible for the extreme thermostability of proteins isolated from microorganisms whose optimum growth temperature is above 80°. Is it an increase in the number of hydrophobic interactions? An increase in the number of ionic interactions? Shorter surface loops? *Longer* surface loops? Improved packing of the protein core? Oligomerization? More secondary structure? Disulfide formation? Tying down the chain termini? In this paper I shall endeavor to show that the answer here, too, is "Yes."

That we are still even asking the question says more about human nature than it does about what the data have told us. It has been fairly obvious for several years that many different factors contribute to the extreme thermostability of any given protein, yet studies continue whose stated goal is to discover a unifying set of rules. It seems hard for some to accept that there is no single factor that dominates. Whether that is because scientists always hope to "discover" something universal, or because it is somehow unsatisfying to conclude that a single great property like extreme thermostability arises from a combination of many different small contributions, is best left for psychologists to ponder; the effect is that studies continue and we have a plethora of new data to examine. It is not the first time this sort of thing has happened in science: two recent examples are the "code" for protein recognition of DNA and the origin of diversity in the immune system. In both cases the search for a single unifying principle led instead to the discovery that several different contributors play a role. The protein–DNA example is particularly relevant to the thermostability issue, because it is now clear that almost every individual protein–DNA complex is *sui generis*. Recognition of, say, an A-T base pair can be done in many different ways and each protein adopts its own strategy. Similarly, we shall see that there are many different ways to make a protein relatively stable to temperatures near the boiling point of water, and although some of them occur frequently, every hyperthermophilic protein employs different mixes. In stabilizing a protein there appears to be, as the saying goes, more than one way to skin a cat. (The origin of this colorful expression, incidentally, is not certain. I'm aware of two possibilities. One is the old British expression "There are more ways of killing a cat than choking it with cream," which implies that whatever was being discussed is foolish, since cats like cream and probably wouldn't choke to death on it. It is conceivable that this expression could have metamorphosed into one indicating that there are more ways than one of accomplishing something. But my preference is for the second possibility, that the original saying was: "More than one way to skin a *catfish*." The meat inside a catfish is tender but the skin is very tough, so over the years many ways have been developed to remove the skin without destroying the meat inside. Why "fish" got dropped I don't know for sure, but because I have a cat I can imagine. . . . By the way, to a friend who was trying to figure out a way to get rid of his cat, Dorothy Parker suggested, "Have you tried curiosity?")

Pace Gertrude Stein

On her deathbed, Gertrude Stein was asked, "What is the answer?" Exasperating to the last, she replied, "What is the question?" Part of the problem with any discussion of thermostability is that there is more than one way to pose the question. For some investigators thermostability means how long a protein survives at some elevated temperature, usually close to 100°, before it is inactivated. For others, it means whether thermal denaturation (usually measured by the loss of catalytic activity or the disappearance of structure as determined spectroscopically) occurs at relatively high temperature. These do not necessarily reflect the same underlying molecular events.

Thermal *denaturation* usually occurs when all or a large fraction of the tertiary structural interactions are disrupted, so an increase in denaturation temperature reflects increased stabilization of the structure. We do not yet understand the onset of thermal denaturation well enough to know if merely "tying down" one or two "hot spots" in a protein is always sufficient to stabilize the overall structure, in which case the increased structural stability may be local, or whether many sites distributed throughout the protein must always be stabilized in order for the protein to survive very high temperatures. Like everything else connected with thermostability, it probably varies from protein to protein, but we can't even be sure of that at present. For the purpose of this discussion, we shall assume that stabilizing interactions can, and do, occur pretty much anywhere in the structure.

Thermal *inactivation* can reflect something very different. For one thing, the loss of activity when a protein is held at elevated temperatures for extended periods of time is usually irreversible; thermal denaturation usually is a reversible process. The reason for this difference is that, as first detailed by Klibanov and associates,[1] there are a number of covalent chemical changes in protein structure that occur on prolonged exposure to high temperatures, in addition to any reversible or irreversible unfolding that takes place. Often these covalent changes prevent the reforming of the active, native structure when the temperature is reduced. Which specific changes are found depends, once again, on the protein being studied, but typical examples include polypeptide chain cleavage,[2] side chain isomerization,[3] and deamidation of asparagine residues.[4] It's easy to see how these chemical processes can often be prevented by simple amino acid replacement: if, for example, the half-life of protein X at 90° is determined by the deamidation of two critical asparagines at a subunit interface, then simply replacing those residues by, say, valine or leucine may greatly increase the stability of X to thermal inactivation. Yet the same sort of substitution may have no effect on the stability of protein Y, whose half-life at elevated temperature is limited by the rate of chain cleavage. Neither will the replacement of asparagine residues with other side chains be likely to have any effect on the denaturation temperature of this protein, for that temperature, at which simple thermal-driven unfolding occurs, should be unrelated to the rate of covalent changes in the molecule. A striking example of this uncoupling is provided by the work of Lebbink *et al.*,[5] who, by combining mutations, engineered a version of glutamate dehydrogenase from *Thermotoga maritima* that had

---

[1] T. J. Ahern and A. M. Klibanov, *Science* **228**, 1280 (1985).
[2] T. J. Ahern and A. M. Klibanov, *Methods Biochem. Anal.* **33**, 91 (1988).
[3] S. J. Tomazic and A. M. Klibanov, *J. Biol. Chem.* **263**, 3086 (1988).
[4] T. J. Ahern, J. I. Casal, G. A. Petsko, and A. M. Klibanov, *Proc. Natl. Acad. Sci. U.S.A* **84**, 675 (1987).
[5] J. H. Lebbink, S. Knapp, J. van der Oost, D. Rice, R. Ladenstein, and W. M. de Vos, *J. Mol. Biol.* **289**, 357 (1999).

30 minutes longer half-life for inactivation at 85° but only a 0.5° higher apparent melting temperature.

Hence, in the search for general or at least frequently employed factors that contribute to protein thermostability, we must first agree on the type of stability we are talking about. For the purpose of this discussion, we shall confine ourselves to thermal unfolding as measured either by loss of structure or loss of activity as the temperature is raised. Since a catalog of interactions in proteins can have all the excitement of the Catalog of Ships in the second half of Book Two of *The Iliad* (that's the part that literature courses all *skip,* remember), I have taken the liberty of trying to enliven the discussion by constructing an imaginary dialog between myself and a skeptic desperately trying to find a single, predominant stabilizing effect.

Is It Increased Hydrogen Bonding? I'll Bet It's Increased Hydrogen Bonding.

Sometimes it is. Comparison of the crystal structure of methionine aminopeptidase from the hyperthermophile *Pyrococcus furiosus* with that of the same enzyme from the mesophile *Escherichia coli* led Tahirov *et al.* to conclude that a major stabilizing factor was an increase in the number of hydrogen bonds between positively charged side chains and neutral oxygens.[6] A similar conclusion was reached by Pfeil *et al.* in their study of ferredoxin from *T. maritima*,[7] and by Macedo-Riberio *et al.*[8,9] Tanner, Hecht, and Krause observe the same thing when they compare their structure of *Thermus aquaticus* glyceraldehyde-3-phosphate dehydrogenase with those from mesophiles.[10] They speculate that the reason such hydrogen bonds, where only one of the participants is charged, are so effective in stabilizing proteins is that they provide electrostatic stabilization without the heavy penalty of increased salt bridges (but see below).

But sometimes it isn't. In contrast to these data, a number of other thermophile/mesophile structure comparisons have found no significant increase in the number of hydrogen bonds in the thermophilic protein. One example is the threefold comparison of *Escherichia coli, Salmonella typhimurium,* and *Thermus thermophilus* 3-isopropylmalate dehydrogenase structures[11] where the number of non-salt-bridge hydrogen bonds didn't change much. The authors of that study make

---

[6] T. H. Tahirov, H. Oki, T. Tsukihara, K. Ogasahara, K. Yitani, K. Ogata, Y. Izu, S. Tsunasawa, and I. Kato, *J. Mol. Biol.* **284,** 101 (1998).

[7] W. Pfeil, U. Gesierich, G. R. Kleemann, and R. J. Sterner, *J. Mol. Biol.* **272,** 591 (1997).

[8] S. Macedo-Ribeiro, B. Darimont, and R. Sterner, *Biol. Chem.* **378,** 331 (1997).

[9] S. Macedo-Ribeiro, B. Darimont, R. Sterner, and R. Huber, *Structure* **4,** 1291 (1996).

[10] J. J. Tanner, R. M. Hecht, and K. L. Krause, *Biochemistry* **35,** 2597 (1996).

[11] G. Wallon, G. Kryger, S. T. Lovett, T. Oshima, D. Ringe, and G. A. Petsko, *J. Mol. Biol.* **266,** 1016 (1997).

the very important point that it is not appropriate to compare numbers of hydrogen bonds between structures that were determined at very different resolutions, because the errors in the precision with which interatomic distances can be determined vary greatly with resolution.

Is It Increased Secondary Structure Formation and Stability?

Sometimes yes. Warren and Petsko examined the amino acid composition of alpha helices in thermophilic proteins and found an increase in the numbers of those amino acids whose presence would be expected to increase helical stability.[12] Of particular interest was an increase in the number of glycine residues. In addition to having a positive delta-s value (Zimm–Bragg helix propagation value) with temperature, glycine is also the most favorable amino acid to form a cap at either the N or C terminus of a helix.[13] Improved capping has been shown by Fersht and co-workers to contribute greatly to the stability of an α helix.[13] And indeed, some thermophilic proteins show more extensive secondary structure, and better capped helices, than their mesophilic counterparts. Examples include the archaeal $O(6)$-methylguanine-DNA methyltransferase[14] and *P. furiosus* methionine aminopeptidase,[6] in both of which the helices are stabilized by interhelical side-chain interactions of the kind predicted earlier[12]; glyceraldehyde-3-phosphate dehydrogenase from *Sulfolobus solfataricus*,[15] phosphoribosylanthranilate isomerase from *T. maritima*,[16] and lactate dehydrogenase from *T. maritima*,[17] in which the number of alpha helices actually increases over that found in their mesophilic counterparts; and improved stabilization by capping such as found in *T. maritima* ferredoxin[9] and indole-3-glycerol phosphate synthase from *S. solfataricus*.[18] Interestingly, much less attention has been paid to β sheets; it is at this time unclear if any changes in sheet number or increased hydrogen bonding in sheets are important for increased thermostability, although one study, of rubredoxin from *P. furiosus*, suggests they may be.[19]

Sometimes no. A number of other studies of specific thermophile/mesophile pairs have found no obvious increase in helix or sheet number, no additional

---

[12] G. L. Warren and G. A. Petsko, *Protein Eng.* **8**, 905 (1995).

[13] Y. Harpaz, N. Elmasry, A. R. Fersht, and K. Henrick, *Proc. Natl. Acad. Sci. U.S.A.* **91**, 311 (1994).

[14] H. Hashimoto, T. Inoue, M. Nishioka, S. Fujiwara, M. Takagi, T. Imanaka, and Y. Kai, *J. Mol. Biol.* **292**, 707 (1999).

[15] M. N. Isupov, T. M. Fleming, A. R. Dalby, G. S. Crowhurst, P. C. Bourne, and J. A. Littlechild, *J. Mol. Biol.* **291**, 651 (1999).

[16] M. Hennig, R. Sterner, K. Kirschner, and J. N. Jansonius, *Biochemistry* **36**, 6009 (1997).

[17] G. Auerbach, R. Ostendorp, L. Prade, I. Korndorfer, T. Dams, R. Huber, and R. Jaenicke, *Structure* **6**, 769 (1998).

[18] M. Hennig, B. Darimont, R. Sterner, K. Kirschner, and J. N. Jansonius, *Structure* **3**, 1295 (1995).

[19] M. W. Day, B. T. Hsu, L. Joshua-Tor, J. B. Park, Z. H. Zhou, M. W. Adams, and D. C. Rees, *Protein Sci.* **1**, 1494 (1992).

stabilizing hydrogen bonds or side-chain interactions in helices or sheets, or better helix capping.[11,20]

It's Better Packing, Isn't It?

Maybe. A decrease in the number of internal cavities has been observed in some cases, such as lactate dehydrogenase[16] and glutamate dehydrogenase[21] from *T. maritima* and *Thermococcus litoralis*.[22]

Maybe not. No such decrease has been found in many other comparisons, including the isopropylmalate dehydrogenase case[11] and the studies of *S. solfataricus* glyceraldehyde-3-phosphate dehydrogenase[15] and the superstable superoxide dismutase (melting temperature 125°!) from *Sulfolobus acidocaldarius*.[23] In fact, a test of this idea was actually carried out on *T. thermophilus* isopropylmalate dehydrogenase: a mutation was made that created a new cavity of 32Å^3 in volume in the interior of the protein, but no decrease in thermostability was observed.[11]

OK, It's Not Exactly Better Packing. It's Really Decreased Surface to Volume Ratio.

It may well be, in some cases. Surface loops are often drastically shortened in hyperthermophilic enzymes; examples include *T. thermophilus* isopropylmalate dehydrogenase,[11] *T. maritima* ferredoxin,[8] and *T. aquaticus* glyceraldehyde-3-phosphate dehydrogenase.[10]

But in other cases, it may well not be. The most dramatic exception seems to be two subtilisin-like proteases from *P. furiosus* and *T. stetteri*.[24] These highly thermostable proteases actually have several extra surface loops compared with their mesophilic counterparts!

Still, there seems to be something to this surface-to-volume ratio idea, because a number of normally monomeric enzymes oligomerize when they are found in hyperthermophilic organisms. Adenylate kinase from *S. acidocaldarius* is a trimer, for example, whereas nearly all other adenylate kinases are monomeric.[25] Perhaps the best evidence that this effect may be important comes from an engineered protein, the repressor of primer (ROP). ROP is normally an all helical homodimeric protein that denatures at 71°. Removal of five amino acids from a surface

---

[20] G. H. Silva, J. Z. Dalgaard, M. Belfort, and P. Van Roey, *J. Mol. Biol.* **286**, 1123 (1999).
[21] S. Knapp, W. M. de Vos, D. Rice, and R. Ladenstein, *J. Mol. Biol.* **267**, 916 (1997).
[22] K. L. Britton, K. S. Yip, S. E. Sedelnikova, T. J. Stillman, M. W. Adams, K. Ma, D. L. Maeder, F. T. Robb, N. Tolliday, C. Vetriani, D. W. Rice, and P. J. Baker, *J. Mol. Biol.* **293**, 1121 (1999).
[23] S. Knapp, S. Kardinahl, N. Hellgren, G. Tibbelin, G. Schafer, and R. Ladenstein, *J. Mol. Biol.* **285**, 689 (1999).
[24] W. G. Voorhorst, A. Warner, W. M. de Vos, and R. J. Siezen, *Protein Eng.* **10**, 905 (1997).
[25] C. Vonrhein, H. Bonisch, G. Schafer, and G. E. Schulz, *J. Mol. Biol.* **282**, 167 (1998).

loop converts the protein to a homotetramer and increases T_m to 101°.[26] Finally, in contrast to what is usually observed for mesophilic proteins, a number of hyperthermophilic proteins have their chain termini tucked back into the body of the protein, which not only would decrease the surface to volume ratio but also would prevent these ends of the chain from serving, as loops might also serve, as "fraying points" where the structure might begin to unravel at high temperatures.[27] Examples include phosphoribosylanthranilate isomerase[16] and ferredoxin[8] from *T. maritima*, and rubredoxin[19] from *P. furiosus*.

Wait, I've Got It; It's More Hydrophobic Residues, Right? Because the Hydrophobic Effect Increases with Increasing Temperature, so....

It makes sense. And, yes, a number of hyperthermostable proteins do show a significant increase in the number of hydrophobic residues, especially in the core of the structure or at subunit interfaces: *T. maritima* lactate dehydrogenase,[16] the hypertherophilic subtilisin-like proteases,[24] *Aquifex pyrophilus* superoxide dismutase,[28] and *S. acidocaldarius* superoxide dismutase[23] are just a few of many examples. However, some proteins show no such increase, and a few actually have more polar water molecules in the core instead (e.g., β-glycosidase from *Thermosphaera aggregans*, ref. 29). And then there's this funny business about aromatic residues. If increased content of hydrophobic residues was a major factor in thermostability, one might expect to see a significant increase in the number of aromatic residues in the core, because they bury more hydrophobic surface area than aliphatic residues do, and also have the opportunity for additional stabilization through aromatic–aromatic interactions. Well, it's like everything else: some hyperthermophilic proteins do have more such interactions,[24] but others do not.[11]

Don't Tell Me It's Increased Rigidity.

OK, I won't tell you. Besides, it's awfully hard to figure out if increased rigidity is a cause of hyperthermostability or an effect. It is certainly true that *most* hyperthermostable proteins are more rigid, at ordinary temperatures, than their mesophilic counterparts,[30] but not *all* of them are, and it's unclear whether this

---

[26] M. W. Lassalle, H. J. Hinz, H. Wenzel, M. Vlassi, M. Kokkinidis, and G. Cesareni, *J. Mol. Biol.* **279**, 987 (1998).
[27] T. Lazaridis, I. Lee, and M. Karplus, *Protein Sci.* **6**, 2589 (1997).
[28] J. H. Lim, Y. G. Tu, Y. S. Han, S. Cho, B. Y. Ahn, S. H. Kim, and Y. Cho, *J. Mol. Biol.* **270**, 259 (1997).
[29] Y. I. Chi, L. A. Martinez-Cruz, J. Jancarik, R. V. Swanson, D. E. Robertson, and S. H. Kim, *FEBS Lett.* **445**, 375 (1999).
[30] P. Zavodszky, J. Kardos, R. Svingor, and G. A. Petsko, *Proc. Natl. Acad. Sci. U.S.A.* **95**, 7406 (1998).

contributes to a high melting temperature or just arises as a consequence of the increased number of different kinds of stabilizing interactions that these proteins all seem to have. The one fact that suggests there may be a causal relationship is the increased number of proline residues in many hyperthermophilic protein sequences.[6,11] Proline reduces the flexibility of the polypeptide chain.

Tang and Dill[31] have attempted a theoretical investigation of this question. Using an HP lattice model, they find a low-temperature point below which large fluctuations are frozen out. They also conclude that proteins having greater stability tend to have fewer large fluctuations, and hence lower overall flexibilities.

Isn't There ANYTHING That's at Least *Common* to All These Proteins?

As a matter of fact, there is. It can't be the *only* stabilizing factor, and may not even be the *most important* stabilizing factor in many cases because of all the other ones already mentioned, but it appears to be observed nearly all the time. It's an increase in the number of ion pairs (salt bridges), especially in networks. Essentially every one of the thermophilic protein structures cited thus far contains an increase in the number of ionic interactions relative to its closest mesophilic homolog. These include intrahelix ion pairs,[10,14] interhelix ion pairs,[14,16] surface ion pairs,[7,29] intersubunit ion pairs,[23,32] and intrasubunit ion pairs,[17,28,33]; any kind of ion pair seems to help, especially when these pairs form networks (e.g., refs. 5, 15, 19). Ogasahara *et al.*,[34] in a very important study with the kind of detail that is sorely needed, used calorimetry to examine the effect of salt on thermal stability of *P. furiosus* methionine aminopeptidase. From this they could establish directly the contribution that the large number of salt bridges in this protein make to its thermostability. It all makes perfect sense.

And yet, even here, there are subtleties. Although it would seem from the above that ion pairs can be put almost anywhere to increase stability, that may not be the case. Lebbink *et al.* tried to enlarge the existing networks in *T. maritima* glutamate dehydrogenase and found that resistance to both thermal denaturation and irreversible thermal inactivation *decreased*. However, combination of destabilizing single mutations often restored stability.[5] From this they conclude that there is a need for a balance of charges at subunit interfaces and high cooperativity between different members of the network. Russell *et al.* found that thermal denaturation in citrate synthase appears to be resisted by intersubunit ion pair networks,

---

[31] K. E. Tang and K. A. Dill, *J. Biomol. Struct. Dyn.* **16**, 397 (1998).

[32] R. J. Russell, U. Gerike, M. J. Danson, D. W. Hough, and G. L. Taylor, *Structure* **6**, 351 (1998).

[33] G. Auerbach, R. Huber, M. Grattinger, K. Zaiss, H. Schurig, R. Jaenicke, and U. Jacob, *Structure* **5**, 1475 (1997).

[34] K. Ogasahara, E. A. Lapshina, M. Sakai, Y. Izu, S. Tsunasawa, I. Kato, and K. Yutani, *Biochemistry* **37**, 5939 (1998).

whereas cold denaturation appears to be resisted by an increase in intramolecular ion pairs.[32] Tanner, Hecht, and Krause concluded that charged residues play a dual role in stabilization by participating not only in salt bridges, but also in charged-neutral hydrogen bonds.[10] And Knapp *et al.* point out that in *T. maritima* glutamate dehydrogenase the number of intersubunit ion pairs is actually *reduced* vis-à-vis the mesophilic enzyme, whereas in *P. furiosus* glutamate dehydrogenase there is a big increase in the size of the intersubunit ion pair network. In both cases the number of intrasubunit ion pairs is increased.[21] We clearly don't understand yet just where some interaction has to be placed in a structure to guarantee that it will increase stability. And as for how all the different kinds of interactions balance out, well. . . .

So is EVERYTHING Important EVERYWHERE?

Potentially, yes. What all of these studies seem to show is that protein stability at hyperthermophilic temperatures arises from a combination of many factors, each of which contributes to a different extent in different proteins, and not all of which need be present. As to where the increased interactions need to be placed in the structure to contribute to stability, the answer seems to be all over. The structural distribution of stability in a thermophilic enzyme has been examined in detail in a paper by Hollien and Marqusee that has just appeared.[35] Using NMR, they determined the native state hydrogen exchange rates for each residue in RNase H from *T. thermophilus*. They found that the general distribution of stability in the thermophilic protein is similar to that of its mesophilic homolog from *E. coli*, with a proportional increase in stability for almost all residues. Consequently, the residue-specific stabilities of the two proteins are remarkably similar under conditions where their global stabilities are the same. From these data they conclude that this enzyme is stabilized in a delocalized fashion, with the stabilizing interactions—and presumably (although their H/D data measure this only indirectly) any decrease in flexibility—being distributed throughout the structure.

Extreme thermostability thus seems to be achieved in nature by distributing many different kinds of additional intramolecular interactions throughout the protein rather than by concentrating just one kind in one or a few places (but there is no indication yet that it would be impossible to achieve it in that simpler way artificially). Although increased ionic interactions and greater compactness seem to be the most frequently observed strategies based on comparison of thermophilic and mesophilic protein structures, in only a few instances have the contributions of these factors been tested experimentally, for example by site-directed mutagenesis of hyperthermophilic proteins. There is still considerable room for further work along those lines.

[35] J. Hollien and S. Marqusee, *Proc. Natl. Acad. Sci. U.S.A.* **96**, 13674 (1999).

As is so often the case in biology, searches for an overarching principle frequently are doomed to failure from the beginning. Evolution can, and does, make use of anything that works. As the somewhat whimsical title of this piece is meant to remind us: there's more than one way to skin a cat (and I have a cat, so believe me, I've thought about this).

This is All Just Your Opinion, Though, Isn't It?

Well, it's my *conclusion* based on the available data. And it's not uniquely my conclusion, either. A number of other scientists have already made the point that the only general conclusion that can be drawn is that there are many different ways to stabilize a protein. Two papers of note that emphasize this are by Daniel, Dines, and Petach,[36] who state that "there is currently no strong evidence that any particular interaction ... plays a more important role in proteins that are stable at 100°C than in those stable at 50°C," and by Jaenicke and Bohm,[37] who conclude that "proteins are individuals that accumulate increments of stabilization; in thermophiles these come from charge clusters, networks of hydrogen bonds, optimization of packing and hydrophobic interactions, each in its own way."

I'm particularly fond of that last quotation. Anyone who has a cat knows all about individuals that go their own way.

---

[36] R. M. Daniel, M. Dines, and H. H. Petach, *Biochem. J.* **317,** 1 (1996).
[37] R. Jaenicke and G. Bohm, *Curr. Opin. Struct. Biol.* **8,** 738 (1998).

Author Index

A

Abelson, J., 424(58), 436
Åberg, A., 216, 223(14), 225(14)
Abola, E., 423
Abramson, R. D., 92
Adam, E., 271, 272(11), 273(11), 276(11), 280
Adams, J. M., 120
Adams, M. W. W., 3, 15, 16, 16(42), 17, 17(40;
  41), 18(7; 8), 23, 27, 27(8), 30, 31, 32, 32(6;
  9; 12; 13; 16–18; 20), 33, 34(6; 30–33; 37),
  36(5; 30; 31; 33), 37(6; 31; 33), 38, 38(19;
  34), 39(30; 31), 40, 40(30–33; 42), 41, 43,
  43(4), 44(4), 45, 46, 48, 48(5; 20–22), 49,
  49(11; 15–19), 50, 50(5), 51(5), 52, 54(5;
  14; 27), 55, 56, 56(1; 7), 57, 57(1), 58,
  59(1), 60(1; 9; 10), 61(1; 4), 62(3; 4), 88,
  93, 114(26), 115(26), 232, 255, 289, 306,
  311, 312(19), 325, 347, 348, 349, 350, 351,
  352, 352(24), 353, 354, 354(5; 10; 21),
  355(5; 21), 356(19), 358, 358(10), 359,
  359(2), 360, 360(42), 361(42), 362(42),
  363, 363(42), 364, 364(42), 365, 365(22),
  366, 366(42), 367, 367(42; 48), 368(9; 10;
  19; 22; 24; 42; 43; 48), 369, 369(10), 370,
  370(10; 20; 22; 24; 42), 371, 371(9; 10; 22;
  24; 43), 372, 372(24), 373, 373(24; 42; 43;
  48), 374, 374(9; 22; 42; 48), 375, 375(9;
  22), 376, 376(22; 42), 377, 377(5; 9; 21;
  22), 378, 378(22), 379(5; 19; 21; 22; 42),
  380, 380(20; 42), 381, 381(19; 24; 59),
  382(19; 42), 383, 383(19; 42), 384,
  384(24), 385(9; 22; 24; 43; 68), 386,
  386(19; 24; 43; 59), 387, 387(20; 24), 388,
  388(10; 19; 24; 68), 389, 392, 393, 423,
  424(1; 2; 37; 39; 67), 431, 434, 435, 435(1),
  437, 439, 444, 473, 474, 475(19), 476(19)
Adinolfi, B. S., 424(66), 437
Adler, E., 447, 448(62), 450(62)
Adman, E. T., 12, 29, 351, 367, 367(6), 372(6),
  379(6), 380(6), 381(6), 382(6), 386(6)
Agback, P., 130
Aguilar, C. F., 434(43), 435

Ahern, T. J., 399, 404(41), 471
Ahgren, A., 216, 223(14), 225(14)
Ahmad, F., 341
Ahn, B. Y., 425(76), 437, 475, 476(28)
Akanuma, S., 284
Akerman, M., 218
Åkesson, B., 84(28), 85
Akhmetzjanov, A. A., 92
Al-Karadaghi, S., 424(66), 437
Alber, M., 27, 368, 374(55), 381(55)
Alcaide, F., 274
Alex, L. A., 425(70), 437
Alilat, M., 159
Aliverti, A., 130, 131(11), 143(11), 144(11)
Alm, A. A., 207
Almagor, A., 326
Almond, E. L., 319
Alonso, J. C., 193
Alseth, I., 99, 102(43)
Altanasov, B., 393
Altschul, S. F., 273
Alvarez, M., 328, 330(1), 448, 450(67)
Amaya, K., 274
Amos, L. A., 424(52), 436
Amster, I. J., 31
Andera, L., 146, 147, 148(11)
Andersson, A., 207
Andersson, I., 429
Andersson, J., 217
Anderton, B. H., 296
Anemüller, S., 8(33), 10, 11(33), 22
Anfinsen, C. B., 63, 144, 393
Ang, D., 296
Angerer, B., 94
Anjum, F., 341
Ankenbauer, W., 94
Anraku, Y., 274
Antranikian, G., 83
Aoki, K., 48
Aono, S., 17, 27, 30, 31, 32(6), 34(6), 37(6), 41,
  46, 48(5), 50(5), 51(5), 54(5), 56, 351,
  354(5), 355(5), 377(5), 379(5)
Aoyama, H., 166

Arakaki, A. K., 40
Araki, H., 259
Archer, M., 19
Archer, S. J., 130
Archontis, G., 392
Argos, P., 328
Armstrong, F. A., 8(34; 37), 9, 10, 11, 11(34)
Armstrong, R. N., 87
Arnheim, N., 115
Arnold, F. H., 284
Arseniev, A. S., 362
Arthur, L., 146(13), 148, 157, 162(13)
Artymiuk, P. J., 325, 424(41), 434, 461
Asada, K., 92, 93, 115(17), 250, 255
Asai, K., 146
Asano, T., 317
Aslund, F., 62, 74
Assman, C., 22
Atomi, H., 261, 262
Atta, M., 217
Atta-asafo-Adjei, E., 21
Aubry, A., 425(82), 437
Auerbach, G., 328, 424(49; 56), 436, 448, 452, 458, 459(86), 462, 463, 463(93), 473, 476, 476(17)
Aujay, M., 48
Ausubel, F. M., 120
Avron, B., 443
Axelsson, K., 70

B

Baas, P. D., 204
Baase, W. A., 142
Baba, S., 48
Babini, E., 379
Backes, H., 334
Backmann, J., 328, 329, 330(1; 6), 333(6), 334(6), 335(11), 336(6), 340, 340(11), 341(6; 11; 30), 448, 450(67), 463
Bacon, D. J., 77
Bader, G., 328, 448, 462, 463, 463(93)
Baek, K., 424(57), 436
Bai, Y., 341, 344, 345, 346(5), 350
Bailey, K. A., 116, 117, 121, 121(4), 123(4)
Bailone, A., 264
Baitin, D. M., 262, 265(14)
Baker, B. M., 330, 411, 414(63), 419(63)

Baker, J. R., 446
Baker, P. J., 325, 424(41), 434, 461, 474
Baker, T. A., 91, 105(7), 108(7), 193
Baldi, M. I., 163
Baldwin, J. E., 429
Baldwin, R. L., 123, 442, 444, 466
Ball, L. J., 130
Ballou, D. P., 21
Balny, C., 316, 327
Balson, D. F., 193
Bambera, R. A., 108
Banci, L., 356
Bannwarth, W., 24
Barbacid, M., 239
Barbé, J., 218, 219(30), 226(30), 227(30)
Barbeyron, T., 193, 195(1)
Bardwell, J. C. A., 63, 75, 79, 87
Barrow, K. D., 307
Bartels, K., 84(26), 85
Bartholmes, P., 467
Bartolucci, S., 62, 63, 66, 74, 76(3; 4), 77, 77(5), 78, 79(4), 80, 82, 87, 130, 132(13), 312, 424(59), 436
Bartunik, H. D., 425(79), 437
Basehore, S. L., 92
Basso, A. L., 306, 314(13)
Batie, C. J., 21
Batista, R., 8(33), 10, 11(33)
Battino, R., 323
Bau, R., 46, 48, 54(14), 424(37), 434
Baucher, M. F., 193, 195(1)
Bauer, H. H., 341
Bauer, K., 120
Bauer, W. R., 151
Bauerle, R., 425(75), 437
Baumann, H., 130, 138(19)
Baumeister, W., 294, 440, 468(18)
Bax, A., 48, 49(16)
Bayer, P., 438
Beard, B., 49
Bearden, A. J., 359
Beardwood, P., 21
Beaucamp, K., 59, 439, 442(6), 447, 447(6), 448, 448(62), 450(62–64; 67), 453(6), 458(6), 459(6), 468(6)
Beaucamp, N., 328, 330(1), 448, 449(69), 450(69)
Beckman, E. J., 316
Becktel, W., 390
Beckwith, J., 63, 75

AUTHOR INDEX

Bedell, J., 130, 393, 400(31), 403(31), 414(31), 419(31), 420(31), 421(31)
Beechem, J. M., 422
Beese, L. S., 92
Beinert, H., 30
Belay, N., 306, 308(18)
Belfort, M., 270, 271(1), 272, 276(15), 278, 278(15), 279(15; 30), 474
Belintsev, B. N., 205
Bell, G. S., 447
Bell, S. D., 159, 161(39), 227, 229
Bell, S. P., 91, 105(7), 108(7)
Belley, R. T., 4
Belly, R., 132
Belova, G. I., 179
Ben-Bassat, A., 120
Benbouzid-Rollet, N., 193, 195(1)
Bendixen, C., 146(13), 148, 157, 162(13)
Benedetti, P., 163
Benner, J., 271, 276(3), 278(3)
Benner, S., 217, 218(18)
Benning, M. M., 40
Benson, F. E., 261
Bentrop, D., 7, 8(31), 27, 29, 351, 354(11), 368, 368(11), 374(55), 381(11; 55), 382, 382(11), 386(11), 388(11)
Berens, C., 334
Bergerat, A., 148, 160, 163, 172, 173, 176, 176(4), 179(4; 7), 206
Berglund, O., 219
Bergman, T., 217
Bergmeyer, H. U., 59
Berk, A. J., 227
Berkessel, A., 312
Bernal, J. D., 443, 444(37)
Bernander, R., 260
Berndt, K. D., 393
Bertini, I., 7, 8(31), 353, 355(17), 356, 357(17; 33; 35), 358(17; 33), 359, 359(17), 361(17; 33), 362(17; 33; 35), 363(35), 364(35), 368(15; 37), 374(17; 33), 377(17; 33; 40), 379, 380(66), 385(15)
Bertini, T. A. J. I., 367
Bertrand, P., 21
Betlach, M. C., 52
Beutler, H.-O., 59
Bevington, P. R., 408
Beynon, R. J., 287
Bidnenko, V., 193
Bill, E., 22

Biltonen, R., 401
Bilwes, A. M., 425(70), 437
Birolo, L., 70
Bishop, D. K., 261
Bjornson, K. P., 155
Björnstedt, A. M., 59
Blacher, R. W., 63, 75, 87(14)
Blair, D., 425(74), 437
Blake, P. R., 46, 48, 48(5), 49(11; 15–17), 50(5), 51(5), 54(5), 56, 388, 423, 424(2)
Blamey, J. M., 15, 17(40; 41), 23, 27, 27(8), 31, 36(5), 38, 41, 351, 359(2), 439
Blanchard, D. K., 39
Blanchard, J. S., 208, 212(18)
Blöchl, E., 205, 302
Blusson, H., 56, 61(5)
Board, P. G., 87
Bocchini, V., 424(66), 437
Bocs, C., 163, 172
Bodgen, J. M., 241
Boenisch, H., 424(60), 436
Bogert, A. M., 148
Bogert, M., 147
Böhm, G., 316, 328, 388, 389, 416(3), 431, 438, 444, 452(47), 454, 456(78), 462, 463(93), 464, 465(100), 478
Bolen, D. W., 334
Bollinger, J. M., 216, 223(15), 225(15)
Bonch-Osmolovskaya, E. A., 262, 265(14)
Bondi, A., 434
Bonetto, V., 145
Bonicel, J. J., 9, 29, 351, 367(7), 380(7), 381(7)
Bönisch, H., 328, 329, 330(6), 333(6), 334(6), 335(11), 336(6), 340(11), 341(6; 11), 463, 474
Bonomi, F., 27
Bonovoisin, J. J., 373
Booker, S., 216, 217, 223(15; 20), 224, 225, 225(15)
Boone, D. R., 306, 308(18)
Boosman, A., 120
Borges, K. M., 147, 148
Borges, N., 306, 307(15), 312
Borsari, M., 379
Boteva, R., 297
Bott, K. F., 165
Boulton, N., 166
Bourenkov, G., 425(79), 437
Bourne, P. C., 425(72), 437, 473, 474(15), 476(15)

Bouthier de la Tour, C., 146, 147, 148(9; 10), 150(10), 155(5), 157(10), 158(6; 10), 159(5), 160, 164, 165, 169(13), 180
Bovier-Lapierre, G. E., 9
Bowman, M. K., 21
Bradford, M. M., 65
Bradley, E. A., 49
Braithwaite, D. K., 91, 249
Braman, J. C., 92, 113, 115, 255
Brönden, C.-I., 75, 83
Brandenburg, K., 332, 335(19), 341(19)
Brandts, J. F., 394, 409, 410(61), 411(61), 439
Brandts, J. M., 394
Branlant, G., 425(82), 437
Brautigam, C. A., 94
Bray, R. C., 19
Bren, K. L., 31, 52, 352, 352(24), 353, 354, 365, 365(22), 366, 368(22; 24), 370(22; 24), 371, 371(22; 24), 372, 372(24), 373(24), 374(22), 375(22), 376, 376(22), 377(22), 378(22), 379(22), 381(24), 384(24), 385(22; 24), 386(24), 387, 387(24), 388(24)
Brendel, V., 261
Brent, R., 120
Brereton, P. S., 30, 31, 32, 33, 34(31–33), 36(31; 33), 37(31; 33), 39(31), 40(31–33), 45, 52, 352, 352(24), 354, 368(24), 370(24), 371, 371(24), 372, 372(24), 373(24), 381(24), 384(24), 385(24), 386(24), 387, 387(24), 388(24)
Breslauer, K. J., 392
Breslow, R., 425(78), 437
Bresnick, E., 241
Breton, J. L., 8(34; 37), 9, 10, 11, 11(34), 367
Brewer, L., 390, 400(11), 403(11)
Brick, P., 259
Briganti, F., 379, 380(66)
Brito, J., 92
Britt, R. D., 21
Britton, K. L., 325, 424(41; 65), 434, 436, 461, 474
Broadhurst, R. W., 130
Brock, K. M., 4, 132
Brock, T. D., 4, 132
Brockman, J. P., 262, 265(13)
Broderick, J., 217, 223(20)
Brodsky, G., 180
Brow, M. A., 92, 101, 109(19), 112(19)
Brown, A. D., 303
Brown, N. C., 102

Brunner, N., 313
Bruschi, M., 7, 9
Brutlag, D., 163
Bryant, F. O., 30, 33, 41, 43, 46, 48(5), 50, 50(5), 51(5), 54(5), 56, 57
Buc, H., 107
Buchanan, B. B., 62(3), 63
Buhler, C., 163, 172, 176, 179(7)
Bujalowski, W., 140
Bujard, H., 24
Bult, C. J., 94, 147, 157, 252, 426
Bundle, D. R., 412, 413(64)
Burden, A. E., 49, 51(32), 54(32)
Burgess, B. K., 40
Burggraf, S., 181, 205, 302, 307
Burley, S. K., 439, 443(10)
Burnett, B., 264
Busse, S. A., 31, 32(16; 17)
Busse, S. C., 359, 360, 367(39), 368(43), 371(43), 373(43), 381, 385(43; 68), 386(43), 388(68)
Bustrov, V. F., 362
Butler, S. L., 326
Butt, J. N., 8(34; 37), 9, 10, 11, 11(34)
Butzow, J. J., 214
Bycroft, M., 443
Byer, R., 280

C

Calderone, T. L., 52
Calendar, R., 178
Calhoun, D. B., 345
Calzolai, L., 31, 32(18; 20), 52, 351, 352, 352(24), 353, 354, 363, 365, 365(22), 366, 367(48), 368(9; 22; 24; 48), 370(20; 22; 24), 371, 371(9; 22; 24), 372, 372(24), 373(24; 48), 374, 374(9; 22; 48), 375, 375(9; 22), 376, 376(22), 377, 377(9; 22), 378, 378(22), 379(22), 380, 380(20), 381(24), 384(24), 385(9; 22; 24), 387, 387(20; 24), 388(24)
Camardella, L., 130, 132(13)
Cambillau, C., 83
Cammack, R., 10, 15, 17(35), 21, 22(61), 30, 351, 354(1), 357(1)
Campbell, J. L., 91, 95(1), 116(1)
Campos, A. P., 8(33), 10, 11(33)
Candussio, A., 446

Cann, I. K. O., 94, 116(37), 249, 254(5), 257(4), 258, 259(5), 260, 260(6; 20)
Cannio, R., 63, 66, 312
Capozzi, F., 359, 377(40), 379
Carballeira, N., 92
Carey, J., 121
Carita, J. C., 9
Carlini, L., 332
Carlow, C. K. S., 271, 276(3), 278(3)
Carmack, C., 207
Caron, P. R., 180
Carot, V., 159
Carredano, E., 21
Carrell, C. J., 21
Carreto, L. S., 306, 310(17), 468
Carrillo, N., 40
Carroll, W., 144
Carter, C. W. J., 427, 429, 429(9)
Casabadan, M. Y., 24
Casal, J. I., 471
Case, D. A., 21, 355, 359(25; 26), 377(25)
Cassuto, E., 162
Cavagnero, S., 49, 54(27)
Cavanagh, J., 353, 362(14), 363(14), 364(14)
Ceccarelli, E. A., 40
Cerchia, L., 130, 132(13)
Certa, U., 24
Cesareni, G., 406, 407(53), 475
Chabrière, E., 16, 16(46), 17
Chakrabarti, P., 427
Chalfoun, D., 223
Chamberlain, A. K., 350
Champoux, J. J., 180
Chan, H. S., 466
Chan, M. K., 424(39), 434, 435
Chan, S. I., 49, 54(27)
Chang, S., 120
Chang, S.-Y., 92, 120
Changeaux, J. P., 348(19), 349
Chao, H.-G., 334
Charbonnier, F., 115
Charon, M.-H., 16, 16(46), 17
Charron, C., 425(82), 437
Chatterjee, A., 279
Chattoraj, D. K., 121
Cheetham, M. E., 296
Chen, B., 27
Chen, C. T., 48(20; 21), 49
Chen, G. C., 136
Chen, J., 48(21), 49

Chen, J.-S., 39, 42
Chen, L., 56, 62(6), 314
Chen, T. L., 355, 359(25), 377(25)
Cheng, H., 40, 353, 356(16), 357(16), 358(16), 361(16), 362(16), 374(16)
Chi, Y. I., 425(77), 437, 475, 476(29)
Chiaraluce, R., 325, 424(41), 434, 461
Chicau, P., 9
Childers, S. E., 446
Cho, C. S., 425(83), 437
Cho, S. J., 425(76), 437, 475, 476(28)
Cho, Y. J., 424(57), 425(76; 83), 436, 437, 475, 476(28)
Choate, W. L., 144
Choi, H.-J., 85, 87
Choi, T., 108
Choli, T., 129(6; 7), 130, 138(6; 7)
Chong, S., 274
Chothia, C., 335
Chow, C., 117, 121(4), 123(4)
Chow, R. L., 294
Chrebet, G., 180
Christensen, J. J., 396
Christova, P., 393
Chung, Y. J., 259
Ciaramella, M., 434(43), 435
Cioni, P., 326
Ciulla, R. A., 304, 306(10), 307, 308(10)
Ciurli, S., 353, 355(17), 357(17), 358(17), 359(17), 361(17), 362(17), 374(17), 377(17)
Clantin, B., 328, 424(5), 426
Clark, A. J., 261, 262, 265(13), 389
Clark, D. S., 316, 318, 319, 319(22), 322, 323, 323(28), 324(22; 26; 28; 29), 325, 326(36), 327, 327(29)
Clark, J. M., 258
Clarke, S., 438, 442(2)
Clayton, R. A., 147, 252
Clever, H. L., 323
Cline, J. F., 21, 91, 113, 115, 255
Clore, G. M., 79, 84(18; 19)
Cohen, A., 425(78), 437
Colas, B., 289
Cole, S. T., 221
Comb, D. G., 271, 276(3), 278(3)
Confalonieri, F., 146, 147, 155(5), 159(5), 160
Connelly, G. P., 345
Connelly, P. R., 392
Connolly, M. L., 434

Conover, R. C., 17, 27, 31, 32, 32(6; 9; 13), 34(6), 37(6), 351, 354, 354(5), 355(5), 377(5), 379(5)
Consalvi, R., 461
Consalvi, V., 325, 424(41), 434
Consonni, R., 130, 131(12), 143(12)
Cornett, D. S., 31
Cosper, N. J., 3(17), 4, 7(17), 8(17), 9, 10(17), 13(17)
Couderc, E., 147, 148
Coulondre, C., 239
Coulson, A. R., 67, 107
Couture, M. M. J., 356, 368(37)
Cowan, J. A., 28
Cowan, N. J., 294
Cowan, S. W., 87
Cox, M. M., 261, 264, 265(1)
Cozzarelli, N. R., 163, 164, 180
Craig, E. A., 294
Cramer, S. P., 48(20; 21), 49
Cramer, W. A., 21
Crane, B. R., 425(71), 437
Creighton, T. E., 63, 75, 79, 81, 87, 87(16), 119
Crick, F. H. C., 151
Crothers, D. M., 127
Crowhurst, G. S., 473, 474(15), 476(15)
Cruz, F. S., 39
Csonka, L. N., 303
Cui, X., 56, 61(4), 62(4)
Culard, F., 160
Cull, M., 93, 116(24), 173
Curmi, P. M. G., 434
Curth, U., 332

D

Da Costa, M. S., 302, 303, 304(8), 306, 307(14–16), 308(16), 310(14; 17), 314, 389, 468
Dabrowski, S., 93
Dahl, K. S., 142
Dahlberg, J. E., 92, 109(19), 112(19)
Dahlquist, F. W., 424(51), 436
Dahm, A., 328, 447
Dalby, A. R., 425(72), 437, 473, 474(15), 476(15)
Daldal, F., 21
Dalgaard, J. Z., 271, 273, 276(14), 279, 279(14), 474

Dams, T., 328, 424(56), 436, 442, 448, 458, 459(86), 462, 463, 463(93), 473, 476(17)
Daniel, R. M., 283, 284, 288, 289, 289(12), 290, 291, 291(3; 14), 292(3), 478
Dankesreiter, A., 442, 450(25)
Danson, M. J., 283, 285, 287, 288, 316, 424(50), 436, 447, 476, 477(32)
Dao Thi, M.-H., 328, 330(1)
Dao-pin, S., 443
Darby, N. J., 63, 75, 79, 81, 87, 87(16)
Darcy, T. J., 118
Darimont, B., 24, 26, 26(11), 28(11), 29, 30(29; 30), 351, 354(11; 12), 367(12), 368(11), 372(12), 379(12), 381(11; 12), 382, 382(11; 12), 383(12), 384(12), 386(11; 12), 388, 388(11; 12), 434(45), 435, 444, 472, 473, 474(8), 475(8)
DasSarma, S., 194
Datta, P., 129, 130(5), 135(5), 140(5), 141(5), 144, 393
Daugelat, S., 274
Dauter, Z., 367
Davidson, E., 21
Davie, P., 129, 130(2), 139(2)
Davis, B., 393
Davis, E. O., 270, 271(1)
Davis, J. L., 178
Davis, R. W., 156
Davy, S. L., 367, 368
Davydov, R., 31
Dawson, R. M. C., 318
Day, E. P., 20(57), 21, 373
Day, M. W., 48, 388, 423, 424(1), 434, 435, 435(1), 444, 473, 475(19), 476(19)
Day, W. W., 388
de Groot, F. M. F., 48(20), 49
De Jongh, H. H., 345
De Kruif, K. G., 341
De la Cruz, A. F. A., 424(51), 436
De Lucia, F., 159
de Massy, B., 160, 172
de Pascale, D., 62, 63, 74, 76(4), 77, 77(5), 78, 79(4), 82, 87, 424(59), 436
de Ropp, J. S., 362, 363(44), 364(44)
de Rosa, M., 130, 132(13), 306, 314(13)
De Vendittis, E., 424(66), 437
de Vos, W. M., 206, 312, 460, 461(89), 462(89; 90), 471, 474, 475(24), 476(5), 477(21)
Dean, F. B., 180

Dean, G. E., 270, 271(1)
Deckert, G., 48, 148, 160(14), 445
Deckut, G., 147
Déclais, A.-C., 146
DeDecker, B. S., 424(47), 434(44), 435
Deits, T. L., 74
del Solar, G., 193
Delong, E. F., 261
Deloughery, C., 252
Demelar, B., 297
Demple, B., 240, 242
Dennis, D. T., 293(1), 294
Derbyshire, V., 93, 103(32), 114(32)
Deronzier, C., 217
DeRosa, M., 133
Dervertarnian, L., 27
Desmarais, D., 304
Deutzmann, R., 452, 458(75), 461(75)
Devoret, R., 264
deVos, W. M., 425(70), 437
Dewan, J. C., 434
Dhawan, I. K., 32
di Cera, E., 390
Di Ruggiero, J., 393
Dias, C., 115
Diaz Orejas, R., 193
Dibbens, J. A., 121
Dickson, D. P. E., 355
DiGate, R. J., 157, 180
Dijk, J., 129, 130(1; 2), 139(1–4)
Dijkstra, K., 63, 79, 87
Dikanov, S. A., 20, 21, 22(56)
Dikiy, A., 7, 379
Dill, K. A., 439, 443(8), 444(8), 466, 476
Dines, M., 284, 291(3), 292(3), 478
Dirr, H. W., 85, 87
DiRuggiero, J., 147, 148, 325
Dixon, D. A., 265, 266(22)
Doan, P. E., 21
Dobrinski, B., 117, 118(8), 121(2; 8), 129, 139(4)
Dobson, C. M., 466
Docampo, R., 39
Dohino, S., 94, 116(41)
Doi, H., 94, 116(40), 249, 251(3), 252(3), 253(3), 256(3), 260
Domaille, P. J., 130

Donaire, A., 31, 354, 356, 356(19), 359, 360, 360(42), 361(42), 362(42), 363, 363(42), 364, 364(42), 366, 366(42), 367, 367(42), 368(19; 37; 42), 370(42), 373(42), 374(42), 376(42), 379(19; 42), 380(42), 381(19), 382(19; 42), 383, 383(19; 42), 384, 386, 386(19), 388(19)
Donato, M. M., 302
Donigian, J. R., 425(78), 437
Doolittle, W. F., 94, 116(36), 260, 445
Dordick, J. S., 316
Doucette-Stamm, L. A., 252
Downey, K. M., 259
Drenth, J., 427
Drezdzon, M. A., 426
Driessen, M. C. P. F., 54
Drlica, K., 160
Drozd, J. W., 288
Drummond, R., 92
Duan, X., 274
Dubendorff, J., 133
Dubois, J., 252
Duderstadt, R. E., 32, 33, 34(32; 33), 36(33), 37(33), 40(32; 33)
Duff, J. L. C., 8(34; 37), 9, 10, 11, 11(34)
Duguet, M., 146, 147, 148, 148(9; 10), 150, 150(10), 151(17), 152(3; 17), 153(3), 154, 154(17), 155, 155(5), 156, 156(29; 30), 157(10; 17), 158(2; 6; 10), 159(5), 160, 160(33), 161(2), 162, 162(2), 164, 180
Dunham, W. R., 20(57), 21, 359
Dunn, J., 133
Dutreix, M., 264
Dutton, P. L., 22
Dwinell, D. A., 32
Dyall-Smith, M. L., 262

E

Easterby, J. S., 287
Eastman, A., 241
Eccleston, E., 46, 48(5), 50(5), 51(5), 54(5), 56
Echols, H., 91, 115(5)
Eckerskorn, C., 313
Eckert, K. A., 105
Eckstein, F., 219
Edgell, D. R., 94, 116(36), 260
Edman, J. C., 63, 75, 87(14)

Edmondson, S., 118, 119(10), 120(10), 129, 130, 130(5), 131, 132, 132(15), 135(5), 139(16), 140(5; 16), 141(5; 16), 142, 143, 144, 329, 335(12), 389, 392(4), 393, 393(4; 5), 400(31), 403(5; 31), 408(4), 414, 414(31), 418(4; 5), 419(4; 31), 420(31), 421(31), 424(54; 63), 436
Eftedal, I., 99, 102(43)
Eftink, M., 403
Eggen, R. I., 460
Eggleston, A. K., 265, 266(22)
Ehrlich, S. D., 162, 193
Eichinger, A., 94
Eichorn, G. L., 214
Eidness, M. K., 51, 424(37), 434
Eidsness, M. K., 46, 48, 48(23), 49, 51(32), 54(14; 32)
Eijsink, V. G. H., 440
Eisenberg, D., 442, 469
Eisenberg, H., 331
Eisenthal, R., 287
Ejchart, A., 27, 368, 374(55), 381(55)
Ekberg, M., 216, 226(13)
Eklund, H., 21, 75, 79, 83, 87, 216, 222(10), 226(13)
Eliasson, R., 217, 218, 219(30), 222(31), 226, 226(30), 227(30; 43)
Elie, C., 146, 147, 155(5), 159(5)
Elkin, C. J., 46, 51(2)
Ellis, L., 63, 75, 87(14)
Ellis, R. J., 293(1; 2), 294
Elmasry, N., 473
Eltis, L. D., 356, 368(37), 379
Empadinhas, N., 314
Engel, A. M., 440, 468(18)
Engel, P. C., 325, 424(41), 434, 461
Engh, R. A., 19, 94
Englander, J. J., 341, 344, 345
Englander, S. W., 49, 341, 342, 344, 345, 346, 346(5), 347, 348, 349, 350, 393
Engström, Y., 218
Eom, S. H., 92, 259
Epp, O., 84(26; 27), 85
Epstein, C. J., 63
Erauso, G., 193, 195(1)
Eriksson, M., 216, 225, 226(13)
Eriksson, S., 70, 218, 220
Eritja, R., 193
Erlich, H. A., 92
Ermler, U., 424(53), 425(79), 436, 437

Espinosa, M., 193
Evans, J. N. S., 313
Evans, T. C., 274
Eyring, H., 404

F

Fabian, H., 340, 341(30)
Faham, S., 32, 373, 381(59), 386(59), 424(67), 437
Fahrney, D. E., 308
Fairbrother, W. J., 353, 362(14), 363(14), 364(14)
Fairman, R., 334
Falke, J. J., 326
Fan, C., 31
Fareleira, P., 56, 62(6)
Faria, A., 9
Faria, T. Q., 307, 310(27)
Farr, A. L., 26
Farr, G. W., 294
Faunt, L., 408
Fedarko, N. S., 304
Fee, J. A., 20(57), 21, 22
Feldman, R. A., 48
Felli, I. C., 356, 368(37)
Ferguson, J. M. C., 424(50), 436
Ferreira, C., 314
Ferreira, S. T., 330
Ferrin, T., 132, 133, 145
Fersht, A. R., 417, 443, 444(40), 466, 473
Fiala, G., 173, 205, 302
Fierke, C. A., 52
Filmer, D., 348(18), 349
Findling, K. L., 20(57), 21
Finnegan, M. G., 32
Finnin, M. S., 425(78), 437
Finnzymes, O., 92
Fisher, H. F., 325
Fitz-Gibbon, S., 48
Flaman, J. M., 115
Flannery, B., 141, 408
Fleischmann, D. R., 147, 252
Fleming, P. J., 434(44), 435
Fleming, T. M., 425(72), 437, 473, 474(15), 476(15)
Flossdorf, J., 406, 407(53)
Foiani, M., 259
Follmann, H., 74, 76(1)

Fong, S. K., 52
Fontecave, M., 215, 216, 217, 217(7), 218, 218(7), 222(31)
Fontecilla-Camps, J.-C., 16, 16(46), 17, 429
Ford, R., 135, 396
Forest, E., 8(34), 9, 10, 11(34)
Forterre, P., 130, 146, 148, 148(9), 150, 151(17), 152(3; 17), 153(3), 154(17), 155, 155(5), 156, 156(29), 157(17), 159(5), 160, 160(33), 162, 163, 164, 165, 166, 169, 169(13), 172, 173, 176, 176(4), 179(4; 7), 193, 195, 195(1; 18), 202(18; 19), 203(18), 205, 206, 207, 212(5; 12), 460
Fourage, L., 289
Frank, J., 316
Franks, F., 439
Frasier, M. S., 109
Frauenfelder, H., 434
Frebourg, T., 115
Freedman, R. B., 70, 71(25), 75, 80
Freire, E., 335, 392, 393(21), 394, 400, 401, 401(44), 405(19), 406, 407(52), 414, 420, 422(68)
Freskgard, P.-O., 284
Frey, M., 29, 351, 367(7), 380(7), 381(7)
Friedman, J. M., 259
Friend, S. H., 115
Fringeli, U. P., 341
Fritsch, E. F., 65, 66(18), 121, 123(23), 129(23), 134
Fsihi, H., 221
Fu, J., 439
Fu, W., 17, 27, 31, 32(6; 12), 34(6), 37(6), 351, 354(5), 355(5), 377(5), 379(5)
Fuentes, E. J., 350
Fuggle, J. C., 48(20), 49
Fujii, T., 3, 3(16), 4, 5, 6(14; 16; 25), 7, 7(24), 8(14; 16), 9, 10, 11(16), 12(14; 16), 13(14; 16; 24; 27), 17(16), 20(24), 166, 434(42), 435
Fujimaki, K., 166
Fujimoto, J., 300
Fujimoto, N., 7, 9, 13(28)
Fujita, K., 92, 115(17), 255
Fujita, N., 278
Fujiwara, S., 240, 243(13), 244(13), 245(13), 246(13), 247, 278, 293, 294, 295, 295(6; 7), 296(6), 300(6), 425(81), 434(46), 435, 437, 473, 476(14)
Fukuyama, K., 29, 351, 367(8), 380(8), 381(8)

Funahashi, T., 48
Fusi, P., 130, 131(11; 12), 143(11; 12), 144(11)

G

Gaasterland, T., 48, 147, 445
Gabrielsen, O. S., 255
Gackstetter, T., 108
Gadelle, D., 160, 172, 173, 176, 176(4), 179(4; 7)
Gagua, A. V., 205
Gaillard, J., 217, 362
Galinski, E. A., 303, 304(6; 8)
Gallay, O., 85
Galperin, M. L., 444
Gamba, D., 259
Gambacorta, A., 133, 306, 314(13)
Gan, Z. R., 74
Gangloff, S., 146(13), 148, 157, 162(13)
Gao, Y.-G., 131, 294, 414, 424(54; 63), 436
Garcia, O., 92
Garcia-Moreno, E. B., 416
Garman, E. F., 429
Gassner, G. T., 21
Gasterland, T., 148, 160(14)
Gaudet, R., 85
Gavish, B., 326
Gay, R., 56, 61(5)
Gayda, J.-P., 21
Gazit, E., 341
Geerling, A. C., 206
Geiger, J. H., 434(44), 435
Geisler, N., 193
Gelfand, D. H., 92, 101, 103, 104(46)
Gellert, M., 147, 148, 149(19), 150(19), 151, 155(22), 156(23), 158(22), 164, 165, 178, 180, 181(2)
Gellet, M., 147
Gentz, R., 24
Genzor, C., 40
George, S. J., 8(34), 9, 10, 11(34), 48(20–22), 49
Georgiadis, M. M., 427
Georgopoulos, C. P., 293(1), 294, 296
Gerez, C., 218, 222(31)
Gerfen, G. J., 217, 223, 223(19), 224(19; 37)
Gerike, U., 476, 477(32)
Gershenson, A., 284
Gervais, A., 160
Gesierich, U., 28, 351, 383(3), 442, 472, 476(7)

Ghosh, G., 424(47), 435
Giacovazzo, C., 429
Gibert, I., 218, 219(30), 226(30), 227(30)
Gibson, D. T., 21
Gibson, J. F., 21
Gill, S. J., 414, 439, 444(16)
Gillespie, S. E., 410
Gilliland, G. L., 87
Gillis, M., 302
Gimble, F. S., 270, 271(1), 274, 279
Ginoza, W., 207
Giori, C., 355, 359(25), 377(25)
Giraldo, R., 193
Girerd, J. J., 355
Giroux, C. N., 173
Giver, L., 284
Glansdorff, N., 328, 424(5), 426
Glusker, J. P., 429
Godde, J. S., 127
Goebel, W., 194
Goette, J., 341
Golay, M. J. E., 136
Golberg, R. F., 63
Goldman, A., 332, 335(19), 341(19), 425(73), 431, 437
Goldman, E., 295
Gomes, C. M., 8(33), 9, 10, 11(33)
Gomez, J., 335
Gomez-Moreno, C., 40
Gonzalez, A., 425(75), 437
Goodfellow, B. J., 353, 356(18), 357(18), 359(18), 361(18), 362(18), 368(18), 374(18)
Goodman, M. F., 91, 115(5)
Goodman, M. R., 115
Goormaghtigh, E., 345
Gootz, T. D., 166
Gordon, A., 135, 396
Gorkovenko, A., 308
Gorst, C. M., 31, 32(18; 20), 38(19), 351, 353, 354, 354(10), 358, 358(10), 359, 360, 360(42), 361(42), 362(42), 363, 363(42), 364, 364(42), 365, 365(22), 366, 366(42), 367, 367(42; 48), 368(9; 10; 22; 42; 48), 369, 369(10), 370, 370(10; 22; 42), 371(9; 10; 22), 373(42; 48), 374, 374(9; 22; 42; 48), 375, 375(9; 22), 376, 376(22; 42), 377(9; 22), 378, 378(22), 379(22; 42), 380(42), 382(42), 383(42), 385(10; 22), 388(10)

Grabarse, W., 425(79), 437
Graham, D. E., 48, 147, 445
Grant, W. D., 306, 314(13)
Gräslund, A., 217
Grättinger, M., 424(49), 436, 442, 450(25), 452, 476
Graumann, P., 464, 465
Graves, M. C., 9
Grayling, R. A., 117, 118, 118(8), 119(10), 120(10; 11), 121, 121(8), 124(3), 389, 393(5), 403(5), 418(5)
Green, M., 187
Griffig, J., 10
Griko, Y., 390, 393, 403(12)
Grindley, N. D. F., 101
Grisa, M., 130, 131(12), 143(12)
Gronenborn, A. M., 79, 84(18; 19)
Gronenborn, B., 193
Gross, M., 316
Grote, M., 129, 139(3; 4)
Growhurst, G. S., 425(72), 437
Gruss, A., 193
Guagliardi, A., 63, 74, 76(3; 4), 79(4), 130, 132(13)
Guerlesquin, F., 7
Guerritore, A., 130, 131(11), 143(11), 144(11)
Guigliarelli, B., 21
Guillot, F., 24
Guinier, A., 135
Guipaud, O., 160, 162, 164, 165, 169, 169(13)
Gupta, R., 129, 130(5), 135(5), 140(5), 141(5), 144, 362, 393
Gurbiel, R. J., 21
Gustafson, C. E., 207
Gutierrez, M. C., 302

H

Habermann-Rottinghaus, S. M., 341
Häckel, M., 399
Hackett, N. R., 194
Hafenbradl, D., 205
Hafezi, R., 40
Hage, A., 27
Hagedoorn, P. L., 54
Hagen, W. R., 8(33), 10, 11(33), 22, 54
Hahn, U., 132, 141(25), 144(25), 340, 341(30), 443, 444(41)
Hahne, S., 216, 223(14), 225(14)

AUTHOR INDEX

Haikawa, Y., 48, 252
Hain, J., 66
Hajdu, J., 429
Hall, D. O., 10, 17(35)
Hamana, H., 389
Hamatake, R. K., 259
Hamiche, A., 159
Hamlin, R., 259
Han, Y. S., 425(76), 437, 475, 476(28)
Hanai, R., 180, 193
Handel, T. M., 350
Hansen, C. J., 92
Hansen, J. L., 259
Hansen, L. D., 396
Hansen, T., 332, 335(19), 341(19), 425(73), 437
Hanson, A. D., 303
Hanson, S. F., 193
Hansson, H., 139, 140(36), 408, 409(59)
Hara, T., 429
Hard, T., 130, 138(19), 139, 140(36), 408, 409(59)
Harder, J., 217, 218, 222(31)
Hare, D. R., 48, 49(11), 388, 423, 424(2)
Harpaz, Y., 473
Harrieder, S., 464
Harris, J. I., 452, 454, 454(72)
Harrison, R. W., 434, 435
Hartl, F.-U., 294, 297
Hartmann, H., 434
Hase, T., 29
Hashimoto, H., 247, 425(81), 437, 473, 476(14)
Hata, Y., 3, 3(16), 4, 6(14; 16), 8(14; 16), 9, 10, 11(16), 12(14; 16), 13(14; 16), 17(16), 434(42), 435
Hatchikian, E. C., 9, 16, 16(46), 17, 29, 351, 367(7), 380(7), 381(7)
Hausner, W., 206
Hayano, T., 87
Hayashi, I., 260, 262
Haynie, D. T., 400, 420
He, H., 259
Hearshen, D. O., 20(57), 21
Hecht, K., 349
Hecht, R. M., 472, 474(10), 476(10), 477(10)
Hedman, B., 48(23), 49
Hedwig, G. R., 399
Hegerl, R., 294
Hei, D., 316, 323, 324(29), 327(29)
Heijna, J. A., 162
Heinemann, U., 443, 444(41)

Heinrich, P., 446
Helbert, M., 289
Helbl, V., 334
Hellgren, N., 425(69), 437, 474, 475(23), 476(23)
Hellman, U., 218, 219(30), 226(30), 227(30)
Heltzel, A., 27, 31, 36(5)
Hemmingsen, S. M., 293(1), 294
Hempstead, S. K., 271, 276(3), 278(3)
Henderson, E., 239
Hendrix, H., 16
Hendrix, R. W., 293(1), 294
Hennecke, H., 62
Hennig, M., 424(40), 434, 442, 463, 464, 473, 474(16), 475(16), 476(16)
Henning, P., 129(6), 130, 138(6)
Henningsen, I., 92
Henrick, K., 473
Hensel, R., 214, 292, 293, 307, 308, 311, 312(21; 29), 313, 314, 328, 389, 447
Herault, D. A., 312
Heremans, K., 316, 432
Hernandez, G., 49
Hester, G., 425(75), 437
Hethke, C., 206
Hicks, P. M., 446
Higuchi, R., 92
Hildebrandt, J. J., 345
Hill, C. P., 425(74), 437
Hille, R., 19, 20(57), 21
Hillen, W., 334
Hiller, R., 49, 342, 347, 348, 349, 350, 393
Hilser, V. J., 335
Hino, Y., 48, 252
Hinz, H.-J., 399, 406, 407(53), 408, 475
Hiraga, S., 164
Hirst, R., 70, 71(25)
Hirst, T. R., 75, 80
Hisano, T., 41
Hoaki, N., 295
Hodges, A. E., 40
Hodges, R. A., 271, 276(10), 278(10)
Hodgson, K. O., 48(23), 49
Hoelle, C. J., 207
Hof, P., 19
Hoffman, B. M., 21, 31, 32(9; 12), 223, 224(37), 354, 354(21), 355(21), 377(21), 379(21)
Hofmann, A., 294, 447, 450(64)
Hogenkamp, H. P. C., 63, 74, 76(2), 80(2)
Hogrefe, H. H., 91, 92, 108, 113, 115, 255

Hol, W. G. J., 427
Holden, H. M., 40
Holler, T. P., 216, 223(15), 225(15)
Holley, W. R., 279
Hollien, J., 477
Hollung, K., 255
Holm, R. H., 30, 375
Holmgren, A., 59, 62, 69, 74, 75, 83
Honig, B., 416, 417(69)
Höög, J.-O., 83
Hoogstraten, R. A., 193
Hopfner, K.-P., 94
Horii, T., 264
Horikawa, H., 48, 252
Horn, G. T., 92
Hornstra, L. J., 271, 274(4), 278(4)
Horovitz, A., 443
Horwich, A. L., 294, 297
Horwitz, P. M., 297
Hosfield, D. J., 425(80), 437
Hosoyama, A., 48
Hoste, B., 302
Hough, D. W., 285, 316, 424(50), 436, 447, 476, 477(32)
Housset, D., 29, 351, 367(7), 380(7), 381(7)
Howard, E. J. B., 56
Howard, J. B., 31, 32(16; 17), 38(19), 46, 48(5), 50(5), 51(5), 52, 54(5), 351, 352, 352(24), 354, 354(10), 358, 358(10), 359, 360, 367(39), 368(10; 24; 43), 369, 369(10), 370, 370(10; 24), 371, 371(10; 24; 43), 372, 372(24), 373(24; 43), 381, 381(24), 384(24), 385(24; 43; 68), 386(24; 43), 387, 387(24), 388(10; 24; 68)
Hsieh, T., 163
Hsu, B. T., 48, 388, 423, 424(1), 435, 435(1), 444, 473, 475(19), 476(19)
Hu, G., 115, 116(62), 258
Hu, Y., 32, 424(67), 435, 437
Huang, C., 132, 133, 145
Huang, H., 31, 32, 46, 48, 54(14), 354, 354(21), 355(21), 377(21), 379(21), 424(37), 434
Huang, M.-M., 115
Huang, W., 84(28), 85, 87
Huang, Y.-H., 48, 49(18; 19)
Huber, H., 9, 48, 101, 181, 187(15), 205, 302, 306, 307(16), 308(16), 389, 448
Huber, J. G., 362

Huber, R., 19, 23, 29, 30(29), 84(26), 85, 87, 94, 146, 148(9), 160, 164, 205, 306, 307, 307(16), 308, 308(16), 351, 354(12), 367(12), 372(12), 379(12), 381(12), 382(12), 383(12), 384(12), 386(12), 388(12), 389, 424(38; 49; 56), 434, 434(45), 435, 436, 438, 440, 440(4), 442(4), 452, 453, 454(76), 458, 459(86), 463, 468(18), 473, 476, 476(17)
Hudepohl, U., 66
Hudson, R. C., 290, 291(14)
Hugenholtz, J., 55, 261
Hughey, R., 271, 276(14), 279(14)
Hung, L. W., 424(61; 64), 436
Hunsmann, G., 24
Hurley, J. K., 40
Hutchins, A., 16
Huth, J. R., 79, 84(18)
Huynh, B. H., 355, 373
Huynh, T. V., 156
Hvidt, A., 346
Hwang, K. Y., 424(57), 425(83), 436, 437
Hynes, G. M., 294

I

Ichiyanagi, K., 278
Ichiye, T., 49
Iggo, R., 115
Ihara, K., 239
Iizuka, T., 3, 4, 4(9), 5(9; 22), 6(9), 7(9), 8(9), 11(9), 15(9), 17(9), 20(22), 22(22)
Ikai, A., 392
Ikeda, A., 55
Ikeda, H., 164
Ikeda, S. H., 41
Ilynia, T. V., 193, 194, 195(15)
Imai, T., 3(15), 4, 6(15), 7(15), 8(15), 10(15), 11, 12(15), 13(15), 17(15), 20, 22(55)
Imamura, M., 249, 250(2), 255
Imamura, R., 164
Imanaka, T., 93, 239, 240, 243(13), 244(13), 245(13), 246(13), 247, 261, 262, 262(8; 9), 268(9), 269(9), 278, 293, 294, 295, 295(6; 7), 296(6), 298, 300(6), 425(81), 434(46), 435, 437, 473, 476(14)
Inatomi, K., 4
Ingelman, M., 79, 87
Innis, M. A., 101

Inoue, H., 93, 166, 247
Inoue, T., 425(81), 437, 473, 476(14)
Isaacs, N. S., 317
Ishii, S., 300
Ishikawa, J. K., 438, 442(2)
Ishimoto, M., 55
Ishino, S., 249, 254, 254(5), 259(5), 260, 262
Ishino, Y., 92, 93, 94, 115(17), 116(37; 40; 41), 249, 250, 250(2), 251(3), 252(3), 253, 253(3), 254, 255, 256(3), 257(4), 258, 260, 260(6; 20), 278, 280
Ishioka, C., 115
Isogai, T., 4, 5(22), 20(22), 22(22)
Isogai, Y., 3, 4(9), 5(9), 6(9), 7(9), 8(9), 11(9), 15(9), 17(9)
Isupov, M. N., 424(55), 425(72; 82), 436, 437, 473, 474(15), 476(15)
Ito, J., 91, 249
Ito, N., 280
Ito, T., 239
Itoh, T., 164, 389
Ivanov, V. I., 142
Ivens, A., 425(75), 437
Iwasaki, T., 3, 3(15; 17), 4, 4(9; 12; 13), 5, 5(9; 12; 13; 22), 6(9; 15), 7(9; 15; 17; 23; 24), 8(9; 15; 17; 23), 9, 10(15; 17), 11(9; 23), 12(15; 23), 13(12; 13; 15; 17; 24), 14(13), 15, 15(9; 12; 13; 20), 16, 16(13), 17(9; 13; 15), 18(12; 13), 19, 19(12), 20, 20(20; 23; 24), 22(20; 22; 55; 56)
Iwaskewycz, B., 316
Iwata, S., 21
Izatt, R. M., 396, 410
Izawa, Y., 295
Izu, Y., 424(48), 436, 472, 476, 476(6)
Izumi, M., 294, 295(7)

J

Jablonski, P., 304
Jack, W. E., 93, 94(25), 100, 101, 109(25), 112(25), 113, 270, 271, 271(1), 276(3; 10), 278(3; 10)
Jackson, A. P., 163, 296
Jackson, S. P., 159, 161(39), 229, 434(44), 435
Jacob, L., 292, 293
Jacob, U., 424(49), 436, 452, 476
Jacobs, W. R., Jr., 274
Jacobsson, A., 242
Jaenicke, R., 316, 328, 330(1), 349, 388, 389, 393, 416(3), 424(38; 49; 56), 431, 434, 436, 438, 439, 440, 440(9), 441(28), 442, 442(6; 9), 443, 443(9), 444, 446, 447, 447(6), 448, 448(62), 449(69), 450(25; 28; 62–64; 67; 69), 451(26), 452, 452(47), 453, 453(3; 6; 74), 454, 454(19; 74; 76), 456, 456(28; 78), 457, 458, 458(6; 75; 82), 459(6; 86), 461(75; 85), 462, 463, 463(93), 464, 465(100; 103), 466, 467, 468, 468(6; 53; 103), 473, 476, 476(17), 478
Jakobsen, K. S., 255
Jancarik, J., 425(77), 437, 475, 476(29)
Janin, J., 335
Jannasch, H. W., 93, 94(31), 100, 109(31), 181, 187(15), 205, 271, 276(3), 278(3), 308
Jansonius, J. N., 424(40), 425(75), 434, 437, 464, 473, 474(16), 475(16), 476(16)
Jansz, H. S., 204
Jarlier, V., 165, 166, 169(18)
Jarvis, L., 132, 133, 145
Jaxel, C., 146, 147, 148, 150, 151(17), 152(3; 17), 153(3), 154(17), 155, 156(29; 30), 157(17), 158(6), 159, 161(39), 162(38)
Jeanthon, C., 218, 219(30), 226(30), 227(30)
Jecht, M., 457
Jeng, M.-F., 345
Jenney, F. E., Jr., 31, 45, 49, 52, 56, 61(4), 62(3; 4), 352(24), 354, 368(24), 370(24), 371, 371(24), 372, 372(24), 373(24), 381(24), 384(24), 385(24), 386(24), 387, 387(24), 388(24), 435
Jeno, P., 425(68), 437
Jensen, L. H., 12, 29, 351, 367, 367(6), 372(6), 379(6), 380(6), 381(6), 382(6), 386(6), 388
Ji, X., 87
Jiang, J., 423
Jiang, Y., 259
Joachimiak, A., 294, 295(5)
Joelson, T., 79, 87
John, R. A., 286
Johnson, B. B., 142
Johnson, C. E., 355
Johnson, K. A., 31, 32, 34(32; 33), 36(33), 37(33), 40(32; 33), 48, 49(18; 19), 115
Johnson, K. S., 150
Johnson, M., 141, 408
Johnson, M. K., 17, 27, 31, 32(6; 9; 12; 13), 33, 34(6), 37(6), 351, 354(5), 355(5), 377(5), 379(5)

Johnson, R. C., 354
Johnson, W. C., 142
Johnston, L. H., 259
Johnston, M. I., 216, 223(15), 225(15)
Jokela, M., 259
Jollie, D. R., 356
Jones, A., 84(26), 85
Jones, S., 83
Jones, T. A., 87
Jordan, A., 215, 216(2), 217, 217(2), 218, 219(2; 30), 220(2), 225(2), 226, 226(30), 227(30; 43)
Jörnvall, H., 83, 145, 217, 218, 222(31)
Josephs, R., 294, 295(5)
Joshua-Tor, L., 48, 388, 423, 424(1), 435, 435(1), 444, 473, 475(19), 476(19)
Jovin, T. M., 101
Joyce, C. M., 91, 93, 101, 103(32), 114(32)
Jrnney, F. E., Jr., 352
Jung, Y. S., 40
Jupin, I., 193
Jurica, M. S., 272, 276(16), 278(16), 279(16)

K

Kagawa, H. K., 294
Kai, Y., 239, 247, 425(81), 437, 473, 476(14)
Kakasu, S., 148, 161(15)
Kakihara, H., 93
Kallenbach, N. R., 346
Kaltoum, H., 146, 147, 148(10), 150(10), 157(10), 158(10), 160, 164, 180
Kamat, S. V., 316
Kamath, U., 141
Kampranis, S. C., 163
Kanai, S., 87, 249, 254(5), 259(5)
Kanaya, S., 261, 262, 262(8; 9), 268(9), 269(9), 294, 295(7)
Kandler, O., 3, 87
Kanei-Ishii, C., 300
Kang, C. H., 49
Kang, J. D., 294
Kang, S. W., 85, 87
Kanodia, S., 308, 313
Karawya, E., 105
Kardinahl, S., 425(69), 437, 474, 475(23), 476(23)
Kardos, J., 340, 349, 475
Karlin, S., 261

Karlsson, M., 216, 223(14), 225(14)
Karpiel, A. B., 21
Karplus, M., 392, 434, 466, 475
Karran, P., 239
Karshikoff, A., 393
Karsten, U., 307
Käslin, S. A., 446
Kath, T., 329, 335(11), 340(11), 341(11)
Kato, I., 92, 93, 94, 115(17), 116(40), 249, 250, 250(2), 251(3), 252(3), 253, 253(3), 255, 256(3), 260, 424(48), 436, 472, 476, 476(6)
Kato, J., 164
Kato, R., 262, 264, 265(15)
Katsube, Y., 29, 351, 367(8), 380(8), 381(8)
Katterle, B., 225
Katti, S. K., 79, 87
Kauppi, B., 21
Kautz, R., 280
Kauzmann, W., 318, 443
Kavina, V., 394
Kawakami, B., 93
Kawarabayasi, Y., 48, 147, 252
Kawasaki, M., 274
Kawate, H., 239
Keeney, S., 173
Keller, M., 48
Kellerer, B., 447, 450(64)
Kelly, R. M., 439, 446
Keltjens, J. T., 313
Kemmink, J., 63, 75, 79, 87, 87(16)
Kemper, E. S., 205
Kendrew, J. C., 428
Kennedy, W. M. P., 79, 84(18; 19)
Kent, T. A., 20(57), 21, 355
Keohavong, P., 115
Kerscher, L., 3, 4(5), 5(5), 6(15), 7, 7(5), 8(5), 9, 10, 10(4; 5), 13(28), 14(4; 5), 15, 15(10; 11), 17(4; 10; 11; 35), 20(4)
Kersters, K., 302
Kessel, M., 325, 393
Kessler, C., 105
Khan, S. A., 193
Khorana, H. G., 194
Kidokoro, S.-I., 401
Kiefer, J. R., 92
Kikuchi, A., 146, 152(4a), 153(4a), 154
Kikuchi, H., 48
Kikuchi, M., 87
Kil, Y. V., 262, 265(14)
Kilpelainen, S., 259

Kim, C., 30
Kim, C.-H., 31
Kim, H. Y., 424(57), 436
Kim, K. K., 424(61; 62; 64), 436
Kim, R., 424(61; 62; 64), 436
Kim, S. H., 424(61; 62; 64), 425(76–77), 436, 437, 475, 476(28; 29)
Kim, S. S., 425(83), 437
Kim, Y., 92, 259
Kimura, J., 129, 130(2), 139(2)
Kimura, M., 129, 130(2), 139(2)
Kindler, B., 332
King, R. J., 307
Kingma, J., 46
Kingston, R. E., 120
Kirkegaard, K., 156, 180
Kirschner, K., 424(40), 425(75), 434, 437, 442, 457, 463, 464, 473, 474(16), 475(16), 476(16)
Kissinger, C. R., 12, 29, 351, 367(6), 372(6), 379(6), 380(6), 381(6), 382(6), 386(6)
Kitabayashi, M., 93
Klaassen, C., 313
Klabunde, T., 274
Klar, A. J., 279
Kleckner, N., 173, 261
Klee, W. A., 144
Kleemann, G. R., 28, 351, 383(3), 442, 463, 472, 476(7)
Kleibl, K., 239
Klein, M. P., 21
Klenk, H. P., 94, 147, 252, 260
Klenow, H., 92
Kletzin, A., 15, 16, 16(42), 32, 38(34), 424(39), 434
Kleywegt, G. J., 87
Klibanov, A. M., 330, 399, 404(41), 471
Klump, H., 325, 393, 424(65), 436
Knaff, D. B., 21
Knapp, S., 130, 138(19), 139, 140(36), 145, 160, 393, 408, 409(59), 425(69; 70), 437, 460, 461(89), 462(89; 90), 471, 474, 475(23), 476(5; 23), 477(21)
Knecht, R., 27
Knittel, T., 99, 102(42)
Knoblich, I. M., 92
Knoechel, T., 425(75), 437
Knowles, J. R., 447, 448(62), 450(62)
Kobrehel, K., 62(3), 63
Koga, Y., 249, 257(4)

Kohda, K., 239, 294, 295(6), 296(6), 300(6)
Kohlhoff, M., 328, 447
Kohlstaedt, L. A., 259
Koike, G., 239, 242(8)
Kok, M., 46
Kokkinidis, M., 406, 407(53), 475
Komiya, H., 427
Komori, K., 249, 253, 254, 254(5), 257(4), 259(5), 260, 278
Kon, T., 3(15), 4, 6(15), 7(15), 8(15), 10(15), 12(15), 13(15), 17(15)
Kong, H., 91, 92(9), 93, 93(9), 94(25; 31), 95, 95(9), 100, 101, 109(9; 25; 31), 112(25), 113, 114(9), 116(9)
König, H., 23, 160, 164, 214, 308, 312(29), 389, 438, 440(4), 442(4)
Koo, H. S., 146
Koonin, E. V., 193, 194, 195(15), 444
Kopylov, V. M., 180
Korkhin, Y., 227, 228
Kornberg, A., 91, 193
Korndorfer, I., 424(38; 56), 434, 436, 453, 454(76), 458, 459(86), 473, 476(17)
Kornuc, J. J., 427
Korolev, S., 92
Kort, R., 460
Kosa, P. F., 159, 161(39), 228, 424(47), 435
Koshland, D. E., 348(18), 349
Kossiakoff, A. A., 340
Kosugi, H., 48
Kosyavkin, S. A., 148, 149(19), 150(19)
Kovalsky, O. I., 155, 180
Kowal, A. T., 17, 27, 31, 32(6), 34(6), 37(6), 351, 354(5), 355(5), 377(5), 379(5)
Kowalczykowski, S. C., 261, 262, 265, 265(13), 266(22)
Kozyavkin, S. A., 147, 151, 155, 156(23), 178, 179, 180, 181, 181(2), 187, 187(14)
Krah, R., 147, 148, 149(19), 150(19), 151, 155(22), 156(23), 158(22)
Kraulis, P. J., 77
Krause, K. L., 472, 474(10), 476(10), 477(10)
Krebs, M. P., 194
Kress, L. F., 388
Kreuwels, M. J. J., 313
Kreuzer, K. N., 164
Kriauciunas, A., 21
Krokan, H. E., 99, 102(43)
Krook, M., 217, 218, 222(31)
Krutzsch, H., 393

Kryger, G., 472, 474(11), 475(11), 476(11)
Krzycki, J. A., 117, 121(2)
Kucera, R. B., 93, 94(25; 31), 100, 101, 109(25; 31), 112(25), 113, 271, 276(3), 278(3)
Kudoh, Y., 48
Kuila, D., 21, 22
Kujo, C., 262, 265(15)
Kukuchi, A., 148, 156(15), 161(15)
Kulms, D., 132
Kulmus, D., 132, 141(25), 144(25)
Kumar, S., 91, 92(9), 93(9), 95, 95(9), 100, 109(9), 114(9), 116(9)
Kunkel, T. A., 105, 115
Kuntz, I. D. J., 434
Kunugi, S., 316, 318
Kur, J., 93
Kuramitsu, S., 262, 264, 265(14; 15)
Kurganov, B. I., 404
Kurigan, J., 63
Kurinov, I. V., 434, 435
Kuriyan, J., 75, 79, 87
Kurr, M., 181, 184, 187(15), 308
Kurtz, D. M., 48(23), 49, 51(32), 54(32)
Kurtz, D. M. J., 424(37), 434
Kurtz, D. M., Jr., 27, 46, 48, 51, 51(2), 54(14)
Kushida, N., 48
Kuszewski, J., 79, 84(19)
Kutyshenko, V., 390, 403(12)
Kyogoku, Y., 280

L

La Mar, G. N., 31, 32, 32(16–18; 20), 37(36), 38(19), 52, 351, 352, 352(24), 353, 354, 354(10), 356(19), 358, 358(10), 359, 360, 360(42), 361(42), 362, 362(42), 363, 363(42; 44), 364, 364(42; 44), 365, 365(22), 366, 366(42), 367, 367(19; 42; 48), 368(9; 10; 19; 22; 24; 42; 43; 48), 369, 369(10), 370, 370(10; 20; 22; 24; 42), 371, 371(9; 10; 22; 24; 43), 372, 372(24), 373(24; 42; 43; 48), 374, 374(2; 22; 42; 48), 375, 375(9; 22), 376, 376(22; 42), 377, 377(9; 22), 378, 378(22), 379(19; 22; 42), 380, 380(20; 42), 381, 381(19; 24), 382(19; 42), 383, 383(19; 42), 384, 384(24), 385(9; 22; 24; 43; 68), 386, 386(19; 24; 43), 387, 387(20; 24), 388(10; 19; 24; 68)
Labedan, B., 169, 460
Lachinat, C., 379
Ladenstein, R., 63, 74, 77, 77(5), 78, 82, 83, 84(26–28), 85, 87, 130, 138(19), 139, 140(36), 145, 160, 393, 408, 409(59), 424(59), 425(49; 70), 436, 437, 460, 461(89), 462(89; 90), 471, 474, 475(23), 476(5; 23), 477(21)
Laderman, K., 393
Laemmli, U. K., 63
LaHaie, E., 21
Lahti, R., 425(73), 437
Laidler, K. J., 283
Lake, J. A., 147, 148, 149(19), 150(19), 178, 179, 180, 181, 181(2)
Lamb, J. D., 410
Lamed, R. J., 298
Lamotte, B., 355, 359(25), 377(25)
Landa, I., 54
Landre, P. A., 92
Lang, D. A., 425(68), 437
Langer, T., 297
Langowski, J., 123
Langridge, R., 132, 133, 145
Langworthy, T. A., 23, 160, 164, 438, 440(4), 442(4)
Lanzer, M., 24
Lanzov, V. A., 262, 265(14; 15)
Lapshina, E. A., 476
Larsen, P. L., 124
Larsson, A., 216, 223(14), 225(14)
Lasken, R. S., 102
Laskowski, M., Sr., 388
Lassalle, M. W., 475
Lau, K., 146
Lauder, L. D., 265, 266(22)
Laue, E. D., 130
Laue, F., 94
Lauerer, G., 302
Laufs, J., 193
Laurents, D. V., 466
Laval, F., 239
Lavoie, T. B., 334
Lawyer, F. C., 92
Lazaridis, T., 392, 475
Lazcano, A., 444
Le Master, D. M., 87
Le Noble, W. J., 317
Lebbink, J., 312, 460, 462(90), 471, 476(5)
Lebowitz, J., 205

Lechmacher, A., 313
Leclere, M. M., 240, 243(13), 244(13), 245(13), 246(13)
Ledbetter, M., 165
Lee, B., 48, 49(17)
Lee, D.-S., 92, 259
Lee, G. E., 117
Lee, G. H., 294
Lee, H.-I., 31, 354, 354(21), 355(21), 377(21), 379(21)
Lee, H.-J., 159, 162(38), 429
Lee, H.-M., 252
Lee, I., 475
Lee, K., 21
Lee, M. Y. W. T., 259
LeGall, J., 19, 26, 48, 49(18), 55, 56, 62(6), 370, 373
Legrain, C., 328, 424(5), 426
LeGrice, S., 24
Lehman, I. R., 261, 265(1)
Leibrock, E., 438
LeMaster, D. M., 49, 79
Lenox, A. L., 48, 147, 445
Lepape, L., 217
Leppanen, V.-M., 332, 335(19), 341(19), 425(73), 437
Lesk, A. M., 335
Lewis, G. N., 390, 400(11), 403(11)
Lewis, M., 393, 429
Lewis, V. A., 294
Li, B. Q., 259
Li, H., 424(58), 436
Li, W.-T., 116, 118, 119(10), 120(10), 389, 393(5), 403(5), 418(5)
Li, Y., 92
Liberek, K., 296
Licht, S., 217, 223, 223(19; 20), 224, 224(19; 37)
Liddington, R. C., 163
Liebl, W., 446, 458, 460, 461(85), 468(54)
Liepinsh, E., 373, 386(58)
Lim, H. M., 159, 162(38)
Lim, J. H., 425(76), 437, 475, 476(28)
Lim, L., 131, 424(63), 436
Lin, D., 423
Lin, H.-J., 48(21), 49
Lin, L.-N., 394, 409, 410(61), 411(61)
Lindahl, T., 207, 239, 240, 242
Linderstrom-Lang, K. U., 348(17), 349
Ling, L. L., 115
Link, T. A., 21, 22

Linn, S., 91, 116(6)
Lipman, D. J., 273
Little, M. C., 101
Littlechild, J. A., 424(55; 72), 425(82), 436, 437, 473, 474(15), 476(15)
Littlefield, O., 227, 228
Liu, L. F., 146, 158, 178
Liu, M.-Y., 56, 62(6), 370
Liu, Z., 325
Ljungdahl, L. G., 26, 55
Lloyd, M. D., 429
Lloyd, S. A., 425(74), 437
Loehle, C., 422
Loferer, H., 62
Loh, S. N., 344
Lohman, T. M., 140, 155
Lohmüller, K., 446
Long, A. M., 259
Lopez-Garcia, P., 130, 160
Lottspeich, L., 292
Lovejoy, A., 91, 108
Loveless, A., 239
Lovenberg, W., 45, 359
Lovett, S. T., 472, 474(11), 475(11), 476(11)
Lowary, P. T., 127
Lowe, J., 424(52), 436
Lowenstein, J. M., 288
Lowry, O. H., 26
Lucchini, G., 259
Luchinat, C., 7, 8(31), 353, 355(17), 356, 357(17; 35), 358(17), 359, 359(17), 361(17), 362(17; 35), 363(35), 364(35), 367, 368(15; 37), 374(17), 377(17; 40), 379, 380(66), 385(15)
Lüdemann, H.-D., 438
Ludlow, J. M., 318, 319, 319(22), 322, 324(22; 26)
Ludwig, H., 316
Lumry, R., 404
Lundbach, T., 130, 138(19)
Lundbäck, T., 139, 140(36), 408, 409(59)
Lundberg, K. S., 93, 114(26), 115(26), 255
Luque, I., 392
Lurz, R., 117, 118(8), 121(2; 8), 124(3), 129, 139(4)
Lustig, A., 463
Lustig, H., 442
Lyamichev, V., 92, 109(19), 112(19)
Lyubarev, A. E., 404
Lyubchenko, Y., 205

M

Ma, K., 38, 40, 40(42), 41, 43(4), 44(4), 55, 56(1), 57(1), 58, 59(1), 60(1; 9; 10), 61(1), 351, 354(10), 358, 358(10), 368(10), 369, 369(10), 370, 370(10), 371(10), 388(10), 474, 4343
Ma, Y., 48(20), 49
Macedo, A. L., 23, 353, 356(18), 357(18), 359(18), 361(18), 362(18), 368(18), 370, 373, 374(18)
Macedo-Ribeiro, S., 29, 30(29; 30), 351, 354(12), 367(12), 372(12), 379(12), 381(12), 382(12), 383(12), 384(12), 386(12), 388, 388(12), 434(45), 435, 444, 472, 474(8), 475(8)
Mach, H., 438
Macinai, R., 7
Macke, T. J., 21
MacPherson, A., 427
Madden, T. L., 273
Maeder, D. L., 316, 424(65), 431, 436, 474
Maekawa, T., 300
Maes, D., 328, 330(1), 448, 450(67)
Magnuson, J. K., 46, 48(5), 50(5), 51(5), 54(5), 56
Makhatadze, G., 392, 393(22), 398, 399
Maki, H., 239, 242(8)
Makimoto, S., 323
Makiniemi, M., 259
Makino, K., 207
Makrides, S. C., 295
Malkin, R., 21
Maltsev, N., 294
Malykh, A., 187
Manaia, C. M., 302, 306, 307(14), 310(14)
Maniatis, T., 65, 66(18), 121, 123(23), 129(23), 134
Mannervik, B., 70, 87
Manning, N., 423
Mao, C., 92
Mao, S. S., 216, 223, 223(15), 225(15)
Marahiel, M. A., 464, 465, 465(103), 468(103)
Marczak, R., 56, 61(5)
Marguet, E., 160, 164, 165, 166, 169(13), 205, 207, 212(5; 12)
Marians, K. J., 157, 180
Marini, F., 259
Markley, J. L., 40, 316, 344, 353, 356, 356(16), 357(16), 358(16), 361(16), 362(16), 374(16), 434
Marks, P. A., 425(78), 437
Marliére, P., 120
Marqusee, S., 350, 477
Marsin, S., 193, 195, 195(1; 18), 202(18; 19), 203(18)
Martial, J. A., 328, 330(1), 448, 450(67)
Martin, B. A., 166
Martin, C., 115
Martin, D. D., 304, 306(10), 308(10)
Martin, J. L., 63, 75, 79, 85(12), 87, 87(12)
Martin, J. M., 297
Martin-Zanca, D., 239
Martinez-Cruz, L. A., 425(77), 437, 475, 476(29)
Martins, L. O., 306, 307, 307(15; 16), 308(16), 310(17; 23), 314, 389, 468
Marugg, J. D., 314
Marz, A., 442
Mason, J. R., 21, 22(61)
Mason, R. P., 39
Masson, P., 316, 327
Masui, R., 262, 264, 265(14; 15)
Mather, M. W., 21
Mathur, E. J., 93, 114(26), 115(26), 255
Matsubara, H., 7, 9, 13(28), 23, 27(1), 29, 30, 55, 351, 367(8), 380(8), 381(8)
Matsueda, G. R., 334
Matsuura, K., 4, 15(20), 22(20)
Matthews, B. W., 439, 440, 441, 441(12; 20), 443, 443(12)
Mattoccia, E., 163
Matussek, K., 313
Maurizot, J. C., 160
Max, L., 434
Maxwell, A., 163
Maxwell, D. P., 193
Mayne, L., 341, 344, 345, 346(5), 350
Mayorga, O. L., 392, 405(19)
Mazel, D., 120
McAfee, J., 129, 130, 130(5), 135(5), 139(16), 140(5; 16), 141(5; 16), 142, 143, 144, 393
McClure, W. R., 101
McCrary, B. S., 130, 131, 132(15), 329, 335(12), 389, 392(4), 393, 393(4), 400(31), 403(31), 408(4), 414, 414(31), 418(4), 419(4; 31), 420(31), 421(31), 424(54; 63), 436
McDonald, C. C., 359

McDonald, J. P., 146(13), 148, 157, 162(13)
McFarlan, S. C., 63, 74, 76(2), 80(2)
McGhee, J., 140, 413
McGovern, K., 75
McHenry, C. S., 93, 116(24), 173
McLendon, G., 379
Medvedkin, V. N., 398
Meiga, G., 48(21), 49
Mendes, J., 7, 8(31; 33), 9, 10, 11(33)
Mendoza, J. A., 297
Menon, A. L., 16, 33, 34(37), 38, 41, 43, 50, 56(7), 57
Menon, N. K., 27
Menon, S., 18
Merckel, M. C., 424(53), 436
Merkle, H. P., 341
Merritt, E. A., 77
Mersha, F. B., 271, 274(4), 278(4)
Messori, L., 379, 380(66)
Meulenberg, C. H. C., 46
Meyer, J., 367
Mian, I. S., 271, 276(14), 279, 279(14)
Michel, H., 21
Michels, P. C., 316, 322, 323(28), 324(28)
Middaugh, C. R., 438
Middleton, P., 355
Miercke, L. J. W., 52
Miguel, C., 314
Mikawa, T., 264
Miklos, D., 294
Mikulik, K., 146, 148(11)
Miller, A., 141
Miller, J. F., 318, 319, 319(22), 322, 324(22; 26)
Miller, J. H., 239
Miller, S., 335, 444
Miller, W., 273
Milne, J. S., 341, 344, 345, 346(5), 350
Mims, W. B., 21
Min, T. P., 49
Minami, Y., 9
Minomura, S., 318
Mirambeau, G., 146, 147, 148, 150, 151(17), 152(3; 17), 153(3), 154(17), 155, 156, 156(29), 157(17), 160(33)
Mirsky, A. E., 443
Mishra, N. C., 124
Mitra, A. K., 52
Mitsuhashi, S., 166
Miyagi, M., 255
Mizuuchi, K., 164, 165

Mol, C. D., 425(80), 437
Mombelli, E., 130, 131(12), 143(12)
Monnanni, R., 379, 380(66)
Monod, J., 348(19), 349
Monty, K. J., 178
Moody, P. C., 452, 456(73)
Moore, A., 259
Moore, D. D., 120
Moore, G. R., 367, 368
Moracci, M., 434(43), 435
Morais Cabral, J. H., 163
Moran, L. S., 271, 276(3), 278(3)
Moras, D., 434(46), 435
More, N., 291
Moreau, V., 115, 165, 166, 169(18)
Morel, P., 162
Moreno, S. M. J., 39
Morgan, H. W., 288, 289(12), 302(4), 303
Morikawa, K., 260, 261, 262, 262(8; 9), 268(9), 269(9), 278, 295
Morikawa, M., 261
Morild, E., 321(25), 322
Moritz, P., 313
Moriyama, H., 3(16), 4, 5, 6(16; 25), 8(16), 11(16), 12(16), 13(16), 17(16), 99, 434(42), 435
Mortenson, L. E., 42
Moscoso, M., 193
Moser, M. J., 271, 276(14), 279, 279(14)
Mostaert, A. S., 307
Mouesca, J.-M., 355, 359(25; 26), 377(25)
Moulinier, L., 434(46), 435
Moulis, J.-M., 362, 367
Moura, I., 19, 48, 49(18), 55, 370, 373
Moura, J. J. G., 19, 23, 27, 48, 49(18), 55, 370, 373
Mous, J., 24
Mozhaev, V. V., 316
Mueller, L., 334
Mueller-Dieckmann, H. J., 424(64), 436
Mukhophyay, G., 121
Mukulik, K., 147
Mukund, S., 23, 27(8), 351, 359(2), 424(39), 434, 439
Mullenbach, G. T., 9
Müller, M., 341
Mulliez, E., 217
Mullis, K. B., 92
Münck, E., 20(57), 21, 30, 31, 32(13), 355, 370
Muniz, R. P. A., 39

Murphy, K. P., 330, 341, 411, 414(63), 419(63)
Murthy, M. R. N., 430
Murzin, A. G., 130
Murzina, N. V., 130
Musgrave, D. R., 160, 302(4), 303
Myambo, K., 92, 101, 120
Myers, J. K., 332, 335(21), 336(21)
Myers, T. W., 103, 104(46)

N

Nadal, M., 146, 147, 148, 150, 151(17), 152(3; 17), 153(3), 154(17), 155, 155(5), 156, 156(29; 30), 157(17), 158(6), 159, 159(5), 160(33), 161(39), 162(38)
Nadeau, J. G., 101
Nagahara, Y., 29
Nagahisa, K., 261, 262(9), 268(9), 269(9)
Nagai, Y., 48
Nakamura, T., 239, 246
Nakasu, S., 146, 152(4a), 153(4a), 154
Nakazawa, H., 48
Narszalek, J., 296
Nash, H. A., 165
Nasheuer, H.-P., 259
Natarajan, K., 28
Natario, V., 239
Naumann, D., 329, 335(11), 340, 340(11), 341(11; 30)
Nazabal, M., 92
Neet, K. E., 330, 331(14), 333(14)
Neff, N., 270, 271(1)
Nelson, C. M., 318, 319(22), 322, 324(22; 26)
Nelson, E. K., 147
Nelson, K. E., 252
Nelson, P. E., 316
Nelson, P. J., 227
Nemethy, G., 348(18), 349
Nemoto, N., 4
Neupert, W., 294
Newman, R. C., Jr., 318
Nicholet, P., 289
Nicholson, H., 443
Nicklen, S., 67, 107
Nicolas, A., 160, 172
Nicolaus, B., 306, 314(13)
Nielson, J. P., 157
Nielson, K. B., 91
Nielson, S. O., 346

Niitsu, M., 389
Niki, H., 164
Nikkola, M., 79, 87
Nikolov, D. B., 228
Nilges, M., 63, 75, 79, 87, 87(16)
Nimmesgern, E., 294
Nishimura, Y., 164
Nishino, T., 19
Nishioka, M., 93, 240, 243(13), 244(13), 245(13), 246(13), 247, 278, 425(81), 437, 473, 476(14)
Nobile, V., 63, 74, 76(3; 4), 79(4)
Noda, H., 392
Nogami, S., 274
Noirot-Gros, M. F., 193
Nojima, H., 392
Noll, K. M., 160, 164, 165, 169(13), 194, 446
Nomura, N., 260
Noodleman, L., 355, 359(25; 26), 377(25)
Nordin, J. H., 439
Nordlund, P., 79, 87, 216
Noren, C. J., 270, 271, 271(1), 274, 276(10), 278(10), 279(23), 280(23)
Novick, R. P., 193
Novotny, J., 334
Nowitzki, S., 3, 4(5), 5(5), 7(5), 8(5), 10(5), 14(5)
Nucci, R., 434(43), 435
Nummelin, H., 425(73), 437
Nunes, O. C., 302, 306, 307(14), 310(14)
Nurse, P., 157
Nwankwo, D. O., 271, 276(3), 278(3)
Nyberg, B., 207
Nycz, C. M., 109

O

O'Dea, M. H., 151, 155(22), 158(22), 164, 165
O'Dell, S. E., 51
O'Donohue, M. F., 159
Obmolova, G., 425(68), 437
Obrien, R., 434(44), 435
Oda, N., 280
Oesterhelt, D., 3, 4(5), 5(5), 6(15), 7, 7(5), 8(5), 9, 10, 10(4; 5), 13(28), 14(4; 5), 15, 15(10; 11), 17(4; 10; 11; 35), 20(4)
Ogasahara, K., 424(48), 436, 472, 476, 476(6)

Ogata, K., 424(48), 436, 472, 476(6)
Ogata, M., 55
Ogawa, H., 261, 264
Ogawa, T., 261, 264
Oguchi, A., 48
Ogura, K., 48
Ohba, M., 4
Ohfuku, Y., 48
Öhman, M., 217
Ohmori, D., 3(15), 4, 6(15), 7(15), 8(15), 10(15), 11, 12(15), 13(15), 17(15)
Ohnishi, T., 21
Ohshima, T., 262, 265(15)
Ohya, Y., 274
Oka, M., 93
Oki, H., 424(48), 436, 472, 476(6)
Okuyama, S., 240
Oldenhuis, R., 46
Olin, B., 87
Oliveberg, M., 417
Ollagnier, S., 217
Ollis, D. L., 259
Olsen, G. J., 3, 10(2), 48, 252, 271, 272(11), 273(11), 276(11)
Olsen, J. G., 147
Olsson, M., 239, 242
Oohara, G., 121
Oozeki, M., 3(16), 4, 6(16), 8(16), 9, 11(16), 12(16), 13(16), 17(16)
Oppermann, U. C., 145
Ormö, M., 216, 223(14), 225(14)
Orr, E., 164, 166
Osborne, M. J., 367, 368
Oscarson, J. L., 410
Oshima, T., 3, 3(15–17), 4, 4(9; 12; 13), 5, 5(9; 12; 13; 22), 6, 6(9; 14–16; 25), 7, 7(9; 15; 17; 23; 24), 8(9; 14–17; 23), 9, 10, 10(15; 17), 11(9; 16; 23), 12(14–16; 23), 13(12–17; 24; 27), 14(13), 15, 15(9; 12; 13; 20), 16, 16(13), 17(9; 13; 15; 16), 18(12; 13), 19, 19(12), 20, 20(22–24), 22(20; 22; 55), 284, 389, 392, 434(42), 435, 472, 474(11), 475(11), 476(11)
Oshishi, T., 21
Oshumi, S., 207
Osipiuk, J., 294
Ostendorp, R., 424(56), 436, 439, 442(6), 446, 447, 447(6), 448(62), 450(62–64), 453(6), 458, 458(6), 459(6; 86), 461(85), 468(6), 473, 476(17)

Ostermann, J., 294
Ota, K., 11
Otomo, T., 280
Otsuka, R., 48
Otting, G., 373, 386(58)
Overbeek, R., 3, 10(2), 48, 294, 445
Oyaizu, H., 6

P

Pace, C. N., 136, 331, 332, 335(21), 336(21), 343, 443, 444(41)
Pace, N. R., 87, 261
Pacheco, I., 8(33), 10, 11(33)
Padmanabhan, S., 131, 424(63), 436
Pagani, S., 27
Page, J., 144
Paitan, Y., 166
Palma, P. N., 23, 373
Palma, R., 434
Palmer, G., 359
Pang, G., 48(21), 49
Panne, D., 280
Papaefthymiou, V., 355, 370
Pappenberger, G., 456
Parales, R. E., 21
Park, D., 261
Park, J. B., 56, 325, 473, 475(19), 476(19)
Park, J.-B., 17, 27, 31, 32, 32(6; 13), 34(6), 37(6), 46, 48, 48(5; 20), 49, 49(11; 15–19), 50(5), 51(5), 54(5), 351, 354(5), 355(5), 377(5), 379(5), 388, 393, 423, 424(1; 2), 435, 435(1), 439
Parrish, R. G., 428
Parthasarathy, S., 430
Pasquo, A., 325, 424(41), 434, 461
Pauling, L., 443
Paulus, H., 271
Pavletich, N. P., 425(78), 437
Pearl, L. H., 434(43), 435
Peck, H. D., Jr., 55
Peck, L. J., 153
Peebles, C. L., 164
Peng, C. Y., 355, 359(26)
Pereira, S. L., 116, 117, 117(1), 124(3), 126(31; 32), 127
Perham, R. N., 442
Perl, D., 464, 465, 465(103), 468(103)

Perler, F. B., 91, 92(9), 93, 93(9), 94(31), 95, 95(9), 100, 109(9; 31), 114(9), 116(9), 270, 271, 271(1), 272(2; 11), 273(11), 274, 274(4; 5), 276(2; 3; 10; 11), 278(2–4; 10), 279(23), 280, 280(5; 23)
Perlman, P. S., 278, 279(30)
Perrakis, A., 429
Persson, A. L., 225
Perutz, M. F., 443, 454(34)
Petach, H. H., 291, 478
Peters, W., 389
Peterson, J., 373
Pétillot, Y., 8(34), 9, 10, 11(34)
Petitdemange, H., 56, 61(5)
Petrach, H. H., 284, 291(3), 292(3)
Petratos, K., 79, 87
Petsko, G. A., 340, 349, 434, 439, 443(10), 469, 471, 472, 473, 474(11), 475, 475(11), 476(11)
Pfanner, N., 294
Pfeifer, F., 10
Pfeil, W., 28, 29, 335, 341, 351, 383(3), 439, 442, 472, 476(7)
Phillips, G. N., 429
Phillips, W. D., 359
Phipps, B. M., 294
Picard, D., 99, 102(42)
Piccioli, M., 7, 8(31), 356, 359, 368(37), 377(40)
Pickering, I. J., 48(22), 49
Pierik, A. J., 22
Pietrokovski, S., 271, 272(12; 13), 273(12; 13), 276(12; 13)
Pieulle, L., 16, 16(46), 17
Pinkau, T. J., 294
Pinsonneault, J. K., 93, 103(32), 114(32)
Pitzer, K. S., 390, 400(11), 403(11)
Place, A. R., 228
Plevani, P., 259
Plotnikov, V., 394
Pochet, S., 120
Poe, M., 359
Polesky, A. H., 101
Pontis, E., 217, 218, 222(31), 226, 227(43)
Portemer, C., 146, 147, 148, 148(9; 10), 150, 150(10), 151(17), 152(17), 154(17), 157(10; 17), 158(10), 160, 164, 180
Porter, D., 207, 212(9)
Porter, R. D., 332
Pospiech, H., 259
Post, C. B., 345

Potekin, S., 398
Pötsch, S., 225
Prade, L., 424(56), 436, 458, 459(86), 473, 476(17)
Prehoda, K. E., 344
Press, W., 141, 408
Prieur, D., 193, 195(1)
Priev, A., 326
Prilusky, J., 423
Prince, R. C., 22, 48(22), 49
Privalov, G., 343, 394, 398, 439
Privalov, P. L., 390, 392, 393, 393(22), 394, 396, 396(14), 398, 400, 403(12; 14), 417(14), 420(14), 439, 444(16)
Prunell, A., 159
Przybyla, A. E., 27
Pugh, B. F., 264

Q

Qiang, B., 271, 276(3), 278(3)
Qin, J., 79, 84(18; 19)
Qiu, L., 130, 132
Quaite-Randall, Q., 294, 295(5)
Querellou, J., 252
Quiocho, F. A., 274
Qureshi, S. A., 229

R

Rabinowitz, J. C., 9
Rachel, R., 205, 302, 440, 457, 458(82), 468(18)
Radding, C. M., 264
Ragsdale, S. W., 18
Raidt, H., 443, 454(34)
Raine, A. R., 130
RajBhandary, U. L., 194
Ramakrishnan, V., 306, 312(19)
Ramaswamy, S., 21, 216, 226(13), 429
Ramos, A., 312
Randall, M., 390, 400(11), 403(11)
Randall, R. J., 26
Rashid, N., 261, 262, 262(8; 9), 268(9), 269(9), 295
Rashtchian, A., 102
Ratcliffe, H. D., 288
Raven, N., 306, 307, 310(27), 312
Reddy, T., 128(8; 9), 130

Redfield, R. R., 144
Reece, R. J., 163
Rees, D. C., 32, 46, 48, 54(14), 373, 381(59), 386(59), 388, 392, 423, 424(1; 37; 39; 67), 427, 429, 431, 434, 435, 435(1), 437, 444, 473, 475(19), 476(19)
Reeve, J. N., 116, 117, 117(1), 118, 118(8), 119(10), 120(10; 11), 121, 121(2; 4; 8), 123(4), 124(3), 126(31; 32), 127, 128(31), 160, 389, 393(5), 403(5), 418(5)
Regalla, M., 9
Regnström, K., 216, 226(13)
Rehaber, V., 440, 454(19), 467(19), 468
Rehrauer, W. M., 265, 266(22)
Reichard, P., 215, 216, 216(2), 217, 217(2), 218, 219(2; 30), 220(2), 222(31), 225(2), 226, 226(30), 227(30; 43)
Reinemer, P., 85, 87
Reinhardt, R., 129, 129(6; 7), 130, 130(1; 2), 138(6; 7), 139(1–4)
Reisler, E., 331
Reiter, W.-D., 66, 156, 160(33)
Remington, S. J., 424(51), 436
Ren, B., 63, 74, 77, 77(5), 78, 82, 84(28), 85, 87, 424(59), 436
Revel-Viravau, V., 166, 169(18)
Revet, B., 159
Reysenbach, A.-L., 302(4), 303
Rhee, S. G., 85, 87
Ricchetti, M., 107
Riccio, R., 306, 314(13)
Rice, D., 460, 461(89), 462(89; 90), 471, 474, 476(5), 477(21)
Rice, D. W., 424(41; 65), 425(70), 434, 436, 437, 461, 474
Rice, P. A., 259
Richards, F. M., 144, 434
Richardson, C. C., 101, 105
Richardson, J. S., 372
Richie, K. A., 46, 49, 51(2; 32), 54(32)
Richon, V. M., 425(78), 437
Riera, J., 216, 217(7), 218(7)
Rietveld, A. W., 330
Rifkind, R. A., 425(78), 437
Ringe, D., 434, 472, 474(11), 475(11), 476(11)
Rinker, K. D., 446
Rishi, V., 341
Ritter, O., 423
Rius, G., 355, 359(25), 377(25)
Rivera, M. C., 181

Robb, F. T., 48, 147, 148, 216, 217(7), 218(7), 228, 316, 325, 326(36), 393, 424(65), 431, 436, 474
Roberts, M. F., 304, 306, 306(10), 307, 308, 308(10; 18), 313, 314
Roberts, R. J., 272, 276(15), 278(15), 279(15)
Roberts, V. A., 40
Robertson, A. D., 408
Robertson, D. E., 21, 304, 306, 308(18), 425(77), 437, 475, 476(29)
Robins, M. J., 225
Robins, P., 240, 242
Robinson, D. K., 408
Robinson, H., 131, 414, 424(54; 63), 436
Robson, R. L., 51
Roca, J., 146, 152(4c), 153(4c)
Roder, H., 346, 350
Rodgers, D. W., 429
Rodrigues-Pousada, C., 55
Roefs, S. P., 341
Rogero, J. R., 345
Rojo, F., 193
Rolfe, M., 180
Romão, M. J., 19
Ron, E. Z., 166
Ronchi, S., 130, 131(11), 143(11), 144(11)
Ronimus, R. S., 302(4), 303
Roovers, M., 328, 424(5), 426
Rosato, A., 353, 356, 368(15; 37), 385(15)
Rösch, P., 23, 27, 29, 351, 354(11), 368, 368(11), 374(55), 381, 381(11; 55), 382, 382(11), 386(11), 388(11)
Rose, G. D., 466
Rose, J. P., 259
Rose, K., 48(23), 49
Rosenberg, A. H., 133, 295
Rosenberg, E., 166
Rosenbrough, N. J., 26
Rossi, M., 62, 63, 66, 74, 76(3; 4), 77, 77(5), 78, 79(4), 82, 87, 130, 132(13), 312, 424(59), 429, 434(43), 435, 436
Roth, M., 29, 351, 367(7), 380(7), 381(7)
Roth, R. A., 63, 75, 87(14)
Rothstein, R., 146(13), 148, 157, 162(13), 180
Rouviere, P., 181
Roy, R., 32, 373, 381(59), 386(59), 424(67), 437
Royer, C. A., 316, 327, 434
Ruddock, L. W., 70, 71(25), 80
Ruger, R., 105
Ruile, P., 446

Ruiz-Echevarria, M. J., 193
Rush, J. D., 355
Russell, A. J., 316
Russell, R. J., 424(50), 431, 436, 447, 476, 477(32)
Rutkat, K., 457, 458(82)
Rutter, W. J., 63, 75, 87(14)
Ruttersmith, L. D., 290, 291(14)
Ruysschaert, J. M., 345
Ryu, S.-E., 85, 87

S

Sacchettini, J. C., 274
Saeki, K., 7, 23, 27(1), 30
Saenger, W., 340, 341(30), 443, 444(41)
Sagner, G., 105
Sahlin, M., 217, 225
Saibil, H., 294
Saiki, R. K., 67, 92
Saito, H., 11
Sakai, M., 48, 476
Sakamoto, Y., 11
Sako, Y., 260
Sakoda, H., 298
Sakumi, K., 239
Sali, A., 466
Samano, V., 225
Sambrook, J., 65, 66(18), 121, 123(23), 129(23), 134
Samejima, K., 389
Sanborn, B. M., 207, 212(9)
Sanchez-Ruiz, J. M., 329, 337(9), 392, 404, 405(19)
Sanderson, I., 434(43), 435
Sandler, S. J., 261, 262, 265(13)
Sandman, K., 116, 117, 117(1), 118, 118(8), 119(10), 120(10; 11), 121(2; 8), 126(31), 127, 128(31), 160, 389, 393(5), 403(5), 418(5)
Sands, R. H., 359
Sandström, J., 366
Sanger, F., 67, 107
Santoro, M. M., 334
Santos, H., 23, 56, 62(6), 302, 303, 304(8), 306, 307, 307(14–17), 308(16), 310(14; 17; 23; 27), 312, 314, 356, 389, 468
Satake, T., 41
Sato, K., 166

Sato, Y., 94, 116(40), 249, 251(3), 252(3), 253(3), 256(3)
Satoh, M., 55
Satow, Y., 274
Sauer, K., 21
Sauer, R. T., 341
Sauer, U., 443
Savelyeva, N. D., 146, 147, 148(11)
Savitsky, A., 136
Sawada, M., 48, 147, 252
Saynovits, M., 21
Scandurra, R., 424(41), 434, 461
Schachman, H. K., 28
Schaefer, G., 424(60), 436
Schäfer, G., 8(34), 9, 10, 11(34), 22, 132, 141(25), 144(25), 328, 329, 330(6), 332, 333(6), 334(6), 335(11; 19), 336(6), 340(11), 341(6; 11; 19), 425(69; 73), 437, 463, 474, 475(23), 476(23)
Schäfer, W., 307, 312(21)
Schaffer, A. A., 273
Schäffer, J., 85
Schägger, H., 36, 134
Scharf, S. J., 92
Scheer, H., 292
Schellman, J., 390, 443
Schellman, J. A., 348(17), 349
Schimpff-Weiland, G., 74, 76(1)
Schindler, T., 464, 465, 465(103), 468(103)
Schleper, C., 261
Schlicht, F., 74, 76(1)
Schmid, F. X., 329, 331(10), 334, 464, 465, 465(103), 468(103)
Schmidt, C. L., 22
Schmidt, P. P., 217
Schmitt, E., 434(46), 435
Schneider, M., 19, 429
Schofield, C. J., 429
Scholtz, J. M., 332, 335(21), 336(21), 403
Scholz, S., 307, 312(21), 314
Schönheit, P., 23
Schoonover, J. R., 21
Schramel, A., 297
Schröder, C., 23
Schröder, K., 464, 465(103), 468(103)
Schroder, K. L., 92
Schultes, V., 349, 393, 452, 453(74), 454(74), 458(75), 461(75), 467
Schultz, C., 340, 341(30)
Schultz, S. C., 259

Schulz, G. E., 424(60), 436, 474
Schumacher, S., 193
Schumann, J., 446, 468(53)
Schurig, H., 328, 424(49), 436, 439, 442, 442(6), 446, 447, 447(6), 448, 448(62), 449(69), 450(25; 62; 63; 69), 452, 453(6), 456, 457, 458, 458(6; 82), 459(6), 461(85), 462, 463(93), 464, 465(100), 468(6), 476
Schurmann, P., 63
Schuster, D. M., 102
Schut, G. J., 38, 41, 43, 50, 56(7), 57
Schut. G. J., 33, 34(37)
Schwander, M., 425(68), 437
Schweiger, A., 349, 393, 452, 453(74), 454(74)
Scopes, R. K., 284, 290, 290(2)
Scott, R. A., 3(17), 4, 7(17), 8(17), 9, 10(17), 13(17), 46, 48, 48(23), 49, 51, 51(32), 54(14; 32), 424(37), 434
Scozzafava, A., 379, 380(66)
Scudiero, D., 239
Searcy, D., 229
Searle, B. G., 48(20; 21), 49
Seckler, R., 467
Sedelnikova, S. E., 325, 424(41), 434, 461, 474
Sedgwick, B., 240
Seely, R. J., 308
Segerer, A., 205, 302
Seidman, J. G., 120
Seitz, E. M., 262, 265(13)
Seki, S., 55
Seki, Y., 55
Sekiguchi, M., 239, 242(8), 246
Sekine, M., 48
Selig, M., 23
Sellman, E., 92
Serrano, L., 443
Serre, L., 29, 351, 367(7), 380(7), 381(7)
Séry, A., 29, 351, 367(7), 380(7), 381(7)
Setlow, J. K., 165
Seville, M., 93, 116(24)
Shadle, S. E., 48(23), 49
Shah, N. N., 318, 319, 319(22), 322, 324(22; 26)
Shand, R. F., 52
Shank, D. D., 101
Shao, Y., 271
Sharma, S., 274
Sharp, R. J., 312
Shaw, W. V., 193
Shen, B. H., 425(80), 437
Shergill, J. K., 21

Sherman, F., 120
Shibata, T., 148, 154, 161(15)
Shikotra, N., 163
Shilling, J., 294
Shima, S., 312, 424(53), 425(79), 436, 437
Shimizu, F., 55
Shinagawa, H., 278
Shinohara, A., 261
Shirotani, I., 318
Shnyrov, V. L., 404
Shoemaker, D. D., 93, 114(26), 115(26), 255
Shore, D., 123
Short, J. M., 48, 93, 114(26), 115(26), 255
Shrader, T. E., 127
Shriver, D. F., 426
Shriver, J. W., 118, 119(10), 120(10), 129, 130, 130(5), 131, 132, 132(15), 135(5), 139(16), 140(5; 16), 141, 141(5; 16), 142, 143, 144, 329, 335(12), 389, 392(4), 393, 393(4; 5), 400(31), 403(5; 31), 408(4), 414, 414(31), 418(4; 5), 419(4; 31), 420(31), 421(31), 424(54; 63), 436
Sieber, V., 465
Siegel, L. M., 178
Sieker, L. C., 12, 29, 351, 367, 367(6), 372(6), 379(6), 380(6), 381(6), 382(6), 386(6), 388
Siezen, R. J., 474, 475(24)
Sigler, P. B., 227, 228, 424(47), 434(44), 435
Sigurskjold, B. W., 412, 413(64)
Silva, D. J., 225
Silva, G. H., 474
Silva, J. L., 316
Silva, Z., 306, 307(15)
Simon, M. I., 424(51), 425(71), 436, 437
Singh, N., 325
Singh, P. B., 130
Singleton, M. R., 424(55), 436
Sinning, I., 87
Sivaraja, M., 21
Sjöberg, B.-M., 83, 215, 216, 217, 217(3), 220, 223(14), 225, 225(14), 226(13)
Skarzynski, T., 452, 456(73)
Skelton, N. J., 353, 362(14), 363(14), 364(14)
Slatko, B. E., 271, 276(3), 278(3)
Slesarev, A. I., 146, 147, 148, 149(19), 150(19), 151, 152(8), 155, 156(23), 178, 179, 180, 181, 181(2), 187, 187(13)
Sleytr, U. B., 160, 164, 438, 440(4), 442(4)
Slupphaug, G., 99, 102(43)
Smeller, L., 432

Smith, C. V., 163
Smith, D. B., 150
Smith, D. L., 345
Smith, D. R., 252
Smith, E. T., 23, 27, 31, 32(9; 12; 16; 17), 36(5), 49, 354, 359, 360, 368(43), 371(43), 373(43), 381, 385(43; 68), 386(43), 388(68)
Smith, J. A., 120
Smith, J. L., 21
Snead, M. A., 48, 445
So, A. G., 259
Soares, D., 116
Sobel, B. E., 45
Sobol, A. G., 362
Söderberg, B. O., 75, 79, 87
Solomon, E. I., 48(23), 49
Solomon, M. J., 124, 127
Soltis, S. M., 429
Sonnenbichler, J., 307, 312(21)
Sonnerstam, U., 79, 87
Soppa, J., 228
Sorge, J. A., 93, 114(26), 115(26), 255
Sosnick, T. R., 350
Sougakoff, W., 165, 166, 169(18)
Sousa, R., 259
Southworth, M. W., 93, 94(31), 100, 109(31), 271, 274, 274(4), 278(4), 280
Spargo, C. A., 109
Spears, P. A., 109
Spicer, L. D., 52
Spies, M., 262, 265(15)
Spikes, D., 165
Spiliotis, E. T., 314
Srivastava, K. K. P., 31, 32(13)
Stalhandske, C. M. V., 3(17), 4, 7(17), 8(17), 9, 10(17), 13(17)
Stalon, V., 328, 424(5), 426
Stankovich, M. T., 40
Staples, C. R., 32, 33, 34(32; 33), 36(33), 37(33), 40(32; 33)
Starich, M. R., 117
Stasiak, A., 261
Staudenbauer, W. L., 164, 166
Steif, C., 406, 407(53)
Steipe, B., 424(38), 434, 453, 454(76)
Steitz, T. A., 91, 92, 94, 101, 259, 262
Stelter, K. O., 147
Stemflinger, I., 446
Stemmer, W. P., 469
Stephens, P. J., 356

Sterner, R., 23, 24, 26, 26(11), 27, 28, 28(11), 29, 30(29; 30), 351, 354(11; 12), 367(12), 368, 368(11), 372(12), 374(55), 379(12), 381(1; 12; 55), 382, 382(11; 12), 383(3; 12), 384(12), 386(11; 12), 388, 388(11; 12), 424(40), 425(68), 434, 434(45), 435, 437, 442, 444, 463, 464, 472, 473, 474(8; 16), 475(8; 16), 476(7; 16)
Sternlicht, H., 294
Sternlicht, M. L., 294
Stetter, K. O., 3, 9, 10(3), 23, 87, 146, 147, 148, 149(19), 150(19), 160, 164, 173, 179, 180, 181, 181(2), 187(15), 205, 294, 302, 306, 307, 307(16), 308, 308(16; 18), 389, 438, 440, 440(4), 442(4), 446, 458, 468(18; 53)
Stevens, L., 286, 287, 288(7)
Stewart, D. E., 49, 55
Stewart, J. W., 120
Steytr, U. B., 23
Sticht, H., 23, 29, 351, 354(11), 368(11), 381, 381(11), 382, 382(11), 386(11), 388(11)
Stillinger, F. H., 444
Stillman, T. J., 325, 424(41; 65), 434, 436, 461, 474
Stoddard, B. L., 272, 276(16), 278(16), 279(16)
Stoffel, S., 92
Stoll, V. S., 208, 212(18)
Story, R. M., 262
Stott, F. J., 130
Stout, C. D., 40, 367
Stowell, M. H. B., 429
Strambini, G. B., 326
Stratagene, 94, 110(33)
Straume, M., 141, 392, 414, 422(68)
Strauss, B., 239
Streett, W. B., 319
Struhl, K., 120
Stubbe, J., 215, 216, 217, 223, 223(15; 19; 20), 224, 224(19; 37), 225, 225(15)
Stüber, D., 24
Studier, F. W., 295
Studier, W., 133
Sturtevant, J., 399, 401(42)
Su, S. Y., 131, 424(63), 436
Sugino, A., 164, 165, 259
Suh, S. W., 92, 259
Sukumar, S., 239
Summers, M. F., 46, 48, 48(5), 49(11; 15–17), 50(5), 51(5), 54(5), 56, 117, 388, 423, 424(2)

Sun, M. M. C., 316, 325, 326(36), 327
Sun, X., 217
Sun, Y., 259
Sung, H. C., 425(83), 437
Surerus, K. K., 31, 32(13), 355, 370
Suryanarayana, T., 128(8; 9), 130
Sussman, J., 423, 425, 430, 431
Suzuki, T., 3(15), 4, 6(15), 7(15), 8(15), 10(15), 12(15), 13(15), 17(15), 164, 207, 459
Svingor, A., 340, 349
Svingor, R., 475
Swack, J. A., 105
Swanson, R. V., 48, 424(51), 425(77), 436, 437, 475, 476(29)
Swartz, P. D., 49
Sweet, R. M., 427, 429, 429(9)
Swint-Kruse, L., 408
Syvaoja, J. E., 259
Szadkowski, H., 442, 463
Szary, N., 389

T

Ta, T. T., 27
Tabor, S., 101, 105
Tafazzul, G., 313
Tahirov, T. H., 424(48), 436, 472, 476(6)
Tainer, J. A., 425(80), 437
Takagi, M., 239, 240, 243(13), 244(13), 245(13), 246(13), 247, 278, 293, 294, 295, 295(6; 7), 296(6), 300(6), 425(81), 437, 473, 476(14)
Takahashi, M., 155, 156(29)
Takamiya, M., 48
Takano, K., 246
Takegi, M., 93
Takenaka, A., 5, 6(25)
Takeya, H., 239, 242(8)
Talfournier, F., 425(82), 437
Tan, C. K., 259
Tanaka, K., 3, 3(16), 4, 4(9), 5, 5(9), 6(9; 14; 16; 25), 7(9), 8(9; 14; 16), 10, 11(9; 16), 12(14; 16), 13(14; 16), 15(9), 17(9; 16), 48, 99
Tanaka, N., 284, 434(42), 435
Tanford, C., 332, 443, 444
Tang, K. E., 476
Taniguchi, Y., 323
Tanner, J. J., 472, 474(10), 476(10), 477(10)
Tarr, G. E., 20(57), 21
Tateishi, S., 264

Tatusov, R. L., 444
Tauer, A., 217, 218(18)
Taylor, G. L., 424(50), 431, 436, 447, 476, 477(32)
Tedeschi, G., 130, 131(11), 143(11), 144(11)
Teixeira, M., 7, 8(31; 33), 9, 10, 11(33), 22, 55
Telenti, A., 274
Telser, J., 31, 32(9; 12), 354, 354(21), 355(21), 377(21), 379(21)
Teng, Q., 31, 32(17; 18; 20), 38(19), 46, 51(2), 52, 351, 352, 352(24), 354, 354(10), 358, 358(10), 363, 367(48), 368(9; 10; 24; 48), 369, 369(10), 370, 370(10; 24), 371, 371(9; 10; 24), 372, 372(24), 373(24; 48), 374, 374(9; 48), 375(9), 377(9), 378, 381, 381(24), 384(24), 385(9; 24; 68), 386(24), 387, 387(24), 388(10; 24; 68)
Terayama, Y., 318
Terazawa, T., 41
Terrell, C. A., 63, 74, 76(2), 80(2)
Terwisscha van Scheltinga, A. C., 429
Tetas, M., 288
Teukolsky, S., 141, 408
Teyssier, C., 160
Thakur, B. R., 316
Thanki, N., 328, 330(1), 448, 450(67)
Thauer, R. K., 312, 424(53), 425(79), 436, 437
Thelande, L., 218
Thelander, L., 220
Then, T., 41
Thi, M. H. D., 448, 450(67)
Thierry, J. C., 434(46), 435
Thilly, W. G., 115
Thoma, R., 425(68), 437, 442, 464
Thomas, C. D., 193
Thomas, J. O., 294
Thomas, T. M., 284, 290, 290(2)
Thomm, M., 23, 160, 164, 206, 438, 440(4), 442(4)
Thompson, A., 429
Thompson, M. J., 469
Thomson, A. J., 8(34; 37), 9, 10, 11, 11(34)
Thorner, J., 270, 271(1), 279
Thornton, J. M., 83
Tibbelin, G., 74, 77, 77(5), 78, 82, 87, 424(59), 425(69), 436, 437, 474, 475(23), 476(23)

Tibellin, G., 63
Tiktopulo, E., 396
Tilly, K., 293(1), 294
Tilton, R. F., 434
Timm, D. E., 330, 331(14), 333(14)
Tinoco, I. J., 142
Tjeng, L. H., 48(21), 49
Tocchini-Valentini, G. P., 163
Toh, H., 87, 93, 249, 250, 254(5), 259(5), 260
Tolliday, N., 325, 326(36), 424(65), 436, 474
Tollin, G., 40
Tolman, C. J., 313
Tomazic, S. J., 471
Tomb, J. F., 147, 252
Tomizawa, J. I., 164
Tomschy, A., 424(38), 434, 453, 454, 454(76), 456(78), 457
Toomey, N. L., 259
Torrents, E., 218, 219(30), 226(30), 227(30)
Tortora, P., 130, 131(11; 12), 143(11; 12), 144(11)
Tosco, A., 70
Totty, N., 240
Toulme, F., 160
Touzel, J. P., 160
Trachsel, H., 157
Trautwein, A. X., 22
Travers, F., 327
Trent, J. D., 294, 295(5)
Tricot, C., 328, 424(5), 426
Trotta, C. R., 424(58), 436
True, A. E., 21
Trumpower, B. L., 21
Truong, Q. C., 165, 166, 169(18)
Trüper, H. G., 41
Trust, T. J., 207
Tse, Y. C., 156, 180
Tsuda, M., 318
Tsukihara, T., 29, 351, 367(8), 380(8), 381(8), 424(48), 436, 472, 476(6)
Tsunasawa, S., 120, 255, 424(48), 436, 472, 476, 476(6)
Tu, Y. G., 475, 476(28)
Tuite, M. F., 75
Turano, P., 356, 357(33), 358(33), 361(33), 362(33), 374(33), 377(33)
Turner, D., 356
Typke, D., 294
Tyrshkin, A. M., 21

U

Uegaki, K., 280
Ueki, T., 4
Uemori, T., 92, 93, 94, 115(17), 116(40), 249, 250, 250(2), 251(3), 252(3), 253(3), 255, 256(3), 260
Ueno, T., 253, 255
Uhlin, T., 79, 87, 216, 226(13)
Uhlin, U., 216, 222(10)
Urbanke, C., 332, 335(19), 341(19)
Ursby, T., 424(66), 437
Urushiyama, A., 3(15), 4, 6(15), 7(15), 8(15), 10(15), 11, 12(15), 13(15), 17(15), 20, 22(55)
Usher, K. C., 424(51), 436
Uyemura, D., 108
Uzawa, T., 4

V

Vakhitov, V. A., 92
Valentine, R. C., 187
Vallini, G., 379, 380(66)
van Alebeek, G.-J. W., 313
Van Beeumen, J., 328, 424(5), 426
Van Cleve, M., 109
van den Bosch, M., 54
van den Burg, B., 440
Van der Donk, W., 224
van der Drift, C., 313
van der Linden, M. P., 46
van der Oost, J., 312, 460, 462(90), 471, 476(5)
van der Vies, S. M., 293(1), 294
van Elp, J., 48(20; 21), 49
Van Roey, P., 474
Vanoni, M., 130, 131(12), 143(12)
Varoutas, P. C., 160, 172
Varshavsky, A., 124, 127
Vaupel, M., 425(79), 437
Veltman, O. R., 440
Venema, G., 440
Venters, R. A., 52
Ventosa, A., 302
Venyaminov, S., 390, 403(12)
Verhagen, M. F., 16, 32, 33, 34(31; 37), 36(31), 37(31), 39(31), 40(31), 45
Verhagen, M. F. J. M., 30, 31, 43, 50, 56, 56(7), 57, 61(4), 62(4), 306, 312(19)

AUTHOR INDEX

Vernède, X., 429
Vessey, K. B., 207
Vetriani, C., 325, 326(36), 424(65), 436, 474
Vetterling, W., 141, 408
Vickery, L. E., 27
Vidugiris, G. J. A., 316, 434
Vieille, C., 388, 389
Viezzoli, M. S., 7
Vihinen, M., 259
Vila, A. J., 356, 357(33), 358(33), 359, 361(33), 362(33), 374(33), 377(33; 40)
Villeret, V., 328, 424(5), 426
Vincent, V., 221
Vincenzini, M., 7
Vinograd, J., 205
Vitoux, B., 425(82), 437
Vlassi, M., 475
Vogels, G. D., 313
Vogt, A.-B., 313
Vogt, G., 328
Volbeda, A., 16, 16(46), 17
Volden, G., 99, 102(43)
Volker, S., 294
Volkin, D. B., 438
Völkl, P., 302
Volkman, B. F., 356
von Hippel, P., 140, 413
von Jagow, G., 22, 36, 134
Vonrhein, C., 424(60), 436, 474
Voorhorst, W. G., 474, 475(24)
Vorholt, J. A., 425(79), 437
Vriend, G., 440
Vuilleumier, S., 417

W

Wada, A., 121, 401
Wada, K., 7, 9, 13(28)
Wagner, G., 346
Wahl, A. F., 259
Wait, R., 306, 307(15)
Wakabayashi, S., 7, 9, 13(28), 55
Wakagi, T., 3, 3(16), 4, 4(9; 12; 13), 5, 5(9; 12; 13), 6(9; 14; 16; 25), 7, 7(9; 24), 8(9; 14; 16), 9, 10, 11(9; 16), 12(14; 16), 13(12–14; 16; 24; 27), 14(13), 15, 15(9; 12; 13), 16, 16(13), 17(9; 13; 16), 18(12; 13), 19(12), 20(24), 434(42), 435
Waksman, G., 92

Walegård, K., 429
Walker, G. T., 101, 109
Walker, J. E., 452, 454(72)
Wall, J. S., 294
Wallis, J. W., 180
Wallon, G., 472, 474(11), 475(11), 476(11)
Walsh, K. A. J., 288, 289(12)
Walter, S., 334
Wampler, J. E., 49
Wand, A. J., 350
Wang, A. H.-J., 131, 414, 424(54; 63), 436
Wang, B.-C., 259
Wang, C., 410
Wang, J., 92, 259, 274, 279(23), 280(23), 344
Wang, J. C., 146, 152(4b), 153, 153(4b), 156, 158, 162, 172, 176, 179(7), 180, 193
Wang, P., 410
Wang, P.-L., 31, 52, 352, 352(24), 354, 356(19), 368(19; 24), 370(24), 371, 371(24), 372, 372(24), 373(24), 379(19), 381(19; 24), 382(19), 383, 383(19), 384, 384(24), 385(24), 386, 386(19; 24), 387, 387(24), 388(19; 24)
Wang, T. S., 91
Wang, T. S.-F., 259
Ware, J., 93, 94(31), 100, 109(31)
Warner, A., 474, 475(24)
Warren, G. L., 473
Warren, P. V, 48, 147, 148, 160(14), 445
Warshel, A., 356
Wassenberg, D., 446
Watenpaugh, K. D., 388
Waterfield, M. D., 240
Waters, M., 454
Watson, R., 205
Weber, G., 316
Weber, P., 406, 407(53)
Weber-Main, A. M., 40
Weigel, J. A., 289, 375
Weinhold, F., 356
Weinstock, G. M., 261
Weiss, R. L., 4, 48, 132, 216, 217(7), 218(7)
Welker, C., 464, 465(100; 103), 468(103)
Wells, W. W., 74
Wendel, A., 84(26; 27), 85
Wendler, I., 24
Wenzel, H., 475
Wernstedt, C., 218, 219(30), 226(30), 227(30)
West, J. A., 307
West, S. C., 261

Westermann, P., 92
Westler, W. M., 356
Wever, I. T., 262
Wheelis, M. L., 3, 87
When, O., 147
Whitby, F. G., 425(74), 437
White, J. H., 151
White, O., 147, 157, 252
White, R. H., 308
Wich, G., 446
Widom, J., 127
Wiegel, J., 88
Wiener, M. C., 429
Wierenga, R. K., 328, 330(1), 427, 448, 450(67)
Wildegger, G., 27, 29, 351, 354(11), 368, 368(11), 374(55), 381(11; 55), 382, 382(11), 386(11), 388(11)
Wilkens, S. J., 356
Willems, J.-P., 223, 224(37)
Williams, K. S., 274
Willison, K., 294
Williston, S., 409, 410(61), 411(61)
Wilmanns, M., 425(68; 75), 437
Wilson, K. S., 367
Wilson, S. H., 105
Wiltholt, B., 46
Winterhalter, C., 446, 468(54)
Wiseman, T., 409, 410(61), 411(61)
Wittershagen, A., 425(79), 437
Wittmann-Liebold, B., 129(6; 7), 130, 138(6; 7)
Woell, S., 328
Woese, C. R., 3, 10(2), 23, 87, 160, 164, 181, 438, 440(4), 442(4)
Wolff, S., 314
Wolffe, A. P., 127
Wonacott, A. J., 452, 456(73)
Wong, J. H., 62(3), 63
Woo, D., 427
Woodruff, W. H., 21
Woods, W. G., 262
Woolford, C., 293(1), 294
Wotkowicz, C., 274
Wray, V., 373
Wrba, A., 349, 393, 446, 452, 453(74), 454(74), 458, 468(53)
Wright, D. J., 109
Wright, G. E., 102
Wu, H. Y., 146
Wuthrich, K., 346

Wüthrich, K., 353, 363(13), 367(13), 373, 386(58)
Wyman, J., 348(19), 349, 414
Wyns, L., 328, 330(1; 6), 333(6), 334(6), 336(6), 341(6), 448, 450(67), 463

X

Xavier, A. V., 55, 56, 62(6), 356
Xavier, K. B., 23
Xia, B., 40, 356
Xie, D., 335
Xu, L., 261
Xu, M.-Q., 271, 274, 274(4; 5), 278(4), 280(5)
Xun, L., 21
Xuong, N. G., 259

Y

Yaffe, M. B., 294
Yagi, T., 55
Yamagishi, A., 4, 6, 284
Yamamoto, S., 48
Yamamoto, T., 300
Yamazaki, J., 48
Yamazaki, T., 280
Yan, Z., 294, 295(6), 296(6), 300(6)
Yang, A.-S., 416, 417(69)
Yang, C.-H., 85, 87
Yang, J. T., 136
Yang, J.-Y., 27
Yang, S.-S., 26
Yang, Y., 74
Yasuda, M., 6
Yasui, K., 154
Yasukawa, T., 300
Yedgar, S., 326
Yeh, A. P., 435
Yeh, T., 363, 367(48), 368(48), 373(48), 374(48)
Yeh, Y.-H., 31, 32(18), 374
Yip, K. S. P., 325, 424(41; 65), 434, 436, 461, 474
Yitani, K., 472, 476(6)
Yokota, H., 424(61; 64), 436
Yoshida, T., 20(57), 21
Young, G. W., 147
Young, P., 217
Young, R. A., 156

Young, W. G., 48, 445
Yu, C.-A., 21
Yu, G. X., 216, 223, 223(15), 225(15)
Yu, J. S., 194
Yu, L. P., 31, 32(16), 359, 360, 368(43), 371(43), 373, 373(43), 381(59), 385(43), 386(43; 59)
Yu, Y. G., 425(76; 83), 437
Yuasa, T., 240, 243(13), 244(13), 245(13), 246(13)
Yutani, K., 424(48), 436, 476

Z

Zaiss, K., 424(49), 436, 442, 451(26), 452, 476
Zaitsev, E. N., 261
Zaitseve, E. M., 261
Zaitzev, D. A., 180
Zak, P., 239
Zale, S. E., 330
Zarembinski, T. I., 424(64), 436
Zavitz, K. H., 157
Zavodszky, P., 340, 349, 393, 452, 453(74), 454(74), 475
Zeelen, J. P., 328, 330(1), 448, 450(67)
Zegar, I., 130, 139(16), 140(16), 141(16), 142, 143
Zeikus, J. D., 298, 388
Zeikus, J. G., 389
Zhang, H., 21
Zhang, J., 273
Zhang, P., 87, 259
Zhang, Q., 3, 4(13), 5(13), 13(13), 14(13), 15(13), 16, 16(13), 17(13), 18(13)
Zhang, S.-H., 259
Zhang, X. X., 410
Zhang, Z., 273, 345
Zhao, Z.-H., 351, 368(9), 371(9), 374(9), 375(9), 377(9), 378, 385(9)
Zheng, D., 294
Zhou, J. H., 363, 367(48), 368(48), 373(48), 374(48)
Zhou, J. Q., 259
Zhou, L., 147, 252
Zhou, Y., 259
Zhou, Z. H., 17, 23, 27, 31, 32, 32(16–18; 20), 34(30; 31), 36(5; 30; 31), 37(31), 38(19), 39(30; 31), 40(30; 31), 41, 45, 48, 48(21; 22), 49, 49(11; 16; 17; 19), 54(27), 347, 348, 349, 350, 351, 353, 354, 354(10), 356(19), 358, 358(10), 359, 360, 360(42), 361(42), 362(42), 363, 363(42), 364, 364(42), 365, 365(22), 366, 366(42), 367, 367(42), 368(10; 19; 22; 42; 43), 369, 369(10), 370, 370(10; 20; 22; 42), 371(10; 22; 43), 373(42; 43), 374, 374(22; 42), 375, 375(22), 376, 376(22; 42), 377, 377(22), 378(22), 379(19; 22; 42), 380, 380(20; 42), 381, 381(19), 382(19; 42), 383, 383(19; 42), 384, 385(22; 43; 68), 386, 386(19; 43), 387(20), 388, 388(10; 19; 68), 393, 423, 424(1; 2), 435, 435(1), 444, 473, 475(19), 476(19)
Zhu, W., 117
Zhurkin, V. B., 142
Zillig, W., 228
Zilling, W., 66
Zinder, S. H., 319
Zipp, A., 318
Zivanovic, Y., 166, 193, 195(1)
ZoBell, C. E., 316
Zollweg, J. A., 319
Zou, Q., 341
Zubay, G., 295
Zuber, H., 458
Zylicz, M., 296

Subject Index

A

ADH, *see* Alcohol dehydrogenase
Alcohol dehydrogenase, protection from irreversible heat denaturation by CpkB
    remaining activity after heat treatment, 299–300
    residual activity at elevated temperature, 298–299
Alpha helix, thermostabilization of proteins, 473–474
Aspartate, compatible solutes in thermophiles and hyperthermophiles, 308
ATP, thermostability, 438

B

Barophiles, *see* Pressure effects, hyperthermophilic enzyme assay
B factor, *see* X-ray crystallography

C

Calorimetry, *see also* Differential scanning calorimetry; Isothermal titration calorimetry
    coupling of folding and binding reactions
        differential scanning calorimetry endotherm, 416
        equilibrium constant for unfolding reaction, 415
        fraction of states in native and unfolded ensembles, 415
        modeling, 414–415
        observed heat of unfolding, 415–416
    free energy change as function of temperature, 390
    ligand binding effects on protein stability
        linkage model, 419
        linkage with pH effects, 419–420
        salt effects, 418
        linkage analysis and global nonlinear regression, 420, 422
    parameters of protein stability, 390, 392
    pH effects on protein stability
        electrostatic interactions, 416
        linkage model, 417
        linkage with ligand binding effects, 419–420
        melting temperature, 417
    state functions, 389–390
CD, *see* Circular dichroism
Chaperonin
    applications of thermostable chaperonins, 295
    ATPase assay, 296–297
    classification, 293
    hyperthermophile function, 294–295
    stress response, 293–294
    Thermococcus kodakaraensis KOD1 protein
        aggregation prevention of denatured rhodanese by CpkA, 297
        alcohol dehydrogenase, protection from irreversible heat denaturation by CpkB
            remaining activity after heat treatment, 299–300
            residual activity at elevated temperature, 298–299
        coexpression of cobyric acid synthase and CpkB for inclusion body solubilization
            electrophoretic analysis, 301
            expression in *Escherichia coli*, 300–301
            plasmids, 300
        purification of recombinant protein in *Escherichia coli*
            ammonium sulfate precipitation, 296
            anion-exchange chromatography, 296
            cell growth and induction, 296
            extract preparation, 296
            heat treatment, 296
            vectors, 295
        subunits, 295
Circular dichroism
    DNA-binding assay for histones, 124
    Sac7d and Sso7d DNA-binding proteins

chemical denaturation, 136–137
cooperative structural transitions induced in
  DNA on binding, 141–143
secondary structure, 136
thermal denaturation, 137–138
Cold shock protein, *Thermotoga maritima*
  folding in mesophiles versus thermophiles, 468
  free energy of unfolding, 464
  homology between species, 464
  melting temperature, 464
  nucleic acid binding, 465
  thermostability factors, 464–465
Compatible solutes, *see* Organic solutes, thermophiles and hyperthermophiles
Conformational rigidity, thermostabilization of proteins, 348–350, 475–477
CpkA, *see* Chaperonin
CpkB, *see* Chaperonin
Csp, *see* Cold shock protein
Cyclic 2,3-bisphosphoglycerate, compatible solutes in thermophiles and hyperthermophiles
  biosynthesis, 313–314
  enzyme thermostabilization, 312
  species distribution, 308

D

DHFR, *see* Dihydrofolate reductase
Differential scanning calorimetry, *see also* Calorimetry
  acquisition of data
    buffers, 396
    running conditions, 397
    sample
      loading, 396–397
      preparation, 396
  advantages for hyperthermophile protein analysis, 393
  enthalpy change determination, 400
  error analysis, 407–408
  ferredoxin from *Thermotoga maritima*, 28–29
  heat capacity determinations
    absolute heat capacity, 398–399
    change in heat capacity, 403–404
    excess heat capacity, 399–400
    linkage effects, 420
  instrumentation
    calibration, 396
    cell cleaning, 395
    design and performance, 394–395
  irreversible versus reversible unfolding of monomer, 404–405
  linkage analysis and global nonlinear regression, 420, 422
  monomeric protein stability analysis, 401
  non-two-state unfolding transitions, 401–403
  nonlinear regression analysis of thermodynamic parameters, 400–401
  oligomer stability analysis
    concentration of protein, 405–406
    equilibrium constant for unfolding, 406
    free energy of unfolding, 406–407
    temperature dependence of unfolded fraction, 406–407
  overview of technique, 393–394
  protein folding reversibility, 329, 393, 397–389
  two-state unfolding transitions, 401
Dihydrofolate reductase, *Thermotoga maritima*
  function, 462
  thermostability factors, 462–463
  unfolding transitions, 462
Di-*myo*-inositol phosphate derivatives, compatible solutes in thermophiles and hyperthermophiles
  biosynthesis, 314
  enzyme thermostabilization, 311–312
  species distribution, 310
  types, 306–307
DNA-binding proteins, *see* Sac7d; Sso7d
DNA gyrase, *Thermotoga maritima*
  applications, 171
  assay of supercoiling
    incubation conditions, 170
    optimization, 170–171
    overview, 169–170
    temperature dependence, 170
    unit definition, 171
  distribution
    bacteria, 165
    hyperthermophiles, 163
  purification
    cell culture and lysis, 166–167
    electrophoretic analysis, 169
    novobiocin–Sepharose chromatography, 167–169
    preparation, 166
    overview, 164, 166

SUBJECT INDEX 513

salt precipitation, 167–168
storage, 168
sign inversion mechanism, 163
subunits, 162–163
Topo IV comparison, 163–164
DNA polymerase, hyperthermophiles
  applications, 92, 116
  assays
    buffers, 96
    error rate measurement, 113–115
    exonuclease assay
      calculations, 113–114
      incubation conditions, 112
      pol II activities, 113–114
      principle, 111–112
      scintallation counting, 112–113
      substrate preparation, 112, 114
    nucleotide incorporation assay
      calculations, 98–99
      formats, 105
      incubation conditions, 98
      nucleotide analog incorporation, 99, 102–103
      overview, 96–97
      pol I activities, 101
      pol II activities, 100
      primed single stranded DNA versus activated DNA templates, 96–97
      reverse transcriptase assay, 103–104
      steady-state kinetic parameters, 104–105
      temperature effects, 97
      template preparation, 97
      thermostability assay, 99
    processivity assay
      calculations, 107
      factors affecting processivity, 105
      incubation conditions, 107
      pol I activities, 101
      pol II activities, 100, 108
      principle, 105–106
      template preparation, 106–107
    strand displacement activity
      electrophoretic analysis, 110–111
      factors affecting activity, 109
      incubation conditions, 110
      principle, 109
      substrate preparation, 110
    temperature, 95
    terminal extendase activity, 115–116
  classification, 91, 249

commercial sources, 95
mechanisms, 94
overview of features, 91
Pol I family
  activities, 92
  homology, 92
  polymerase chain reaction application, 92
  proteolytic fragments, 92
  types, 92
Pol II, *see also* Pol D
  editing capacity, 93–94
  homologs, 93–94
  types, 93
Pol III family, 92–93
stability, 95
DNA thermostability
  melting temperature of hyperthermophile DNA, 205
  thermodegradation
    agarose gel assays
      alkaline gel electrophoresis, 209
      calcium effects, 213
      depurination evaluation, 208–209
      ethidium bromide staining, 209
      glutamate effects, 214
      incubation conditions, 208
      kinetics analysis, 209, 211
      magnesium effects, 212–213
      migration of products, 211–212
      neutral gel electrophoresis, 209
      potassium effects, 214–215
      zinc effects, 214
    relevance in denaturation studies, 208
    salt effects, 207–208, 212
    steps, 207
    topologically open versus closed DNA, 205–206
DNA topoisomerase
  classification, 172, 180–181
  function, 179–180
DNA topoisomerase I, *see* Reverse gyrase
DNA topoisomerase IIA, *see* DNA gyrase; *Thermotoga maritima*
DNA topoisomerase V, *Methanopyrus kandleri*
  applications
    accuracy increasing for automated sequencing, 188–190
    sequencing at high temperatures, 187–188
    signal intensity increasing for radioactive sequencing, 188

ThermoFidelase sequencing protocols, 190–192
assay, 181
comparison with other type IB topoisomerases, 184
DNA unlinking activity
  electron microscopy of unlinked DNA, 187
  mechanism, 185–186
function, 192
overview of features, 181
purification
  ammonium sulfate precipitation, 182–183
  cell culture and lysis, 182
  gel filtration, 183–184
  heparin chromatography, 183
  phosphocellulose chromatography, 183
  polyethyleneimine precipitation, 182
DNA topoisomerase VI
assay for decatenation activity
  gel electrophoresis, 177
  incubation conditions, 176–177
  specificity of reaction, 177
  stabilizers, 177–178
  unit definition, 177
classification of topoisomerases, 172, 180
function, 172–173
properties of *Pyrococcus furiosus* and *Sulfolobus shibatae* enzymes
  kinetic parameters, 179
  reaction specifcity, 179
  sizes, 178–179
purfication from *Pyrococcus furiosus*
  ammonium sulfate precipitation, 173–174
  cell culture and lysis, 173
  heparin affinity chromatography, 174–175
  hydrophobic interaction chromatography, 174, 176
  polyethyleneimine precipitation, 173
  storage, 176
  sucrose density gradient centrifugation, 175–176, 179
unknotting assay, 178
DSC, *see* Differential scanning calorimetry

E

Electrophoretic mobility shift assay
DNA-binding assays for histones
  gel mobility acceleration in agarose, 121
  gel mobility retardation in polyacrylamide, 121–123
  overview, 120–121
Rep proteins, 200
EMSA, *see* Electrophoretic mobility shift assay
Enolase, *Thermotoga maritima*
  kinetic parameters, 457
  quaternary structure, 457
  thermostability factors, 457

F

Ferredoxin, *Pyrococcus furiosus*
assays
  coupled assay, 39–40
  direct assays
    metronidazole assay, 39
    pyruvate ferredoxin oxidoreductase assay, 38–39
  electrophoretic analysis, 36–37
iron–sulfur cluster
  forms, 30–32, 37, 354, 368
  structure, 32–33, 351
nuclear magnetic resonance, *see* Nuclear magnetic resonance
purification
  native protein
    anion-exchange chromatography, 34
    cell culture, 33
    extract preparation, 34
    gel filtration, 34
    hydroxyapatite chromatography, 34
    overview, 33–34
  recombinant protein from *Escherichia coli*
    anion-exchange chromatography, 35
    apoprotein removal from mutants, 37
    cell culture, 35
    extract preparation, 35
    gel filtration, 35–36
    mutant purification, 36
    vectors, 34–35
redox states, 37–38, 40
sequence homology between species, 351–352
stability, 31, 33
Ferredoxin, *Sulfolobus*
  cluster type and structure, 7–8, 11–12
  crystallization, 6
  crystal structure, 7, 10–12

electron paramagnetic resonance, 3, 8
function in hyperthermophiles, 3
isolated zinc site, 12–13
phylogenetic analysis, 10
purification
   anion-exchange chromatography, 5
   cell culture, 4
   extraction, 4–5
   gel filtration, 5
   hydroxylapatite chromatography, 5
   yield, 5
redox potential, 3
sequence alignment with other ferredoxins, 7, 9
X-ray absorption analysis, 13
Ferredoxin, *Thermoplasma acidophilum*
cluster type and structure, 7–8, 11–12
electron paramagnetic resonance, 3, 8
function in hyperthermophiles, 3
isolated zinc site, 12–13
phylogenetic analysis, 10
purification, 6–7
redox potential, 3
sequence alignment with other ferredoxins, 7, 9
X-ray absorption analysis, 13
Ferredoxin, *Thermotoga maritima*
apoenzyme preparation and characterization, 27–28
crystal structure, 29–30
differential scanning calorimetry and thermostability, 28–29
extinction coefficient determination, 27
function, 23
iron content determination, 27
nuclear magnetic resonance, *see* Nuclear magnetic resonance
purification of recombinant protein from *Escherichia coli*
   absorbance monitoring, 25
   anion-exchange chromatography, 25
   extract preparation, 25
   gel filtration, 26
   gene cloning, 24
   heat treatment, 25
   hydroxylapatite chromatography, 25
   large-scale expression, 25
   test expression, 24
   yield, 26
redox potential, 23

sequence homology between species, 351–352
size determination, 26
Ferredoxin:NADP oxidoreductase, *Pyrococcus furiosus*
assays
   ferredoxin-dependent reduction of elemental sulfur and polysulfide, 42
   ferredoxin-dependent reduction of NADP, 41
   NADH-dependent reduction of benzyl viologen, 42
   NADPH-dependent reduction of elemental sulfur and polysulfide, 41–42
electron paramagnetic spectroscopy, 44
function, 40–41
purification
   anion-exchange chromatography, 43–44
   Blue Sepharose chromatography, 43
   cell culture, 43
   extract preparation, 43
   gel filtration, 43–44
reaction specificity, 45
size, 44
substrate specificity, 44–45
FNOR, *see* Ferredoxin:NADP oxidoreductase
Fourier transform infrared spectroscopy
β-aggregation detection, 341
hydrogen exchange of thermostable proteins
   data acquisition, 341
   deuterium exchange conditions, 340–341
   principle, 340
FTIR, *see* Fourier transform infrared spectroscopy

G

GAPDH, *see* Glyceraldehyde-3-phosphate dehydrogenase
Gibbs–Helmholtz equation, 342, 390
Glutamate, compatible solutes in thermophiles and hyperthermophiles, 306
Glutamate dehydrogenase, *Pyrococcus furiosus*
glycerol stabilization, 326
melting temperature, 325
pressure stabilization, 325
thermodenaturation model, 325–326
Glutamate dehydrogenase, *Thermotoga maritima*
function, 460

mutant studies of thermostability, 471–472, 476–477
structural comparison between species, 460
thermostability factors, 460–462
X-ray crystallography, 460–462
Glyceraldehyde-3-phosphate dehydrogenase, *Thermotoga maritima*
  domain studies of thermostability, 456–457
  folding, temperature effects on mechanism, 466–467
  function, 452
  mutation analysis of thermostability, 454, 456
  physiochemical properties compared with mesophiles, 452–453
  sequence homology between species, 452
  X-ray crystallography, 452–454
GroEL, Chaperonin

H

Heat shock protein 60 (HSP60), *see* Chaperonin
Histones, Archaea
  dimerization, 117
  DNA-binding assays
    DNA circular dichroism, 124
    DNA circularization assay, 123
    gel mobility acceleration in agarose, 121
    gel mobility retardation in polyacrylamide, 121–123
    overview, 120–121
  gene abundance, 117
  purification
    native HMf or HMt
      extract preparation, 118
      heat treatment, 118–119
      heparin affinity chromatography, 119
      protein assay, 119
      storage, 119
    nucleosomal preparations, histone isolation, 126–127
    recombinant proteins from *Escherichia coli*
      ammonium sulfate precipitation, 120
      coexpression of processing enzymes, 118
      extract preparation, 119
      growth conditions for processing enhancement, 120
      vectors, 119
Hydrogenase, *see* Pressure effects, hyperthermophilic enzyme assay

Hydrogen bond, thermostabilization of proteins, 472–473
Hydrogen exchange, protein unfolding
  conformational flexibility and stability, 348–350
  detection techniques, 344–345
  Fourier transform infrared spectroscopy of thermostable proteins
    data acquisition, 341
    deuterium exchange conditions, 340–341
    principle, 340
  hydrogen bond breaking, 344
  nuclear magnetic resonance
    denaturant titration, 346–347
    deuterium exchange, 345–346
    exchange rate calculations, 345–346
    nitrogen-15 spectra, 345
    pH effects, 346
    temperature effects, 347–348
  principle, 344
Hydrophobic interactions, thermostabilization of proteins, 475
Hydroxyproline, compatible solutes in thermophiles and hyperthermophiles, 308

I

Inteins, hyperthermophiles
  adenosylcobalamin-dependent enzyme from *Pyrococcus furiosus*, 221–222
  catalytic splicing of exteins, 271, 273–274
  definition, 270
  discovery, 271
  distribution between species, 276–278
  expression in heterologous proteins and organisms, 274, 276
  homing endonuclease domain and intein mobility, 278–280
  identification criteria, 271–272
  protein engineering applications, 280
  sequence motifs, 272–273
  Vent polymerase, 276
Iron–sulfur cluster
  ferredoxin, *Pyrococcus furiosus*
    forms, 30–32, 37, 354, 368
    structure, 32–33, 351
  nuclear magnetic resonance of hyperthermophilic ferredoxins, *see* Nuclear magnetic resonance

SUBJECT INDEX

2-oxoacid:ferredoxin oxidoreductase, *Sulfolobus*, 15
red iron–sulfur flavoprotein, *Sulfolobus*, 19
sulredoxin, *Sulfolobus*, 20–21
L-Isoaspartyl methyltransferase, function in *Thermotoga maritima*, 438
Isothermal titration calorimetry, *see also* Calorimetry
  acquisition of data
    experimental design, 411
    running conditions, 411
    sample
      loading, 411
      preparation, 411
  instrumentation
    calibration, 410
    cell cleaning, 409–410
    design and performance, 409
  ligand concentration calculations, 412–413
  linkage analysis and global nonlinear regression, 420, 422
  molar heat change, 413
  nonlinear regression analysis, 413
  overview of technique, 409
  Sac7d binding studies, 413–414
  software for data analysis, 411–412
ITC, *see* Isothermal titration calorimetry

J

Johnson–Eyring equation, 322

L

Lactate dehydrogenase, *Thermotoga maritima*
  function, 458
  sequence homology between species, 458
  X-ray crystallography, thermostability analysis, 458–459
LDH, *see* Lactate dehydrogenase

M

Mannosylglycerate, compatible solutes in thermophiles and hyperthermophiles
  biosynthesis, 314–315
  enzyme thermostabilization, 312
  forms, 307
  species distribution, 310–311
McGhee–von Hippel equation, 140, 413

Methanopyrus kandleri DNA topoisomerase V, *see* DNA topoisomerase V, *Methanopyrus kandleri*
Methanothermus fervidus histones, *see* Histones, Archaea
Methanothermus thermoautotrophicum histones, *see* Histones, Archaea
Methylguanine methyltransferase, hyperthermophiles
  assay
    incubation conditions, 242
    principle, 240
    substrate preparation, 241–242
    thermostability assay, 246
    unit definition, 242
  in vivo assay in *Escherichia coli* mutant strain, 246
  DNA repair, 239
  gene cloning, 243
  purification of recombinant *Thermococcus kodakaraensis* KOD1 enzyme in *Escherichia coli*
    anion-exchange chromatography, 245
    crystallization, 246–248
    extract preparation, 245
    gel filtration, 245
    sequence homology between species, 243–244
  species distribution, 239–240
MGMT, *see* Methylguanine methyltransferase
Michaelis constant (K_m), variation with temperature, 289–290

N

NAD(P)H:rubredoxin oxidoreductase, *Pyrococcus furiosus*
  absorption spectroscopy, 59–60
  assays
    benzyl viologen reduction, 56–57
    electron acceptor specificity, 57
    rubredoxin assay, 56
  features compared with other anaerobic bacteria enzymes, 56, 61–62
  flavin characterization, 60
  function, 55–56
  purification
    anion-exchange chromatography, 57–58
    Blue Sepharose chromatography, 58
    cell culture, 57

extract preparation, 57
gel filtration, 58
yield, 59
size, 59
stability, 60
substrate specificity, 60–61
temperature dependence, 61
NMR, see Nuclear magnetic resonance
NROR, see NAD(P)H:rubredoxin oxidoreductase
Nuclear magnetic resonance
  ferredoxins from hyperthermophiles
    cubane cluster magnetic properties
      four-iron clusters, 355
      spin coupling overview, 354
      three-iron clusters, 355–356
    magnetization transfer between reduced and oxidized forms, 365–366
    molecular structure determination, 368, 380–381
    Pyrococcus furiosus enzyme
      cluster architecture and ligand mutation effects on structure, 387
      disulfide bond cleavage effects on structure, 385
      disulfide bond orientational isomerism in Fd_A, 371–373
      disulfide bond orientation effects on Fd_A structure, 385–386
      electron exchange rate of cluster, 380
      forms of iron–sulfur cluster, 354, 368
      oxidized four-iron cluster, 374–375
      oxidized three-iron cluster, 373
      redox-active disulfide bond cycling, 369–370
      reduced four-iron cluster, 375–379
      sequence-specific assignments, 367–368
      structural modeling, 385
    reference spectra, 362
    relaxation time acquisition, 362
    sample preparation, 361–362
    spatial proximity effects, 364–365
    spectral parameters, cluster paramagnetism effects
      hyperfine shifts, 357–359
      relaxation, 356–357
      temperature behavior of cluster ligand contact shifts, 359, 361
    spin connectivity, 362–363
    temperature variation in studies, 353

Thermococcus literalis enzyme
  electron exchange rate of cluster, 379–380
  oxidized four-iron cluster, 374
  oxidized three-iron cluster, 373
  reduced four-iron cluster, 376–377
  sequence-specific assignments, 367–368
  structure of oxidized four-iron cluster form, 382–383
  thermostability implications, 387–389
Thermotoga maritima enzyme
  oxidized four-iron cluster, 374
  structure of oxidized four-iron cluster form, 381–382
hydrogen exchange in hyperthermophilic proteins
  denaturant titration, 346–347
  deuterium exchange, 345–346
  exchange rate calculations, 345–346
  nitrogen-15 spectra, 345
  pH effects, 346
  temperature effects, 347–348
three-dimensional structure elucidation
  Sac7d, 130–131
  Sso7d, 130–131
Nucleosome, Archaea
  formation assays, see Histones, Archaea
  mapping
    end mapping with restriction enzymes, 128–129
    microccocal nuclease digestion, 127
    overview, 127
    template synthesis, 127
  purification
    DNA isolation, 127
    formaldehyde treatment, 124
    gel electrophoresis, 126
    gel filtration, 124
    histone isolation, 126–127
    microccocal nuclease digestion, 124
    overview, 124
  structural overview, 117

O

Organic solutes, thermophiles and hyperthermophiles
  aspartate, 308
  compatible solutes, overview, 303–304
  cyclic 2,3-bisphosphoglycerate

biosynthesis, 313–314
enzyme thermostabilization, 312
species distribution, 308
di-*myo*-inositol phosphate derivatives
biosynthesis, 314
enzyme thermostabilization, 311–312
species distribution, 310
types, 306–307
distribution by species, 306–309
glutamate, 306
halotolerance, 302, 315
hydroxyproline, 308
mannosylglycerate
biosynthesis, 314–315
enzyme thermostabilization, 312
forms, 307
species distribution, 310–311
osmotic adaptation, 303–304, 315
phylogenetic distribution, 310–311
tetracarboxyhexane, 308
trehalose, 304, 306, 308
unique compounds, 304–305, 315
uptake versus *de novo* synthesis in
hyperthermophiles, 308, 310
Osmotic adaptation, *see* Organic solutes,
thermophiles and hyperthermophiles
2-Oxoacid:ferredoxin oxidoreductase,
Sulfolobus
aerobic species enzyme comparison, 16–17
assay, 14–15
catalytic mechanism, 17–18
ferredoxin electron acceptors, 15, 17
iron–sulfur cluster, 15
phylogenetic analysis, 15–16
purification
anion-exchange chromatography, 13
glycerol density gradient centrifugation, 14
hydrophobic interaction chromatography, 14
hydroxylapatite chromatography, 14
yield, 14
stability, 14
structure, 15
substrate specificity, 15

P

PDI, *see* Protein disulfide oxidoreductase,
Pyrococcus furiosus
PGK, *see* Phosphoglycerate kinase

pGT5 replication initiator protein Rep75,
see Rep75
Phosphoglycerate kinase, *Thermotoga maritima*
function, 446–447
fusion partner, *see* Triose-phosphate
isomerase
gene regulation and fusion, 447
structure, 447–448
thermostability of bifunctional complex
versus monofunctional enzyme
chemical denaturation studies, 449
physiochemical data, 450
thermodynamic analysis, 449–450
unfolding transition state, 450–451
X-ray crystallography, 452
Phosphoribosylanthranilate isomerase,
Thermotoga maritima
function, 463
thermostability factors, 464
X-ray crystallography, 464
Pol D, *Methanococcus jannaschii*
activators and inhibitors, 258
purification of recombinant enzyme from
Escherichia coli, 257
sequence homology between species,
258–260
Pol D, *Pyrococcus furiosus*
accessory proteins, 260
activators and inhibitors, 254
classification, 249
exonuclease activity, 255–257
fractionation of polymerase activities, 250
function, 260
gene cloning, 251–252
pH optimum, 255
polymerase chain reaction applications, 258
primer–template preference, 255
purification of recombinant enzyme from
Escherichia coli
ammonium sulfate precipitation, 252
anion-exchange chromatography, 253
cell growth and induction, 252
extract preparation, 252
gel filtration, 253
heat treatment, 252
heparin affinity chromatography, 253
polyethyleneimine precipitation, 252
yield, 253
sequence homology between species,
258–260

subunits, 253–254
temperature optimum, 255
PRAI, *see* Phosphoribosylanthranilate isomerase
Pressure effects, hyperthermophilic enzyme assay
　barophiles, 316, 327
　buffer selection, 317–318
　food processing applications, 316–317
　hydrogenase assay of methyl viologen reduction
　　data analysis, 321–322
　　incubation conditions and operation, 320–321
　　instrumentation with fiber optic probe, 318–320
　protease assay
　　data analysis, 323–324
　　incubation conditions and operation, 323
　　instrumentation, 322–323
　　substrate, 322
　thermostability measurements under high pressure
　　glutamate dehydrogenase from *Pyrococcus furiosus*
　　　denaturation model, 325–326
　　　glycerol stabilization, 326
　　　melting temperature, 325
　　　pressure stabilization, 325
　　instrumentation, 324–325
　　pressure-stabilized enzymes, 326–327
　　sampling, 325
Protein disulfide oxidoreductase, *Pyrococcus furiosus*
　assays
　　insulin reductase activity, 69
　　peptide oxidation
　　　fluorescence assay, 70–71
　　　reversed-phase high-performance liquid chromatography assay, 72
　　　synthetic peptide, 70
　　thioltransferase activity, 69–70
　crystal structure
　　active sites, 80–82
　　crystal packing contacts and substrate binding, 83–84
　　crystallization, 75
　　monomer structure, 77–80
　　thioredoxin fold, 75, 78–79, 84–85, 87

　　zinc-binding site and dimer formation, 82–83
　discovery, 63
　evolution, 84–85, 87–88
　function of superfamily members, 62–63, 74–75
　gene cloning
　　chromosomal DNA isolation, 65–66
　　colony hybridization, 66–67
　　gene bank construction, 66
　　sequencing, 67–68
　homology with *Sulfolobus solfataricus* protein, 76
　isoelectric point, 72
　N-terminal sequence analysis, 65
　purification
　　native protein
　　　anion-exchange chromatography, 64
　　　extract preparation, 64
　　　gel filtration, 64–65
　　　reversed-phase high-performance liquid chromatography, 65
　　　yield, 64
　　recombinant protein from *Escherichia coli*
　　　cell growth and induction, 68
　　　extract preparation, 68
　　　gel filtration, 69
　　　heat treatment, 69
　　　vector construction, 68
　sequence motifs, 76
　size, 73
　stability, 72–73
　substrate specificity, 73, 76
Protein folding, *see also* Calorimetry
　concentration dependence of temperature denaturation profile, 329
　denaturant half-concentration C_m
　　concentration dependence, 337–338
　　determination, 334
　　temperature dependence, 336
　detection and denaturation techniques, 329–330
　enthalpy change on unfolding determination, 336
　equilibrium states between folding and unfolding, 329–331
　free energy extrapolation toward zero denaturant concentration, 332–333
　heat capacity change on unfolding determination, 335–336

SUBJECT INDEX

hydrogen exchange studies, see Hydrogen exchange, protein unfolding
kinetics of thermally induced isothermic loss of native structure, 341
mechanistic effects of temperature, 466–468
melting temperature
  concentration dependence, 337–338
  determination from temperature dependence of denaturant half-concentration, 336
monomer association contribution to overall stability, 330
m-value determination, 335
oligomeric proteins, concentration normalization of free energy function in unfolding studies, 338–339
packing, thermostabilization of proteins, 474
residual structure in thermodynamics calculation, 335–336
reversibility monitoring, 328–329, 393
surface-to-volume ratio, thermostabilization of proteins, 474
thermal stability as function of temperature, 342–343
thermophilic versus mesophilic proteins, 468
unfolding isotherm analysis
  equilibrium constant calculation, 334
  free energy determination, 334
  nontransitional regions, 333–334
Pyrococcus abyssi Rep proteins, see Rep50; Rep75
Pyrococcus furiosus
  DNA topoisomerase VI, see DNA topoisomerase VI
  ferredoxin, see Ferredoxin, *Pyrococcus furiosus*
  ferredoxin:NADP oxidoreductase, see Ferredoxin:NADP oxidoreductase, *Pyrococcus furiosus*
  glutamate dehydrogenase, see Glutamate dehydrogenase, *Pyrococcus furiosus*
  NAD(P)H:rubredoxin oxidoreductase, see NAD(P)H:rubredoxin oxidoreductase, *Pyrococcus furiosus*
  pfu, see DNA polymerase, hyperthermophiles
  Pol D, see Pol D, *Pyrococcus furiosus*
  protein disulfide oxidoreductase, see Protein disulfide oxidoreductase, *Pyrococcus furiosus*
  ribonucleotide reductase, see Ribonucleotide reductase
  RNA polymerase, see RNA polymerase
  rubredoxin, see Rubredoxin, *Pyrococcus furiosus*
  TATA binding protein, see TATA binding protein
  transcription factor B, see Transcription factor B

R

Rad51, see RecA/Rad51
RecA/Rad51
  function, 261, 270
  genes in hyperthermophiles, 261
  phylogenetic analysis, 262–263
  Thermococcus kodakaraensis KOD1 homolog
    aggregation, 268
    ATPase activity
      assay, 269
      products, 269–270
    complementation of ultraviolet resistance in *Escherichia coli*, 264–266
    DNase activity
      assay, 268
      metal ion effects, 268–269
      temperature optimum, 269
    function, 262, 264
    mutant studies, 265–266
    purification of recombinant protein from *Escherichia coli*, 266
    sequence analysis, 262
    size, 262, 266–267
Red iron–sulfur flavoprotein, *Sulfolobus*
  function, 18
  iron–sulfur cluster, 19
  purification
    anion-exchange chromatography, 18
    glycerol density gradient centrifugation, 19
    hydrophobic interaction chromatography, 18
    hydroxylapatite chromatography, 18–19
    yield, 19
  structure, 19
Rep50
  assays
    electrophoretic mobility shift assay, 200
    nicking–closing reaction, 200
    nucleotidyl-terminal transferase, 200

oligonucleotide substrates, 199–200
comparison with Rep75, 203–204
purification of recombinant *Pyrococcus abyssi* protein from *Escherichia coli,* 199
Rep75
  assays
    electrophoretic mobility shift assay, 200
    nicking–closing reaction, 200–201
    nucleotidyl-terminal transferase, 200
    oligonucleotide substrates, 199–200
  common features of Rep proteins, 193–195
  comparison with Rep50, 203–204
  divalent cation requirements, 202–203
  nucleotidyl-terminal transferase activity, 195, 201–202
  pGT5 replication initiation, 193, 195
  purification of recombinant *Pyrococcus abyssi* protein from *Escherichia coli*
    cell growth and induction, 196
    electrophoretic analysis, 198
    gene cloning, 195–196
    inclusion body purification and renaturation, 197–198
    overview, 195
    phosphocellulose chromatography, 197–198
  temperature optima of activities, 202
  thermostability, 198
Reverse gyrase
  ATPase activity assay, 154–155
  ATP-dependent linking number increase assay, 152–154
  equation, 151
  mechanism, 152
  distribution in prokaryotes, 146–148, 160–161
  DNA cleavage specificity assay, 155–156
  domains, 146, 162
  function
    chromatin remodeling, 162
    DNA topology in hyperthermophiles, 159–161
    overview, 148
    recombination, 162
    renaturase activity, 161–162
    repressor activity, 162
  noncovalent stoichiometric binding of DNA, 156
  physiochemical properties from hyperthermophiles, 151

positive supercoiling mechanistic models, 158–159
purification from hyperthermophiles
  native proteins
    ammonium sulfate fractionation, 149
    extract preparation, 148
    heparin affinity chromatography, 149
    hydrophobic interaction chromatography, 149
    phosphocellulose chromatography, 149
    variations, 149–150
  recombinant protein from *Escherichia coli,* 150–151
  sequences from hyperthermophiles, 156–158
R factor, *see* X-ray crystallography
Ribonucleotide reductase
  adenosylcobalamin-dependent enzyme from *Pyrococcus furiosus*
    acetate stimulation, 218
    assay, 219–220
    catalytic mechanism, 224–225
    cofactor thermoprotection, 227
    kinetic parameters, 220
    purification
      ammonium sulfate fractionation, 219
      dATP-Sepharose chromatography, 219
      extract preparation, 219
      hydrophobic interaction chromatography, 219
      overview, 218–219
    sequence analysis
      homology with other ribonucleotide reductases, 223
      size, 220–221
      splicing variants, 221–222
    allosteric regulation, 225–227
    classification
      class Ia, 216
      class Ib, 217
      class II, 217
      class III, 217
      hyperthermophile distribution of classes, 218
    function, 215–216
RNA polymerase
  preinitiation complex
    components, 227–228
    reconstitution, 238–239
  purification

SUBJECT INDEX

epitope selection for immunoaffinity
  chromatography, 233
Pyrococcus furiosus enzyme
  conventional chromatography, 232–233
  immunoaffinity chromatography, 233, 235
Sulfolobus acidocaldarius enzyme
  ammonium sulfate precipitation, 230
  anion-exchange chromatography, 230
  cell culture, 228–229
  extract preparation, 230
  gel filtration, 232
  heparin affinity chromatography, 230
Sulfolobus shibatae enzyme, 232
RNR, *see* Ribonucleotide reductase
Rubredoxin, *Pyrococcus furiosus*
  function, 55
  hydrogen exchange
    denaturant effects, 346–347
    temperature effects, 347–348
  metal coordination, 48–49
  purification
    native protein
      anion-exchange chromatography, 50–51
      cell culture, 50
      color in detection, 49–50
      extract preparation, 50
      gel filtration, 51
      hydroxyapatite chromatography, 50–51
      yield, 51
    recombinant protein from *Escherichia coli*
      anion-exchange chromatography, 53–54
      cell culture, 51–52
      extract preparation, 52–53
      gel filtration, 53
      hydroxyapatite chromatography, 53
      metal incorporation, 51, 54
      nitrogen-15 labeling, 52
      N-terminal processing, 51
      vectors, 52
      yield, 54
  related proteins, 46
  sequence homology between species, 46–48
  size, 54
  three-dimensional structure, 48–49, 54

S

Sac7d
  circular dichroism
    chemical denaturation, 136–137
    cooperative structural transitions induced in DNA on binding, 141–143
    secondary structure, 136
    thermal denaturation, 137–138
  DNA-binding assays, overview, 138–139
  DNA thermal denaturation stabilization studies, 143
  extinction coefficient determination, 135
  function of family members, 129–130
  isothermal titration calorimetry binding studies, 413–414, 420
  lysine monomethylation, 138
  nomenclature, 131–132
  purification
    extract preparation, 134
    gel filtration, 134
    ion-exchange chromatography, 134
    recombinant protein expression in *Escherichia coli*, 133–134
    Sulfolobus culture, 132–133
    yield, 134
  ribonuclease assays
    gel filtration of protein, 144–145
    incubation conditions, 144
    overview, 143
    principle, 144
    ribonuclease A contamination, 145
  size, 134–135
  three-dimensional structure
    nuclear magnetic resonance, 130–131
    X-ray crystallography, 131
  tryptophan fluorescence DNA-binding assay
    binding constant determination, 140–141
    site size determination, 140–141
    titration conditions, 139–140
    overview, 135–136
Sso7d
  circular dichroism
    chemical denaturation, 136–137
    cooperative structural transitions induced in DNA on binding, 141–143
    secondary structure, 136
    thermal denaturation, 137–138
  DNA-binding assays, overview, 138–139
  DNA thermal denaturation stabilization studies, 143
  extinction coefficient determination, 135
  function, 129–130

lysine monomethylation, 138
nomenclature, 131–132
purification
  extract preparation, 134
  gel filtration, 134
  ion-exchange chromatography, 134
  recombinant protein expression in
    Escherichia coli, 133–134
  Sulfolobus culture, 132–133
  yield, 134
ribonuclease assays
  gel filtration of protein, 144–145
  incubation conditions, 144
  overview, 143
  principle, 144
  ribonuclease A contamination, 145
size, 134–135
three-diminensional structure
  nuclear magnetic resonance, 130–131
  X-ray crystallography, 131
tryptophan fluorescence
  DNA-binding assay
    binding constant determination, 140–141
    site size determination, 140–141
    titration conditions, 139–140
  overview, 135–136
Sulfolobus
  DNA topoisomerase VI, *see* DNA topoisomerase VI
  DNA-binding proteins, *see* Sac7d; Sso7d
  ferredoxin, *see* Ferredoxin, *Sulfolobus*
  2-oxoacid:ferredoxin oxidoreductase, *see* 2-Oxoacid:ferredoxin oxidoreductase, *Sulfolobus*
  red iron–sulfur flavoprotein, *see* Red iron–sulfur flavoprotein, *Sulfolobus*see
  reverse gyrase, *see* Reverse gyrase
  RNA polymerase, *see* RNA polymerase
  sulredoxin, *see* Sulredoxin, *Sulfolobus*
  TATA binding protein, *see* TATA binding protein
  transcription factor B, *see* Transcription factor B
Sulredoxin, *Sulfolobus*
  discovery, 20
  iron–sulfur cluster, 20–21
  purification, 20
  redox-linked ionization, 22
  redox potential, 21
  structure, 20–22

Surface-to-volume ratio, thermostabilization of proteins, 474

T

Taq, *see* DNA polymerase, hyperthermophiles
TATA binding protein
  preinitiation complex
    components, 227–228
    reconstitution, 238–239
  purification
    Pyrococcus furiosus recombinant protein in *Escherichia coli*
      ammonium sulfate precipitation, 237
      cell growth and lysis, 236
      gel filtration, 237
      hydroxyapatite chromatography, 237
      vector, 236
    Sulfolobus acidocaldarius recombinant protein in *Escherichia coli*
      cation-exchange chromatography, 235–236
      cell growth and lysis, 235
      hydroxyapatite chromatography, 236
      vector, 235
TBP, *see* TATA binding protein
Tetracarboxyhexane, compatible solutes in thermophiles and hyperthermophiles, 308
TFB, *see* Transcription factor B
Thermococcus kodakaraensis KOD1
  chaperonin, *see* Chaperonin
  methylguanine methyltransferase, *see* Methylguanine methyltransferase
  RecA/Rad51 homolog, *see* RecA/Rad51
Thermoplasma acidophilum ferredoxin, *see* Ferredoxin, *Thermoplasma acidophilum*
Thermostability, DNA, *see* DNA thermostability

Thermostability, hyperthermophilic enzyme assay
  applications of enzymes, 283
  Arrhenius activation energy and catalytic activity with temperature, 283, 290–291
  buffer selection, 286–288
  calorimetry, *see* Differential scanning calorimetry; Isothermal titration calorimetry
  cofactor thermostability, 288–289
  continuous versus discontinuous assays, 284–285

crystallography studies, *see* X-ray crystallography
denaturation versus degradation, 291–292
denaturation versus inactivation, 470–471
duration of assay and inactivation, 283–284
equipment, 285–286
intensive versus extensive parameters, 328
K_m variation with temperature, 289–290
pressure effects under high temperature, *see* Pressure effects, hyperthermophilic enzyme assay
renaturation by lowering of temperature, 292
safety, 285–286
stabilizers, 293
substrate thermostability, 285, 288–289, 438
temperature control, 286
temperature optimum versus denaturation, 284
thermodynamic analysis, *see* Protein folding
Thermotoga maritima
  DNA gyrase, *see* DNA gyrase, *Thermotoga maritima*
  ferredoxin, *see* Ferredoxin, *Thermotoga maritima*
  thermostability of proteins
    amino acid composition, 444–445
    cold shock protein, 464–465
    dihydrofolate reductase, 462–463
    enolase, 457
    experimental approaches for study, 440–441
    folding and stability, 465–468
    Gibbs free energy of stabilization, 441–442
    glutamate dehydrogenase, 460–462
    glyceraldehyde-3-phosphate dehydrogenase, 452–454, 456–457
    homologous protein comparisons, 445–446
    intrinsic versus extrinsic factors, 440
    lactate dehydrogenase, 458–459
    phosphoglycerate kinase, 446–452
    phosphoribosylanthranilate isomerase, 463–464
    stabilizing interactions, 443–444, 468–469
    temperature effects on folding mechanism, 466–468
    temperature optima for activity, 442
    triose-phosphate isomerase, 446–448

Thioredoxin fold
  protein distribution and evolution, 75, 84–85, 87
  protein disulfide oxidoreductase from *Pyrococcus furiosus,* 78–79
TIM, *see* Triose-phosphate isomerase
Topoisomerase, *see* specific DNA topoisomerases
Transcription factor B
  preinitiation complex
    components, 227–228
    reconstitution, 238–239
  purification
    Pyrococcus furiosus recombinant protein in *Escherichia coli,* 238
    Sulfolobus acidocaldarius recombinant protein in *Escherichia coli*
      cell growth and lysis, 237
      gel filtration, 238
      nickel affinity chromatography, 237–238
      vector, 237
Trehalose, compatible solutes in thermophiles and hyperthermophiles, 304, 306, 308
Triose-phosphate isomerase, *Thermotoga maritima*
  function, 446–447
  fusion partner, *see* Phosphoglycerate kinase
  gene regulation and fusion, 447
  structure and thermostability of bifunctional complex, 447–448

U

Unfolding isotherm, *see* Protein folding

V

Vent, *see* DNA polymerase, hyperthermophiles

X

X-ray crystallography
  B factors, 430
  cryocrystallography of oxygen-sensitive proteins, 429
  crystallization of oxygen-sensitive proteins, 427–428
  expression and purification of hyperthermophilic proteins

SUBJECT INDEX

anaerobic handling, 426–427
recombinant proteins, 425–426
ferredoxin
  Sulfolobus, 7, 10–12
  Thermotoga maritima, 29–30
glutamate dehydrogenase from *Thermotoga maritima*, 460–462
lactate dehydrogenase from *Thermotoga maritima*, 458–459
overview of steps for structure elucidation, 423
phosphoglycerate kinase from *Thermotoga maritima*, 452
phosphoribosylanthranilate isomerase from *Thermotoga maritima*, 464
prospects for hyperthermophilic proteins, 436–437
Protein Data Bank entries from hyperthermophiles, 423–425
protein disulfide oxidoreductase from *Pyrococcus furiosus*
  active sites, 80–82
  crystal packing contacts and substrate binding, 83–84
  crystallization, 75
  monomer structure, 77–80
  thioredoxin fold, 75, 78–79, 84–85, 87
  zinc-binding site and dimer formation, 82–83
R factors, 430–431
resolution, 430
rubredoxin from *Pyrococcus furiosus*, 48–49, 54
Sac7d, 131
Sso7d, 131
thermal expansion of hyperthermophilic proteins
  coefficient of thermal expansion, 432
  temperature-dependent data, 434–435
  volume of activation for protein unfolding, 434
thermostability analysis, interaction types in hyperthermophiles, 431–433, 443–445, 476–478

Z

Zinc-containing ferredoxin, *see* Ferredoxin, *Sulfolobus*; Ferredoxin, *Thermoplasma acidophilum*